HOME REHAB HANDBOOK

HOME REHAB HANDBOOK

Steven Winter Associates

Michael J. Crosbie, Editor

McGraw-Hill

New York Chicago San Francisco Lisbon London
Madrid Mexico City Milan New Delhi San Juan Seoul
Singapore Sydney Toronto

Cataloging-in-Publication Data is on file with the Library of Congress.

McGraw-Hill

A Division of The ***McGraw-Hill*** *Companies*

1 2 3 4 5 6 7 8 9 0 AGK/AGK 0 7 6 5 4 3 2 1

ISBN 0-07-137777-8

The sponsoring editor for this book was Larry Hager, the editing supervisor was Caroline Levine, and the production supervisor was Sherri Souffrance.

Printed and bound by Quebecor/Kingsport.

This book is printed on recycled, acid-free paper containing a minimum of 50 percent recycled, de-inked fiber.

Contents

Preface xi

CHAPTER 1 FOUNDATIONS 1

1.1 Foundation Design and Engineering 3
Existing Foundation Overview 3
Basement Floors 4
Crawl Space Floors 5
1.2 Permanent Wood and Prefabricated Foundations 7
Permanent Wood Foundation Systems 7
Prefabricated Foundations 8
1.3 Drainage 12
Surface and Subsurface Drainage 12
Foundation Drainage 15
1.4 Dampproofing and Waterproofing 19
1.5 Insulation 24
Insulating Below-Grade Walls 24
Insulating Crawl Spaces 27
Insulating Slabs 28
1.6 Ventilation 32
Ventilating Basement Spaces 32
Ventilating Crawl Spaces 35
Ventilation for Soil Gases 36
1.7 Shoring, Underpinning, and Repair 41
1.8 Crack Repair, Coatings, and Finishes 48
Cracks in Walls and Slabs 48
Coatings and Finishes 52

CHAPTER 2 EXTERIOR WALLS 57

2.1 Design and Engineering 57
Exterior Wall Overview 57
Wood-Frame Seismic Resistance 58
Wood-Frame Wind Resistance 62
Reinforcing Existing Masonry Wall Construction 64
Moisture Deterioration 66
Mitigating Insect Damage 68
Fire Damage to Wood Framing 70
2.2 Masonry and Brick Veneer 70
Clean Existing Masonry Walls 72
Apply Colorless Protective Coatings 74
Repoint Existing Walls 74
Repair Existing Masonry Walls 75
2.3 Sheathing 79

2.4 Vapor Retarders and Air Infiltration Barriers 83
Vapor Retarders 83
Air Infiltration Barriers 86
2.5 Insulation 90
2.6 Vinyl Siding 95
2.7 Metal Siding 99
2.8 Wood Shingles and Shakes 102
2.9 Solid Wood Siding 106
2.10 Hardboard Siding 111
2.11 Engineered Wood Siding 114
2.12 Plywood Panel Siding 117
2.13 Fiber-Cement Siding 120
2.14 EIFS and Stucco 124
Exterior Insulation and Finish Systems 124
Stucco 127
2.15 Exterior Trim 129
2.16 Sealants and Caulks 131
2.17 Paint and Other Finishes 135

CHAPTER 3 ROOFS 141

3.1 Design and Engineering 143
Roof Systems Overview 143
Typical Framing Errors 144
Wood-Frame Wind Resistance 145
Wood-Frame Seismic Resistance 146
Structural Decay 147
Fire Damage 147
3.2 Sheathing 150
3.3 Flashing 153
3.4 Underlayments and Moisture Barriers 161
3.5 Insulation 163
3.6 Wood Shingles and Shakes 171
3.7 Asphalt Shingles 175
3.8 Low-Slope Roofing 179
3.9 Metal Roofing 184
3.10 Slate 190
3.11 Clay, Concrete, Fiber-Cement, Composite Tile 194
3.12 Gutter and Leader Systems 199

CHAPTER 4 WINDOWS AND DOORS 209

4.1 Existing Window and Door Overview 211
Evaluations of Existing Conditions, Options, and Selection 211
Window and Door Types 212
Ratings and Standards 214
Installation 216
Costs and Benefits 217
4.2 Glazing 218
4.3 Window Frames and Replacement Units 224
4.4 Storm Windows and Screens 230

4.5 Skylights 234
4.6 Doors and Frames 236
Primary Entry Doors 236
Garage and Bulkhead Doors 239
Storm and Screen Doors 240
Interior Doors 241
4.7 Casing and Trim 244
4.8 Hardware 248
4.9 Flashing 251
4.10 Caulking and Weatherstripping 254
Caulks and Sealants 254
Weatherstripping 265
4.11 Shutters and Awnings 261

CHAPTER 5 PARTITIONS, CEILINGS, FLOORS, AND STAIRS 263

5.1 Floor and Ceiling Structure 265
Overview 265
Common Structural Problems 265
Alternatives to Solid Lumber for Floor Framing 270
Moisture Deterioration 275
Fire Damage to Floor Framing 277
Sound Control 278
5.2 Subflooring and Underlayments 280
Damaged Subflooring 280
Floor Squeaks 281
Underlayments 284
5.3 Finish Flooring 287
Wood Flooring 287
Vinyl Sheet Flooring and Tile 291
Ceramic Tile 295
Carpet and Rugs 296
5.4 Finish Walls and Ceilings 299
Plaster and Drywall 299
Paints and Wall Coverings 306
Moldings and Trim 308
5.5 Stairs 311
Repairing Treads and Risers 311
Replacing Treads and Risers 313
Sagging Carriages 314
Damaged or Broken Balusters 315
Prefabricated Stairs 317
Attic Ladders 318

CHAPTER 6 KITCHENS AND BATHS 321

6.1 Kitchens and Baths Overview 323
6.2 Cabinets 326
Surface Maintenance and Repairs 326
Hardware to Maximize Access and Function 327
Accessories to Maximize Access and Improve Storage 332
Replace or Add Cabinetry 336

6.3 Countertops 340
Surface Maintenance and Repairs 340
Improve Indoor Air Quality 342
Improve Backsplash and Countertop Seams 343
Maximize Access and Function 345
Improving Functional Countertop Space 346
Enhance Backsplash with Surface Materials 348
6.4 Appliances 350
Replace Outdated or Nonfunctioning Appliance 350
Improve Access and Function within Work Centers 353
Upgrade Appliance Appearance 357
Improve Resource and Energy Efficiency of Existing Appliances 359
Install Reslource- and Energy-Efficient Appliances 360
6.5 Sinks and Lavatories 363
Surface Maintenance and Repairs 363
Water Purification 365
Water Conservation 366
Maximize Access and Function 367
6.6 Tubs and Showers 370
Surface Maintenance and Repairs 370
Moisture Control 373
Maximize Access and Function 375
Water Conservation 378
6.7 Toilets and Bidets 380
Maintenance and Repairs 380
Water Conservation 381
Maximise Access and Function 383

CHAPTER 7 ELECTRICAL AND ELECTRONIC SYSTEMS 387

7.1 Electrical and Electronic Systems Overview 389
7.2 Service Panels 390
7.3 Wiring and Receptacles 397
Wiring Overview 397
Aluminum Wiring 399
Receptacles 401
7.4 Lighting and Controls 404
Interior Lighting 404
Exterior Lighting 406
Controls 408
7.5 Electric Baseboard Heating 413
7.6 Phone, Computer, and TV Cabling 416
7.7 Security Systems 419
Intrusion and Alarm Systems 419
Smoke Detectors 422
Carbon Monoxide Detectors 424
Lightning Protection 425
Surge Protection 428
Garage Door Openers 430

CHAPTER 8 HVAC AND PLUMBING 435

8.1 HVAC Design and Engineering 437
HVAC Systems Overview 437
Replacement System Sizing 438
Energy Sources 439
8.2 Distribution Systems 442
8.3 Heating 451
8.4 Cooling 457
8.5 Heat Pumps 463
8.6 Indoor Air Quality 468
8.7 Controls 475
8.8 Fireplaces and Chimneys 480
8.9 Domestic Hot-Water Heating 488
8.10 Plumbing Design and Engineering 496
8.11 Water Supply and Distribution Systems 498
8.12 Drain, Waste, and Vent Systems 502
8.13 Fuel Supply Systems 509
8.14 Appliance Vents and Exhausts 511
8.15 Fire Protection Systems 514

CHAPTER 9 SITE WORK 517

9.1 Decks, Porches, and Fences 519
Deck and Porch Strructure 519
Repair and Replace Wood Decking Materials 522
Preservative-Treated Wood 525
Stairs and Handrails 527
Wood Fences and Retaining Walls 529
9.2 Paved Driveways, Walks, Patios, and Masonry Walls 534
Paved Driveways, Walks, and Patios 534
Masonry Walls 540
9.3 Underground Construction 544
Wells 544
On-Site Wastewater Treatment 545
Water and Sewer Lines 548
Underground Storage Tanks 551
9.4 Landscaping 553
Landscape Care 553
Energy-Efficient and Sustainable Landscaping 557

FURTHER READING 561

HOME REHAB TECHNIQUES CHECKLIST 597

PRODUCT INFORMATION 621

PROESSIONAL ORGANIZATIONS 677

Index 691

PREFACE

Home Rehab Handbook is a guide for architects, designers, builders, contractors, and homeowners to the innovations and state-of-the-art practices in home rehabilitation. As is too often the case, innovative techniques, materials, technologies, and products are slow to make their way into accepted practice. Such innovations will not advance unless the industry is made aware of them and they are tested. *Home Rehab Handbook* is intended to accelerate this process by informing professionals in the home design and construction industry about such innovations and state-of-the-art practices.

Home Rehab Handbook also fills the need for a comprehensive publication to make the design and construction industry aware of innovative and cost-saving developments in housing rehabilitation. Professional trade magazines, conferences, and trade shows offer some dissemination of this information, but they are rarely focused exclusively on housing rehabilitation, as this book is, nor are they comprehensive.

The focus of this handbook is on housing rehabilitation, which is different from home improvement. Rehabilitate means "to restore to good condition," not necessarily to improve to a state that is significantly different than the original. This is a fine line, but it distinguishes this publication from "home improvement" books written for the amateur. *Home Rehab Handbook* focuses on building technology, materials, components, and techniques rather than "projects" such as adding a new room, converting a garage into a den, or finishing an attic. Nor is *Home Rehab Handbook* intended to be a "diagnostic" tool; a number of such books are already available to the industry.

The content for this book has been gathered from professionals in the housing rehabilitation field: manufacturers and suppliers of innovative technologies, materials, components, tools, and equipment; trade shows, conferences, reports, and publications considering such issues; trade organizations; and building research centers. Drafts of the content were reviewed by professional members of the National Association of Home Builders' Remodelors Council, and the National Association of the Remodeling Industry (NARI).

A variety of excellent resources exists for information on home rehabilitation. Monthly publications of interest include the *Energy Design Update, Environmental Building News, Journal of Light Construction*, *Home Energy, Old House Journal*, *This Old House*, *Remodeling*, and *Traditional Builder*.

Helpful information is also accessible via the Internet. Most equipment manufacturers and monthly magazines have Web sites where specific product information and past articles can be retrieved.

This book is divided into nine chapters. Each one is devoted to distinct elements of the house, and within each is a range of issues that is common to that element of home rehabilitation work. Chapter 8, HVAC and Plumbing, for example, covers topics from new piping materials for the repair of an existing plumbing system to the criteria for selection of an entirely new central heating system. Each chapter addresses a wide range of techniques, materials, and tools, and recommendations based on regional differences around the country. Throughout *Home Rehab Handbook*, special attention is given to issues related to energy efficiency, accessible design, and sustainability.

Home Rehab Handbook is written and presented in a format intended for easy use. Drawings, photos, and other graphic materials supplement written descriptions of a broad range of items: state-of-the-art and innovative building technology, products, materials, components, construction and management techniques, tools, equipment—virtually any and all items that make housing rehabilitation more efficient in terms of cost and time. While the content focuses on present technologies and techniques that are currently part of the house-building industry, *Home Rehab Handbook* also includes information on materials, products, and procedures from other construction sectors (such as commercial, industrial, and institutional) that are relevant to housing rehabilitation.

The information within each chapter is organized in different sections according to rehab subjects, and under headings that make this book easy to understand. "Essential Knowledge" gives the reader a basic overview of the important issues related to the section heading. Next, "Techniques, Materials, Tools" presents state-of-the-art and innovative approaches to accomplishing the work. Each entry is explained in detail, including its advantages and disadvantages. This makes it easy for readers to compare approaches and choose the one that is most applicable to their particular project. By design, the "Techniques, Materials, Tools" section is an overview, not a detailed description of implementation. The section on "Further Reading," at the end of the book, lists the valuable resources relevant to the chapter subjects that readers can go to for more detailed information. "Product Information" provides names and addresses of manufacturers of products, materials, systems, and components mentioned in the text so that more information can be obtained. By virtue of their being listed here, such products are not necessarily recommended; their existence and availability are being brought to the reader's attention. New products should be carefully evaluated in the field with regard to their performance. The product lists are not necessarily comprehensive, and we encourage readers to bring new materials and products to our attention to be included in later editions of *Home Rehab Handbook.* Also included is a "Home Rehab Techniques Checklist" that lists the various approaches to rehab work as explained in the book. The checklist can be used as a handy tool in considering all the options in completing a rehab project. The section on "Professional Organizations" lists trade groups, manufacturer representatives, and research centers where readers can obtain additional information.

Research for this publication was originally conducted for the U.S. Department of Housing and Urban Development's Office of Policy Development and Research, which produced a series of guidebooks on the subject. HUD's David Engel, William Freeborne, and Nelson Carbonell provided valuable guidance. Staff members of Steven Winter Associates, Inc. who contributed to this book include Jeff Bellows, William Bobenhausen, Harold Bravo, Christine Bruncati, Donald Clem, Catherine Coombs, Michael J. Crosbie, Christopher Demeter, Masaki Furukawa, Jeff A. Goldberg, Harold Grice, Dianne Griffiths, Alexander Grinnell, Deneé Hayes, Pawan Kumar, Debra Lombard, Roque Rey, Paul Romano, Peter A. Stratton, Gordon Tully, Adrian Tuluca, Christoph Weigel, Steven Winter, and William Zoeller. Mary Jo Peterson and Terry McBride of Mary Jo Peterson, Inc., contributed to Chapter 6. Michele L. Trombley assisted in the preparation of the manuscript.

Michael J. Crosbie
Editor

CH. 1

HOME REHAB HANDBOOK

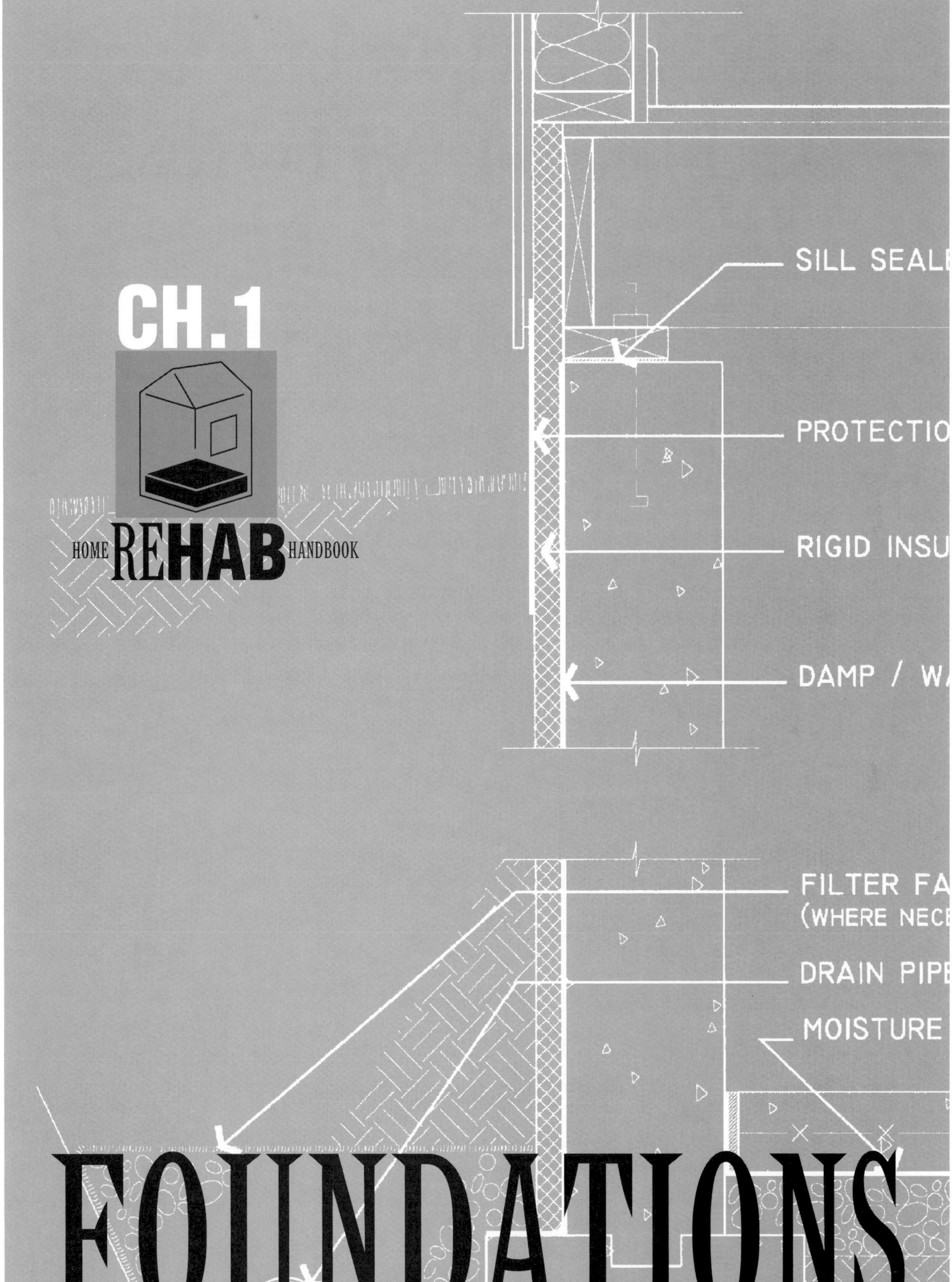

FOUNDATIONS

SILL SEALE

PROTECTIO

RIGID INSU

DAMP / W

FILTER FA
(WHERE NECE

DRAIN PIPE

MOISTURE

Chapter 1
FOUNDATIONS

1.1. FOUNDATION DESIGN AND ENGINEERING
1.2. PERMANENT WOOD AND PREFABRICATED FOUNDATIONS
1.3. DRAINAGE
1.4. DAMPPROOFING AND WATERPROOFING
1.5. INSULATION
1.6. VENTILATION
1.7. SHORING, UNDERPINNING, AND REPAIR
1.8. CRACK REPAIR, COATINGS, AND FINISHES

1.1 FOUNDATION DESIGN AND ENGINEERING

EXISTING FOUNDATION OVERVIEW

The great majority of residential foundations are made of either concrete or concrete block. Concrete dominates in most areas of the United States, especially in the south, southwest, and west. Concrete block is more common in the industrialized states of the northeast and the north central United States.

Foundations of older houses built more than 60 years ago, particularly those in rural areas, are often made of stone. Within the last 15 years prefabricated foundations of pressure-treated wood or concrete have become more popular. Recent foundation developments have included various combinations of expanded polystyrene (EPS) or extruded polystyrene (XPS) and rigid sheathing materials as well as systems that combine concrete and polystyrene forms, commonly referred to as insulating concrete form systems. Such systems are not commonly used in rehab work and thus are not included in *Home Rehab Handbook*. Detailed information on the different types of insulating forms is available from the Insulating Concrete Forms Association.

Building codes typically require that new construction, where it replaces existing construction, be installed in accordance with current codes. Because a variety of codes—regional, state, and municipal—are in force throughout the United States and these codes are now in the process of being revised and consolidated into a single code, it is incumbent upon the rehab architect and contractor to carefully research and understand the pertinent code issues as they relate to foundation design and construction.

Building codes typically referencing American Concrete Institute (ACI) standards set minimum requirements for the size, strength, composition, reinforcement, and installation of concrete and concrete block foundation walls and footings. Local building inspectors responding to construction problems they have encountered, as well as to the prevailing practice in their areas, will sometimes require thicker or more heavily reinforced walls than those required by code minimums. However, if engineering calculations prove otherwise, thinner walls or less reinforcement may be allowed. Production builders in many areas of the country have historically preferred to build thicker concrete and concrete block foundation walls with as little reinforcement as possible. This means that the footings and foundation walls of many existing houses are often unreinforced, contributing, in some instances, to settlement and cracking problems. When replacing concrete walls, it is important to analyze the soil conditions carefully and to comply with local code requirements, adding reinforcement to compensate for poor drainage soil or subsurface water conditions.

The design of concrete foundation walls is well covered by building codes and the ACI standards referenced in the codes, as well as in publications by the Portland Cement Association and the

National Association of Home Builders. The design of concrete block foundation walls is likewise covered by building codes, the ACI standards referenced in the codes, as well as by publications by the Concrete Masonry Association (NCMA).

BASEMENT FLOORS

ESSENTIAL KNOWLEDGE

Most houses with basements built in the past century have concrete basement floor slabs, typically 4" thick. A number of these floors may be in poor repair due to deterioration, heaving from expansive soils, or settlement due to poor initial soil compaction or water-related soil displacement. In the event of severe cracking, spalling, or other distress, it may be advisable to replace sections of, or the entire, floor slab. If the floor slab is in poor condition but retains substantial structural integrity, a new unbonded floor can be poured over the existing one. If the existing floor is structurally sound, but uneven or moderately deteriorated, it is possible to pour a thin, self-bonding topping that can provide a smooth finish that is suitable as a finished surface or an underlayment for floor coverings such as tile or carpet.

TECHNIQUES, MATERIALS, TOOLS

1. PROVIDE A NEW FLOOR SLAB OR REPLACE A PORTION OF THE EXISTING SLAB.

If the existing basement floor is earth, it must be excavated to a point no deeper than the bottom of the wall footing. The subgrade should be prepared to provide uniform support. Slabs supported by expansive soils should be designed to withstand or accommodate swelling and shrinkage of the subgrade. A minimum of 4" of gravel, crushed stone, or coarse sand should be placed under the slab. Where possible, this granular base should cover the top of the footing by several inches so that the slab can settle somewhat without being restrained by the footing—which could lead to cracking. Moisture retarders, usually polyethylene sheeting, are typically required by code or local conditions in areas with poor draining soils and high water tables, to help prevent moisture migration through the slab. (The term *moisture retarder* is used instead of *vapor barrier* for on-grade or underslab conditions because this layer inhibits the migration of groundwater as well as vapor. True "barriers" are rarely achieved in conventional construction; thus the term *retarder* is used.) The ACI, however, recommends eliminating the moisture retarder where local ground conditions and codes permit because the polyethylene sheet retards the curing of the bottom surface of the concrete and can aggravate slab edge curling, drying, and plastic shrinkage-cracking problems. More guidance can be found in the reference documents listed in Further Reading.

ADVANTAGES: This repair will provide a usable, finished basement space.

DISADVANTAGES: Excavation and pouring a of new slab is potentially costly and difficult to undertake without sufficient access.

2. POUR A NEW FULLY BONDED FLOOR SLAB OVER THE EXISTING SLAB.

If the surface of the existing slab is clean, sound, and of good quality but needs to be leveled to serve as a base for a finished floor, a 1"- to -2" thick (or thicker) overlay topping can be poured on the existing slab. Conventional, low-slump concrete with a high sand content and small aggregate (maximum size 3/8"), with or without the use of latex admixtures, can be used. In addition, fast-drying, specially formulated proprietary portland cement-based thin toppings and underlayments are available. These products are often referred to as self-leveling and are primarily used to provide a level floor surface and to repair floors that have deteriorated or spalled. Toppings provide a finished floor surface; underlayments require a floor covering material such as tile or carpet. Gypsum self-leveling toppings work satisfactorily in the absence of moisture and are not normally specified for basement slabs where there may be moisture problems.

ADVANTAGES: This is a relatively inexpensive repair that can contribute significantly to the appearance and use of a basement. Self-leveling mixtures are typically rapid-setting and designed to reduce shrinkage and cracking.

DISADVANTAGES: Self-leveling mixtures cannot be used where there is a possibility of a significant water problem. In general, cracks in the base slab can be expected to be transmitted through the new slab. This repair will raise the floor height and reduce headroom.

3. PROVIDE A NEW UNBONDED FLOOR SLAB OVER THE EXISTING SLAB.

When the existing floor slab is not in suitable condition for the application of a bonded overlay, a new unbonded 4" slab can be applied. The existing slab should be swept clean, and badly worn areas or holes should be filled with a cement-sand mortar to provide a reasonably flat base. A layer of polyethylene sheeting serves as a bond-breaker as well as a moisture retarder. Nonstructural welded wire reinforcement is typically recommended to help distribute shrinkage stresses and to minimize the size of cracks (reinforcement should be placed at the slab center and have sufficient topping to prevent spalling). Polyethylene or nylon fiber reinforcement can also be added to the concrete to help control (but not eliminate) cracks. In areas with extremely expansive soils or an exceptionally high water table, an additional slab reinforcement may be advisable.

ADVANTAGES: This is a permanent fix that can add value to the house without the need to remove the existing slab. A new unbonded floor will act independently of the existing slab and may prevent cracks from reappearing on the new surface.

DISADVANTAGES: A new unbonded floor is more expensive than a topping; will raise the floor height; and may require significant time to cure, due to use of nonabsorbtive polyethylene sheeting under the slab.

CRAWL SPACE FLOORS

ESSENTIAL KNOWLEDGE

Many crawl spaces in older houses do not have permanent, hard-surface floors. Typically the ground has been left exposed, frequently resulting in excessive moisture, odors, vermin, and insects. It is difficult and costly to place conventional concrete in the confined areas of existing crawl spaces, and because the grade of the crawl space may slope, conventional concrete mixtures will not provide uniform coverage. New concrete mixtures using lightweight aggregates have been introduced specifically for this use.

TECHNIQUES, MATERIALS, TOOLS

POUR A CRAWL SPACE FLOOR SLAB OF LIGHTWEIGHT CONCRETE.

A specialty concrete product using Zonolite or Vermiculite aggregate has been developed by Neutocrete that can be pumped from lightweight mobile mixing equipment and quickly installed. The fast-drying mixture with a consistency of heavy shaving cream will adhere to almost any surface, including walls. Trowel-finished to an average depth of 3" over a polyethylene moisture retarder, this material is cured for light traffic in 7 days.

ADVANTAGES: Lightweight concrete can conform to sloping and irregular surfaces, is less expensive and disruptive than conventionally pumped concrete, can be pumped through narrow openings and hard-to-reach areas, dries fast, and reduces moisture and soil gas problems.

DISADVANTAGES: Lightweight concrete cannot sustain heavy traffic or heavy objects.

1.2 PERMANENT WOOD AND PREFABRICATED FOUNDATIONS

PERMANENT WOOD FOUNDATION SYSTEMS

ESSENTIAL KNOWLEDGE

Permanent wood foundations have been used in over 200,000 homes over the past quarter-century and offer an alternative to concrete or masonry systems. In rehab work, permanent wood foundations might be suitable if a large section of an existing foundation wall is damaged beyond repair and must be replaced. They should also be used when replacing damaged portions of an existing permanent wood foundation. The basic elements of permanent wood foundations include a 2x footing plate resting on crushed stone, on top of which a 2x4, 2x6, or 2x8 (depending on number of stories, stud spacing, and backfill height) stud wall is constructed and sheathed with plywood. All wood must be pressure-treated with either ammoniacal copper arsenate (ACA), ammoniacal copper zinc arsenate (ACZA), or chromated copper arsenate (CCA). All fasteners should be made of galvanized or stainless steel, and the exterior of the plywood sheathing should be covered with 6-mil polyethylene sheeting for drainage.

TECHNIQUES, MATERIALS, TOOLS

REPLACE PORTION OF EXISTING FOUNDATION WITH PERMANENT WOOD FOUNDATION SYSTEM.

Once the source of damage has been corrected, the damaged part of the existing foundation wall should be removed so that the new foundation wall can mate with the existing wall. The ground must be excavated to the level of the existing footings. The new foundation wall can be supported on treated wood footing plates and crushed stone footings (Fig. 1.2.1) or can rest on the existing footings, if feasible. According to guidelines developed by the Southern Pine Council, the new permanent wood foundation sections are connected to the existing foundation using lead expansion shields with 1/2"-diameter galvanized lag screws for concrete or masonry foundations. If mating the section to an existing permanent wood foundation, you may need additional studs in the older section to fasten the new portion with lag screws.

ADVANTAGES: Work can be conducted in cold weather that limits concrete or masonry construction, giving more flexibility to the construction schedule. Permanent wood foundations can be installed without the use of heavy equipment and in areas of the existing building that have limited access and are approved by all model code agencies.

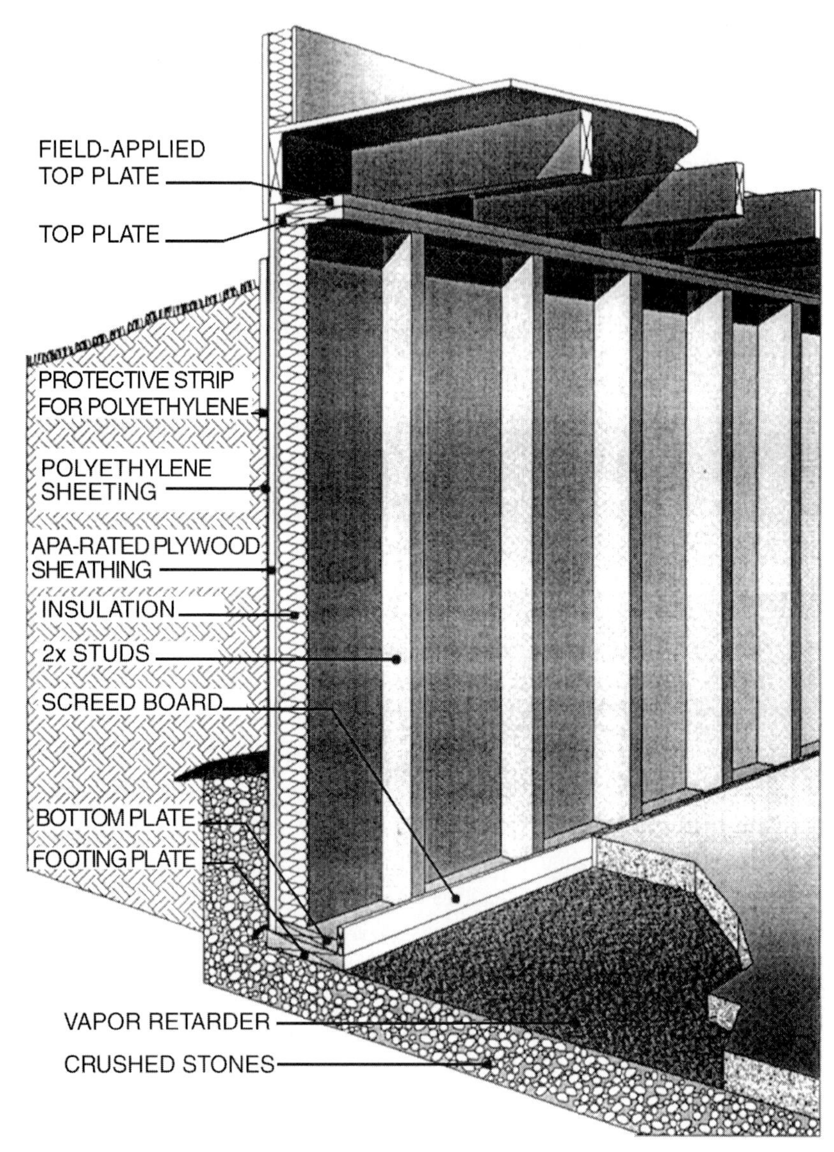

FIGURE 1.2.1 PERMANENT WOOD FOUNDATION

DISADVANTAGES: Significant excavation may be required, depending on the extent of foundation wall to be replaced.

PREFABRICATED FOUNDATIONS

ESSENTIAL KNOWLEDGE

The two major types of prefabricated foundation systems are structural insulated panels (SIPs) and precast concrete panel systems. While these systems are used primarily for new house construction, they may be used to replace damaged sections of existing foundation walls. The SIP foundation systems are virtually the same in detail as the panels used for walls and roofs and offer an advantage over precast systems in that the components can be easily installed by one or two people without heavy equipment. Precast concrete panels must be craned into place. As is the case for permanent wood foundation systems, both SIPs and precast panel systems can be installed in cold weather, thus permitting flexibility in the project schedule.

TECHNIQUES, MATERIALS, TOOLS

1. REPLACE DAMAGED FOUNDATION SECTIONS WITH A SIP FOUNDATION SYSTEM.

In concept, a SIP foundation system is identical to a permanent wood foundation system (Fig. 1.2.2). SIPs are commonly 4'x 8' or 4'x 9' EPS foam core panels with 1/2" pressure-preservative treated plywood exterior sheathing and 7/16" plywood or oriented strand board (OSB) interior sheathing. The interior sheathing is normally required to be covered with a fire barrier such as gypsum board, although some recently developed proprietary coating systems such as AFM Corporation's Firefinish™ meet the requirement of a 15-minute thermal barrier. The ground must be excavated to the level of the

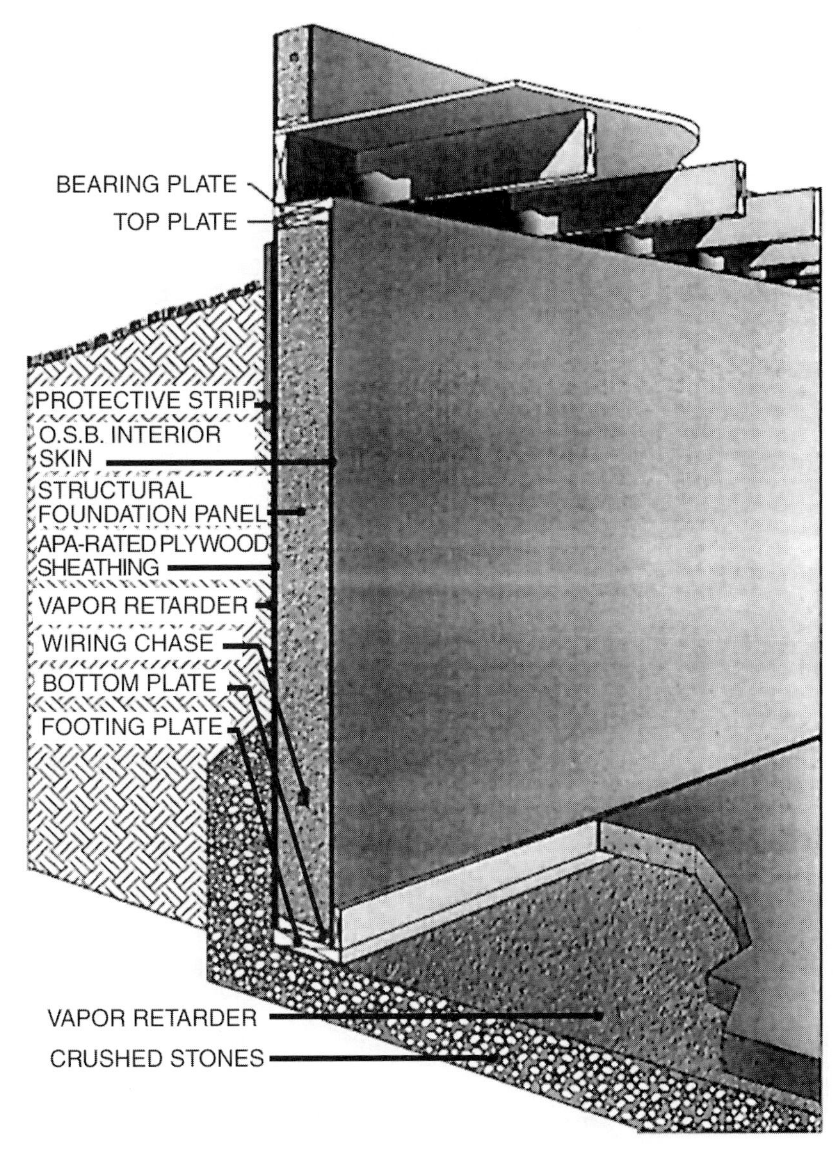

FIGURE 1.2.2 STRUCTURAL INSULATED PANEL FOUNDATION WALL

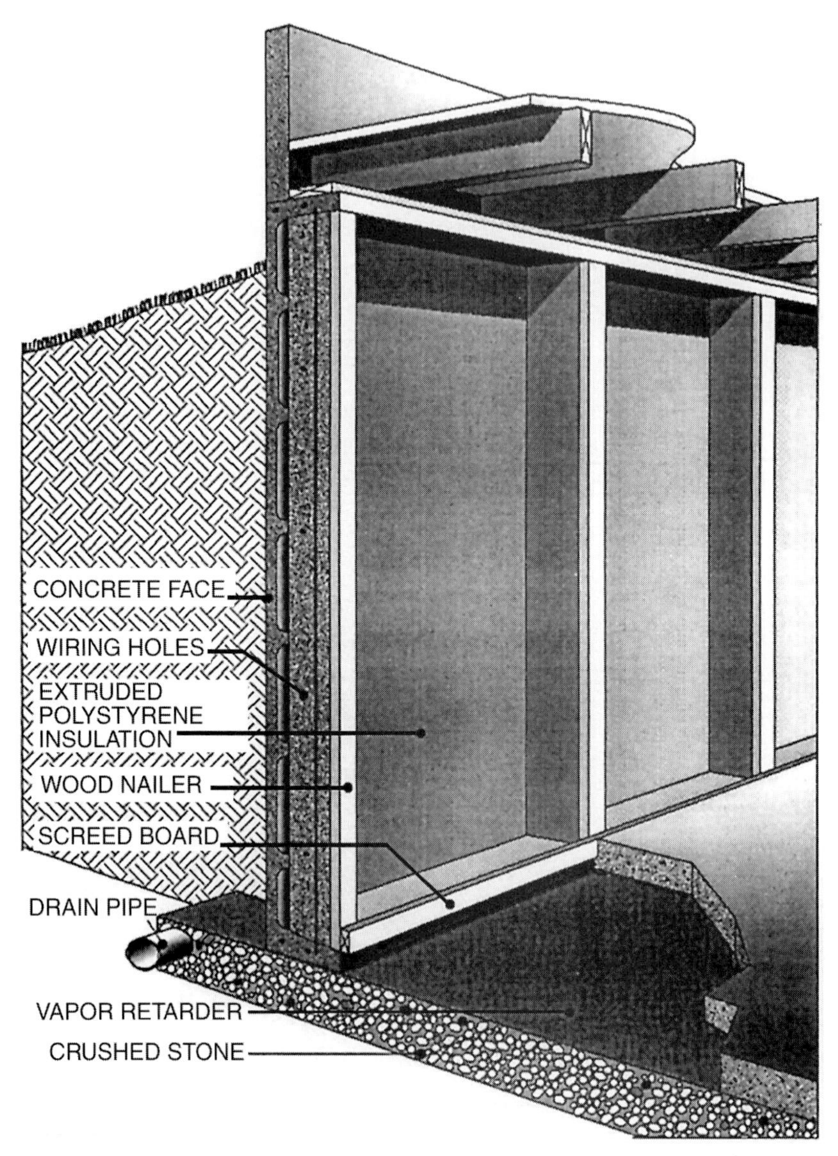

FIGURE 1.2.3 PRECAST CONCRETE WALL

existing footings. The SIPs are prefabricated and arrive at the site ready for installation on a 2x footing plate over a crushed stone, gravel, or concrete footing. The SIPs mate with the existing foundation wall with a 2x splice that is fastened with lag screws into the existing concrete or masonry wall.

ADVANTAGES: Work can be conducted in cold weather that limits concrete or masonry construction, giving more flexibility to the construction schedule. SIP foundations can be built without heavy equipment and in areas of the existing building that have limited access. SIPs deliver good thermal performance.

DISADVANTAGES: Standard panel sizes may not accommodate the portion of the wall to be replaced. Siginificant excavation may be required depending on the extent of foundation wall to be replaced.

2. REPLACE DAMAGED FOUNDATION SECTIONS WITH PRECAST CONCRETE PANEL SYSTEM.

Precast concrete foundation walls, made to order, can be craned into place after the affected section of the existing foundation has been excavated and the damaged portion of the wall is removed (Fig.

1.2.3). Precast sections are bolted together and may contain holes or notches in the ribs for plumbing or electrical conduits. Superior Walls of America and Kistner Concrete Products, Inc., offer a precast stud wall system that uses lightweight concrete to make the handling of the sections easier, but a crane is necessary to place them. The panels are made with 5000 pounds per square inch (psi) concrete, and the manufacturers state that they usually do not need waterproofing, except at the joints. Insulated concrete wall systems, which are placed on gravel or crushed stone footings, incorporate varying thicknesses of polystyrene insulation with an R-5 per inch rating. Additional insulation can be placed in the cavity between the studs. The inside edge of the concrete studs, which are 24" on-center, have factory-installed nailers to accept a finish material. The studs also have predrilled holes for plumbing and electrical conduit. The panels can be bolted to the edge of existing foundation walls.

ADVANTAGES: A concrete foundation wall can be replaced in cold weather; systems include insulation and furring for finished walls.

DISADVANTAGES: Installation requires heavy machinery, such as a crane, and significant excavation, which may not be feasible in affected foundation wall areas with limited access.

1.3 DRAINAGE

SURFACE AND SUBSURFACE DRAINAGE

ESSENTIAL KNOWLEDGE

Poor surface and subsurface drainage can lead to ponding of water around the house, leakage of groundwater through the basement or crawl space walls, and structural damage to the foundation from the buildup of hydrostatic pressure and the freeze-thaw action of water on the foundation system. Successful drainage requires surface water to be led away from buildings by appropriate grading. The water can be dispersed slowly over the landscape or led offsite through underground gravel drainage ways or piping. Surface runoff is usually not a problem in low-density developments (one to two houses per acre) with porous soils and vegetation that allows water to percolate into the ground. Higher densities, hard surfaces such as roofs and pavement, and poor soil conditions necessitate drainage systems. Successful drainage requires that houses be built above surrounding groundwater tables and be protected from groundwater migrating through the soils adjacent to the foundation.

TECHNIQUES, MATERIALS, TOOLS

The most successful techniques for improving surface and subsurface drainage around foundations include:

1. GRADE AWAY FROM THE HOUSE.
Ground around the foundation should slope away a minimum of 10% for a distance of 8 to 12 ft (codes variously state minimums of 4% to 8.3% for 6 to 8 ft, which is not sufficient, according to many experts).

ADVANTAGES: Grading is easily monitored and maintained, allows for the filtration of water-borne pollutants from the land; recharges groundwater tables and aquifers; and is a natural-appearing technique that requires little or no maintenance.

DISADVANTAGES: Depending on natural slope, site, and soil conditions, achieving natural runoff may be expensive or not possible. Grading may be costly to achieve depending on the existing slope and character of the building's perimeter.

2. PROVIDE A "GROUND ROOF" AROUND THE PERIMETER OF THE HOUSE.
This entails the placement of an impervious layer of clay or bentonite under top soil adjacent to the foundation to act as a "ground roof." This layer directs water away from the foundation. Sod in roll form can act as a relatively impervious layer as it is grown on clay soil.

ADVANTAGES: This remedy has a natural appearance and promotes good drainage.

DISADVANTAGES: This remedy may be difficult and costly to achieve depending on the slope around the house and the amount of earth-moving necessary.

3. CREATE SWALES TO CHANNEL WATER AWAY FROM FOUNDATION.
Furrows can be cut into the existing slope to lead surface runoff water away from the foundation walls (Fig. 1.3.1)

ADVANTAGES: This is a natural-appearing technique that requires little or no maintenance.

DISADVANTAGES: This technique may be costly to achieve depending on the existing slope and size of the perimeter.

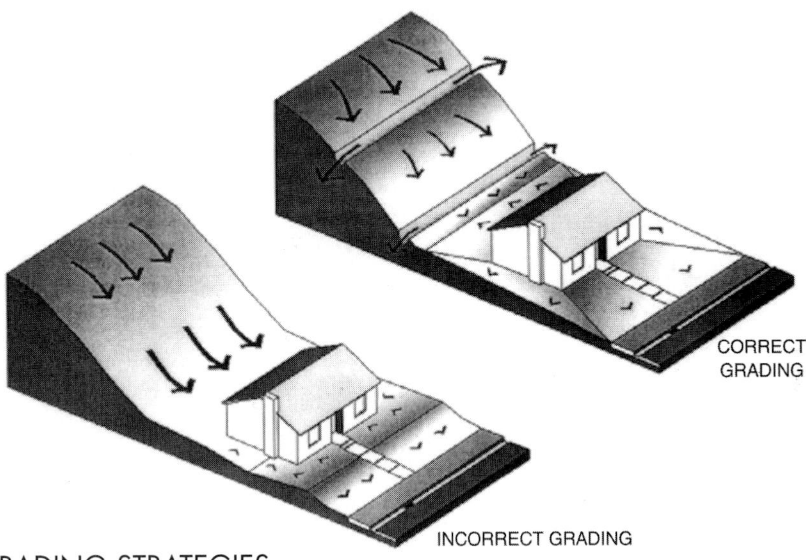

FIGURE 1.3.1 SITE GRADING STRATEGIES

4. TERRACE SLOPE TO REDUCE WATER FLOW.
Where slopes are steep and directed at the foundation wall, they can be terraced to slow and reduce the flow of water (Fig. 1.3.1).

ADVANTAGES: This technique requires little or no maintenance.

DISADVANTAGES: This technique may be costly to achieve depending on the height and character of the existing slope.

5. PROVIDE AND MAINTAIN ROOF GUTTERS AND LEADERS.
Clean gutters and leaders (downspouts) that direct water away from the foundation by means of leader extensions, splash blocks, or underground drain lines is the first and most cost-effective line of defense against water-related problems.

ADVANTAGES: This is a relatively low-cost technique to support good foundation drainage.

DISADVANTAGES: This technique requires diligence in keeping gutters and leaders clean and splash blocks in place.

6. PROVIDE TRENCH OR SOIL STRIP DRAINS.
Trench or strip drains placed between the slope and the foundation can intercept and redirect the flow of water from uphill slopes (Figs. 1.3.2 and 1.3.3).

ADVANTAGES: This type of drain preserves the natural appearance of the slope.

DISADVANTAGES: These drains can be costly depending on their size and need to be monitored and maintained to prevent silting and clogging.

FIGURE 1.3.2 TRENCH DRAIN

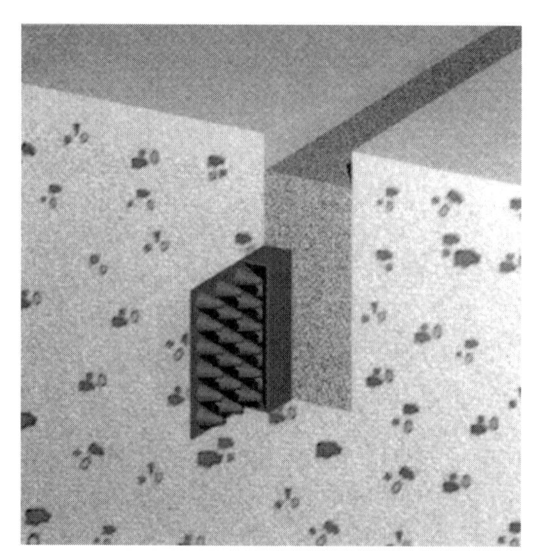

FIGURE 1.3.3 STRIP DRAIN

7. PROVIDE GOOD DRAINAGE UNDER BASEMENT WINDOW AREAWAYS.

Particularly at low points around the foundation, areaways can fill up with plant material, debris, and water, causing leaks around basement windows. Keep them clean and well drained with a gravel pit that extends down to the foundation drain or leads via a drain pipe to a separate gravel (French) drain.

ADVANTAGES: This is a relatively low-cost and effective technique.

DISADVANTAGES: Maintenance is required.

8. REPLACE BACKFILL ADJACENT TO HOUSE WITH FREE-DRAINING MATERIAL.

A possible remedy when backfill against the house is overly compacted and nonporous.

ADVANTAGES: This technique may be cost-effective in some instances, depending on the amount of soil to be replaced.

DISADVANTAGES: This is a disruptive technique requiring removal of shrub and plant material. Existing footing drains may not be adequately sized or functioning properly.

FOUNDATION DRAINAGE

ESSENTIAL KNOWLEDGE

Wet basements can be an indication of drainage problems around foundations. Most codes require that new housing with habitable basements have an approved foundation drainage system (except in locations with well-drained soils). Houses built within the last 40 years probably have drainage systems similar to current techniques but which may have ceased to properly function due to blockage or a breach in the system. Without inspecting a drainage outfall or discharge pipe for flow rate, it is virtually impossible to determine actual conditions.

Corrugated, flexible polyethylene piping has replaced concrete and clay foundation drainage pipe as the industry standard. Some manufacturers, such as Hancor, sell a nonwoven filter fabric "sock" that fits over the drain. This may be satisfactory in some areas, but where there are significant fines in the soil that migrate with water, it is preferable to encase the gravel area around the drain tile, as the greater surface area will extend the life of the filter fabric. Local conditions and codes should be consulted. Considerable interest has been generated in wall drainage systems, called *sheet* or *geocomposite* drains, as they have been successfully used on a large scale in Canada. They are increasingly seen as a cost-effective addition to dampproofing and water proofing techniques in the United States, as they are less expensive than corresponding aggregate drainage material. There are four basic types of sheet drainage systems.

TECHNIQUES, MATERIALS, TOOLS

If site drainage and roof drainage deficiencies have been corrected but problems persist, there are four options to consider.

1. INSTALL AN INTERIOR PERIMETER BASEBOARD "GUTTER" DRAINAGE SYSTEM.

A perimeter baseboard drainage system can be placed on top of the floor slab or be cut into the juncture of the wall and the floor and concreted in place (Fig. 1.3.4). The system picks up water draining from wall cracks, the cores of concrete block walls, and at the floor-wall intersection, and directs it to a sump pump. Such packaged systems include WaterGuard™ by Basement Systems, Inc., and Basement De-Watering (Systems)™. Baseboard systems should be sloped 1/8" to 1/4" per foot to induce flow.

ADVANTAGES: These systems range from relatively simple and inexpensive to moderately expensive. In combination with the sump pump, water penetration through the wall-slab juncture is reduced.

DISADVANTAGES: Although these systems collect leakage, they do not remedy the wall dampness or drainage problems per se, as the walls remain moist and the exposed perimeter drains may be considered unsightly in finished spaces. Mold and mildew can appear on wet walls and in the gutter itself.

2. INSTALL A SUMP PUMP.

A sump pump is used to lower the groundwater table to a point below the basement slab. It is used when water problems persist and exterior drainage systems are not possible or practicable. If soil gases such as radon or methane exist, sump pumps should be covered and vented to the outdoors. An elaboration of the sump pump solution is to install a radial drainage pipe system, such as that offered by Sanford Irrigation, under the slab (Fig. 1.3.5) that directs water to the sump. To ensure sump operation in the event of a power outage, water-driven emergency back up pumps, such as Home Guard from HiLo Industries, provide continuous operation with a minimum of 22 psi pressure from a municipal water system.

ADVANTAGES: This is the least-cost alternative to lowering the water table and reducing water problems in the basement.

DISADVANTAGES: This may not be an adequate solution if a major and continuous groundwater problem exists. Sump pump operation is an added household expense, and may be disrupted by power outages.

3. REPLACE THE BASEMENT SLAB AND INSTALL INTERIOR FOUNDATION DRAINS.

A technique to be used when a sump pump alone is not adequate, and where the ground beneath the floor slab is relatively impervious and greater drainage area is required. The existing slab perimeter must be removed and new drains installed (Fig. 1.3.6), which direct water to a sump pump.

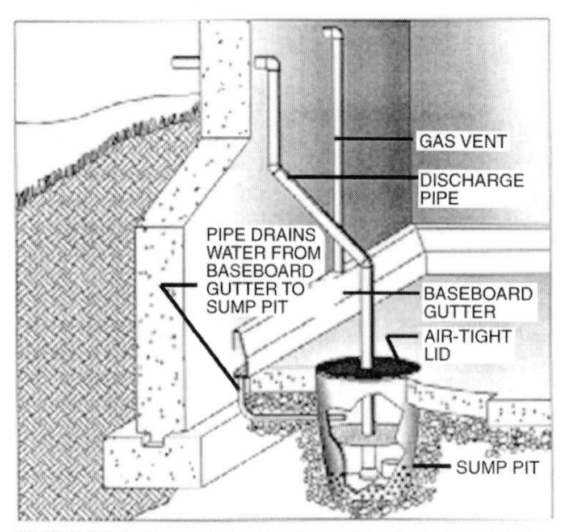

EXPOSED SYSTEM LEADING TO SUMP PUMP

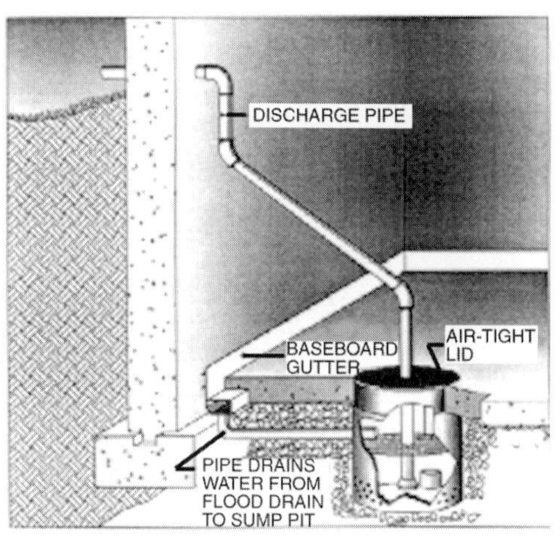

PARTIALLY EXPOSED SYSTEM LEADING TO SUMP PUMP

FIGURE 1.3.4 INTERIOR BASEBOARD DRAINAGE SYSTEMS

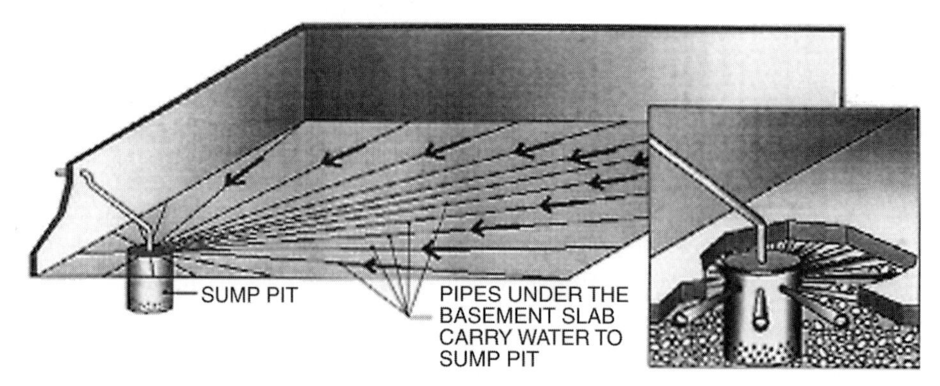

UNDERGROUND BASEMENT DRAINAGE SUMP SYSTEM

FIGURE 1.3.5 INTERIOR RADIAL SUMP SYSTEM WITH SUMP PUMP

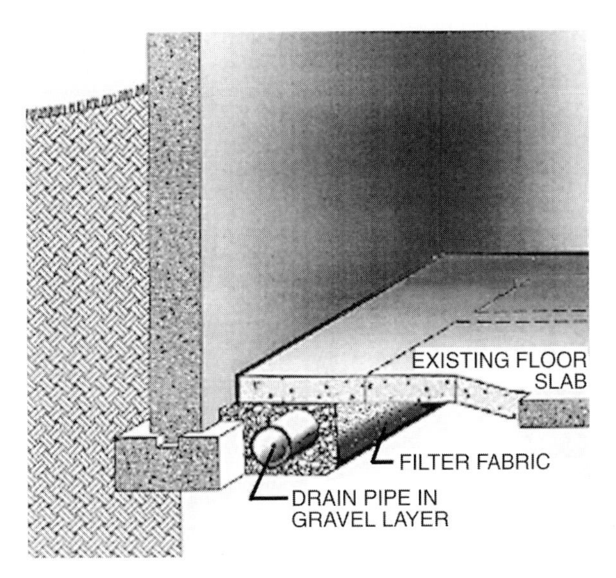

FIGURE 1.3.6 INTERIOR FOUNDATION DRAIN

ADVANTAGES: This technique eliminates the need to remove soil from against the outside of the foundation wall and the need to disturb the foundation.

DISADVANTAGES: This technique entails removal of at least portions of the existing slab and pouring a new slab, which may be difficult and expensive, based on access to the basement space. These systems must be carefully designed and built by experienced professionals and may not work where there is permanently high groundwater.

4. EXCAVATE THE EXTERIOR FOUNDATION WALL AND INSTALL AND REPAIR OR REPLACE THE EXISTING DRAINAGE SYSTEM.

Soil is removed from against the basement foundation wall, and new drainage material is installed next to the wall, which diverts water to new 4"- diameter (or larger) drainage pipes or other approved system at the base of the wall. In poorly draining soils, a sheet drainage material can be used to enhance the movement of water to the foundation drain and reduce hydrostatic pressure against the outside of the wall.

There are four common types of drainage materials.

1. Dimpled polyethylene or polystyrene sheets used in conjunction with a filter fabric to form a continuous drainage channel. The material is installed with the dimples and filter fabric away from the wall to assure good adherence to it. The sheets, typically used with a dampproofing or waterproofing membrane, can also be used for underslab drainage. One manufacturer, American Wick Drain Corp., offers a sheet drain that connects to the foundation drain system (Fig. 1.3.7).

2. Formed polyethylene sheets without filter fabric that orient the dimples toward the wall, providing a continuous 1/4" drainage space (Fig. 1.3.8). This system, widely used in Canada, does not require dampproofing to be applied to the foundation wall, and keeps backfill soil and moisture from resting against the wall. The system is sold under the name of System Platon and is manufactured by the Armtec, Ltd.

3. Matting of various types that forms a wall drainage system to be used with applied waterproofing and dampproofing systems. Products include: WARM-N-DRI, a 6-lb-density fiberglass mat, used in combination with a sprayed TUFF-N-DRI polymer modified asphalt, not only drains the channel water, but also has an R-rating of approximately 4 per inch when dry—offered by Koch Materials Co.; Roxul Drainboard, a mineral wool blanket manufactured by Roxul, Inc.; Enkadrain drainage matting of geotextile fabric heat-bonded to a three-dimensional, high-density polyethylene core, which allows water

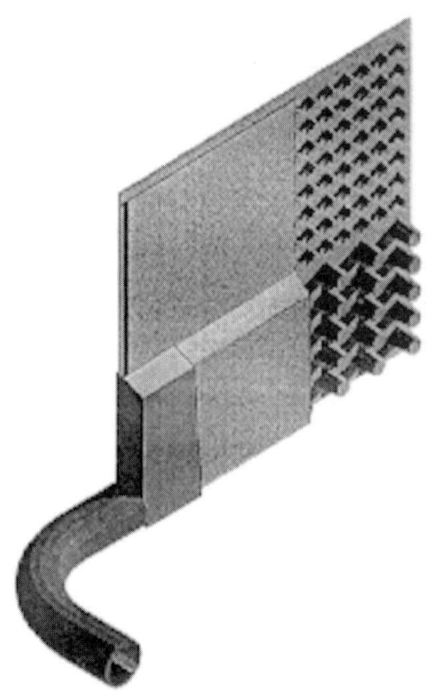

FIGURE 1.3.7 AMERICAN WICK DRAIN

FIGURE 1.3.8 SYSTEM PLATON

to seep into the core while it holds back adjacent soil—manufactured by Akzo Nobel Geosynthetics.

4. Drainage systems using grooved extruded polystyrene boards to enhance drainage against foundation walls. Among these products is Owens Corning's INSUL-DRAIN™, which has a network of drainage channels cut into one side of the board which allows groundwater to drain away from the foundation wall. Dow Chemical's Styrofoam THERMADRY board also has grooves to encourage drainage.

ADVANTAGES: With a properly drained foundation wall, the possibility of leakage through basement walls is reduced significantly and interior wall finishes can be applied with less concern for water damage.

DISADVANTAGES: This work requires removal of the existing plant material and soil, which can be expensive. This solution may be costly, but it is the best strategy for certain drainage problems.

1.4 DAMPPROOFING AND WATERPROOFING

ESSENTIAL KNOWLEDGE

Dampness in a basement can render it virtually useless, even for storage. Sources of moisture in basements include the intrusion of groundwater through cracks in walls and slabs and their junctures, wicking of moisture through concrete by capillary action, and the infiltration of moist air from outside (this last source is discussed in Section 1.6). Dampness can also be a health hazard. Recent studies sponsored by the Canada Mortgage and Housing Corporation (CMHC), Health Canada, and the National Research Council of Canada as well as Harvard University have established that moisture can lead to increases in a number of biological contaminents including mold, dust mites, and bacterial toxins; some of the molds produce toxins that are harmful to humans.

Building codes generally require that all below-grade basement walls be dampproofed or waterproofed. Dampproofing is used in the absence of hydrostatic pressure and is intended to protect the interior surface of foundation walls from water vapor diffusion, caused by temperature and humidity differentials, and the wicking of moisture through the wall by capillary action.

Waterproofing is designed to bridge nonstructural cracks up to a maximum of 1/16". It is required when hydrostatic pressure exists or is likely to occur due to the presence of groundwater at elevations above the basement slab. In such cases, both floors and walls have to be waterproofed and designed to withstand hydrostatic pressure. Waterproofing is usually applied from the top of the footing to not less than 12" above the maximum elevation of the water table. Under some codes, the remainder of the wall can be dampproofed, but it is usually better to continue waterproofing all the way to grade rather than changing materials. Although it is substantially more expensive than dampproofing (sometimes by a factor of 10 or more), waterproofing is increasingly being specified in lieu of dampproofing in mid- to high-end housing because of its static, nonstructural crack-spanning capability and because basements, now more commonly used for living space, are expected to be dry. Materials and applications that come with written performance guarantees are preferred.

TECHNIQUES, MATERIALS, TOOLS

There are a number of techniques and materials used for effective dampproofing and waterproofing. The following alternatives consider dampproofing first, and then waterproofing.

Dampproofing materials for walls include those complying with American Society for Testing and Materials (ASTM) C887, materials permitted for waterproofing, and other approved materials. There are a number of approaches to dampproofing.

1. APPLY CRYSTALLIZATION PRODUCTS OR CEMENTITIOUS COATINGS TO FOUNDATION WALL INTERIOR.

A variety of coatings are sold as interior (referred to as the *negative* side) waterproofing products, including sealers and paints. Some of these products may work in the absence of hydrostatic pressure, but most interior coating systems are not long-term solutions and will fail under sustained pressure and substrate movement. Products such as Bonsal's Sure Coat™ and Thoro Systems' ThoroSeal™ are represented as being able to prevent water intrusion even under pressure. Crystallization products such as Xypex, Vandex, and Permaquick combine proprietary chemicals with sand and cement. This slurry produces crystallization growth in concrete or masonry substrates that fill open pores and is represented as being able to block the passage of water.

ADVANTAGES: These products may protect the interior surface of the wall from water wicking from both the exterior wall surface and the foundation base, can be relatively inexpensive solutions compared to exterior work; are usually easy to apply;. can be applied to damp concrete, allowing water vapor to pass through; and are nontoxic.

DISADVANTAGES: These products are often not effective, especially on active cracks and those greater than 1/64". They do not prevent groundwater from penetrating the concrete masonry unit (CMU) wythe on the exterior side and becoming trapped inside the wall which is a serious problem in cold climates. The finish may not be acceptable and may require paneling or drywall covering. Crystallization products are not effective over control joints.

2. APPLY CEMENTITIOUS COATINGS TO FOUNDATION WALL EXTERIOR.

Commonly known as *parging*, these coatings are primarily used for concrete block walls. They should be combined with acrylic modifiers (made by companies such as Bonsal, Thoro Systems, and Sonneborn).

ADVANTAGES: This system has been widely used for many years and may be cost-effective in locations with well-draining soils. It protects from moisture wicking from outside the wall, does not require completely dry substrate or protection boards, can be combined with polyester reinforcing mesh to enhance performance, and uses nontoxic materials.

DISADVANTAGES: These coating have no crack-spanning capabilities. There have been cracking and spalling problems with some cementitious coatings over time due to water migration behind the substrate and freeze-thaw cycling. Excavation is required.

3. APPLY ASPHALTIC COATINGS WITHOUT MODIFIERS TO FOUNDATION WALL EXTERIOR.

Asphaltic dampproofing products come in a variety of formulations—solvent- or emulsion-based—with or without fibers to give them body. Hot-applied coatings are necessary in cold weather when emulsionbased coatings may freeze. Both hot- and cold-applied coatings have varying consistencies and formulations, allowing them to be troweled, rolled, brushed, or sprayed. Cold-applied coatings are becoming more common and can be applied to slightly damp surfaces. Solvent-based materials cannot be applied until concrete has cured sufficiently to meet the manufacturer's recommendations.

ADVANTAGES: These coatings are everywhere, are economical, and are easy to install with little skill. They are cost-effective when serious water problems do not exist and are used almost exclusively in many regions for starter and low-end housing.

DISADVANTAGES: Materials can lose their limited elasticity due to aging and freeze-thaw cycling. They will not span cracks, are unattractive when exposed, require protection board and excavation.

4. APPLY ACRYLIC OR OTHER APPROVED POLYMER SEALERS TO FOUNDATION WALL EXTERIOR OR SLAB SURFACE.

Some recently developed polymer-based materials are available that have model code approvals for use as dampproofing. One of these products, Poly-Wall™, is an attractive gray, nonasphaltic coating that can be painted to match any color.

ADVANTAGES: These sealers are fast-drying, easy to apply by spraying, and can also be used on the interior of foundation wall to reduce or eliminate moisture migration in the absence of hydrostatic pressure. Exterior use of these sealers will qualify as waterproofing when used in conjunction with rubberized asphalt sheet membrane accessory component.

DISADVANTAGES: Without additional membrane, these sealers will not span cracks. They are solvent-based, require a curing period prior to application of sealer on new concrete, are more expensive than other dampproofing products, are normally used in higher-end housing, and require excavation.

5. APPLY POLYETHYLENE SHEET BELOW FLOOR SLAB.

Polyethylene sheet is generally required by codes to be not less than 6- mil thick, with joints lapped not less than 6" (12" is preferred). It should be sealed with double-sided asphaltic tape such as that manufactured by Reef Industries, Inc., prior to pour to assure continuity of the barrier, especially in clay soils where capillary action is severe. Fiber-reinforced or high-density polyethylene should be considered. These materials cost a little more, but an undamaged barrier is key to the slab's performance. Dampproofing systems that come with written material and performance warranties are preferred.

ADVANTAGES: If properly installed, polethylene sheet provides a continuous, impervious barrier to moisture migrating from below the slab. It can be installed over cracked slabs, with new slab poured on top.

DISADVANTAGES: Care must be taken not to puncture the sheet so it will maintain good performance. There are a number of approaches to exterior (positive side) waterproofing. Systems are usually required to meet ASTM C836, which sets elastomeric characteristics.

1. APPLY AN ASPHALTIC-BASED PRODUCT TO THE FOUNDATION WALL EXTERIOR.

These hot- or cold-applied products are similar to those used for dampproofing, but are used in two or more coats in combination with polyester, fiberglass, or other types of fabric membranes.

ADVANTAGES: This technique has been used extensively in the past and has a proven record. It is still preferred and is economical in some regions where roofing contractors use similar systems.

DISADVANTAGES: Asphaltic-based product is less environmentally friendly than other products, does not bridge cracks or weather as well as some other systems, more labor intensive to install than some of the new one-coat products, and requires backfill protection layer. Hot systems are difficult to apply on vertical surfaces due to the weight of the fabric membrane and viscosity of asphalt.

2. APPLY A RUBBERIZED ASPHALT COATING TO THE FOUNDATION WALL EXTERIOR OR SLAB SURFACE.

These products originated in pipeline protection and commercial building markets, but are increasingly used in residential projects. They are considered higher performing than conventional asphalt products and have become lower in cost in recent years. They are made and distributed nationwide by such manufacturers as W.R. Grace, Sonneborn, and Koch Materials Co.

ADVANTAGES: These products often come with material and performance guarantees. They have excellent elastomeric qualities and can span static, nonstructural cracks. They can be spray-applied or installed with self-adhering rolls that eliminate the need for seam adhesive. Sheet material guarantees consistent thickness, unlike spray applications. Spray applications require less skill.

DISADVANTAGES: These products are more expensive than nonmodified asphalt products. Spray applied solvent-based materials should not be applied over fresh concrete. A backfill protection layer is required.

3. APPLY AN ASPHALT-MODIFIED URETHANE COATING TO THE FOUNDATION WALL EXTERIOR.

Asphalt is used as a filler with urethane to lessen the cost of the material, without sacrificing performance. Products have typically been used in commercial applications but are becoming more common in residential work. They are distributed nationally by Sonneborn and applied by brush or spray.

ADVANTAGES: This coating has good elongation and warranties are available.

DISADVANTAGES: A backfill protection layer is required. This coating is more expensive than nonmodified asphalt coatings.

4. APPLY A URETHANE COATING SYSTEM TO THE FOUNDATION WALL EXTERIOR.

These systems have the highest elastomeric capabilities of fluid-applied membranes. Urethane systems are available in one- or two-component materials. Typically black in color, urethanes are solvent-based, requiring substrates to be completely dry to avoid membrane blistering. They are available nationally through Karnak, Sonneborn, Mameco, and GACO Western.

ADVANTAGES: Fluid-applied systems are self-flashing, allowing seamless covering of complicated joints and protrusions. They have high performance, and are easy to apply. One-coat systems are available. They have good resistance to chemicals and alkaline conditions on masonry substrates and warranties are available.

DISADVANTAGES: The coating is solvent-based. The materials cost can be as much as three times that of the least-expensive waterproofing system, but the coating is economical to apply. Skilled applicators are required.

5. APPLY A RUBBER-BASED COATING TO THE FOUNDATION WALL EXTERIOR.

A number of rubber-derived materials—including neoprene, butyl, and hypalon—are used in high-end waterproofing systems. These materials have excellent elastomeric capabilities, but less than that of urethane. They are resistant to most chemicals likely to be encountered below grade. Sheet materials are also available, including EPDM and butyl systems where significant hydrostatic pressure is evident. Used primarily on horizontal surfaces such as slabs. Most manufacturers have experience in commercial applications, and a few new wall waterproofing products, such as Rub-R-Wall™ and Composeal™, have been developed specifically for residential applications.

ADVANTAGES: This coating has high performance and is resistant to chemicals. Warranties are available.

DISADVANTAGES: This coating can be expensive. It is likely to be cost-effective only in high-end housing or when serious foundation water problems are encountered. Sheet systems are not self-flashing at protrusions and changes in plane. A backfill protection layer and skilled applicator are required. Some products are highly flammable and toxic.

6. APPLY A CLAY-BASED WATERPROOFING SYSTEM TO THE FOUNDATION WALL EXTERIOR.

Natural clay systems, commonly referred to as bentonite, typically contain 85% to 90% clay and a maximum of 15% natural sediments such as volcanic ash. When exposed to water, bentonite swells 10% to 15% above its dry volume and becomes impervious to water. It is available in panels, sheets, and in combination with urethane-, rubber-, and asphalt-based products to enhance performance. It is also available in mat form with a textile backing similar to carpet. It is not widely used in residential construction, but is recommended by some architects and contractors who have used it. A widely distributed product is Volclay Waterproofing, manufactured by American Colloid Co.

ADVANTAGES: This is an excellent waterproofing material, can be applied to recently poured concrete, is not toxic or harmful to the environment, and is useful in underslab waterproofing. Minimal substrate preparation is required.

DISADVANTAGES: These products require careful application. Hydration and swelling must occur in confined space after backfilling for waterproofing properties to be effective. If space is too confined, bentonite can swell with enough force to raise a floor slab or crack concrete. It is subject to premature hydration if not properly protected. Wetting and drying the material will cause it to lose its waterproofing properties. The material should not be installed where free-flowing groundwater occurs that could wash clay away from substrate. These products are not resistant to chemicals.

1.5 INSULATION

INSULATING BELOW-GRADE WALLS

ESSENTIAL KNOWLEDGE

Insulation is an important element in controlling thermal comfort of basement spaces. There are two basic approaches to insulating existing concrete or masonry foundation walls: applying insulation to the exterior or interior of the wall. Research has shown that there is not a significant difference in energy savings.

The recent widespread use of foam plastic insulation below grade has increased awareness that foam plastic in contact with the soil can allow undetected access by insects including termites and carpenter ants through the insulation to cellulose (food) material in the structure above. In response to this problem, the *1977 North Carolina State Building Code*, Vol. III, *Residential*, requires inspection and treatment gaps in exterior foundation insulation. In addition, in construction where wood is used as a structural member, the *1977 Standard Building Code* and the proposed *CABO One and Two-Family Dwelling Code* prohibit the use of foam plastics on the exterior face or under interior or exterior foundation walls or slab foundations located below grade in areas where the probability of termite infestation is "very heavy." A reference map indicating these areas includes most of California, eastern Texas, Louisiana, Mississippi, Alabama, Georgia, Florida, and South Carolina. A CABO exception is "buildings where the structural members of walls, floors, ceilings, and roofs are entirely of non-combustible materials or pressure-preservative-treated wood."

TECHNIQUES, MATERIALS, TOOLS

1. APPLY EXTERIOR INSULATION.

Covering the upper half only is cost-effective in regions with low heating degree day (HDD) requirements (below 2,000 HDD or in areas with up to 2,400 HDD and low fuel costs). Covering the entire wall (Fig. 1.5.1) is cost-effective at varying levels of insulation from R-10 to R-15 depending on regional fuel costs.

ADVANTAGES: Exterior applications can provide continuous insulation with no thermal bridges; protect and maintain waterproofing and dampproofing at moderate temperatures, thereby extending its life; reduce or eliminate interior moisture condensation problems, since the wall is warm; do not affect interior finishes or reduce usable floor area; will allow inspection of interior walls, sills, and rim joists for insect infestation and mold.

DISADVANTAGES: Some codes now prohibit the use of or require that the insulation be discontinuous so that insect pathways can be detected. Excavation of the foundation wall is required for installation.

2. INTERIOR INSULATION COVERING THE ENTIRE WALL FROM FLOOR TO CEILING.

Insulation is applied to the interior wall surface, either as rigid insulation board in combination with furring strips to receive drywall or paneling and accommodate electrical wiring, or in batt form between furring studs. Dow Chemical and Owens Corning have introduced insulation panels with slotted grooves to accept wood furring strips.

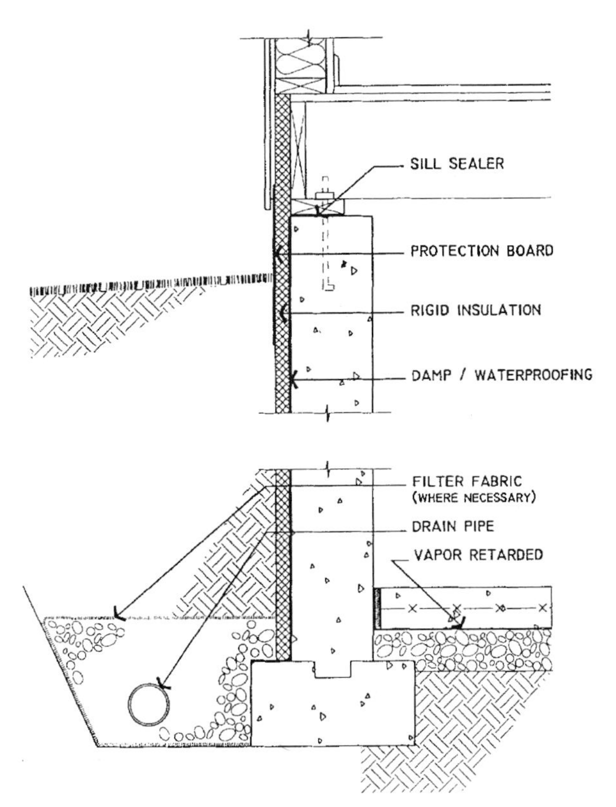

FIGURE 1.5.1 CONCRETE BASEMENT WALL WITH EXTERIOR INSULATION

ADVANTAGES: Interior applied insulation (Fig. 1.5.2) is easier to install than exterior applications and can be fairly inexpensive to install in batt form.

DISADVANTAGES: Interior surfaces of basement walls should be dry, as application of insulation can inhibit detection of insects and mold within furred wall cavity. Most rigid foam insulation requires a 15-minute fire-resistant thermal barrier (usually drywall) to slow the spread of fire to the insulation.

XPS board has been the material of choice for the exterior insulation of foundation walls due to its superior compressive strength, impermeability, and durability compared to EPS, mineral wool, and polyisocyanurate products. XPS is made by Owens Corning, Tenneco Building Products, and Dow Chemical with a nominal density of 2 $lb/ft.^3$ EPS, however, is also promoted in some areas, due to its lower cost. One manufacturer, AFM Corp., has a product called Perform Guard that is treated with borate to discourage termite and carpenter ant infestation. Some EPS manufacturers suggest using 1.5- or 2.0-lb-density board on exterior applications in lieu of 1-lb-density board commonly used for interior and roof insulation, since at higher densities EPS is more durable and easily handled. Mineral-fiber or fiberglass boards, such as those used in Koch Industries' TUFF-N-DRY system, have performed satisfactorily. EPS and polyisocyanurates can be used for interior applications, along with insulating batts. Polyisocyanurate boards easily absorb water and are damaged with freeze-thaw cycling, so their use should be avoided where water is a problem.

Relatively new developments include vertically grooved extruded polystyrene boards to enhance drainage against foundation walls. For example, Owens Corning's INSUL-DRAIN™ has a network of drainage channels cut into one side of the board which allows groundwater to drain away from the foundation wall. Tongue and-groove edges permit easier board alignment and help seal joints between the boards. Dow chemical's Styrofoam THERMADRY™ is also made with drainage channels (see Chapter 4).

As mentioned earlier, foam board insulation provides a potential pathway for insect (particularly termite) infiltration, into the house. One way to detect this, according to a technical bulletin by LiteForm International, is to provide a "vision strip" by removing a continuous strip of insulation (exposing the concrete) around the entire perimeter of the foundation (Fig. 1.5.3) prior to covering

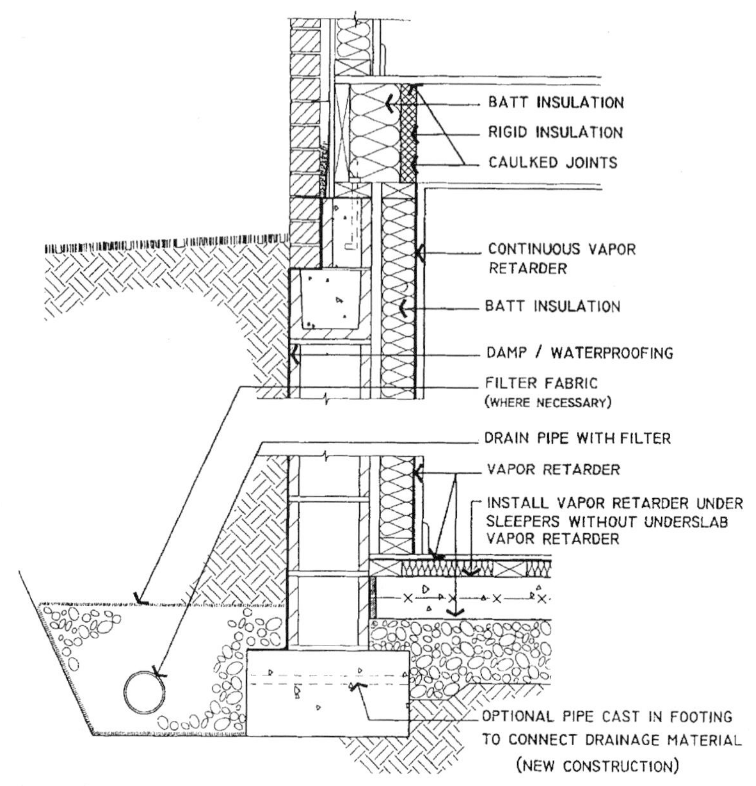

FIGURE 1.5.2 CONCRETE MASONRY BASEMENT WALL WITH INTERIOR INSULATION AND WOOD FLOOR ON SLEEPERS

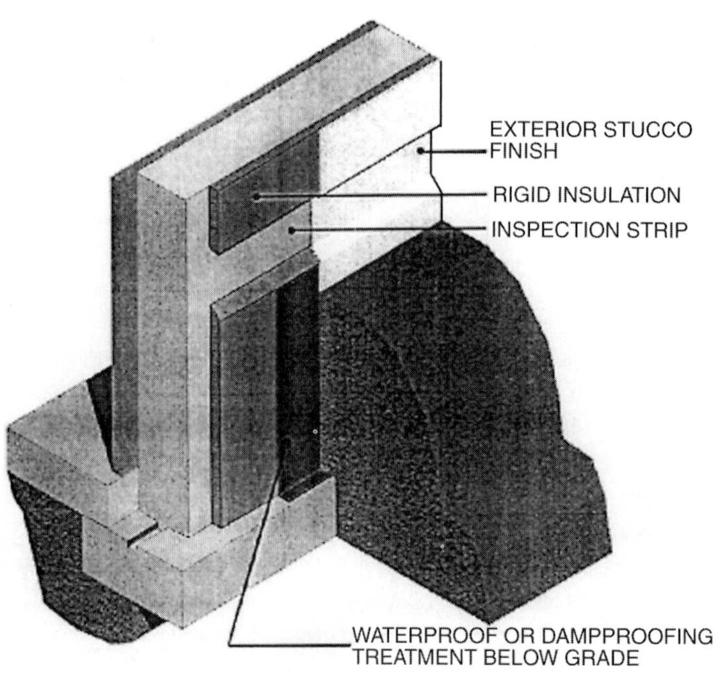

FIGURE 1.5.3 INSULATION INSPECTION STRIP

or finishing the walls. The strips should be approximately 12" above grade level and approximately 8" wide. The bottom edge of the cut should be trimmed so that it slopes away from the wall at 45° to shed moisture. It should be noted that this breach in the insulation seriously reduces the overall R-value of the system, and may cause condensation to form on the interior of the wall. The bulletin also recommends a 20- to 22-gauge continuous metal flashing strip (termite shield) be placed at the top of the foundation wall before the wood sill plate is attached.

The flashing should overlap the plate by 1" and turn down to shed moisture. Termite shields have long been recommended as a way of stopping termite tunnels. Recently, their use has decreased because they interrupt exterior finishes and complicate the overlap of sheathing and siding at the juncture of the foundation. When they are used, termite shields are often poorly installed, damaged, or bent in such a way that they do not provide a continuous barrier.

Some code agencies have created restrictions against the use of foam products below grade. As this requirement continues to be researched, it is important to contact local officials for current use requirements. In addition, most foam product manufacturers are researching the use of insect-inhibiting chemical additives in their products. Codes require the protection of insulation board on foundation walls above grade level. DFI Pultruded Composites, Inc., among others, makes a rigid, 1/16"-thick fiberglass panel that covers the insulation. The panels come in 12" or 24" widths and have a 1" lip that goes over the top of the insulation board.

INSULATING CRAWL SPACES

ESSENTIAL KNOWLEDGE

Crawl spaces are insulated to protect plumbing and ducting that may run through the space and to mitigate the effect of temperature swings on the living spaces above. Crawl spaces are either vented or unvented, and insulation strategies vary in each case.

TECHNIQUES, MATERIALS, TOOLS

There are two basic approaches to insulating crawl spaces.

1. INSULATE VENTED CRAWL SPACES.

This approach places batt insulation between the joists, in the underside of the first floor, to protect the living space.

ADVANTAGES: This is often the least costly method of crawl space insulation, and is accomplished with mineral wool or fiberglass batts. It eliminates the need to insulate crawl space walls (required in unvented spaces) and protects pipes and ducts that run in the plenum space between floor joists.

DISADVANTAGES: This insulation may be difficult to install if there is limited access to the crawl space. It also may be difficult to seal the underside of the floor adequately with a vapor barrier due to multiple penetrations. Ducts and pipes below the floor must be insulated.

2. INSULATE UNVENTED CRAWL SPACES.

Preferred by many researchers, this strategy of treating the crawl space as a shallow basement permits the crawl space to act as an insulated and conditioned plenum space through which pipes and ducts can be run without freezing. It also reduces or eliminates odors, insects, pests, dirt, and debris.

Crawl space walls can be insulated from inside or outside in fashions similar to that for basement walls (Figs. 1.5.1, 1.5.2, 1.5.4, 1.5.5, 1.5.6).

ADVANTAGES: Exterior rigid insulation on crawl space walls can be relatively easy to apply to the foundation surface and may be the only insulation alternative where crawl spaces are inaccessible; reduces moisture condensation problems; and allows inspection of crawl space interior walls (if accessible), sills, and rim joists for insect infestation. Interior insulation in the form of boards or batts can be applied without excavation; can be adhered with mastic to the foundation wall or attached to the top of the wall and draped down.

DISADVANTAGES: Insulating unvented crawl spaces can be more costly than insulating a comparable vented crawl space. It can be difficult to install interior insulation in crawl spaces with limited access; it is harder to inspect insulation-covered walls for insect infestation.

Insulation materials and techniques were discussed earlier in this section.

INSULATING SLABS

ESSENTIAL KNOWLEDGE

There are two basic types of slab-on-grade foundations: those with conventional footings and concrete or masonry stem walls (Fig. 1.5.7) and those with shallow perimeter grade beams (Figs. 1.5.8 and 1.5.9). In rehab work, insulation can be added only to the exterior without disturbing the slab floor.

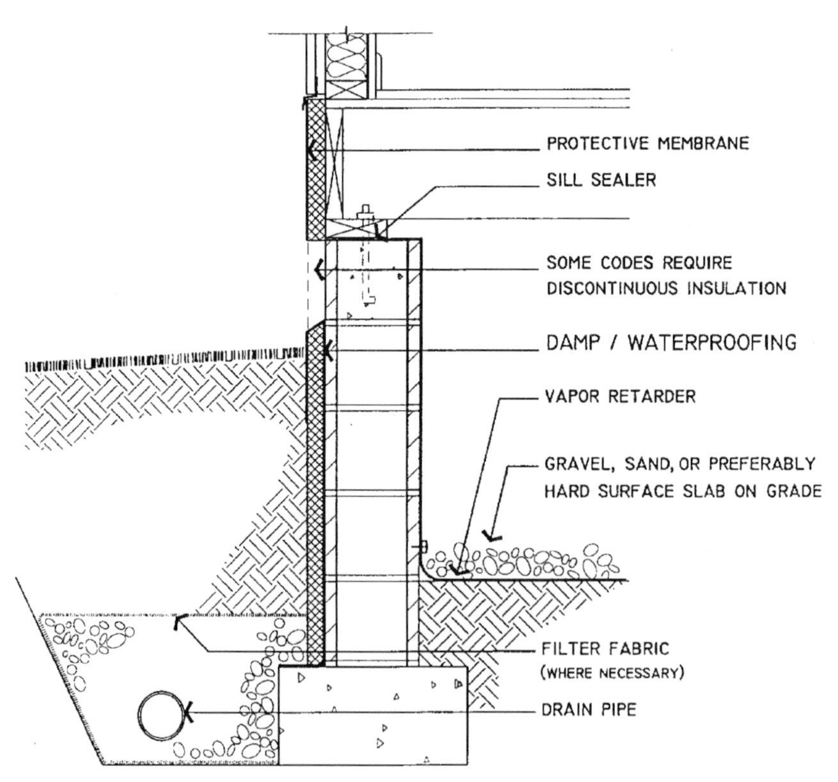

FIGURE 1.5.4 CONCRETE MASONRY CRAWL SPACE WALL WITH EXTERIOR INSULATION

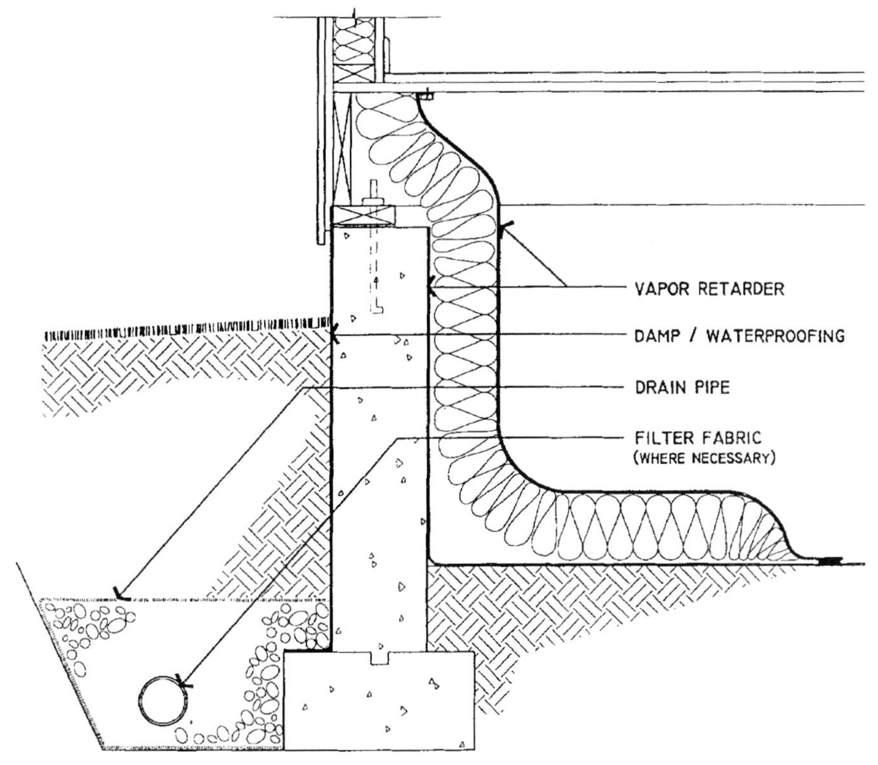

FIGURE 1.5.5 CONCRETE CRAWL SPACE WALL WITH INTERIOR INSULATION

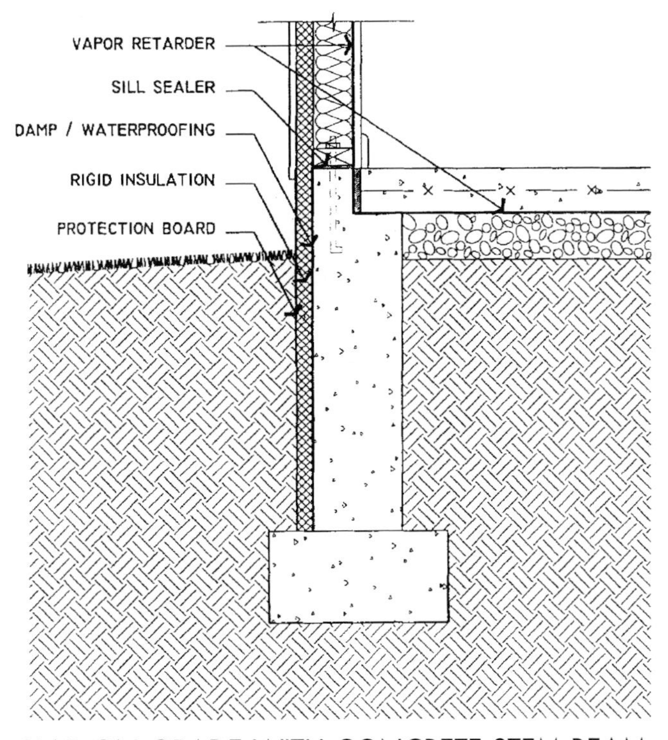

FIGURE 1.5.6 SLAB ON GRADE WITH CONCRETE STEM BEAM

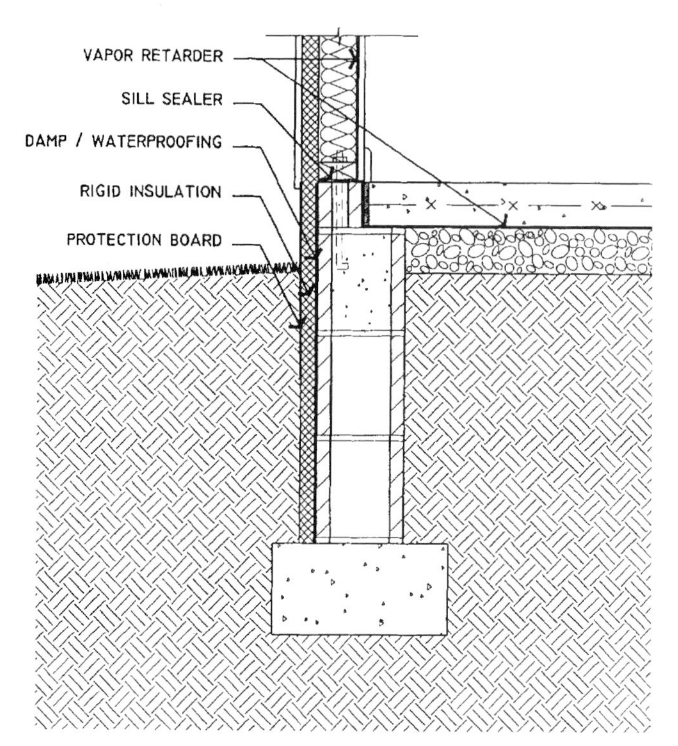

FIGURE 1.5.7 SLAB ON GRADE WITH MASONRY STEM WALL

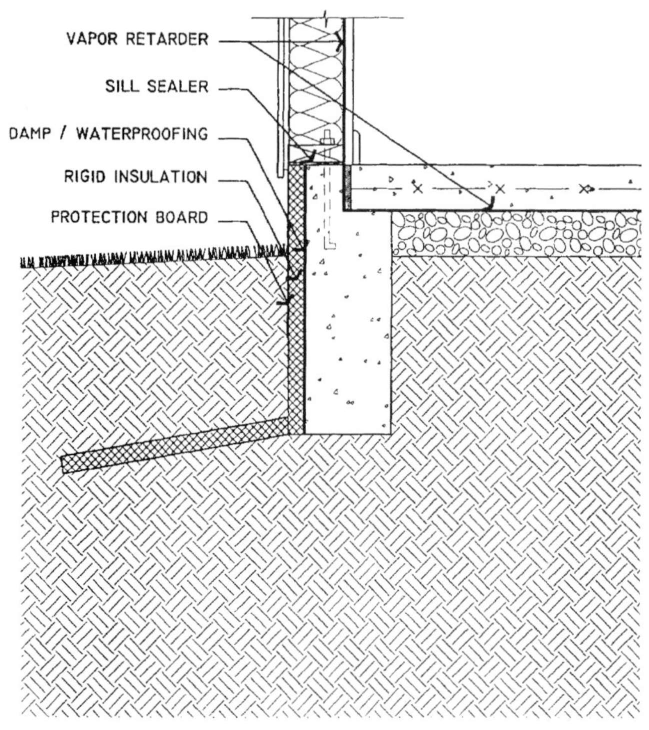

FIGURE 1.5.8 SLAB ON GRADE WITH FROST-PROTECTED SHALLOW FOUNDATION SYSTEM

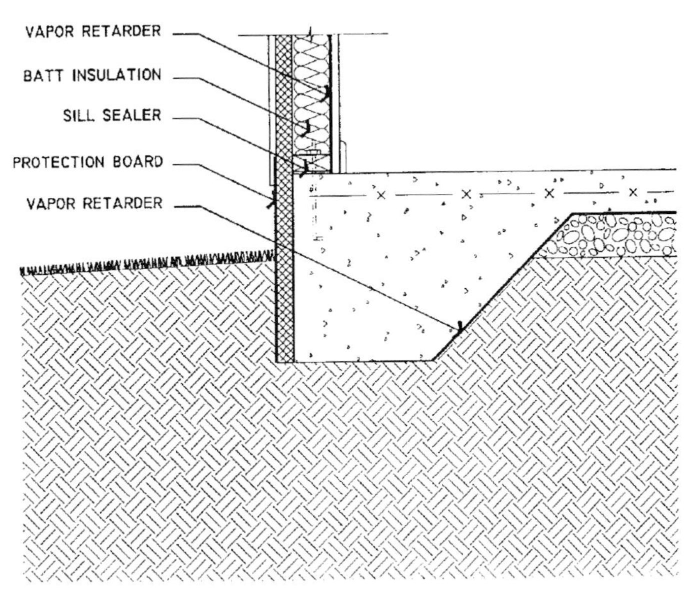

FIGURE 1.5.9 SLAB ON GRADE WITH PERIMETER "TURN DOWN"

TECHNIQUES, MATERIALS, TOOLS

There are three basic approaches to insulating slabs:

1. EXTERIOR INSULATION FOR SLABS WITH STEM WALLS.

The stem wall depth will typically vary with the frost line. Full-depth insulation is recommended in all areas of the country. An alternate insulation strategy is to insulate in accordance with the National Association of HomeFull Builders (NAHB) *Frost Protected Shallow Foundation* recommendations.

2. EXTERIOR INSULATION FOR SLABS WITH GRADE BEAMS.

This type of slab is typically used in the west, southwest, and south. Full-depth insulation is recommended in areas with moderate or high fuel costs and heating degree days in excess of 2,000.

3. INSULATING BASEMENT SLABS FROM INSIDE.

While not cost-effective in terms of energy savings versus first cost, basement slab insulation makes finished floors more comfortable. The traditional method (Fig. 1.5.2) is to seal the existing floor with a vapor barrier; fasten pressure-treated wood nailers or sleepers to the floor; cover with another vapor barrier (to prevent moist air from condensing in the floor cavity); install sheathing (plywood or OSB); and finish with wood flooring or resilient tile. An alternate method is to float the new floor over loosely applied insulation and subfloor.

Insulation materials and techniques were discussed earlier in this section.

1.6 VENTILATION

VENTILATING BASEMENT SPACES

ESSENTIAL KNOWLEDGE

Basement air can suffer from soil gases, mold, and mildew. Soil gases can penetrate the basement through cracks in foundation walls and slabs and through unsealed penetrations. Mold and mildew growth is encouraged by cool surface temperatures, high relative humidity, and by the presence of moisture in walls and floors. Basements are often used to store solvents and fuel tanks, which outgas volatile organic compounds (VOCs). The quality of the basement air also affects the quality of air in the entire house. If the house uses a forced-air system for heating or cooling, basement air is induced into the return ducts and at the fan housing and is redistributed to other spaces in the house. If the house is heated with a hydronic system or with electric baseboards, the basement air can still migrate to other areas. This type of air movement can be reduced through sealing and caulking but is unlikely to be eliminated.

If the basement is used as a living space, it especially needs to be properly ventilated. Without proper ventilation, indoor pollutants can reach high concentrations. Even if the basement is mostly used for storage or mechanical equipment, ventilation is important because complete separation of basement air from house air is not feasible. Basement ventilation is usually provided directly through operable windows, vents, or fans, or indirectly through air infiltration. Air infiltration is not a good method of ventilation because the rate of infiltration varies with wind and temperature and cannot be relied upon to provide sufficient outside air at all times. The best solution is to have an airtight basement that is well-ventilated.

For basements with boilers or furnaces, good ventilation also helps reduce the potential for depressurization. When a basement furnace or boiler pulls air from the rest of the house for combustion and exhausts it out the flue, a low-pressure zone is created in the basement, which can pull soil gases into the space. If the equipment is placed in an area open to the rest of the basement, depressurization will usually be negligible; if the equipment is enclosed in an airtight room, depressurization is more significant. The basement can also have more than one combustion appliance (a furnace plus a hot water heater). The effect of these appliances firing at the same time is additive. For air systems, return ducts usually pull air from the basement, causing depressurization. Ducts should be sealed and taped, but some leakage will remain. A separate supply of outside air to the combustion area is beneficial. The basement can also be pressurized by the air system to keep soil gases out.

Since ventilation has many positive effects, it is important that it be provided effectively. In many residences, supply of outside air to the combustion equipment is achieved by windows left open in the basement space that contains the furnace or boiler, or by a vent with insect screening installed in the foundation wall. These methods will induce cold air into the basement during winter and moist air during summer, increasing the energy cost. Additionally, the window that provides combustion air could be closed accidentally. For this reason, windows are not recommended as a means to provide combustion air.

Combustion air can be ducted in close proximity to the boiler or furnace. This method is not that different from creating a screened opening in the wall. The best solution is to purchase equipment that is certified as direct-ventilation (sealed-combustion) and that draws the outside air directly into the combustion chamber. This method raises the energy efficiency of the heating system and increases the safety of operation. If sealed combustion is not feasible, it is preferable to provide a fan that brings in air to the combustion equipment only when this equipment operates.

TECHNIQUES, MATERIALS, TOOLS

There are a number of approaches to providing adequate ventilation to basements and sufficient combustion air for a furnace or boiler, without creating low-pressure zones that draw soil gases.

1. PROVIDE DIRECT-VENTILATION (SEALED-COMBUSTION) BOILERS OR FURNACES.

This type of mechanical equipment provides outside air for combustion through a duct that leads directly from a wall vent to the combustion chamber of the boiler or furnace.

ADVANTAGES: Warm air from the house is not used for combustion and is exhausted through the chimney. The risk of incomplete combustion and of flue gas backflow into the house is eliminated.

DISADVANTAGES: There is an extra cost to purchase a sealed-combustion boiler or furnace.

2. PROVIDE A FAN WITH A MOTORIZED DAMPER TO INDUCE OUTSIDE AIR INTO THE BOILER OR FURNACE ROOM ONLY WHEN THE EQUIPMENT FIRES.

This method provides combustion air and is similar to the sealed-combustion technique, except that the combustion air is not ducted directly to the equipment and has to cross the boiler or furnace room. When the equipment does not fire, the damper closes over the fan and does not allow cold air to seep in.

ADVANTAGES: Warm air from the house is not used for combustion and is exhausted through the chimney. The risk of incomplete combustion and of flue gas backflow into the house is eliminated.

DISADVANTAGES: Using this method is more costly and complex than using a window or vent.

3. PROVIDE A SCREENED, OPEN VENT TO ALLOW OUTSIDE AIR INTO THE BASEMENT NEAR THE FURNACE OR BOILER.

The constantly opened vent allows outside air into the boiler and furnace area of the basement at all times (operable windows are not a recommended alternative, because they might be closed). To mitigate the intrusion of cold or moist air throughout the basement, the furnace or boiler should be enclosed in its own room.

ADVANTAGE: This is a low-tech, low-cost solution.

DISADVANTAGES: The equipment must be located next to an exterior wall with the vent opening above grade; if located close to the exterior grade level, the vent will require maintenance to keep it free from debris or snow. The air in the boiler and furnace room can get cold, increasing the heat loss from the equipment and therefore increasing the fuel use. Also, it is possible (even if not likely) that under strong wind conditions a supply vent placed on the leeward side could *exhaust* air, rather than admit it.

4. SEPARATE THE BASEMENT AIR FROM THE AIR IN THE REST OF THE HOUSE.

Separation of basement air from house air is desirable when the basement is not used for living space. The purpose is to minimize mixing basement air, which could be polluted by dust, mold, or VOCs, with

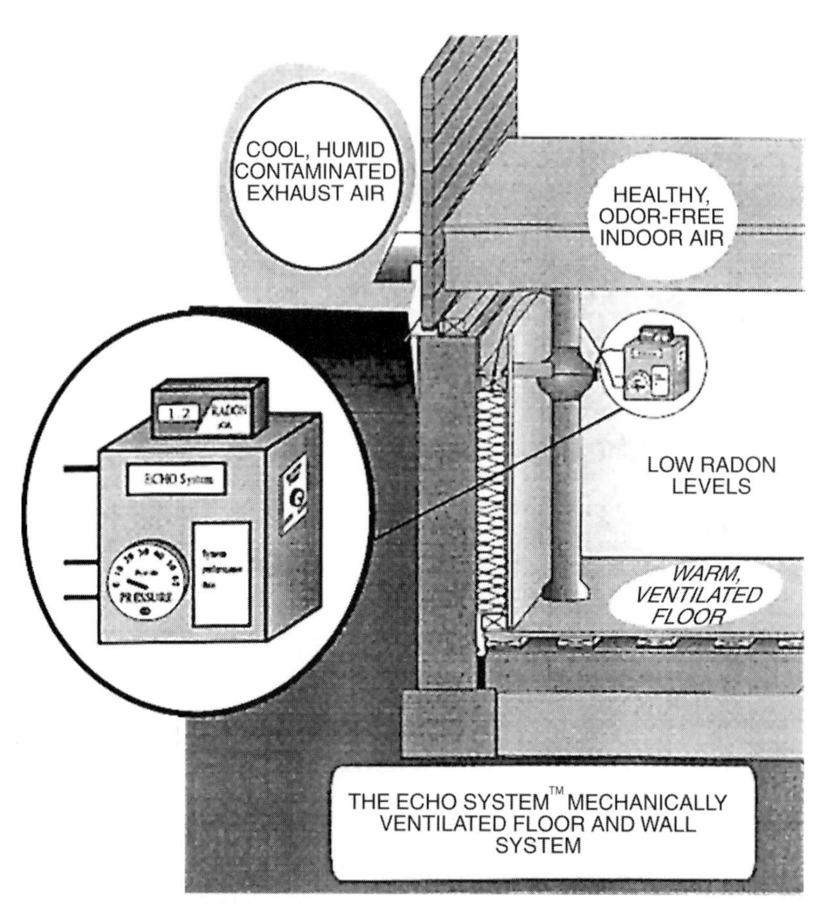

FIGURE 1.6.1 ECHO SYSTEM PERFORMANCE CHARACTERISTICS

conditioned air for the house. The basement space is separated from the rest of the house by doors and continuous floor. Basement ventilation is provided by adequately sized windows or mechanical intake and exhaust fans. The basement is heated with a hydronic system, electric baseboard heat, or with a separate air heating system that keeps the air in the basement from mixing with that of the rest of the house. Containers that can outgas VOCs are stored in an airtight, enclosed room. This room should have continuous exhaust to the outdoors. Fan housings and ducts in the basement that serve the house must be sealed and taped to avoid drawing basement air into the rest of the house.

ADVANTAGES: The basement is well-ventilated, and the air in the rest of the house is cleaner.

DISADVANTAGES: If a heated air system is used for the house, there will be a significant added cost for a separate, decoupled air system for the basement (or for a separate hydronic or electric heat system). Extra costs are incurred to carefully seal fan housings and ducts in the basement that serve the living area.

5. TREAT THE BASEMENT AS CONTIGUOUS, CONDITIONED SPACE OPEN TO THE REST OF THE HOUSE.

This method is recommended when the basement is maintained to the same standards of cleanliness as a living space. Separate ventilation for the basement can be provided with windows or mechanical intake and exhaust systems. The basement can also be ventilated using outside air brought in by the air system that serves the house. If an air system is used for the entire house, ducts located in the basement must be well-sealed to avoid pressure imbalances that could lead to the suction of soil gases. An additional step in avoiding soil gas seepage is to pressurize the basement. This condition can be achieved by supplying outside air at the upper floors and by exhausting at the basement level.

ADVANTAGES: Proper ventilation can be provided in the basement at all times.

DISADVANTAGES: There may be discomfort at times as warm air rises and the upper spaces overheat, while the basement remains cooler. Adjustable dampers in the ducts leading to each floor, or mechanical destratification ducting, will remedy this problem at an added cost. Also, tight basement walls will reduce the air infiltration rate in the basement and will reduce the airflow from the basement into other spaces.

6. PROVIDE A VENTILATED "ROOM WITHIN A ROOM."

When walls are continuously moist from water penetration, or when radon cannot be satisfactorily controlled with Environmental Protection Agency (EPA)-recommended techniques (see Ventilation for Soil Gases later in this section), rooms or areas within the basement space can be built that have continuous air exhaust under the floor and in the wall cavity between the room's walls and the basement walls. This innovative technology, called the ECHO system (Fig. 1.6.1) is sold by Indoor Air Technologies, Inc. Ideally, the furnace, boiler, or domestic hot water heater nearby should have sealed combustion, with a ducted outside air supply. More important, the return ducts and fan housing need to be well-sealed to avoid air being drawn from the basement, creating negative pressure that can reduce the effectiveness of this system.

ADVANTAGES: This method several aspects of basement air quality. It is effective when exterior source problems cannot be eliminated. The insulation installed on the interior walls (not on the foundation walls) remains dry even if the basement walls leak.

DISADVANTAGES: Monitoring of equipment and conditions in ventilated wall spaces is required. This method can be costly.

VENTILATING CRAWL SPACES

ESSENTIAL KNOWLEDGE

The venting of many existing crawl spaces is inadequate, and moisture and odor problems are frequently apparent. Moisture encourages mold, mildew, and wood rot, and degrades insulation R-value. There is currently much debate about the ratio of ventilation area to floor area. Building codes typically require that crawl spaces be provided with ventilation openings in the surrounding walls with a net free area of 1 ft^2 for each 150 ft^2 of crawl space area, or reduced to 1/1,500 of the crawl space area when an approved moisture retarder is used over the ground surface. Natural ventilation is useful during dry winter months, but can actually increase the moisture in the crawl space during humid winter spells and during the cooling season, in areas with humid summers. Crawl spaces that are not ventilated with foundation vents must be equipped with a mechanical ventilation system conforming to an approved mechanical code.

TECHNIQUES, MATERIALS, TOOLS

There are a number of approaches to ventilating crawl spaces and mitigating moisture problems.

1. NATURALLY VENTILATE THROUGH REQUIRED-SIZED OPENINGS IN THE FOUNDATION WALL.

Ventilation can be achieved through the existing foundation wall by removing portions of the block or concrete wall and installing ventilation grilles. The floor above the crawl space must be insulated, as temperatures inside the crawl space will be seasonal (see Insulating Crawl space in Section 1.5).

Special care must be taken to insulate water and waste piping located below the floor insulation. Dirt floors are a major source of moisture in crawl spaces and should be covered with a polyethylene moisture retarder. A concrete slab, or other protective surface, can be placed over the polyethylene to protect the sheet against damage.

ADVANTAGES: Compared to mechanical ventilation, this approach is relatively inexpensive in terms of operating costs and requires little maintenance.

DISADVANTAGES: This approach requires insulation of the floor above the crawl space, which may be difficult or impossible to achieve. Insulation must be tightly packed between floor joists; any air spaces between the insulation and the joists promote air convection, which short circuits the insulation. Removing portions of the existing foundation walls may not be possible or practicable, such as with rubble stone walls or concrete walls with rebar. Natural ventilation may not always provide enough airflow to remove moisture and may induce moisture in summer and during humid winter periods. Removable covers for the grilles can address this problem, but add a maintenance cost.

2. MECHANICALLY VENTILATE THE UNCONDITIONED CRAWL SPACE.
This option is similar to the one discussed in option 1 (natural ventilation), except that the ventilation is mechanically provided. There are ventilation systems on the market, such as the CellarSaver by Tamarack Technologies, that can be adjusted to activate at a certain humidity level. They also have insulated dampers that close when the fans are not in use.

ADVANTAGES: This approach ensures a minimum ventilation rate even when there is no breeze. Dampers, manual or automatic, close the crawl space to humid outside air when the fans are off.

DISADVANTAGES: This approach has the same construction drawbacks as option 1. There is an additional cost of operating and maintaining the fans.

3. MECHANICALLY VENTILATE THE SEMICONDITIONED CRAWL SPACE.
This strategy essentially treats the crawl space as a basement space, ventilating it similarly through mechanical means. The crawl space is mechanically heated in addition to being ventilated. The foundation walls are insulated (see Insulating Crawl Spaces in Section 1.5).

ADVANTAGES: This method ensures that ventilation is timely and sufficient. When the crawl space is properly insulated, this technique saves money in fuel costs and extends the life of the floor structure.

DISADVANTAGES: There is a higher initial cost. Access to the crawl space is required for installation and to observe the condition of the space and equipment.

VENTILATION FOR SOIL GASES

ESSENTIAL KNOWLEDGE

In the past few years, awareness of the dangers of soil gases, especially radon, has grown. According to the U.S. EPA, 14,000 cancer deaths in the United States are caused by radon each year. The EPA estimates that one out of every 15 houses has elevated radon levels, with concentrations heavier in some regions than in others. A colorless, odorless, radioactive gas, radon is most often found in basements, entering through crawl spaces, dirt floors, cracked foundation floors and walls, and openings around drain pipes and sump pumps.

TECHNIQUES, MATERIALS, TOOLS

The EPA has released details of techniques for radon abatement in basements and crawl spaces. Consideration should also be given to the mechanical ventilation system's effect on radon control.

1. PROVIDE SUBSLAB VENTILATION OF SOIL GASES IN BASEMENT SPACES.

Vent pipes penetrating the slab can run vertically through the house, and be vented out the roof or a side wall. The vent pipe has a continuously operating exhaust fan that maintains negative pressure. All cracks and penetrations through the slab, and junctures between the slab and the walls, should be sealed with caulking or grout. New slabs should be poured on a 4" layer of gravel or sand, overlaid with a gas-retarder membrane (Figs. 1.6.2 and 1.6.3).

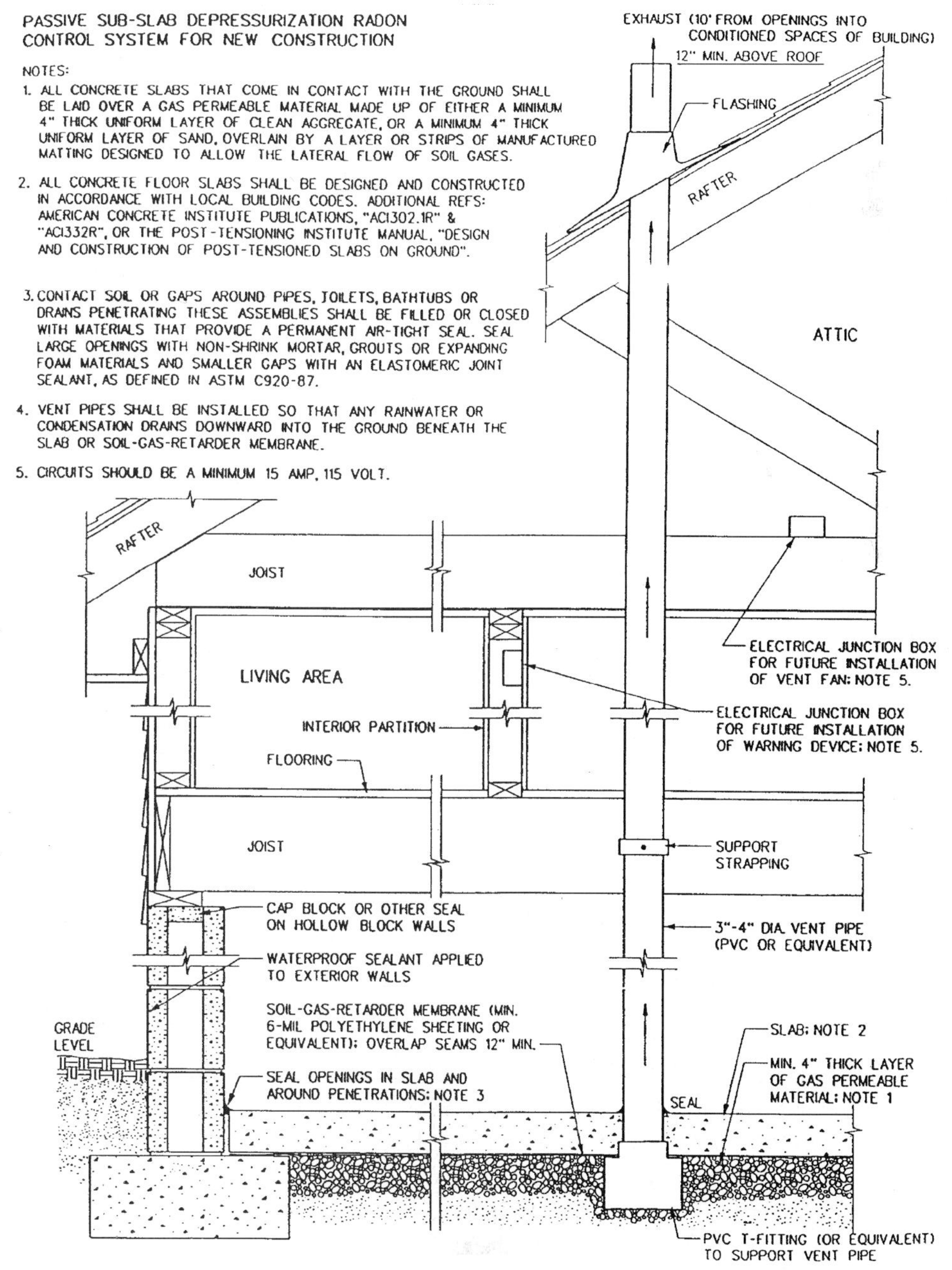

FIGURE 1.6.2 PASSIVE SUBSLAB DEPRESSURIZATION RADON CONTROL SYSTEM

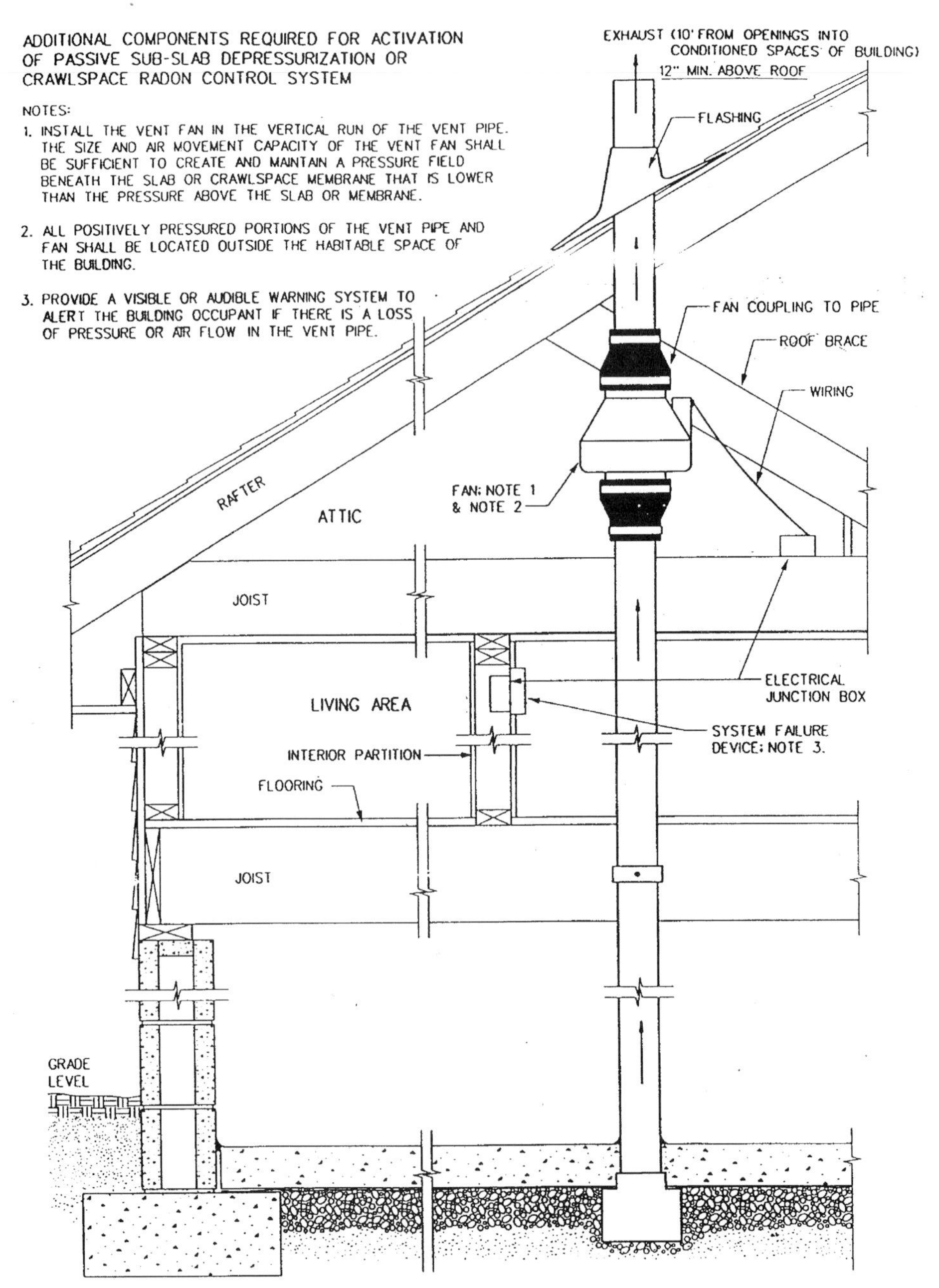

FIGURE 1.6.3 ADDITIONAL COMPONENTS REQUIRED FOR ACTIVATION OF DEPRESSURIZATION RADON CONTROL SYSTEM

ADVANTAGES: This is an approved technique, endorsed by the EPA, that effectively exhausts gases below the slab.

DISADVANTAGES: This technique can be costly, depending on the size of the basement and the configuration of the house. It is less effective when the basement walls are made of hollow core (CMUs) since radon could find points of entry into such walls. In cold climates, there is the possibility of frost heaving due to the cold air introduced under the slab.

2. PROVIDE FOUNDATION WALL DEPRESSURIZATION.

The foundation wall is depressurized through weeping tiles (drain tiles). A continuously operating exhaust fan maintains negative pressure. All cracks and penetrations at the wall should be sealed with caulking or grout.

ADVANTAGES: This method is relatively easy to implement.

DISADVANTAGES: The area protected by this technique is limited. The fan must operate at all times. This method is ineffective if weeping tiles are connected to the sewer system or to roof leaders.

3. PROVIDE BASEMENT HEATING AND VENTILATION, SEPARATE BASEMENT AIR FROM THAT OF THE REST OF THE HOUSE, AND PRESSURIZE THE SPACE TO KEEP OUT SOIL GASES.

This approach, similar to that discussed in Ventilating Basement Spaces earlier in this section, is useful to supplement the EPA-recommended radon mitigation techniques described earlier. Pressurization is achieved via a dedicated air system for the basement, with an outdoor air intake. This technique ensures that the basement is always at a positive pressure with respect to air spaces in the basement walls and in the cavities below the slab. This approach is most useful in basements with CMU walls with hollow cores, through which radon can enter. If a furnace, boiler, or domestic hot water heater is located in the basement, combustion air must be ducted to a sealed appliance, or outside air must be supplied using a fan with a damper (the fan operates only when the appliance fires). Open windows or screened vents are not recommended because they can compromise positive pressure. If the house has an air system, fan housings and ducts must be sealed and lapped to avoid entraining basement air.

ADVANTAGES: This method ensures good ventilation and has increased effectiveness in keeping radon out.

DISADVANTAGES: There is a significant cost required to achieve a tight basement-house separation; a significant cost in supplying a dedicated air system, such as a furnace with an outside air supply; and an additional cost for a ducted outside air system for combustion (although such a system is a good idea even if the basement has no radon problems; see Ventilating Basement Spaces earlier in this section).

4. PROVIDE VENTILATION OF SOIL GASES FROM CRAWL SPACES.

Where a crawl space slab exists, the technique is similar to that described in Ventilating Basement Spaces earlier in this section. For dirt-floor crawl spaces (Fig. 1.6.4) perforated pipe should be laid over the soil, parallel to the house's long dimension, and should extend no closer than 6 ft to the foundation wall. A T-fitting joins the perforated pipe to an exhaust stack that extends through the roof. The pipe and crawl space floor are then covered with a gas-retarder membrane, and sealed against the foundation wall and all vertical penetrations.

ADVANTAGES: This is an approved technique sanctioned by the EPA that effectively exhausts soil gases.

DISADVANTAGES: This method can be costly, depending on the size of the crawl space and the configuration of the house.

5. PROVIDE A VENTILATED "ROOM WITHIN A ROOM".

Refer to option 6 in Ventilating Basement Spaces earlier in this section. The ECHO system described there is also effective in reducing radon penetration into the basement and other occupied spaces.

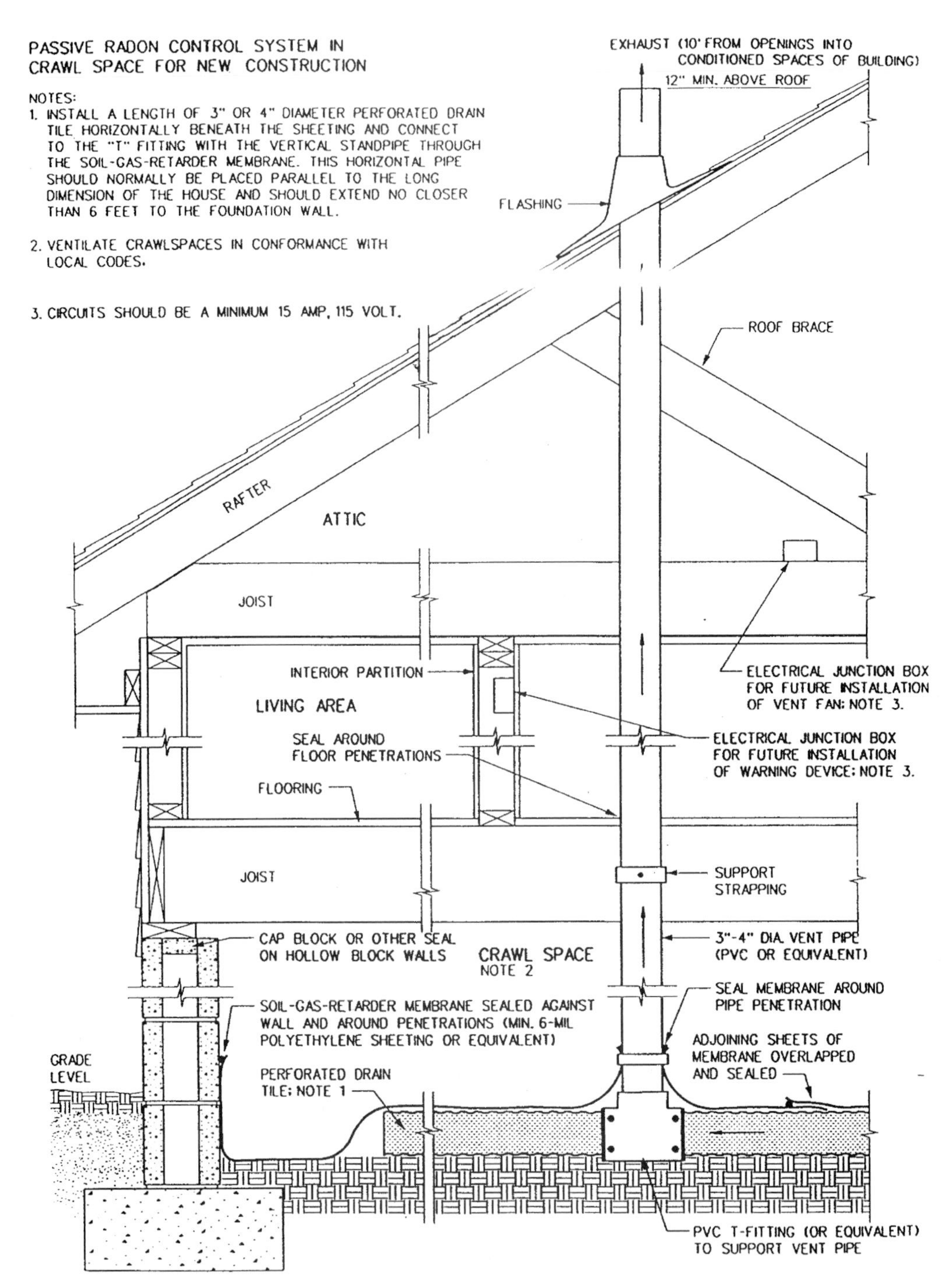

FIGURE 1.6.4 PASSIVE RADON CONTROL SYSTEM IN CRAWL SPACE FOR NEW CONSTRUCTION

1.7 SHORING, UNDERPINNING, AND REPAIR

ESSENTIAL KNOWLEDGE

When a foundation settles or is displaced laterally, cracks appear, water may intrude, and the building's basic structure is compromised. Foundation walls may crack, settle, move upwards from freeze-thaw cycling or expansive soils, or move sideways. The primary causes of such foundation failures are problematic sites; poor soil conditions; seismic activity; water migration through soils or flooding; poor foundation design, engineering, or construction materials. Problematic site conditions include steep slopes beyond the soil's natural angle of repose; lack of terracing under fill; sites with high water tables, springs, or underground streams; and sites adjacent to bodies of water or within flood plains. Poor soil conditions include highly expansive soils; organic (peat or discarded plant material) or silty soils; and poorly compacted fill that was not layered when it was placed. Seismic activity can damage inadequately reinforced brick, concrete, concrete block, or stone foundations. Water-related damage includes the migration of fines under foundations, decreased soil cohesiveness (which causes settlement), and hydrostatic pressure against foundation walls and floors (which can lead to cracks and displacement). Poor design and engineering may result in improper materials, under-reinforcement, or lack of attention to many of the conditions mentioned here.

Foundation settlement usually happens over time, and is not a dramatic event; there is usually time to prepare a proper fix. The cause of the settlement should be carefully and deliberately researched, as remedial work is expensive and care should be taken to understand the nature of the problem. A structural engineer or a soils (geotechnical) engineer who specializes in such work should be consulted.

TECHNIQUES, MATERIALS, TOOLS

There are a number of approaches to stabilizing and repositioning foundations, which are applicable to all types, including stem walls, piers with grade beams, and slabs on grade. Some of these systems are commonly used to repair commercial or industrial building, but they can also be used for residential multifamily and single-family housing rehab. Repairs should be undertaken by experienced contractors under the direction of a professional engineer.

1. STABILIZE AND UNDERPIN SETTLED FOUNDATION WITH REINFORCED CONCRETE PIERS.

This process involves auguring a pier excavation next to the settled portion, in some instances (such as with granular soils) inserting round cardboard forms, placing reinforcing steel, and pouring a concrete pier including an elbow or cap under the existing footing. Concrete piers can bear on rock or

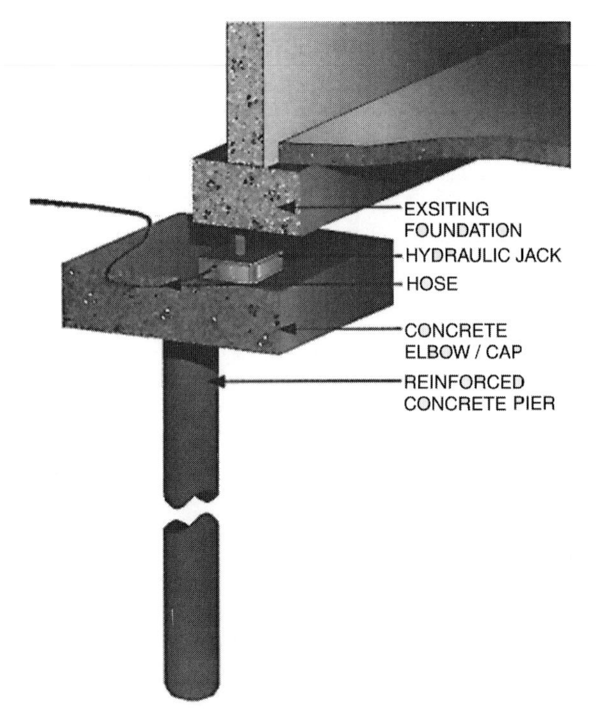

FIGURE 1.7.1 REINFORCED CONCRETE PIER

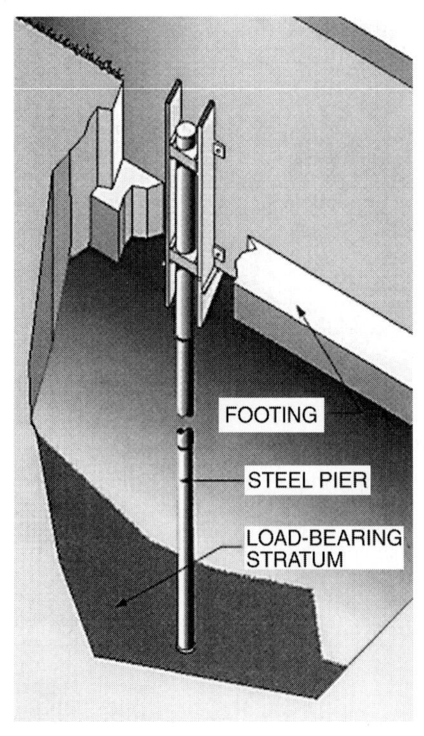

FIGURE 1.7.2 MINI-PIER

can be designed as friction piers. Once the concrete is cured, a hydraulic jack can be set between the newly formed elbow or cap and the underside of the footing (Fig. 1.7.1). The jack distributes the load to the new pier and raises the settled portion. Shims are then inserted to hold the raised foundation in place so that the jack can be removed and the joint packed with a no-slump mixture of sand and cement (drypack). The number of piers is determined by the type of foundation and the size of the settled portion. If the loads are small and the soil's capacity is good, the jack can be eliminated and the loads transferred through the shims directly.

ADVANTAGES: This is a conservative approach, with a proven track record.

DISADVANTAGES: Concrete piers are limited in depth in sandy soils that will collapse the auger hole and may not work as friction piers in soil with poor bearing capacity. This method requires that the concrete cure sufficiently before load is applied. It is difficult to raise the foundation more than a few inches. This method is more disruptive to landscaped areas adjacent to the building than mini or helical piers.

2. STABILIZE OR RAISE SETTLED FOUNDATION WITH STEEL MINI-PIERS.

Usually hydraulically driven, mini-piers range from 2" to 8" in diameter. The smaller sizes, 2" to 4", are typically used in residential projects. The piers are set adjacent to the interior or exterior face of the footing at intervals of 6 to 8 ft, depending on the condition and reinforcement of the footing (Fig. 1.7.2). They are then hydraulically or pneumatically installed through a drive frame mounted on the foundation wall to rock or an adequate bearing stratum. The building's dead weight, plus the weight of the soil adhering to the footing and foundation wall, provide the resistance and reaction during installation of the mini-piers. The piers are usually driven to depths of 15 to 25 ft, and can go as deep as 150 ft, but are increasingly uneconomical and difficult to drive over 60 ft. A number of companies including Atlas Systems, Inc., and Perma-Jack hold patents on mini-pier product installation systems. Another variant, called the Mini-Pile™ system, developed by Heywood Baker, Inc., employs an oversized point that concentrates resistance at the end of the pipe pile. The pile is hollow, and if bedrock is not encountered, a grout bulb can be pressure injected at the end of the pipe to supply the necessary bearing. There are two attachment systems—one used in expansive soil attaches to the foundation by means of a steel bracket; the other employs a 12"-thick reinforced concrete pile cap and jacking pad that is used to seat a hydraulic jack that is in turn used to raise the footing. The space between the surface of the jacking pad and the footing underside is then filled with concrete to complete the pilecap.

ADVANTAGES: This is a time-proven, economical, and stable method, where the bearing layer is within 50 to 60 ft of grade. No heavy equipment is required. Mini-piers can be installed in limited-access areas and can be used to raise slabs.

DISADVANTAGES: Mini-piers can crack the foundation and slab if used incorrectly. Economical use requires the bearing layer to be relatively near to grade and 2 to 3 ft^2 access pits, which can disturb the foundation perimeter if multiple pits short distances apart are used.

3. STABILIZE OR RAISE SETTLED FOUNDATION WITH HELICAL PIERS.

This is a patented system that screws a steel shaft with a helical plate into stable soil. The steel shaft attaches to the footing and foundation wall by means of a bracket. The rotation of the helical plate by a hydraulic jack will eventually cause enough resistance to stabilize the foundation or raise it, if necessary (Fig. 1.7.3). Installers of the helical pier system are trained and licensed by such companies as A.B. Chance Co. and Atlas Systems, Inc., which provide geotechnical support. Warranties are available.

ADVANTAGES: This is a proven, versatile, technically sound system. It is economical and not as disruptive as some other methods; no heavy equipment is required, it can be installed in limited-access

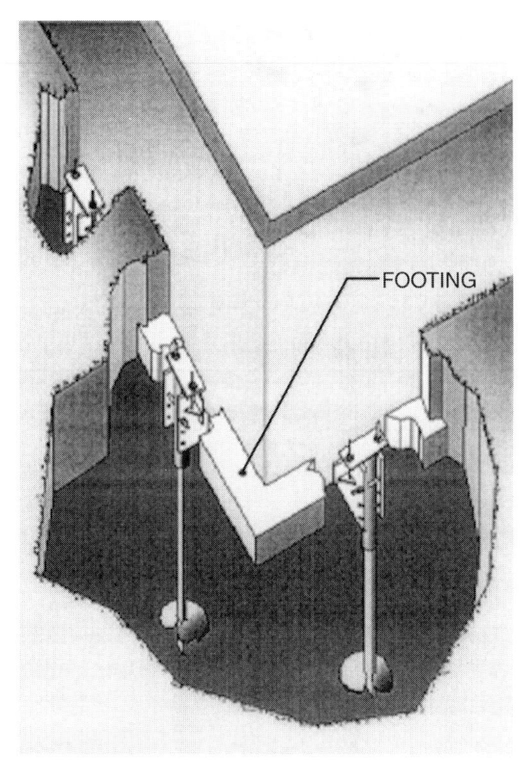

FIGURE 1.7.3 HELICAL PIER

areas; and it can be used for shallow pier foundations, as a wall anchor to pull back and hold basement walls, and to raise slabs. Hydraulic jack torque can be measured, thereby determining actual bearing capacities of helical piers.

DISADVANTAGES: This method will not work in soils of limited bearing capacity. The building perimeter, including walks and planting, may be disturbed if on-center spacing of piers is close. Notching the footing to apply the bracket is required.

4. STABILIZE OR RAISE SETTLED FOUNDATION BY "PRESSURE GROUTING" OR "MUD GROUTING."

This technique pressure-injects a slurry of cement, water, and sand into the soil to stabilize the foundation or to raise the basement slab. This method is difficult to control, and the location of the grout and its effect are impossible to determine. It is more commonly used for slab stabilization and less for foundation shoring, where other techniques are considered more reliable and cost-effective.

ADVANTAGES: This is an economical solution, if it works.

DISADVANTAGES: It is often not effective, is difficult to control and evaluate, and may crack slabs. Grout may travel well beyond the desired treatment area.

5. STABILIZE OR RAISE FOUNDATION WITH COMPACTION GROUTING.

Compaction grouting involves the pressure injection of a very low slump grout through a buried pipe to displace and compact soils. Major uses include densifying loose and poorly compacted soils; filling voids including large sink holes; preventing liquefication of soils due to seismic activity by densifying the soil beyond the liquefication threshold; releveling settled structures; and using compaction grout bulbs as structural elements for mini-piles or underpinning. The key to the system's success is that the pumped grout remains local to the area around the buried pipe and does not migrate, as less controlled forms of pressure grouting do.

ADVANTAGES: This method is relatively controllable, and there are a wide variety of applications. Compaction grouting remains in a mass, can displace poor soils, and can be used to raise foundations and slabs. This method is cost-effective primarily for large projects such as townhouses and multifamily buildings.

DISADVANTAGES: This method is too expensive for some single-family house foundation repairs, requires access by heavy machinery, and is not recommended in saturated clay soils or in soils containing high amounts of organic material.

6. STABILIZE, WATERPROOF, OR RAISE FOUNDATION OR SLAB WITH VARIOUS HIGHLY SPECIALIZED GROUTING TECHNIQUES.

A variety of other specialized grouting techniques exist to stabilize foundations in larger buildings, such as multifamily housing. Commonly used by foundation repair specialists around the country, these techniques include *jet grouting*—high-pressure liquid grouting to form a soil cement called soilcrete used for sophisticated underpinning work; *chemical grouting*—injecting sodium silicate into sandy soils to stabilize them and make akin to sandstone; *urethane grouting*—to waterproof under slabs and difficult-to-access spaces; *vibro-compaction*—uses probe-type vibration to densify granular soils; *injection systems*—for expansive soils, injecting potassium, lime, and water to limit movement.

ADVANTAGES: These are sophisticated site-specific techniques that can solve difficult problems.

DISADVANTAGES: They are expensive and not normally used in residential rehab.

7. UNDERPIN MASONRY OR CONCRETE STEM WALL WITH ENLARGED FOOTING.

If the settlement of a foundation wall is localized, it may be possible to underpin the foundation directly by increasing the footing size in order to distribute the wall loads over a greater area. This process involves jacking a portion of the house to take the load off the foundation wall, excavating a portion of the footing (in alternating 3' to 6' sections, for example), adding steel reinforcement, and filling the footing excavation with concrete. After the concrete has cured sufficiently (usually 7 days) the intermediate sections can be excavated and the process repeated. Alternatively, if the wall is seriously deteriorated, a new wall and footing can be erected alongside the existing wall.

ADVANTAGES: This is a relatively inexpensive, expedient local repair if the affected area of the foundation is restricted to a relatively small perimeter.

DISADVANTAGES: Work should be designed by an engineer so that the new footing is sized and reinforced for the existing soil conditions. This is a dangerous procedure that should be undertaken by experts. It may not work in silty or expansive soils as the new footing mass may continue to settle. It is not cost-effective over long wall lengths.

8. BUTTRESS STONE FOUNDATION WALLS.

Freeze-thaw action, expansive soil, migrating water, and hydrostatic pressure can all raise havoc with stone foundation walls, particularly those laid without mortar. In some cases walls may be repointed or stabilized from either inside or outside. In other cases, the pointing may not be cost-effective and may not be easily waterproofed due to irregular surfaces. A repair that can both stabilize and waterproof stone walls is shown in Fig. 1.7.4. In this case, a new concrete surfacing is poured into the cavity between plywood forms and the existing wall face, consolidating the wall and providing a smooth surface on which to apply waterproofing.

ADVANTAGES: This method allows the existing wall to remain and reinforces and waterproofs the wall.

DISADVANTAGES: Excavation and formwork are required. This method can be expensive.

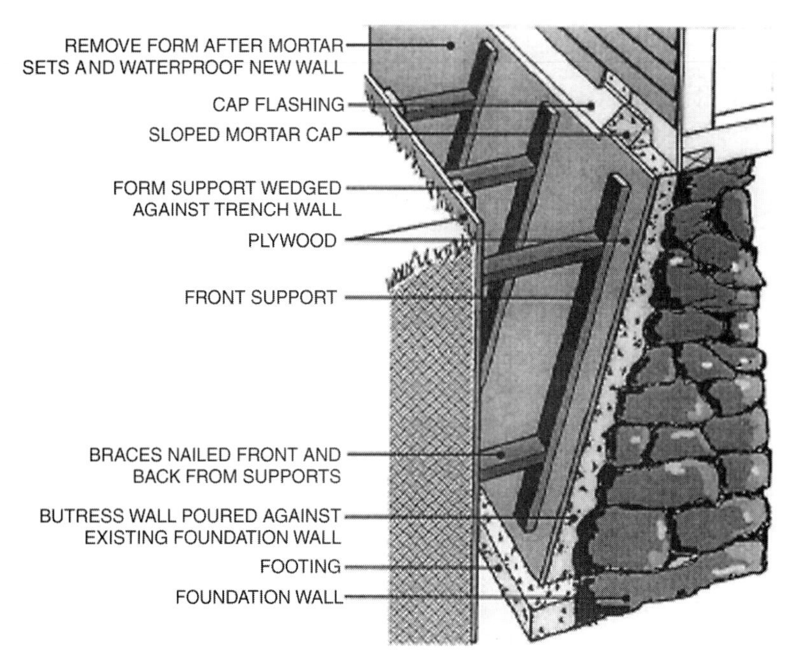

FIGURE 1.7.4 BUTTRESSING STONE FOUNDATION WALL

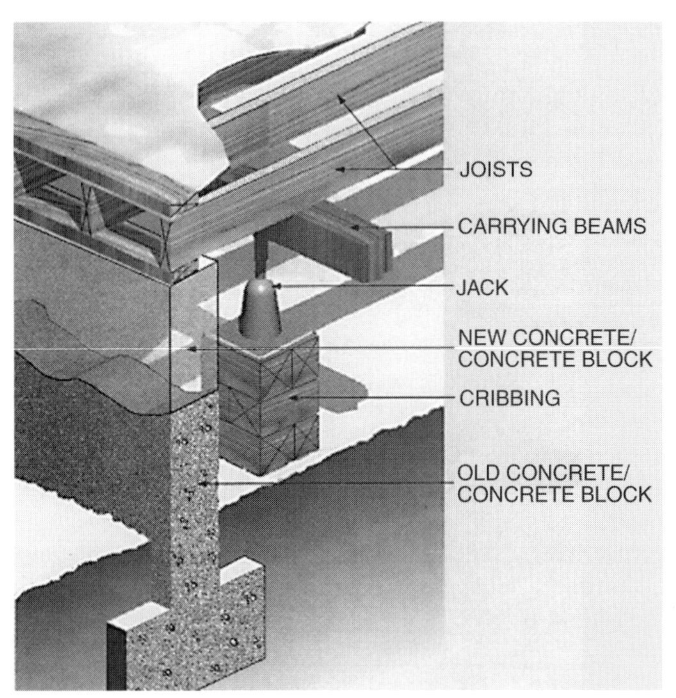

FIGURE 1.7.5 JACKING A WALL WITH PERPENDICULAR JOISTS

9. REPLACE THE FOUNDATION.

Extensive settlement; unsightly, cracked, or displaced foundation walls; rotting sill plates; or the need to gain additional height in a basement may necessitate the replacement of all or part of the foundation. When only one foundation wall is to be replaced, platform framing offers several options. If the first floor framing is perpendicular to the wall, it may be possible to place a supporting girder just inside the exterior wall, parallel to it (Fig. 1.7.5). In structures where the joists run parallel to the exterior wall, needle beams can be placed perpendicular and under the wall to

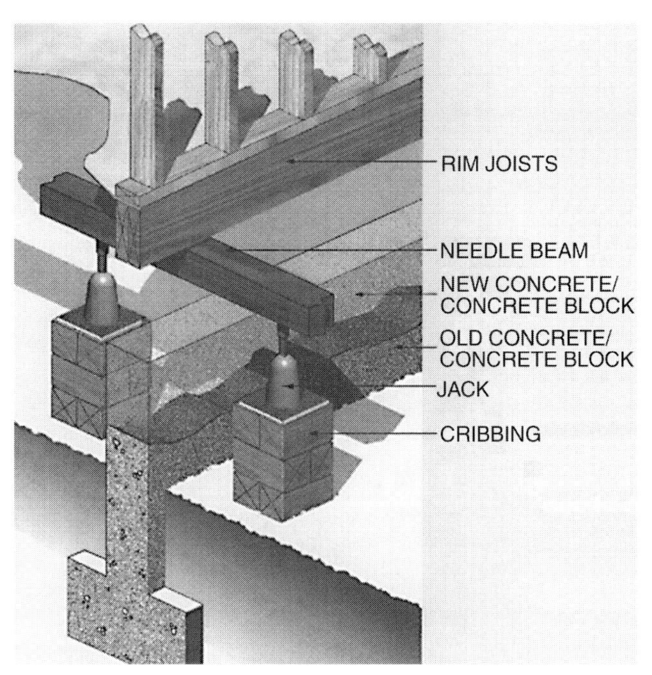

FIGURE 1.7.6 JACKING A WALL WITH PARALLEL JOISTS

support it while the foundation is replaced (Fig. 1.7.6). If the entire house is being raised, a system of carrying beams, cribbing, and jacks should be designed to support the critical load-carrying members of the house. An alternative to using cribbing and jacks is to install helical piers to provide the footing for a column support to temporarily underpin portions of the structure. Professional guidance should be sought.

ADVANTAGES: This method provides the opportunity to build new, dry living and storage space while fixing a number of basic problems.

DISADVANTAGES: This method can be expensive, requires changes to all plumbing and electrical service connections, and requires skills beyond those of most house-building contractors.

1.8 CRACK REPAIR, COATINGS, AND FINISHES

CRACKS IN WALLS AND SLABS

ESSENTIAL KNOWLEDGE

Basement wall and floor cracks can be caused by a variety of factors, including drying shrinkage, thermal contraction, restraint (internal or external) to shortening, subgrade settlement, and applied loads, including those due to expansive soils, hydrostatic pressure from high water tables, and in some parts of the country, seismic activity. Factors that can contribute to cracking problems include inadequate reinforcement in concrete, concrete block, and stone walls, incorrect construction or lack of construction and control joints, and improperly mixed or placed mortar or concrete. Stationary hairline cracks (those 1/16" and less in width) are usually nonstructural and are primarily cosmetic concerns unless they are associated with moisture problems. Cracks 1/8" in width and greater may also be nonstructural, but may reflect significant wall settlement or displacement. Repair should not be undertaken until the principal cause of the cracking has been determined (see Chapter 8).

TECHNIQUES, MATERIALS, TOOLS

1. REPAIR CRACKED AND OUTWARDLY DISPLACED FOUNDATION WALL BY EXTERIOR JACKING.

If a section of a foundation wall has been outwardly displaced beyond the plane of the adjacent walls, it may be possible, rather than rebuilding that section, to move it back into place. This can be accomplished by excavating in front of and to the side of the affected section and jacking the displaced wall by means of hydraulic ram jacks (Fig 1.8.1). Care must be taken to provide a jacking pad of sufficient area so that the jack doesn't punch through the wall. It is also important to shore up the excavation against which the wall is jacked and to provide drainage from the excavation in the event of rain.

ADVANTAGES: This method is less expensive than completely replacing portions of the wall, allows visual inspection of wall conditions, and provides an opportunity to waterproof the wall and inspect drainage.

DISADVANTAGES: Wall conditions may not permit jacking. The wall may not be aligned when jacking is complete. Cracks will require patching, grouting, and waterproofing. This technique will disturb grounds and plantings adjacent to the affected wall portion.

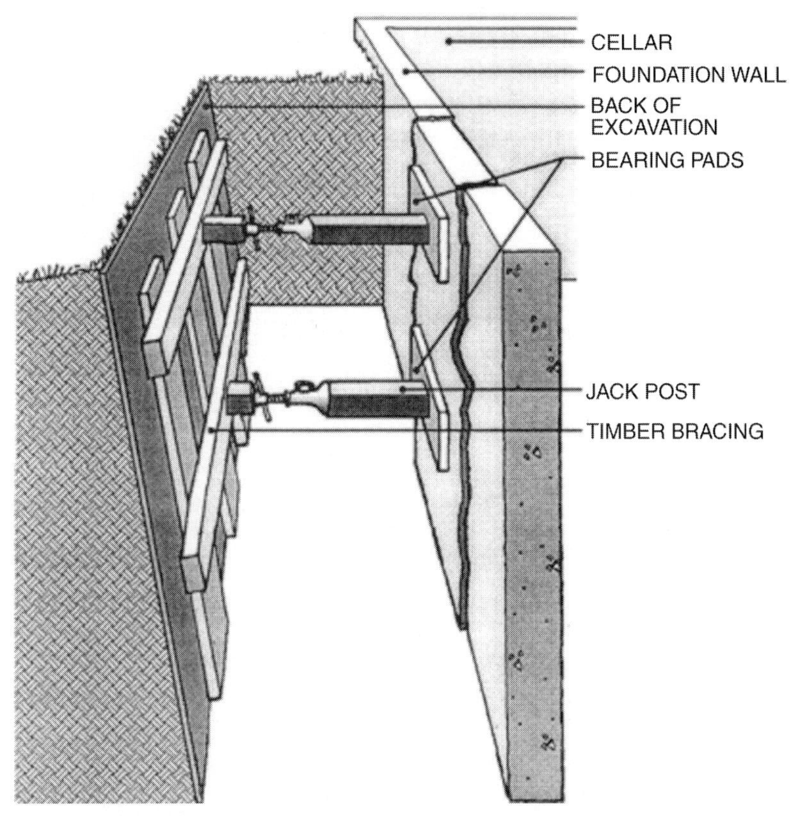

FIGURE 1.8.1 JACKING CRACKED FOUNDATION WALL

2. STABILIZE CRACKED AND INWARDLY BOWED FOUNDATION WALL BY USE OF "EARTH ANCHOR" AND WALL PLATE.

If an upper section of the foundation wall has been inwardly displaced, usually at the midpoint of the wall, it may be possible to stabilize or straighten the wall with an "earth anchor" (Fig. 1.8.2). Typically installed in an exterior trench approximately 6'0" on-center horizontally and below the average frost depth. The Grip-Tite wall anchor system utilizes an anchor rod that is pushed through the soil by means of a hammer drill/rotary hammer from a reinforced wall plate through a 1¹/8" hole in the foundation wall to an earth anchor plate set at a distance of 7 to 8 ft from the face of the exterior wall. The anchor rod has a nut and washer that is tightened with a torque wrench against the wall plate, usually during dry seasons. The wall can be brought back to a more vertical position in a process that takes, according to the manufacturer, 1 to 3 years. An offset trencher or narrow bucket backhoe are occasionally used to excavate 12"-wide trenches alongside the wall to relieve soil pressure and to allow for faster vertical realignment.

ADVANTAGES: This method stabilizes and straightens walls without the need for a complete excavation along the length of the exterior wall. In some instances, it can reduce the size of wall cracks and bring the wall to a more vertical position. It is less expensive than wall replacement or sidewall excavations and can prevent further wall displacement.

DISADVANTAGES: This method is not capable of bringing the wall back to a fully vertical position, something which could only be accomplished if the soil on the outside of the wall was removed. Exterior excavation for the wall plate is required. Wall anchor plates are exposed on the inside wall surface.

3. STABILIZE CRACKED AND INWARDLY BOWED FOUNDATION WALL BY USE OF HELICAL SCREW ANCHOR.

An alternative technique to the "earth anchor" includes the use of a helical screw anchor. Installed at intervals along a wall depending on the job site conditions, helical screw anchors are similar in con-

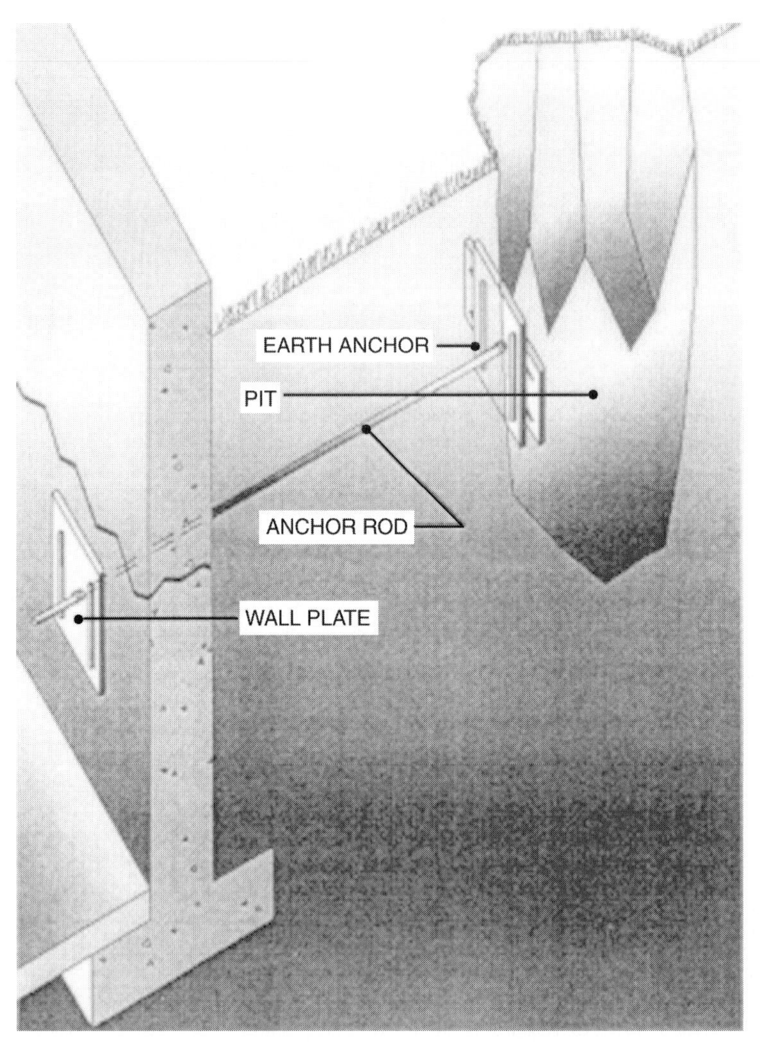

FIGURE 1.8.2 EARTH ANCHOR

cept to earth anchors described earlier, except that the helical plate is a part of the anchor rod itself and is screwed into the soil bank rather than placed in a pit (Fig. 1.8.3). Once the rod develops the appropriate resistance, a plate is placed on the inside of the wall and a nut is tightened with a torque wrench to secure and possibly straighten the wall. The two largest manufacturers of helical systems are A.B. Chance Co. and Atlas Systems, Inc. Their anchors are installed by certified dealers throughout the United States.

ADVANTAGES: This method stabilizes and may straighten walls without the need for continuous excavations. It may reduce the size of wall cracks by bringing walls to a more vertical position. It is less expensive and disruptive than wall replacement or continuous excavation.

DISADVANTAGES: This method may not straighten walls significantly. The interior anchor plate is exposed.

4. REPAIR WALL CRACKS WITH CONVENTIONAL GROUTING TECHNIQUES.

If cracks are not expanding in size and water penetration is not an issue, conventional grout repairs may be possible. The limitations of grouting systems are inherent in their material consistency as fairly stiff; dry mixes typically used for repair work cannot easily flow into cracks, especially thin ones. Most cracks must be chipped open on the surface for the patch to be effective. Even in this case, however,

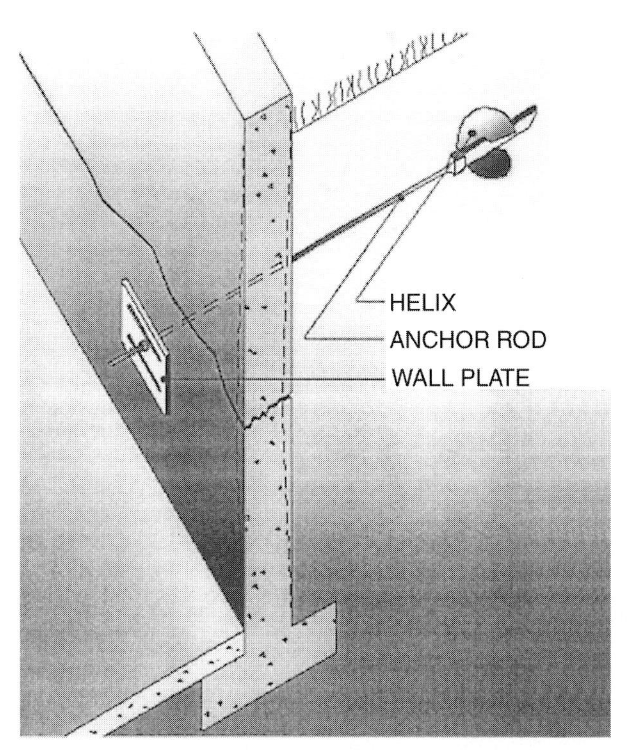

FIGURE 1.8.3 HELICAL WALL ANCHOR

the patch depth is only 1" to 2" and is more cosmetic than structural in nature. A number of grouting materials are available through a wide variety of suppliers, including conventional portland cement and sand, with or without acrylic admixtures to enhance the bond to existing material, various proprietary low slump dry pack mixtures, and fast-acting hydraulic cements that set up fast and can be used to plug holes and wide cracks where water seepage is a problem. In addition, fiber-reinforced cements, called surface-bonding cements, are designed to strengthen block walls and can even be used on both surfaces to form a structural skin on walls laid "dry" without mortar.

ADVANTAGES: These are inexpensive repairs that do not require a high degree of skill to complete.

DISADVANTAGES: These are essentially nonstructural surface repairs. Except for hydraulic cements, these products will not usually prevent water penetration. This method will not work where there is crack movement.

5. REPAIR CRACKED WALL WITH EPOXY INJECTIONS.

Epoxy injections can be used to restore structural soundness when foundations have been stabilized and when cracks are not growing in size. Cracks as narrow as 0.002" can be bonded by the injection method. The typical technique involves drilling holes at close intervals along the crack and installing tubes (ports) in the holes into which epoxy is injected under pressure (Fig. 1.8.4).

ADVANTAGES: This is a proven and widely used technique. It can create a structural bond that is stronger than the concrete itself. Epoxy grout materials are available from a wide variety of manufacturers.

DISADVANTAGES: This method will not prevent the wall from cracking adjacent to the repair if further movement occurs. It cannot be used if cracks are leaking water. While moist cracks can be injected, water or contaminants in the crack will reduce the effectiveness of the repair. It works best with concrete rather than concrete block walls, because the cores in the block reduce the contact area. The method is and expensive and requires skilled applicators.

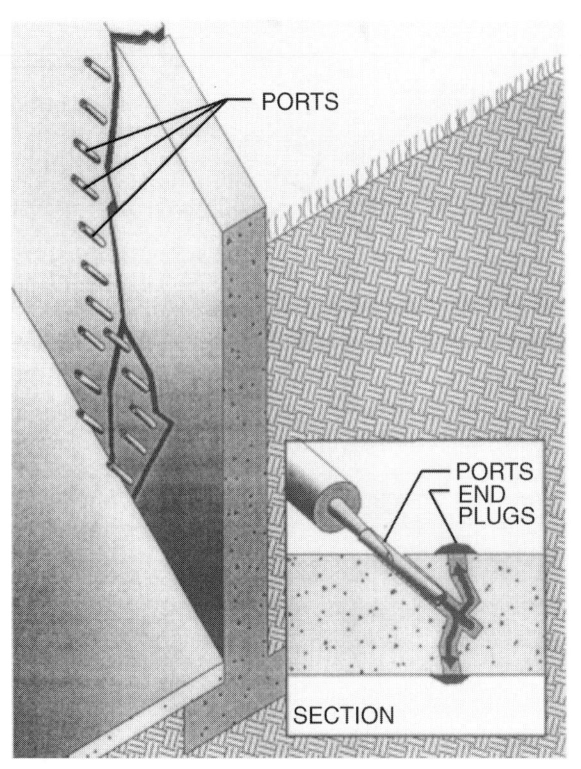

FIGURE 1.8.4 EPOXY OR URETHANE GROUTING

6. REPAIR CRACKED WALL WITH URETHANE INJECTIONS.

Urethane grouts are used to stop water infiltration and to fill cracks in foundation walls when there is substantial seepage through the wall. Urethane grouts are also flexible, which makes them suitable for repairing cracks with limited movement (up to 1/8" with some grouts). Urethane grouts are not structural (as epoxy grouts are) but will serve to help bond materials together. Grouts are either hydrophilic (which chemically react with water and can dehydrate in the absence of water) or hydrophobic (which do not dehydrate). Both grouts expand to create a dense, closed-cell foam through which water will not pass. Specific applications should be reviewed with the manufacturers, some of which include the 3M Co., Green Mountain International, Prime Resins, and the DeNeef Co. The application of urethane grout is similar to that of epoxy, pumped through a series of injection ports that are drilled to intercept the crack. Urethane sealant manufacturers also have developed fast-acting formulations that are applied directly to holes in concrete to stop leaks.

ADVANTAGES: These are state-of-the-art products used in critical situations where water and crack movement conditions exist. They were developed for commercial projects such as municipal water and sewer lines, tunnels, and other large-scale projects.

DISADVANTAGES: These products are more expensive than other kinds of products and are non-structural. Skilled applicators and careful attention to installation instructions are required. This method is not recommended for concrete block walls, as expansion of urethane may crack block.

COATINGS AND FINISHES

ESSENTIAL KNOWLEDGE

Surface treatments for concrete and concrete block walls repel water, which helps to control exterior surface staining as well as the migration of water through the wall, which can lead to efflorescence, staining, mold, and other problems on interior surfaces. In general, exterior surface treatments should allow for vapor transmission to ensure that humidity within the wall can escape. Treatments that are impermeable to water vapor tend to fail by blistering and peeling when moisture builds up behind the exterior surfaces. (Special water-retarding coatings were discussed in Section 1.4.)

Typical interior coatings include a wide variety of latex paints that use polyvinyl acetate as a binder. These paints are not subject to weathering and are less expensive than exterior vinyl acrylic paints. Most exterior coatings can be used on interior applications. As is the case with all paints, manufacturers' recommendations as to surface preparations and applications should be followed carefully. Manufacturers estimate that up to 80% of paint failures are due to improper surface preparation and application.

Clear treatments, which can be used to enhance the water resistance of walls without significantly altering their appearance, are made with either silicone, silicate, or acrylic resins. Epoxy paints are used in some interior wall and floor commercial applications, such as in schools and hospitals, where corrosive chemicals are present or high traffic is anticipated but are not normally used in residential applications.

TECHNIQUES, MATERIALS, TOOLS

1. APPLY CEMENTITIOUS COATINGS.

Cementitious coatings include conventional parging and stucco coatings, and specially formulated cement-based products that include latex binders and other additives that enhance performance. Included in this category are surface bonding cements meeting ASTM C887 requirements that contain glass fiber reinforcements that can strengthen walls.

ADVANTAGES: These coatings can change the color and texture of existing walls and can enhance water resistance. Fiber reinforcements enhance strength.

DISADVANTAGES: These coatings are subject to cracking and delamination due to weathering and expansion and contraction of walls.

2. APPLY ELASTOMERIC COATINGS.

A number of manufacturers produce elastomeric polymer coatings that are formulated to bridge static cracks up to 1/16". These products are often used as the final coating on exterior insulation and finish systems (EIFS) and can be applied directly on concrete and concrete block walls. These coatings are also used in conjunction with other brush- or trowel-applied elastomeric underlayment products that fill cracks up to 1/4".

ADVANTAGES: These coatings can change the color and texture of existing walls and can cover and protect exterior surfaces subject to deterioration from expansion and contraction.

DISADVANTAGES: These coatings are significantly more expensive than other coatings.

3. APPLY PAINT COATINGS.

The currently preferred exterior paints are latex, breathable, water-based paints. Latex paints typically come with acrylic or polyvinyl acetate binders. Paints with all-acrylic binders are recommended for exterior use as they have the greatest durability and flexibility. Most paint manufacturers offer a variety of crack fillers that can be applied before the finish paint to smooth out surface irregularities in walls. Alkyd paints use chemical compounds made from vegetable oil and synthetic resins as binders and are more difficult to apply and clean up than water-based latex paints. They are also adversely affected by alkaline surfaces, especially new concrete and concrete block construction. Recently developed acrylic paints have a number of superior characteristics to alkyd paints, including easier workability and better breathability, flexibility, color retention, resistance to chalking, and resistance to mildew.

ADVANTAGES: These coatings can alter color and surface texture with use of aggregate additives and can increase water resistance of wall. They are relatively economical.

DISADVANTAGES: These coatings will not bridge active cracks. On-going maintenance and renewal is required.

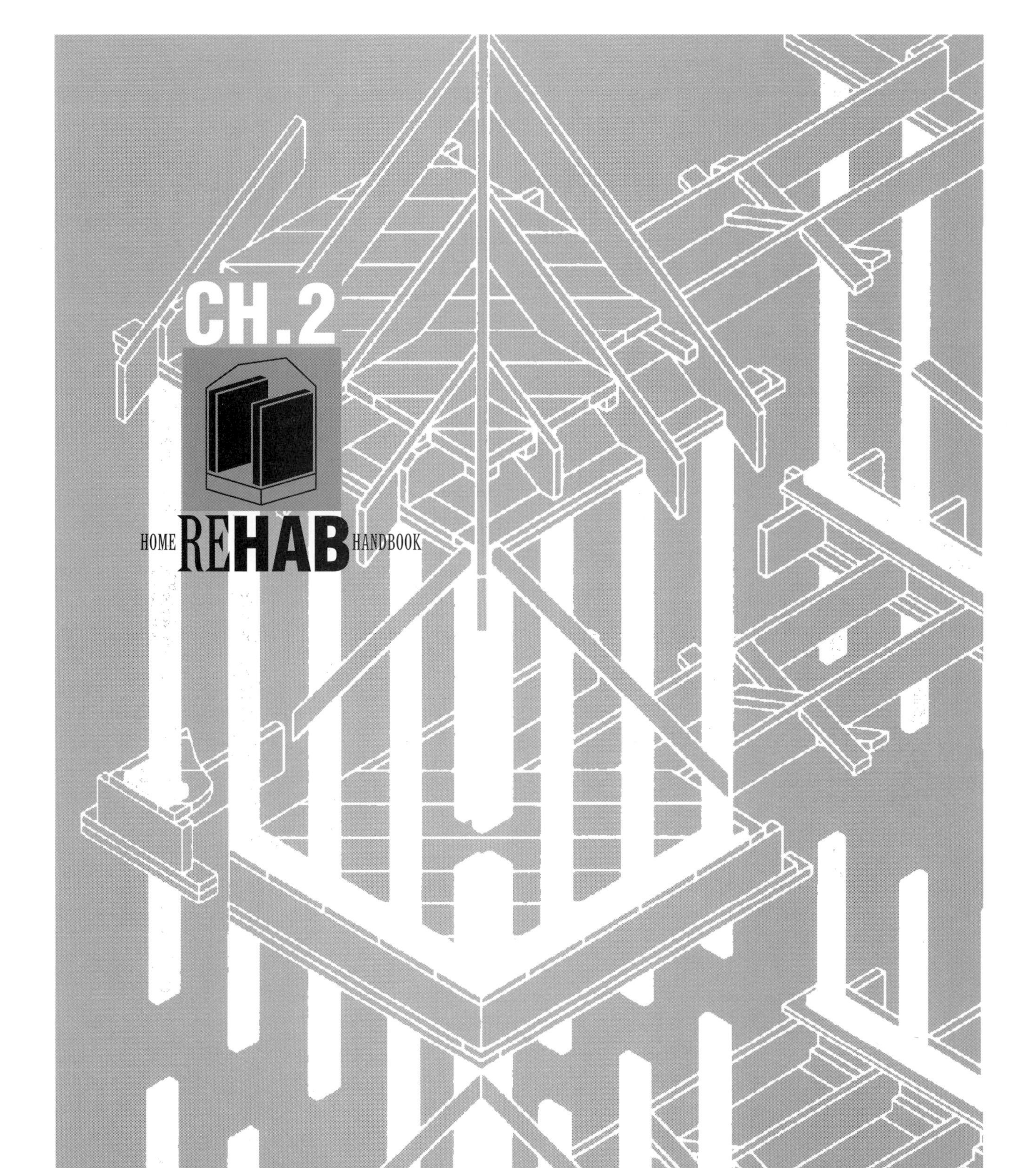

EXTERIOR WALLS

Chapter 2
EXTERIOR WALLS

2.1. DESIGN AND ENGINEERING
2.2. MASONRY AND BRICK VENEER
2.3. SHEATHING
2.4. VAPOR RETARDERS AND AIR INFILTRATION BARRIERS
2.5. INSULATION
2.6. VINYL SIDING
2.7. METAL SIDING
2.8. WOOD SHINGLES AND SHAKES
2.9. SOLID WOOD SIDING
2.10. HARDBOARD SIDING
2.11. ENGINEERED WOOD SIDING
2.12. PLYWOOD PANEL SIDING
2.13. FIBER-CEMENT SIDING
2.14. EIFS AND STUCCO
2.15. EXTERIOR TRIM
2.16. SEALANTS AND CAULKS
2.17. PAINT AND OTHER FINISHES

2.1 DESIGN AND ENGINEERING

EXTERIOR WALL OVERVIEW

From the time of the first European settlers in North America, the predominant wall framing system for houses was timber (with wood exterior cladding). Also popular, but to a lesser degree than wood framing, was masonry construction (most commonly brick or stone). Other exterior wall systems less widely used included log construction and adobe.

In the first half of the nineteenth century, the introduction of machine-sawn lumber and factory-made nails led to lighter structural systems (Fig. 2.1.1,), including braced-frame construction (which combines timber framing and infill studs) and balloon framing, which replaced heavy timber columns and girts (beams) with lightweight framing members that ran continuously from the foundation to the roof. By the beginning of World War II, balloon framing had largely been replaced with platform framing, which uses shorter framing pieces and gains lateral stability from the floor platform. This system prevails today in both stick-built and prefabricated housing.

Because it has been the dominant building material, the rehabilitation of wood-frame systems will be given the most attention in this guide. Masonry systems such as stone, brick, and concrete block will be addressed briefly. Additional recommendations for remedial work will be addressed in the individual chapters that discuss specific wall materials and application systems. Steel framing has not been used extensively in residential rehab, except occasionally for interior non-load-bearing

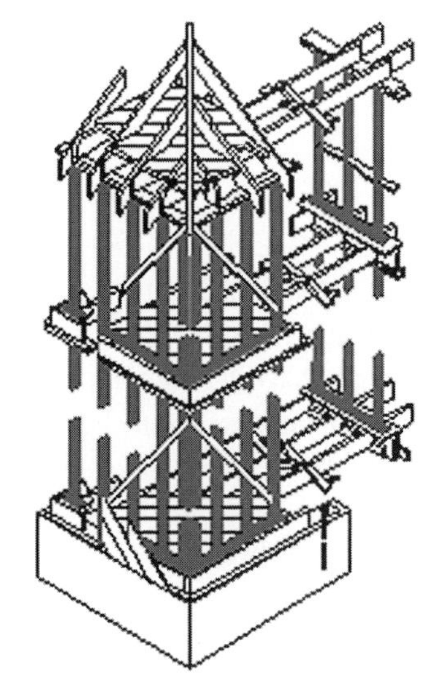

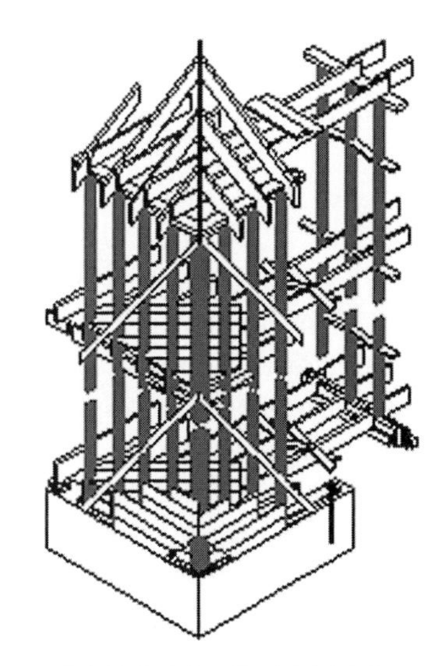

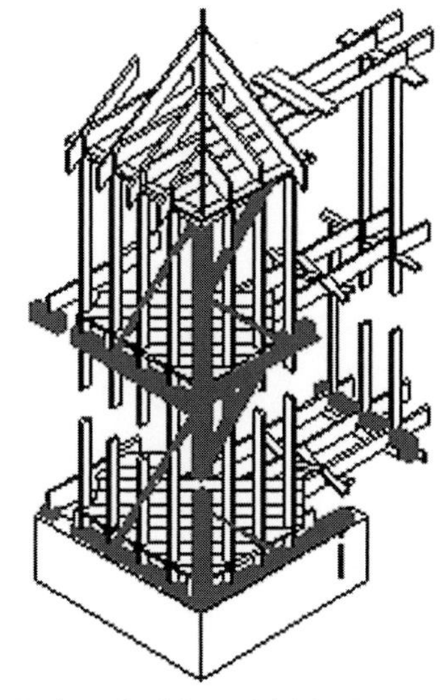

FIGURE 2.1.1 BRACED FRAMING BALLOON FRAMING PLATFORM FRAMING

partitions. Structural insulated panels (SIPs) have also not been used much in residential rehab work, because it is rare that large sections of walls are completely replaced.

Older, pre-code-complying domestic structures employed carpenters' rules of thumb, and buildings were, for the most part, strong, resilient, and adequate for normal conditions. When buildings fail structurally, which is infrequent, it is typically due to one or more of the following causes: inadequate design, earthquakes, storms and high winds, fire, insect damage, and structural deterioration caused by moisture. This chapter will outline some of the causes of structural failure and recommended remedial approaches and reference sources to be contacted for additional recommendations.

WOOD-FRAME SEISMIC RESISTANCE

ESSENTIAL KNOWLEDGE

A house's load-bearing walls and columns transmit live and dead loads from the roof to the foundation, which in turn distributes these loads to the ground. Resistance is also needed to lateral forces from wind and seismic occurrences, which can cause racking and displace buildings from their foundations. These loads are taken into account in the design of newer code-complying buildings, but for houses constructed before state and local code enforcement it is likely that they were not specifically addressed. For instance, the use of anchor bolts was not uniformly enforced until the late 1950s, and seismic requirements were not developed and enforced until the early 1960s.

The most serious structural damage to wood-frame houses in seismic areas results from insufficient anchoring of the frame to the foundation, and the collapse of "cripple walls" in crawl spaces. Local municipalities, working on their own and with code agencies, have developed prescriptive standards that are accepted by local building departments and insurance providers. Typical standards, such as those approved by the city of San Leandro, California, are described below. Other municipalities may reference the *Uniform Code for Building Conservation* (UCBC), 1994 edition, or 1997 for seismic requirements. Codes are evolving, may vary among municipalities, and should be researched carefully.

TECHNIQUES, MATERIALS, TOOLS

1. ATTACH A SILL TO THE FOUNDATION WITH ANCHOR BOLTS.

Unreinforced brick and block foundations are problematic because anchor bolts are difficult to drill and install properly, and the mortar may not be strong enough to hold the wall together in an earthquake. Crumbling, cracked, or porous concrete cannot hold mechanical anchors and will tend to shear adjacent to epoxied anchor bolts (inadequate or substandard foundations should be rebuilt or replaced to current code standards). Reinforced concrete foundations are preferable, but they are not typical in older homes. If the foundation is adequate and there is sufficient height in the crawl space to use an impact or rotary drill, the easiest method of attachment of a sill is by means of an expansion bolt or an anchor bolt epoxied into the foundation (Fig. 2.1.2). Sills should be bolted at a maximum of 6' intervals with bolts located within 12" of each joint or step in the sill, but not less than 9" from the end of a sill board. In addition to, or in lieu of, conventional anchor bolts, special hold-down brackets are often installed at shear walls or at wall openings. These hold-downs secure the studs or post through the bottom plate into the foundations (Fig. 2.1.3). Specific reinforcement requirements will depend on individual site and building code requirements and should be reviewed with a structural engineer.

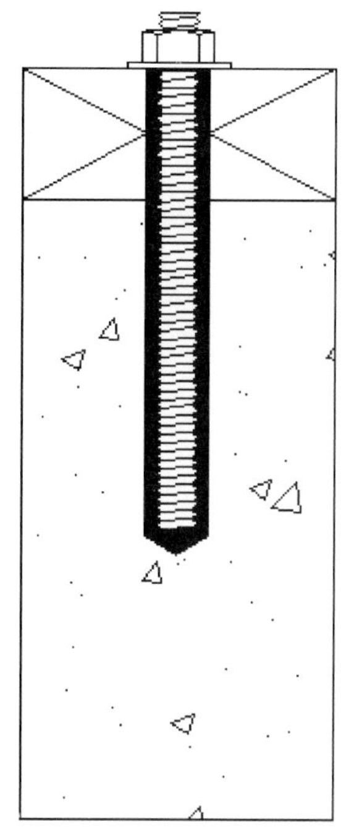

FIGURE 2.1.2 EPOXIED ANCHOR BOLT EXPANSION ANCHOR BOLT

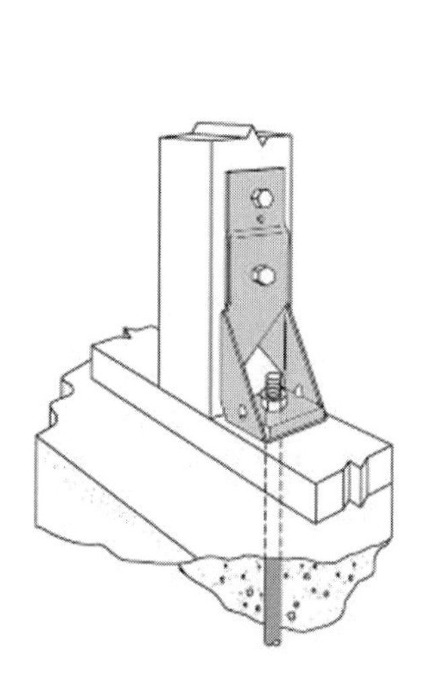

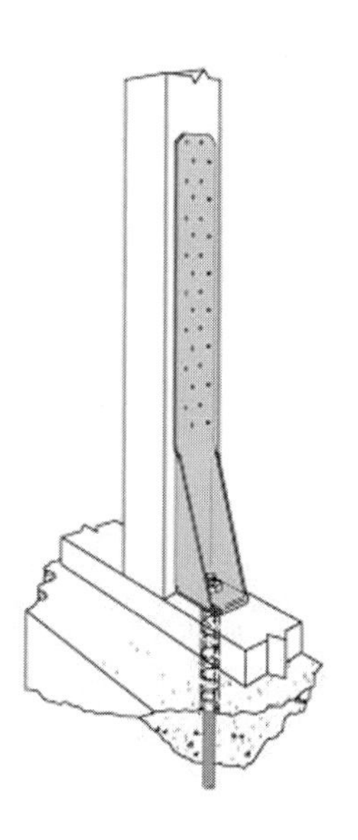

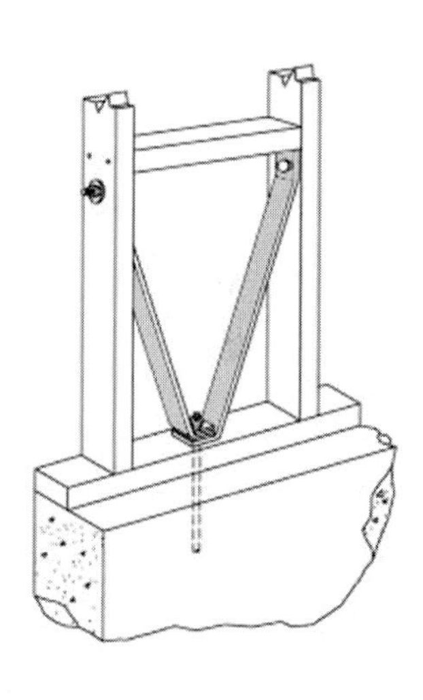

FIGURE 2.1.3 TYPICAL HOLD-DOWNS

ADVANTAGES: This technique provides the simplest, most positive connection.

DISADVANTAGES: This technique may not be possible where there is insufficient headroom to drill.

2. ATTACH A SILL, JOIST, OR STUD TO THE FOUNDATION WITH SIDE BRACKETS OR STRAPS.

Where it is not possible to install anchor bolts because of insufficient headroom, a variety of fasteners has been specially developed to affix frames to foundations. Anchors are available from manufacturers such as Simpson Strong-Tie Co., Inc., among others. Typical products include straps and plates designed for attachment of plates and joists to the face of foundations and mud sills (Fig. 2.1.4).

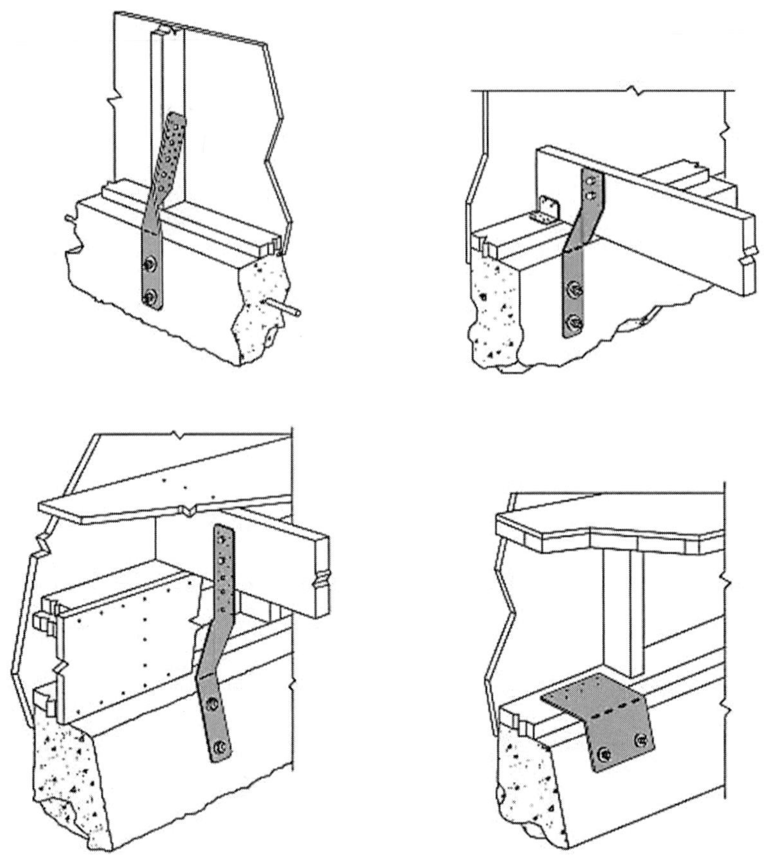

FIGURE 2.1.4 TYPICAL ANCHORS

ADVANTAGES: Side brackets can connect walls to foundations in areas with limited headroom.

DISADVANTAGES: Side brackets do not provide as strong or direct a connection as anchor bolts.

3. REINFORCE CRIPPLE WALLS WITH PLYWOOD OR OSB.

The lateral forces of an earthquake are concentrated on the interface of the foundation and the wood frame. Cripple walls are inherently weak connections and have to be reinforced to prevent buckling and collapse. This is easily accomplished with plywood or oriented strand board (OSB) structural sheathing used as a rigid diaphragm connecting the top and bottom plates with the studs (Fig. 2.1.5). The selection of the proper fastener type and spacing is critical. For crawl spaces that are not accessible, most municipalities have standards for the application of plywood or alternative structural sheathing to the outside of the crawl space. Consultation with a licensed engineer is recommended.

ADVANTAGES: This is an inexpensive and effective remedy.

DISADVANTAGES: An accessible crawl space is required, otherwise existing siding and sheathing have to be removed and new sheathing has to be applied to the building exterior.

4. PROVIDE SECURE LOAD PATH FROM ROOF TO FOUNDATION.

Engineers recommend that a continuous *load path* or *hold-down path* be created with metal connectors or sheathing so the walls, floors, and roof act together as a structural unit. This is accomplished by providing, in addition to the wall-foundation connection, a secure load path between the walls and floors (in platform construction) and between the walls and the roof. Typical floor-to-floor connectors include bolted hold-downs with threaded rods or straps designed specifically for that use. The choice would depend on job-site conditions and loading requirements (Fig. 2.1.6). Another mate-

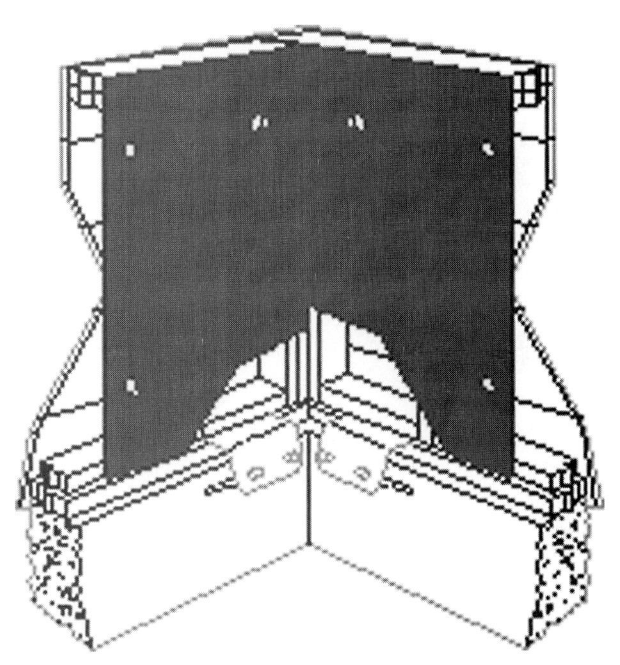

FIGURE 2.1.5 PLYWOOD REINFORCEMENT

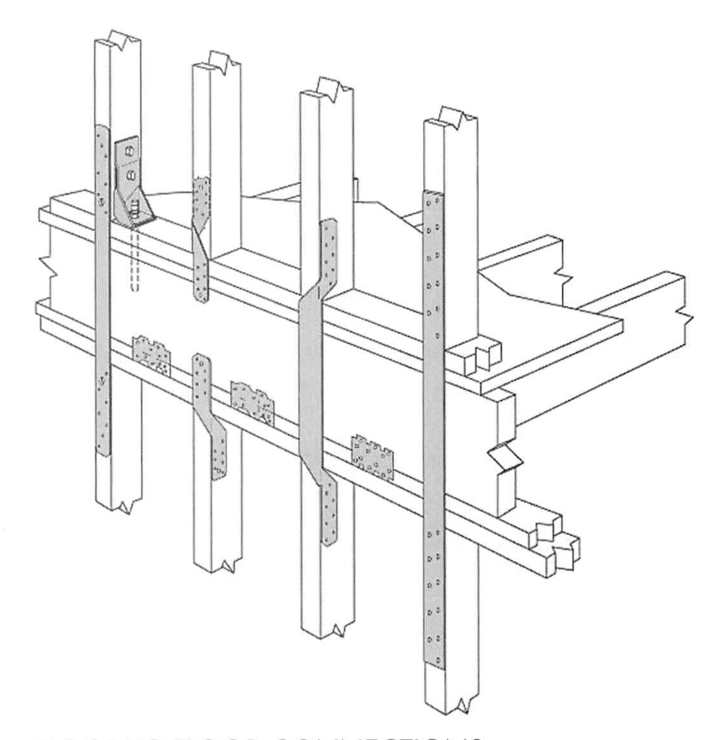

FIGURE 2.1.6 FLOOR-TO-FLOOR CONNECTIONS

rial used to tie building components together is plywood or OSB sheathing which provides a hold-down path and resists shear or racking forces. The lower portion of the plywood sheathing should connect the lower to the upper floor and be nailed into the bottom plate, the wall studs, the top plate, and the second floor rim joist, the upper portion should connect the rim joists, bottom plate, and studs to the top wall plate (Fig. 2.1.7). In areas of high seismic probability, engineers may specify a top grade of plywood, Structural #1, in lieu of regular rated sheathing. Consultation with a licensed engineer is recommended. Wall-roof connections in seismic areas are reviewed in Chapter 3.

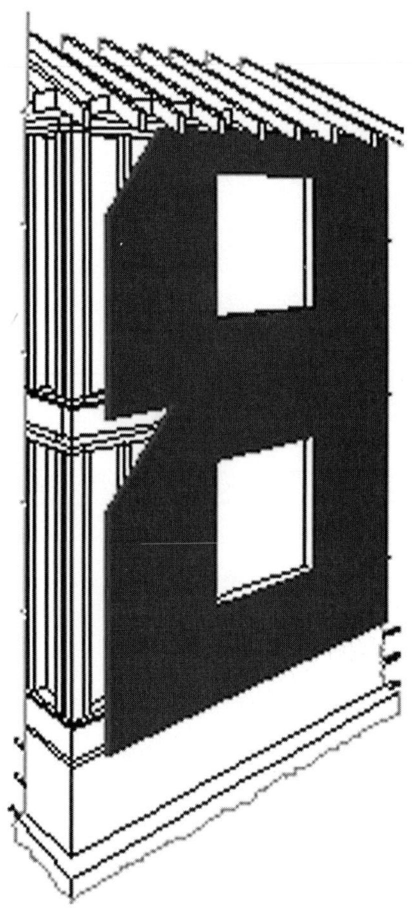

FIGURE 2.1.7 SHEATHING RESISTS SHEAR AND RACKING

ADVANTAGES: This method provides structural continuity to the entire house.

DISADVANTAGES: This method is costly, and requires removal of siding and possibly sheathing.

WOOD-FRAME WIND RESISTANCE

ESSENTIAL KNOWLEDGE

Exterior walls, in combination with interior shear walls that brace them, resist perpendicular and lateral loads and uplift forces generated by high winds (which can tear off roofs and porches). The increasing amount of damage caused by hurricanes to the Gulf and Atlantic coastal states and the mounting costs to repair and replace existing structures demonstrate the need to better design, build, and retrofit houses for wind resistance.

While code requirements regarding wind resistance are extensive for new home construction, there are relatively few requirements for rehab of existing houses (unless the work is extensive enough to warrant bringing the entire building up to code). Exceptions to this occur in some municipalities, such as Dade County, Florida, which have specific requirements for rehabbing existing buildings. Rehab guidelines regarding wind resistance being considered by local, state, and model code agencies include the following recommendations. A key aspect of these considerations is whether to provide prescriptive "fixes" for simple building configurations or whether to require that a professional engineer or architect prescribe the specific details. The latter is recommended for complicated building geometries.

TECHNIQUES, MATERIALS, TOOLS

1. REINFORCE CONNECTIONS OF WOOD-FRAME WALLS TO FOUNDATIONS.

Depending on the type of foundation and access possible, there are a number of anchor bolts, straps, or threaded-rod connections that can be used to reinforce the connection of the frame to the foundations. Many of these connectors are similar to those used in seismic areas. In northern areas, crawl spaces might provide accessibility, while foundations in southern regions are slab on grade or, along the coast, pilings or piers. Typical rehab options for reinforcement are shown in Figs. 2.1.3 and 2.1.4.

ADVANTAGES: These are relatively simple, cost-effective remedies.

DISADVANTAGES: These remedies may involve removing exterior or interior finishes.

2. REINFORCE WOOD-FRAME WALLS FOR SHEAR RESISTANCE.

Winds cause lateral forces on buildings that can, in severe cases, displace and collapse the building walls. Resistance to these forces is provided by exterior and interior shear walls that brace the building's structure and transfer loads to the floors and foundations. The most effective shear walls are made of plywood or OSB. Alternative systems include other structurally approved sheathings, let-in wood bracing, metal strapping, T-bracing, or special stud connectors. For buildings that have qualifying structural sheathings, a cost-effective retrofit is to increase the number of fasteners or connectors from the sheathing to the studs or to add resistance with foamed-in-place adhesives such as Foam Seal® products. This can be accomplished with the removal and replacement of the siding.

ADVANTAGES: This method is a simple way of providing shear resistance.

DISADVANTAGES: Removal of siding will be required, and removal of sheathing may be required if it is not structurally adequate.

3. REINFORCE CONNECTIONS OF WOOD-FRAME WALLS TO FIRST FLOOR.

Connections are simple to make in new construction but difficult in rehab work unless the siding is removed to expose the wall sheathing. Metal straps similar to those used in seismic areas provide structural continuity from one building component to another (Figs. 2.1.5 and 2.1.6).

ADVANTAGES: This method provides a continuous load path.

DISADVANTAGES: Removal of siding and sheathing may be required.

4. REINFORCE CONNECTIONS OF WOOD-FRAME WALLS TO ROOF TRUSSES AND RAFTERS.

The connection of the exterior walls to the roof structure is the key element in transferring wind loads to the building frame and in preventing uplift forces from tearing off the roof. While new houses in high-wind areas are required to have metal connectors, older houses most likely do not have them. The simplest connection is made from the outside after the soffit is removed (Fig. 2.1.8). This juncture can be reinforced from the inside, but the top portions of the wall as well as portions of the ceiling at the wall have to be removed for access. See Chapter 3 for further discussion.

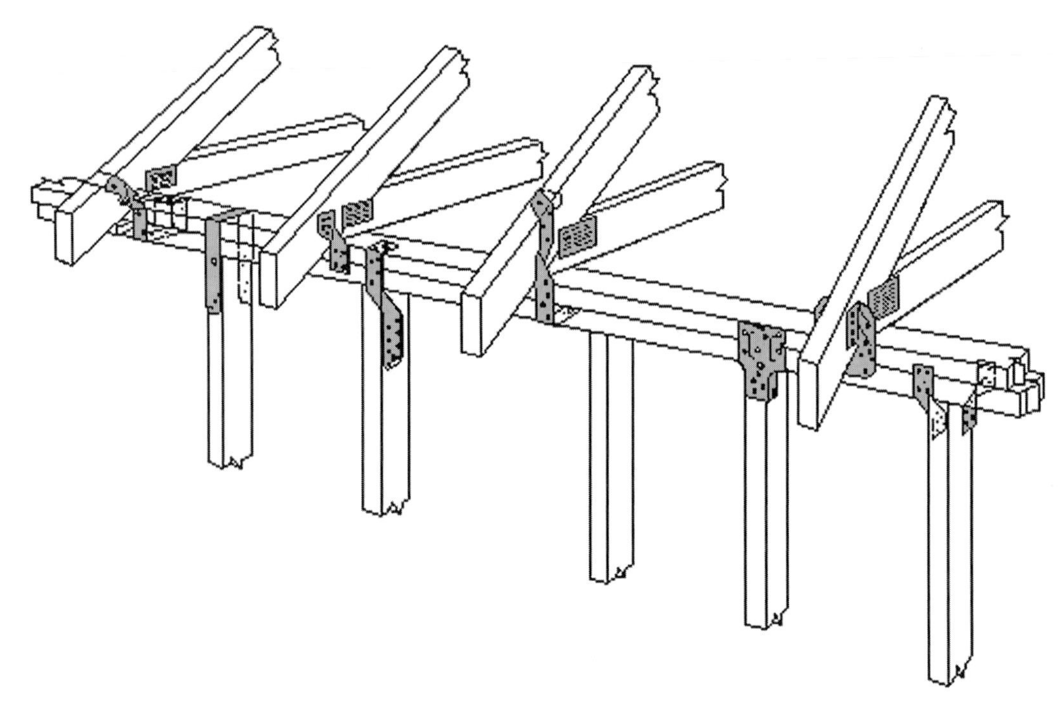

FIGURE 2.1.8 WALL-ROOF CONNECTORS

ADVANTAGES: This is an effective means of providing uplift resistance.

DISADVANTAGES: Removal of soffit material will be required.

5. REINFORCE CONNECTIONS OF WOOD-FRAME WALLS TO ROOF OVERHANGS.

The most vulnerable portion of a building for wind uplift is the connection of roof overhangs and walls. The typical connection of the wall to the ladder overhang (Fig. 2.1.9) is inadequate if the overhang exceeds 1' in depth, and uplift forces can lead to separation of the ladder from the wall. The preferred detail on new or repaired construction is to use *lookouts* tied to the top of the exterior wall and anchored back to the adjoining truss or rafter (Fig. 2.1.10).

ADVANTAGES: This method is an effective means of providing uplift resistance.

DISADVANTAGES: Require removal of soffit and modification to the gable end will be required.

REINFORCING EXISTING MASONRY WALL CONSTRUCTION

ESSENTIAL KNOWLEDGE

It is often difficult to reinforce existing masonry walls for seismic or high-wind resistance. Reinforcement strategies should be developed for individual buildings on a case-by-case basis by a licensed professional. Masonry buildings generally perform well in high wind as long as they are rein-

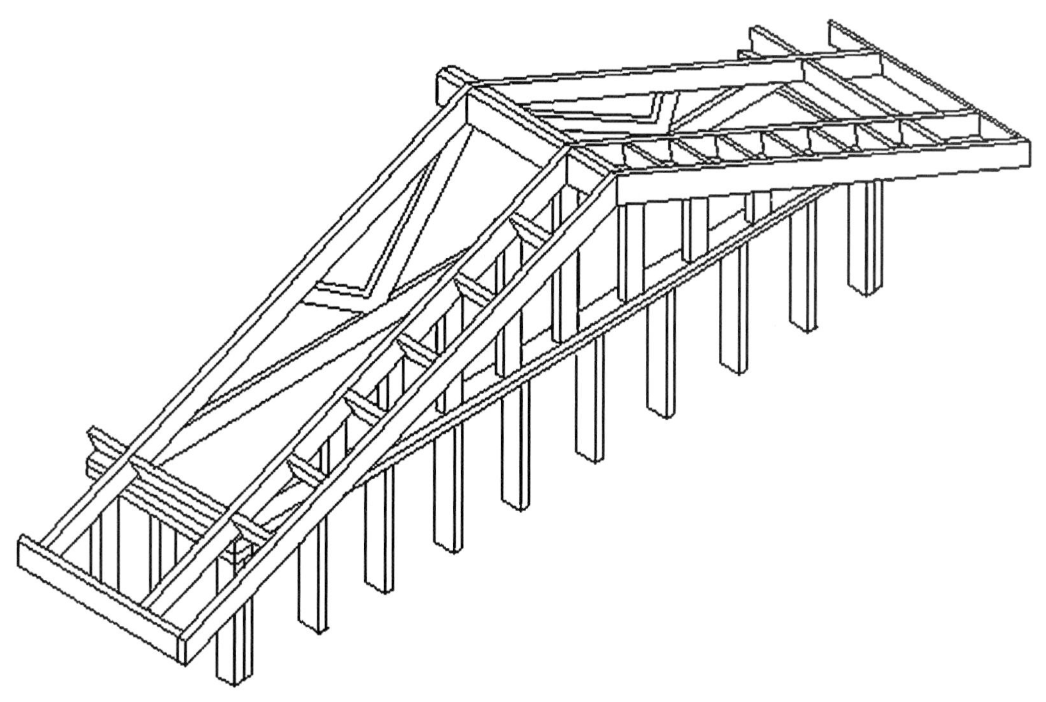

FIGURE 2.1.9 SHALLOW LADDER OVERHANG

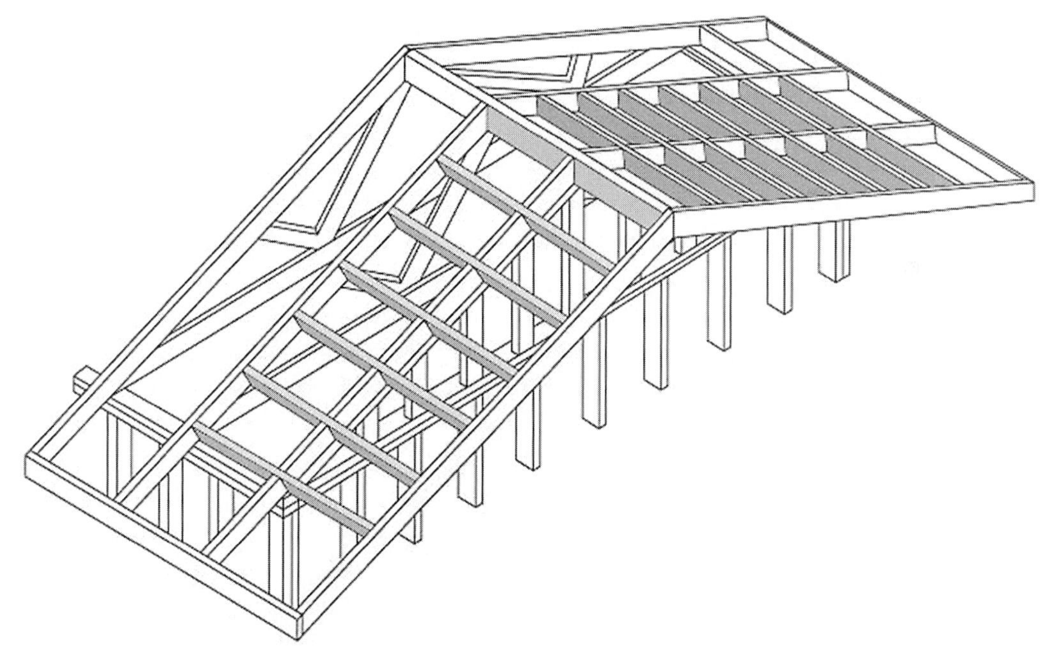

FIGURE 2.1.10 DEEP OVERHANGS SUPPORTED BY LOOKOUTS

forced in accordance with codes and as long as the connections to the roof structure are adequate to prevent uplift failure. Unreinforced masonry buildings perform poorly in seismic areas. The connections to roof structures and secondary structures such as porches are similar in concept to those used for wood-frame construction but are adapted to masonry. Typical masonry-to-roof connectors are illustrated in Fig. 2.1.11.

ADVANTAGES: This is an effective means of providing uplift resistance.

DISADVANTAGES: Removal of soffit material will be required.

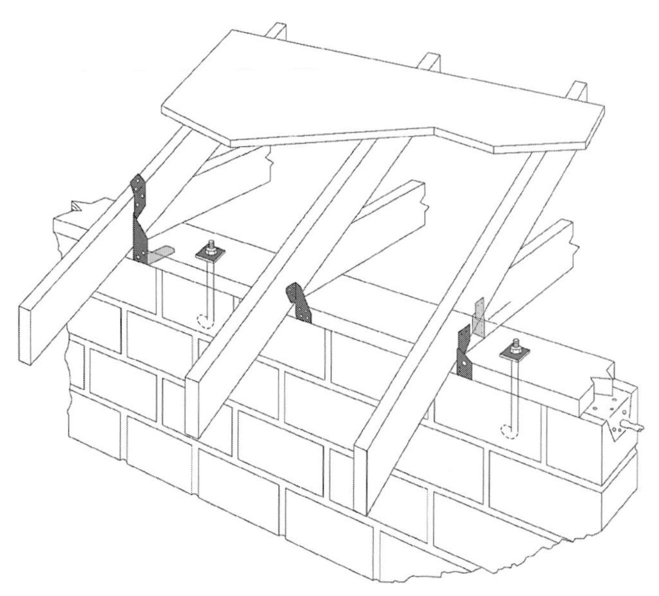

FIGURE 2.1.11 MASONRY-TO-ROOF CONNECTORS

MOISTURE DETERIORATION

ESSENTIAL KNOWLEDGE

Water absorbed by structural wood framing can raise the wood framing's moisture content, reduce its compressive and tensile strength, ultimately cause rot and decay, and also attract termites. The most critical points of the building envelope susceptible to leakage are tears or gaps in the roofing material; flashings and penetrations of the roof plane; roof-wall connections; wall penetrations such as windows and doors; rainwater penetration through siding materials; and wall-foundation connections. Roofing water-related problems are discussed in Chapter 3, wall-penetration leaks at wall openings are covered in Chapter 4, and water penetrations through wall materials are covered in other chapters.

Wall-foundation junctures are particularly critical because runoff from roofs and walls collects on the ground at that location. If the sill and floor assembly are not sufficiently elevated, rot will occur. Older timber-framed and balloon-framed structures with sill plates that rest on a few courses of stone (or, occasionally, directly on grade) are at greatest risk. Platform-framed houses that have foundation walls with the sill a minimum of 6" to 8" above grade (to comply with code minimums) are less susceptible to rot and decay from moisture. However, unless the grade below the siding is sufficiently sloped away and kept clear of debris and plantings, moisture can wick up through the siding and cause decay. Rot and decay cannot progress in the absence of moisture.

Sills can be inspected from inside the building in the crawl space or from outside by removing a portion of the siding and sheathing. The condition of the wood can be checked with a sharp object such as a screwdriver or pocket knife. Sound wood will split into fibrous splinters, while decayed wood will separate into small chunks of a dark brown, black, or gray color. Decay can also be revealed by rapping the surface of the wood member; a dull, hollow sound frequently indicates decay below the surface.

Decayed sills can be replaced with full-sized members, partially replaced with built-up lumber, or stabilized with structural epoxy conservation techniques.

TECHNIQUES, MATERIALS, TOOLS

1. REPAIR SILL WITH BUILT-UP LUMBER.

If the sill is a heavy timber section (4x6 to 8x8) and the wall studs are 3" or 4" wide, the house is most likely of post-and-beam construction. The roof and floor loads are transferred by means of beams (girts) to the columns and the studs between columns carry very little weight. Accordingly, it is possible to temporarily support the building at its bearing columns and replace sections of the sill below. If the building is balloon-framed or platform-framed, with individual studs carrying the load, the wall has to be supported along its length. Once the load is taken by the shoring, deteriorated sections of the sill can be removed with a reciprocating saw and a mallet and chisel. Pressure-treated sections of lumber can be scabbed (spliced) into the affected area (using APA-approved gap-filling adhesives) and fastened with galvanized drywall screws, spikes, or other rust-resistant fasteners.

ADVANTAGES: This is a relatively simple fix for sills requiring isolated repairs.

DISADVANTAGES: An accurate assessment of the sill condition is required; sections of rotting sill may be overlooked.

2. REPLACE LARGE SECTIONS OR THE ENTIRE SILL.

If significant decay runs the length of the sill, the sill should be replaced in its entirety. The exterior wall can be supported by jacking timbers placed next to the plate, running perpendicular to the joists (Fig. 2.1.12). Sections of the sill can be cut with a chain saw or reciprocating saw between joists that

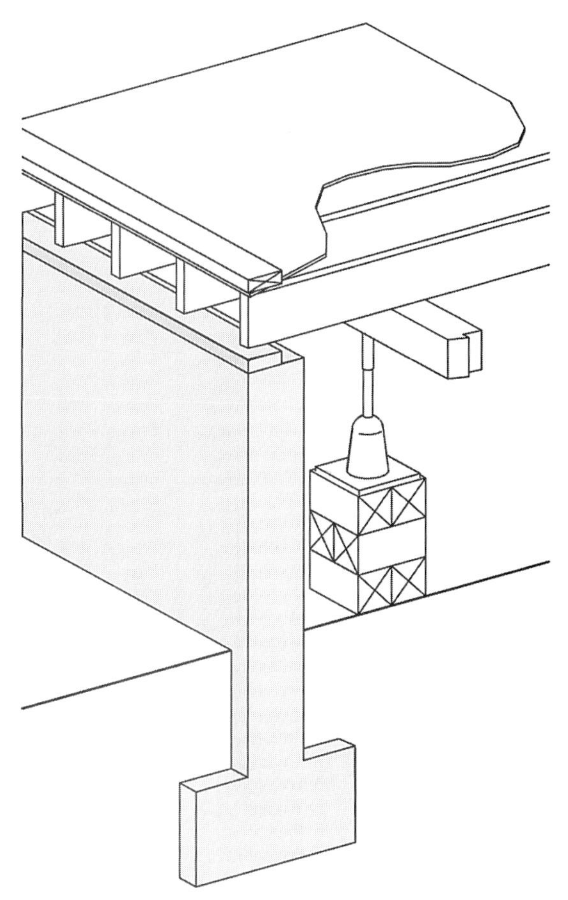

FIGURE 2.1.12 JACKING FLOOR JOISTS

frame into the sill. A new sill is placed on top of the foundation. If the floor joists do not rest directly on top of the sill, they can be hung from the new sill with joist hangers or, if the ends of the joists are not decayed, they can be mortised into the new sill. Replacing a sill is much easier with stud-framed houses, as individual joists rest on top of the sill. Because each stud of a load-bearing wall carries a relatively small portion of the load, the process of supporting the floor joists while removing the rotted sill or rim joist is relatively simple. Once the new sill is anchored to the foundation, the grade next to the wall should be sloped to provide drainage away from the wall.

ADVANTAGES: Replacing a major portion or the entire sill is a way to comprehensively address the problems of decay and may be more cost-effective than a series of small, interim repairs.

DISADVANTAGES: This method is costly and might require extensive exterior sheathing and siding repairs.

3. REPAIR PORTIONS OF THE FOUNDATION OR SUPPORTING COLUMNS USING EPOXY TECHNIQUES.

Small portions of the foundation or columns that support the structure above can be reconstituted and consolidated using liquid epoxy or epoxy putty (Fig. 2.1.13). This is particularly appropriate if the building is of historic significance.

ADVANTAGES: The existing structure can be repaired without being removed.

DISADVANTAGES: This technique is time-consuming and not practical for large areas of work.

MITIGATING INSECT DAMAGE

ESSENTIAL KNOWLEDGE

Destructive insects include termites, carpenter ants, and wood-boring beetles (Fig. 2.1.14). Termites access aboveground wood through cracks in foundation walls or slabs or build tubes from the ground

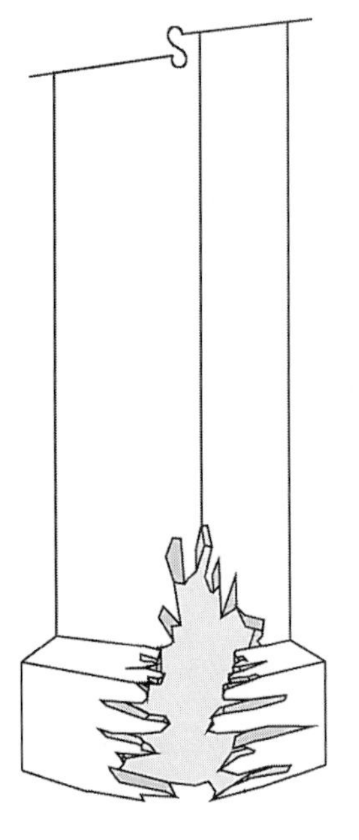

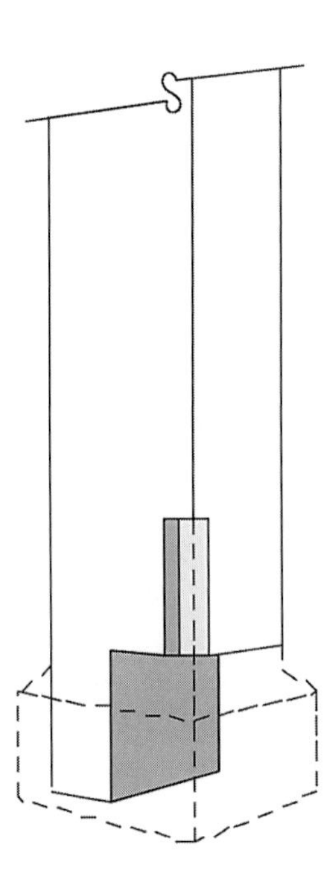

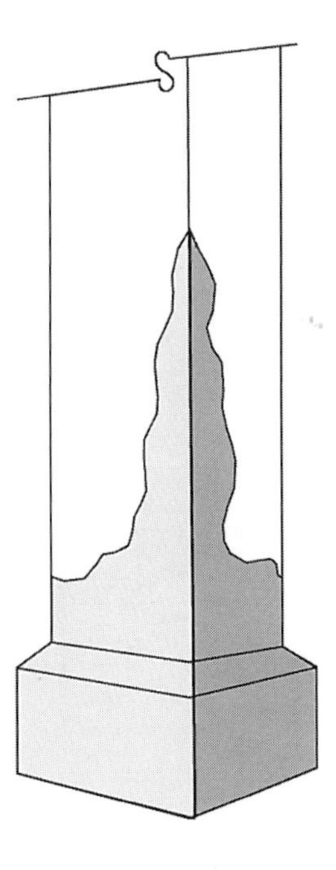

FIGURE 2.1.13 EPOXY REPLACEMENT

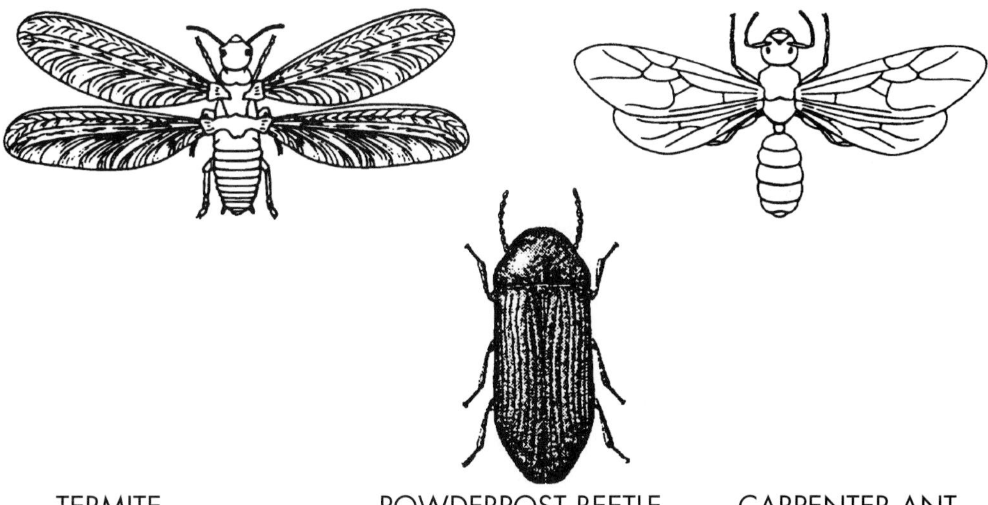

FIGURE 2.1.14 TERMITE POWDERPOST BEETLE CARPENTER ANT

up to the wood above to provide the necessary moist environment. Having infested the wood, they can live there indefinitely with no ground contact if the moisture level is adequate. Sources of moisture include roof leaks, condensation, or plumbing problems. Termites in crawl spaces may build free-hanging tubes from wood members to the ground. Evidence of termite infestation includes the presence of mud tubes, damaged wood, active swarms of winged termites or large numbers of discarded wings, or evidence of conducive conditions (e.g., moisture, wood-to-ground contact, inadequate ventilation, settlement cracks, and other likely entry points). Detection tools include visual inspection, probing of the wood surface with a screwdriver or pocket knife, sounding (tapping) of the affected area with a hard object such as the handle of the screwdriver, the use of a moisture meter to detect likely environments, infestations, listening devices, electronic gas (methane) detectors, and fiber-optic devices to inspect areas otherwise inaccessible to visual inspections. Many species of termites prefer wood that has been previously invaded by fungi. To guard against drying, termites consume wood only until the outer shell remains, leaving the wood looking intact. Wood that has been tunneled by termites will sound hollow when tapped with a solid object. Termites will attack all types of wood including redwood, cypress, and junipers if the wood has aged and the chemicals that provide termite-resistance (alcohols, oils, gums, and resins) have leached out. Termites have been known to penetrate and damage many noncellulosic materials such as drywall, plaster, stucco, and plastics. Items damaged include some softer metals (e.g., lead, copper, aluminum). Insulation board used as a substrate for EIF systems has often been riddled by termites leading to the prohibition of foam plastic within 8" of the ground in states with very heavy infestation of termites.

Ants are the most reported pests in many parts of the country and range from the arctic to the tropics. Carpenter ants, the major group that damages buildings, tunnel through wood but do not consume it for food. Main colonies of carpenter ants, given their name because they typically dwell in and excavate wood, are located in trees surrounding the infested structures. Inside buildings, satellite colonies will nest in a variety of voids including walls, hollow doors, behind appliances, in floor cavities and attic rafter spaces, under kitchen cabinets, and bathroom fixtures. Carpenter ant infestations can be located by their very presence or by piles of *frass*, pieces of dead ants and other insects mixed in with bits of wood.

There is a variety of wood-boring beetles. Among the best known and most destructive are the powderpost, roundhead, and flathead beetles. Some beetles attack both hardwood and softwoods, usually limiting their feeding to the sapwood portion. Their presence is usually indicated by frass and a number of exit holes in wood, although the number of holes does not necessarily indicate the activity or severity of infestation. The potential for damage is greatest when the infestations are new and the number of exit holes is low. Beetles are typically introduced into structures in building materials that have been infested at lumberyard stockpiles, although they may also enter homes in finished wood products such as flooring, paneling, furniture, and firewood. Dead tree limbs may serve as a habitat from which flying adults may enter the house.

TECHNIQUES, MATERIALS, TOOLS

MITIGATE AND CONTROL INSECTS.

Mitigation methods to reduce the likelihood of termite infestations include removal of all cellulous material on or beneath the soil adjacent to structures or in crawl spaces; eliminating details that allow materials to continue from the exterior wall into the soil; providing adequate clearance between grade and structural members to allow access and inspection of termite tunnels; eliminating dirt-filled porches, steps, and similar raised attachments; providing termite shields; and using pressure-preservative-treated lumber. Treatment strategies include the use of liquid termiticides, termiticidal foams that fill cracks and gaps in materials and can be injected into soils, borate insecticides, and termite baits. The variety of treatment products is widespread and new products are being developed. The effectiveness of individual products, especially new ones, may vary depending on local soil and building-related conditions. Local pest control experts should be contacted for site-specific recommendations. Certain treatments may be prohibited by local authorities.

Mitigation methods for subterranean termites are of little benefit against drywood termites since infestations originate from *swarmers* entering through vents, cracks, or other openings. Drywood termites live within the affected wood. Their presence is indicated by piles of fecal pellets or discarded wings adjacent to holes or joints in the affected wood. Methods to control drywood termites include the use of pressure-preservative-treated wood, sprayed insecticides, silica aerogel dust (a desiccant), borate dust, fumigation, heat treatment, and drill-and-treat with termiticides. Some new, and as yet unproven, strategies include the use of microwave equipment and electrocution. Local pest control experts should be contacted for site-specific recommendations. Ant mitigation methods include reducing moisture conditions, sealing cracks in the building's exterior, trimming trees away from the house roofs, keeping log piles away from houses, and eliminating wood-to-soil contact. Treatment strategies include a variety of sprayed insecticides, baits, and soil treatment. Treatment of affected houses for wood-boring beetles includes removing infested wood, use of sprayed insecticides, fumigation moisture reduction, and use of borate sprays.

ADVANTAGES: Soil treatment and elimination of cellulose in backfill can be effective deterrents.

DISADVANTAGES: New chemical treatments need replenishing sooner and are less effective than those approved in the past. Treatments can be costly.

FIRE DAMAGE TO WOOD FRAMING

ESSENTIAL KNOWLEDGE

Damage from fire can range from the total loss of a building and its contents to minor inconvenience from smoke odors. The process of determining the restoration requirements of a fire-damaged building varies considerably with the building location and extent of damage. Insurance adjusters often make settlement offers based on their own evaluation of restoration needs, although they may employ consultants on more complex projects. Recommendations may also originate from local fire marshals, building department officials, contractors, consulting engineers, industrial hygienists, public adjusters, and architects hired by the building owner. Unless the damage is limited, the restoration process can be complicated, involving structural, electrical, HVAC, and plumbing systems, as well as building finishes. In addition, significant health and comfort issues arise from the residual smoke, combustion gases, moisture from fire department hoses, and the existence of products containing asbestos. For these reasons the selection of a restoration contractor who is experienced and knowledgeable in current techniques is critical. At least one national association, the Association of Specialists in Cleaning and Restoration (ASCR) manages training and certification programs and publishes a restoration guideline.

TECHNIQUES, MATERIALS, TOOLS

FIRE-DAMAGE RESTORATION.

The first step in a restoration project is to assess the damage to the wall structure. In 2x4 construction, significantly charred members are generally removed in their entirety. Heavy timber construction can remain (according to the American Society of Civil Engineers), once the char is removed and if the remaining section is still structurally adequate (after a reduction in size factor of 1/4" on all sides). Char is removed by scraping and abrasive blasting. It should generally be removed because it holds odors, although encapsulating coatings inhibit their transmittal. New construction, replacing the damaged construction, should meet codes for new construction. Smoke-damaged materials should be cleaned and deodorized as necessary. The use of ozone generators, sometimes used to remove odors and contaminants, is controversial and considered by a number of specialists to be ineffective and potentially dangerous (see Further Reading).

ADVANTAGES: Restoration can be relatively effective in removing odors when fire damage is slight.

DISADVANTAGES: Restoration after serious fires is expensive and complicated. Finding competent restoration professions is sometimes difficult.

2.2 MASONRY AND BRICK VENEER

Brick and stone masonry are among the oldest, long-lasting, and most versatile materials. Throughout the United States many brick homes, centuries old, continue to perform well. In many regions brick is the predominant building material because of its low maintenance, noncombustibility, availability, moisture resistance, and aesthetic appeal.

Any corrective work should be preceded by a careful visual assessment of the wall's conditions to determine overall patterns of deterioration and distress so that underlying problems can be appraised and corrected. Some common problems include foundation displacement (see Chapter 1), water penetration into the wall assembly, inappropriate material choices, poor construction practices, stresses caused by expansion and contraction due to temperature changes, shrinkage of the wood structural walls, and routine aging of the masonry facing and joints.

This section focuses on repair and rehabilitation of brick masonry, primarily clay and concrete brick veneer wall construction, as this is the most common construction type, although many of the recommendations apply also to stone, concrete block, and solid brick construction. Topics include cleaning, protective coatings, repointing, and repair.

CLEAN EXISTING MASONRY WALLS

ESSENTIAL KNOWLEDGE

The decision to clean a masonry veneer facade requires careful consideration because the cleaning process may remove weathered material as well as accumulated dirt. It might be justified if the dirt and pollutants (such as that from acidic rain, efflorescence, bird droppings, and deteriorated paint) are having a harmful effect on the wall; however, "lightening up" the facade for cosmetic reasons may not be prudent. The benefits of cleaning should be weighed against the possibility of adverse affects on the masonry surfaces and mortar joints, as well as on flashing, windows, and other elements.

The least invasive cleaning is usually recommended; improper cleaning or overcleaning can cause irreparable damage. Cleaning strategies for historic buildings should be reviewed with a restoration professional. All cleaning procedures and materials should be tested prior to the start of the project. For large jobs, the test area should be a minimum of 20 ft^2. Before chemical cleaners are used, the wall should be saturated with water to avoid staining by heavy concentrations of cleaning agents. A waiting period of at least one week after finishing the test area is recommended in order to judge the results of the cleaning procedure, especially if chemical agents are used. For recommendations on cleaning specific stains see Further Reading.

TECHNIQUES, MATERIALS, TOOLS

1. CLEAN WITH BRUSH BY HAND.

This technique employs a variety of cleaning agents including water, detergents, proprietary cleaners, and acid. Efflorescence can often be removed by dry brushing, with pressurized water, or with proprietary cleaners. Dirt can be removed with water or with a detergent solution such as trisodium phosphate and laundry detergents dissolved in water. Many stains can be removed with conventional kitchen cleaners. Other stains, resulting from leaching of salts or coloring agents within the brick, require acid cleaners, which should be used very carefully in diluted form. Some acids, such as hydrochloric acid (muriatic acid), can seriously degrade mortar. Acid can also discolor lighter masonry surfaces and damage metal, glass, marble, terra cotta, limestone, and cast stone surfaces, and can leave a white film that is difficult to remove. Walls treated with acid must be thoroughly flushed with water after cleaning.

ADVANTAGES: This is the easiest and most conservative approach, employs the widest variety of cleaning options, and allows for the most cost-effective approach to be used. The appropriateness of cleaning strategies can be confirmed prior to large-scale application.

DISADVANTAGES: This method is appropriate only for relatively small areas, is time-consuming, and requires direct access to wall surfaces.

2. CLEAN WITH PRESSURIZED WATER.

Useful for covering large areas, pressurized cleaning may be accomplished with low- or moderate-pressure water, steam, or water in combination with detergents or other cleaning solutions. Walls should be saturated with water prior to cleaning and completely flushed after cleaning. Care should be taken with acidic compounds as discussed above.

ADVANTAGES: This method is cost-effective for large areas and can reach heights of 100'.

DISADVANTAGES: Specialized equipment is required. Nozzle pressures in excess of 700 psi may damage brick, especially sand-finished material, and erode mortar joints. Certain stains may not be removed as effectively as brush cleaning by hand. Disposal of water run off may be a problem. Excess water can bring soluble salts from within the masonry to the surface. This technique cannot be used during periods of freezing weather. Steam cleaning with or without chemicals may be useful in removing paint and embedded grime, but requires careful analysis, testing, and experienced professionals. This method can be costly.

3. CLEAN BY ABRASIVE BLASTING.

Abrasive blasting, usually with sand, is often considered a means of last resort, and in many cases is prohibited because it can erode ornamental details and destroy or scar brick and stone faces. Less abrasive and softer aggregates than sand, such as glass and plastic beads, and organic matter, such as finely ground nut shells, wheat starch, peach and apricot pits, and cherrystones, are sometimes used on small sections of decorative brick, stone, or metal elements.

ADVANTAGES: This method will clean when other techniques will not.

DISADVANTAGES: This method is potentially destructive and should be used only after careful analysis and testing. It may be prohibited.

APPLY COLORLESS PROTECTIVE COATINGS

ESSENTIAL KNOWLEDGE

Colorless coatings are sometimes considered for masonry walls in order to enhance water resistance or to repel graffiti. The chemicals used fall within two classifications: films and penetrates. These two have significantly different physical properties and performance. Whereas these coatings may have some limited usefulness, in many cases they provide little or no advantages, require frequent replacement, and can have adverse effects. Brick manufacturers should be contacted for recommendations on the use of colorless coatings.

TECHNIQUES, MATERIALS, TOOLS

1. APPLY FILM COATING TO MASONRY WALL.

Products include acrylics, stearates, mineral gum waxes, urethane, and silicone resins. The large molecular size of these materials prevents them from penetrating into the masonry.

ADVANTAGES: Film coatings can reduce the absorption of some bricks and can bridge hairline cracks. They can keep surfaces clean and help prevent graffiti from penetrating into the masonry surface.

DISADVANTAGES: Film coatings can inhibit evaporation of water within the masonry through the exterior face, which can cause the coating to cloud or spall under some freeze-thaw conditions. This method is generally not recommended in such environments. Sheen or gloss may darken material. Film coatings are vulnerable to cracking due to thermal fluctuations. Urethane often breaks down under ultraviolet (UV) light. Silicones do not chemically bond with substrate, and as a result have a short life.

2. APPLY PENETRATING COATING TO MASONRY WALL.

These coatings typically penetrate into the masonry to depths up to 3/8", due to their small molecular structure. Materials include silane and siloxanes that chemically bond with silica- or aluminum-content materials to make them water repellent. These coatings are not generally necessary on new walls or brick veneer walls with drainage cavities.

ADVANTAGES: These coatings can last up to 10 years, decrease absorption, and increase water repellence. Unlike film coatings, penetrating coatings allow walls to breathe. Siloxanes have been shown to be effective on some multiwythe brick barrier walls where water penetration is a problem.

DISADVANTAGES: Penetrating coatings may have a limited lifetime and cannot be applied over film coatings. They may react with other building materials, can kill vegetation and emit harmful vapors, and will not seal cracks in mortar joints.

REPOINT EXISTING WALLS

ESSENTIAL KNOWLEDGE

While the service life of many types of brick can exceed 100 years, the longevity of mortar joints, depending on the exposure, is closer to 25 years, according to the Brick Industries Association (BIA).

At some point the mortar joint will fail, allowing water to enter the wall cavity. Conditions that require repointing may include mortar erosion more than 1/4", crumbling mortar, and hairline cracks in the mortar and between the mortar and brick.

TECHNIQUES, MATERIALS, TOOLS

REPOINT EXISTING WALL.

Visual observation in combination with a light scraping with a metal tool can detect most deficiencies. Other conditions requiring repairs beyond repointing are discussed in Section 3.4. Where repointing work is undertaken on houses of special architectural or historical significance, advice should be sought from a preservation specialist. Portland cement mortar was not used before the beginning of the twentieth century. To avoid serious brick damage the compressive strength of the repointing mortar should be similar to or weaker than that of the original mortar. If it is not, dead loads and stresses from the expansion and contraction of the brick can transfer loads through the new mortar into the brick and can spall and crack the brick face. Mortars used in more recent construction include types N and O (Table 2.2.1).

TABLE 2.2.1 MORTAR TYPES AND INGREDIENTS BY VOLUME

Type	Cement	Hydrated Lime	Sand
N	1	1	6
O	1	2	9
K	1	4	15

ADVANTAGES: Repointing walls can stabilize deterioration, strengthen walls, and provide weather-tightness.

DISADVANTAGES: Repointing walls can be costly and may require scaffolding. Skilled and thorough mechanics are required.

REPAIR EXISTING MASONRY WALLS

ESSENTIAL KNOWLEDGE

There are certain conditions where repointing alone is not effective and replacement of a portion of a wall may be required. Some of these include:

- *Wall cracking associated with thermal movement.* Such cracks are cyclical and will open and close with thermal swings. These cracks may gradually expand as dislodged mortar accumulates in the crack after each cycle. The cracks should be cleaned and protected with flexible sealants. Remortaring cyclical cracks will prevent them from closing and may lead to further cracking. In some instances, the masonry may need to be cut and expansion joints installed.

- *Wall cracking associated with moisture penetration and caused by freeze-thaw cycles and corrosion.* Examples include cracking around sills, cornices, eaves, parapets, joints between dissimilar materials, and other elements subject to water penetration and freezing; cracking around clogged or nonfunctioning weep holes at lintels and at the base of brick veneer cavity walls. A number of companies, including Mortar Net™ and Heckman Building Products, Inc., make plastic mesh products for cavity wall construction that suspend mortar droppings above the weep holes, thereby reducing the chance of blocking them with mortar debris (Fig. 2.2.1). Mortar Net™ also makes a vertical insert between bricks that acts as a continuous weep (Fig. 2.2.2).

■ *Wall cracking associated with failure of structural elements.* Aboveground examples include cracking or displacement of brick over openings resulting from deflection or failure of lintels or the deterioration of mortar joints in masonry arches; cracking from outward displacement of sloped roofs due to lack or failure of collar ties; bulging and cracking of walls caused by deteriorated or inadequate wall ties; cracking due to inadequately supported point loads; cracking due to ground tremors, nearby construction, or heavy traffic. Light gauge (22 or 24 ga) corrugated wall ties typically used in residential construction are not recommended, according to the Brick Industry Association, for three reasons: (1) the tie shape allows water to flow more freely to the interior of the wall, (2) they are susceptible to corrosion and, (3) they have poor strength to transfer loads between the brick wythes and the building structure. Adjustable ties, similar to those recommended for use with metal studs, (Fig. 2.2.3) are preferred. Serious structural problems require a professional engineer's assistance in determining appropriate corrective measures.

■ *Deteriorated masonry*. A number of factors, in addition to structural distress, can contribute to the deterioration of a masonry wall, including weathering effects of rain, UV light, temperature

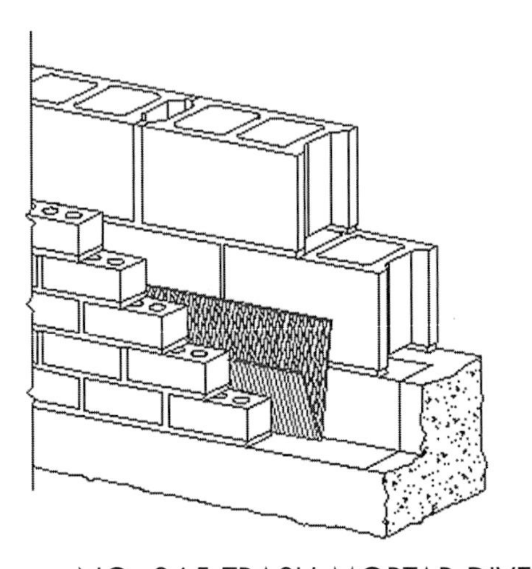

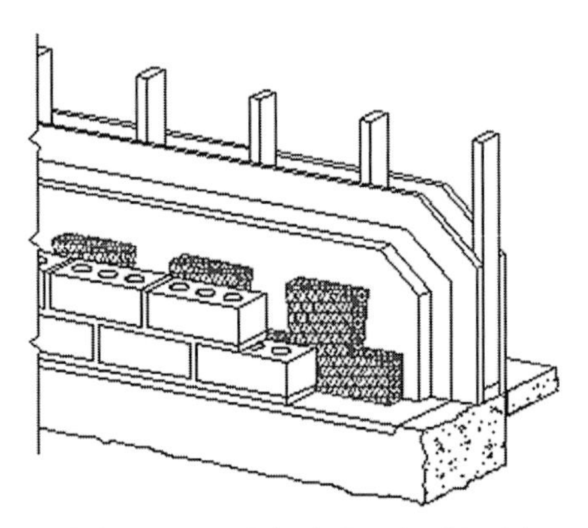

FIGURE 2.2.1 NO. 365 TRASH MORTAR DIVERTER CAVITY DRAINAGE SYSTEM

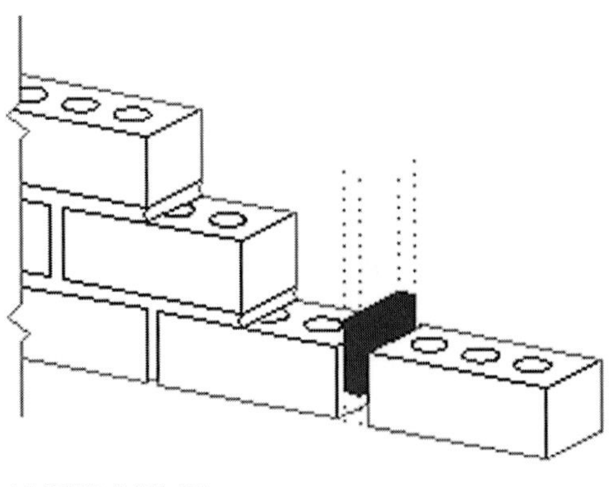

FIGURE 2.2.2 WEEP VENT

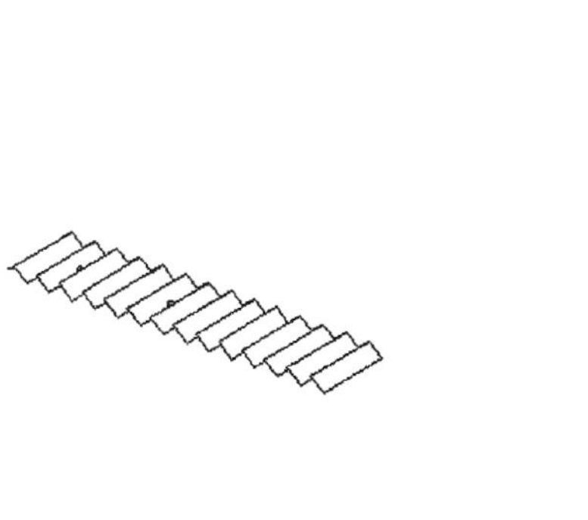

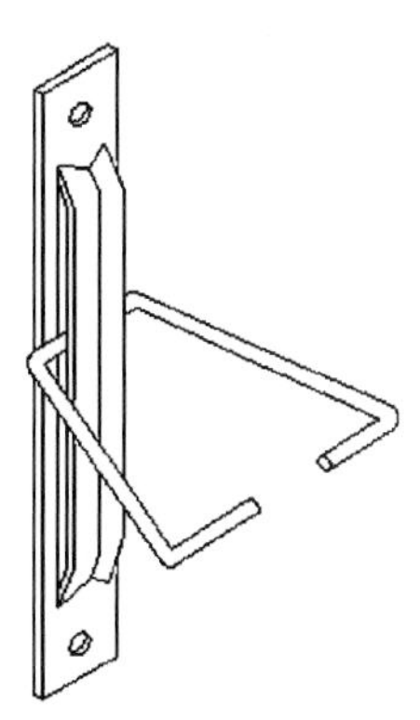

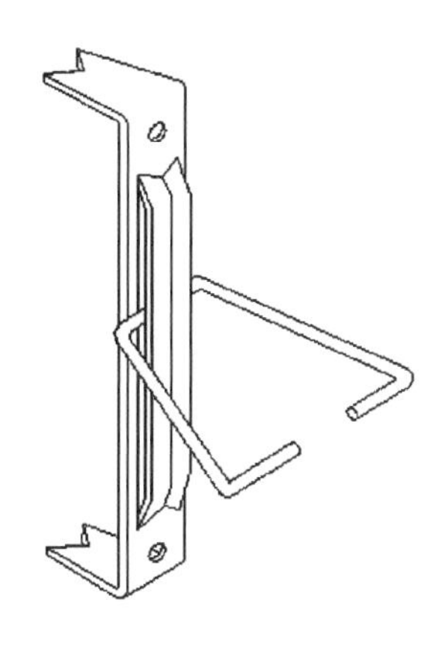

FIGURE 2.2.3 CONVENTIONAL CORRUGATED TIE PREFERRED WIRE TIES

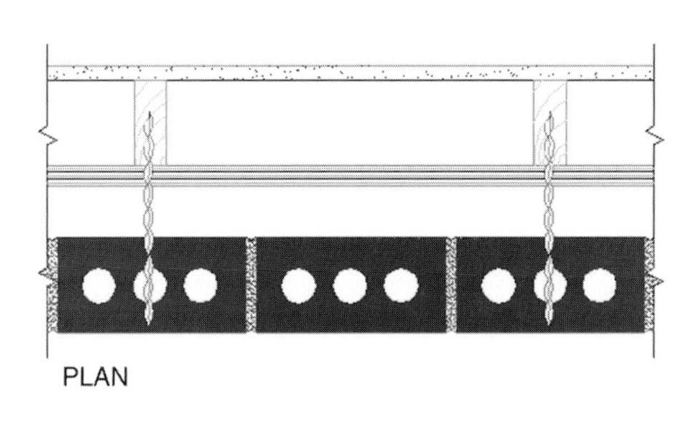

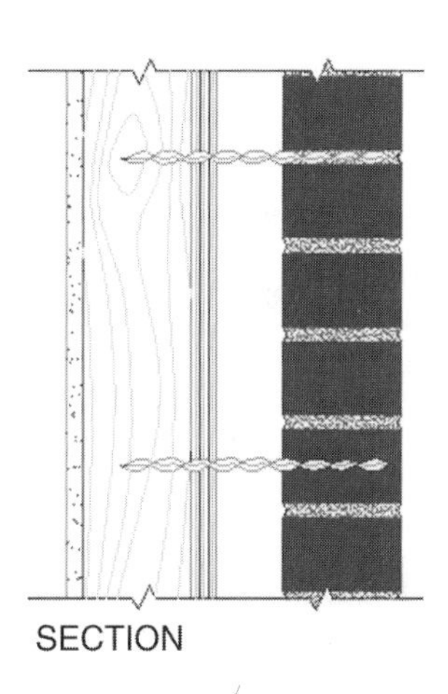

FIGURE 2.2.4 STABILIZATION TIES

changes, as well as the effects of chemicals in the air or ground, inappropriate cleaning or coatings, and erosion from faulty leaders and downspouts.

TECHNIQUES, MATERIALS, TOOLS

REPAIR MASONRY WALL.

If individual bricks or masonry units, or sections of the masonry wall, require replacement, they may be removed relatively easily by cutting out the units to be replaced and replacing them with new material. The arch action of masonry walls can often prevent adjacent sections from collapsing if the area to be removed is small. Alternatively, in the event that the masonry ties are missing or deteriorated, a number of companies make stabilization systems that connect exterior brick wythes with backup walls (Fig. 2.2.4). These are either mechanically attached pins or ties that are drilled directly through the veneer into its substrate, or ties that are anchored to the substrate with epoxy cements (see Product Information). Cracks in masonry walls can also be stitched and bonded (Fig. 2.2.5). Consultation with a professional engineer is advised if the affected wall area is significantly large, or if the brick failure is due to an underlying structural problem.

ADVANTAGES: The repair or replacement of brick on low structures is relatively simple and cost-effective.

DISADVANTAGES: Replaced brick and mortar will not match the color of the existing wall. The replacement of brick on high walls will require scaffolding and is costly.

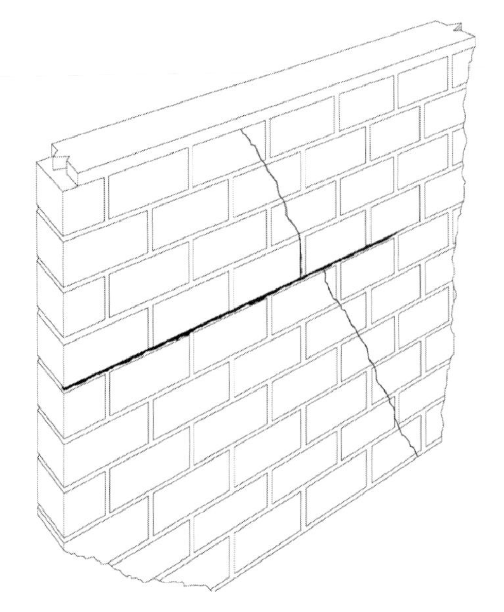

1. REPAIR CRACKED MASONRY

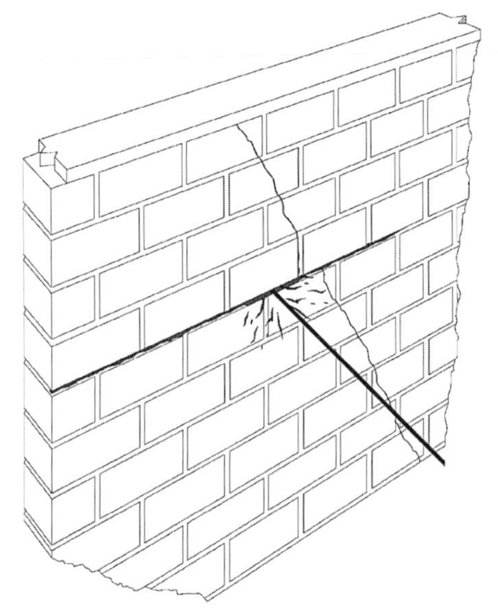

2. CLEAN OUT SLOTS WITH BLOW PUMP AND FLUSH WITH WATER

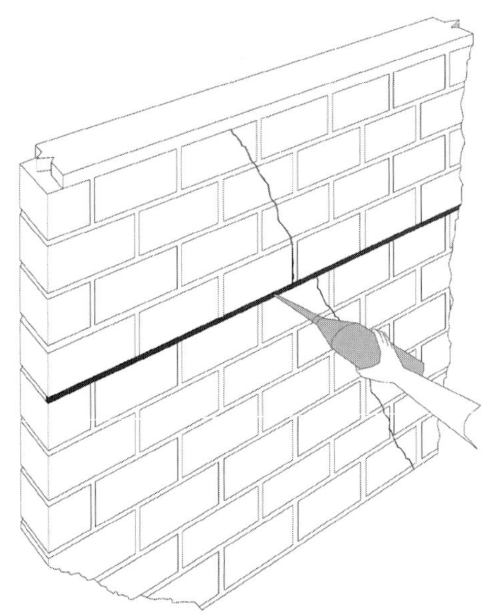

3. USING A POINTING GUN INJECT A BEAD OF HELIBOND TO THE BACK OF THE SLOT. PUSH A STAINLESS-STEEL ROD INTO THE GROUT TO OBTAIN GOOD COVERAGE. INJECT HELIBOND OVER THE EXPOSED BAR UNTIL COVERED

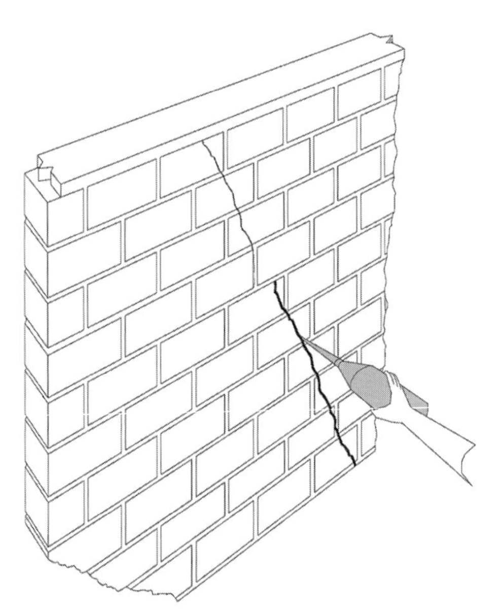

4. SEAL THE VERTICAL CRACK WITH A WEATHERPROOF FILLER

FIGURE 2.2.5 CRACK STITCHING AND BONDING

2.3 SHEATHING

ESSENTIAL KNOWLEDGE

Exterior wall sheathing serves a number of functions. It provides rigidity and shear resistance to the main framing elements; it is part of the barrier system that keeps out the destructive effects of moisture; it provides varying degrees of insulation; and it can serve as the nail base for exterior siding.

Until the mid-1960s, when plywood was introduced, the dominant sheathing material was 1x3 and 1x4 wood boards, typically nailed diagonally to the stud frame. Today the most common wall sheathing materials are OSB and plywood, which together account for approximately 55% of the sheathing market, with slightly more plywood sold than OSB (other siding products are described below). It is important to recognize that APA The Engineered Wood Association (APA) does not differentiate between plywood and OSB under its APA Rated Sheathing program. OSB prices have dropped by half over the past several years, and in some areas of the country it is about half the price of comparable plywood. OSB is fast becoming the most common sheathing material.

If the finished siding has been well maintained, the wall sheathing should not have deteriorated. Exceptions to this would be deterioration from moisture trapped behind the finished siding. Moisture entry is due to improperly flashed or caulked joints between the siding and openings such as doors and windows; inadequate or poorly fabricated flashing at the wall-roof juncture; water driven by high winds between siding material during rainstorms; moisture penetration through mortar joints in brick veneer walls; and the lack of, or improperly lapped, moisture retarder such as building paper or housewrap behind the siding. Much of the research into sheathing failure suggests that, particularly in the case of EIFS, it should be assumed that moisture will penetrate the finished siding and that provisions should be made to allow the system to be self-draining between the finish material and the sheathing by means of furring strips, drainage channels, plastic matting, or other devices.

TECHNIQUES, MATERIALS, TOOLS

1. REPAIR EXISTING WALL SHEATHING.

The specific sheathing repair will depend on the location and extent of damage and the type of sheathing encountered. There is very little possibility of consolidating existing sheathing material. Replacement is necessary if the material is unsound and can no longer function as intended. Replacement of sheathing will require removal and replacement of siding as well.

ADVANTAGES: Localized repairs of sheathing are cost-effective if the damage is limited.

DISADVANTAGES: Localized repairs will only mask the problems if they are widespread and result in a patchwork of new siding. If the problems are widespread, the affected sheathing should be replaced in its entirety and new siding installed.

2. REPLACE EXISTING SHEATHING WITH ORIENTED STRAND BOARD.

Introduced in the early 1980s, OSB (made with rectangular-shaped wood strands cross-oriented in layers for better structural performance) has replaced particleboard, flakeboard, chipboard, and waferboard as the most popular alternative to plywood sheathing. OSB utilizes a variety of fast-growing wood species, including aspen, southern yellow pine, poplar, birch, and mixed hardwoods, with waterproof phenolic resin or polyisocyanate binders. Available in varying thicknesses, it typically comes in 4x 8 sheets, but can also be custom ordered in lengths up to 24' and in widths up to 12'.

ADVANTAGES: OSB has excellent shear resistance, dimensional stability, and bond durability under normal conditions. It is increasingly popular, readily available nationally, and economical (significantly less expensive than plywood). OSB is recognized by all three model code agencies and can serve as a nail base for a variety of siding material.

DISADVANTAGES: The edges of OSB can swell if subject to continuous wetting. OSB has a relatively low R-factor of 1.25/inch.

3. REPLACE EXISTING SHEATHING WITH PLYWOOD.

Structural plywood is generally identified in terms of the veneer grades (A-B, B-C, C-D) used on the face and back of the panels or sheets, or by a term suggesting the panel's intended use (APA Rated Sheathing). Veneer grades define appearance in terms of natural unrepaired growth characteristics and allowable number and size of repairs that may be made during manufacture. According to the APA-The Engineered Wood Association, the minimum face veneer grade permitted in exterior plywood is C. D-grade veneer is used in panels intended for interior use or in applications protected from the weather. CDX (exterior adhesive) panels should not be used in applications where the plywood is permanently exposed to weather or moisture. According to APA, the CDX plywood is not a recognized grade designation. A better designation is (DOC) PS1-95 (a prescriptive standard that references plywood only). (DOC) PS2-92 is a performance standard that allows the supplier to submit either plywood or OSB. Plywood sheathing also comes in a structural—1 grade, a higher performing grade than normal sheathing, which is used for shear walls in seismic areas.

ADVANTAGES: Plywood has long been considered the highest-performing sheathing. It has excellent structural and weathering characteristics and can be used as a nail base for virtually any type of siding, including both cements and synthetic stuccos, and brick veneer applications.

DISADVANTAGES: Plywood is considerably more expensive than similar structural sheathing such as OSB, and is thus losing market share. It has a relatively low R-factor of 1.25/inch.

4. REPLACE EXISTING SHEATHING WITH FIBERBOARD SHEATHING.

Fiberboard sheathing has a 40-year track record in the construction industry and is in wide use throughout the country. Known under a variety of names, such as *blackboard*, grayboard, and *buffalo board*, the material is made from recycled newspaper, wood fiber, and other cellulose products, held together with a binder. It is available in regular density, which requires additional corner bracing, and high density with sufficient racking strength to be used without additional bracing (unless required by local code officials).

ADVANTAGES: Fiberboard sheathing is historically less expensive than most other sheathing material. It has a higher R-value (approximately. 2.4/inch) than OSB, gypsum sheathing, and paper board sheathing. It is vapor-permeable and can be used as a substrate for a variety of finish materials including stucco and wood.

DISADVANTAGES: Fiberboard sheathing has a lower R-value than insulated sheathings. It cannot be used as a nail base for aluminum and vinyl clapboard siding (siding must be nailed to studs or furring strips). It is becoming less cost-competitive with OSB in some areas.

5. REPLACE EXISTING SHEATHING WITH GYPSUM SHEATHING.

Gypsum sheathing has been in use for many years as a substrate for stucco, brick veneer, and a variety of other siding materials where fire-rated assemblies are required by code officials. There is a variety of different types of gypsum sheathing available as both nonfire-rated and fire-rated.

5.1. PAPER-FACED GYPSUM SHEATHING.

This type of gypsum sheating has been in use for over 30 years. It combines a wax-treated water-resistant gypsum core with a water-repellant paper facing.

ADVANTAGES: It is economical and widely available in 4'×8' and 2'×8' sheets of varying thickness. It is relatively inexpensive sheathing for brick veneer, stucco, and EIFS. It provides fire ratings for a variety of assemblies.

DISADVANTAGES: Standard paper-faced gypsum cannot be left exposed for more than 4 weeks before applications of finish material. There have been problems with delamination of the paper face when used under EIFS. It requires careful handling, as edges are subject to breakage.

5.2. GLASS MAT-FACED GYPSUM SHEATHING.

A product recently developed by Georgia Pacific, Dens-Glass Gold™ combines inorganic glass mats embedded with a water resistant and silicone-treated gypsum core and an alkali-resistant surface treatment. The product is more water-resistant and generally performs better than paper-faced gypsum sheathing, particularly as a substrate to EIFS and brick veneer. Available in 1/2" and 5/8" thicknesses and lengths of 8', 9', 10', and greater on special order.

ADVANTAGES: This sheathing is resistant to wicking, moisture penetration, and delamination. It can be installed and exposed up to 6 months before application of finish siding, has a superior performance compared to paper-faced gypsum, can be used as a substrate for a wide variety of siding applications but not as a nail-base, does not require additional bracing for normal applications, and is as lightweight and easy to handle as paper-faced sheathing.

DISADVANTAGES: This sheathing costs up to 50% more than paper-faced gypsum sheathing. Its R-value for 1/2" thickness is 0.56 and for 5/8" thickness is 0.67, which is considerably less than for insulative sheathing. This sheathing cannot be used as a nail base for siding.

5.3. GYPSUM SHEATHING MADE WITH A NON-PAPER-FACED BLEND OF CELLULOSE FIBER AND GYPSUM.

Developed recently as a high-performing alternative to paper-faced boards by Louisiana Pacific, FiberBond™ fiber-reinforced wall sheathings are made from recycled newsprint, perlite, and gypsum, with a special water-resistant face treatment.

ADVANTAGES: This sheathing is stronger and more moisture resistant than paper-faced boards. Structural wall bracing provides superior resistance to screw withdrawal and can be used as a substrate for EIFS, brick veneer, and a variety of other claddings. It has a higher impact strength than other gypsum sheathings, harder edges and ends, is available in up to 12' lengths, and uses recycled materials.

DISADVANTAGES: This sheathing is somewhat heavier than other 4'×8' sheets of gypsum sheathing (paper-faced: 56 lb, Dens-Glass: 64 lb, FiberBond: 72 lb). Its price is comparable to Dens-Glass, but it is significantly more expensive than paper-faced sheathing. Its unfinished exposure is limited to 60 days and it cannot be used as a nail base for sidings.

6. REPLACE EXISTING SHEATHING WITH PAPERBOARD SHEATHING.

In use for over 60 years, paperboard sheathing is a code-approved, low-cost alternative to the other structural sheathings, and has found a considerable following among large home builders for new construction. Available from Simplex Products Division (Thermo-Ply™) and other manufacturers, in thicknesses from 0.078" to 0.137", it can be obtained in sheets up to 80" wide and 16' long, with both reflective foil surfaces and nonreflective surface. Vapor-permeable sheathing is under development. It is often available to builders with their own private label.

ADVANTAGES: Paperboard sheathing does not require additional shear bracing. It is recognized as structural sheathing by national model codes. Paperboard sheathing is less expensive than other sheathing alternatives, and has excellent air infiltration resistance due to overlapping joints.

DISADVANTAGES: The thinness of the material makes it difficult to use in small-scale rehab projects as infill for thicker sheathing products. It is not as strong as OSB or plywood. This sheathing material has an R-value of 0.2, but is claimed to be greater with reflective surface and air space, but less than other insulating sheathing. Adjustments to wood window trim detailing due to thinness of material may be required. It cannot be used as a nailbase for siding products.

7. REPLACE EXISTING SHEATHING WITH FIBER-CEMENT SHEATHING.

A number of fiber-cement sheathing products are available as structural sheathing underlayments. These products range from 30-year-old cement and wood fiber products such as Wonderboard, to high-tech fiber-cement products such as Hardiboard™ and Eternit™, which perform well in high-moisture locations. As such, they are frequently used as underlayments for thin brick, tile, and EIFS.

ADVANTAGES: This sheathing performs well in high-moisture locations. It is resistant to face delamination and is noncombustible, strong, and rigid.

DISADVANTAGES: This sheathing is more costly than gypsum board and other types of sheathing and is not typically used for siding systems other than for EIFS.

8. REPLACE EXISTING SHEATHING WITH FOAM INSULATING SHEATHING.

With increased energy conservation mandated by state and model energy codes, and an increased awareness by the public of possible cost savings and environmental benefits, the use of insulating sheathings including polyisocyanurate (ISO), extruded polystyrene (XPS), and molded expanded polystyrene (EPS) has grown steadily. This is especially true with steel construction, which has potentially large heat losses through thermal bridging. Foam insulating sheathings generally are not structural and require structural sheathing underlayment, such as OSB, or other approved form of shear bracing; they require a 15-minute fire-rated barrier (usually gypsum) when used on the interior of habitable residential spaces. Foam insulating sheathings are discussed in Chapter 6. See Product Information for a list of suppliers.

ADVANTAGES: Foam insulating sheathing provides the most energy-saving method of providing insulation on the outside of walls with R-values up to 7.7/inch for ISO insulation material. It provides a thermal break and can also be used in cavity wall construction and as a substrate to stucco and EIFS.

DISADVANTAGES: Most foam sheathings are not structural sheathings. Applications of many siding products over foam sheathings require special nailing provisions (see individual siding manufacturers' specifications). Thicknesses of 1" and over present attachment problems to existing or new framing if not adequately addressed.

2.4. VAPOR RETARDERS AND AIR INFILTRATION BARRIERS

VAPOR RETARDERS

ESSENTIAL KNOWLEDGE

Vapor retarders first appeared in building construction in the 1920s. Early theories held that moisture vapor will migrate from a region of high concentration toward a region of low concentration along a linear path. The amount of moisture transfer is dependent on the differences in concentration and the vapor permeability of the membrane separating the two regions. This is the theory of vapor diffusion, which viewed the flow of moisture vapor directly analogous to the conductive flow of thermal energy. In this theory, air movement, and the moisture propelled by it, were not considered to be major factors. In the early 1950s, Canadian research found that air movement was the primary mechanism of moisture vapor migration. Without active air infiltration control, vapor retarder barriers become ineffective.

Current theory on vapor retarders indicates that both air infiltration and direct diffusion play significant roles in the transfer of moisture vapor and, therefore, both must be accounted for. Effective vapor retarders must have a water vapor permeability not exceeding 1.0 grain per hour per square foot per inch of mercury vapor pressure difference (referred to as 1.0 perm) and must be installed in such a manner as to prevent air leaks at joints and laps.

Although the issue of what makes a vapor retarder effective is generally settled, controversy still remains as to where to install it, if at all. From this standpoint, the authority on the subject is the 1997 *ASHRAE Handbook of Fundamentals*, which has more to say on the topic than any of the model codes. In what is defined as heating climates (4,000 heating degree days, base 65°F, or more), vapor retarders belong on the interior side of the insulation. In warm, humid, cooling climates (Florida and the Gulf Coast) where moisture vapor transfer conditions are effectively reversed, vapor retarders are best placed close to the exterior.

In mixed climates (not fitting either of the above definitions), the vapor retarder should be placed to protect against the more serious condensation condition, summer or winter. If in a mixed climate the winter indoor relative humidity is kept below 35%, a vapor retarder at the interior side of the insulation is usually not required, and an exterior vapor retarder strategy is most effective. Where winter interior humidity is not controlled or if a humidifier is used, an interior vapor retarder is most useful.

Vapor retarders should never be placed on both sides of a wall. Where a vapor retarder is employed, the opposite wall surface must provide a permeable surface to allow drying to occur. Thus, in hot, humid, cooling climates, where a vapor retarder is employed at the exterior, the interior wall surfaces should be permeable. No vapor retarder paints, kraft-faced insulation, or vinyl wall coverings should be used. Conversely, in northern heating climates, with interior vapor retarders, the exterior wall coverings should be vapor permeable.

The primary purpose for installing a vapor retarder in residential rehabilitation is to minimize moisture vapor migration into a wall or roof assembly where it has the potential to deposit condensate when the dew point is reached. The resulting water in liquid form may cause decay in structural wood framing, wood-based sheathing materials, and interior gypsum board or plaster wall coverings. The prolonged presence of moisture will also encourage and facilitate mold and mildew growth, raising potential serious health concerns for the homes' occupants.

TECHNIQUES, MATERIALS, TOOLS

Vapor retarders can be classified into two major groups: flexible or coatings. Metal foils, laminated foils, treated paper, and plastic films are flexible sheet goods, while paint, semifluid mastic, and hot melt are coatings. In typical residential construction and rehabilitation, the commonly used materials are exterior- or interior-applied plastic films, interior-applied foil-faced products, interior treated paper-faced products, and interior paint coatings.

1. APPLY A VAPOR RETARDER PAINT COATING.

A relatively new product on the market suitable for interior applications is vapor retarder paint. Produced by several manufacturers, including Sherwin-Williams and Glidden, vapor retarder paints are available as interior latex primers, typically with a perm rating of approximately 0.7. These primers are formulated to behave much like standard latex interior primers, in terms of consistency, coverage, and application. They are tintable and suitable for use over new gypsum board or previously painted surfaces. As with standard interior primers, normal prep work is needed, and stained areas will require a stain-hiding primer prior to application. The cost per gallon of the vapor barrier primers is generally competitive with standard interior primers.

ADVANTAGES: Vapor retarder primers are the simplest application in situations where existing wallboard or plaster surfaces are not to be significantly disturbed. Where interior primers are used, the vapor retarder function comes at virtually no additional cost. You can effectively upgrade the vapor transmission performance of an exterior frame wall with no more effort and cost than a new primer and finish coat paint application.

DISADVANTAGES: This method is appropriate for interior wall surface applications only. With the vapor retarder at the inside surface of the wall assembly, damage to the paint can compromise retarding ability. If required prepriming prep work is inadequate, the primer coat vapor retarder effectiveness will be diminished. To be fully effective, all penetrations and material intersections at the interior surface of the wall must be caulked or otherwise sealed.

2. INSTALL TREATED PAPER OR FOIL VAPOR RETARDERS.

For residential rehabilitation purposes, treated paper and foil vapor barriers usually take the form of kraft and foil-faced batt installation. In a situation where interior wall finish has been removed and new exterior wall insulation is to be installed, kraft or foil-faced batts are cost-effective and do provide an adequate to marginal vapor barrier. The amount of unsealed edge is significant and does provide a path for moisture vapor migration. To improve effectiveness, the kraft or foil flanges can be installed over the face of the studs and lapped instead of stapled to the inner stud faces (Fig. 2.4.1). Convenient and

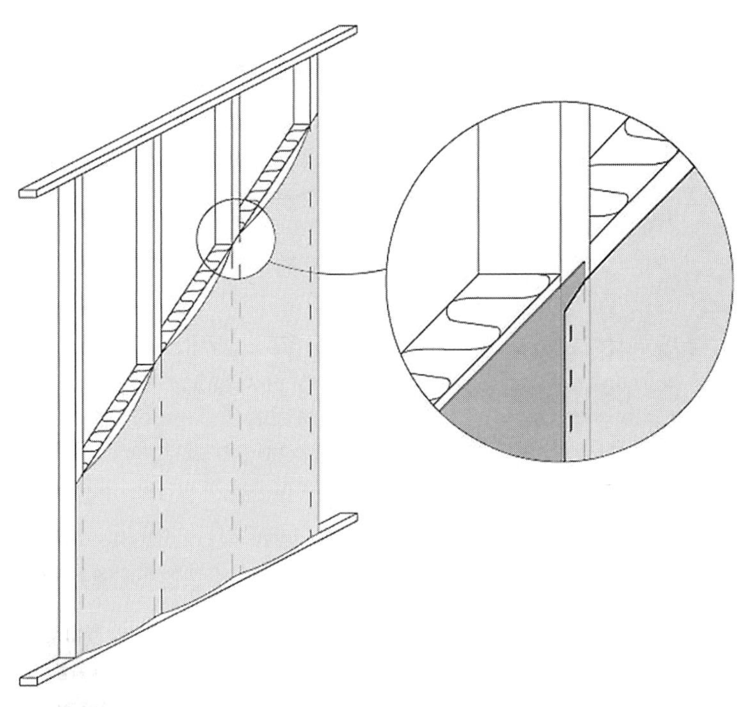

FIGURE 2.4.1 LAPPED FLANGE

cost-effective, kraft and foil batt insulation facings do have limitations, and their use as a primary vapor barrier should be limited to applications where vapor barrier performance is not critical, such as in mixed, nonhumid climates. In heating climates with 4,000 degree days or more, a more continuous vapor barrier surface should be considered.

ADVANTAGES: This is the most cost-effective interior vapor retarder strategy where exterior wall framing is exposed and new insulation is to be installed. It saves labor costs since the fiberglass batt insulation and vapor retarder are installed in one step.

DISADVANTAGES: Installation requires that walls are stripped to rough framing and that fiberglass batt insulation be installed. The number of joints and edges inherent in this system make for a functionally marginal vapor retarder, but sufficient for mixed climates or where indoor humidity is controlled in heating climates. Performance can be improved by installing faced batts with flanges attached to the narrow face of studs and lapped.

3. INSTALL CLEAR POLYETHYLENE VAPOR RETARDERS.

Most plastic barrier films are either clear polyethylene, black polyethylene, cross-laminated polyethylene, or reinforced polyethylene. The most basic of these materials, clear polyethylene, is also the most economical. Available in 4-, 6-, and 10-mil thicknesses, it is best suited for interior wall applications over framing and insulation. As clear poly's content is up to 80% reprocessed material, it is also an environmentally sustainable choice. The high recycled content comes at a cost: Its quality can be uneven and it generally has poor tear and puncture resistance. Clear polyethylene should never be used for exterior applications or for applications with more than limited exposure to sunlight. Clear polyethylene is available in widths of 4' up to 32' in 100'-long rolls. As with all polyethylene vapor retarders, for horizontal application over wood framing, staples are most often used. For maximum effectiveness, joints should be kept to a minimum and seams should be lapped and taped.

ADVANTAGES: Clear polyethylene is relatively inexpensive and easy to install. In more severe heating climates, the use of interior polyethylene films is most effective and is practical where interior finish surfaces are removed. Since polyethylene is transparent, attachments to framing members are simplified, as is the installation of wallboard material over the polyethylene, because the studs are visible.

DISADVANTAGES: Clear polyethylene has limited tear and puncture resistance and must be installed with care to avoid damage. All penetrations such as electrical junction boxes must be taped and sealed to ensure effectiveness. Clear polyethylene can be used only in instances where wall finishes and surfaces have been removed, fully exposing the wall framing.

4. INSTALL BLACK POLYETHYLENE VAPOR RETARDERS.

Black polyethylene is nearly identical to clear polyethylene, except for the addition of carbon black to the composition as a UV inhibitor. This permits the use of the polyethylene where some limited exposure to sunlight is required, such as at exterior wall surfaces. Black polyethylene strength characteristics are similar to clear polyethylene, with low tear and puncture resistance.

ADVANTAGES: For exterior wall surface applications in hot, humid, cooling climates, black UV-protected polyethylene films can provide superior vapor retarder performance.

DISADVANTAGES: Black polyethylene has limited tear and puncture resistance. Unreinforced black polyethylene must be installed with care to avoid damage. Its opaque nature makes installation more difficult by obscuring underlying framing, sheathing, and other components. Joints and seams must be lapped and taped for full effectiveness. Installation is limited to conditions where siding has been fully removed and attachment directly to exterior sheathing can be made.

5. INSTALL CROSS-LAMINATED POLYETHYLENE OR FIBER-REINFORCED POLYETHYLENE VAPOR RETARDERS.

Compared with standard polyethylene, high-density cross-laminated polyethylene and fiber-reinforced polyethylene are both specialty products manufactured for applications where higher strength is required. For retrofitting over rough, irregular surfaces, such as solid board sheathing, both products would be less susceptible to tearing or puncture by lifted nail heads, splinters, or exposed sharp corner edges. Either product would also be appropriate where rough handling and adverse site conditions are expected.

ADVANTAGES: Stronger than standard polyethylene, reinforced and laminated material can withstand more adverse site conditions and rough handling. The reinforced and laminated products are typically rated for limited UV exposure for exterior use and situations where the installation of siding and coverings is delayed. Black reinforced and laminated polyethylene can be used as the required weather barrier under exterior siding and cladding.

DISADVANTAGES: These products have a higher initial cost compared to standard black polyethylene. Application is limited to conditions where siding and exterior wall coverings have been removed. Seams must be lapped and sealed for full effectiveness.

AIR INFILTRATION BARRIERS

ESSENTIAL KNOWLEDGE

Air infiltration barriers, or "housewraps," as they are known in the industry, have grown in popularity since their appearance in the 1970s in the wake of the energy crisis. DuPont, one of the first companies to introduce such a product, came out with Tyvek™ in the late 1970s. Today there is a variety of similar products that reduce air infiltration and improve energy performance.

The primary attribute of housewraps is their ability to operate as air infiltration barriers while not forming an impervious vapor barrier. When placed over the exterior surface of the wall sheathing, the material allows moisture vapor to escape from the frame-wall cavity while reducing convective air movement in the insulation, thereby helping to maintain the composite R-value of the wall. The greater the exterior air movement, the greater the benefit.

The 10 biggest selling housewrap products fall into one of two basic categories: perforated and nonperforated. Perforated products are either woven polyethylene, woven polypropylene, spun bonded polypropylene, or laminated polypropylene film. These materials are more impervious to moisture vapor migration than nonperforated wraps, and thus are provided with microperforations to allow vapor migration and diminish their vapor-retarding properties. With the exception of the polyethylene films, all the perforated housewraps are further coated with either polyethylene or polypropylene for added air infiltration resistance. In contrast, nonperforated housewraps are either spun-bonded polyethylene or fiber-mesh-reinforced polyolefin. The structure of these materials allows water vapor to pass through, while inhibiting air infiltration.

In addition to their primary functions as air infiltration barriers and water vapor transmitters, some (but not all) of the major housewrap brands are code approved as substitutes for required moisture protection barriers. To gain national code approval as a substitute for No. 15 felt, the product manufacturer must apply to each of the three major model building codes, or CABO, and supply specific testing data on water penetration resistance. With code recognition, the product can be used under all siding applications, including stucco and masonry veneer. Currently, at least four products are listed by all three model codes as acceptable moisture protection barriers: Amowrap, Pinkwrap, R-Wrap, and Tyvek. Tyvek also produces a product, StuccoWrap™, that is specifically intended for use with traditional and synthetic stucco and is code listed for that application. Other housewraps are acceptable to some codes as weather-resistant barriers. Before a particular product is used as a weather barrier, its approval should be verified with the governing code.

In addition to air leakage resistance, permeance, and moisture resistance, two other material characteristics are worth considering: UV sunlight resistance and strength. All major housewrap brands have a manufacturer's rated UV exposure time ranging from 120 days to more than a year. Some products are manufactured with antioxidants and UV stabilizers, while others are naturally more resistant by their composition. In the field, however, covering the housewrap as quickly as practicable is recommended, as some UV degradation will occur even over a short period, and other unrelated damage to the membrane can be avoided.

The strength of the housewrap can be critical, as wind conditions or adverse job site handling can tear or puncture the material during and after installation. Even small holes can negatively affect overall performance. The inherent strengths of housewrap can be judged on three levels: tensile strength, tear strength, and burst strength. Respectively, these are the material's ability to withstand damage from pulling and stretching; withstand tearing at nail and staple locations; and withstand separation of material fibers, fabrics, or films. Unfortunately, testing procedures and standards vary between manufacturers, so product comparison is difficult. Generally, the spun-bonded products have good tensile and burst strength but tear easily; woven and fiber-reinforced products have good tear and burst strength, but are susceptible to diagonal tensile loading; laminated film products tend to be weakest of all and can lose strength significantly, making a tight installation more difficult.

Although the wide variety of housewrap products with varying performance characteristics may appear confusing, they offer a wide selection for any particular job. In northern heating climates, where interior vapor barriers are the norm, a highly moisture-vapor permeable housewrap may be required. In hot, humid, cooling climates, where an interior vapor barrier is not required, a housewrap with a low-air leakage rate may be preferred. In low-wind environments, a low-strength material may be selected. A particularly cost conscious choice would be laminated film.

TECHNIQUES, MATERIALS, TOOLS

INSTALL HOUSEWRAP OVER NEW OR EXISTING SHEATHING.

For rehab applications, housewraps will generally be placed over existing solid board sheathing, plywood, or OSB, or over new plywood or OSB where the existing sheathing needs replacement. Housewraps come in rolls of varying widths, with 9'-0" being the standard. Other widths are available, depending on the manufacturer, including 1'-6", 3'-0", and 4'-6". Roll lengths vary from 60' to 200'. Some custom sizes and lengths are available. Material thickness varies somewhat but is irrelevant in terms of application. Beginning at an outside corner, hold the roll of housewrap vertically and unroll the material across the face of the sheathing for a short distance. Make sure the roll remains plumb and that the bottom edge of the housewrap extends over the foundation by 2". The application should start at an outside corner extending around the starting point corner by 6" (Fig. 2.4.2).

Manufacturers specify acceptable fasteners, typically large head nails, nails with plastic washers, or large crown staples. Fastener edge and field spacing patterns are also specified. Housewrap sheets are installed shingle-style, from the bottom up. Horizontal laps should be a minimum of 2"; vertical laps of 6" are acceptable (Figs. 2.4.3 and 2.4.4). To be fully effective in their primary role as air infiltration barriers, all seams and edges must be taped or caulked. While some manufacturers mar-

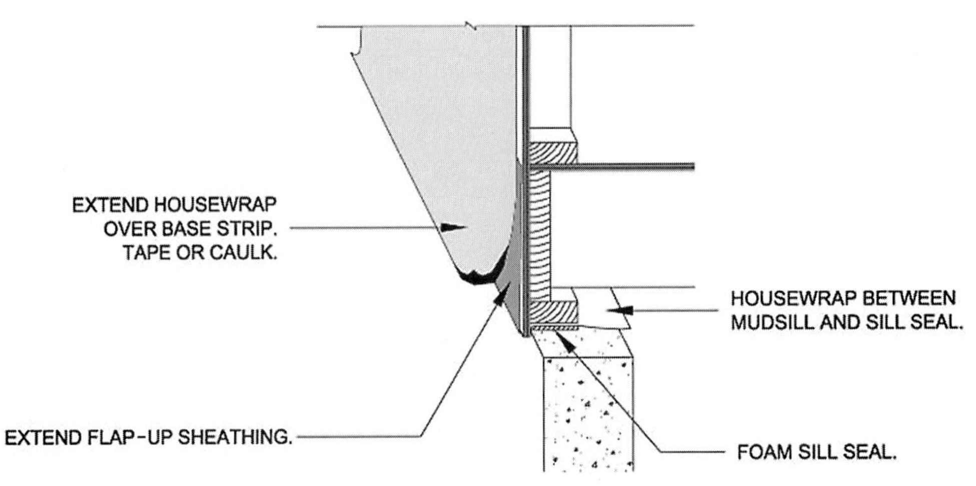

FIGURE 2.4.2 HEAD LAP

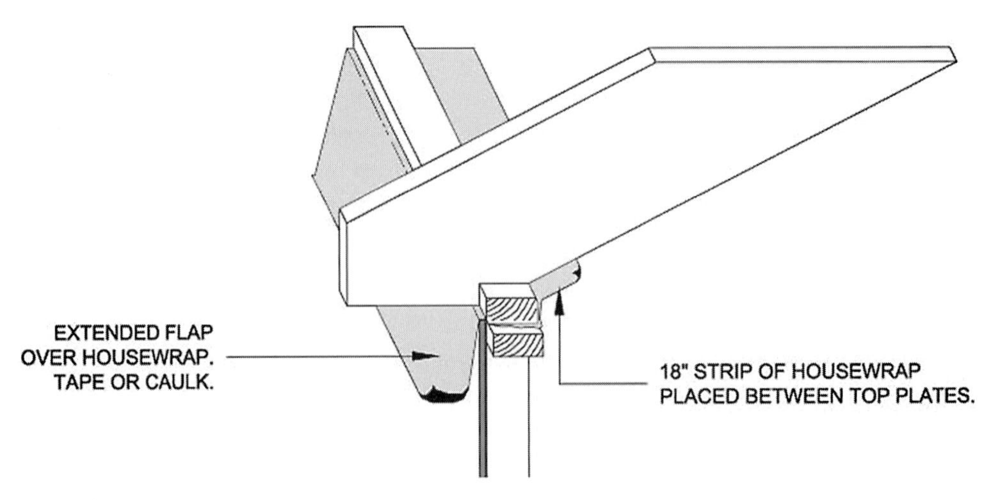

FIGURE 2.4.3 SILL LAP

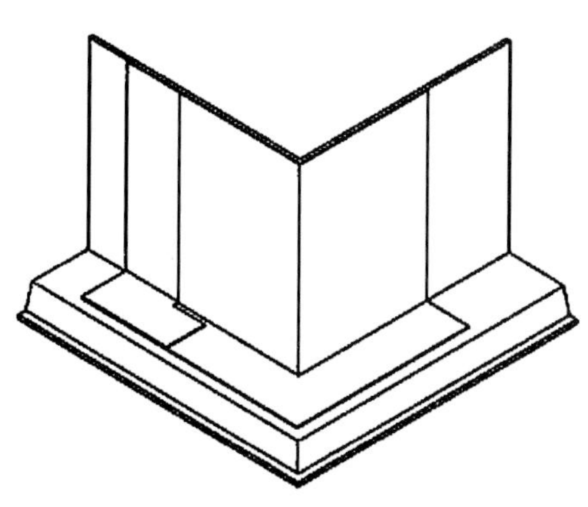

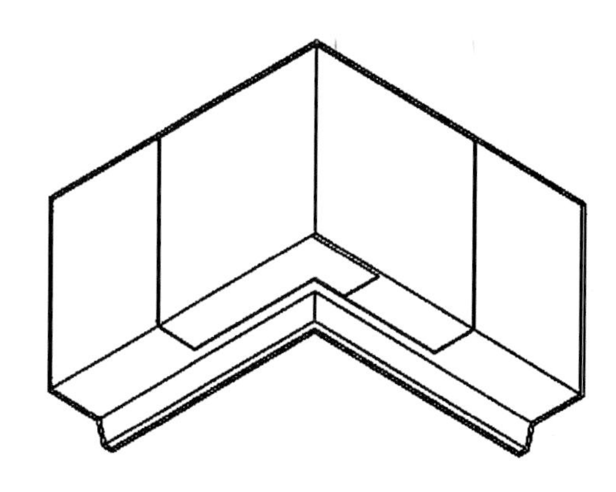

FIGURE 2.4.4 OUTSIDE CORNER EDGE FLASHING

ket products for this purpose, others provide information outlining the performance requirements for approved products.

ADVANTAGES: This is a relatively low cost, lightweight, easily installed energy conservation and moisture control product. It is especially effective in mixed and northern heating climates where unchecked air infiltration can significantly degrade house energy performance and occupant comfort. It is beneficial in limiting airborne moisture vapor transmission into the wall cavity by limiting air movement, while allowing moisture in the cavity to be expelled. Some products can be used as a code-approved substitute for building felt.

DISADVANTAGES: The initial cost of these products is slightly more than building felt. Availability of some products may be limited. These products have an inferior performance as a weather barrier compared with building felt. Nail penetrations in housewrap are not self-sealing, as they tend to be in felts. Housewraps are not selective vapor-permeable membranes: moisture vapor will pass through in both directions. As water-absorptive siding materials such as wood and brick veneer dry, moisture in vapor form can be forced through housewrap into sheathing and insulation. Less-vapor-permeable building felt can better withstand reverse vapor migration. Some recent studies appear to indicate that surfactants, a class of substances found in wood, stucco, soap, and detergents, can decrease the natural surface tension of water and allow it to pass through housewraps wetting the underlying materials. According to anectdotal field observations, this process is most likely to occur in regions with heavy rainfall and when unprimed wood siding is placed in direct contact with the housewrap.

2.5 INSULATION

ESSENTIAL KNOWLEDGE

Insulation is one element in a tightly knit construction system intended to improve indoor comfort and reduce energy consumption. In rehab work, installing insulation or improving existing insulation levels will be critical in providing comfort. Insulation should never be applied without considering its effect on other aspects of construction. Some factors to consider when evaluating wall insulation are density and compressibility, air leakage, moisture control, fire safety, and wall construction in existing homes.

Each type of insulation has a density at which its R-value per inch is greatest, but reaching this density is not always cost-effective. For 3½"-thick fiberglass batts, an R-13 batt contains 40% more material, and an R-15 batt 180% more material, than an R-11 batt (Fig. 2.5.1).

To achieve a desired overall R-value for dry blown-in insulation, and to prevent settlement, the installed density must be above a recommended minimum. For convenience in comparing estimates or monitoring the installation, have the bidder or installer calculate the number of bags required to achieve the required density.

Unless insulation completely fills all the wall cavities, air leakage can bypass the insulation and create a risk of condensation. Reducing air leakage is an inseparable part of insulating: you should not do one without doing the other. Typical locations for air leakage through walls are at the sill, the wall plates, vertical corners, around openings, and at electrical devices.

Before filling stud cavities of older homes with blown-in or foamed-in-place retrofit insulation, explore the construction. Stud cavities are often interrupted by blocking halfway up the wall, or in girtframe construction, by full-depth diagonal corner braces. The outside walls may be "backplastered," where a hidden layer of plaster creates two parallel cavities within each stud space, neither deep enough to receive loose-fill insulation. After insulating, an infrared camera scan of the wall will show cavities that have not been fully insulated.

Avoid deliberately ventilating walls, since any convective airflow within an outside wall risks condensation within the wall and compromises its R-value. Ventilation passages behind the exterior finish are called for when extreme interior humidity is expected and no vapor retarder can be applied or where wood siding is applied directly over exterior foam insulation.

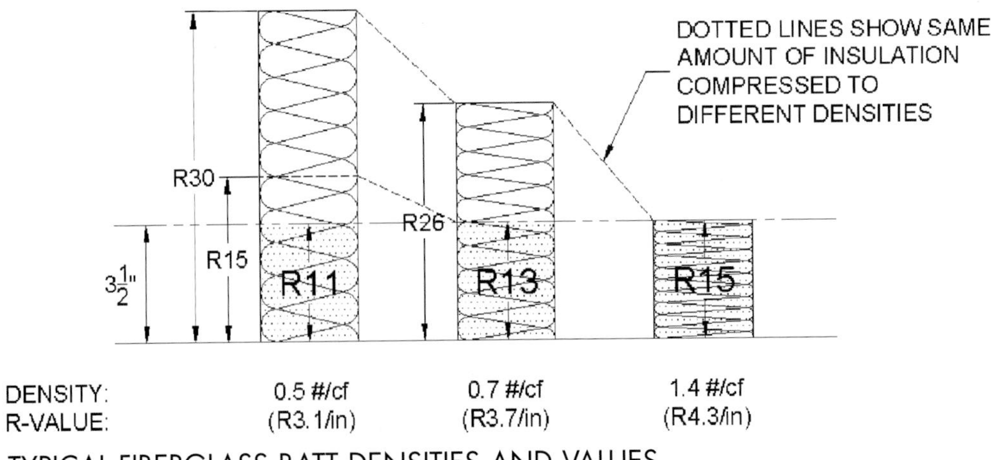

FIGURE 2.5.1 TYPICAL FIBERGLASS BATT DENSITIES AND VALUES

TECHNIQUES, MATERIALS, TOOLS

Of the innumerable possible combinations of insulating materials and wall configurations, the following list covers those in common use and uncommon systems that are recommended.

1. INSTALL BATT INSULATION.

Fiberglass insulation is available in batt form, typically sized 93" long to fit within the stud space of an 8' wall, or in continuous rolls. Both forms are here referred to as *batt insulation.* It is available in many thicknesses, densities, and in widths to fit framing at 16" and 24" centers. Unfaced batts can easily be cut to fit into odd-sized spaces and are preferred where a continuous membrane vapor retarder is installed. Residential batts are available faced with kraft-paper and aluminum foil, and commercial batts with a flame-resistant foil facing are available. All have extended tabs on the facings to secure them in place. When properly applied, the facings create a partial vapor retarder. Only unfaced or fire-retardant-faced batts can be left exposed in attics or occupied spaces. If not accurately cut around wiring and other obstacles, faced batts create large air cavities that compromise their effectiveness. Tabs can be inset-stapled to the sides of the studs, or face-stapled to the inner face (Fig. 2.5.2). Face stapling is preferred because it creates a better vapor retarder and avoids the air cavity left between the facing and the wall finish when inset stapling. Unless this cavity is carefully sealed at the top and bottom, it can compromise the wall's airtightness and R-value. Staples into stud faces must be fully set to avoid interfering with drywall installation. In a three-sided wall cavity, friction-fit unfaced batts, covered with a separate vapor retarder, will typically result in a more effective installation than will stapled faced batts. In an open wall, the facings are usually necessary for attachment.

ADVANTAGES: Batt insulation is an economical, flexible, and well-known product. It provides a dependable thickness of uniform density and does not settle, if properly installed. Faced batts can insulate an open stud wall.

DISADVANTAGES: Careful installation of batt insulation is required to avoid gaps and consequent convective losses. Glass fibers can be irritating if touched or inhaled.

2. INSTALL ENCAPSULATED FIBERGLASS INSULATION.

Fiberglass insulation is available in rolls or batts, encapsulated with kraft paper or plastic to reduce mechanical irritation to installers. These can be used in any installation where batts are appropriate. One face is extended to form attachment tabs. The faces on sound control batts do not have a vapor

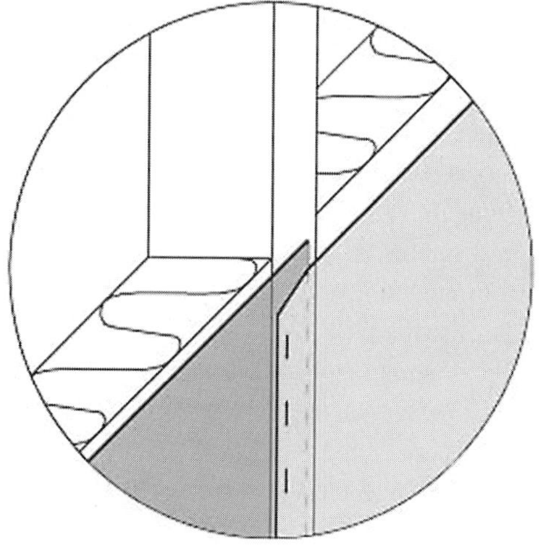

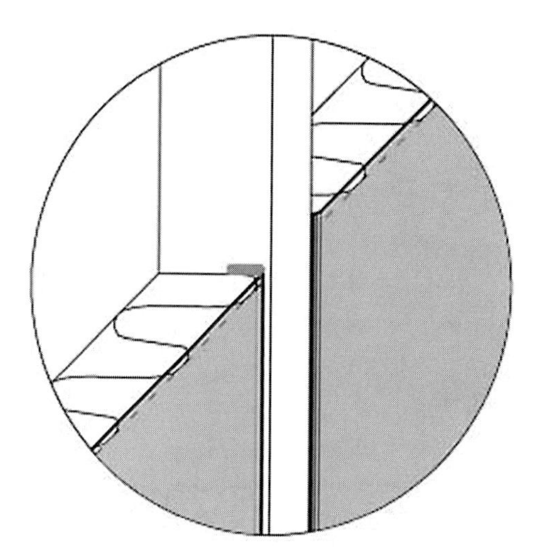

FIGURE 2.5.2 FACE STAPLED INSET STAPLED

retarder; and some exterior wall batts have a polyethylene vapor retarder on the flange side. Some encapsulated batts have a Class A fire-rating and can be left exposed if allowed by local codes. Owens Corning markets Miraflex, made from loose, virtually itch-free glass fibers with no binder, contained in a plastic sleeve. Cutting encapsulated batts around obstructions is possible, but exposes the fiberglass. Cutting Miraflex releases the fibers; the product is therefore intended primarily as attic floor insulation.

ADVANTAGES: Encapsulated fiberglass insulation is ideal for rehab contractors. It can be left exposed (check local codes) and some products can be flange-attached with or without a vapor retarder.

DISADVANTAGES: Encapsulated fiberglass insulation is more costly than regular batt insulation. Cutting encapsulated batts exposes the fiberglass, and cutting Miraflex releases the fibers.

3. INSTALL BLOWN-IN LOOSE-FILL INSULATION INTO CLOSED STUD SPACES.

Loose-fill insulation (fiberglass, cellulose, or mineral wool) can be blown into closed stud spaces through openings formed by drilling holes through the interior or exterior finish, or by removing strips of interior or exterior finish, at the top and bottom of each rafter space. Careful installation is required because material can bridge over electric lines and other obstructions, causing voids and later settlement. Beware of blocking; blow into cavities above and below it. At least a moderate amount of pressure is required to produce sufficient density to inhibit settlement. Fiberglass blown into a 2'×4' stud cavity at a density of about 1.5 pounds per cubic feet (pcf) produces R-13 without excess pressure on finishes. In a "dense-pack" installation of cellulose, dry material is applied at high velocity through a narrow tube inserted through a single hole at the top and extended to the bottom of the cavity. The tube is gradually withdrawn, compacting the material to a density of 3 to 3½ pcf. Stabilized cellulose includes an adhesive, and Fiberized™ cellulose is made in strands instead of chunks; both processes are claimed to inhibit or prevent settlement. Insulation packed into and filling wall cavities suppresses air movement within the cavity, does not create a vapor retarder, but may substantially improve fire safety.

ADVANTAGES: If the cavities are completely filled under sufficient pressure, this method provides superior insulating performance without settlement, greatly reducing air circulation within the walls, and may improve fire safety.

DISADVANTAGES: Some types of installation may leave voids and/or settle after installation. Blown-in materials do not form a vapor retarder and form only a partial air barrier.

4. INSTALL BLOWN-IN OR SPRAYED-ON INSULATION INTO OPEN STUD SPACES.

An inexpensive material can be applied as a membrane over open studs to form see-through cavities, within which any blown-in insulation can be applied under moderate pressure. In Ark-Seal's Blown-in-Blanket® (BIBS) system, fiberglass mixed with some water and adhesive is blown through slits cut in a tightly stretched nylon netting. In Par-Pac's Dry-Pac Wall System™, dry cellulose is blown at 3 pcf density into a cavity closed by a reinforced polyethylene vapor retarder (Fig. 2.5.3). The material is installed from the bottom up to minimize voids and settlement. All such installations will cause the membrane to bulge out; make sure this bellying does not interfere with drywall installation. Various types of polyurethane- and polyisocyanurate-based insulations, such as Icynene, can also be spray-applied into stud cavities. A thin layer of such material can form an air-barrier skin, over which cheaper material can be placed. Excess material must be scraped off, and windows and electrical devices protected or cleaned. Water is mixed with cellulose in a wet-spray application. The water combines with the starch in the cellulose to form a natural adhesive, which holds the material in place. Excess material must be scraped off, but can be reused. The material must dry out before a finish is applied; exces-

FIGURE 2.5.3 BIBS SYSTEM

sive water can prevent drying and generate rot or mildew. High-density insulation filling cavities may improve fire safety.

ADVANTAGES: This method fills the cavities without settling. Visual inspection is possible to ensure filled cavities. Air circulation within walls is greatly reduced and fire safety may be improved.

DISADVANTAGES: Bellying of the interior membrane may interfere with drywall installation. Sprayed-on foam products are more expensive than batt or loose-fill installations, and are messy processes, requiring cleanup and protection. Excess water in wet-spray applications may lead to rot and mildew.

5. INSTALL RIGID WALL INSULATION.

A 3/4" to 1" layer of rigid insulation, typically polyisocyanurate, molded expanded polystyrene, or extruded polystyrene, is a widely used adjunct to cavity insulation. Where cavity insulation cannot be installed, rigid foam may be the only way to insulate a wall. It is preferably applied on the outside of the framing, keeping the framing warm enough in cold weather to prevent condensation within the walls, and inhibiting thermal short-circuits through the studs. It is also useful on the outside as a base for cement stucco or an EIFS, although the latter should be part of an engineered system that provides interior drainage. A layer of foam is essential in conjunction with conventional steel framing to prevent surface condensation. In Gulf Coast climates, a layer of foam behind the interior finish is preferred over an exterior layer or a layer on both sides of the studs, especially with steel framing. Since foam cannot be relied upon to resist racking, it must be applied over structural sheathing or in parallel with a system of wall shear bracing. Structural sheathing separated from the framing by an outside layer of foam may not meet code racking requirements (consult with a structural engineer). Celotex makes a structural polyisocyanurate foam sheathing that, when glued and nailed to the framing, acts as racking bracing. Wood siding should not be applied directly to foam insulation.

ADVANTAGES: Rigid wall insulation isolates framing to minimize or eliminate internal condensation and reduce cold bridging through framing. It can add more than its rated R-value to a wall assembly.

DISADVANTAGES: Rigid wall insulation is more costly per R-value than fiberglass insulation. If substituted for exterior sheathing, it requires other measures to create racking resistance in the structure. It should not be used with an EIFS except as part of an engineered system with interior drainage. Expanded polystyrene cannot support one-coat cement stucco over more than 16" stud spacing.

6. INSTALL A RADIANT BARRIER.

While radiant barriers and coatings are commonplace in high-temperature industrial applications (typically 500° F or more), they are marginally effective at ordinary temperatures. To be cost-effective in building applications, they must have a very low incremental cost (from \$0.02 to \$0.10/ft^2,depending on the application). Clean, shiny aluminum foil facing a 3/4" or deeper air space can create a radiant barrier. Foil-faced insulation held back and inset-stapled creates only a marginally effective radiant barrier because the insulation bulges into the air space, and because the cavity can create heat loss through convection: It is always more effective to fill the cavity with insulation. If an air space is present for other reasons (for example, the cavity between sheathing and brick veneer), facing the air space with foil-faced sheathing will add thermal resistance if the material remains clean. A new form of radiant barrier is Radiance™ paint, which contains aluminum dust.

ADVANTAGES: This is an easy way to add insulating value at brick cavity walls.

DISADVANTAGES: This method is marginally cost-effective, but unlikely to be effective if exposed to dirt or condensation. It is not cost-effective in cold climates, except possibly if Radiance™ paint is used. It is seldom or never cost-effective if the air space is deliberately "stolen" from conventional insulation. Foil may create a vapor barrier where one is not desired.

7. INSTALL A STRUCTURAL INSULATED PANEL WALL.

Rigid foam insulation adhered to structural skins can create a structural insulated panel (SIP). SIPs provide a combination of structure and insulation. Depending upon the design, the panels may be self-supporting or may be a non-load-bearing exterior skin applied over a post-and-beam frame. Very tight joints are crucial, since a small amount of air leakage through a joint is guaranteed to create destructive condensation at the most critical structural point. Panels are typically fabricated to order and delivered to the site for quick erection. Acoustical tightness is readily noticeable.

ADVANTAGES: This method creates a high-R wall that can be load bearing and resists racking. It provides excellent acoustical resistance and allows a high level of prefabrication and fast on-site erection.

DISADVANTAGES: This method is not yet in common use, and therefore is more expensive than ordinary framing. Careful air-sealing at all joints is required.

2.6 VINYL SIDING

ESSENTIAL KNOWLEDGE

Introduced in the 1960s, vinyl has become the leading wall cladding material for siding (60% to 70% of the national re-siding market, and 40% to 50% of the national new siding market). Regionally, its most prevalent use is east of the Mississippi. Originally considered a relatively low-performing product that had fading and cold-weather cracking problems, vinyl siding has evolved into a high-performing product with good weatherability and a level of detailing and finish that, in the most sophisticated and innovative examples, comes fairly close in appearance to the wood siding products that it emulates. It is essential, however, that vinyl siding be allowed to expand and contract freely. It is estimated that over 90% of the problems with vinyl siding are caused by a lack of sufficient clearance between the nail head and the nailing slot or between the siding and trim.

Vinyl siding is made of polyvinyl chloride (PVC) resins with inorganic color pigments, UV stabilizers, and various plasticizers. It has a typical exposure of 8" or 10" plus a fastening tab (hem) and is commonly available in 12' lengths, although longer lengths are available from some manufacturers. Physical characteristics are established by ASTM D3679, which sets the minimum thickness at 0.035". Thickness is not the only indicator of performance. The specific PVC formulation, siding profile, and attachment details are also important characteristics. The industry is reviewing ways to set multiple performance requirements, but at this time siding thickness is the primary means of differentiating products.

There are three general classifications for vinyl siding based on thickness, with subclassifications for low- and high-end products.(see Table 2.6.1). Prices vary accordingly, with some super premium products costing up to twice the cost of super economy. Most of the material sold is in the economy range. Industry consensus holds that material below 0.040" may be too thin to conceal uneven substrates. The super premium products, at 40% thicker, will not necessarily last 40% longer than standard products, although they will be straighter, less wavy, more resistant to impact damage, and may be architecturally more distinctive. Most vinyl siding is sold as a "commodity" product, in standard economy styles and finishes. However, some fairly recent evolutions and innovations in vinyl siding have been introduced largely in higher-end products. Many of these reflect small niche markets today, but the market is becoming more selective and quality conscious.

■ *Product formulations.* Most manufacturers have continued to refine existing formulas to develop better weathering and nonfading characteristics. Nonfading warranties have been introduced that extend the warranty period. New premium resins are being developed that will allow darker-colored panels to perform as well as lighter ones.

■ *Finishes.* A number of manufacturers have developed low-gloss finishes that replicate sanded, sealed, and painted cedar. Some closely resemble the texture of cedar clapboard and the color of stained siding. Simulated plain and scalloped cedar shingles and decorative cedar panels, such as Certainteed's Cedar Impressions™, made from polypropylene, are also available. Extended

TABLE 2.6.1

Super Economy	Economy	Standard	Premium	Super Premium
0.035" to < 0.040"	0.040" to < 0.042"	0.042" to < 0.044"	0.044" to < 0.048"	0.048" +

warranties are available against peeling, blistering, rotting, flaking, chipping, cracking, corroding, and excessive fading.

■ *Profiles and reinforcements.* High-end products, such as Wolverine's Super Premium Portfolio HP™, provide thicknesses up to 0.055", deeper reveals, reinforced nailing hems, and stronger locking profiles. Wolverine offers a fiberglass reinforcement bar in its Benchmark™ series that overlaps adjacent panels and provides increased rigidity. Wolverine has recently developed a flexible nail hem on its Millennium™ series that eliminates the conventional slotted hem, making fastening faster and simpler, reducing expansion and contraction problems, and allowing the use of stapling as well as nailing. Certainteed, Heartland, and Alside also have developed reinforced interlocks (Fig. 2.6.1).

■ *Insulated siding products.* Progressive Foam Products manufactures an insulated contoured underlayment, ThermoWall®, that is designed to go over existing siding products, provide a rigid base for new vinyl siding and add an R-value of up to 4.2 to the existing wall. The underlayment panels (Fig. 2.6.2), 20" high, 48" long, and minimum 1/2" thick, are profile-specific and have been developed for over 750

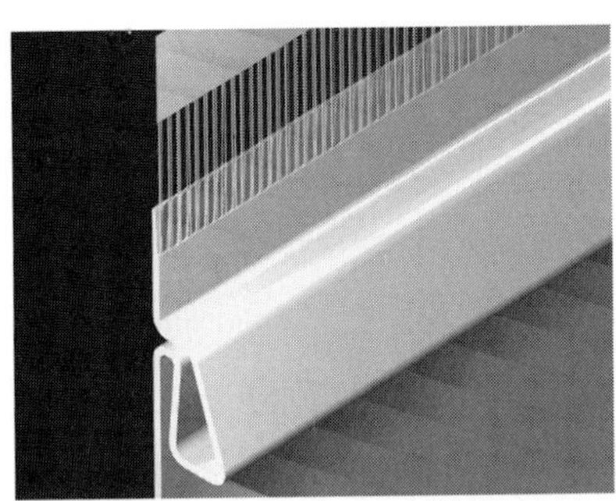

WOLVERINE® MILLENNIUM™ SIDING

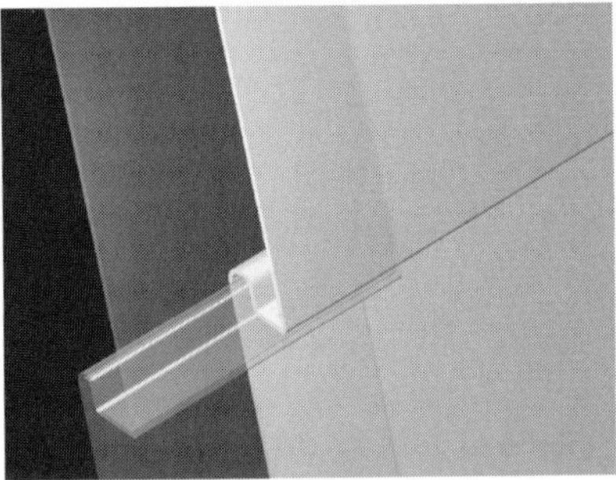

ALSIDE CENTERLOCK™

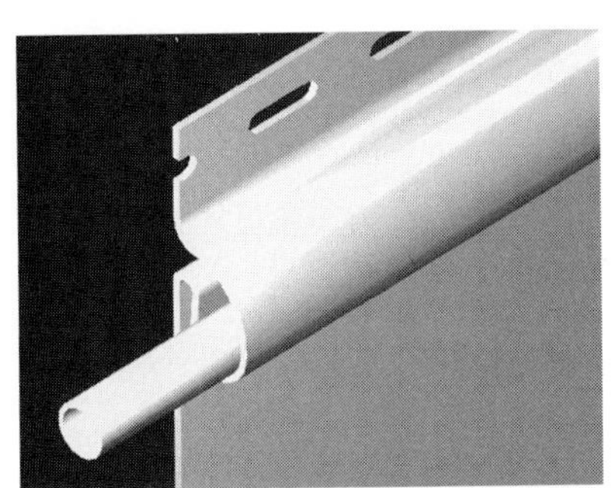

BENCHMARK™ LAPLOCK™

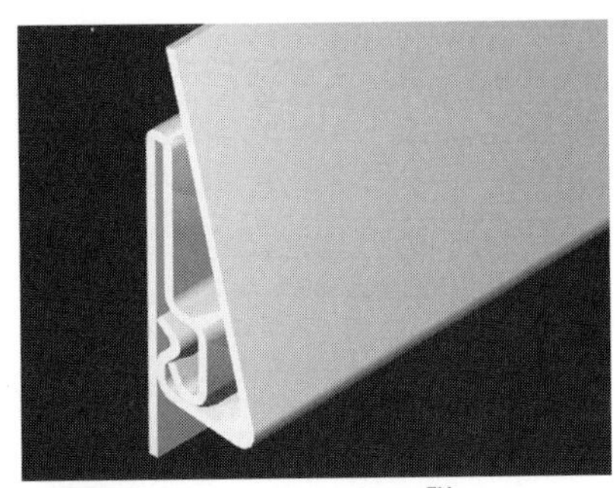

CERTAINTEED CENTILOCK™

WOLVERINE® GRIPLOCK™

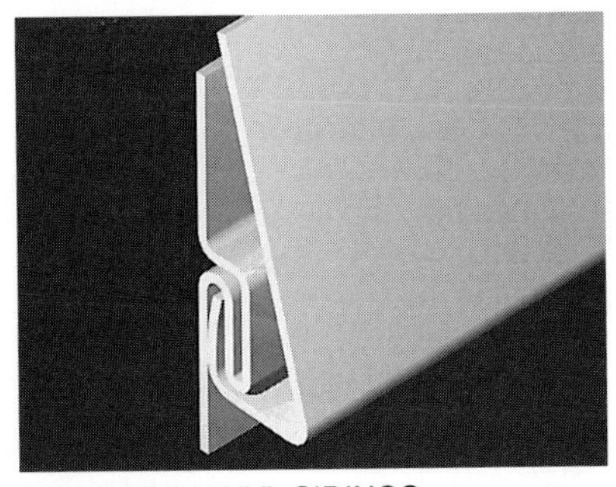

STANDARD VINYL SIDINGS

FIGURE 2.6.1 VINYL SIDING LOCKING PROFILES

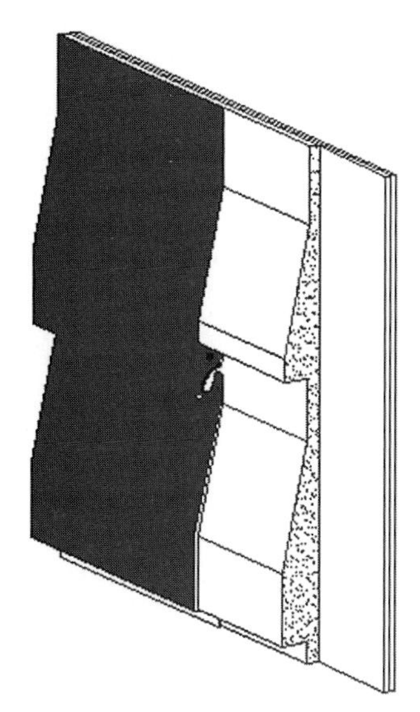

FIGURE 2.6.2 THERMOWALL®

existing vinyl siding products from different manufacturers. Progressive has helped develop a four-course profile for VIPCO (a division of Crane Plastics) that is laminated to a vinyl siding panel approximately 12' long x 16" high, sold under the name of TechWall™.

■ *Trim and accessory panels.* A number of manufacturers produce wide window, door, and corner trim with reveals that eliminate the standard J-channel and make the joining of material appear closer to that of wood siding. Note that siding cannot be butted directly against wood trim without use of a J-channel. Vinyl siding requires trim on outside and inside corners.

TECHNIQUES, MATERIALS, TOOLS

1. REMOVE STAINS FROM EXISTING VINYL SIDING.

Some vinyl siding, which might appear to need replacement, can be rehabilitated by careful cleaning. Vinyl siding will fade over time, but maintenance and cleaning will prolong the service life and appearance of the material. Vinyl siding is easily cleaned with a variety of approved cleaners that are formulated for specific staining problems. The Vinyl Siding Institute has prepared a comprehensive list of those cleaners that will remove most stains (see Further Reading).

ADVANTAGES: This is a low-cost approach to vinyl siding rehab.

DISADVANTAGES: Cleaning might not eliminate all staining and will not eliminate fading.

2. REPAIR EXISTING VINYL SIDING.

Sections of vinyl siding that are buckled, dented, cracked, stained, or otherwise damaged can be easily replaced by means of a *Zip* tool that slips behind the bottom of the siding panel above the damaged panel, allowing access to the damaged panel for replacement (Fig. 2.6.3). Instructions are provided by the Vinyl Siding Institute and individual manufacturers (see Further Reading).

ADVANTAGES: This method allows portions of vinyl siding to be replaced without complete re-siding.

DISADVANTAGES: New vinyl replacement siding will not match weathered vinyl siding. Small discolored sections can be painted with an all-acrylic paint, although the Vinyl Siding Institute does not specifically endorse painting.

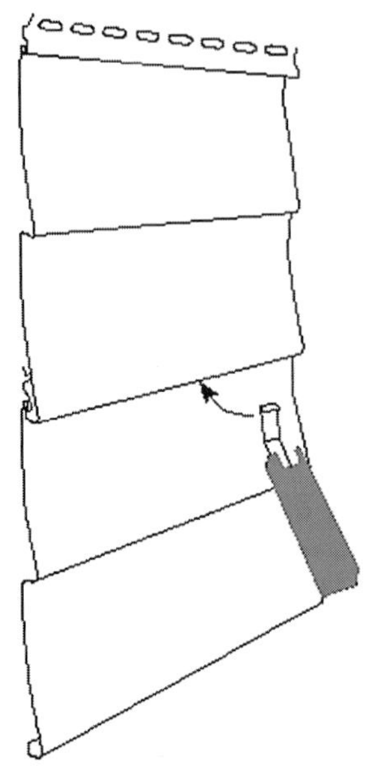

FIGURE 2.6.3 ZIP TOOL

3. REPLACE AND COVER EXISTING SIDING WITH NEW VINYL SIDING.

Existing siding can be prepared in three ways:

1. Strip off existing siding that has deteriorated to the point where it cannot be used as a substrate. This will assure the straightest and flattest application and will allow inspection of the sheathing and insulation, which can be replaced if necessary. Install new vinyl siding as per manufacturer's installation requirements. Note that vinyl siding requires a smooth, even, rigid substrate such as plywood, wood composition, rigid foam insulation, or fiber sheathing. It cannot be installed directly to structural framing or it will sag and deflect between framing members. (Failure to establish a smooth, solid substrate may constitute a misapplication under the terms of the warranty.)
2. Apply rigid or semirigid (e.g, FoamCore™ by International Paper) sheathing to existing siding to provide a smooth substrate. Nail securely through old siding and into framing members. Flash as necessary around projections and openings.
3. Apply vertical furring strips to old siding to strengthen and straighten uneven surfaces. Then apply rigid sheathing as described in example 1 above. Vinyl siding cannot be applied directly over furring. Residing over existing material will require jamb and trim extensions.

Vinyl siding should be applied with corrosion-resistant nails (aluminum or galvanized) with a minimum 3/4" penetration into wood or wood composition substrate. When foam sheathing is used directly over studs, nails must penetrate studs by at least 3/4". Nails should be driven so that the heads are 1/16" to 1/8" away from the slotted nailing tab to allow for shingle movement. Pneumatic staplers or nailers can be used but can bind siding more easily (especially staples) than hand nailing, unless used by an experienced installer. Individual manufacturers' installation guidelines should be followed carefully.

ADVANTAGES: Better quality vinyl siding replicates wood siding appearance. New formulations, textures, colors, and details are now available that allow more choice and improve the appearance of most architectural styles.Vinyl siding is relatively low maintenance product.

DISADVANTAGES: Vinyl siding, especially the thinner products, might appear wavy and will reflect the irregularity of some substrates. It expands and contracts more than other siding materials. Thinner gages are susceptible to "oil canning" and may become brittle over time. Overdriven or improperly placed fasteners can resist siding movement and cause buckling. Colors, especially dark ones, will fade over extended periods. Vinyl siding is not weatherproof and requires a weather barrier.

2.7 METAL SIDING

ESSENTIAL KNOWLEDGE

Aluminum and steel siding gained great popularity in the 1950s and 1960s as the most durable and cost-effective materials for replacing or covering up old, deteriorating siding. It frequently contained thin foam inserts which, manufacturers claimed, greatly enhanced the insulating qualities of the material. While the new siding may have been effective in reducing air infiltration, the thinness of the insulation and the great thermal conductivity of metal made this siding a poor performer in terms of energy.

The use of aluminum and steel siding has drastically declined in recent years with the emergence and popularity of vinyl and fiber cement siding products, to the point where metal siding now accounts for only about 1% of the new siding market. Primary markets include the midwest (particularly for steel siding) where cladding must resist hailstorms and temperature extremes. In metropolitan areas where local codes require noncombustible building materials, metal siding is an obvious choice. The plethora of existing metal siding makes it a prime candidate for repair or replacement in rehab work.

TECHNIQUES, MATERIALS, TOOLS

1. MAINTAIN METAL SIDING.

Under normal conditions, metal siding will require only an occasional washdown with a garden hose and a soft bristle brush. If the siding is moderately dirty, use a solution of ⅓ cup of a mild cloth-washing detergent to a gallon of water. For heavier dirt and stains the nonabrasive detergent can be mixed with ⅔ cup trisodium phosphate (Soilax, or Spic-N-Span for example) to a gallon of water. Use mineral spirits to remove caulking compounds, tar, and similar substances. Clean from bottom to the top. Rinse thoroughly. Avoid abrasive cleaners and strong solvents.

ADVANTAGES: Maintenance is simple and effective.

DISADVANTAGES: This method will not remove fading and severe caulking.

2. REPAIR METAL SIDING.

The replacement of metal panels is relatively simple: (1) Cut the damaged panel along its center with a utility knife or metal shears. Remove and discard the bottom section; (2) Cut and remove the top lock or a new panel; (3) Apply a heavy bead of Gutterseal (Alcoa or other brand) the full length of the defective panel; (4) Install the new panel over the Gutterseal. Apply pressure with the palm of the hand. Do not nail the panel (Fig. 2.7.1).

ADVANTAGES: Repair is a relatively simple procedure.

DISADVANTAGES: The color of the replacement will not match original.

3. REPLACE EXISTING SIDING WITH STEEL SIDING.

Considered by some in the industry to be one of the highest-performing siding products on the market because of its resistance to cracking, bending, high winds, and high temperatures, steel continues to have a strong niche market. Steel can be placed directly over existing siding materials, over

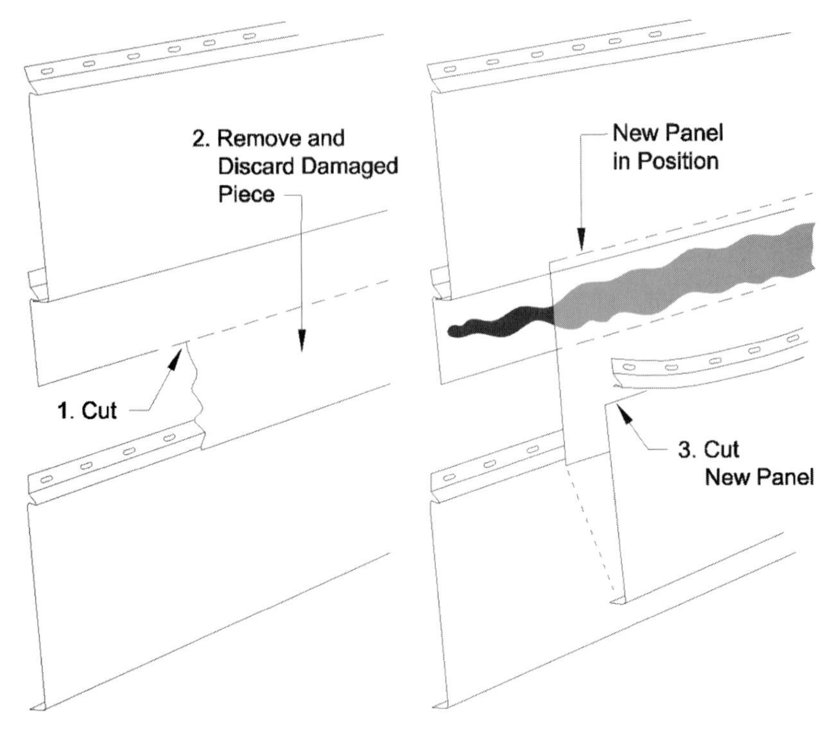

FIGURE 2.7.1 REPLACEMENT OF SIDING

FIGURE 2.7.2 ON-SITE ROLL FORMING

insulated sheathing on top of existing material, or attached to furring strips placed on masonry or uneven walls. It typically comes in 12' lengths and panel widths of double 4" or 5" exposures, with a PVC or acrylic finish. A number of companies, such as ABC Seamless, provide seamless steel siding through franchised installers. It is roll-formed on site to eliminate vertical joints (Fig. 2.7.2).

ADVANTAGES: Steel siding is one of the strongest and most damage-resistant siding products available. It lays flat and straight against most substrates; its color finish is warranted against fading and peeling; it is noncombustible; it can be touched up with paint.

DISADVANTAGES: Steel siding is approximately double the cost of standard vinyl siding and 30% more expensive than aluminum siding. Fewer profiles, styles, colors, trim, and accessories than for vinyl siding are available.

4. REPLACE EXISTING SIDING WITH ALUMINUM SIDING.

Aluminum continues to have some niche market appeal, although its use is diminishing. Aluminum siding is typically available in 12' lengths and panel widths of double 4" or 5" exposures, with a PVC or acrylic finish. The most popular color is white, but a limited color palette is available. Like steel siding, aluminum can be applied directly over wood-sided walls that are sound and straight, over insulated siding, and over furring strips.

ADVANTAGES: Aluminum siding lays straight over most substrates is less likely to show waviness than vinyl siding is lightweight, and is noncombustible, durable, and easy to clean.

DISADVANTAGES: Aluminum siding is more costly than vinyl, dents relatively easily, and comes in limited styles and colors.

2.8 WOOD SHINGLES AND SHAKES

ESSENTIAL KNOWLEDGE

Wood shingles and shakes (thicker versions of shingles) have been used for siding for more than 300 years. Today, most of this material is milled in Canada and is made of western red cedar, eastern white cedar, or Alaskan yellow cedar. Pressure-preservative-treated southern yellow pine is also used for shakes on a limited, regional basis. Western red cedar weathers a darker gray than the other two materials. A few mills make redwood shingles, but these shingles are not as popular as cedar shingles because they weather considerably darker. Cedar shingles and shakes are warranted against material defects for a minimum of 20 years (30 years if pressure-treated with chromated copper arsenate CCA) by members of the Cedar Shake and Shingle Bureau (CSSB).

Western red cedar shingles are available in a variety of grades, including No. 1, BLUE LABEL, (100% heartwood, 100% clear, 100% end grain) and No. 2, RED LABEL, with some flat grain and limited sapwood. Other grades are available for secondary structures, economy installations, and undercoursing. No. 1 is the preferred grade for both roofing and siding, but No. 2 grade is also used for siding because the weathering conditions are not as extreme as roofing.

Eastern white cedar shingles (increasingly coming from small, second growth trees) are not available in 100% edge grain and are graded by knot content. Grade A, BLUE LABEL, is all heartwood with no imperfections; Grade B, RED LABEL, allows imperfections such as knots on nonexposed parts and has a recommended maximum exposure of 6". Grades A and B are recommended for siding. Grade C, BLACK LABEL is an economy grade that allows sound knots on exposed portions and has a rustic appearance. Grade D is a utility grade for underlayment. Eastern white cedar shingles are available prefinished from Sovebec, Inc. (eastern Canada's largest consortium of white cedar mills) in a tailored rebutted and squared configuration with a peroxide bleaching agent and latex stain called Ultra Bleach, which accentuates and accelerates the silver gray weathered appearance (see Product Information).

Western red cedar shakes are available in a variety of textures and finishes including machine grooved, handsplit face and resawn back, taper sawn on both sides (resembling an extra-thick shingle), taper split by hand on both sides, and straight-split by machine both sides. See the *Cedar Shake and Shingle Bureau Design and Application Manual* for detailed specifications on cedar siding.

TECHNIQUES, MATERIALS, TOOLS

1. REPLACE INDIVIDUAL CEDAR SHINGLES.

If a small number of individual shingles are badly curled, cracked, or missing, they can be removed and replaced relatively easily. Cut nails holding damaged shingles with a hack saw blade. Split shingles with a chisel and remove pieces. Cut a new shingle to fit with a 1/8" to 1/4" clearance on each side.

ADVANTAGES: This is an inexpensive way to repair existing siding.

DISADVANTAGES: This method will not work with a large area of defective shingles.

2. RE-SIDE WITH NEW CEDAR SHAKES AND SHINGLES.

Certain types of existing siding including vertical wood siding or paneling, and existing wood clapboard that are flat and in sound condition, can be left in place and new shingles applied over them. Shingles can be applied over beveled siding by filling in the low points of the wall with low-grade timber strips (called "horse feathers") and thereby increasing the potential nailing surface, or by nailing the shingles or shakes to the high points of the bevels of each course of the old wall (Fig. 2.8.1). However, many shingle manufacturers recommend the use of furring strips or a plastic mesh product such as Cedar Breather™ be used to allow for air circulation and to reduce the potential of excessive moisture build-up behind the shingle (Fig. 2.8.2). If the existing siding is stucco or masonry, or if the surface is uneven, horizontal, or a combination of horizontal and vertical, furring is necessary to flush out the wall, allow for air circulation, and to provide a nailing surface for the new siding (Fig. 2.8.3). If the existing siding is substantially deteriorated, removal

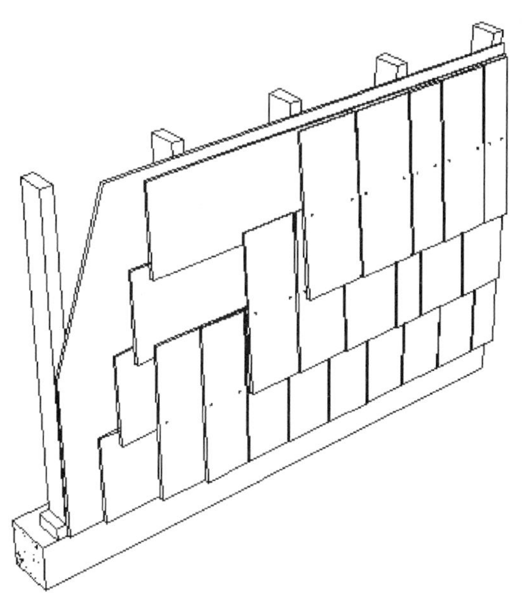

FIGURE 2.8.1 BEVELED SIDING DETAIL

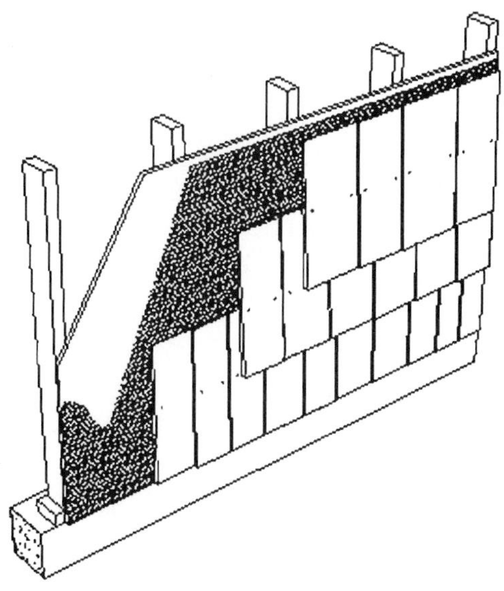

FIGURE 2.8.2 CEDAR BREATHER

of the shingles allows for the inspection, removal, and reinstallation of existing insulation, sheathing, flashing, caulking, building paper, or housewrap as necessary. If the sheathing is nonstructural or foam, new shingles should be fastened to furring that is laid over the sheathing. Furring is typically 1x3 or 1x4 material (Fig. 2.8.4). Wherever possible, butt lines should align with tops or bottoms of windows or other openings for appearance. Shingle exposure should be consistent. Corners can be butted against corner boards or laced together on outside and inside corners, or mitered on outside corners (Fig. 2.8.5).

ADVANTAGES: Cedar is an attractive, natural material for both traditional and contemporary buildings. It can be treated with a variety of coatings and preservatives or left to weather naturally, and it

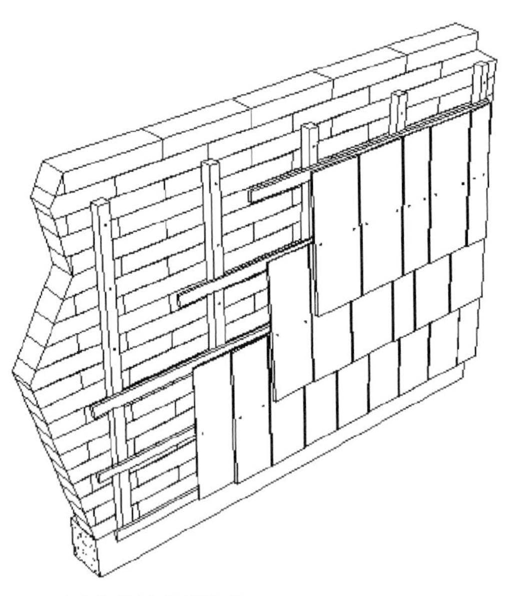

FIGURE 2.8.3 MASONRY DETAIL

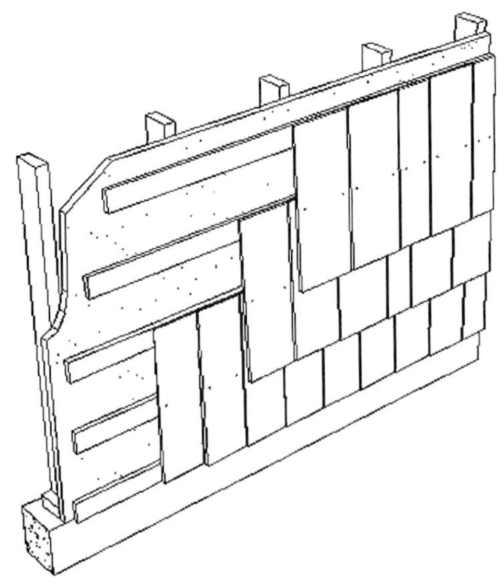

FIGURE 2.8.4 STUCCO DETAIL

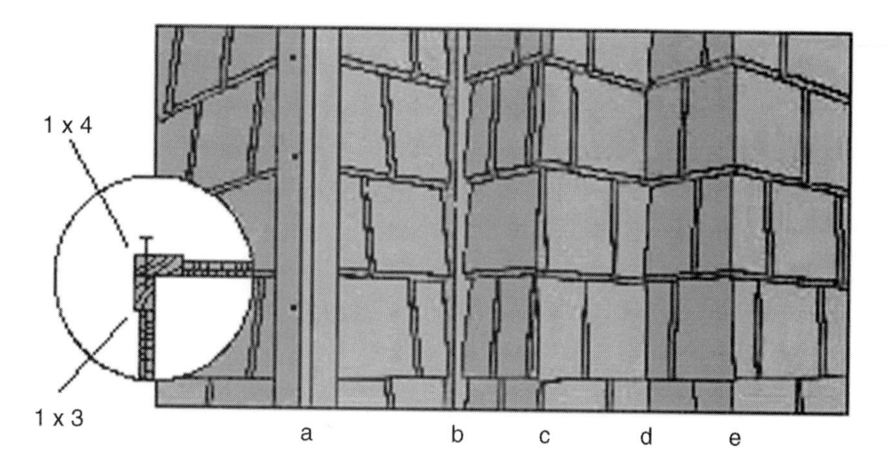

FIGURE 2.8.5 CORNER DETAILS

is a sustainable material that comes from renewable sources. Re-siding over existing wall eliminates the time and expense of removing existing siding.

DISADVANTAGES: Cedar is more costly than other siding material. It can weather unevenly in certain conditions, it is combustible unless pressure-treated with fire retardants; and some warping and cupping will occur.

2.9 SOLID WOOD SIDING

ESSENTIAL KNOWLEDGE

Solid wood siding (particularly beveled siding such as clapboard) has been popular in the United States for the past 300 years. The use of wood siding has increased somewhat recently, according to the Western Red Cedar Lumber Association. However, its use is expected to decline as lumber becomes more costly, as quality materials become harder to obtain, and as other similar looking and less-expensive materials become more popular (such as vinyl and fiber-cement siding).

The large majority of solid wood siding comes from the western United States and Canada and is made of western red cedar. A small portion is also made from western redwood, and is available from individual mills. Solid wood siding from other softwoods such as vertical grain western Douglas fir is occasionally available, but in decreasing amounts. Some eastern mills produce siding from softwoods such as white pine and spruce. Wood siding is usually treated with a protective coating such as a semitransparent or opaque stain or paint. Individual industry associations have their own finishing recommendations (see Further Reading).

Most bevel siding (the most popular form of solid wood siding) is made from resawn lumber (1"-thick boards that are sawn from logs and then cut diagonally on a band saw. Depending on where the 1"-thick boards are cut, resawn lumber can be vertical grain, curved grain, or flat grain. Quartersawn (also known as radial sawn) siding is made from cuts radiating out from the center of a log, with the result that each piece is vertical grain (Fig. 2.9.1). Vertical grain siding is better at resisting warping and twisting. Curved and flat grain siding tends to twist opposite to the curve of the growth ring. Vertical grain siding also takes and holds paint better than flat grain because the maximum number of grain

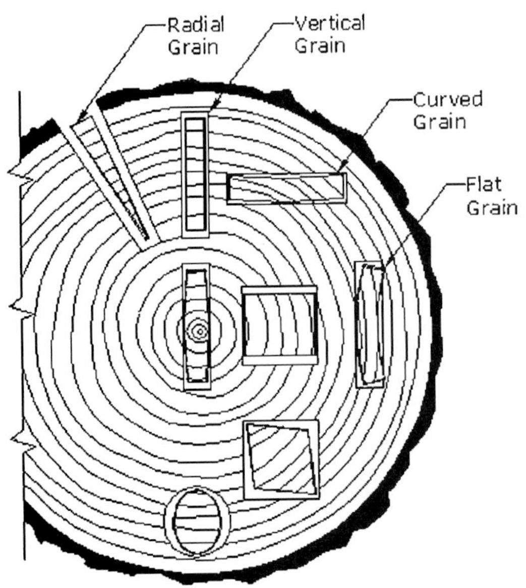

FIGURE 2.9.1 GRAIN CONFIGURATION AND SHRINKAGE CHARACTERISTICS

surfaces, which act as capillaries absorbing paint or stain for better adhesion, is exposed. Since vertical grain is more stable than flat grain, the paint is also less stressed by twisting.

TECHNIQUES, MATERIALS, TOOLS

1. REPAIR OR REPLACE DAMAGED PIECES OF EXISTING BEVEL WOOD SIDING.

Individual pieces of bevel wood siding can easily be removed and replaced by cutting the nails holding the damaged sections with a hacksaw blade, and then by wedging and sawing the damaged sections out. New material can then be installed to butt existing material and fastened into place.

ADVANTAGES: This repair is the most economical.

DISADVANTAGES: This repair will not work where there are large areas of deteriorated material, which will have to be stripped down to the sheathing or structure.

2. REPLACE DETERIORATED OR DAMAGED SIDING WITH NEW WESTERN RED CEDAR BEVEL SIDING.

Resawn western red cedar bevel siding has either two textured sides, or one textured and one smooth sanded side, depending on its grade. Clear bevel siding results in the highest quality appearance with no visible knots. Bevel siding is available in plain bevel and rabbeted bevel (Fig. 2.9.2). Plain bevel is the most popular. Rabbeted bevel lays flatter on the wall and is somewhat easier to nail, but it has a less pronounced shadow line. Bevel siding ranges in exposure from 4" to 12", with a butt thickness ranging from 1/2" to 7/8". Red cedar bevel siding is available in six traditional grades: Clear V.G. Heart (smooth face, all vertical grain), A Clear (mixed grain), Rustic (some knots, sawn texture), B Clear (more knots and other growth characteristics), Select Knotty (significant sound and tight knots), Quality Knotty (considerable knots); and Architect Knotty Bevel Siding (comes with a variety of knots and other growth characteristics and is factory primed). Bevel siding is also available finger-jointed in lengths up to 16' in clear and knotty grades and in smooth, resawn, and combed (multiple-grooved) textures.

A number of mills have arrangements with paint and stain companies, including Olympic, Cabot, and Sherwin Williams, which will provide 5-year warranties for factory-applied primer coat, increased to 15 years after the field application of an approved topcoat. If the siding is recoated before the 15-year warranty is over, it can be extended another 15 years for up to 30 years. Although

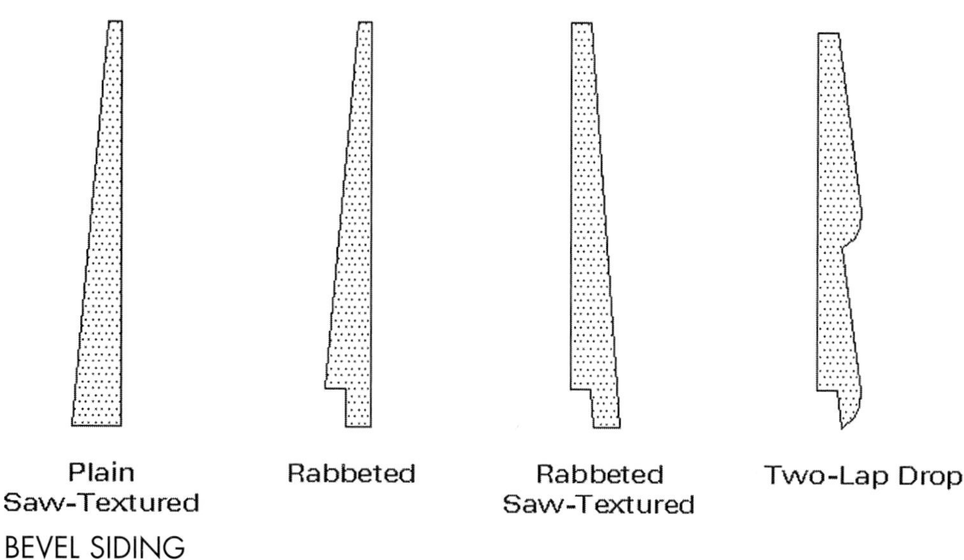

FIGURE 2.9.2 BEVEL SIDING

some lower grades are used in conjunction with some extended warranty painting programs, vertical grain cedar will perform the best over an extended period (see Product Information).

Bevel siding can be applied directly, over building paper or housewrap to solid siding or sheathing, or to insulating sheathing if special application requirements for the insulating sheathing are adhered to (see Further Reading). It can also be applied to furring over masonry (on uneven walls or to allow for the ventilation of the inside face of the cedar). Face nailing of bevel siding is recommended as blind nailing will split the thin, feathered portion of the shingle (Fig. 2.9.3). Stainless-steel nails are recommended, especially in areas near the ocean. Hot-dipped galvanized or aluminum nails are acceptable in less corrosive environments. A variety of inside and outside corner details is possible (Fig. 2.9.4). Mitered corners are usually caulked and can separate if the wood is not properly seasoned and knots have been allowed to get wet prior to application. Corner boards are a good alternative. Nonhardening caulks are recommended, including polyurethane, polysulfide, or latex-silicone. For more information on installation refer to *Installing Cedar Siding* (see Further Reading).

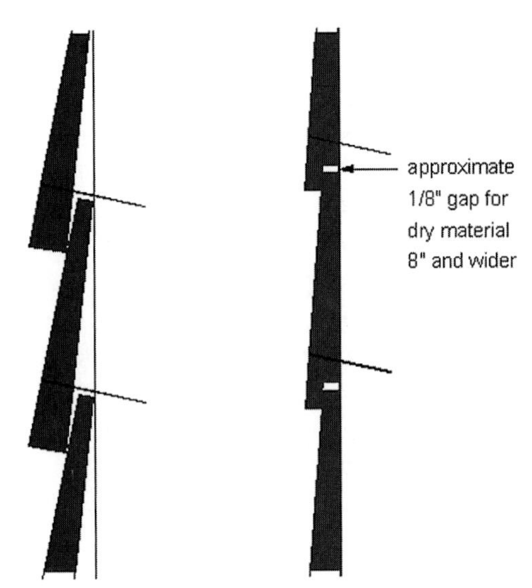

FIGURE 2.9.3 FACE NAILING OF BEVEL SIDING

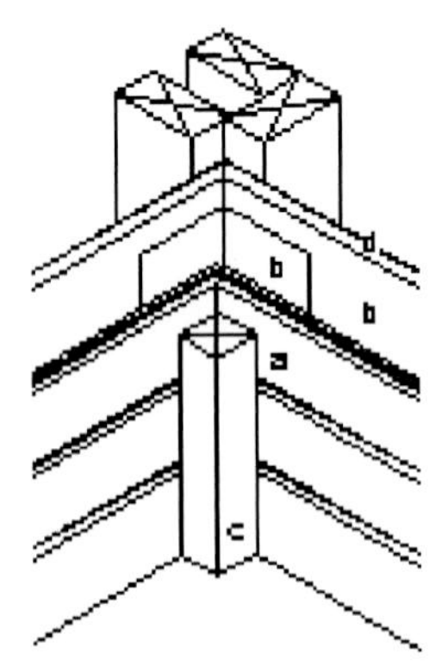

a. Horizontal Siding
b. Building Paper
c. Trim
d. Sheathing

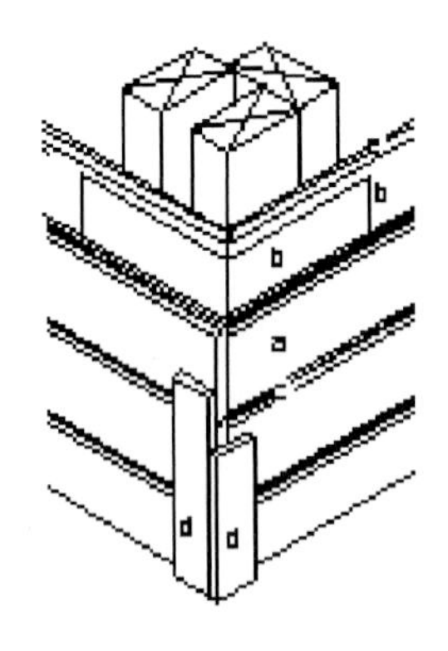

a. Horizontal Siding
b. Building Paper
c. Caulk
d. Corner Boards
e. Sheathing

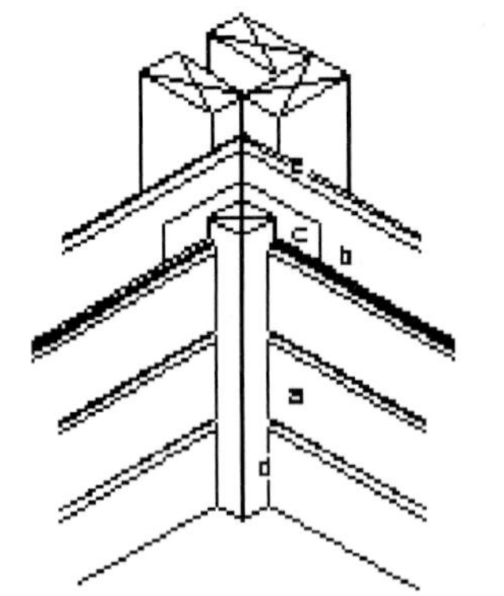

a. Horizontal Siding
b. Building Paper
c. Metal Flashing or Building Paper
d. Use Caulk at Inside Corner or Open Joint. Provide Metal Flashing Behind Joint if Caulk is not used.

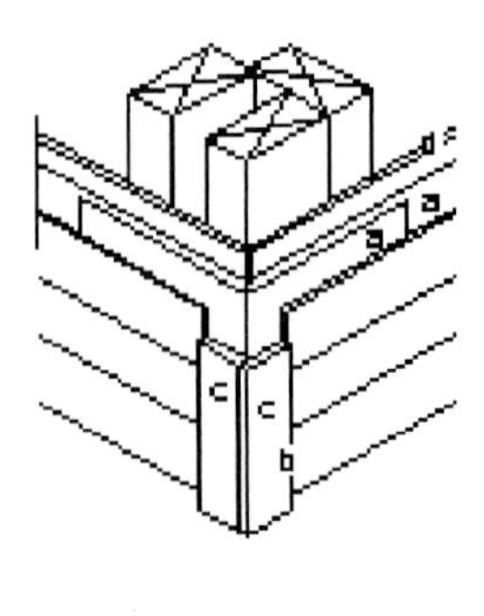

a. Building Paper
b. Caulk
c. Corner Boards
d. Sheathing

FIGURE 2.9.4 CORNER DETAILS

ADVANTAGES: Western red cedar bevel siding has natural decay resistance, dimensional stability, workability, and paintability. Up to 30-year warranties available from paint companies make painted bevel siding competitive with other siding products. Western red cedar bevel siding is available in vertical grain and in a variety of sizes, lengths (up to 16'), and finishes.

DISADVANTAGES: Western red cedar beval siding will deteriorate and discolor over time if not maintained properly. It expands and contracts with changes in humidity and will shrink over time, causing vertical joints (which should be beveled) to open. It may cup and twist, especially if flat grain is used instead of vertical grain. It is a combustible material and is more costly than other siding products.

3. REPLACE DETERIORATED OR DAMAGED SIDING WITH NEW WESTERN REDWOOD BEVEL SIDING.

Redwood bevel siding has all of the favorable attributes of cedar bevel siding, but it is less available nationally and usually is more costly than cedar. It comes in a variety of grades including Clear All Heart (all heartwood and free of knots), Clear (some sapwood and some small, tight knots), B Heart (a limited number of tight knots and other characteristics), and B Grade (similar to B Heart except that it permits sapwood as well as heartwood). All these grades are available as certified kiln dried, which is recommended for top performance and minimal shrinkage. Clear or Clear All Heart may be ordered either flat grain or vertical grain (recommended for best performance). Rustic grades are also available. Redwood bevel siding comes plain and rabbeted, smooth and rough sawn, and in a full range of sizes from ½" × 3 ½" to ¾" × 10". Thinner widths are less likely to shrink and split than the wider widths. Redwood takes paint well, but individual mills haven't been as aggressive as the cedar mills in developing warranty programs with the leading paint companies. Redwood bevel siding is applied and finished in the same manner as cedar.

ADVANTAGES: Redwood bevel siding has natural decay resistance, dimensional stability, paintability, and workability. It is available in a variety of finishes, grains, sizes, and styles.

DISADVANTAGES: Redwood bevel siding will deteriorate and discolor over time if not maintained properly. It expands and contracts with changes in humidity and will shrink over time, causing vertical joints (which should be beveled) to open. It may cup and twist, especially if flat grain is used instead of vertical grain. It is a combustible material and is more costly than other siding products.

4. REPLACE DETERIORATED OR DAMAGED SIDING WITH NEW QUARTERSAWN SPRUCE OR PINE BEVEL SIDING.

Quartersawn eastern spruce or pine bevel siding is available from the Granville Mfg. Co. The bevel siding comes in 1x Clear (clear, unmarked, or exposed section of siding), 2x Clear (up to one knot or blemish per piece), and Cottage (not more than three knots or blemishes, no loose knots). It is available in sizes from 7/16" × 4" to 6" and up to 6' long. The material comes unpainted, primed, or painted.

ADVANTAGES: Vertical grain spruce or pine bevel siding resists twisting and warping and accepts paint well. Authentic manufacturing meets all colonial reproduction specifications. It is less costly than western red cedar or redwood siding and has a long lifespan.

DISADVANTAGES: Spruce or pine bevel siding requires maintenance and repainting and is available only in relatively short lengths (which shrink less). It will not match resawn siding. It is available only in 2½" to 4" exposures and is a combustible material.

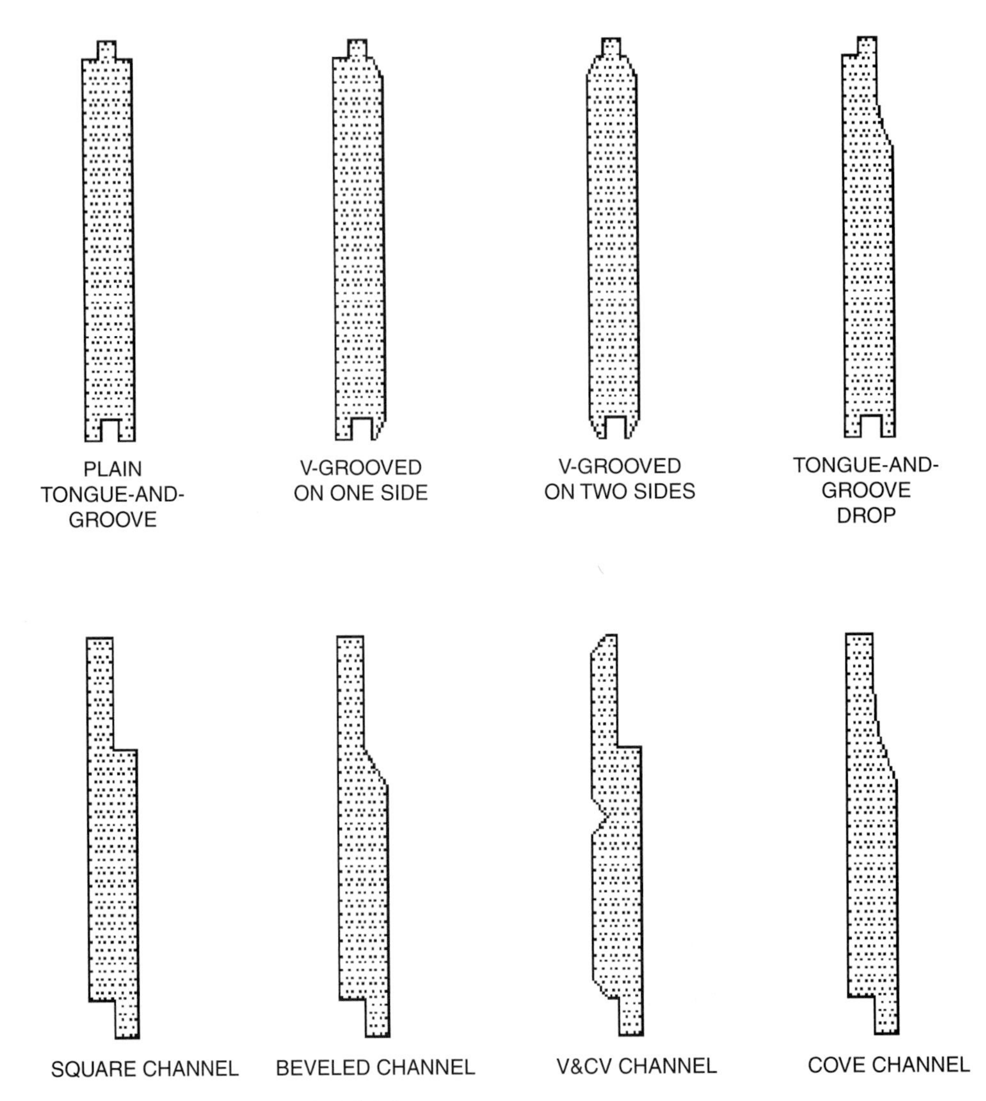

FIGURE. 2.9.5 NONBEVEL WOOD SIDING

5. REPLACE DETERIORATED OR DAMAGED SIDING WITH NONBEVEL SOLID WOOD SIDING.

Nonbevel wood siding types include board and batten, drop siding, shiplap, tongue-and-groove, and variations of these types (Fig. 2.9.5). Many, including tongue-and-groove and shiplap, can be installed horizontally in the same manner as bevel siding, and vertically with blocking between studs, or over horizontal furring strips. The grades of the siding products are similar to those for bevel siding. For detailed sizes and specifications see Further Reading and Product Information.

ADVANTAGES: A wide variety of distinctive, historically correct products for both traditional and contemporary use, are available from individual suppliers with vertical grain and extended painting warranties.

DISADVANTAGES: Periodic maintenance is required. These products are more costly than other siding products and are combustible.

2.10 HARDBOARD SIDING

ESSENTIAL KNOWLEDGE

Hardboard siding was first developed by William Mason in 1922. The product is made of wood chips converted into fibers, combined with natural and synthetic binders, bonded under heat and pressure. The early success of this material led to a proliferation of hardboard manufacturers in the early 1950s when the original patents expired. More than a score of companies were producing hardboard siding by the 1970s when the EPA's newly declared environmental restrictions on the dumping of wastewater from hardboard's wet manufacturing process led to the closing of a number of plants that did not have the space or resources to make required changes. By the mid-1980s hardboard siding accounted for about 30% of the U.S. residential siding market. Today, hardboard siding's market share has declined to approximately 15%. The number of current manufacturers has been reduced to six. The constricted market is due to the growth in use of other materials such as vinyl siding, exterior insulation and finish systems, and fiber-cement siding.

There has been considerable publicity about class action lawsuits against certain manufacturers regarding the material's performance when exposed to moisture. Manufacturers have countered that, while lapses in quality control may have occurred, hardboard has a long history of good performance and that the majority of problems is due to inadequate field supervision (particularly regarding incorrect flashing, caulking, nailings, and painting) by installers around windows, doors, deck terminations, and other potential moisture entry points, and the lack of sufficient maintenance on the part of homeowners.

The majority of the class action lawsuits has been settled, and the major manufacturers continue to produce and sell hardboard siding with limited 25- to 30-year warranties. Some products are undergoing continued design improvements; others have changed relatively little over the years. Masonite Corp., for example, has developed a new lap siding product called HiddenRIDGE™ with concealed fasteners and interlocking design that speeds installation and provides level alignment (Fig. 2.10.1).

TECHNIQUES, MATERIALS, TOOLS

1. REPAIR EXISTING HARDBOARD LAP SIDING.

Small sections of damaged or deteriorated lap siding can be cut out and replaced with matching profiles relatively easily with conventional carpentry tools, including handsaws and power saws. Deteriorated sheathing should be replaced as required. Panel siding sections can be repaired as well, but with greater difficulty.

ADVANTAGES: This repair is the most economical.

DISADVANTAGES: Large areas of damaged material have to be stripped down to sheathing or structure. Repaired sections may not exactly match existing siding.

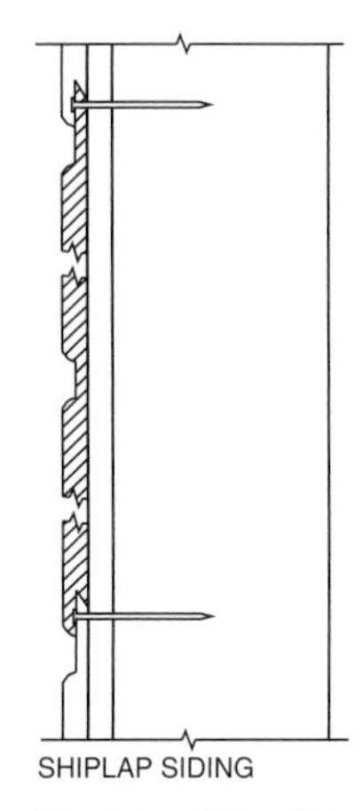

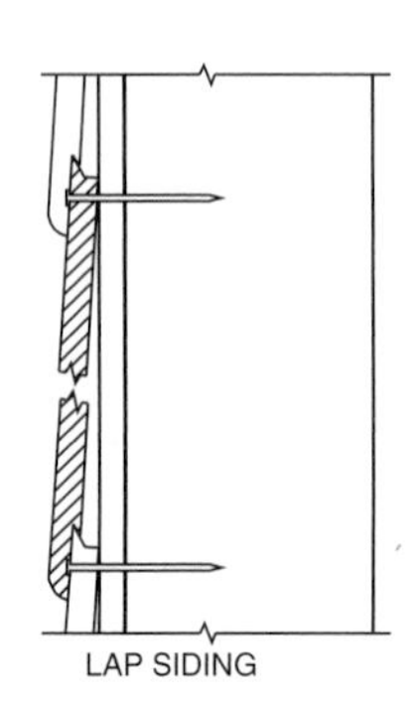

FIGURE 2.10.1 HiddenRIDGE

2. REPLACE EXISTING SIDING WITH NEW HARDBOARD LAP SIDING.
Lap siding is available in 7/16" and 1/2" thicknesses and in a wide variety of configurations including shiplap siding 12" and 16" wide in lengths up to 16' with varying profiles, as well as conventional 6", 8"-, or 12"-wide lap siding made to look like beveled siding (Fig. 2.10.2). Most products are preprimed and require two coats (4 dry mils) of field painting. Masonite offers a completely prefinished, Colorlock siding that comes with a 15-year limited warranty on the finish. Lap siding can be applied over existing sound and level siding, over solid or insulating sheathing, directly over studs 16" on-center (o.c.) with an approved water-resistant barrier, or over furring strips. Staples, t-nails, or bugle head nails are not recommended. Hardboard, as all materials, will expand and contract with temperature and humidity variations. A minimum 3/16" space is recommended between the siding and windows, door frames, and corner boards and 1/16" between vertical butt joints, which must fall on studs. All joints must be caulked with nonhardening, mildew-resistant exterior grade sealant.

ADVANTAGES: Hardboard lap siding has been used successfully for over 50 years and remains the preferred siding material in some markets. It is available in a wide variety of profiles, textures, and styles that accurately simulate the appearance of wood products. It is easily worked and handled, is a consistent product with no surface defects, is available in lengths up to 16', and is less costly than solid-wood siding products.

DISADVANTAGES: Hardboard lap siding is more susceptible to moisture-related problems than other siding materials. It requires careful storage and protection during construction and should not be applied over wet sheathing.Careful attention to fastening, caulking, and painting recommendations is required. Long lengths of siding may buckle due to expansion. Periodic inspection and maintenance of caulking and painting is required.

3. REPLACE EXISTING SIDING WITH NEW HARDBOARD PANEL SIDING.
Of the hardboard siding sold, panel siding represents about a third. It is available in 4'x8' and 4'x9' square-edge panels (for board and batten applications) and shiplap-edge panels. The panels come in a variety of textures and groove configurations that simulate vertically applied wood boards. The material is also available without grooves in a stucco appearance. Temple-Inland offers a 16"x48" panel called Shadowround that simulates scalloped shingles. Hardboard panels may be applied to sheathed or unsheathed walls with studs no more than 24" o.c. or over sound, flat existing siding. Panel edges must fall on and be nailed to framing members. Horizontal joints must have adequate blocking and be overlapped a minimum of 1" or be provided with Z-shaped preformed flashing. The material is available preprimed for field painting.

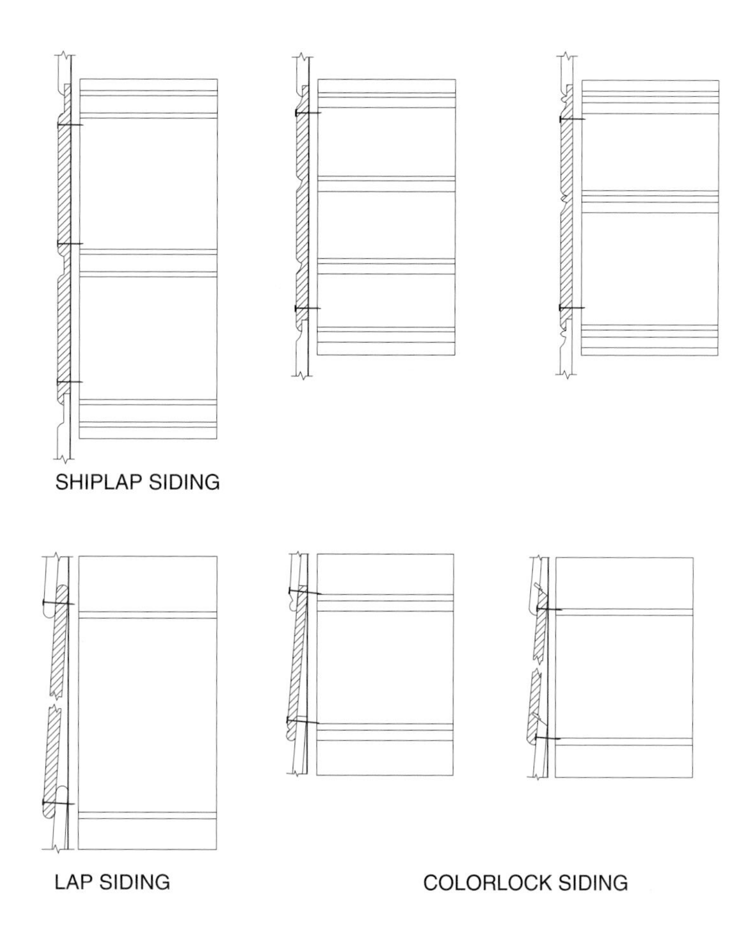

FIGURE 2.10.2

ADVANTAGES: Hardboard panel siding is a consistent material without knots, raised grain, checks, or other surface defects. It takes paint well and has been used successfully for many years when properly installed and maintained. It can provide shear resistance when installed directly over stud and is available in a variety of attractive patterns and finishes. It is easily worked and handled.

DISADVANTAGES: Hardboard panel siding requires careful storage and protection during construction. Careful attention to fastening, caulking, and painting requirements and periodic inspection and maintenance of caulking and painting are required.

2.11 ENGINEERED WOOD SIDING

ESSENTIAL KNOWLEDGE

Louisiana-Pacific (L-P) (the largest producer of oriented strand board) introduced Inner-Seal™ OSB siding in the mid-1980s as an economical alternative to conventional wood siding products that had become more costly due to rising lumber prices. While the material proved popular with builders, some performance problems arose, particularly in humid environments such as the northwest and southeast. Where nail holes, material surfaces, and edges were not adequately painted or sealed, water penetrated, causing the OSB siding to swell and expand. Some OSB siding deteriorated due to rot, fungus, and invasion by insects, in some cases causing damage to sheathing beneath. A number of class action lawsuits were brought against the manufacturer, and L-P has settled these claims.

Since then, L-P has reengineered the material and has reintroduced it as a "treated engineered wood product" instead of an OSB product, offering bevel-edged siding in 16' lengths with 6", 8", 9 1/2", and 12" widths, and in 4'×8' to 4'×16' panels. The composition and production process have been substantially revised. The binder has been changed from a phenolic to methylene diphenyl diisocyanate (MDI), which is more water resistant and provides a stronger bond to the wood flakes, therefore reducing the swelling and expansion. Powdered zinc borate has been included to prevent rot, fungal growth, and insect-caused deterioriation. A prefinished resin-saturated paper embossed with a pronounced wood grain is thermal-fused to the wood substrate, and a proprietary sealant treatment is applied to ends and edges. The siding is prefinished with an all-acrylic primer. Fascia and trim products are also available.

TECHNIQUES, MATERIALS, TOOLS

1. REPLACE DAMAGED OR DETERIORATED SIDING WITH NEW L-P SMART LAP™ SIDING.

Depending on the most cost-effective approach, engineered wood lap siding can be installed directly to studs with a weather barrier (Fig. 2.11.1), directly to nailable sheathing, over "fanfold" insulating sheathing; or to furring strips over masonry or irregular surfaces (siding should not come in contact with masonry). If a rainscreen or drainage channel behind the siding is desired, the siding can be installed over vertical furring strips at a maximum spacing of 24" o.c. Sections of existing Inner-Seal™ OSB siding that have swelled, edge cracked, or otherwise deteriorated can be cut out and replaced with new sections of L-P engineered wood siding, which will closely match the appearance of the old OSB siding. For complete instructions refer to *Application Instructions for Smart Panel™ and Smart Lap™ Siding*.

ADVANTAGES: This siding is lightweight, strong, and easily worked with conventional tools. It has the appearance of textured painted cedar siding, joints at the trim similar to wood bevel siding, and lays flat. It is dimensionally stable; will not warp or cup; is free from knotholes; and is resistant to end-checking and splitting, shrinkage, and buckling. It has a 25-year prorated limited warranty against fungal degradation, cracking, peeling, separating, chipping, flaking, or rupturing of the resin-impregnated surface overlay. It is environmentally sound and made from a renewable resource.

DISADVANTAGES: This siding requires repainting and recaulking over time. It is a new product with a limited performance history, is a combustible material, and is currently not available in smooth patterns.

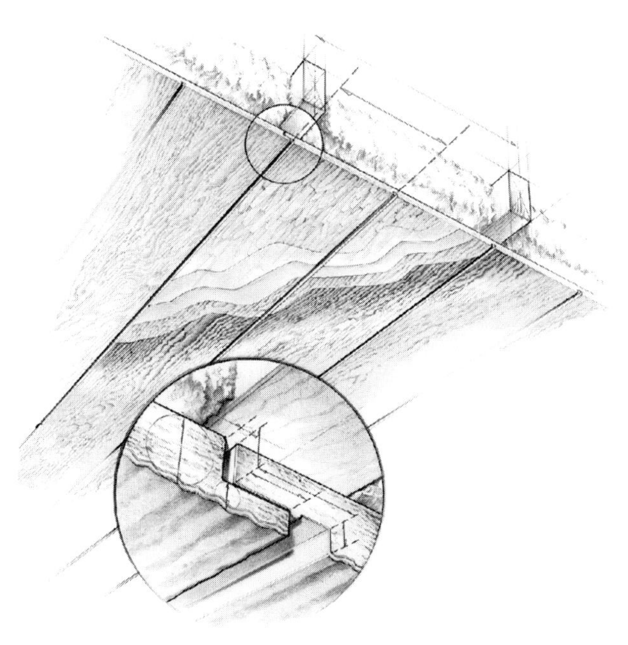

FIGURE 2.11.1 SMART LAP™

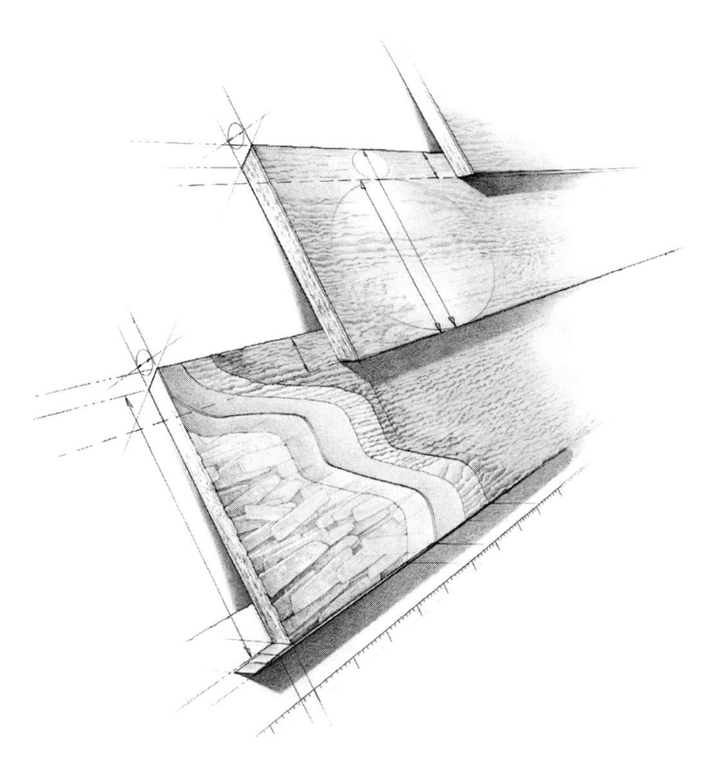

FIGURE 2.11.2 SMART PANEL™

2. REPLACE DAMAGED OR DETERIORATED SIDING WITH NEW L-P SMART PANEL™ AND EZ PANEL™ ENGINEERED WOOD PANELS.

These panels are available in a variety of thicknesses and patterns simulating vertically grooved plywood siding (Fig. 2.11.2). Panels have shiplapped edges and may be used directly over studs, over structural or nonstructural sheathing, over existing sound wood siding, and on furring strips over masonry and uneven walls. Panels should not contact masonry surfaces, and vertically grooved panels should not be applied horizontally. Nails must be stain and corrosion resistant and have a minimum 1/4" head. Nails driven below the surface must be sealed. All joints must be caulked with nonhardening paintable sealant with a service life of at least 25 years. Horizontal trim should not be placed over grooved siding without proper flashing. If flashing is not practicable, space trim away from siding so that moisture is not trapped between siding and trim. For detailed installation and painting requirements refer to *Application Instructions for Smart Panel™ & EZ Panel™ Siding*.

ADVANTAGES: Smart Panel™ is less costly than plywood, can act as a shear wall, and does not require sheathing. It comes primed and can be painted any color, has a natural-looking wood grain texture, and is easily handled and installed with conventional tools. It has a 25-year prorated limited warranty against fungal degradation, cracking, peeling, separating, chipping, flaking, or rupturing of the resin-impregnated surface overlay. There is a comparable 5-year warranty on EZ Panel™, which is a utility grade.

DISADVANTAGES: Engineered panels require regular repainting and recaulking. These panels are a new product with limited performance history and are made of a combustible material.

2.12 PLYWOOD PANEL SIDING

ESSENTIAL KNOWLEDGE

Plywood panel siding has been a popular material, especially in the northwest, north central, and northeastern states, since the early 1950s. Its use has declined in the last decade, however. Further market erosion is expected in the next decade as other siding products, such as vinyl siding and fiber cement, continue to increase in popularity.

As a material, plywood panel siding has changed very little in terms of specifications or product configuration, except that a variety of special veneer facings is now available that increases paintability and extends the life of the siding. These products include Simpson Forest Products' Guardian™ siding made with a medium-density overlaid resin-impregnated craft paper, Roseburg Forest Products' Breckenridge Siding™ made with "okoume" hardwood surface overlay imported from New Zealand, and Stimson Lumber Company's Duratemp™ made with a hardboard surface overlay. As an alternative to special surface overlays, many manufacturers offer preprimed plywood siding, although the great majority of siding is still sold as unprimed.

TECHNIQUES, MATERIALS, TOOLS

1. REPAIR EXISTING PLYWOOD SIDING.

Sections of damaged or deteriorated plywood panels can be cut out and repaired with conventional carpentry tools if there is sheathing below, but the most typical repair would be to replace the individual panel.

ADVANTAGES: This is the most economical repair.

DISADVANTAGES: Repair of individual panels is often not practical, especially if the panels are fastened directly to the structure.

2. REPLACE EXISTING SIDING WITH NEW PLYWOOD PANEL SIDING.

Made from Douglas fir or (increasingly) Southern pine veneers, plywood siding is available in 4'x8', 4'x9', and 4'x10' panels in varying thicknesses, including $^{11}/_{32}$", $^{15}/_{32}$", and $^{19}/_{32}$". Panels are available as smooth-faced or textured, grooved or ungrooved, square or shiplap-edged. Siding patterns include channel-grooved, brushed, overlaid, Texture 1-11, reverse board-and-batten, rough sawn, and kerfed rough sawn (Fig. 2.12.1). Battens can be applied for a board-and-batten appearance. Depending on the depth of the grooves, the panels are designed to be attached directly to framing members 16" o.c. or 24" o.c., or over existing flat siding, insulating or solid sheathing, or furring strips against masonry or uneven walls. Plywood siding must be primed and finish painted or stained within 30 days of installation. Unless properly maintained, the surface ply will degrade due to UV light and weathering effects and will become brittle, cracked, and eroded.

ADVANTAGES: Plywood panel siding has a long history of successful applications. It can be applied directly to studs without sheathing, can provide shear resistance, and is easily worked and erected

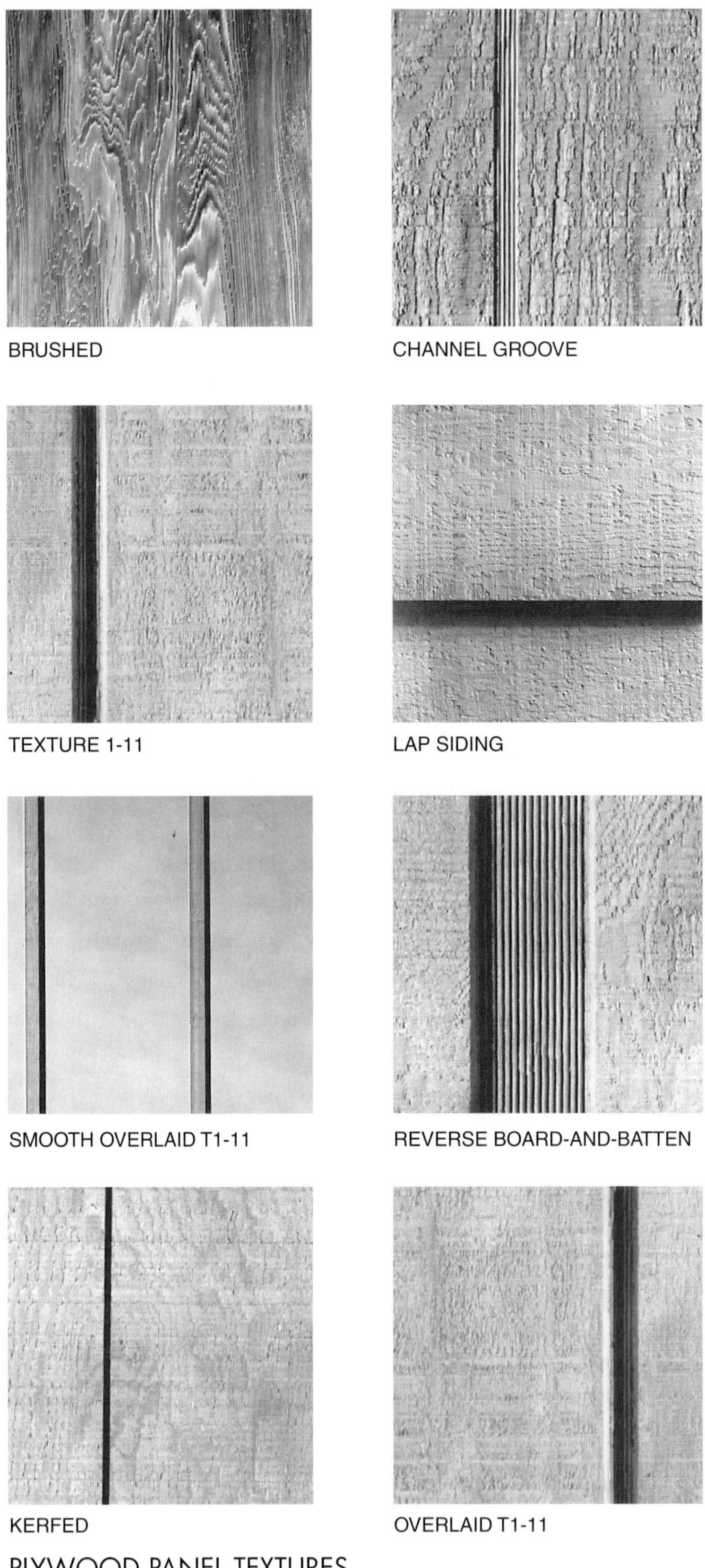

FIGURE 2.12.1 PLYWOOD PANEL TEXTURES

with conventional tools. One panel covers a large surface area. It is dimensionally stable and available in a variety of textures and styles. Products with special overlaps have surfaces that, after painting, will not split, check, or crack. Limited 25-year warranties are available.

DISADVANTAGES: Vertical grooves impart a distinctive, contemporary appearance that may not be suitable given the existing design of the house to be rehabbed. Use of plywood panel siding is generally confined to northern regions of the United States. Conventional plywood without overlays requires careful attention to initial painting and staining recommendations. The material is combustible. Ongoing maintenance is required.

2.13 FIBER-CEMENT SIDING

ESSENTIAL KNOWLEDGE

Fiber-cement siding, in its present form, is a relative newcomer to the U.S. homebuilding market, having been introduced in the late 1980s. Asbestos-cement siding, an earlier incarnation of the material, had been used extensively in the United States as well as in Europe, throughout the twentieth century. The use of asbestos in the United States was discontinued in the 1970s. Currently fiber-cement siding products are composed of portland cement, sand, clay (in some products), and specially treated wood. Today's products are thicker, less brittle, and easier to cut and work with than asbestos materials. They are also available in a wider variety of products such as backer board, lap siding, panel siding, trim, soffits, and fascias.

Fiber-cement siding has generated a great deal of interest among builders and homeowners because of its strength and impact, and rot and fire resistance. Expectations are that the use of fiber-cement products and the development of new product types, such as fiber-cement shingle siding, will increase dramatically in the next several years.

Fiber-cement siding can be cut and drilled with conventional woodworking tools (although some installers use diamond-tipped masonry blades for cutting) or scored with special shearing tools and broken much like paper-faced drywall. Fiber-cement panels are available either preprimed or unpainted, depending on the individual manufacturer. Paint adheres very well to the material's slightly textured and porous surface. The material itself is not affected by intermittent wetting, but it will discolor and stain unless painted.

TECHNIQUES, MATERIALS, TOOLS

1. REPAIR EXISTING FIBER-CEMENT SIDING.

Dented, cracked, or otherwise distressed siding can be repaired with the use of a latex-modified cementitious patching compound available from fiber-cement manufacturers or from specialty product manufacturers (see Product Information). Damaged sections can be cut out with handsaws or power saws and new sections installed as necessary. Joints between new and old materials should be primed and caulked.

ADVANTAGES: Fiber-cement siding can be repaired in much the same manner as solid wood siding.

DISADVANTAGES: Some of the wood grain and stucco patterns have changed somewhat. Adjoining new and old sections may not match exactly.

2. REPLACE EXISTING SIDING WITH FIBER-CEMENT LAP SIDING.

Fiber-cement lap siding is typically available in a variety of textures, widths from 6" to 12", lengths up to 12', and in thicknesses between 5/16" and 7/16" (Fig. 2.13.1). It can be installed directly to studs with a suitable weather barrier such as housewrap or building paper, or over solid or insulating sheathing, existing solid wood siding, or furring strips applied to uneven walls or existing masonry surfaces. Fiber-cement lap siding can be used with conventional wood, fiber-cement, hardboard, or vinyl trim. Lap siding is fastened by means of stainless or galvanized-steel nails or by means of screws with corrosion-resis-

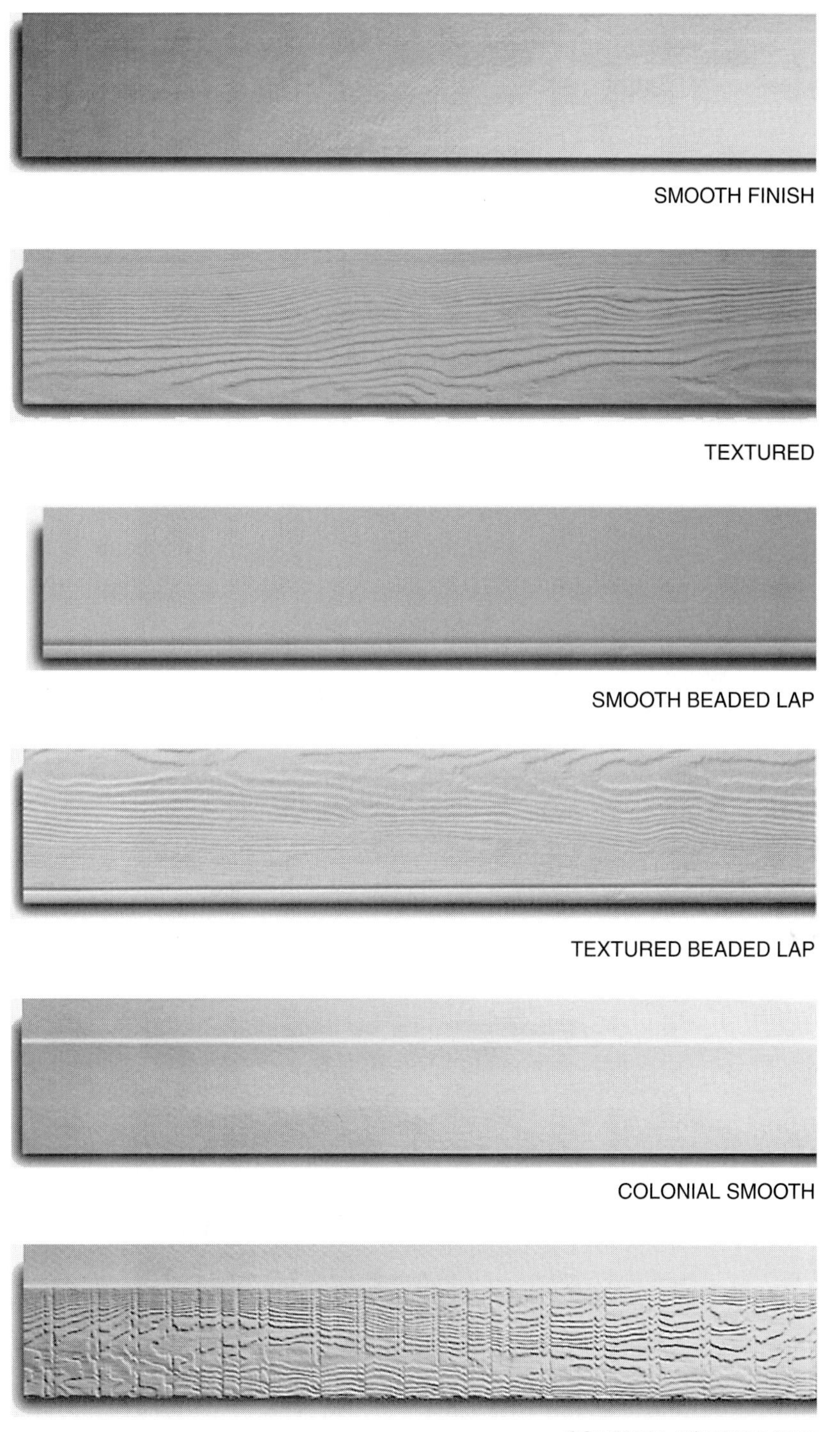

FIGURE 2.13.1 LAP SIDING TEXTURES

tant coatings. Staples are not recommended. Refer to individual manufacturer's installation manuals for specific recommendations.

ADVANTAGES: Smooth lap siding is close to wood in appearance when painted. It performs well in high-humidity environments, will not rot, is termite resistant, is noncombustible, has good impact resistance, and can be painted or stained. Up to 50-year limited product warranty against manufacturing defects are available.

DISADVANTAGES: The wood grain is somewhat more pronounced than that of rough-sawn cedar. The surface texture can be distracting in appearance at joints where the grain does not align. Periodic painting and caulking are required; however, less frequent painting is required than for wood. The variety of siding patterns is less than with other siding materials. The material is heavier than wood, and cutting of the material with saws produces silica dust. The brittleness of the material results in significant wear on tools.

3. REPLACE EXISTING SIDING WITH FIBER-CEMENT PANEL SIDING.

Fiber-cement panels are typically 4'×8', 4'×9', and 4'×10', and 5/16" in thickness. A variety of textures is available, depending on the manufacturer (Fig. 2.13.2). Installed vertically to studs directly, the

SMOOTH FINISH

STUCCO FINISH

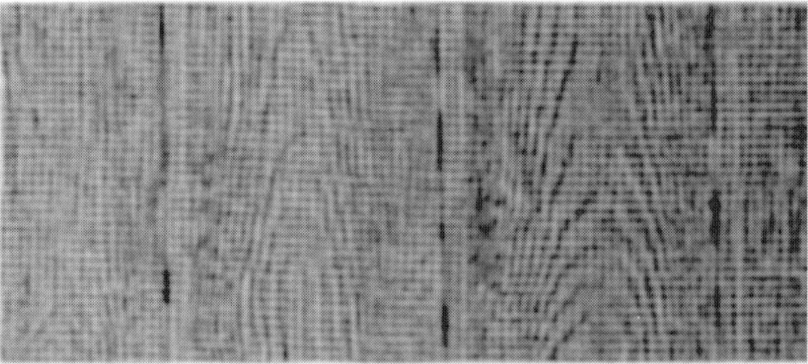

CEDAR TEXTURE WITH 5/8" GROOVES, 8" ON-CENTER

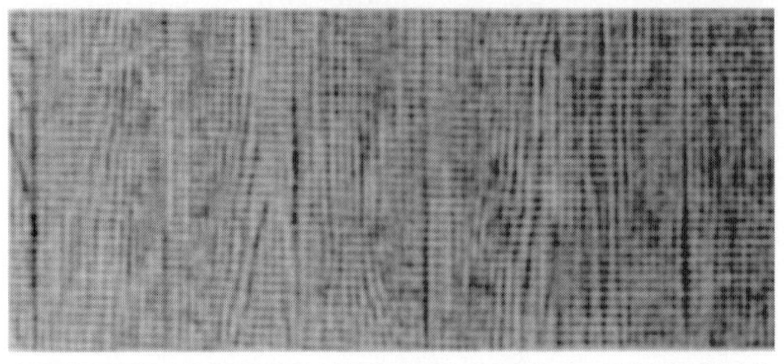

CEDAR TEXTURE WITH MOLDED GROOVES AT 4" INTERVALS

FIGURE 2.13.2 PANEL SIDING TEXTURES

panels can act as shear wall bracing. They can also be installed over solid sheathing, existing flush solid wood siding, or over furring strips against masonry or uneven walls. If required by code, a weather barrier must be installed. Corrosion-resistant nails or screws may be used. Staples are not recommended. Joints should fall on structural framing members and are typically caulked or covered with a batten strip. Horizontal joints are flashed with metal Z flashing.

ADVANTAGES: Fiber-cement panel siding can be installed quickly, covering a large area, and performs well in high-humidity environments. It can provide shear bracing, is termite resistant, and noncombustible, and has good impact resistance. It can be painted or stained. Up to a 50-year limited product warranty against manufacturing defects is available.

DISADVANTAGES: The textures do not accurately reflect materials they simulate. The material is heavier and more brittle than most siding materials and requires special tools to cut and install. Periodic painting and caulking are required. Cutting of the material with saws produces silica dust. The variety of panel patterns and textures is less than for other types of panel materials.

2.14 EIFS AND STUCCO

EXTERIOR INSULATION AND FINISH SYSTEMS

ESSENTIAL KNOWLEDGE

Developed in Europe in the 1950s, and introduced into the United States by Dryvit Systems, Inc., in the early 1970s, exterior insulation and finish systems (EIFS), which are sometimes called synthetic stucco systems, have largely replaced conventional three-coat portland cement stucco systems. The industry's association, the EIFS Industry Members Association (EIMA) estimates that EIFS currently account for about 3.5% of the residential wall market in the United States.

The system's attributes include the benefits of insulation outside the structure (a significant thermal break advantage), reduced air infiltration due to the monolithic nature of the finished membrane, and great design flexibility. While many of the thousands of buildings clad with EIFS have had few, if any, problems, poor EIF performance has been documented on individual projects on an ongoing basis, particularly in climates with severe temperature swings and high moisture levels. Most recently, attention has focused on large-scale repairs to houses in Wilmington, North Carolina, following the discovery that moisture had entered the EIFS barrier cladding systems through the juncture between the EIFS and windows or doors, at deck terminations, at roof-wall connections, and at windowsills. In the absence of drainageways, moisture trapped behind the EIFS cladding caused deterioration of the substrate. The adverse publicity and class action lawsuits that followed have led to a curtailment in the use of barrier EIFSs in some states, although not nationally.

Concern about the use of EIFS among the public, the insurance industry, building officials, and manufacturers has also led to the development and promotion of new EIFS products that incorporate drainageways and moisture barrier membranes behind the insulation boards that allow infiltrating water to drain out (Fig. 2.14.1). The EIFS industry remains fragmented (EIMA represents only 8 of the 30 to 40 EIFS manufacturers, although its members produce an estimated 85% to 90% of the systems sold),

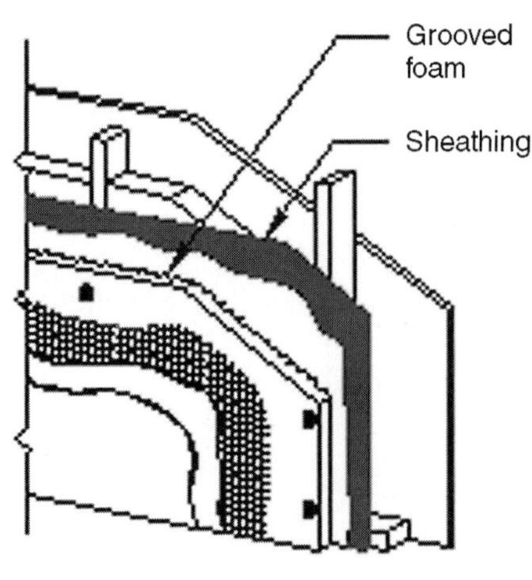

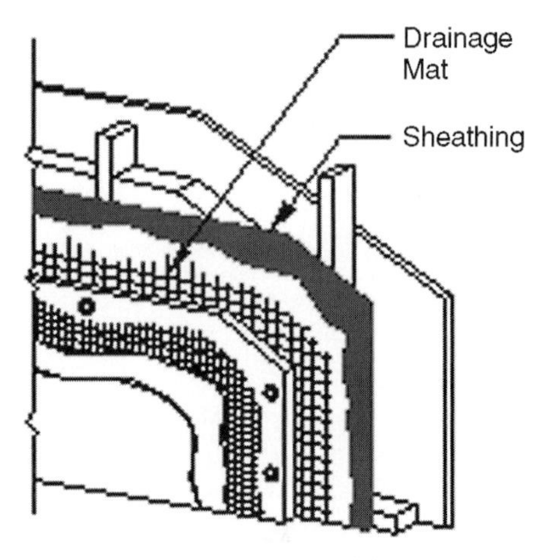

FIGURE 2.14.1 GROOVED INSULATION BOARD DRAINAGE LATH OR MAT

and EIFS installation systems and details vary among manufacturers. However, the ASTM has developed recommended installation details and specifications that are part of a new ASTM standard (ASTM C1397) for barrier-type systems. Water-managed systems have yet to be addressed by ASTM. EIMA is working with the NAHB Research Center to develop a third-party applicator certification program.

TECHNIQUES, MATERIALS, TOOLS

There are two major classifications of EIFS: barrier systems (which depend on the integrity of the EIFS surface, flashing, and sealants to prevent entry of water)and drainage systems (which employ a variety of drainage techniques to allow moisture, that may have entered, to exit the system).

1. REPAIR EXISTING EIFS.

The repair of small damaged areas of an EIFS, where the substrate is sound, is relatively simple. The affected area is cut out and the system is rebuilt in much the same way as it was originally installed. If the damage was caused by water infiltration from deficient flashing, the flashing should be repaired or replaced before corrective work begins. If the sealant has failed, it should be removed and replaced or covered with caulking products designed for this remedial work. Dow Corning, among other manufacturers, has developed a variety of restoration recommendations and remedial products, including a preformed silicone extrusion (Dow Corning 123 silicone seal; see Fig. 2.14.2) that is designed to span failed sealant joints and weatherseals at the perimeter of windows and other wall penetrations (see Further Reading). If the deterioration due to faulty installation or water entry is extensive, the EIFS may have to be removed in its entirety and, in some cases, the sheathing, trim, windows, and structure will have to be removed as well.

ADVANTAGES: Repairs are simple and work well on small damaged areas where significant water penetration and substrate deterioration has not occurred.

DISADVANTAGES: Serious water entry problems can require extensive removal and rebuilding work. Repaired areas will probably not match the color of adjacent existing areas and may require a new color coat.

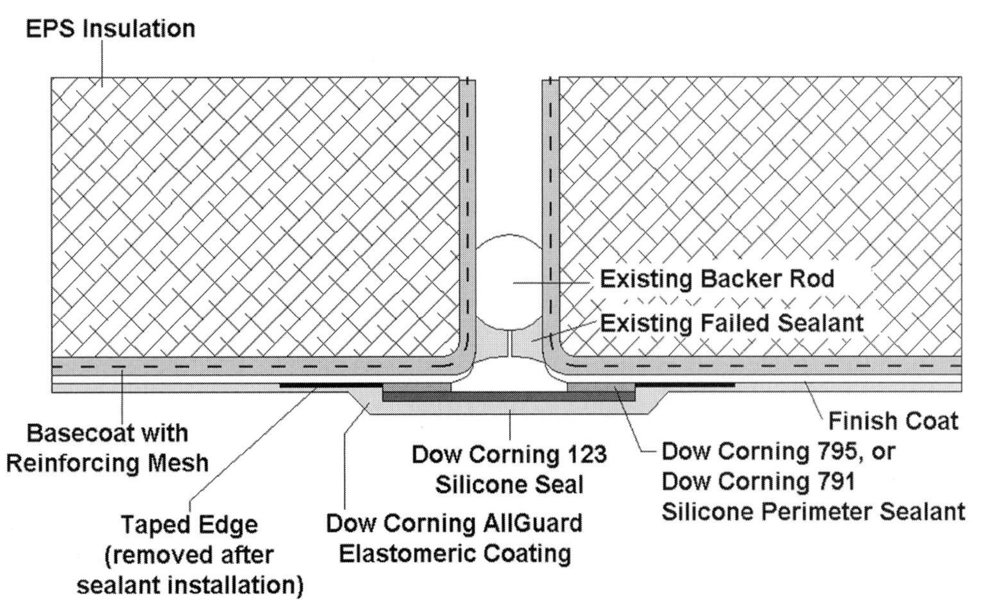

FIGURE 2.14.2 SILICONE SEAL

2. INSTALL AN EIFS BARRIER SYSTEM.

Barrier systems have been the basic industry standard until recently, and EIMA and many manufacturers maintain that when properly detailed and installed they will perform satisfactorily. Some manufacturers, however, disagree. US Gypsum offers only draining, water-managed systems, and Senergy, Inc., will not authorize the installation of barrier systems on wood-frame construction (all manufacturers agree that barrier systems work over masonry block or concrete substrates that are relatively unaffected by moisture). An EIFS typically consists of insulation board made of polystyrene (usually expanded) or polyisocyanurate foam, which is secured to the exterior wall surface (usually DensGlass Gold®, plywood, OSB, or fiber-cement board) with a specially formulated adhesive or with mechanical fasteners. The foam provides insulation and allows the coating to flex during temperature swings. Some manufacturers provide an EIFS without the insulation for installation over masonry or directly to sheathing when insulation is not required (Fig. 2.14.3). A water-resistant base coat is applied to the top of the insulation and reinforced with glass-fiber mesh for added strength. A final coat, typically acrylic, similar to a thickened acrylic paint with a fine aggregate, is applied as the finish surface. Some manufacturers provide elastomeric coatings, which are softer and more flexible than the coatings typically used.

ADVANTAGES: An EIFS barrier system is more effective than (although often used in conjunction with) comparable between-the-studs insulation, especially over metal studs. It provides an important thermal break, reduces air infiltration (EIMA claims up to a 55% reduction over standard masonry or wood construction), and provides attractive exteriors and design flexibility through a wide variety of colors and textures. It can be fashioned into a variety of shapes and sizes to produce decorative details such as cornices, quoins, keystones, arches, columns, reveals, shadow lines, and special moldings. It is relatively easy to clean and repair small areas.

DISADVANTAGES: Proper detailing and the choice of an experienced applicator is critical. Some manufacturers have significantly less experience than others and have more limited technical support staff. Some manufacturers have model code research reports; others do not. Many residential contractors do not have the skills and experience of commercial applicators. The specification control and inspection, typical with commercial work, is frequently absent on residential projects. Warranties vary considerably and should be studied and compared carefully. Insurance and code agencies may not allow barrier systems without certification by design professionals as to their proper detailing. Some states, such as North Carolina, and model codes [Uniform Building Code] (UBC) do not allow the use of barrier systems with wood-frame construction. The use of barrier systems should be carefully monitored on a case-by-case basis, as the codes are changing. Extreme heat, dryness, cold (especially freeze-thaw), and moisture conditions affect the system's performance. Proper flashing, high-performance sealants, and weather barriers are essential to good performance, but sealants

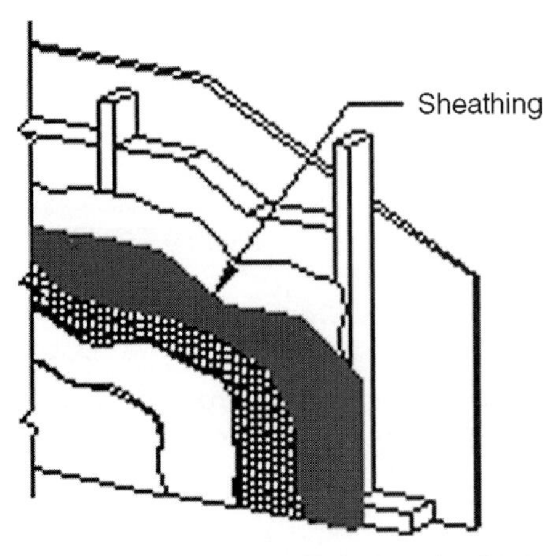

FIGURE 2.14.3 AN EIFS WITHOUT INSULATION

should not be relied upon in the absence of other weather barriers. Careful selection of windows and flashing detailing is required to assure the wall integrity is not compromised by water weeping into the wall cavity at the sill and the sill-jamb interface. Termites and carpenter ants can tunnel into foam plastic and use it as a habitat.

3. INSTALL AN EIFS MOISTURE DRAINAGE SYSTEM.

Moisture drainage systems, also called water-managed or rain-screen systems, have been used in commercial applications since the early 1990s and are currently becoming the system of choice for housing among architects, designers, builders, and code officials. They are similar to barrier systems except that they employ a drainageway behind the insulation either in the form of vertical grooves cut in the insulation board, vertical furring strips, or a woven fabric drainage mat or other drainage system (Fig. 2.14.1). Proper flashings, weather barriers, and sealant details remain critical as water should still be kept out of the system. EIFS should stop well above grade to restrict insect access and allow for inspection.

ADVANTAGES: This sytem has all the advantages of a barrier EIFS with the additional protection of drainageways to evacuate moisture from the system. It is perceived by most builders, architects, and designers as an improved system and is recommended or required by increasing numbers of insurance companies and code agencies.

DISADVANTAGES: This system generally requires mechanical fastening of the foam insulation board, which is more labor intensive than using adhesive applications. More attention and cost are required for secondary weather barriers and accessories. Careful attention to system details, including housewrap overlaps to avoid the channeling of moisture behind these weather barriers, particularly at windows, doors, and other openings, is required. Termites and carpenter ants can tunnel into foam plastic backing and use it as a habitat. There has been limited testing and performance data on the various drainage systems employed. Some building researchers and experts remain unconvinced that this system solves all the performance problems with barrier systems.

STUCCO

ESSENTIAL KNOWLEDGE

Portland cement stucco is a traditional finish material that has been in use in North America for over 300 years. Currently installed as a three-coat system (scratch, brown, and color coat) it is sometimes abbreviated to a two-coat system, particularly over masonry.

TECHNIQUES, MATERIALS, TOOLS

1. PATCH EXISTING STUCCO.

Hairline cracks are very difficult to patch without making the repair more noticeable than the defect. Some installers recommend "dusting" with stucco. Cracks between 1/8" to 1/4" can be repaired by scraping out the topcoats to expose the scratch coat. An acrylic bonding agent can be used to help bond the repair coats. On even larger cracks or holes, a self-adhesive fiberglass mesh can be used to strengthen the repair. Some large holes will require the repair of the building paper behind the wire mesh. For recoloring old stucco a "fog" coat of cement, color pigment, and lime, but no sand, is sometimes used. Acrylic additives increase the ease of application and cover. Elastomeric coatings can also be used to seal and recolor old stucco.

ADVANTAGES: Stucco repair is relatively easy and cost-effective.

DISADVANTAGES: If large areas are affected, removal may be more practical.

2. INSTALL A STUCCO EXTERIOR WALL FINISH.

Stucco is installed in the northeastern and north central states over gypsum sheathing using dimple or expanded metal diamond lath that holds the plaster away from the sheathing allowing for better cement bond, for expansion and contraction with temperature changes, and for the creation of a drainageway. Over masonry substrates, stucco is typically applied directly to the substrate without lath. In hotter climates in the south and southwest, stucco is often installed over paper-backed lath without sheathing material. This requires adequate bracing of the walls to prevent shear cracks. Stucco can be colored and scored to appear like brick, stone, and other materials.

ADVANTAGES: Stucco is a proven material that stands up well to moisture in cold climates. It can have integral color or can be coated with elastomeric coatings. It can have a variety of finishes and is easily repaired.

DISADVANTAGES: Experienced applicators are required. Fogging to prevent excessive fast curing and hairline cracking is required. The integral color can fade, especially if a dark color is chosen. Stucco can be stained at grade by rain-splattered earth. Reinforcement at all openings and periodic expansion joints is required. Stucco is a poor insulator.

2.15 EXTERIOR TRIM

ESSENTIAL KNOWLEDGE

Trim has always been an important element in the appearance of houses, as well as a key weather and waterproofing component. As a finishing element around doors, windows, porches, roof edges, at corners, and at other building features, trim provides a decorative element and scale. There is a wide variety of material available for use as trim other than traditional solid wood, such as laminated wood lumber, engineered wood, wood-thermoplastic composites, fiber cement, and polymers. These new materials have the appearance of wood trim but promise longer service life. The major threats to trim are UV radiation, water damage, snow and ice, mold, rot, and insect infestation (all except UV radiation are related to moisture). Materials should be selected, detailed, and installed to limit the effects of exposure.

TECHNIQUES, MATERIALS, TOOLS

1. REPAIR EXISTING WOOD TRIM WITH EPOXY FILLER.

The decision whether to repair or replace the existing trim will depend on its condition and whether the building is historically significant (if so, follow the U.S. National Park Services' *Guidelines for Rehabilitating Historic Buildings* referenced in Further Reading). Before repair or replacement, the conditions that caused damage to the trim should be corrected, if possible, and the decision to repair or replace can then be addressed. Most wood, even if it is seriously decayed, can be reconstituted by means of liquid epoxy consolidants that impregnate the wood fibers and harden into a mass that can be sawn, planed, drilled, nailed, sanded, glued, and painted. Most of these epoxy materials have weatherability as good as or better than wood, and work well at sills, thresholds, and other parts of the building that cannot be easily replaced. Epoxy putties are also available that work in conjunction with liquid epoxy to rebuild missing sections of decorative features and trim (Fig. 2.15.1).

ADVANTAGES*:* Restoration of damaged trim material may be less disruptive and less expensive than replacement. The repair helps maintain the historical integrity of the trim.

DISADVANTAGES*:* Careful application is required. This repair may not be cost-effective for the average rehabilitation project and may not be practicable if deterioration is extensive.

FIGURE 2.15.1 USE OF EPOXY PUTTY TO REBUILD TRIM

2. INSTALL NEW TRIM.

If a building's trim has deteriorated to the extent that it is unsightly or does not function as intended, and if epoxy consolidation is not cost-effective, the trim and any deteriorated substrate should be removed and replaced. If deterioration was caused by a lack of flashing (especially at window or door heads), sealants, or poor detailing, those deficiencies should be corrected or the condition will be repeated. The choice of trim material has expanded considerably beyond that of solid wood. Trim options include:

■ *Solid Wood.* For trim and fascias, solid wood is still the favorite material in many parts of the country due to its ease of application and general availability. Wood species include white and (to a lesser extent) southern yellow pine, imported pine species, and locally available fir including Douglas fir and Hem-fir. Heartwood grades of western red cedar and redwood are more expensive, but considered more dimensionally stable and resistant to decay because of their natural extractives. Cedar often comes rough-sawn. Redwood is available in wide boards and is often used for fascias. Clear, vertical grain, all-heart wood material takes paint better, is more stable, and lasts longer than other grades. Finger-jointed trim is increasingly used and is available in larger sizes because it is comprised of glued sections of material. Back priming of all solid wood trim is recommended to protect it from moisture and to keep it from warping. Finished sides are best protected with two coats of paint or stain.

■ *Laminated Veneer Lumber (LVL).* At least one company, South Coast Lumber, makes LVL trim products. Its ClearLam™ product is made of Douglas fir core veneers and older face veneers glued together with phenolic adhesives and sprayed with a preservative to protect the trim in the field. The face is overlaid with a phenolic-based medium-density overlay (MDO) sheet that eliminates face checking and serves as an excellent substrate for paint. All edges are fully coated with an elastomeric edge coating and primer. Easy to cut, nail, and install, ClearLam™ can be used for fascia, corner boards, and window and door trim, and is dimensionally stable without knots, checks, or cracks.

■ *Engineeered Wood Trim.* Engineered trim is a composite of wood fibers and resins. It resembles hardwood, but has added waxes, resins, and oils to give it better weather resistance. Engineered wood can be used for corner boards, fascias, rake boards, soffits, and door and window trim. It is relatively inexpensive, uniform, consistent product, tht is smoother and straighter than regular wood.

■ *Wood-Thermoplastic Trim.* Wood-thermoplastic trim is a relatively new composite product made from thermoplastic resins and wood fiber. This material is exceptionally durable and is becoming popular for exterior decking applications and as windowsills and door jambs. Several companies make limited sizes of flat stock for trim and extruded brickmold.

■ *Fiber-Cement Trim.* Fiber-cement trim is available from manufacturers of fiber-cement siding products and is generally used in conjunction with those materials, although it need not be. It is used as fascias, rake boards, corner boards, soffits, and window and door trim and is available in smooth and wood grain finishes, primed or unprimed. This material takes paint well and is available with a 50-year warranty against warping, cracking, and delamination.

■ *Polymer Trim.* Polymer trim, made from high-density polyurethane, is cost-effective in replicating the appearance of heavily decorative trim elements such as columns, railings, balusters, brackets, trellises, pediments, and shutters.

2.16 SEALANTS AND CAULKS

ESSENTIAL KNOWLEDGE

Sealants and caulks are the first line of defense, serving as a barrier to both water and air infiltration. However, not all joints are meant to be caulked; some provide an exit for air or moisture trapped within the wall assembly. In low-rise residential structures sealants and (to a lesser extent) caulks are used as elements of a weather barrier system that includes the exterior finish material, drainage planes (building paper or housewrap), ice and weathershield membranes, and flashing. Sealants and caulks are typically used at expansion joints, joints between dissimilar materials, joints at window and door openings, at the juncture of siding and trim, and at flashing. Several factors should be considered in order to achieve satisfactory performance of both sealants and caulks.

- *Material selection.* Materials must have the proper physical characteristics for the specific application, including elasticity (the ability of a sealant to return to its original profile), elongation (the ability of a sealant to stretch, as expressed as a positive or negative percentage), adhesion (bonding between the sealant and adjacent materials), durability, paintability, and compatibility with substrate and adjacent materials.

- *Weatherability.* Sealants vary with respect to weathering characteristics. Indications of weathering include hardening of the material, chalking or discoloration, alligatoring, wrinkling, bubbling, sagging, erosion, or softening of the sealant surface. Sealants and caulks showing these characteristics should be replaced.

- *Joint design.* Sealant manufacturers recommend that sealants should adhere to only two surfaces by use of polyethylene backer rods or bond breaker tape. Three-surface adhesion will lead to cohesive failure (tearing). Narrow (less than 3/8") or excessively deep sealant joints that exceed a depth-to-width ratio of 1 to 2 may not allow for the proper compression, elongation, or adhesion of sealants. Small joints prove to be the most difficult to seal because the smallest movement can represent a significant percentage of expansion. Interior applications typically do not require nearly the same degree of elongation because the temperature is maintained within a narrow range.

- *Installation.* The leading cause of sealant failure is improper installation, elements of which include improper priming or cleaning of the substrate; installation over incompatible coatings, materials, or contaminants (including existing sealants and lubricants); installation during periods of excessive cold or heat, rain, or dampness. Such failures are prime reasons for replacement of sealants and caulks in rehab work.

The distinction between caulk and sealant, terms often used interchangeably, is essentially the ability to conform to movement. Caulk typically provides for less movement but is easier to work and is used for interior applications, while a sealant is used for exterior purposes. Caulking usually refers to latex sealing compounds that meet ASTM C834 *Standard Specification for Latex Sealing Compounds* while sealants usually refer to ASTM C920 *Standard Specifications for Elastomeric*

Joint Compounds. Manufacturers' instructions and technical assistance should be closely followed. The selection of caulk should be guided by knowledge of the materials that are to be adhered and the material properties that are most critical, such as elongation, durability, or ease of installation.

Sealants are continuously evolving with new formulations for lower cost, ease of installation, adherence, flexibility, and durability. Newer formulations have allowed greater range of uses, but no one product is ideal in all these respects. It is estimated that there are over 300 sealant manufacturers. Some make their own sealants while others sell sealants manufactured by others under their "private label." The most common types are listed here.

- Latex and oil-based sealants, generally referred to as caulks with low flexibility and relatively poor durability, are low cost, easy to work, and suitable for interior applications not exposed to prolonged moisture.

- Acrylic latex, sometimes referred to as rubberized latex, is a more durable and elastic variation suitable for interior and exterior applications. Small amounts of silicone emulsions are frequently added to enhance performance somewhat.

- Butyl rubber is commonly employed in insulated window assemblies and between layers of metal flashings because of its good adhesion qualities, ability to resist water and temperature extremes, and because it remains tacky. It has only moderate flexibility and is difficult to install.

- Kraton, a primerless, solvent-based, synthetic rubber, has become popular as a general-purpose sealant that adheres to most common substrates. Because it is solvent-based, it may shrink slightly.

- Silicone is used extensively in curtain wall, EIFS, and glazing applications. It is the most elastic and durable sealant, but not generally paintable. It is difficult to remove and not suitable for porous materials in some formulations. It has minimal shrinkage.

- Polyurethanes have excellent movement and durability characteristics, but the flexibility degrades over time, particularly in direct sunlight, and they are difficult to apply and clean up.

TABLE 2.16.1 TYPES OF CAULKING MATERIALS

Base Type	Retail $ (per 10 oz)	Est. Life (years)	Uses	Clean Up
Oil	1.00–2.00	1–3	Most dry surfaces*	Paint thinner
Polyvinyl acetate	1.50–2.00	1–3	Indoor surfaces only*	Water
Styrene rubber	2.00–2.50	3–10	Most dry surfaces*	Paint thinner
Butyl	2.50–3.00	4–10	Masonry and metal**	Paint thinner
Acrylic latex	2.00–4.00	5–20+	Most dry surfaces*	Water
Kraton	5.00–7.50	10–15	Most dry surfaces*	Paint thinner
Polyurethane	4.50–10.00	15–20+	Masonry**	Acetone, MEK
Silicone	4.00–7.00	20+	Glass, aluminum* (not for masonry)	Paint thinner, naphtha, toluene

** Wood, drywall, aluminum, e.g., for gaps in wood frames around perimeter of house, plumbing penetrations, gaps in rough openings around windows and doors, boots around supply and return HVAC grills, seal between bottom plates and subfloor.*
*** Gaps in masonry construction.*

All sealants require surface preparation and appropriate primers as directed by the manufacturer. Sealants are only able to provide for movement in two directions; if the sealant contacts a third surface, it will detach from the surface with the least adhesion. Sealants typically are applied with half the width adhered to either side of the opening in an hourglass shape (Fig. 2.16.1). The width of the opening is exposed on one side and must be prevented from adhering to materials along its other side with a nonadhering surface referred to as a bond breaker or backer material. The bond breaker material also serves to shape and support the profile of the sealant and as a secondary barrier.

TECHNIQUES, MATERIALS, TOOLS

1. PREPARE SURFACE, REMOVE EXISTING SEALANTS.

All surfaces must be sound, clean, dry, and free of frost, dirt, oil release agents, loose particles, efflorescence, old sealants, and other foreign substances that impair the adhesion bond. On impervious surfaces, such as glass, metals, or paints, sealant manufacturers may recommend a commercial grade solvent cleaner such as Xyol, toluene, or alcohol, or may produce one themselves. For porous surfaces such as cement board, concrete, concrete block, old brick, and stone, joints can be cleaned by cutting, scraping, sandblasting, saw cutting, or grinding. Remaining loose dust and particles should be removed by dusting with a stiff nonmetallic brush, vacuuming, or blowing with oil-free compressed air. Some sealants require a primer prior to application; some do not. Follow the manufacturer's recommendations.

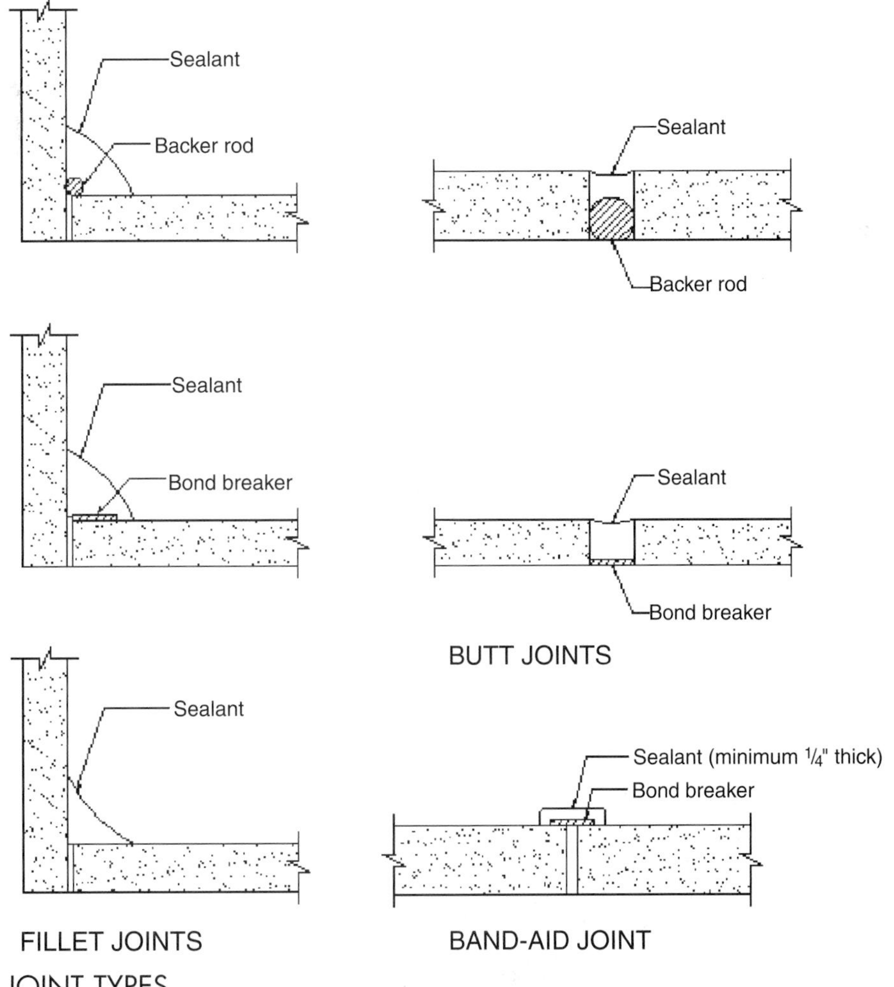

FIGURE 2.16.1 JOINT TYPES

ADVANTAGES: Proper preparation will help ensure maximum life of sealant performance.

DISADVANTAGES: Removing contaminants and old sealants, especially silicones, is time-consuming and expensive.

2. INSTALL SEALANT.

Proper installation of sealant is absolutely critical to performance. The methods of preparation and installation vary among manufacturers, and it is best to consult their literature for instructions. Sealants are available in essentially four types: preformed, tube, cartridge, bulk. Typically, the easier the installation method, the lower the anticipated performance. For this reason materials available in tubes tend to be water-soluble caulk materials suitable for interior repairs. Materials designed for exterior purposes, specifically windows and doors, require higher-performing materials and larger quantities. These materials generally require some form of mechanical means of applying the sealant. The traditional hand-operated gun provides a relatively simple device with convenient cartridges. Larger bulk-loading guns are also available to provide for economy in the packaging of materials or when two-part sealants are to be combined on site. Applications of large amounts of sealant are ideally suited for power-assisted equipment. Traditionally this was pneumatically driven, requiring an air compressor. New equipment developed specifically for this purpose is either electrically or battery driven. With the requirements of application being so specific, these power tools and preformed materials provide for consistency of application.

ADVANTAGES: New automated tools and new materials provide for greater ease of installation and performance than ever before. The wide variety of materials is suitable for virtually any condition.

DISADVANTAGES: The improper selection of sealants has the potential of damaging or discoloring adjacent materials.

2.17 PAINT AND OTHER FINISHES

ESSENTIAL KNOWLEDGE

Paints, stains, and other coatings protect wood from the deteriorating effects of moisture and UV radiation. Areas exposed to the greatest amount of sun and wind-driven rain deteriorate the fastest—typically the south and west exposures, and the higher portions of the building.

The performance of paints and stains on wood and wood-based composite products (plywood, oriented strand board, laminated beams, etc.) is affected by the wide range of properties between and within wood species. Understanding the physical characteristics of various wood species contributes to appropriate paint and stain selection. Varying properties affect the performance of finishes.

- *Density.* High-density woods (southern pine, Douglas fir, oak) tend to swell, cup, and check more than low-density, "light" woods (redwood, cypress, western red cedar) causing stresses in film coatings that can lead to cracking and flaking.

- *Grain characteristics.* Vertical-grained woods (western red cedar, redwood) have excellent paint-holding characteristics because of their narrow bands. Flat-grained woods (southern pine, Douglas fir) have dense, wide bands and hold paint less well, especially if smooth finished.

- *Texture.* Some hardwoods (oak, ash) have large pores that cause pinholes to form in the finish. Other hardwoods (yellow poplar, magnolia, and cottonwood) have smaller pores and good paintability. Paint and penetrating stains will last longer on rough-sawn lumber and plywood than on smooth surfaces because, in order to achieve the proper coverage, the paint buildup is necessarily greater. Smooth surfaces of some species, including western red cedar, may exhibit a condition known as *mill glaze* created during the planing or drying process. This condition can inhibit the adherance of solid body stains.

- *Knots, extractives, and other irregularities.* Knots absorb finish differently than the surrounding wood. Pitch (resin), oils, and other extractives can leach out of wood and cause staining. Better grades of wood have fewer defects and are preferable for painting.

- *Growth ring orientation.* Flat-grained, softwood lumber (typically used in most grades) shrinks and swells to a greater extent than vertical-grained lumber. Edge-grained softwoods (available at a premium price) cup less and hold paint better than flat-grained wood of the same species (Fig. 2.17.1).

Most residential paints and stains are classified as oil- or alkyd-based or latex-based (which includes acrylic). Oil- or alkyd-based paints contain inorganic pigments suspended in a natural oil such as linseed, or synthetic resin (alkyd), and usually a solvent such as mineral spirits (paint thinner), toluene, or xylol (all petroleum distillates). These paints cure by reacting with oxygen to form a polymeric film. Latex paints contain inorganic pigments, petroleum-based solvents, and various latex resins, but the solvent is mostly water. The curing of both paints releases volatile

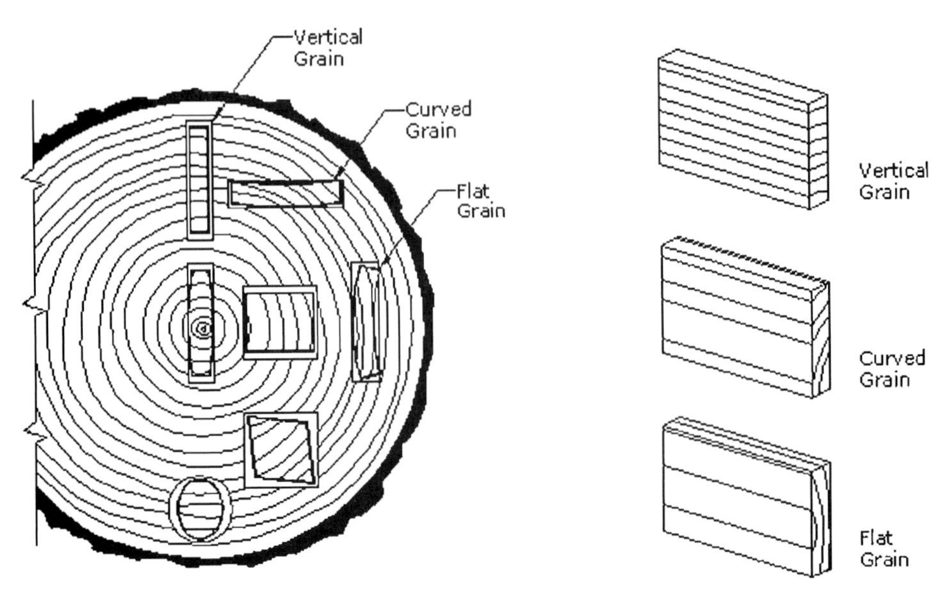

FIGURE 2.17.1 GROWTH RING AND GRAIN ORIENTATION

organic compounds (VOCs), but the amount is much less in latex paints (7% to 9%, compared to as much as 50% for alkyds). New paints on the market release very low or no VOCs.

Modern exterior latex paints, especially the all-acrylic paints, are generally considered to perform better than oil- or alkyd-based coatings. Even though alkyd paints provide a more permeable barrier to moisture, acrylic paints are faster drying, more elastic, hold color better, tend not to crack as much, and are easier to work with and clean up. In general, latex paints and solid-color stains can be applied over either oil- or latex-based finishes. Oil-based coatings should only be applied over oil-based finishes and not latex products. Better-quality paints contain a greater amount of pigment by weight, cover better, last longer, and are more cost-effective than lower-quality paints.

TECHNIQUES, MATERIALS, TOOLS

1. MAINTAIN EXISTING COATED SURFACES.

Painted, stained, or treated surfaces must be protected from UV radiation, rain, dirt, and mold and mildew to perform properly. Roof overhangs can help; however, soffits and sidewalls under wide overhangs will still collect dirt and water-soluble salts (which can interfere with the adhesion of new paints). Dirt, salts, and chalk (individual pigment particles from weathered paint) can be removed by scrubbing with nonmetallic bristle brushes and water. For stubborn stains, a nonammoniated detergent can be added. Mold and mildew can be removed by scrubbing with a mixture of one or two parts of bleach to a gallon of water. Surfaces should be rinsed thoroughly with clean water prior to refinishing. All landscaping should be protected.

ADVANTAGES: This method is cost-effective, and will increase the service life of coatings.

DISADVANTAGES: Surfaces will eventually need refinishing.

2. PREPARE PREVIOUSLY COATED SURFACES.

Film-forming paints and solid-color stains can fail by cracking, flaking, or peeling. Such failures are typically caused by moisture penetration, painting over weathered wood, prolonged weathering, too much time between application of primer and top-coat applications, and chalked, mildewed, or dirty surfaces that were insufficiently cleaned prior to coating. Blistering, another common failure mode, can be caused by high temperature or moisture. Temperature blisters are caused by rapid increases in temperature soon after painting or by poor-quality paint. Moisture blisters can occur anytime excessive moisture penetrates the surface edges or the back side of the painted material. The source of the moisture should be eliminated prior to refinishing. Deteriorated coatings can be removed by scrubbing, scraping, sanding, heat, chemical strippers, or pressure washing. Scrubbing is discussed in technique 1; scraping is best done with long handled professional scrapers; sanding is best done with orbital or siding sanders equipped with tungsten carbide abrasive disks (less likely to clog than conventional sandpaper); electrically heated paint removers can be used to soften and strip oil- or alkyd-based paints; chemical strippers such as lye and trisodium phosphate (TSP) contain caustic solutions and should be used with care. Naturalizing and sanding is essential. Refer to chemical manufacturers' recommendations and directions. Wet sandblasting and high-pressure water sprays are also used, but sandblasting can easily erode and destroy materials. Pressure washing can be effective for large areas, but should also be used carefully as it can also damage material. Dry sandblasting should never be used as it is too destructive. Paint can be removed with an open-flame blowtorch, but this should be left to professionals as the danger from fire is constant and lead paint can give off noxious fumes. The Occupational Safety and Health Administration (OSHA) and Department of Housing and Urban Development (HUD), and state and local health departments should be contacted for recommendations and requirements affecting the removal of lead paint (for houses built or painted prior to 1978).

ADVANTAGES: Proper preparation will make painting easier and enhance the coating performance.

DISADVANTAGES: These techniques are time-consuming and expensive.

3. APPLY PAINT TO NEW OR EXISTING WOOD OR WOOD-BASED COMPOSITE MATERIALS.

To achieve maximum paint life on new wood, the U.S. Forest Products Laboratory recommends that new wood be initially treated with a paintable water-repellant preservative (especially at openings such as windows and doors where water can collect on horizontal surfaces such as sills, and at lap, butt, and end joints of siding where the edge grain is exposed). On existing painted surfaces, sanding is particularly important in order to feather the edges of the existing paint to allow for uniform coverage where new and old coatings abut. If the existing paint is not feathered, the new paint will fail first (Fig. 2.17.2). High-quality stain-blocking acrylic latex primers are recommended. Alkyd- or oil-based primers are recommended for woods with water-soluble extractives such as redwood and western red

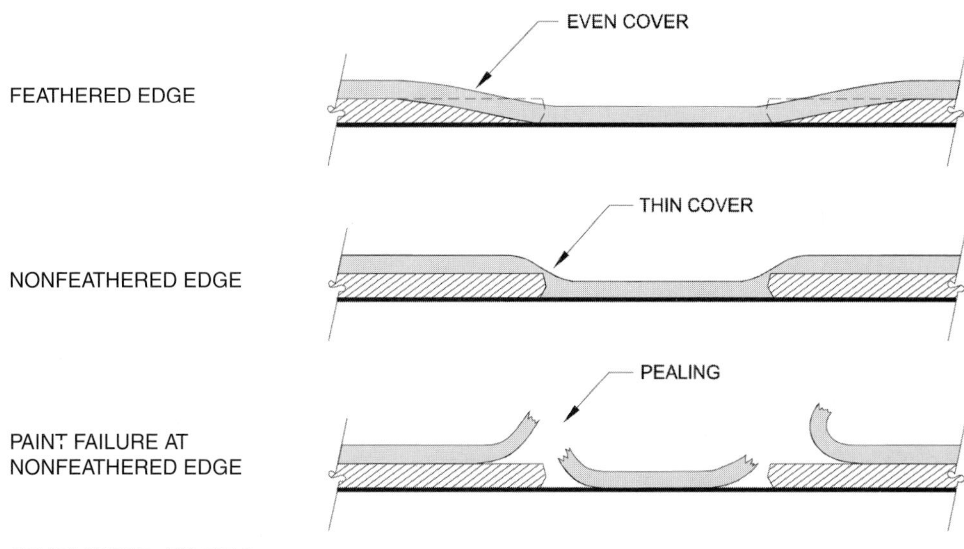

FIGURE 2.17.2 FEATHERED EDGES

cedar and are still preferred by many professional painters for new wood. Raw wood should be primed within a few weeks of installation as a longer delay can cause loss of adhesion. Two coats of good-quality all-acrylic house paint over the primer are recommended, especially on southern and western exposures. A one-coat acrylic house paint over a properly applied primer should last 4 to 5 years; two coats can last twice as long. As with all remedial work, the various product manufacturers' recommendations for paint selection, surface preparation, and paint application should be followed. Most paint failures occur when recommendations are not followed.

ADVANTAGES: Paint protects materials from weathering and deterioration due to the effect of rain and sun and can dramatically alter the appearance of a house. It has the highest percentage of solids of conventional wood finishes and can conceal surface defects and discolorations. New formulations of VOC-free paints are odorless, fast drying, can be applied at lower temperatures than conventional latex paints, and adhere well, including over alkyd paints.

DISADVANTAGES: Paint is not a preservative and will not protect a substrate from decay if moisture penetrates the surface or is absorbed from behind. Paint will fail if applied incorrectly or without proper preparations. Paint is not recommended for horizontal exposed surfaces such as wood decking, as water can get behind paint film and lead to decay. Periodic maintenance and repainting is required. Some new formulations dry so quickly that it is sometimes difficult to keep a wet paint edge. Alkyd paints are not recommended for use directly on masonry or other alkaline surfaces except over an alkali-resisting primer or sealer, or over an unprimed galvanized metal surface. Too frequent refinishing, especially with oil-based paints and solid-color stains, can lead to a thick coating buildup and subsequent cracking and peeling.

4. APPLY OIL-BASED PENETRATING STAINS TO NEW OR EXISTING WOOD OR WOOD-BASED MATERIAL.

Oil-based penetrating stains use linseed oil or alkyd formulas to seal and protect the wood substrate. Stains typically contain fungicides, water repellants, UV blockers, and other additives. These stains are available in varying degrees of pigment densities, from virtually clear and semitransparent formulations designed to reveal and enhance the grain, to increasingly opaque coatings including semisolid and solid stains that hide the grain, but allow the substrate texture to be expressed. The greater the amount of pigment, the greater the hiding power and UV protection. Solid-color stains are usually applied over primers; transparent stains are not.

ADVANTAGES: These are time-proven, popular coatings that protect and enhance the appearance of substrates. They resist blistering, cracking, and peeling better than paints (especially the more transparent formulations); can be applied over a wide variety of new solid sidings, including shingles, shakes, plywoods, and trim; can be used over some previously stained wood of an equally or less dense pigment formulation.

DISADVANTAGES: These coatings are not recommended for use over OSB, cementitious, or hardboard products (especially those with medium-density overlay surfaces). They may not penetrate and may cause unwanted gloss and blotchiness when used over existing unweathered penetrating stains. They are not recommended for decking unless specially formulated for that use. Transparent and semitransparent finishes require more frequent reapplication than more solid stains, especially for smooth-faced woods.

5. APPLY SOLID-COLOR ACRYLIC STAINS TO NEW OR EXISTING WOOD OR WOOD-BASED MATERIAL.

The paint industry's interest in developing water-based low-VOC emitting stains has led a number of manufacturers to develop newly formulated, all-acrylic stains for siding products. These flat stains are not penetrating stains, but form protective films. They are thinner coatings than acrylic paints and are designed to resist cracking, blistering, and peeling while retaining the texture of the substrate materials. Used with primers, these stains are more flexible and have better color retention than oil-based stains. They can be used over previously treated oil-based penetrating stains, trim, and a variety of other materials, including primed metal, cured masonry, plywood, medium-density overlaid plywood and hardboard, cementitious siding, and stucco.

ADVANTAGES: These stains protect and provide strong color accents to substrate materials and are less likely to peel than paints.

DISADVANTAGES: These stains are not as long-lasting or easily cleaned as all-acrylic paints.

6. APPLY SPECIALTY COATINGS.

Many specialty coatings are available that complement paint and stain products.

- Paints designed specifically for masonry.
- Tinted and untinted bleaching oils that contain oxides to accelerate the weathering process.
- Slightly tinted, clear, oil-based finishes that retard the normal gray weathering of wood and impart a slight reddish-brown tint.
- Wood cleaners that are formulated to remove mildew, mold, algae, and dirt using chemicals such as sodium hydroxide.
- Wood "brighteners" that are formulated to remove tanniss bleed in extractive-prone woods, such as cedar and redwood.

ADVANTAGES: These products are unique and potentially useful for special needs.

DISADVANTAGES: It is difficult to anticipate the results of these products without careful research and inspection of their use on existing buildings or without examination of samples.

CH.3

HOME REHAB HANDBOOK

ROOFS

Chapter 3
ROOFS

3.1. DESIGN AND ENGINEERING
3.2. SHEATHING
3.3. FLASHING
3.4. UNDERLAYMENTS AND MOISTURE BARRIERS
3.5. INSULATION
3.6. WOOD SHINGLES AND SHAKES
3.7. ASPHALT SHINGLES
3.8. LOW-SLOPE ROOFING
3.9. METAL ROOFING
3.10. SLATE
3.11. CLAY, CONCRETE, FIBER-CEMENT, AND COMPOSITE TILE
3.12. GUTTER AND LEADER SYSTEMS

3.1 DESIGN AND ENGINEERING

ROOF SYSTEMS OVERVIEW

The performance of a building's roof is key to the integrity of the structure and the comfort and well-being of the occupants. Roof failures run the gamut from catastrophic structural failure from earthquakes, fire, snowstorms, tornadoes, and hurricanes to major leaks caused by falling tree limbs and the intrusion of wind-driven rain under roof shingles or tiles. Damage is also caused by deterioration of roof sheathing and saturation of insulation from ice damming and wind-blown moisture into attic spaces through soffit, gable-end, and ridge vents. Minor leaks due to improper caulking or flashing at roof penetrations, or roof-wall intersections are also common.

This chapter will provide an overview and reference resource for information about roofing systems, subsystems, and materials; a review of current theory in terms of performance of these systems; and a discussion of existing or new materials, techniques, or components that have recently been improved or that represent totally new product lines.

Since this book is primarily about rehabilitation, structural issues will involve retrofitting buildings to protect against catastrophic failure, from such phenomena as wind and earthquakes. Basic framing errors and their correction are outlined, but not discussed in detail. They are treated in a number of publications published by the *Journal of Light Construction* and other sources (see Further Reading).

Protective strategies are undergoing considerable review, and a number of issues remain controversial, including the effectiveness of air infiltration and vapor retarders and the desirability of ventilating attic spaces in cooling climates. As there are currently limited conclusive test data, the differing concepts will be presented, when relevant, to inform the reader of the issues involved. Issues regarding insulation alternatives are also presented to inform the reader with an understanding that some of the current strategies have yet to receive consensus.

In the fairly recent past, steep-sloped residential roofing selection was relatively simple: The choice was between three-tab asphalt shingles, cedar shingles or shakes, clay and possibly concrete roofing tiles, or, in a smaller number of cases, slate. Today, these same basic choices are available, but these materials have evolved considerably, particularly in the development of high-profile laminated asphalt shingles, new tile shapes and colors, and new protective treatments for shingles and shakes. In addition, a whole new set of materials has come into mainstream use including metal shingles, shakes, and tiles; fiber-cement and plastic profiles that simulate wood and slate; and new materials and detailing for conventional systems such as standing seam metal roofing. Low-slope roofing systems have also evolved with the increased use of modified bitumen membranes and single-ply roofing systems. This guideline will review the attributes of steep- and low-sloped systems and materials with the intent of outlining some of the apparent advantages and disadvantages. These attributes are not necessarily comprehensive, and readers are advised to undertake their own research of individual products and their respective warranties. Detailed price comparisons have been avoided because of the fluctuating nature of prices and wide geographic variance.

Recommendations regarding the removal and disposal of asbestos roofing tiles have not been included because they are available from the Environmental Protection Agency's (EPA's) Asbestos Information Hotline: 800-438-2474 as well as from industry sources such as the National Roofing Contractors Association, which publishes a guide titled *A Practical Guide for Handling Asbestos-Containing Roofing Material*. Information regarding the removal of lead-based paint is also available from the EPA (800-532-3394).

TYPICAL FRAMING ERRORS

ESSENTIAL KNOWLEDGE

One of the most telling aspects of a roof's condition is the condition of the ridge. Swayback (sagging) roofs can indicate a number of problems including lack of or an insufficiently sized structural ridge beam, insufficiently sized roof rafters, missing or misplaced collar ties, improper rafter heel bearing, lack of solid blocking at the heel, inadequate rafter bracing or rafter bracing that is not transferred to a structural member such as a bearing wall or column, or interior column displacement due to settled footings. Common problems with trusses include trusses bearing on interior partitions when they are designed to span to the sidewalls, truss members cut to accommodate ductwork or equipment in the attic space, and lack of lateral bracing of top and bottom truss cords.

TECHNIQUES, MATERIALS, TOOLS

REPAIR FRAMING.

The repair of some sloped roofing problems can be accomplished by placing column supports or a supporting beam under the ridge board or beam and jacking it up to take most of the load off the rafter heel–top plate connection. When this is accomplished, the walls can be pulled together; collar ties, ceiling joists, tension members, or intermediate supports added; and rafters refastened, repaired, replaced, or rebraced. In some instances, inadequately sized roof sheathing may have sagged between rafters, resulting in a wavy roof surface. If this condition exists, new sheathing is required.

Most roofing problems, particularly on sloped roofs with multiple, complicated intersections, can only be identified after a thorough, systematic analysis of the roof's structural components and a documentation of systems and possible causes. Because of the great variety of defective conditions and possible causes, it is impossible to provide generalized repair guidelines.

If the structure has excessive deformation; if termite, fire, or dryrot damage is severe; or if the materials used were inferior, it may be wise to gut the structure and reframe it, or demolish all or portions of the structure and rebuild. Low-sloped roof problems are frequently caused by undersized rafters or rafters and sheathing that have deteriorated from rot from roof leaks or moisture condensation. Typical corrective work includes adding new rafters (sistering) adjacent to deficient ones. Project-specific structural repairs should be determined by job conditions and should be reviewed with a licensed professional engineer or architect.

ADVANTAGES: This method retains as much of the existing structure as possible and does not necessarily require the replacement of the finish roofing material. This is the least obtrusive repair and is generally cost-effective.

DISADVANTAGES: Jacking of the ridge may not eliminate roof swayback or wall bulge at the top plate. Removal of interior wall and ceiling finishes and more extensive reconstruction may be required. Correcting extensive fire- or moisture-related problems may require considerable demolition, reframing, and new materials.

WOOD-FRAME WIND RESISTANCE

ESSENTIAL KNOWLEDGE

The increased number and severity of recent catastrophic wind storms and hurricanes has underlined the need for homeowners to assess the structural condition of their houses. Inspections of houses in the aftermath of hurricane Andrew in 1992 indicated that much of the existing housing was not designed to resist high winds or such provisions were not constructed properly. It is important to understand that a building's roofs and walls resist wind loads in a complex manner distributing gravity loads and acting as shear walls and diaphragms to resist and distribute lateral wind loads. In high winds they have to resist very significant uplift forces as well. All the structural elements should act to provide a continuous load transfer path from the roof structures to the walls and floors and into the foundation. This only works when all the elements are well connected, which is not usually the case. Analyses of roof failures indicate typical failures at wall-to-roof connections, roof sheathing-to-roof rafter or truss connections; and the bracing of rafters or trusses to each other to prevent progressive "dominoing." Other failures include the loss of roofing material which adds to flying debris, breaking windows, and allowing water to penetrate into interior spaces.

A recently published brochure from the Institute for Business and Home Safety (IBHS) (a research and communications arm of the insurance industry), titled *Is Your Home Protected from Hurricane Disaster?—A Homeowners Guide to Hurricane Retrofit,* represents one of the most current guidelines on retrofit recommendations. Recommendations include the inspection of the roof components, addition of wall-roof tie downs, refastening of roof sheathing, the refastening or replacement of finished roofing products, the reinforcement of rafter-truss sheathing connections with adhesives, and the reinforcement of gable and wall connections (see Further Reading).

TECHNIQUES, MATERIALS, TOOLS

1. REINFORCE EXISTING STRUCTURE WITH METAL CONNECTORS, STRAPS, AND ADDITIONAL FASTENERS.

One of the recommendations of the IBHS is to remove the existing finish roofing material and underlayment, remove the bottom row of roof sheathing, and install new fasteners at each truss-rafter wall connection. The size and model of the connector will depend on the field condition and uplift resistance requirement. Simpson Strong-Tie Company, Inc., has prescriptive guidelines for simple building types, but in general the advice of a registered professional engineer should be sought. Typical fastener types are shown in Figs. 3.1.1 and 3.1.2. Once the connectors are in place, the roof sheathing can be reinstalled, the entire sheathing reinforced with additional fasteners, and new wind-resistant finished roofing applied. As an alternate to removing the roof sheathing, connectors can often be installed by removing the building's soffits. Currently, there are no connectors available that can be installed from existing attic spaces, but some manufacturers are researching possible design solutions.

ADVANTAGES: Rafter-truss connectors have proven to be very effective against uplift forces. In conjunction with reinforcing the sheathing fastening, this is probably the most recommended step to strengthen the building's structure.

DISADVANTAGES: Removal of a portion of the roof sheathing or soffit is required.

2. REINFORCE EXISTING ROOF SHEATHING TO THE RAFTER-TRUSS CONNECTION WITH ADHESIVES.

A promising technology that is currently under development is the use of adhesives to connect building components such as roof sheathing and rafters and trusses. One of the pioneers in this field, ITW Foamseal, Inc., manufactures a structural urethane foam adhesive that is widely used to attach gypsum ceilings to trusses in the manufactured housing and modular industries. Foamseal has developed a product, SF 2100, that is currently being installed in a number of homes in coastal South Carolina. Foamseal has also applied for acceptance of this product in a number of Florida counties.

ADVANTAGES: This reinforcement forms a continuous fillet weld that attaches and stiffens the components of the entire roof structure. It can also be applied to joints in the sheathing to prevent the intrusion of water. The applicator does not need to have physical contact with the joint as the spray has a range of approximately 10'(other construction adhesives would require that the adhesive gun be in contact with the joint). This is a cost-effective technique to connect the roof framing and sheathing and is considerably less costly than removing sheathing and applying metal connectors.

DISADVANTAGES: This reinforcement does not address the connection of the roof truss and rafter with the top of the wall plate. The stiffening of the roof sheathing may cause stress and load transfers to other structural members with difficult-to-determine results. The services of a professional engineer may be required to review the application.

3. REINFORCE EXISTING ROOF-TO-WALL CONNECTION WITH KEVLAR STRAPS.

A new material development, Millibar V 220, distributed by New Necessities, Inc. utilizes 0.050"-thick by 3"-wide Kevlar straps that can be run from the foundation, up the sidewall over the roof, and back down the other side to resist windloads due to hurricanes, tornadoes, and severe storms.

ADVANTAGES: Kevlar straps are extremely strong with a 525,000-psi tensile strength. They will not corrode or rot and are impervious to heat, cold salt air, and water. They are versatile, can be used as needed, are easily fastened, and create a continuous load path.

DISADVANTAGES: This is a supplemental restraining system used in conjunction with conventional fasteners, straps, and connectors. Removal of roofing and siding material and project-specific engineering are required. This system can be costly.

WOOD-FRAME SEISMIC RESISTANCE

ESSENTIAL KNOWLEDGE

Strengthening existing residential buildings to resist earthquakes is focused primarily on the strengthening of connections of the structure to the foundations and, to a lesser extent, the provision of shear walls and the connection of walls to floors. The existing literature and prescriptive recommendations for the seismic retrofit of residential buildings makes very little mention of roof-to-wall connections except to recommend that a continuous load path be provided where possible from the roof to the foundations. An engineer designing a new residential building (especially with heavy roof tiles) would, however, most likely specify some wall-roof connector to resist lateral forces induced by ground motion. Wall-roof reinforcement requirements should be considered as part of an overall study of structural performance.

TECHNIQUES, MATERIALS, TOOLS

REINFORCE EXISTING WALL-ROOF CONNECTIONS

The connections available to provide resistance to seismic forces are those also used to resist high winds. Whereas they are effective in strengthening the building frame, they are not normally prescribed for existing housing. Individual structures, however, have unique requirements, and the assistance of a professional engineer is required to make building-specific recommendations.

ADVANTAGES: The building envelope is strengthened.

DISADVANTAGES: This method is not usually as cost-effective as reinforcing foundations and shear walls.

STRUCTURAL DECAY

ESSENTIAL KNOWLEDGE

Water absorbed by structural wood framing can raise its moisture content, reduce its compressive and tensile strength, and ultimately allow for rot, decay, and the corrosion of metal fasteners and truss plates. Fungi attack cellulose in the cell walls of roof framing members when three conditions exist: moisture, the presence of air, and a temperature range between 68°F and 86°F (20°C to 30°C). The best way to prevent decay is to eliminate moisture.

TECHNIQUES, MATERIALS, TOOLS

CONTROL MOISTURE INTRUSION.

The most effective ways of reducing moisture in attic spaces are to prevent warm moist air from rising through openings in ceilings of occupied spaces; maintain weathertight roofs and walls; prevent rain and snow from infiltrating through ridge, soffit, gable end, or other venting devices; and provide sufficient ventilation of attic spaces. The process of rehabbing buildings affords the opportunity to install the necessary weather barriers and moisture retarders. These issues and implementation alternatives are discussed throughout the various chapters of this handbook, in particular Section 3.5.

FIRE DAMAGE

ESSENTIAL KNOWLEDGE

Damage from fire can range from the total loss of a building and its contents to minor inconvenience from smoke odors. The process of determining the restoration requirements of a fire-damaged building varies considerably with the building's location and extent of damage. Insurance adjusters often make settlement offers based on their own evaluation of restoration needs, although they may employ consultants on more complex projects. Recommendations may also originate from local fire marshalls, building department officials, contractors, or consulting engineers, industrial hygienists, public adjusters, and architects hired by the building owner. Unless the damage is limited, the restoration process can be complex involving structural, electrical, HVAC,

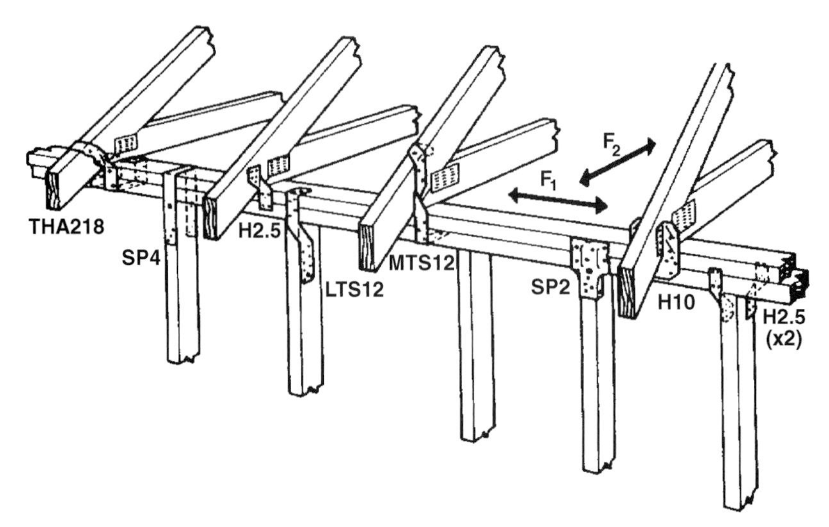

FIGURE 3.1.1 TYPICAL STUD WALL–TO–TRUSS CONNECTORS

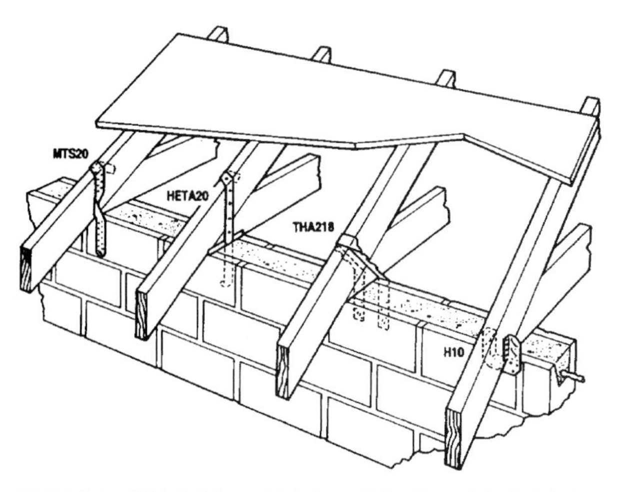

FIGURE 3.1.2 TYPICAL MASONRY WALL–TO–TRUSS CONNECTORS

plumbing systems, as well as building finishes. In addition, significant health and comfort issues arise from the residual smoke, combustion gases, moisture from fire department hoses, and the existence of products containing asbestos. For this reason the selection of a restoration contractor who is experienced and knowledgeable in current techniques is critical. At least one national association, the Association of Specialists in Cleaning and Restoration (ASCR), manages training and certification programs, and publishes a restoration guideline.

TECHNIQUES, MATERIALS, TOOLS

RESTORE FIRE-DAMAGED ROOF ELEMENTS.

The first step in a restoration project is to assess the damage to the building's structure, systems, and finishes. In 2x4 construction, significantly charred members are generally removed in their entirety. Heavy timber construction can remain if, according to the American Society of Civil Engineers, once the char is removed, the remaining section is still structurally adequate (after a reduction in size factor of ¼" on all sides). Char is removed by scraping and abrasive blasting. It should be removed because it holds odors, although encapsulating coatings will inhibit their transmittal. Sheathing materials, especially charred or unsound fire-retardant-treated material, should also be removed. New construction, replacing the damaged elements, should meet codes for new construction. Smoke-damaged materials should be cleaned and deodorized as necessary. Water-damaged materials, such as insulation, should be replaced when the damage is irreversible.

ADVANTAGES: This method allows for rehabilitation of fire-damaged buildings.

DISADVANTAGES: This method may not be cost-effective for severely damaged structures.

3.2 SHEATHING

ESSENTIAL KNOWLEDGE

Roof sheathing serves a number of functions. It is a key element in the barrier system that keeps out moisture; it serves as a nail base and support of roofing materials; it connects and braces the individual roof trusses or rafters; and it provides a diaphragm that, in combination with shear walls, stiffens the entire house against lateral forces from wind and earthquakes.

Roof sheathing in older houses (before the 1950s) is typically 1x boards, either tongue-and-groove, butt-edged, or spaced, that are laid perpendicular to the rafters. Houses built since the mid-1950s most likely have plywood, particleboard, or oriented strand board (OSB) sheathing thinner than 3/4". Exceptions to this are post-and-beam houses with rafters at 4'-0" or greater spacing, with 2x tongue-and-groove decking, 1 1/8" plywood or OSB, or other structural sheathing material such as Homasote, that is designed for longer spans than typical wood structural panel sheathing.

Plywood and OSB are today's leading roof sheathing materials, together accounting for over 95% of the market. Currently, slightly more plywood is used than OSB, but the latter material is expected to dominate the market within the next few years, and already has in some regions. It is important to understand that APA-The Engineered Wood Association (APA) does not differentiate between plywood and OSB under its APA Rated Sheathing program. Nor is APA the only quality assurance agency that certifies that plywood or OSB meet the voluntary product standards that have been set by the National Institute of Standards and Technology (NIST), to which the industry generally conforms. The two product performance standards that are in use are *Voluntary Product Standard DOC PS 1-95 for Construction and Industrial Plywood*, and *Voluntary Product Standard DOC PS 2-92 for the Performance Standard for Wood-Based Structural-Use Panels*. DOC PS 1-95 refers only to plywood; DOC PS2-92 allows the flexibility for the supplier to obtain plywood or OSB if there is no preference. DOC PS 1-95 is a prescriptive manufacturing standard; DOC PS2-92 is a performance standard. The performance requirements for wood structural panels set forth by both standards are the same. Many in the industry refer to colloquial terms such as "CDX" plywood. This is not a recognized grade in the product standards. A better specification for this item would be (under the APA designation) Rated Sheathing, Exposure 1, DOC PS 1-95, for plywood only, or Rated Sheathing Exposure 1, DOC PS 2-92, when there is no preference between plywood or OSB. The reference to a product performance standard should be evident in the stamp of the third-party grading agency that appears on individual panels. Exposure 1 panels are panels that have a waterproof bond and are designed for prolonged exposure prior to application of final roofing material.

Fire-retardant-treated (FRT) plywood sheathing, with a flame spread rating no greater than 25 when tested for 30 minutes in accordance with ASTM E84, is available from a number of producers. FRT OSB is not available because the wood swells unpredictably and a reliable impregnation process has yet to be developed. FRT plywood sheathing is not normally used in detached one- and two-family housing, although it is used in multifamily housing and occasionally in attached single-family housing. The surface of the FRT plywood is designed to char at high temperatures in order to prevent sustained burning. The use of high amounts of ammonium phosphate in products during the 1980s led to a chemical reaction caused by high attic temperatures (150°F–170°F), induced by solar radiation and the presence of moisture in the air. This chemical reaction led to substantial wood degradation of FRT roof sheathing on a number of housing projects, requiring, in many cases, the complete removal of sheathing and finished roofing. Fortunately, according to the U.S. Forest Products Laboratory, most of the deficient material has been replaced and the new standards developed by the

American Wood Preservers' Association and ASTM have led to new formulations which have largely controlled the problem.

TECHNIQUES, MATERIALS, TOOLS

1. REPAIR EXISTING ROOF SHEATHING.

If the finished roofing material has been well maintained, the sheathing should not have deteriorated. Exceptions to this would be damage from fire; inadequate roof structure; hail; airborne debris from high winds; localized problems including rot caused by leaks at improperly flashed roof penetrations such as at vent stacks, skylights, and chimneys; damage to the leading edge of the roof from ice damming; deterioration of the sheathing from condensation caused by inadequate venting of moist air from the attic space; swelling of sheathing edges caused by inadequate protection from rain during original construction; and deterioration of FRT sheathing caused by excessive heat and humidity.

The condition of the existing roof sheathing can be assessed by visual and physical means. Visual signs of deterioration include obvious delamination or deterioration, the existence of water stains, dark patches, mold spores, insect holes, and charring of the sheathing and roof structure. Physical assessment employs probes and soundings to determine the presence of soft, crumbling, split, swollen, or otherwise degraded material.

Sheathing that cannot function well as a nail base and support for roof finishes and as a brace to the roof structure should be replaced. If the deteriorated area is small, it may be able to be patched without extensive reroofing or treated with a fungicide in the event it is caused by mold. For the most part, however, deterioration, if it exists, is likely to be widespread and may require partial or complete removal of the sheathing and the existing finish roofing as well. The specific removal requirements and techniques will depend on individual on-site conditions.

ADVANTAGES: These repairs are cost-effective when the affected area is relatively small and the remaining sheathing is sound.

DISADVANTAGES: If the deterioration is widespread, the building's structure will be compromised. In this instance, the affected sheathing should be replaced.

2. REPLACE DAMAGED EXISTING ROOF SHEATHING WITH PLYWOOD.

Plywood is made of thin sheets of wood veneer (or plies) arranged in cross-laminated layers to form a panel. Plywood always has an odd number of plies consisting of three or more layers. Plywood is available in a wide variety of thicknesses, from 5/16" to 1 1/8", with span capabilities from 12" to 60" (with edge support). The APA has recommended span and load tables for various combinations of panel thicknesses and grades.

ADVANTAGES: Plywood has been the material of choice for many years. It has excellent structural and durability characteristics and is used as a nail base for almost every type of finished roofing application.

DISADVANTAGES: Plywood is more expensive than OSB and becoming more so. It is expected that plywood will continue to lose market share to OSB.

3. REPLACE DAMAGED EXISTING SHEATHING WITH OSB.

Made with rectangular-shaped wood strands cross-oriented in layers for better structural performance, OSB has replaced particleboard, flakeboard, chipboard, and waferboard as the most popular alternative to plywood sheathing. It typically comes in 4' × 8' panels (1.25 × 2.5 meter sizes are available from some manufacturers), but can also be custom ordered in lengths up to 24' and widths up to 12'.

OSB can be manufactured with square or tongue-and-groove edges. OSB is available in the same thicknesses as plywood, from 5/16" to 1 1/8".

ADVANTAGES: OSB is available nationwide, and is significantly less expensive than plywood and becoming more so. It is an excellent nail base for a wide variety of finish roofing materials. Smaller, younger trees and fast-growing species that were previously underutilized are used. OSB has the same waterproof adhesives as plywood.

DISADVANTAGES: The edges of OSB can swell when exposed to moisture. The APA recommends sanding edges down again (this has no structural effect on the panel).

4. REPLACE DAMAGED EXISTING ROOF SHEATHING WITH TONGUE-AND-GROOVE WOOD DECKING.

Some post-and-beam structures utilize exposed 2x4 and 2x6 tongue-and-groove roof decking in conjunction with widely spaced exposed roof rafters or trusses. Tongue-and-groove decking can span over 8' ft depending on species and loading conditions. Exposed decking systems typically have roof insulation on top of the roof (see Section 3.5).

ADVANTAGES: Wood decking is cost-effective with widely spaced trusses and provides an attractive finished ceiling. It provides a nail base for insulation or batten systems used to support finished roofing.

DISADVANTAGES: Wood decking is generally more expensive than other framing or sheathing alternatives. A substantial thickness of insulation above the sheathing is needed to reach the required R-value with attendant problems of fastening and venting the finished roofing products.

5. REPLACE DAMAGED EXISTING ROOF SHEATHING WITH FIBERBOARD-SHEATHING OR DECKING.

Fiberboard roof sheathing products have been in existence since 1908. The dominant manufacturer, Homasote Company, makes a structural sheathing, Easy-ply®, in thicknesses from 1" to 2 1/16" designed to span from 16" to 48" with an R-value of 2.4 for 1" to 5 for 2 1/16" thickness. Easy-ply® has a class C fire rating. A specially treated Firestall® has a class A rating.

ADVANTAGES: Fiberboard decking is a better insulator than wood decking. It can be obtained with a class A rating, prefinished, and as tongue-and-groove. It has a long history of use, is cost-effective on 48" spans, water-resistant, and resource-conserving since it is made from recycled paper with wax emulsion.

DISADVANTAGES: Fiberboard decking is more costly than some competing products such as OSB on short spans. The R-factor is not significantly better than wood. Tongue-and-groove 2 3/8" decking with an R-value of 5 still needs rigid insulation on top. Fiberboard is not as strong as OSB or plywood.

3.3 FLASHING

ESSENTIAL KNOWLEDGE

Residential roofing is typically made up of a multitude of materials and surfaces whose primary task is to maintain a barrier between the interior and the weather. The most pervasive and difficult weather element to control is water. Roof flashing is usually the last line of defense in the battle against water penetration.

Flashing forms the intersections and terminations of roofing systems and surfaces, to thwart water penetration. The most common locations for roof flashing are at valleys, chimneys, roof penetrations, eaves, rakes, skylights, ridges, and roof-to-wall intersections. Flashing must be configured to resist the three mechanisms of water penetration: gravity, surface tension, and wind pressure. To achieve this, flashing can be lapped shingle style, soldered or sealed to function as a continuous surface, or configured with a noncontinuous profile to defeat water surface tension. Flashing materials must be durable, low in maintenance requirements, weather resistant, able to accommodate movement, and be compatible with adjacent materials. Common modes of failure include exposure to salt air, excessive heat, acid rain, heavy snows, and scouring winds.

Traditional materials and methods of installing flashing produce some of the longest lasting of building systems components. Those methods do, however, require experience and are time-consuming. Newer membrane materials and modern sealants are available that complement time-tested techniques, but, regardless of the methods and materials employed, the basic principles of roof flashing must still be adhered to, and the three water penetration mechanisms must be overcome.

TECHNIQUES, MATERIALS, TOOLS

Roof flashing materials can be classified into two primary groups: membrane and sheet metal. Ice and water barriers and roll roofing are membranes. The most typical sheet-metal flashing materials are aluminum, copper, lead-coated copper, lead, stainless steel, galvanized steel, zinc, and Galvalume™. Both sheet-metal and membrane flashing are available unformed or, for some particular applications, in preformed configurations.

1. REPAIR EXISTING FLASHING.

Small areas of loose, bent, split, corroded, or otherwise deficient flashing can often be reinstalled, permanently patched with similar material, or can be replaced with new flashing. Asphaltic patching material is adequate for asphaltic roofs, but is not recommended for metal flashing because it will break down from ultraviolet (UV) radiation exposure and movement of the metal. It is also unsightly and an indication of poor maintenance. If large sections of flashing have deteriorated or have become loose or disengaged, it is time to remove the roofing material and install new membranes and flashing.

ADVANTAGES: Repairs are cost-effective over small areas.

DISADVANTAGES: If large areas need replacement, or if serious leaks develop, postponing replacement may cause damage to the building's structural elements or finishes.

2. INSTALL NEW COPPER OR LEAD-COATED COPPER FLASHING.

Copper flashing is one of the most durable of roof flashing materials. It is also one of the more expensive. Typical applications include chimney flashing; valley flashing on tile, wood shake, or slate roofs; base, step, and counter flashing at roof-to-sidewall intersections; and exposed or concealed ridge and hip flashing. Fabricating and installing copper flashing is a learned skill, at one time reserved for experienced craftspeople. Although the material is relatively soft and malleable, the techniques required to plan, cut, shape, fabricate, and install complicated shapes take practice and patience. Copper can be fabricated into rigid, continuous custom shapes such as chimney crickets or other special configurations. For step flashing roof-to-sidewall applications, copper is available in standard 5" × 7" pieces. The Copper Development Association (CDA) and individual manufacturers have excellent manuals of recommended flashing details. The CDA maintains that acid rain and the acid from red cedar shingles are not a problem when water is not allowed to stand and when cant strips are used to raise the shingles off the copper surface (see Fig. 3.3.1 for recommended details). The CDA recommends a minimum of 16-oz. plain or lead-coated copper be used for valley flashings and 20 oz. when slate or tile is used for the roof material. Lead-coated copper is considered the premiere flashing material in northern and maritime climates because it combines copper's durability with lead's resistance to acid rain and characteristic of not staining adjacent materials.

ADVANTAGES: With proper installation, copper flasing can be one of the most durable of all exterior building components. Soldering joints and intersections is relatively easy, allowing for the formation of permanent, three-dimensional, continuous shapes. Copper is a relatively pliant metal, is easy to form and work, and ages to a familiar protective green patina. At least one manufacturer, Revere Copper Products, Inc., makes a prepatinated copper flashing for those unwilling to wait.

DISADVANTAGES: Copper is a relatively costly material. Labor costs are higher than for other flashing materials. Unless roofs are properly detailed with overhangs and drip edges, rainwater runoff from copper flashing can stain adjacent materials.

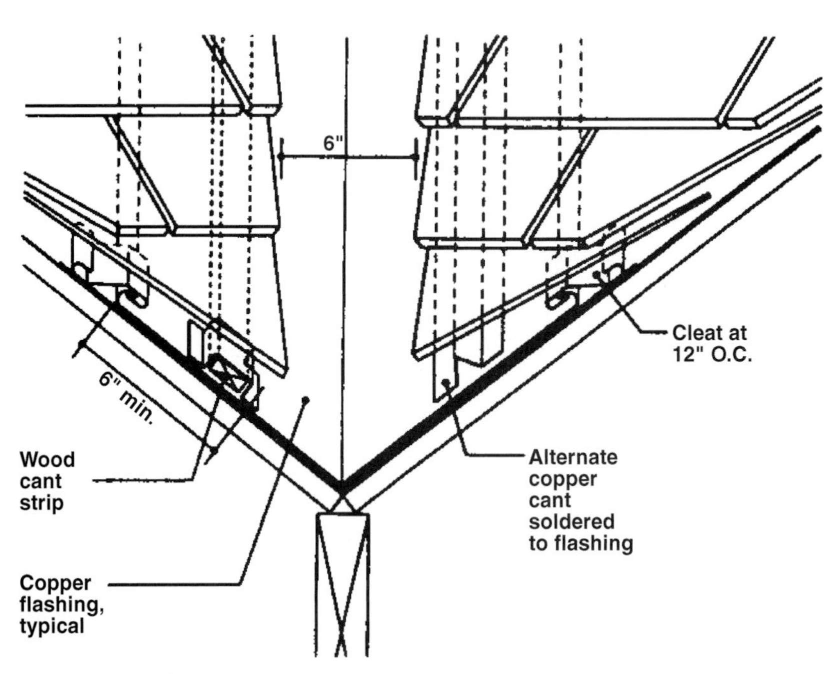

FIGURE 3.3.1 A. OPEN VALLEY

The detail shows a typical open flashing for a shingle or slate roof. Two different cants are illustrated. The cant strip can also be constructed as shown in Figure 3.3.1D. The shingles or slate must lap the flashing at least 6".

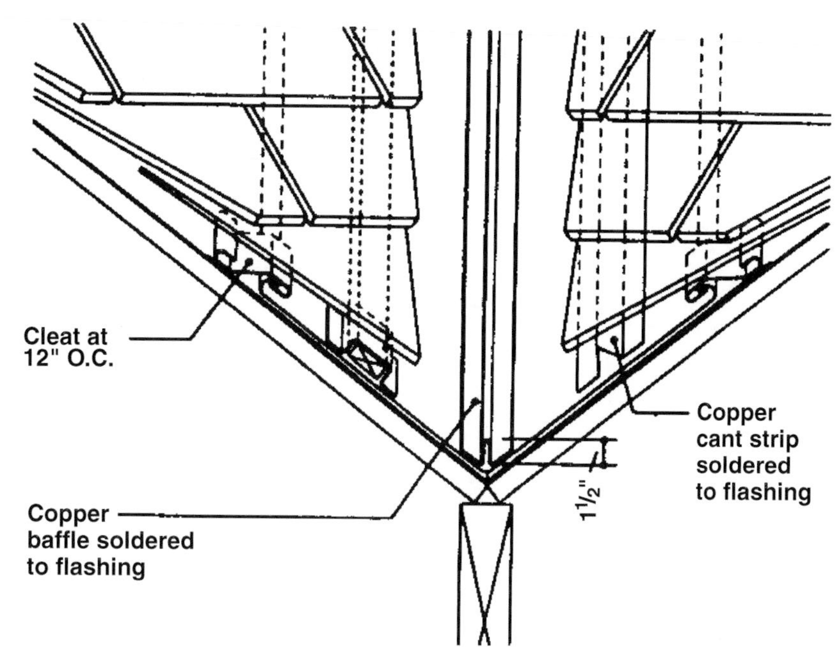

FIGURE 3.3.1 B. EQUAL SLOPES—UNEQUAL WATER FLOW

Where unequal water flow is expected, a baffle, 1½" high, should be installed as shown to prevent water of higher velocity from forcing its way past the opposite edge of the valley flashing. The baffle can also be constructed as shown in Figure 3.3.1D.

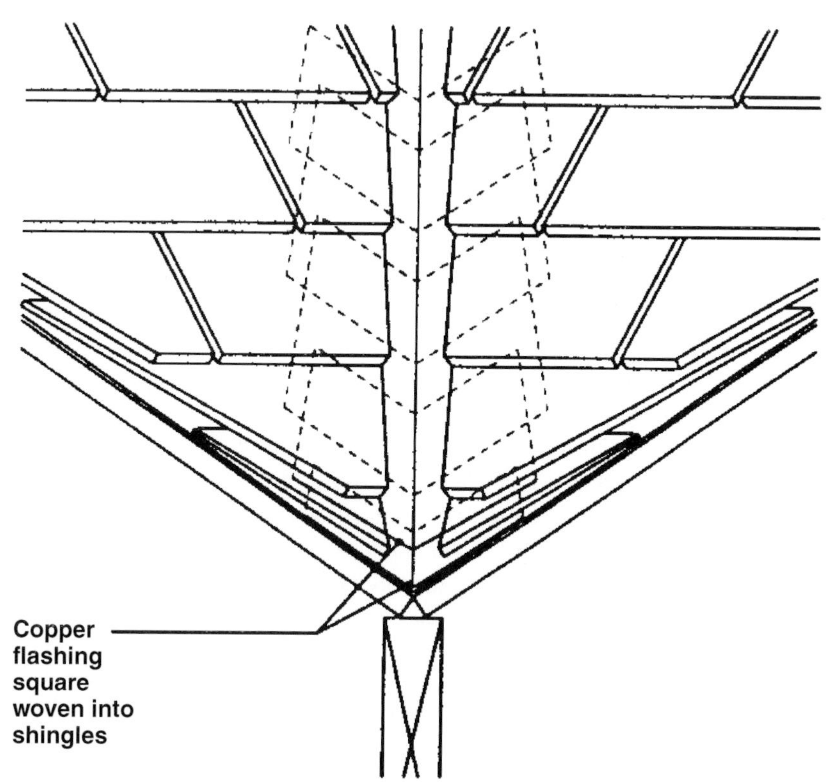

FIGURE 3.3.1 C. CLOSED VALLEY

Intersecting roofs using a closed valley must have the same slopes so that the shingle butts line up at the valley intersection. For roof pitches of 6" or more per foot the flashing extends at least 9" under the roof covering on each side. For roof pitches less than 6" per foot the flashing extends at least 12".

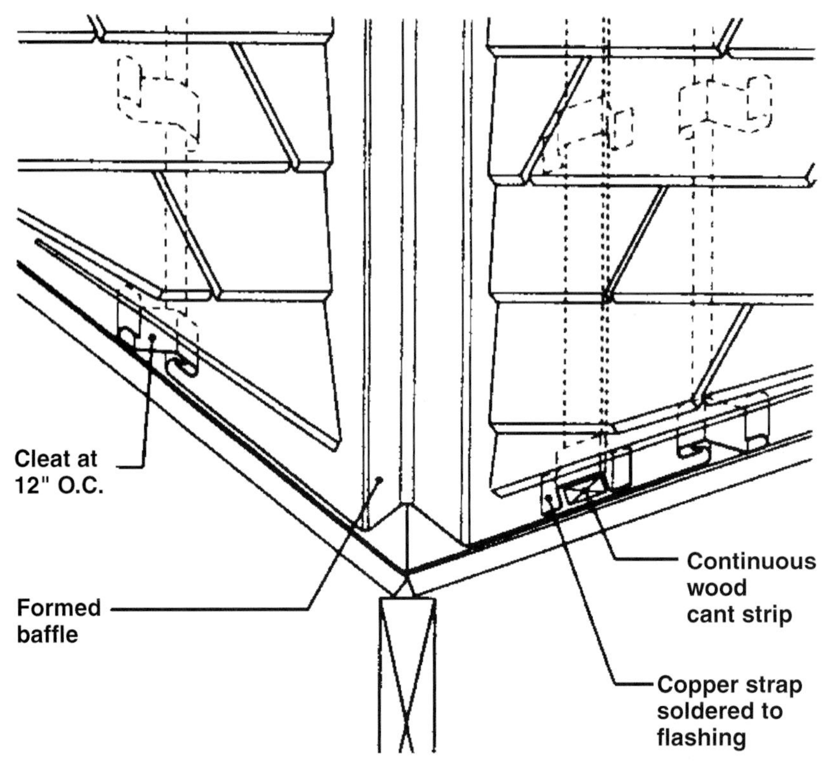

FIGURE 3.3.1 D. UNEQUAL SLOPES

This condition requires a baffle for the same reason as Figure 3.3.1B. It can be constructed as shown in either Figure 3.3.1B or this detail. This detail also shows a different cant strip. Other methods of raising the shingles away from the copper are shown in Figre 3.3.1A and B.

3. INSTALL NEW ALUMINUM FLASHING.

Aluminum is a versatile and durable material appropriate for many roof flashing applications including chimney flashing, valley flashing, step flashing, base flashing, and counter flashing. Aluminum comes in rolls, sheets, and preformed shapes for specific applications such as drip edges. Aluminum comes in standard (mill) finish and factory paint finishes with white and bronze the most common. Aluminum may be left exposed to the elements with or without an applied finish, but anodized or painted material will last substantially longer. Coil stock 24" wide is available in about 30 standard colors for aluminum siding and trim cover applications. Mill or paint finished aluminum readily accepts field-applied paint giving flexibility to desired finishes. It is available in thickness from 0.016" to 0.032". Thicker material will last longer in salty, acidic, or polluted environments. Unlike copper, field soldering of aluminum is not practical. In order to create watertight seams, joints must be lapped, mechanically fastened, and caulked when necessary with high-performance sealants. Fasten aluminum flashing to the framing with aluminum nails to avoid galvanic action between dissimilar materials (such as the aluminum flashing and steel nails), which can cause deterioration.

ADVANTAGES: Aluminum is in the mid-cost range of metal flashing materials. It is a soft, workable material that forms easily and holds its shape well. Preformed aluminum drip edges and other related components speed installation and assure effective results.

DISADVANTAGES: Uncoated aluminum should not be placed in direct contact with concrete, mortar, or other cement-containing materials including fiber-cement siding and trim. The alkalinity of those materials will corrode bare aluminum. Creating watertight joints and laps requires experience, skill, and high-quality sealants.

4. INSTALL NEW GALVANIZED-STEEL FLASHING.

Galvanized-steel flashing is an economical material made by coating sheet steel with a layer of zinc alloy, either through electroplating or hot dipping. Typical gauges are 32 ga (0.010") for roll product and 28 ga (0.015") for preformed shapes. Roll flashing is available in widths up to 48", usually in 50' lengths. Galvanized nails should be used to attach galvanized sheet steel to the frame structure, as dissimilar metal fasteners can cause corrosion. Contact with green lumber and treated lumber should be avoided, because the chemicals in treated lumber can react with the steel, and green lumber has high moisture levels which can lead to rusting. Attachments must be mechanical, such as crimping, or through the use of sealants, or both. Choose only high-performing exterior-grade sealants. Typical applications include valley flashing, base flashing, counter flashing, and chimney flashing. Common preformed shapes include J flashing for roof-to-sidewall intersections, drip edges, 90° bent base flashing, and shingle base flashing. These are normally available in 10' lengths.

ADVANTAGES: Galvanized-steel flashing has the lowest first cost of all metal flashing. It is a rigid material able to be formed into permanent three-dimensional shapes. and will accept paint in order to disguise the raw silver sheen at exposed applications. However, the thin oil coating on the steel must be removed before painting, and primers designed for galvanized steel should be used. Preformed shapes and the 10' lengths make for quick, effective installation.

DISADVANTAGES: In its raw, unpainted form galvanized steel is the least durable of the metal flashing materials. In harsh environments, corrosion may occur in less than 15 years. Galvanized steel is a stiff, relatively nonmalleable material and is somewhat difficult to work with and form. It cannot be field soldered when used to form custom fabrications as the required acid wash and heating process will damage the coating, exposing raw steel. It is not cost-effective when used with long-lasting roofing materials such as slate and tile.

5. INSTALL NEW GALVALUME™ SHEET METAL FLASHING.

Galvalume™ is a product similar to galvanized steel developed by Bethlehem Steel approximately 25 years ago. Where galvanized steel has a hot-dip coating of zinc, Galvalume™ is hot-dipped with an alloy consisting of 55% aluminum and 45% zinc by volume. By weight, aluminum makes up 80% of the coating. Galvalume™ is available in rolls 24" to 48" in width. Mechanical fasteners and sealants are typically required for field fabrications. Choose only high-performing, exterior grade sealants.

ADVANTAGES: An economical material similar in cost to galvanized steel, Galvalume™ is considerably more durable, with approximately twice the service life of galvanized steel. Galvalume™ readily accepts paint finishes, further increasing durability and is a rigid material that can be formed into permanent three-dimensional shapes.

DISADVANTAGES: Galvalume™ should not be allowed to contact concrete foundations, masonry chimneys, or cement board siding, because these materials hold moisture which can lead to deterioration of the flashings. Contact with green lumber, treated lumber, copper, or lead should also be avoided, because of reasons mentioned in option 4, and because of the contact of dissimilar metals which can lead to destructive galvanic action. It is generally available in coil form only. Preformed shapes such as drip edges are not generally available. The stiffness and rigidity of the material makes site braking and fabrication difficult; shop prefabrication is more practical. Galvalume™ should not be field soldered as the required acid wash and heat process will damage the coating, exposing raw steel.

6. INSTALL NEW STAINLESS-STEEL FLASHING.

Stainless steel is a very durable flashing material particularly suited for harsh, corrosive environments. It is available in rolls of 18" and 24" widths as well as sheet stock. Typical gauges are 18 ga and 24 ga. Typical applications include valley flashing, base flashing, and counter flashing. Stainless steel is generally not available in preformed shapes. Material costs for stainless-steel sheet stock is roughly comparable with copper, making it a premium priced material. Use of stainless steel nails is recommended for installation.

ADVANTAGES: Of the most common roof flashing materials, stainless steel is the most durable and least affected by environmental corrosives such as acid rain or salt spray. Strong and rigid, stainless steel can be fabricated into complex shapes. It is a monolithic material which, when cut, maintains its corrosion resistance at the exposed edge. Stainless steel is not affected by contact with masonry mortar or concrete and can be field soldered using a special solder after acid etching of the surfaces to be adhered.

DISADVANTAGES: Although material costs for stainless-steel flashing are comparable to copper, labor and installation are usually higher. Stainless steel is a very stiff, rigid material. It is difficult, and often impractical, to bend and shape stainless steel using a typical roofer's brake. Shop prefabrication is often required.

7. INSTALL RHEINZINK FLASHING.

Rheinzink is a metal flashing and roofing material comprising 99.99% high-grade zinc with 1% copper and 1% titanium alloys. Zinc has been used as flashing in Europe since the early 1800s. Rheinzink was introduced into the United States in 1992. It handles and performs much like copper, but develops a natural blue-gray, gray-green color through weathering, or it can be obtained prepatinated. It is available in sheet or roll form in 0.027" (24 ga) and 0.031" (22 ga) thickness for flashing use.

ADVANTAGES: Rheinzink is easily worked into complex shapes and can be soldered. The manufacturer claims it has a lifespan of 100 years+ if properly detailed. It has a self-healing patina and can be prepatinated. Rheinzink is competitive in price with copper and less expensive than lead-coated copper. It is an inert material, will not leach chemicals, is recyclable, and will not stain adjacent materials. Extensive specification and detailing information is available.

DISADVANTAGES: The underside of the metal requires ventilation to allow protective patination to develop. If the underside is allowed to stay damp, white rust and corrosion can reduce service life severely. Proper detailing is required—the underside must be protected by bituminous sheet material against alkaline influence (e.g., fresh concrete or mortar), acid reacting antifreeze agents, and harmful influence of wood preservative. Contact with copper should be avoided.

8. INSTALL NEW LEAD FLASHING.

Lead is one of the oldest flashing materials. It is durable and malleable, making it a favorite for use as cap flashings, in complex intersections, and with materials that have complicated profiles such as clay and concrete tiles. It is available in rolls from 6" to 20" wide, typically in 2.5-lb/ft.^{2} (0.0391") and 3-lb/ft^2 (0.0468") weights. The use of lead has fallen off as a flashing material because of the concern over lead poisoning. Whereas lead can be used safely with appropriate handling techniques, precautions should be taken to avoid inhaling lead dust or fumes and to avoid hand-to-mouth transfer when eating or smoking. More detailed suggestions are available from the Lead Industries Association (see Further Reading).

ADVANTAGES: Lead is an inert material relatively unaffected by salt and acid rain. It is easily formed into complex shapes, is easily cut and soldered, and has a long service life.

DISADVANTAGES: Lead is a relatively soft material that can tear, especially at right angle cuts. It can fatigue when fastened on all sides, must be used with caution, and requires special handling procedures.

9. INSTALL NEW ROLL ROOF FLASHING.

Mineral surface roll roofing is an economical flashing material for some roof valley installations. When installed in conjunction with asphalt roofing shingles, either three-tab or laminated, roll roofing valley flashing can be a quick, relatively durable installation with the ability to remain functional for 15 to 25 years depending on location, solar orientation, and the quality of the material and installation. The recommended installation involves two layers. The base layer, a minimum of 18" wide, is applied mineral surface down, and the surface layer, 36" minimum width, is installed mineral surface up (Fig. 3.3.2). The material comes in roll widths of 36" and 18", with 36" being most common. Mineral surface roll roofing is available in colors to approximate asphalt shingle colors allowing for a more continuous appearance. The valleys may be installed continuously with no need for laps and seams as expansion and contraction lengthwise is not a large factor. Recommended fasteners are hot-dipped galvanized or aluminum roofing nails with minimum 12-ga shanks and 3/8" heads. The roll roof valley flashing should be considered a part of the roofing system and should be replaced when routine reroofing is required.

ADVANTAGES: Mineral surface roll roofing is an economical material which offers quick, simple installation, with the ability to last the life of the asphalt shingle roof system. No special skills or tools are required for installation.

DISADVANTAGES: Mineral surface roll roofing is appropriate for use in valleys only and with asphalt roofing products. Roll roofing will tend to shrink somewhat in width causing the material to lift from the base of the valley, leaving a void beneath it, which is susceptible to puncture (Fig. 3.3.3). Valley flashing of this type should never be walked on, and other impacts should be protected against. Roll roofing is not self-sealing and is increasingly being replaced by ice and water barriers in colder climates.

10. INSTALL NEW ICE AND WATER BARRIER MEMBRANE.

Ice and water barriers are relative newcomers to the world of flashing materials. Initially intended to combat ice damming at the eaves of sloped roofs and inhibit damage from wind-driven rain, they are now also used in conjunction with more traditional flashing materials as an additional line of defense against water entry at chimneys, valleys, skylights, and other roof penetrations. Some manufacturers also recommend full surface coverage below the roof shingles at low slope applications. Ice and

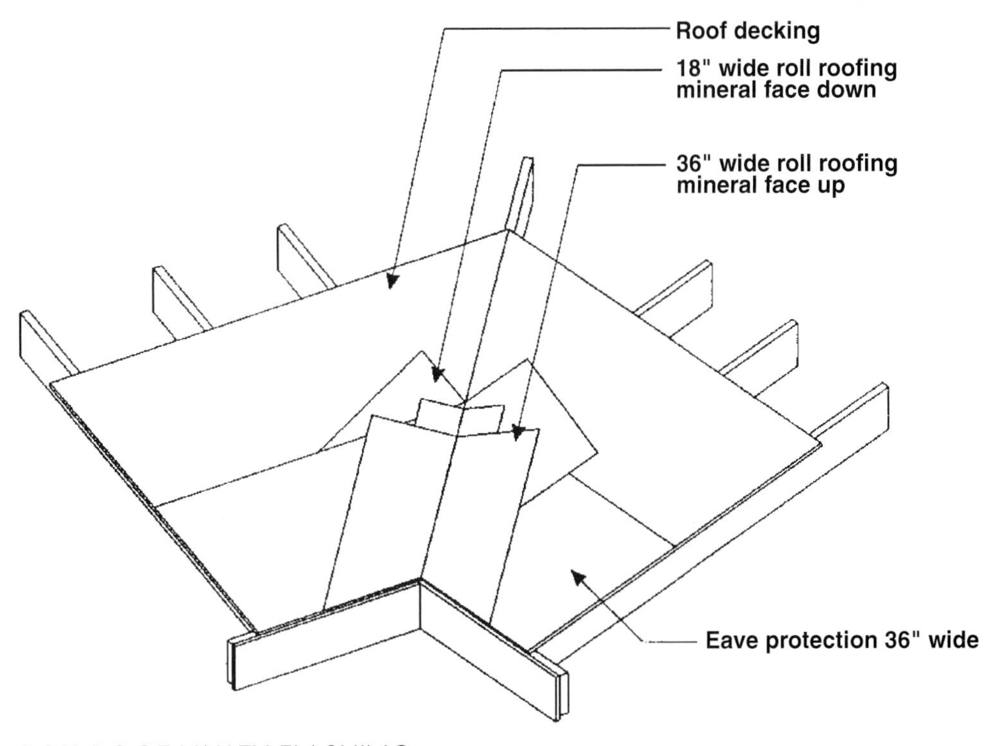

FIGURE 3.3.2 ROLL ROOF VALLEY FLASHING

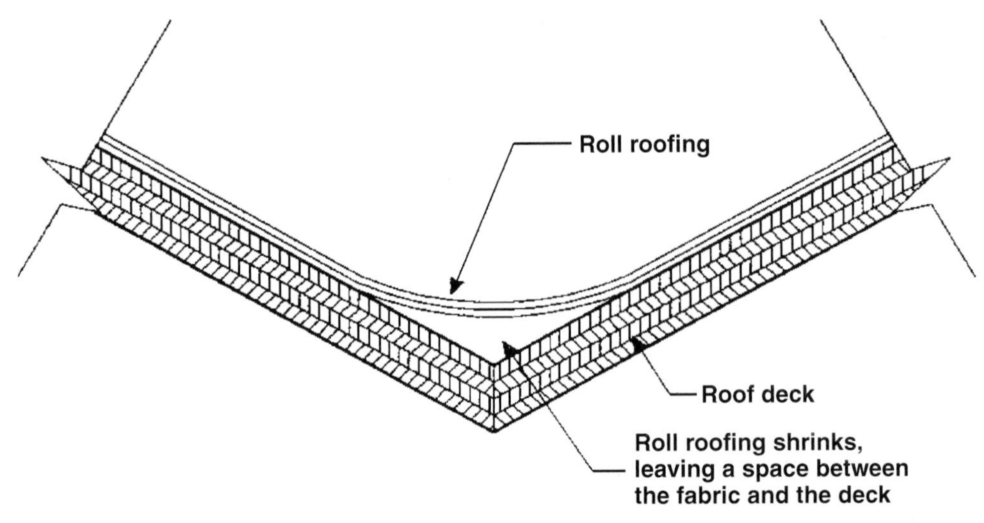

FIGURE 3.3.3 VALLEY CONSTRUCTED OF ROLL ROOFING

water barriers can be installed under all standard sloped roof materials. The materials are easy to work with even for the first-time installer. Ice and water barriers are generally self-adhering and applied directly to roof decking after the removal of the release paper. Manufacturers claim that they self-seal around nail and fastener penetrations. The most common material compositions are modified bitumen with a granular surface, modified bitumen with a polyethylene surface, and rubberized asphalt with a polyethylene surface. Some membranes are available in roll form, for miscellaneous flashing uses (see Section 3.4).

ADVANTAGES: Ice and water barriers are relatively inexpensive protection against water penetration from ice damming. They can provide an additional measure of protection against wind-blown rain in high-wind areas.

DISADVANTAGES: Some types of ice and water barriers should not be installed in hot climates, because of heat degradation, or under dark roofs such as slate, because high summer roof temperatures (up to 180°F) can cause the membrane to bleed. Ice and water barriers can conceal poor detailing and work quality which may cause later problems. The material can degrade when exposed to UV light.

3.4 UNDERLAYMENTS AND MOISTURE BARRIERS

ESSENTIAL KNOWLEDGE

The primary purpose of roofing underlayments is to provide a secondary protective barrier beneath the roofing material. Typically this material has been 15# asphalt-impregnated felt (now called No. 15) for lightweight material such as wood or asphalt shingles, and one or more layers of 30# (No. 30) asphalt-impregnated felt for heavier roofing material such as slate, clay, or concrete tiles. These materials are still in use. However, there is an increasing awareness among manufacturers and roofing consultants of the importance of reinforced underlayments in areas of severe snow or wind-driven rain. In hurricane-prone regions, it is recommended that roofing be designed with the assumption that the first layer will be breached.

Increasing numbers of manufacturers are recommending the use of "special applications" for severe conditions—what have generically become known as ice and water barriers. The pioneer in the development of these products in the early 1980s was W.R. Grace & Company. In the past several years most other asphalt roofing material manufacturers have introduced similar products. Local code requirements and manufacturers' specifications should be consulted prior to selection of a specific product.

TECHNIQUES, MATERIALS, TOOLS

1. REPLACE EXISTING ROOFING FELTS WITH NEW ASPHALT-SATURATED FELTS.
Asphalt-saturated roofing felt underlayments are made from recycled paper and wood products and are essentially the same as the felts used in built-up roofing, except that they are perforated. Typically installed in No. 15 or No. 30 weights (No. 15 weighs approximately 13 lb per square, and No. 30 approximately twice that), roofing felts are installed from the bottom up in the same fashion as shingles so that water does not penetrate lower sheets. In locales with more severe weather, two layers of No. 15 felt or one layer of No. 30 felt are often recommended at building eaves 36" inside of the exterior wall to protect against ice damming. The felt edges are sealed with roofing mastic, with the second layer embedded in mastic. *The Residential Steep-Slope Roofing Material Guide*, produced by the National Roofing Contractors Association, offers detailed underlayment recommendations.

ADVANTAGES: No. 15 roofing felt, when properly applied, is usually satisfactory for lightweight roofing products in areas without severe exposure. It can be doubled up or No. 30 can be used to provide extra protection in areas of more severe exposure.

DISADVANTAGES: No. 15 roofing felt can tear or buckle when subject to rain and wind prior to installation of roofing and may require replacement. It can be torn by roof traffic in the course of installing finish roofing. It is thinner, not self-sealing, less water resistant than ice and water barriers, and will deteriorate over time.

2. REPLACE EXISTING ROOFING FELTS WITH NEW REINFORCED UNDERLAYMENT.

Reinforced roofing underlayments have been developed relatively recently that are less susceptible to tearing than conventional roofing felts. One product, Typar® 30, made of spun-bonded polypropylene, has been used for over 10 years in Canada and the United States as an underlayment for concrete and clay tile in place of No. 30 felt. The material, a moisture and water barrier, is also promoted as an underlayment for metal roofing and other roofing products.

ADVANTAGES: Reinforced underlayment is stronger than roofing felt; is tear resistant; resists wind blow off; will not rot, decay, or become brittle; is lightweight and easy to use; is water resistant; is pliable and flexible at low temperatures; and conforms to irregularities. It can be used for reroofing and has a low flame-spread rating.

DISADVANTAGES: Reinforcement underlayment creates a vapor barrier. The roof must be ventilated to avoid condensation problems. Reinforced underlayment is not self-sealing. National code approvals are pending, and local code approval may not exist.

3. REPLACE EXISTING ROOFING FELTS WITH NEW ICE AND WATER BARRIER.

Ice and water barriers are made with fiberglass and rubberized asphalt, typically styrene-butadienestyrene (SBS) formulations. These materials are fairly recent developments and have high-performance characteristics in terms of elongation, resistance to tear, and longevity. They come in one or two configurations: with a fine mineral granule or sand coating and with a top polyethylene sheet embossed to make it skid-resistant. They generally have a self-adhesive backing that adheres to the sheathing substrate. These products, particularly Grace's Ice and Water Shield®, have a strong following among users. Designed to be installed at the eaves a minimum of 36" inside of the exterior walls, ice and water barriers are also effective in providing extra protection under flashing at valleys as well as at overhangs, skylights, dormers, and vents and chimney flashing. Grace has recently developed a product called Vycor™ Ultra made of butyl rubber that is specifically designed to be used in high-temperature applications such as the hot desert climates of the southwest. Some roofing consultants caution against covering the entire roof, unless it is properly vented. The ice and water barrier can trap moisture in the attic space.

ADVANTAGES: This is a cost-effective, proven way of providing extra protection against potential leaks. When properly installed, it is reported to be effective against ice damming and wind-driven rain penetration under roofing materials. It is self-adhering; seals around nails; is resistant to cracking, drying out, or rot; and is frequently used as a roofing membrane under low-slope roofs.

DISADVANTAGES: This material is more costly than roofing felts. Poor work quality can be disguised, which may lead to later problems. Ice and water barriers are vapor barriers and can cause condensation problems without proper attic or roof ventilation.

3.5 INSULATION

ESSENTIAL KNOWLEDGE

Insulation is one element in a tightly knit construction system intended to improve indoor comfort and reduce energy consumption. Insulation should never be applied without considering its effect on other aspects of construction. Some factors to consider when evaluating roof and ceiling insulation are density and compressibility, air leakage, moisture control, and fire safety.

Each type of insulation has a density at which its R-value per inch is greatest, but reaching this density is not cost-effective. For 3½"-thick fiberglass batts, an R-13 batt contains 40% more material, and an R-15 batt 180% more material, than an R-11 batt (Fig. 3.5.1). Achieving the maximum possible R-value for a 3½" fiberglass batt requires packing in the equivalent of eight R-11 batts.

Some blown-in insulation tends to settle, reducing its R-value. To achieve a desired overall R-value for blown-in insulation, specify the R-value or depth as measured after settlement. The required bag count per net 1,000 ft^2 to achieve a given settled R-value is listed on the bag or can be obtained from the manufacturer. Monitoring the installed bag count is a convenient way to ensure a good installation.

Air leakage can bypass roof and ceiling insulation and create a risk of condensation. Air leakage into attics and through cathedral ceilings typically occurs around vent piping, ductwork, wiring, bath fans, skylights, down lights, and attic stairs, all of which should be carefully sealed or gasketed. Avoid down lights in attic ceilings; if present, they should be of the IC (Insulation Contact) type, and ideally be of the more expensive airtight design. Interior finishes with many joints, such as tongue-and-groove wood decking, can also contribute to air leakage in a cathedral ceiling; an air barrier is recommended behind the planking.

Because the most commonly used roofing material, asphalt shingles, forms an airtight and vaportight surface, moist air can accumulate in an unvented attic or within a cathedral ceiling during cold weather. One way to prevent this problem is to install a vapor retarder at the ceiling, in conjunction with careful air sealing, and thereby inhibit moisture from leaking into the attic or ceiling assembly.

Since vapor retarders and gap sealing are seldom perfect, an additional precaution to prevent condensation, usually required by code, is to provide ventilation into the attic or between the insulation and roof surface in a cathedral ceiling. This is most commonly achieved using *ventilation*

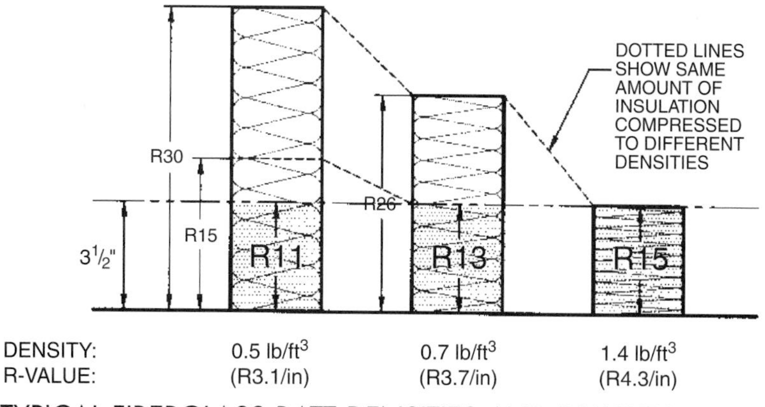

FIGURE 3.5.1 TYPICAL FIBERGLASS BATT DENSITIES AND R-VALUES

baffles—U-shaped channels made from polystyrene foam or cardboard, stapled to the underside of the roof sheathing (Figs. 3.5.2 and 3.5.3). These baffles ensure a space for ventilation air between the insulation and the roof sheathing and are effective at controlling condensation even when they don't communicate with eave or ridge vents (as long as a ceiling vapor retarder is used). In addition, if ventilation is required between insulation and roofing to protect the warranty on roofing, roofing paper, or roof decking, ventilation baffles can provide this ventilation. The need for such ventilation is being studied.

Some sort of air barrier and insulation is desirable around a flue or chimney where it penetrates a floor or roof, but the material used must be rated to resist high temperatures. While fiberglass is noncombustible, at suitably high temperatures the binder holding the fibers together outgasses and the fibers fall apart. For this reason, mineral wool or other "fire-safing" material made for the purpose should be used to seal around high-temperature surfaces.

TECHNIQUES, MATERIALS, TOOLS

Of the innumerable possible combinations of insulating materials and roof and ceiling configurations, the following list covers those in common use and the uncommon systems that are recommended.

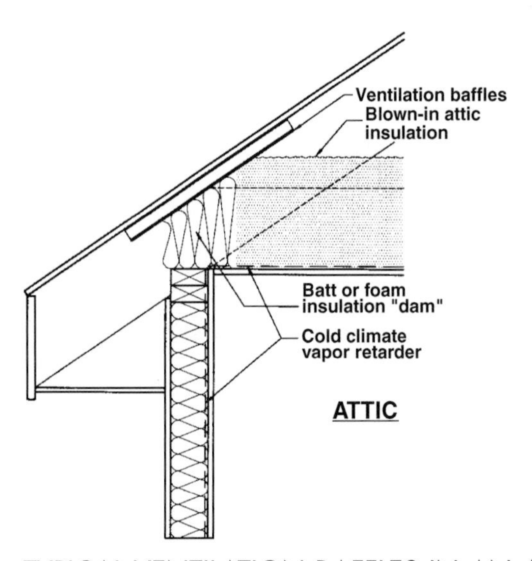

FIGURE 3.5.2 TYPICAL VENTILATION BAFFLES IN AN ATTIC

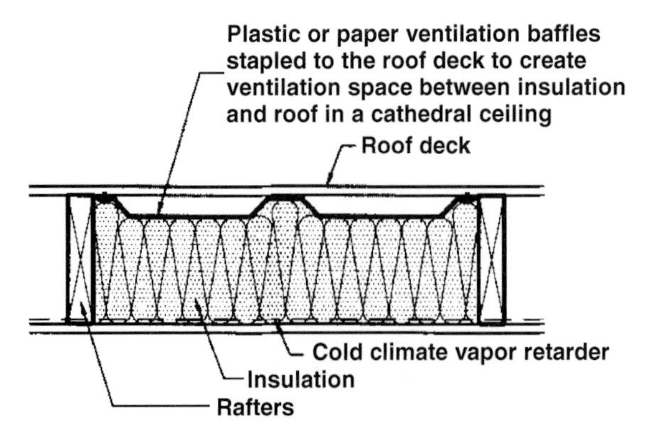

FIGURE 3.5.3 TYPICAL VENTILATION BAFFLES IN A CATHEDRAL CEILING

1. INSTALL BATT INSULATION AT CEILING LEVEL.

Where the insulation is deeper than the ceiling framing members, a gap-free installation requires two layers of batts: a lower one the depth of the ceiling joist or bottom truss chord, plus a continuous upper layer (Fig. 3.5.4). Unfaced fiberglass batts installed from below in new construction may need to be temporarily supported. This is best done with fishing line stretched against the ceiling framing and stapled in place. Faced batts, face- or inset-stapled from below in new construction, or installed from above over existing ceilings, can create a partial cold-weather vapor retarder, as can special vapor-barrier paint or primer applied to the finished ceiling. To create a complete cold-weather vapor retarder and air barrier, a membrane must be installed against the framing in new construction or below the existing ceiling and covered with new ceiling material. In hot, humid climates, a vapor retarder should never be installed at ceiling level, but one is desirable on the attic side of the insulation to inhibit condensation at the ceiling, especially in tightly sealed homes. Applying fire-retardant-faced batts (used in commercial projects) facing up with all joints taped may be a feasible way to achieve an attic-side vapor-retarder.

ADVANTAGES: This method avoids the special equipment needed to blow in loose-fill insulation. Batt insulation does not settle and avoids the cold-weather convective loss in low-density blown fiberglass is avoided. Faced batts form a partial vapor retarder when properly installed.

DISADVANTAGES: Batt insulation depends for its effectiveness upon careful installation to avoid gaps and consequent convective losses. Exposed fiberglass can be irritating if touched or inhaled. This material is typically more costly than blown-in insulation except for small areas.

2. INSTALL LOOSE-FILL INSULATION AT CEILING LEVEL.

Cellulose, mineral wool (either slag wool or rock wool), and fiberglass can be blown onto a ceiling from above. Small blowers can be rented by the day for rehab projects, or the work can be done by a specialty contractor. Cellulose and mineral wool installations typically are less costly than fiberglass, but prices vary by locality. Care must be taken to keep insulation 3" away from down light cans that are not of the IC (zero-clearance) type (Fig. 3.5.5). At eaves, ventilation baffles or other measures are required to prevent the insulation from blocking required ventilation (Fig. 3.5.2). If an attic floor is desired, its support framing must be deep enough for the desired thickness of insulation. A higher density is more effective at inhibiting air circulation through the insulation. Cellulose can be blown in at up to 2 pounds per cubic foot (pcf) density, mineral wool up to 1.5 pcf, and fiberglass up to 1 pcf. Weight limits for gypsum board ceilings are 1.3 pounds per square foot (psf) for 1/2" board supported every 24"; 2.2 psf for 5/8" board supported every 24"; and 2.2 psf for 1/2" board supported

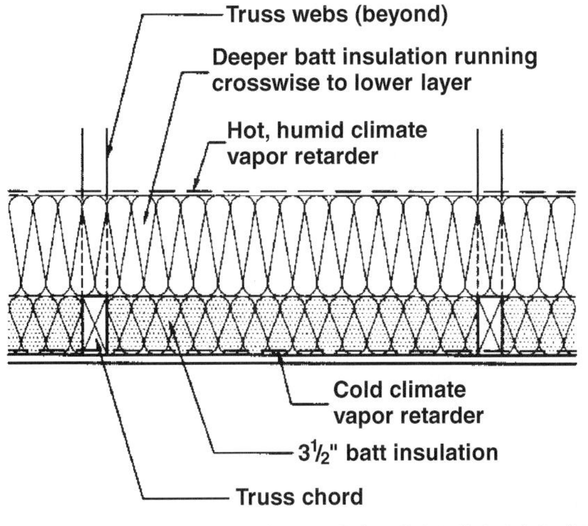

FIGURE 3.5.4 TWO-LAYER BATT APPLICATION IN AN ATTIC

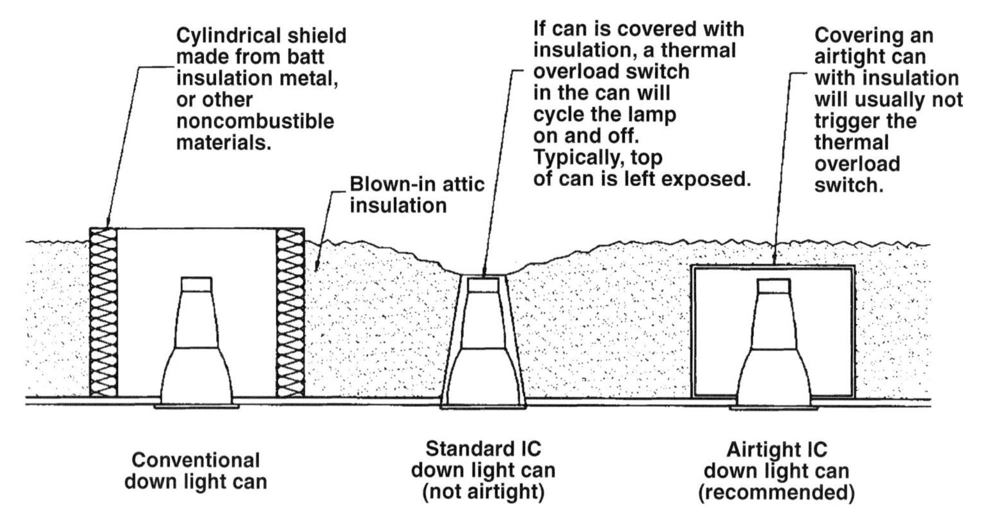

FIGURE 3.5.5 LOOSE-FILL INSULATION APPLICATION AT ATTIC

every 16". During very cold weather, warm air can circulate by convection within low-density fiberglass (1/2 pcf), reducing its R-value. This effect can be inhibited with an added layer of batts, denser loose-fill material, or a layer of blown-in blanket system (BIBS) insulation; check with local energy codes for requirements. For small projects, vermiculite and perlite loose-fill insulation is available for hand pouring if rental blowing equipment is unavailable. No loose-fill material creates a vapor retarder or air barrier.

ADVANTAGES: Loose-fill insulation can be blown in from one or more central points. It is variable in thickness; is usually the least costly form of attic insulation; and forms a complete blanket of insulation, filling irregular cavities.

DISADVANTAGES: Loose-fill insulation does not form a vapor retarder or an effective air barrier. Some lower-density insulation will settle, reducing its R-value accordingly. Low-density blown fiberglass loses some effectiveness during very cold weather because of convection within the insulation.

3. INSTALL BATT INSULATION AT ROOF LEVEL IN CATHEDRAL CEILINGS.

Cathedral ceilings can be insulated with fiberglass batts, which are typically installed against ventilation baffles (Fig. 3.5.3). Faced batts provide a partial vapor retarder if the flanges are face- or inset-stapled to the framing. Unfaced batts need a separate (preferably) 6-mil polyethylene vapor retarder to hold the batts in place and provide a continuous vapor retarder. Stretched and stapled fishing line can be used to temporarily support thin (R-11 through R-19) batts, which are cut 1" narrower than thicker batts and which are less likely to stay in place by friction. At unfinished insulated attic ceilings requiring a vapor retarder, foil-faced or kraft-paper-faced batts cannot be left exposed to the attic space; use fire-retardant-faced batts, or (if allowed by local code) poly-faced encapsulated batts or a separate poly vapor retarder.

ADVANTAGES: This is the least costly way to insulate an open rafter space. Faced batts create a partial vapor retarder when properly installed. Properly installed insulation at roof level saves energy by keeping all ductwork and down lights within the conditioned envelope.

DISADVANTAGES: Batt insulation depends for its effectiveness upon careful installation to avoid gaps and consequent convective losses. A finished ceiling is required.

4. INSTALL ENCAPSULATED FIBERGLASS INSULATION.

Fiberglass insulation is available in rolls or batts, encapsulated with kraft paper or plastic to reduce mechanical irritation to installers. These can be used in any installation where batts are appropriate.

One face is extended to form attachment tabs. The faces on sound-control batts do not have a vapor retarder; some underfloor batts are vapor retardant on all sides other than the flanged face, and could be useful as cold-climate attic insulation installed from above; some exterior wall batts have a polyethylene vapor retarder on the flange side. Some encapsulated batts have a Class A fire rating and can be left exposed if allowed by local codes. Owens Corning markets Miraflex, made from loose, virtually itch-free glass fibers with no binder, contained in a plastic sleeve. Cutting encapsulated batts around obstructions is possible, but this exposes the fiberglass.

ADVANTAGES: Fiberglass insulation is ideal for rehab contractors. It can be left exposed (check local codes). Flange-attached with or without a vapor retarder on some products. Unique underfloor product with vapor retarder opposite the flanged face. Fibers released from cut Miraflex can fill in small voids around obstructions.

DISADVANTAGES: Fiberglass insulation is more expensive. Cutting encapsulated batts exposes the fiberglass, and cutting Miraflex releases the fibers.

5. INSTALL BLOWN-IN LOOSE-FILL INSULATION INTO CLOSED RAFTER SPACES AT ROOF LEVEL.

Loose-fill insulation (fiberglass, cellulose, or mineral wool) can be blown into closed rafter spaces in cathedral ceilings through openings formed by drilling holes through the interior finish, or by removing strips of interior finish, at the upper and lower ends of each rafter space. If roof ventilation is required by code or to protect roof material warranties, the interior finish must be removed to install ventilation baffles, resulting in an open joist cavity (see option 3). Careful installation is required because material can bridge over wiring and other obstructions, causing voids and later settlement. At least a moderate amount of pressure is required to produce sufficient density to inhibit settlement. Fiberglass is typically blown into a cavity at a density of about 1.5 pcf without creating excess pressure on finishes. In a dense-pack installation of cellulose, dry material is applied at high velocity through a narrow tube inserted through a single hole at the top and extending to the bottom of the cavity. The tube is gradually withdrawn, compacting the material to a density of 3 to 3½ pcf. Stabilized™ cellulose includes an adhesive, and Fiberized™ cellulose is made in strands instead of chunks; both processes are claimed to inhibit or prevent settlement. Insulation packed into and filling wall cavities suppresses air movement within the cavity.

ADVANTAGES: If the cavities are completely filled under sufficient pressure, loose-fill insulation provides superior insulating performance without settlement. Properly installed insulation at roof level saves energy by keeping all ductwork and down lights within the conditioned envelope.

DISADVANTAGES: Some types of installation may leave voids and/or settle after installation. Blown-in materials do not form a vapor retarder, form only a partial air barrier, and are very difficult to ventilate.

6. INSTALL BLOWN-IN OR SPRAYED-ON INSULATION INTO OPEN RAFTER SPACES AT ROOF LEVEL.

An inexpensive material can be applied as a membrane over open rafters to form see-through cavities, within which any blown-in insulation can be applied under moderate pressure. In Ark-Seal's Blow-in-Blanket System, fiberglass mixed with some water and adhesive is blown through slits cut in a tightly stretched nylon netting. In Par-Pac's Dry-Pac system, dry cellulose is blown at 3-pcf density into a cavity closed by a reinforced polyethylene vapor retarder. The material is installed from the bottom up to minimize settlement. All such installations will cause the membrane to bulge out; make sure this bellying does not interfere with drywall installation. Various types of polyurethane- and polyisocyanurate-based insulations, such as Icynene, can also be spray-applied into the rafter cavities of cathedral ceilings. A thin layer of such material can form an air-barrier skin over which cheaper material can be placed. Excess material must be scraped off, and windows and electrical devices protected or cleaned.

If roof ventilation is required by code or to protect roof material warranties, ventilation baffles must be installed.

ADVANTAGES: These types of insulation fill the cavities without settling. Visual inspection is possible to ensure filled cavities. Properly installed insulation at roof level saves energy by keeping all ductwork and down lights within the conditioned envelope.

DISADVANTAGES: Bellying of the interior membrane may interfere with drywall installation. Sprayed-on foam products are more expensive than batt or loose-fill installations and are messy processes, requiring cleanup and protection.

7. INSTALL RIGID INSULATION BELOW THE ROOF STRUCTURE.

To increase the R-value of a cathedral ceiling, it may be more cost-effective to add a layer of rigid foam insulation under the rafters than to deepen the rafters and add more cavity insulation. In most climates, it pays to add foam to 2x6 or shallower rafters and not to add foam to 2x10 or deeper rafters, with 2x8 rafters an in-between case. If the roof framing is steel, it is necessary to add a layer of insulation between the framing and the ceiling to prevent interior condensation and ghosting in cold climates and to minimize condensation on the framing in hot, humid climates. An added layer of foam can improve the R-value of the assembly by more than the R-value of the foam, since it suppresses cold bridges through framing members. In cold climates, foil-faced rigid insulation can provide a useful vapor retarder. Do not use a vapor-retarding foam in hot, humid climates, as any retarder should be on the warm side of the insulation.

ADVANTAGES: This method is a useful way to raise the R-value of a cathedral ceiling and is an essential adjunct to normal insulation at steel framing. If the insulation panel is foil-faced, it can create an excellent vapor retarder in cold climates. Properly installed insulation at roof level saves energy by keeping all ductwork and down lights within the conditioned envelope.

DISADVANTAGES: This method is usually not cost-effective relative to deeper framing in new construction. Foam plastics cannot be left exposed in an attic or facing living space.

8. INSTALL RIGID INSULATION ABOVE THE ROOF STRUCTURE.

Installing rigid insulation above the roof structure is the only practical way to insulate an exposed plank-and-beam roof and can be used under low-slope built-up roof (BUR) membranes, modified bitumen (MB) membranes, and flexible membrane systems (single-ply roofing systems). Preformed insulation boards provide resistance to heat flow through roof decks and serve as a base for the roofing membrane. Most types of rigid insulation can be adhered to roof decks with hot or cold asphalt or adhesives, can be mechanically attached, and in some cases, can be ballasted with gravel or crushed stone. Industry associations recommend that a fiberboard overlay or venting base sheet be used over foam products when used under hot-applied BUR or MB systems. Roofing insulation should not be applied over wet roof decks or decks with moisture in the existing membrane or insulation. When multiple layers are used (which is usually recommended) the joints of the top layer should be staggered and offset from the layer below. For specific recommendations, refer to the relevant industry association's guidelines or individual roofing product manufacturer's specifications (see Further Reading and Product Information). A variety of roof insulation products is available:

- *Polyisocyanurate (ISO)*. Composed of a plastic foam manufactured from the reaction of an isocyanate and a polyol formed into a board with a glass-reinforced cellulose felt face sheet on both sides. ISO is available in varying sizes, thicknesses from 1" to 5", and a nominal density of 2 pcf. It is available tapered to provide the necessary slope for drainage. ISO is the dominant roofing insulation because it has high thermal resistance.

■ *Extruded polystyrene (XPS).* Formed by the expansion of a blowing agent in a plastic (polystyrene) polymer in a heated die extruder; available in a variety of sizes, thicknesses, and in densities of 1.5 to 3.0 pcf. XPS is relatively unaffected by occasional immersion in water and can be used in a protected membrane roof (PMR) (at one time referred to as an IRMA roof) in which the insulation is above the membrane, thereby reducing thermal shock. PMR systems require gravel or crushed rock ballast. XPS should not be installed with coal tar pitch (except in protected membrane applications) or solvent-based materials.

■ *Molded expanded polystyrene (EPS).* Formed in a large mold by the expansion of a blowing agent and plastic (polystyrene) polymer under heat and pressure and typically cut by hot wires into a variety of sizes and thicknesses in densities of 1.0 to 2.0 pcf. It is less costly than XPS or ISO. Do not install EPS with coal tar pitch or highly solvent extended mastics.

■ *Fiberglass insulation.* Composed of fiberglass bonded with resin with a glass-reinforced asphalt and kraft paper facer on the top side. It has a stable thermal value and is dimensionally stable. It is available in a variety of sizes, with thicknesses typically from 3/4" to 3". Fibrous glass roof insulation is an excellent recover board due to its ability to conform to irregularities.

■ *Fiberboard insulation.* Composed of wood, cellulosic, or vegetable fibers bonded together to form tough, rigid insulating panels suitable for use as a general-purpose roof insulation under BUR and MBM systems and over foam insulation products. It is available in a variety of sizes and thickness from 1/2" to 2". Its R-factors are considerably below that of foam insulating products. Boards can be adversely affected by moisture and require that a venting base sheet be used to separate boards from existing roof surfaces when the likelihood of moisture is present.

■ *Expanded perlite.* A rigid board composed of expanded siliceous ore particles, cellulose fibers, starch, and other synthetic binders. It has been used as a base for low-slope roofing systems since 1950. It is fire-and water-resistant, is an excellent base for bituminous roofing materials, has a constant insulating value, and is available in thicknesses from 1/2" to 4" and in a variety of sizes. Expanded perlite is available in tapered form and is often used for crickets, saddles, cant strips, and tapered edge strips.

■ *Cellular glass.* A rigid, high-density board material composed of glass foam blown with hydrogen sulfide. Available in sizes 2' to 4' in length and 1" to 4" in thickness.

■ *Phenolic foam.* This product performed unsatisfactorily and is no longer produced in the United States.

■ *Composite insulation boards.* Composed of perlite fiberboard or OSB typically laminated to ISO. These boards add protective and/or nail base material to ISO products.

9. INSTALL A RADIANT BARRIER.

While radiant barriers and coatings are commonplace in high-temperature industrial applications (typically 500°F or more), they are marginally effective at ordinary temperatures. To be cost-effective in building applications, they must have a very low incremental cost (approximately from $0.02 to $0.10 per square foot, depending on the application). Radiant barriers are most effective at roof level, facing down toward the attic. These can be sheets of foil-faced plastic draped over the rafters before sheathing or stapled to the underside of existing rafters. Louisiana-Pacific makes TechShield™ foil-faced OSB sheathing (formerly KoolPly). Because of elevated daytime temperatures, confirm that warranties on the sheathing, building paper, and roofing remain valid before these products are placed above a radiant barrier. A new form of radiant barrier is Radiance paint, which contains aluminum dust.

ADVANTAGES: Down-looking radiant barriers at roof level may be cost-effective when air conditioning ductwork is located in the attic.

DISADVANTAGES: Radiant barriers are marginally cost-effective at ordinary temperatures. They are not effective in cold climates, except possibly Radiance paint.

10. INSTALL A STRUCTURAL INSULATED PANEL ROOF.

Rigid foam insulation adhered to structural skins can create a *structural insulated panel* (SIP). SIPs provide a combination of structure and insulation. Depending upon the configuration, the panels may or may not be supported by beams, trusses, or purlins. Very tight joints are crucial, since a small amount of air leakage through a joint is guaranteed to create destructive condensation at the most critical structural point. SIPs are typically better-insulated and cheaper than structural decking with foam insulation above it. If ventilation under the roofing is required by code or to protect the roofing warranty, SIPs can be made with ventilation cavities at extra cost.

ADVANTAGES: A high-R roof can be achieved with an integrated structural product. Properly installed insulation at roof level saves energy by keeping all ductwork and down lights within the conditioned envelope.

DISADVANTAGES: This product currently is more costly than ordinary framing. Careful air-sealing at all joints and often a subsidiary support structure are required. Ventilation between roofing and the insulation adds to the cost.

3.6 WOOD SHINGLES AND SHAKES

ESSENTIAL KNOWLEDGE

Wood shingles and shakes (thicker, more rustic versions of shingles) have been used for roofing in this country for more than 300 years. Originating in England, the technique of making shingles and shakes involved the radial cutting of large trees, originally including oak, white pine, hemlock, and other softwoods. Today, shingles and shakes are milled largely in Canada and are made mostly from western red cedar, eastern white cedar, and Alaskan yellow cedar. Pressure-preservative-treated southern yellow pine is also used for shakes. Western red cedar weathers a darker gray than eastern white cedar or Alaskan yellow cedar. One or two mills make redwood shingles, but they are not as popular as cedar because they weather considerably darker (although they perform well). Warranted by the members of the Cedar Shake and Shingle Bureau (CSSB) against material defects for a minimum of 20 years, wood shingles and shakes can last longer in colder climates, depending on the building site, local climate, shingle grade, and installation. Wood shingles and shakes perform less well in hot, arid climates. Pressure-preservative-treated (CCA) shingles perform better than untreated shingles in the hot, humid southeastern states. In areas where solid sheathing is required for fire, seismic, high-wind, or snow conditions, direct attachment to solid sheathing is acceptable, especially if the shakes or shingles are pressure-treated with preservatives or pressure impregnated with fire-retardant polymers.

Shingles and shakes can be applied over existing roofing materials with the use of furring strips, but the preferred application is directly over sheathing. For optimum service life, the CSSB recommends that cedar shakes and shingles be attached over spaced sheathing such as 1x6 boards (Figs. 3.6.1 and 3.6.2) or horizontal furring strips over solid sheathing or insulating sheathing (Fig. 3.6.3). Alternatively, shingles can be installed over a plastic mesh product such as Cedar Breather™. The use of spaced sheathing in combination with soffit and ridge vents will allow the shingles and shakes to "breathe" and will reduce excessive moisture. Horizontal furring will also allow for air circulation under the shingles and shakes, prolonging their life.

Wood shingles generally perform better in high-wind locations than some other roof coverings such as asphalt shingles, because they are rigid and do not curl. Testing with normal two-nail fastenings show resistance to 120+ mile per hour (mph) winds. Under severe conditions, heavier than normal roofing felt underlayment, such as two layers of 15# felt, one layer of 30# felt, or a self-sealing reinforced fabric such as an ice and water barrier, is recommended at eaves, roof protrusions, valleys, and other critical areas.

TECHNIQUES, MATERIALS, TOOLS

1. REPAIR EXISTING CEDAR SHINGLES.

It is relatively easy to judge the condition of wood shingles and shakes, although alternate exposures of the roof may wear differently. Very mossy shingles can sometimes be cleaned of moss or lichen with

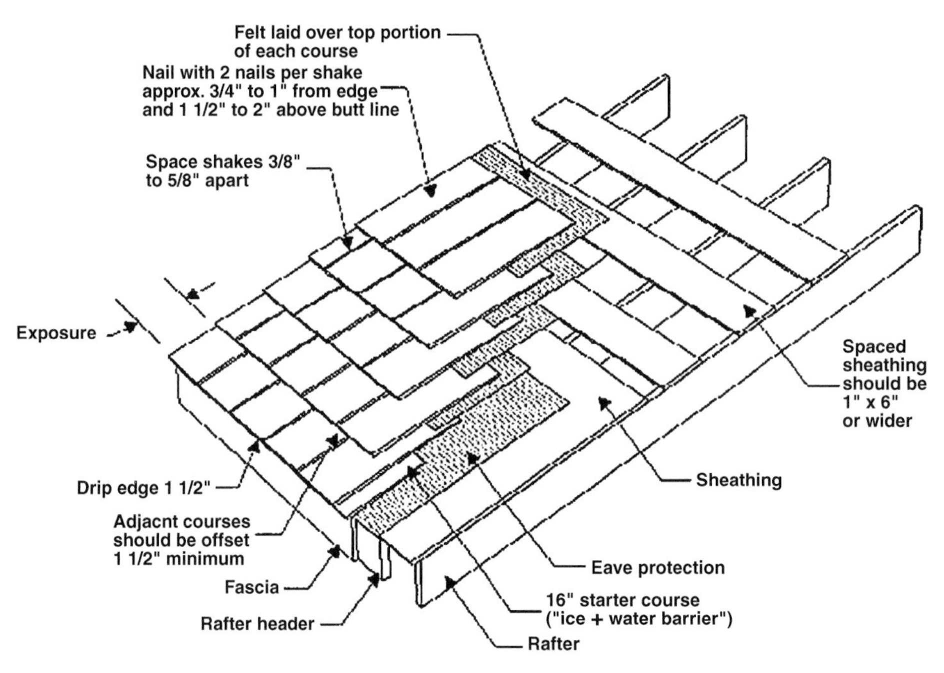

FIGURE 3.6.1 SHAKE APPLICATION

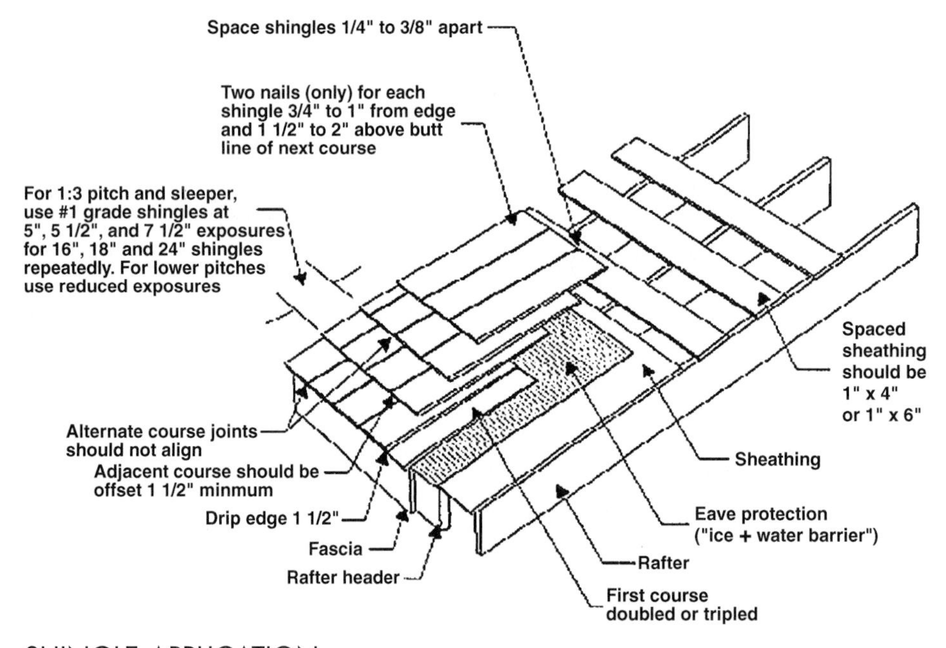

FIGURE 3.6.2 SHINGLE APPLICATION

the use of pressure washing; however, pressure washing can erode shingles. The CSSB has a list of products and techniques that can assist in shingle maintenance. When shingles or shakes begin to split and curl excessively, or become soft, spongy, or crumbly, they should be replaced. Individual cracked, split, or defective shingles or shakes are easily removed and replaced using simple tools such as pry bars.

ADVANTAGES: Repairs are easy to make and are cost-effective over small areas.

DISADVANTAGES: If large areas need replacement, or if serious leaks develop, postponing replacement can cause damage to the building's structural elements or finishes.

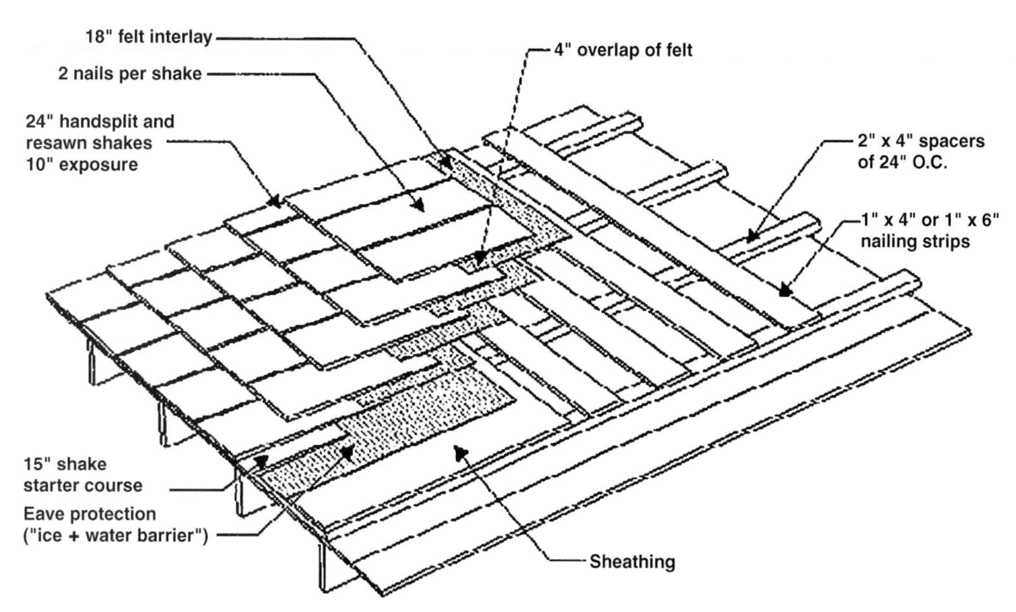

FIGURE 3.6.3 APPLICATION OVER SOLID OR INSULATED SHEATHING

2. INSTALL NEW CEDAR SHINGLES.

Cedar shingles are available in four grades. No. 1 BLUE LABEL with 100% heartwood, 100% clear, and 100% edge (vertical) grain. This recommended grade will have the longest life and is least likely to curl and split due to the content of vertical grain and heartwood. No. 1 shingles are available natural or pressure impregnated with fire retardants for Class A, B, and C roof requirements, or they can be obtained with a 0.40 CCA pressure preservative treatment warranted for 30 years against rot and fungal decay by the treating company. Pressure fire-retardant treatments and pressure preservative treatments cannot be combined. No. 2 RED LABEL is satisfactory for some less exposed siting situations but has significant amounts of flat grain and will not perform as well as No. 1. Since the installation costs are the same, No. 1 or premium shingles will be more cost-effective in the long-term. No. 3 BLACK LABEL is a utility grade for economy applications and secondary buildings. No. 4 UNDER COURSING is used for under coursing, or shimming. (*Note*: A number of suppliers are not members of CSSB and their grading specifications may vary somewhat.)

ADVANTAGES: Cedar shingles are preferred for their natural appearance for both traditional and modern houses, often simulated but never duplicated and have a long history of satisfactory performance. They can be treated with preservative, shingle oil, stain, CCA, or fire-retardant, or can be left to weather naturally. They can be obtained in a variety of butt configurations. Western red cedar shingles are warranted for 20 to 30 years (depending on the product) against material defects by CSSB members and independent suppliers or their treatment applicators. Available with premanufactured hip and ridge caps, wood shingles and shakes are more resistant to damage from high wind and hail than asphalt shingles. Being made from a renewable resource, cedar shingles are a sustainable choice.

DISADVANTAGES: Cedar shingles are more expensive than most other roofing materials. They can weather unevenly under certain conditions and are more combustible, unless treated, than most other roofing products. Eastern white cedar is normally available only in random lengths with only 50% to 66% vertical grain, requiring the expense of culling out unacceptable pieces.

3. INSTALL NEW CEDAR SHAKES.

There are four basic types of cedar shakes available: No. 1 Handsplit & Resawn is the most popular, with hand-split faces and sawn backs. No. 1 Tapersawn are sawn on both sides and resemble thick butt shingles (also available in No. 2 and No. 3 utility grades). No. 1 Tapersplit shakes are produced largely by hand and are popular in preservation applications in a three-ply system, which does not require building paper in snow-free areas. No. 1 Straight Split shakes are the same thickness throughout and are used primarily on exterior walls, but occasionally on roofs. Shakes are thicker than shingles and are warranted against product defects (but not installation) for 25 years. They are used primarily when a more rustic or rugged roof character is desired than that possible with shingles.

ADVANTAGES: The material imparts deep shadows and strong texture to the roof surface. It is a classic, natural material for both traditional and contemporary buildings. Cedar shakes are much copied, but never replicated. They can be treated with preservative, stain, CCA, or fire-retardant, or can be left to weather naturally. Treated heavy shakes are warranted for 50 years by some manufacturers. They are available with premanufactured hip and ridge caps. Being made from a renewable resource, cedar shakes are a sustainable choice.

DISADVANTAGES: Cedar shakes are more expensive than other roofing types including shingles. They can weather unevenly and are more combustible than other roofing materials unless treated with fire retardant. Some warping and cupping will occur.

4. INSTALL NEW SOUTHERN YELLOW PINE SHAKES.

Southern yellow pine shakes have been available since the 1970s, but were not actively promoted until the late 1980s. Initial interest outstripped manufacturing capabilities, and the capacity remains limited to a relatively small market within the eastern and midwestern states. Available taper sawn in No. 1 and No. 2 grades in lengths of 18" and 24" with a butt thickness of 13/16", southern yellow pine shakes are heavier, thicker, and denser than cedar shingles. All shakes are pressure treated with CCA preservative and are warranted against rot and decay by the treating agency for 30 years. The material weathers a natural gray. Southern yellow pine shakes can be applied on solid sheathing because they are treated, but venting by means of furring strips is recommended. Shakes can be applied over existing wood or composition shingles, but application over sheathing is preferred.

ADVANTAGES: Southern yellow pine shakes provide an attractive, heavily textured roof used in high-end custom housing and historic restoration projects. The material is warranted for 30 years against decay, rot, and termites and performs better than other lighter materials in high winds and hail storms. They cost approximately the same as untreated cedar shakes. Being made from a renewable resource, pine shakes are a sustainable choice.

DISADVANTAGES: Southern yellow pine shakes are heavier than cedar shakes, so the roof structure may require strengthening. They are more expensive than some other roofing products and cannot be treated with fire retardant when treated with CCA. Some warping and cupping will occur.

3.7 ASPHALT SHINGLES

ESSENTIAL KNOWLEDGE

Asphalt shingles were introduced into the roofing market in the late 1800s as a by-product of making tar and asphalt-impregnated felts for flat roofs. Early shingles contained up to 33% cotton or wool fibers derived from rags, hence the term *rag felt*. From the early 1940s to the late 1970s asphalt shingle mat was comprised of organic cellulose fibers derived from recycled waste paper and/or wood fiber. Although organic mat is still available, the dominant base material since the late 1970s has been inorganic (fiberglass) mat. According to recent sales information from the Asphalt Roofing Manufacturing Association (ARMA), fiberglass mat shingles comprise approximately 82% of the residential asphalt shingle roofing market, and organic mat shingles (still popular in the midwest and northeast because they are more flexible and considered easier to install in cold weather) 18%. ARMA estimates that asphalt shingles represent 80% to 85% of the total residential roofing market.

The fiberglass mat typically used in the asphalt shingle industry is lighter than organic mat. Fiberglass mat doesn't necessarily perform better but does allow shingles to meet Class A fire resistance ratings, while organic mat only meets Class C. Organic mat is presaturated with asphalt and then coated with a mineral-stabilized material (limestone, slate, fly ash, or traprock). Inorganic (fiberglass) mat is coated with mineral-stabilized material but not presaturated. The mineral-stabilized coating fills the voids between individual mats while at the same time providing increased resistance to fire and weather.

The typical asphalt shingle has been for many years a 36" × 12" three-tab strip shingle. The cutouts between the tabs create the illusion of individual shingles (Fig. 3.7.1). Within the last 5 or 10 years, architectural or laminated shingles that simulate wood and slate shingles or shakes have become increasingly popular for higher-end housing. These shingles are made of two or even three layers laminated into a single unit. The multiple laminate reinforces the impression of individual shingles, and the protective granules are toned to simulate weathered material. Dark-colored granules are added to create the impression of shadows, thereby enhancing the three-dimensional, high-definition effect (Fig. 3.7.2). Dark-colored granules are also added to some three-tab shingles, such as Owens Corning's Prominence® series to simulate the appearance of premium shingles (Fig. 3.7.3). Other recent developments include the increasing use of larger-sized metric shingles, longer multitabbed shingles, and distinctively styled, nonrectangular, diamond, and hexagonal (scalloped) shingles (Fig. 3.7.4).

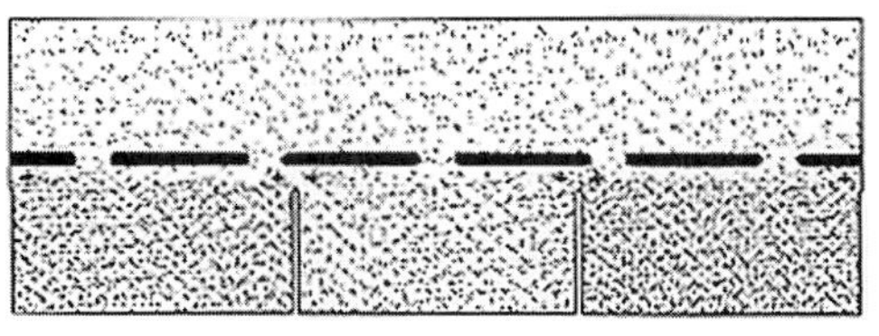

FIGURE 3.7.1 THREE-TAB SHINGLE

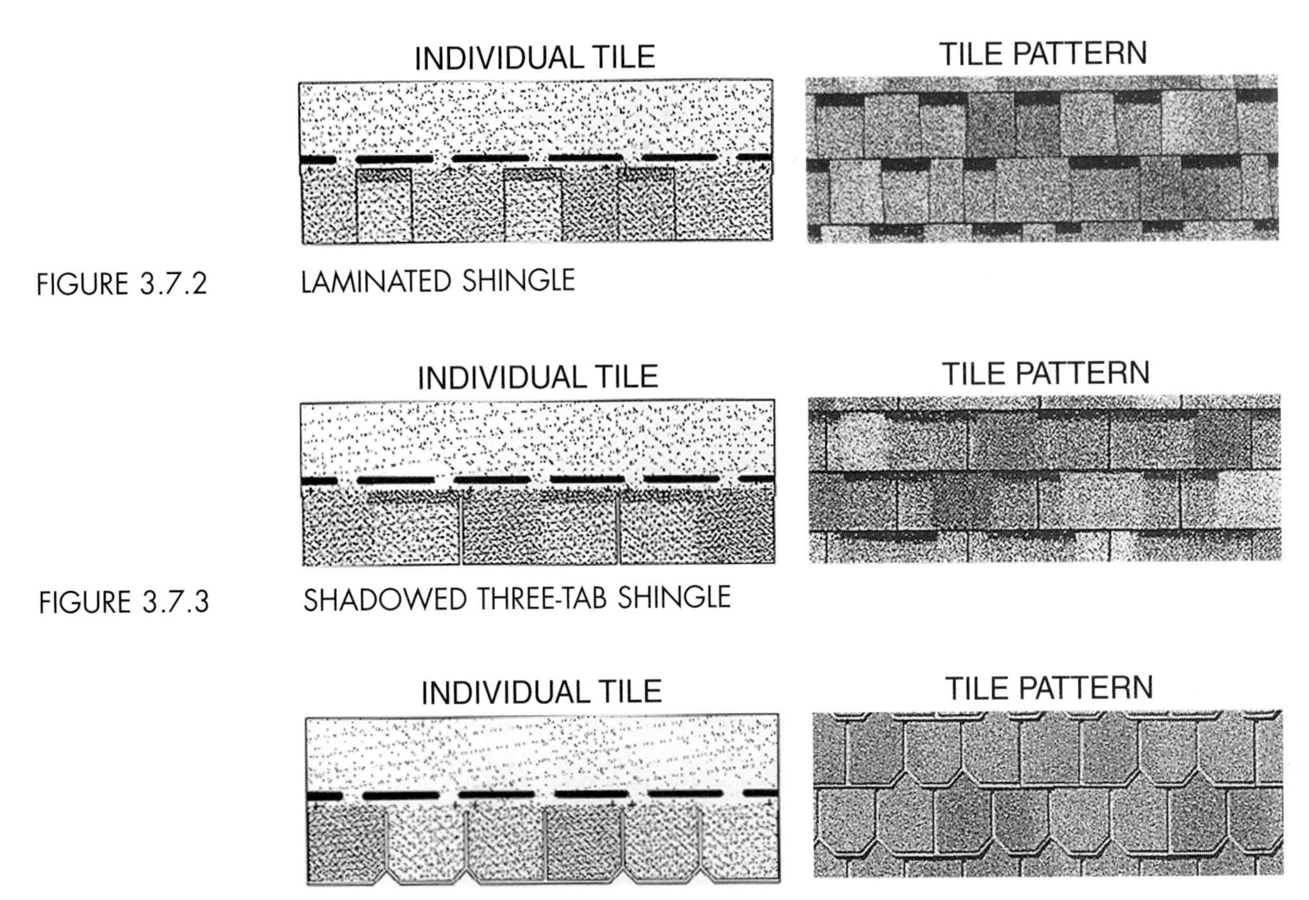

FIGURE 3.7.2 LAMINATED SHINGLE

FIGURE 3.7.3 SHADOWED THREE-TAB SHINGLE

FIGURE 3.7.4 SHAPED SHINGLE

The proliferation of different asphalt roofing shingle types and styles has made the selection of these materials difficult. In the recent past shingles were categorized by weight, such as 235, 240, or 280 lb. The weight was generally related to service life. Currently shingles are classified by warranty duration, such as 20, 25, 30, or 40 year. There is no direct relationship between base mat thickness, shingle weight, performance, and warranty. Furthermore, asphalt coatings, the type of fillers, mat thickness, and shingle weight vary from one manufacturer to another, making it difficult to estimate relative performance. Most of the laminated shingles have been on the market for only the past 5 to 10 years, and there is no long-term history of their performance (the Insurance Institute for Property Loss Prevention assumes that the effective life of an average asphalt shingle is 17 years). There is also no applicable ASTM accelerated wear test that is appropriate for asphalt shingles. Manufacturer warranties are apt to differ in terms of what is warranted (material, labor, or both), against what defect, for what period (prorated or not), with what exclusions (rain, hail, and other variables), or whether the warranty is transferrable.

In response to widespread reports of product failure in the early 1990s, the Midwest Roofing Contractors Association (MRCA) ran shear strength tests on a number of shingle types (asphalt fiberglass, asphalt organic, asphalt laminated fiberglass, and asphalt laminated organic shingles) in accordance with Section 8.1.2 of ASTM D3462, *Standard Specification for Asphalt Shingles Made from Glass Felt and Surfaced with Mineral Granules*. All but two of the 20- to 25-year warranted three-tab asphalt fiberglass shingles, and all the 25- and 30-year warranted asphalt laminated fiberglass shingles failed to meet the test. The ARMA maintains that tensile strength, tensile elongation, and shingle flexibility are better indicators of potential resistance to shingle splitting than tear strength and is working through ASTM to improve the D3462 standard to more accurately reflect shingle performance. ARMA maintains that some shingles that don't meet D3462 perform adequately. Most manufacturers currently produce shingles that meet ASTM D3462, and this standard is increasingly referenced by model code agencies. However, some manufacturers, responding to requests for inexpensive products, make "commodity" 20- and 30-year shingles that do not meet this standard. These shingles, of varying quality, are often sold through discount wholesale and retail outlet stores that supply very price sensitive markets. Until the standards are revised, MRCA and some other industry representatives recommend purchasing organic or fiberglass shingles that are certified as meeting the tear-strength requirements of ASTM D3462.

A recent development in asphalt shingles is algae-retardant surfacing. Manufacturers have in the past experimented with zinc granules to retard algae growth, but these sometimes resulted in white patches and uneven staining. At least one manufacturer, 3M, that provides granules for roofing manufacturers has developed ceramic-coated granules that release copper ions over a 10-year period to help reduce the darkening of roofs from algae in hot, humid climates. For the additional cost of $150 to $300 per house, 3M claims that roofs will not require as frequent cleaning, which can remove surface granules. It is too early to evaluate the long-term effectiveness of this material, but there are indications that in the short term it helps retard algae discoloration.

If installing asphalt roofs in hurricane-prone regions, note that manufacturers generally do not warrant their products for wind speeds greater than 80 mph. Asphalt shingles have frequently performed poorly in high winds and can be a significant source (along with other roofing products) of wind-blown debris. Such performance, manufacturers point out, is frequently attributed to improper application, substrates, or fasteners (staples are not recommended). The industry and some regional standards (such as the Blue Sky guidelines developed by the town of Southern Shores, North Carolina) recommend doubling up on the weight of shingle underlayments, using polymer-modified asphalt underlayment membranes (such as Ice & Water Shield™ or WinterGuard™) instead of, or in addition to, roofing felts, and increasing the nailing pattern from four to six per shingle and the nail size to 1 1/4" #12 nails with 3/8" head diameter. Shingles at roof boundaries such as rakes, ridges, hips, and valleys should be secured with hand-tabbing of an asphaltic roof cement. Existing shingles should be removed prior to reroofing and the existing sheathing inspected for deterioration. Some municipalities, such as Metro Dade County, Florida, have special requirements, such as allowing only shingles that pass 110-mph testing (but which are not typically warranted by manufacturers beyond 80 mph).

Partly in response to cold-temperature and high-wind performance requirements, some manufacturers produce shingles derived from recent flat-roof technology developments, with different base mat materials and configurations, as well as modified asphalt formulas. Malarkey Roofing Co. (with distribution largely in the northwest) offers a shingle that has a base mat of fiberglass, polyester, and styrene-butadiene-styrene modified rubberized asphalt that it claims increases shingle flexibility to resist tearing and cracking under normal as well as low temperatures. Malarkey makes a three-tab and laminated shingle that is warranted against material failure in up to 100-mph winds. Another manufacturer, IKO, produces a plastic polymer-modified asphalt shingle (atactic polypropylene) with similar enhanced performance claims.

TECHNIQUES, MATERIALS, TOOLS

1. REPAIR EXISTING ASPHALT SHINGLES.

Isolated small holes or cracks in shingles can be temporarily repaired by troweling on plastic roofing cement. Curled shingles can often be cemented back in place. Individual shingles that are badly damaged can be replaced by slipping a pry bar under the damaged shingle, removing the nails, sliding the shingle out, replacing the shingle, and applying roofing cement to the new nail heads and the bottom edge of the new shingle and the one above. If a large number of shingles exhibit excessive drying out, curling, loss of protective granules, cracking, or other deterioration, or if there is evidence of significant leaks that are not due to faulty flashing, then a complete shingle replacement is likely to be required.

ADVANTAGES: Repairs are easy to make and are cost-effective over small areas.

DISADVANTAGES: If large areas need replacement, or if serious leaks develop, postponing replacement may cause damage to the building's structural elements or finishes.

2. INSTALL NEW ASPHALT SHINGLES.

A critical decision will be whether or not to remove the existing shingles or apply the new shingles over the existing layer. According to ARMA, in many cases it is not necessary to tear off the old shingles if the roof has only one layer of shingles, is laying flat, and the decking is in good condition. In this instance, the existing asphalt shingles can provide a secondary moisture barrier and the cost of removing and disposing of the old shingles will be eliminated. The existing shingles will probably have to be removed, however, if an inspection of the roof sheathing and substrate reveals significant rotting or warping of sheathing members, if there is more than one existing layer of asphalt shingles on the roof, if the roof structure shows signs of sagging along the ridge or truss lines, or if the condition of the existing shingles is so rough and distorted that new roof shingles would not lay flat. If the roof does not look straight and feel solid, it should be inspected for structural adequacy by a professional engineer or architect. The advice of an experienced roofing consultant or contractor can assist in the determination of the appropriate installations. If there is any serious doubt, a complete removal will lead to the most satisfactory application, as it allows a complete assessment of the condition of the existing roof and will provide the base for a level and flat installation. In any case, adequate roof ventilation should be provided (see Further Reading). Ideally, asphalt shingles should be installed only when the temperature ranges between 40°F and 85°F. At temperatures lower than 40°F shingles become brittle, crack easily, and are hard to cut. At ground-level temperatures of 85°F and above, roof temperatures can be in the mid-100s and the granular material is easily disfigured and scuffed by handling and walking on the roof's surface. In hot climates, roofers usually begin work at daybreak and quit early. In colder climates during the winter, the seal tab adhesives may not set up initially, especially on light, reflective roofs. Seal tabs may also not set up properly on very steep roofs with slopes over 12 on 12. These installations often require additional applications of roofing cement (see Further Reading for cold weather and high slope application recommendations). Application procedures are critical: incorrect nailing above the seal tab line or overdriven nails can lead to product failure. Shingle manufacturers and ARMA recommend against directly applying asphalt shingles on insulation or radiant barrier decks (see Further Reading).

ADVANTAGES: Available in a wide range of types, colors, and patterns, asphalt shingles are by far the most popular residential roofing material. Under most conditions asphalt shingles perform satisfactorily. Laminated shingles can dramatically enhance the building's appearance, especially those with higher roof pitches. New asphalt shingles can be applied over one layer of flat existing asphalt shingles, but reroofing directly over sheathing is best. Asphalt shingles are a relatively economical material with low first-costs compared to other materials and are easily installed.

DISADVANTAGES: Confusing claims and warranties, and the proliferation of material types and specifications, make it difficult to compare and evaluate different manufacturers' products. Competing products are not necessarily equal, and warranties, specifications, and testing data should be carefully examined and compared. Some asphalt shingle products may perform unsatisfactorily, especially in hot, arid climates where thermal shock conditions (high heat with rapid cooling from thunderstorms) occur. Roofs may darken or stain from excessive moisture or humid conditions. Being a product of nonrenewable fossil materials, asphalt shingles are not the best choice from a sustainability standpoint.

3.8 LOW-SLOPE ROOFING

ESSENTIAL KNOWLEDGE

All roofing systems are subject to leaks, but low-slope roofs do not shed water as effectively as high-slope roofs and are more susceptible to damage from water infiltration. Some of the causes of water leaks can be mitigated by careful maintenance, but others, such as those caused by deficient design and installation; long-term weathering and UV exposure; extreme weather conditions including snow, hail, high winds, and drenching rains (that raise the water level above base flashing height); structural deficiencies and changes; and excessive thermal expansion and contraction, cannot. The causes of roof leaks are often difficult to determine accurately, and leaks frequently result in damage at some distance from their source.

An inspection for roof leaks should begin in the building's interior, noting all signs of moisture infiltration, such as stained or deteriorated roof decks, structure, ceilings, and walls. Before corrective work is undertaken, roofs should be inspected by a qualified roofing consultant or contractor to document general problem areas beginning with the obvious ones. These include obstructed or nonperforming drains, scuppers, gutters, or leaders; the existence of foreign matter such as tree limbs, debris, leaves, and pine needles that block drains and puncture roof membranes; displaced ballast or walkway pavers; wear and tear on roofs used for recreation; and uneven, settled, or depressed roof areas that prevent proper drainage and lead to ponding. If the roof is under a manufacturer's warranty, the roofing membrane manufacturer should be contacted as soon as possible about any roof leaks. Upon completion of a general survey, a more detailed inspection should address the condition of each roof component, including copings, caps, and counter flashings; gravel stops, roof edging, and fascia; base flashing, pitch pockets, and boots at roof penetrations; deteriorated sealants; and roof membranes, including unbonded or unsealed seams, and the existence of ridges, blisters, wrinkles, worn spots, holes, or deteriorated areas. The National Roofing Contractors Association (NRCA), ARMA, and the Single Ply Roofing Institute (SPRI) have jointly produced a useful checklist of items to be surveyed (see Further Reading).

The low-slope roofing materials discussed below include built-up roof membranes, modified bitumen membranes, and thermoset and thermoplastic single-ply membranes.

TECHNIQUES, MATERIALS, TOOLS

1. REPAIR EXISTING BUILT-UP ROOFING (BUR) MEMBRANE.

Spot repairs address deficient or degraded conditions at isolated locations in the roof membrane or base flashings. Most repairs of punctured, cracked, blistered, wrinkled, or otherwise distressed areas involve similar repair strategies including the removal of debris, contaminants, or aggregates; checking for water damage to the insulation or decking; removal and replacement if necessary; cutting out of damaged section of membrane; priming the membrane and installation of new plies (to match the number removed) in hot bitumen or cold-applied adhesive. A variety of resaturants and liquid coatings exist that extend the life of existing surface coatings. Some contain pigments that reflect solar radiation. The NRCA and the ARMA have both produced a number of comprehensive manuals of maintenance procedures and recommended repairs (see Further Reading). Permanent repairs should be undertaken by a professional roofing contractor.

ADVANTAGES: Spot repair can be cost-effective in increasing the service life of both new and worn roofs.

DISADVANTAGES: As with all roofing systems, if roof problems are widespread, or if serious water problems exist, spot repairs may not be effective.

2. REPLACE EXISTING ROOF WITH BUILT-UP ROOFING MEMBRANE.

Built-up roofing (BUR) systems comprise multiple overlapping layers of roofing felt coated with asphalt or coal tar pitch (Fig. 3.8.1). BUR systems are classified into two major categories: asphalt systems and coal tar systems. Asphalt is derived from the refining of crude oil; coal tar is derived from the refining of coal. Although they are both used in conjunction with plies of roofing felt, with fiberglass or polyester fabric reinforcement, they have quite different properties and are not necessarily interchangeable. Asphalt-based products should typically be used for the repair of asphalt BURs, and coal tar products for the repair of coal tar BURs, with the exception that asphalt-based products are routinely used for the construction of base flashings in coal tar BURs. BURs have been in use for over 150 years; many 50-year-old and older roofs are still in service. Coal tar roofing pitch comes in three grades or types that have varying viscosities. Type I is the most prevalent. Coal tar is used basically for low slopes and is not recommended for slopes greater than 1/4" in 12". In warm climates this slope factor is reduced to 1/8" in 12". Coal tar roofing pitch is the only material with true cold-flow properties which give it self-healing characteristics beginning at surface temperatures of approximately 60°F (experienced during winter months due to solar radiation). Asphalt is also available in a variety of types according to their viscosity and softening points. Type I asphalt's softening point is 135°F and is used only on roofs with no appreciable slope. Type III asphalt, called *steep asphalt*, can be installed on roofs with a maximum slope of 3 in 12. Its softening point is between 185°F and 205°F. Type III is the most commonly used asphalt in BUR because of its in-service softening point. Asphalt BURs are available in both conventional hot-applied and cold-process roof systems which mix asphalt with petroleum distillates, polymers, fibers, and fillers. Cold-process roof systems are usually brush-applied or applied by commercial airless spray equipment, instead of being hand-mopped or applied by mechanical asphalt spreaders. They normally require fewer plies. Undesirable fumes and the danger of hot spills are said to be eliminated.

FIGURE 3.8.1 BUILT-UP ROOFING BEING INSTALLED

ADVANTAGES: BUR systems, properly designed and applied, have a long history of successful use and can perform well for many years. Cold-process systems can minimize undesirable fumes and are economical. BUR systems can be relatively easily maintained and repaired and can be more forgiving of installation errors, due to the redundancy of multiple plies, than other systems.

DISADVANTAGES: Built-up roofing systems, as with all roofing systems, depend on close adherence to specifications and require careful attention to work quality during installion for their performance. Air pockets must be limited and application of the proper amount of adhesives is critical. Aggregate surfaces on coal tar and some asphalt BUR systems make identification of sources of leaks difficult.

3. REPAIR EXISTING MODIFIED BITUMEN (MB) MEMBRANE

The ARMA, NRCA, and SPRI's recommended repair strategies for MB membranes include inspection of the roof membrane for water infiltration; removal and replacement of damaged or wet insulation or deteriorated decking; removal of debris, contaminants, ballast, aggregate, or loose granules; priming of the membrane surface; patching surface with a similar material 8" wider in all dimensions with 3"-radius corners; installation of the pach in hot asphalt [if styrene-butadiene-styrene (SBS) but not atactic polypropylene (APP)], in cold adhesive, or by heat welding in accordance with the roofing manufacturer's recommendations (see Further Reading).

ADVANTAGES: This method can increase the service life of new and worn roofs and is cost-effective.

DISADVANTAGES: If roof problems are widespread or if serious leaks persist, localized repairs may not be effective and more general repairs or replacement may be necessary.

4. REPLACE EXISTING ROOF WITH MODIFIED BITUMEN (MB) MEMBRANE

Modified bitumen membranes are made from asphalt or coal tar pitch bitumens modified with chemicals (polymers) to provide enhanced weatherability, flexibility, tensile strength, and resistance to flow at high temperatures. The most common polymers used to modify asphalt bitumen are APP and SBS. Other modifiers include SBR, EIP, and SEBS. Coal tar pitch bitumen can be modified with Tardyne® polymers to create an MB membrane that is compatible with coal tar BUR systems. However, most MB membranes combine asphalt, reinforcing fabric such as polyester or fiberglass, and APP or SBS polymers. Most MB membranes are supplied in rolls covering 100 ft^2 (one "square"), installed in multiple plies. Although they differ significantly from single-ply thermoset and thermoplastic membranes, MB membranes are sometimes grouped with those systems into a new classification, *flexible membrane systems*.

APP systems are normally applied with an open-flame torch process, but are also installed with hot-air welding equipment and can be obtained in a peel-and-stick sheet form or in rolls compatible with cold-process adhesives. SBS systems are normally installed with hot asphalt moppings, but can also be installed with open-flame, hot-air self-adhesive, and cold-process adhesive. A variety of surfacings is available, including asphalt cutback aluminum coatings, ceramic granule or mineral surfacing (similar to asphalt composition shingles), and occasionally aggregate toppings. MB membranes with factory-applied ceramic granules are more expensive than unsurfaced sheets but are cost-effective from a life-cycle view. The life expectancy of roofs can be increased with additional layers of MB membrane (or smooth-surfaced sheets) in the roof's field or at flashings. Currently MB systems are frequently combined with BURs into a hybrid roof, i.e., two plies of BUR and an MB cap sheet for a high-tensile-strength (BUR), highly flexible (MB) system. Coatings, such as reflective aluminum roof coatings, can also increase service life, although they require periodic renewal.

ADVANTAGES: MB membranes are available in a variety of configurations and surface coatings and are an increasingly popular alternative to BUR systems. Installation and detection of possible roof leaks are easier than for aggregate-surfaced BUR systems. They can accommodate some roof traffic.

DISADVANTAGES: Installers must be familiar with its use. Careful attention to installation details, particularly seaming, is required. The life expectancy may be greatly reduced by ponding water.

Products containing new polymer types should be carefully researched due to limited history of satisfactory use.

5. REPAIR EXISTING THERMOSET AND THERMOPLASTIC SINGLE-PLY ROOFING MEMBRANE.

Emergency patching of single-ply roofing systems can be made with duct tape, roofing tape, polyethylene sheets, wood blocking, or butyl or polyurethane sealants, depending on size and severity of damage. Plastic roofing cement and fabric patches are acceptable if other means are not possible, but roofing cement may deteriorate some types of single-ply membranes [particularly polyvinyl chloride (PVC)] and possibly insulation as well. Avoid the use of liquid or pourable asphalt repair products. Single-ply roofing systems require permanent repairs that are compatible with the individual materials and systems chosen. An excellent source of permanent repair guidelines is provided in ARMA, NRCA, and SPRI's *Repair Manual for Low-Slope Membrane Roof Systems.* Permanent repairs should be undertaken with professional roofing contractors experienced with the systems at hand. Roofing manufacturers should be consulted to confirm the appropriateness of individual repair techniques (see Further Reading).

ADVANTAGES: Permanent spot repairs can be relatively simply made and are cost-effective in increasing the service life of both new and worn roofs.

DISADVANTAGES: If roof problems are widespread, or if serious water problems exist, spot repairs may not be effective.

6. REPLACE EXISTING ROOFS WITH NEW THERMOSET AND THERMOPLASTIC SINGLE-PLY ROOFING MEMBRANES.

Single-ply roofing membranes, currently promoted as "flexible membranes," were first used in the United States during the late 1950s on such architecturally significant projects as Dulles Airport. Widespread commercial applications were developed in the 1970s spurred on by oil shortages and increases in the cost of petroleum-based products (such as BUR) caused by the Arab oil embargo. Currently, according to the SPRI, flexible membranes comprise approximately 55% of the total commercial roofing market. While their use is largely in commercial projects, they are also used on low-sloped multifamily and townhouse projects, both for rehabilitation and new construction.

In addition to modified bitumen, previously discussed, flexible membranes are grouped into thermoset and thermoplastic categories (Table 3.8.1). Thermoset membranes include EPDM and CSPE, first introduced in 1951 under the trade name Hypalon and neoprene (also known as chloroprene rubber). In thermoset materials, polymers are cross-linked during the manufacturing process for EPDM and during exposure to heat and light (curing) in the case of CSPE. Seaming of thermoset materials has traditionally been accomplished by means of liquid adhesives, or more recently by the use of specially formulated tape. Newer formulations of CSPE are seamed by means of heat welding.

Thermoplastic membranes include PVC; PVC blends; and alloys such as CPA, EIP, and NBP. PIB and TPO are based on polypropylene and ethylene propylene polymers. Thermoplastic membranes differ from thermosets in that there is no cross-linking or vulcanizing—they can be repeatedly softened by heating and will reharden when cooled. This allows them to be seamed together by special heat welding equipment which is an increasingly efficient, economical, and popular form of sealing that provides more uniform attachment than that usually attained by the use of solvents. Flexible membrane roofing systems can be attached to the deck and roof insulation with mechanical fasteners; adhesives including special glues; hot asphalt or cold adhesive; or loosely laid with gravel ballast or concrete pavers. Each installation has its advantages and disadvantages which should be carefully researched.

Some flexible membranes such as PVCs, EPDMs, and CSPEs have been successfully used for over 30 years. Some are new formulations of existing products (CPAs, EIPs, and NBPs). Others including

TABLE 3.8.1 POLYMERS USED IN FLEXIBLE ROOFING MEMBRANES

Polymer Category	Abbreviation	Generic Classification
Butadiene acrylonitrile polymers	NBP	Thermoplastic
Chlorinated polyethylene polymers	CPE	Thermoplastic
Chlorosulfonated polyethylene polymers	CSPE	Thermoset
Copolymer alloys	CPA	Thermoplastic
Epichlorohydrine polymers	ECH	Thermoset
Ethylene propylene polymers	EPDM	Thermoset
Ethylene interpolymers	EIP	Thermoplastic
Isobutylene—isoprene polymers	PIB, IIR	Thermoplastic
APP modified bitumen	MB	Modified bitumen
SBS modified bitumen	MB	Modified bitumen
Polyvinyl chloride polymers	PVC	Thermoplastic
Thermoplastic polyolefin	TPO	Thermoplastic

Certain materials may be derived by combining two or more of the polymers shown.

TPOs are quite recent developments, and even newer products and formulations are continuously being developed. A comparison of individual products is made difficult because of the complexity and large number of competing products, the lack of comprehensive qualitative comparative data, and continuing product evolution. Some data is available from SPRI (in particular its reference manual titled *Flexible Membrane Roofing: A Professional Guide to Specifications*) and individual product manufacturers, but an informed choice will require an evaluation of comparative data including product history (successes and failures); weatherability; chemical and flame spread resistance; slope characteristics; tear and puncture repairability; ability to resist roof traffic wear; compatibility with associated materials (flashings and insulation); color and roll size; warranties; installation choices, procedures, and details; availability of reinforcing screens; life-cycle costs; and availability of experienced applicators (sometimes the most critical determinant of all).

ADVANTAGES: Single-ply roofing membranes are available in a broad range of products, colors, surfacings, application methods, and engineering and design solutions. They can be applied to complex roof profiles and can accommodate building movement, are lightweight and can be used over existing roofing, and are easy to install and inspect. There is continuous research and development of systems in use and new systems. They have a history of satisfactory performance.

DISADVANTAGES: It is difficult to obtain accurate comparative data between systems. Very careful quality control during product installation is required since poor application procedures have led to a number of failures, particularly at seams between rolls of materials. Some newer materials have a limited history of successful use.

3.9 METAL ROOFING

ESSENTIAL KNOWLEDGE

Long a roofing choice in many parts of the world, metal was introduced into the United States in the eighteenth century as a craft industry using small sheets of lead or copper fabricated on-site into flat-seamed roofs. By the beginning of the twentieth century, factory-formed sheets of flat and corrugated galvanized steel became available that were economical and did not require a full supporting roof deck. Used primarily for rural housing and utility buildings, steel roofing was initially considered a somewhat inferior product that often leaked, had a limited life, and wasn't very attractive. Copper was too expensive for all but the most costly housing and institutional projects.

In the years since World War II a number of factors have transformed residential metal roofing: the use of zinc-aluminum alloys has lead to better corrosion resistance; improvement in coating systems has reduced fading and chalking; better-quality sealants have increased weathertightness; and the use of fastening clips has allowed for increased roof movement. In addition, the development of sophisticated site roll-forming equipment has allowed for longer, site-formed sections of roof panels, and the development of economical metal tile products that simulate other traditional roofing materials such as slate, clay tiles, wood shingles, and shakes has increased the range of roofing options.

TECHNIQUES, MATERIALS, TOOLS

1. REPAIR EXISTING METAL ROOFS.

All metal roofs and their fastenings will eventually fail due to the effects of normal weathering, severe wind and hail storms, corrosive airborne pollutants, rapid expansion and contraction, and UV radiation. Localized problems can be repaired by replacing sections of damaged panels with patches of similar materials (applied with a sealant and screws if steel or aluminum, soldered if zinc or copper), or, preferably, by replacing the entire affected panel. Asphaltic patches or acrylic "elastomeric" repairs can be unsightly and are not normally long-term solutions. Some acrylic or polyurethane systems are available to recoat entire roofs, but many of these systems have limited lifetimes (1 to 5 years), are relatively unattractive, and generally are appropriate primarily for low-slope applications where the roof is not visible.

ADVANTAGES: Repairs can be cost-effective if limited to a few localized areas.

DISADVANTAGES: Repairs are not cost-effective if large sections of roof or flashings have deteriorated and the structure or interior finishes are at risk. Replacement panels will not match the color of existing panels.

2. REPLACE EXISTING ROOF WITH NEW STANDING SEAM METAL ROOF.

Standing seam roofing has become increasingly popular for both multifamily and high-end, single-family housing, because it performs well and adds texture and color to simple roofs and continuity to complicated roofing profiles and elevations. It is available in preformed sheets and site-formed panels up to 100' long in a wide variety of materials including aluminum, copper, galvanized steel, Galvalume, Zincalume, painted steel, stainless steel, and zinc. Aluminum is sometimes specified for dry maritime areas where there is little rainfall to wash off the salt. Copper and zinc, the longest lasting materials, are sometimes specified for very high-end projects. Zincalume and Galvalume are also specified where their color characteristics are desired, but the dominant standing seam roofing material is prepainted steel. Prepainted steel roofing is made from cold-rolled coils of steel of varying thicknesses from 0.013" to 0.024", coated in a continuous hot-dipped process with zinc or zinc-aluminum alloys such as Zincalume and Galvalume (similar products). The coils are then finished with various paint systems including polyvinyl fluoride products such as Kynar 500 and Hylar 5000 (similar products) or siliconized modified polyester (SMP). Polyvinyl fluorides with at least 70% resin content (some have less) are warranted to retain 85% of their original color for 20 years, SMPs will retain 80% of their color for 10 years (both have lifetime warranties against chipping or peeling).

Metal roofing systems are fabricated either at manufacturing facilities dedicated to that purpose, at the roofing contractor's own sheet-metal facility, or at the job site using portable roll-forming equipment. Standing seam roofing systems are typically divided into two categories: architectural and structural. Architectural systems are usually installed on solid sheathing at slopes greater than 3 in12 (although lower slopes can be accommodated) and are designed to shed water rapidly over the surface of the panels so the seams do not have to be watertight. Structural systems have more pronounced profiles, may be of heavier gauge, may use sealants applied to the seams, and generally span between purlins. Because they are for commercial applications, structural systems are not covered here. Traditional architectural standing seams include overlap (bent flat to form a flat seam) (Fig. 3.9.1), single-lock (Fig. 3.9.2), and doublelock (Fig. 3.9.3). Single-and double-locked seams are mechanically crimped with electric crimping tools. Double-locked seams are the most watertight of the residential systems. For very low slopes, butyl tape or sealant is sometimes used in the seam to increase the watertightness. Architectural panels typically have ribs 3/4" to 1 1/2" high, are 12" to 16"

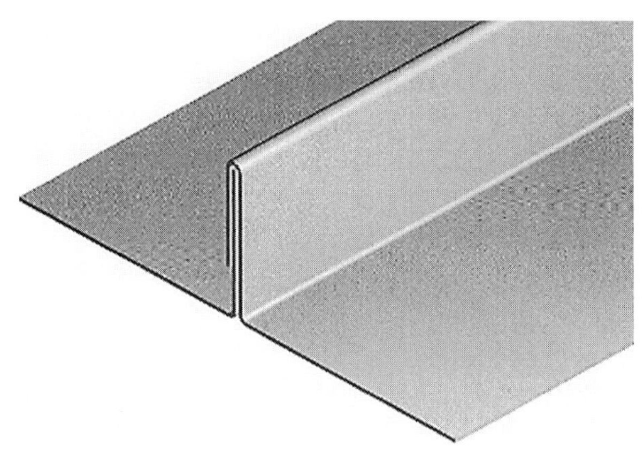

FIGURE 3.9.1 OVERLAP SEAM

FIGURE 3.9.2 SINGLE-LOCK SEAM

wide (wider panels can allow oil-canning and substrate unevenness to show), and have clips at 36" o.c. to allow the roofing to "float" and compensate for expansion and contraction. Some systems use continuous clips, and some less costly systems fasten directly to the substrate (which can cause fastener backout or tearing of the metal at the fastener penetration from expansion and contraction). Architectural panels are available in a number of gauges including 28 ga, sometimes used for utility buildings; 26 ga (typically available only with siliconized modified polyester coatings), used in some cool climates such as the Pacific Northwest; and 24 ga, the most typical gauge, specified in hot, sunny, dry climates where maximum fade resistance is required (Kynar 500 or Hylar 5000 are unavailable from many suppliers in 26 ga material).

Two more recently developed standing seam systems include the snap-on cap and the snap-lock. Snap-on cap systems (sometimes referred to as the mansard system because they were developed for high slopes) use a simple cap (in one system with a continuous vinyl weather seal) to cover the joint between panels (Fig. 3.9.4). A wider snap-on batten system is also available to provide a more pronounced rib (Fig. 3.9.5). Snap-lock systems employ specially formed ribs that snap and lock together (Figs. 3.9.6 and 3.9.7). Snap-lock and snap-on cap systems are less weathertight than double-locked systems and are usually restricted to slopes over 3 in 12. The NRCA recommends a highly water resistant underlayment be used under snap-on or snap-lock systems where leak-free performance is required. The NRCA also recommends that ventilation be provided between the underside of metal

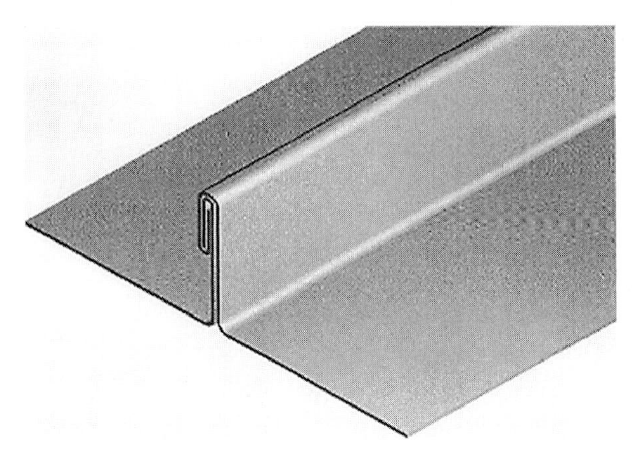

FIGURE 3.9.3 DOUBLE-LOCK SEAM

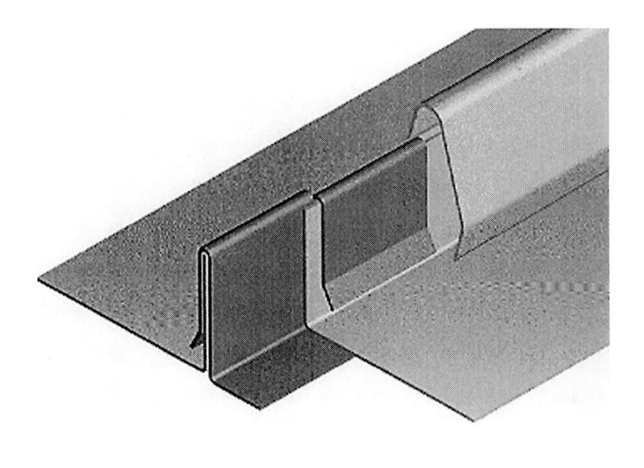

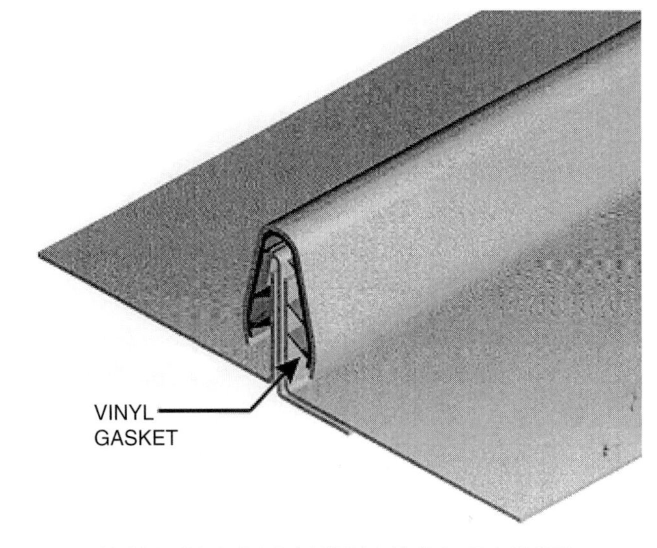

FIGURE 3.9.4 SNAP-ON CAP WITH CLIP SNAP-ON CAP WITH VINYL GASKET

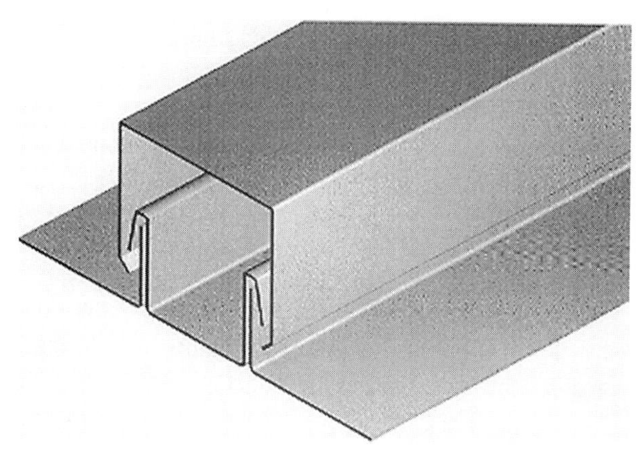

FIGURE 3.9.5 SNAP-ON BATTEN

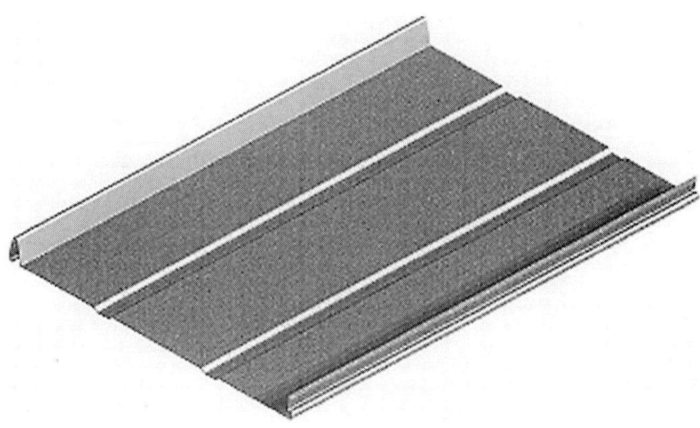

FIGURE 3.9.6 SNAP-LOCK SYSTEM

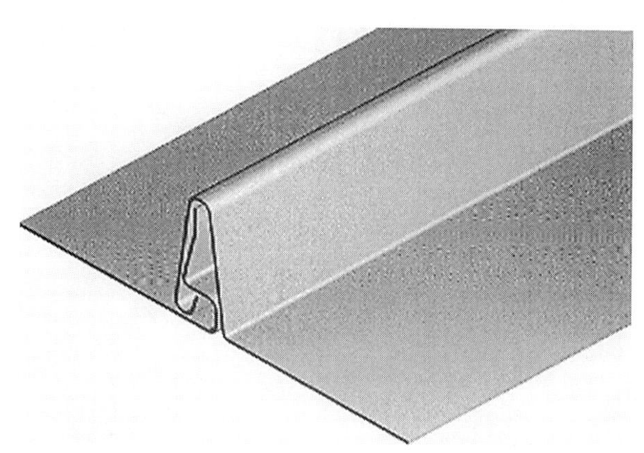

FIGURE 3.9.7 SNAP-LOCK

panels and the substrate, due to the great potential for condensation on the underside of the panel (the panels are an excellent thermal conductor and when temperatures fall below the dew point, condensation can occur on the panel underside).

ADVANTAGES: Standing seam roofing is attractive, and is available in a great variety of colors and materials. It is easy to install, in some instances, over existing roofing; is relatively long lasting, and noncombustible; resistant to decay, discoloration, and mildew; is wind resistant and water, snow, and ice shedding; is lightweight; and can be structural and placed over spaced purlins. It is potentially recyclable.

DISADVANTAGES: Standing seam roofing has a higher first-cost than asphalt shingles and some other roofing products. Zinc-aluminum alloys may not be warranted within 1/4 mile of the coast. Steel is a vapor barrier and can trap moisture leading to condensation if not ventilated. Careful detailing is required to mitigate condensation, excessive expansion and contraction, and uplift in high-wind areas. Some installers may not be qualified. Warranties vary considerably and should be carefully reviewed. Roof warranties with longer duration (5 to 10 years) are preferred.

3. REPLACE EXISTING ROOFS WITH METAL SHAKES AND TILES.

Metal roll-formed shakes and S-tiles were introduced more than 20 years ago. They have generated considerable interest recently because of their noncombustibility, light weight, and generally good performance in high-wind areas. Typically made of 24-ga galvanized or Galvalume steel with smooth polyvinyl fluoride coatings to simulate tile or with textured granules set in acrylic resins to simulate shakes. Metal tiles are available in small panels approximately 16" × 46" (Figs. 3.9.8 and 3.9.9) or in large panels up to 40' long (Figs. 3.9.10 and 3.9.11). Large panels can be placed on battens over solid sheathing, on battens on existing roofs, or over purlins spanning between roof rafters or trusses.

ADVANTAGES: These products are lightweight and strong due to the molded surface; are noncombustible; are wind, mildew, and rot resistant; are available in a variety of textures, profiles, and colors; and are economical, considering ease and speed of installations. Long-term material warranties exist.

DISADVANTAGES: Tiles may not closely resemble the materials they simulate. Ceiling insulation to reduce sound transmission is required.

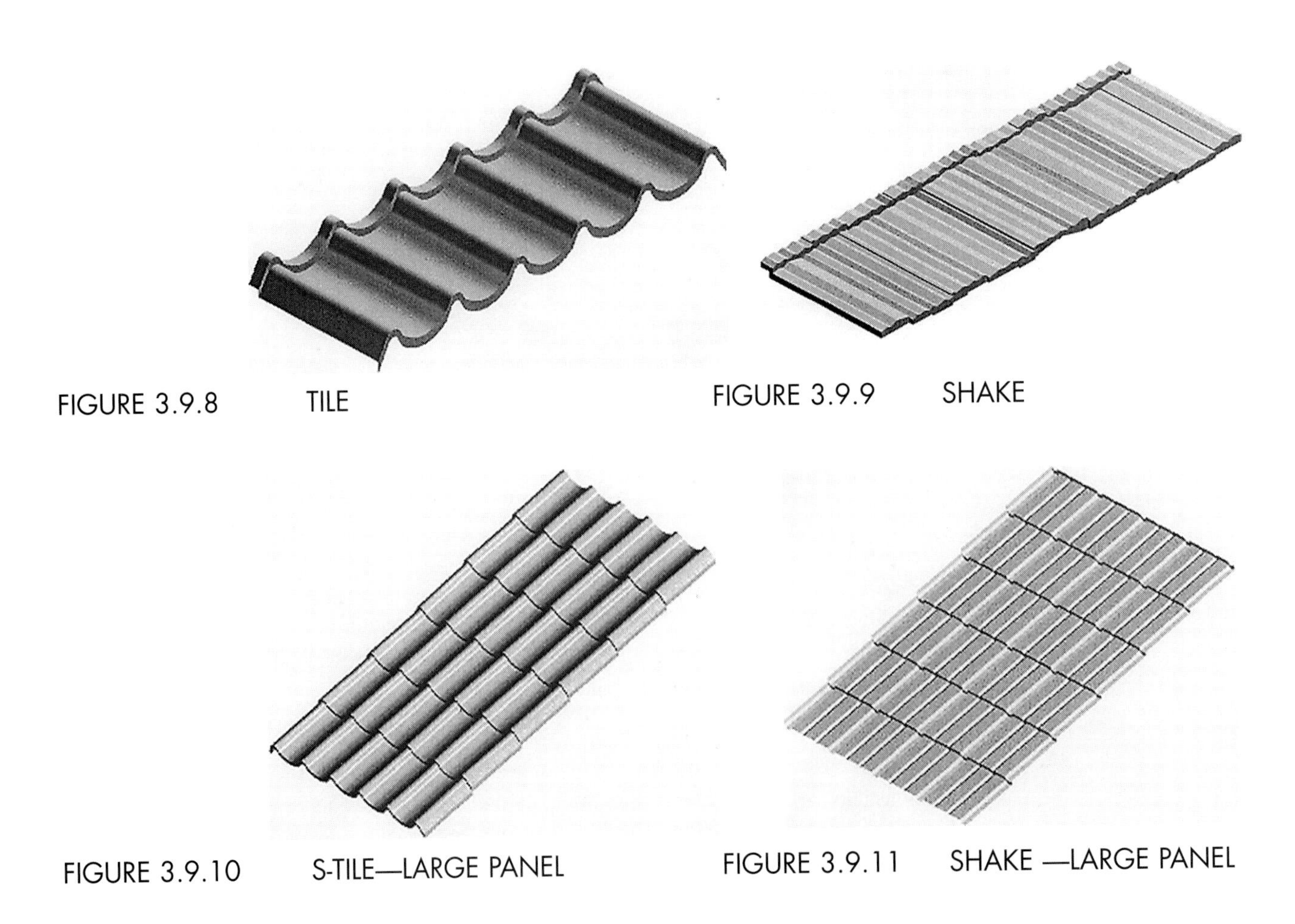

FIGURE 3.9.8 TILE

FIGURE 3.9.9 SHAKE

FIGURE 3.9.10 S-TILE—LARGE PANEL

FIGURE 3.9.11 SHAKE —LARGE PANEL

4. REPLACE EXISTING ROOF WITH FLAT METAL SHINGLES.

Several manufacturers make flat specialty shingle panels, including copper panels by Revere Copper Products, Inc., and Vail Metal Systems; a Victorian shingle made of prepainted or plain Galvalume steel by the Berridge Manufacturing Co.; and a diamond-shaped tile, Castle Top, made of aluminum, copper, zinc, and prepainted steel by Atas International, Inc. The copper shingles are nominally 12" × 48" panels embossed to look like individual shingles (Figs. 3.9.12 and 3.9.13). Vail Metal System's product was designed to function well in high-snow and high-wind environments. It is also available in prepainted metal. Berridge's Victorian shingle was developed for restoration applications (Fig. 3.9.14). All shingles are installed over conventional roof sheathing.

ADVANTAGES: Flat metal shingles have a distinctive appearance; have high wind and weather resistance; are noncombustible and durable; and will not decay, peel, or rot. They are lightweight, and easy to install.

DISADVANTAGES: Flat metal shingles are more costly than other materials (copper is more than twice the cost of prepainted metal), and take longer to install than some other metal roofing products.

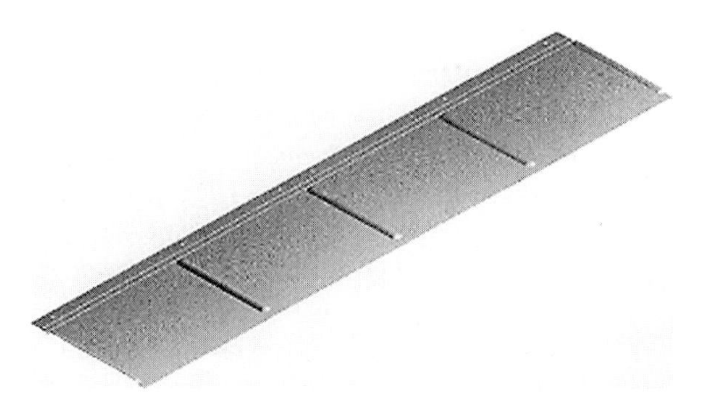

FIGURE 3.9.12 COPPER SHINGLE PANEL

FIGURE 3.9.13 COPPER SHINGLE PANELS

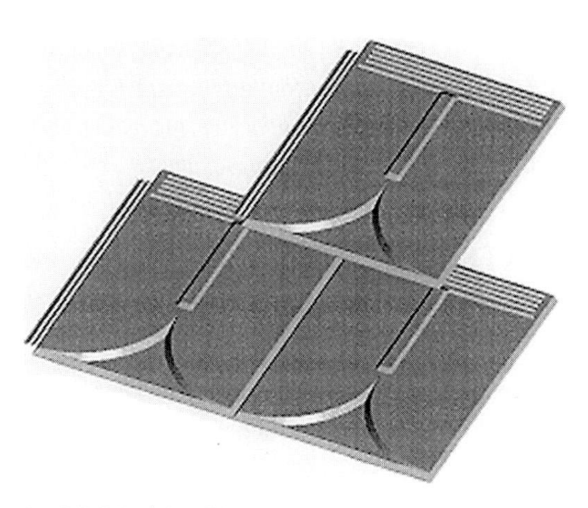

FIGURE 3.9.14 VICTORIAN TILE

3.10 SLATE

ESSENTIAL KNOWLEDGE

Originally introduced into this country from England in the eigtheenth century, slate has been a favorite roofing material for all types of buildings for more than 200 years. Its use peaked around the turn of the nineteenth century, when wood shingles were scarce because of clear-cutting of eastern forests, and a noncombustible material was sought for the fast-growing cities of the east and midwest. Demand abated after the 1929 depression with the fall-off in housing starts and introduction of lower-cost asbestos and asphalt composition shingles. According to the NRCA, slate's share of the current residential roofing market is approximately 5% for new construction and 3% for reroofing.

The most common types of slate were formed millions of years ago from marine deposits of clay and sand. Less-common types were formed from layers of volcanic ash. Slate is typically quarried in the United States in Virginia, Pennsylvania, New York, and Vermont along the Appalachian Mountain chain, as well as in Canada and Newfoundland. U.S. quarried slate roofs have a service life from 60 years for Pennsylvania soft vein slate (hard vein has not been quarried since the 1950s) to more than 125 years for New York and Vermont slate, to more than 175 years for Virginia Buckingham slate. Weathering of some slates' high calcium carbonate content can discolor slate; high iron oxide concentrations can lead to permanent discoloration due to the effect of acid rain. Some manufacturers will provide long-term warranties against delamination and softening due to freeze-thaw cycling.

In the last decade the increased interest in slate for use in high-end housing and restoration projects has led to the reactivation of some dormant quarries and the importation of tile from Spain, England, Newfoundland, Canada, and increasingly from countries with low labor costs such as Brazil, China, and India. The quality of imported tiles varies considerably depending on the specific quarry in the country of origin. Slate is graded largely according to hardness under ASTM C406-89. The grade normally specified for housing is S1. The other grades, S2 and S3, are typically specified for utility buildings. When specifying imported tile, it is advisable to order from a reputable U.S. distributor to avoid substandard material. Slate is also classified into two categories: weathering, which can change color as the slate is exposed to the elements, and nonweathering (unfading), which will not discolor significantly.

TECHNIQUES, MATERIALS, TOOLS

1. REPAIR EXISTING SLATE ROOFS.

As a rule, most slate roofs last about 75 to 100 years. A great number of slate roofs built in the early 1900s are likely to need reroofing due to deterioration of the slate, the flashings, the fasteners, or a combination of all three. The condition of slate roofs can be preliminarily assessed from the ground with the use of binoculars. Critical areas to assess for missing, chipped, cracked, or slipped slates include the ridge, each horizontal row, valleys, and where the roof changes direction. When evaluating a slate roof, note previously repaired areas. Some of these old repairs may not have been done correctly and may need to be reworked. Fasteners and flashings should also be surveyed. Check flashings at valleys, chimneys, dormers, vent pipes, and other roofing protrusions. Check also for the condition of gutters and leaders (on a rainy day for best results). Inspect the attic for structural distress, water stains, rot, or other indications of problems. If there are only a few damaged slates, and if the roof is less than 50 years old, it is likely that replacement slates are available, either salvaged or new, that can match the color of the existing roof. Slate replacement is usually a project for an experienced roofer, especially if the damaged slates are in the roof's field—slate can be easily broken by roof traffic, and the surface is dangerous to walk on. Inexperienced contractors may recommend the removal of a slate roof when, in fact, it is still serviceable. In some instances, even when substantial areas have suffered from neglect, slate can be removed, flashings and underlayment repaired, and the salvaged material reinstalled with new slate.

ADVANTAGES: Repairs are cost-effective over small areas. In some instances tiles can be removed and reused.

DISADVANTAGES: Experienced contractors are required. It is difficult to estimate costs in advance of doing repair work. Repairs over large areas can be very costly. Existing slates may be difficult to match. If large areas need replacement, or if serious leaks develop, postponing replacement may cause damage to the building's structural elements or finishes.

2. REPLACE EXISTING SLATE ROOF WITH NEW SLATE.

Slate is available in a range of colors from gray, blue-gray, black, green, deep purple, red, and variations of these shades, depending on where the material was quarried. Typically available in 1/4" and 3/8" thickness, slates can be obtained in up to 2" thickness for high-end housing where deep shadow lines are desired. The material is heavy (weighing approximately 7 to 8.5 lb/ft^2) and brittle, requiring careful handling. It is installed by hand using copper or stainless-steel roofing nails on minimum 3/4" roof sheathing over trusses or rafters at 16" o.c. Slate can be installed directly over sheathing, on spaced sheathing (with a moisture barrier), or on battens over sheathing. Most manufacturers supply installation details based on industry standards developed in the 1920s by the now defunct National Slate Association. A new, comprehensive and authoritative guide to installation techniques and materials, *The Slate Book*, is well regarded by industry representatives (see Further Reading). Slate is available in a number of patterns, sizes, shapes, and thickness. A standard layout is shown in Fig. 3.10.1. and a custom "French" or hexagonal shape in Fig. 3.10.2. The standard layout ensures that a single slate is overlapped by two other slates (Fig. 3.10.1). An economy installation employing an absolute minimum overlap and asphaltic underlayments interlaced between slates is sometime encountered (Fig. 3.10.3). This system depends on the underlayment lasting the life of the roof, which is usually unrealistic. An important detail in the design of slate roofs is the use of a cant strip at the eaves to elevate the starter course into the same plane as the rest of the slate on the roof. Without the cant strip the slate will not lie properly and will break. New slate products include a ridge vent

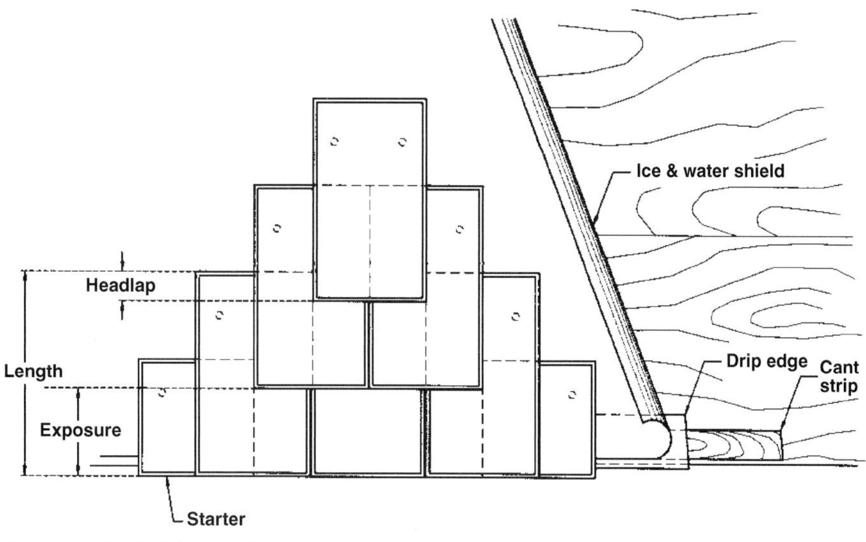

FIGURE 3.10.1 STANDARD LAYOUT

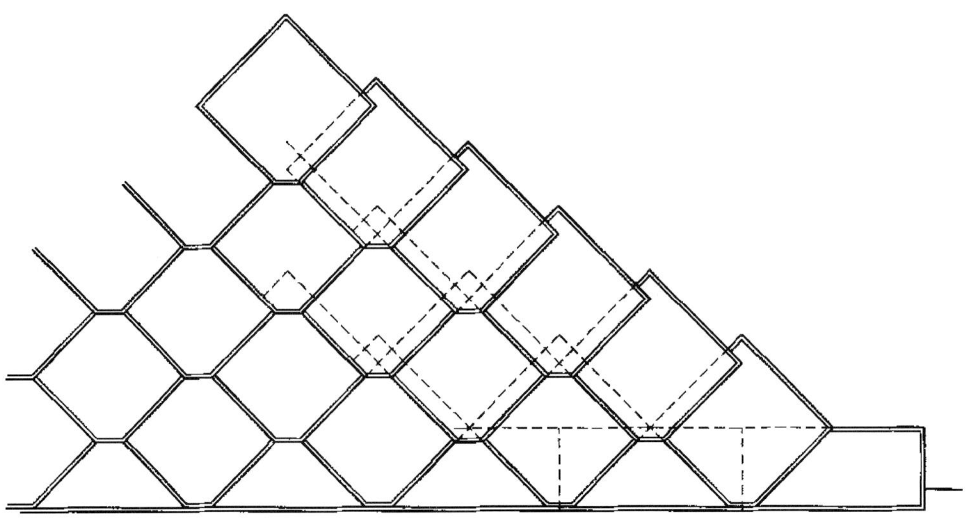

FIGURE 3.10.2 "FRENCH" OR HEXAGONAL LAYOUT

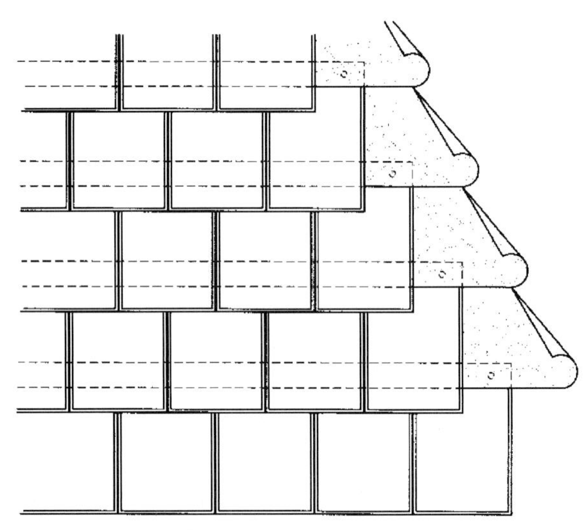

FIGURE 3.10.3 ECONOMY INSTALLATION

designed especially for slate (Top Slate™), manufactured by Peterson Aluminum (Fig. 3.10.4), and a new attachment and ventilation system, Fastrack for Slate™, promoted by Slate International, Inc. (Fig. 3.10.5).

ADVANTAGES: Very long service life makes slate cost-effective on a life-cycle basis, considering the cost of repeated stripping, removal, and reinstallation of other materials with shorter life spans. Slate will permanently add to the quality and character of a house and is low maintenance under normal conditions. Individual broken slates are relatively easily replaced by experienced roofers. Noncombustible, impervious to rot, mold, and mildew.

DISADVANTAGES: The initial material and labor cost is high. Installation prices vary widely depending on availability of qualified roofers and competitive bidding environment. Experienced installers are required. The roof structure may require strengthening to support slate. Some imported slate may be of lesser quality than domestic material, without warranties against fading, discoloration, or delamination. Slate is heavy and brittle, and sharp edges make handling difficult. Corners of slate tiles can be easily chipped. The slate color on existing roofs may be difficult to match.

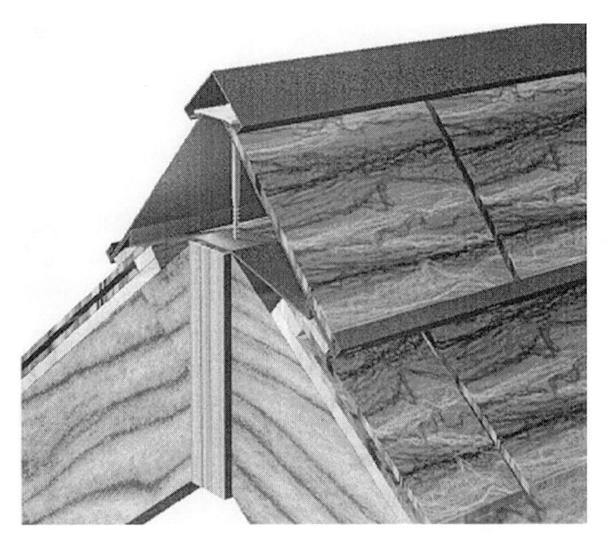

FIGURE 3.10.4 TOP SLATE™ RIDGE VENT

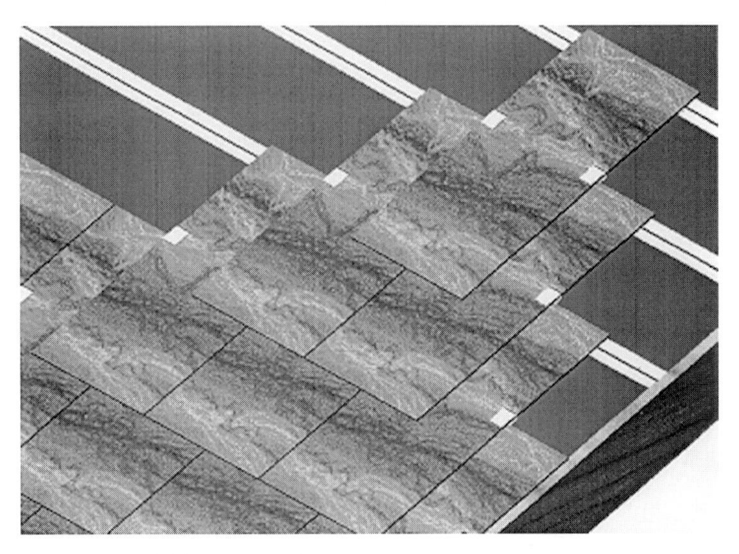

FIGURE 3.10.5 FASTRACK™ FOR SLATE

3.11 CLAY, CONCRETE, FIBER-CEMENT, COMPOSITE TILE

ESSENTIAL KNOWLEDGE

Tile roofing accommodates various building traditions and climatic conditions, and now accounts for over 8% of the residential steep-slope roofing market in the United States for new construction and about 3% for reroofing. Concrete tile, used mostly in the southeast, the southwest, and California, now dominates the tile roofing market due primarily to its lower cost. Concrete tile does not have to be baked in a kiln as does clay tile, and the materials for production are more readily available. Fiber-cement and composite shingles are also gaining acceptance because of their lighter weight and lower cost, although some manufacturers have not found these products profitable and have discontinued them.

Roofing tile is differentiated by its shape and composition. Other important characteristics are breaking strength, absorption or porosity, resistance to freeze-thaw cycles, joining methods, and installation details. Clay and concrete roofing tiles are generally available in two shapes: profile and flat. Profile tile can be further divided into pan and cover, S-tile, and interlocking (Fig. 3.11.1). Flat tile can be subdivided into interlocking and noninterlocking. Interlocking is preferred in areas with heavy rains and snow, because it is inherently a more weathertight system.

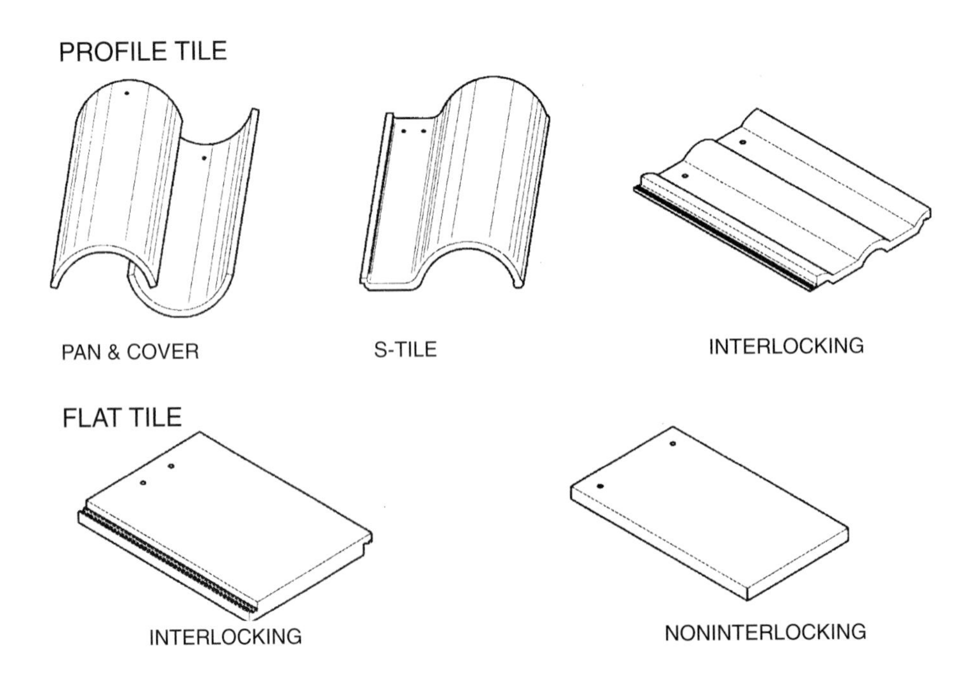

FIGURE 3.11.1

The porosity of roof tiles is very important in climates with a repetitive freeze-thaw cycle. The more porous a roof tile is, the more water it will absorb. Water that freezes in a roof tile can cause the material to spall or crack. According to ASTM Standard C1167 there are three grades of resistance to frost: Grade 1 "Provides resistance to severe frost action"; Grade 2 "Provides resistance to moderate frost action"; Grade 3 "Provides negligible resistance to any frost action." The selection of a tile should reflect an appropriate grade. Some fiber-cement tiles have delaminated in climates with severe freeze-thaw cycles, and most manufacturers will not warrant their products in these areas. The best indication of a roof tile's durability is a service record of similar tile used in a comparable climate.

The roof installation method should be based upon roof slope, type of roof deck or substrate, climate, seismic considerations, local building codes, and manufacturer's recommendations. Methods include lug-hung, bedding in mortar, nails, screws, wire, clips, and adhesives. Nails are the least expensive and most widely used installation system. Adhesives outperform mortar bedding in hurricane-force winds and are used extensively in southern Florida. They cost the same as two nails and a hurricane clip, but installation is faster. Adhesives are new on the market so little field knowledge of their durability exists.

TECHNIQUES, MATERIALS, TOOLS

1. REPLACE EXISTING CLAY OR CONCRETE TILES.

The condition of tile roofs can be preliminarily assessed from the ground with the use of binoculars. Critical areas to assess for missing, cracked, or slipped tiles and missing mortar include the ridge, each horizontal row, valleys, and where the roof changes direction. Fasteners and flashings should also be surveyed. Check flashings at valleys, chimneys, dormers, vent pipes, and other roofing protrusions. Check also for the condition of gutters and leaders (on a rainy day for best results). Faulty gutters and leaders will direct water onto roof and wall surfaces which were not designed for high water flows. These areas should be inspected carefully for damage and deterioration. Inspect the attic for water stains, rot, or other indications of problems. If there are only a few damaged tiles, and if the roof is less than 100 years old, it is likely that replacement tiles are available and individual tiles can be replaced. This is usually a project for an experienced roofer, especially if the damaged tiles are in the roof's field, because tiles can be easily broken from roof traffic and the surface is dangerous to walk on. The isolated replacement of a few tiles is relatively easy, but for more extensive work, a contractor experienced in the installation and maintenance of tile should be engaged (suppliers of roofing tile can frequently be a good referral source). Inexperienced contractors may recommend the removal of a tile roof when, in fact, it is still serviceable. In some instances, even when substantial areas have suffered from neglect, tiles can be removed, flashings and underlayment repaired, and the tiles reinstalled with new tiles placed in areas where they are less visible.

ADVANTAGES: Repairs require more skill than other roof materials, but are cost-effective over small areas. In some instances tiles can be removed and reused.

DISADVANTAGES: Experienced contractors are required. Costs are difficult to estimate in advance of doing repair work. Repairs over large areas can be very expensive. Existing tile may be difficult to match.

2. INSTALL NEW CLAY ROOF TILE.

Because of the long service life of clay tile, it is recommended that a high-performance underlayment be installed. Clay roof tiles come in a variety of colors in both earth tones and glazed primary colors. Unlike some other roofing products, clay roof tiles maintain their color over time. Celadon Ceramic Slate produces a clay tile that weighs 580 lb per square. This product looks like slate and costs around $220 to $260 per square. Costs are a function of tile size and the manufacturer's production volume and processes. On a per-square basis, smaller tiles cost more to purchase and install than large tiles.

ADVANTAGES: Clay roof tiles are a traditional roofing material. They come in a variety of types and styles (Fig. 3.11.2). They are perceived as a high-end, quality construction material; are long lasting and virtually maintenance free; are fireproof and impervious to insects and rot; will withstand hurricane winds better than other roofing products; and many resist the effects of freeze-thaw cycles.

DISADVANTAGES: Clay roof tiles are more difficult to install than other roofing products. They are not recommended for roofs with less than a 4-in-12 pitch. They are heavy and can weigh anywhere from 520 to 1,250 lb per square. Additional roof supports may be needed for existing roofs before clay roof tile is installed. Clay roofing tile has a high first cost ranging from $180 to $1,800 per square. Because of the limited number of clay roofing tile manufacturers, the material may have to be shipped long distances, increasing costs dramatically.

3. INSTALL NEW CONCRETE ROOF TILE.

Concrete tile makes up about 6% of the U.S. residential steep-slope roofing market, with California representing the largest market. Studies show that concrete tile life-cycle costs are significantly less than other roofing products. Because of the long service, it is recommended that a high-performance underlayment be installed. The southeast is particularly vulnerable to the underlayment cracking due to high levels of heat and humidity. Concrete tile comes in a variety of shapes, colors, and textures, simulating clay roof tile (Fig. 3.11.3) as well as wood shakes and slate. The material typically weighs 1,000

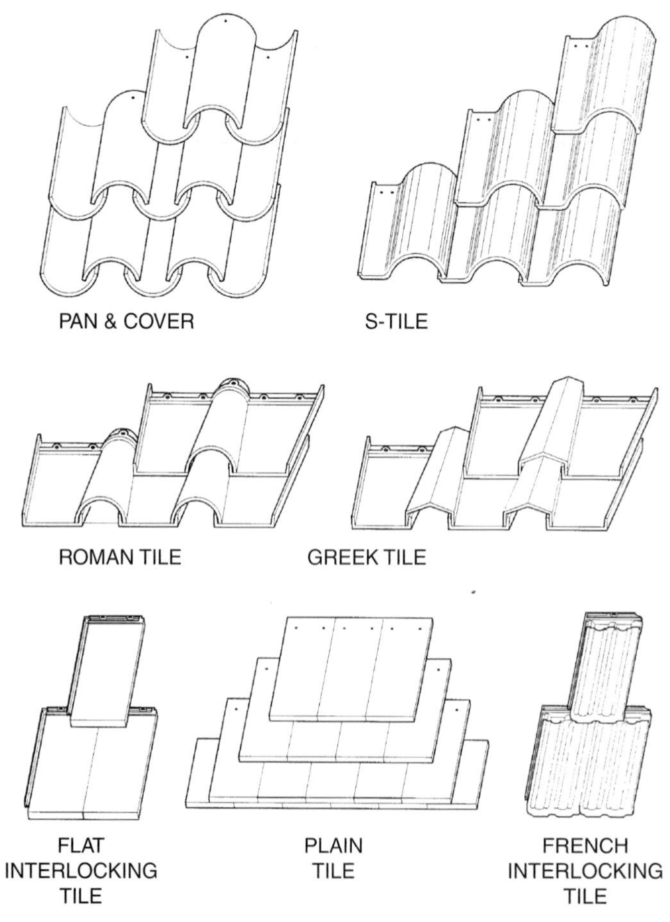

FIGURE 3.11.2 CLAY TILES

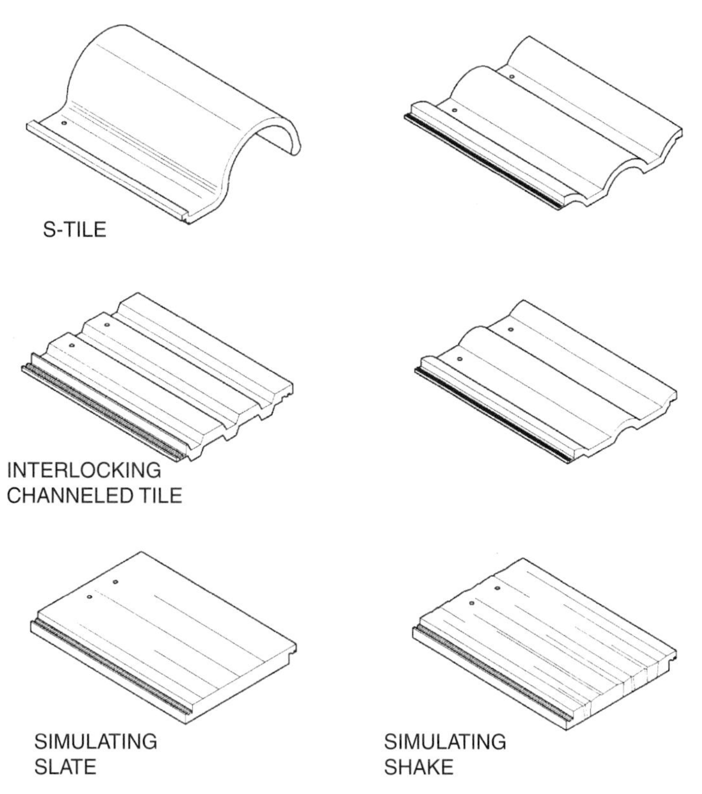

FIGURE 3.11.3 CONCRETE TILES

lb per square, although Westile's FeatherStone® product weighs 690 lb per square (costs 30% more than its standard tile).

ADVANTAGES: Concrete tile is generally less expensive than clay tile. It is long lasting, estimated 40 to 50 years in California and the southwest, and 20 to 30 years in the southeast. It is fireproof and can withstand insects, rot, and hurricane winds better than other roofing products. Some products are designed to resist the effects of freeze-thaw cycles.

DISADVANTAGES: Concrete tile is more difficult to install than other roofing products. It is not recommended for roofs with less than a 4-in-12 pitch. It has a high first cost, three times the cost of asphalt shingles and 150% that of cedar roofing. Because of the limited number of concrete tile manufacturers, the material may have to be shipped, raising the cost dramatically. Existing roofs may require additional support. Sealant-covered roof tile will change from a glossy to a mat finish as the sealant wears off.

4. INSTALL NEW FIBER-CEMENT ROOF TILES.

Fiber-cement shingles are made from a blend of portland cement, synthetic or natural wood fibers, and, on occasion, lightweight aggregate. The material can simulate natural slate and wood shakes. One product, Cembrit B7, is made in 24" × 93" panels that simulate clay barrel tiles. There is conflicting information about which manufacturing process produces the most durable product. Some manufacturers claim that the steam-curing autoclave process increases the strength and durability of fiber cement while greatly reducing water absorption. Others claim this creates a more brittle product than air-cured fiber cement, which is more likely to break from seismic stresses or when walked on. Air-cured products are pressed to increase density and therefore reduce absorptivity. Fiber-cement roofing is available with a manufacturer's warranty of from 25 to 60 years. Some manufacturers are finding that fiber-cement production is not profitable. GAF Materials Corporation has recently discontinued sales of its roofing product, UltraSlate®, and American Cemwood Corporation has also discontinued its fiber-cement roofing products, Cascade® and Royal® shakes and Pacific® slates.

ADVANTAGES: Fiber-cement shingles are lighter and can be less costly than clay or concrete roofing tile. Some products can be installed over asphalt shingles. Fiber-cement shingles will not rot, split, or curl. The material can be cut with a saw and fastened with conventional fasteners. These shingles simulate quite accurately the texture of shakes and especially slate.

DISADVANTAGES: Fiber-cement shingles are more porous than cement and clay tiles and may crack or delaminate in repeated freeze-thaw cycles. The appropriateness of specific products for use in cold climates should be confirmed with the manufacturer. Repeated dry-wet cycling could also cause some products to deteriorate. Fiber-cement shingles may appear different from the products they simulate. Some fiber-cement products are susceptible to breakage from foot traffic and should not be used for recover applications. They are not recommended for slopes with less than 4-in-12 pitch.

5. INSTALL NEW COMPOSITE ROOF SHINGLES.

In a process parallel to the development of fiber-cement shingles, some manufacturers now make products that are composites of various materials. One such product from Owens Corning, MiraVista™ shake shingle, is made of slate and clay reinforced with fiberglass and bonded under pressure with polymeric resin.

ADVANTAGES: These shingles are configured with a rough texture, an uneven shape, and detailing that is quite similar in appearance to wood shakes. They are lightweight and available in natural, nonfading colors that will weather to a natural looking weathered tone. Lightweight, Class A rating shingles, will not absorb water. The shingles install like wood shakes and are competitively priced with fiber-cement shingles.

DISADVANTAGES: These shingles are a new product without a history of durability. They may not be available nationally and may appear different from the products they simulate.

6. INSTALL NEW PLASTIC COMPOSITE ROOF TILES.

A number of new plastic composite products have become available recently, including Eco-Shake®, made from recycled pallet wood, shower curtains, and plastic bottles; Authentic Roof Shakes® made from 100% postconsumer waste; and Perfect Choice® 21" × 40" molded panels of Noryl plastic made to look like 13 individual hand-split shakes.

ADVANTAGES: These tiles are very lightweight and easy and quick to install with conventional fasteners. Their appearance is reasonably close to shakes and tiles. They are wind and hail resistant, and their recycled content makes them environmentally sound.

DISADVANTAGES: Most materials have class C ratings and are new and untested. Special cutting equipment may be required. These tiles may not have model code approvals and may appear different from the products they simulate.

3.12 GUTTER AND LEADER SYSTEMS

ESSENTIAL KNOWLEDGE

The purpose of gutter and leader (downspout) systems is to collect rainwater from the roof and direct it away from the building foundation by means of leader extensions, splash blocks, or underground drain lines. An improperly functioning gutter and leader system can contribute to water and ice backing up against fascias and under roof shingles, can damage soffits, and can discolor or deteriorate siding materials. Faulty gutters and leaders can also lead to soil erosion adjacent to buildings and serious water and foundation displacement problems in basements and crawl spaces. Unfortunately, gutter systems take more abuse from extreme weather conditions, particularly ice and snow, than any other component of the building envelope. They are also subject to damage from ladders and being stepped on, as well as from falling tree limbs and debris.

Gutters should be sloped a minimum of 1" for every 40' of run. Standing water may indicate a sagging or incorrectly pitched gutter. Gutters are often sized according to the roof area they drain. Five-inch-wide K-style gutters are the residential industry's standard. Six-inch-wide K-style gutters are used for larger roofs. Half-round gutters are typically sized 1" wider than K-style to provide the equivalent capacity. Therefore, 6" half-round gutters are equivalent to 5" K-style. Four-inch-wide K-style and 5" half-round gutters (with equivalent capacity) are rarely used except for small roofs. Wider gutters may be required for certain hard surface roofing materials, such as slate and tile, or used on steeply pitched roofs, to prevent water from shooting over the gutter. Gutters should be positioned tight against roofing materials and the fascia. In heavy snow areas, snow guards should be used to prevent gutter tear-off.

Vertical leaders (downspouts) are used to capture and distribute rainwater to storm drainage systems or, by means of splash blocks, to areas away from the building's foundation walls to prevent the buildup of water in the soil and possible resulting structural or basement moisture problems. Leaders are typically rectangular and of the same material as the gutter to prevent destructive galvanic actions. Connections between gutters and leaders and leaders and storm drains require continuing maintenance to assure the drain is free of leaves and debris and the connection has not become loose. Other maintenance points include the connections of leaders to the building. Leader diameters are sized according to the roof area they drain. A rule of thumb used in the industry is that a 2" × 3" leader will suffice for a 600 ft^2 roof and a 3" × 4" leader for a 1,200 ft^2 roof. The typical leader size for a 5" K-style gutter is 2" × 3", but 3" × 4" is preferable because it is less likely to become clogged and is easier to clean out.

Before 1960, most gutters were made of wood or metal in a "half-round" shape. During the 1960s, roll-formed metal gutter technology was introduced that allowed metal gutters to be made lighter and less expensively. Initially available primarily in galvanized steel, roll-formed gutters are now available in copper, aluminum, galvanized steel, and painted steel. Gutter profiles include half-round and square, but the K-style (also called formed, Ogee, or OG) (Figs. 3.12.1 to 3.12.3) predominates because it is visually compatible with recently built housing and adds interest to simple fascia details. Recent developments in truck-mounted roll-forming equipment have allowed gutter installers to form continuous, seamless metal gutters to fit site-measured field dimensions and individual roof profiles. Other recent developments include the use of vinyl (PVC) snap-together gutter systems used primarily

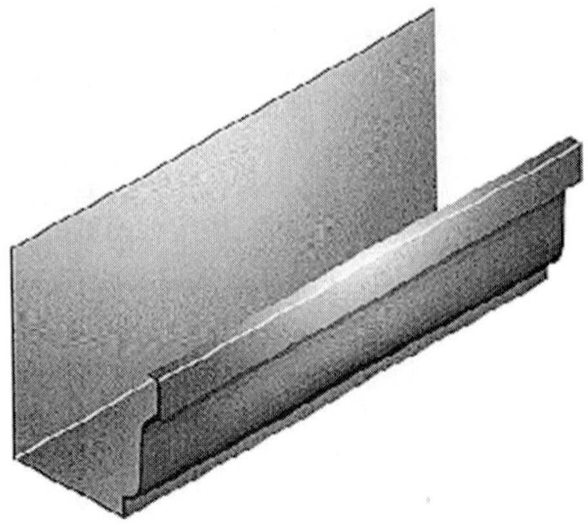

FIGURE 3.12.1 SQUARE BOX GUTTER

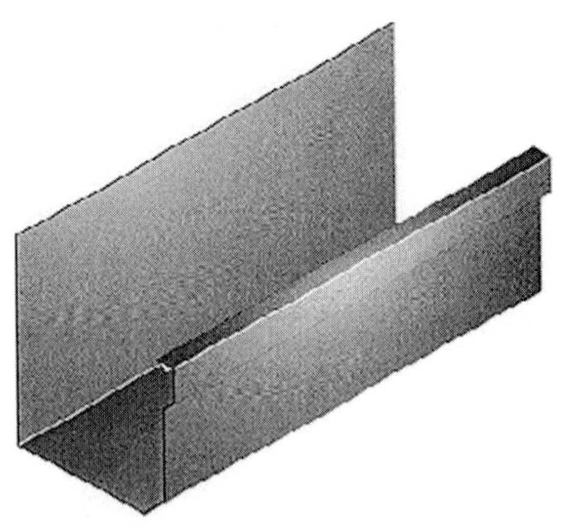

FIGURE 3.12.2 SQUARE BOX GUTTER

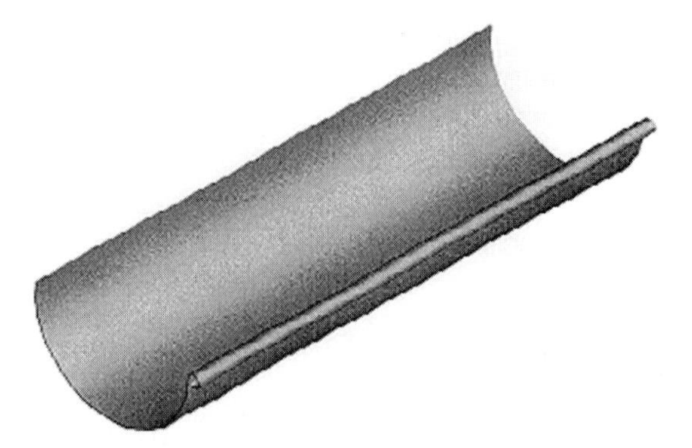

FIGURE 3.12.3 HALF-ROUND GUTTER

in the rehab and do-it-yourself market because of their ease of installation. Another relatively new product, Rainhandler™, is claimed to eliminate the need for gutters by deflecting rain away from the building (Fig. 3.12.4). Each gutter system has its own special characteristics and uses.

TECHNIQUES, MATERIALS, TOOLS

1. REPAIR EXISTING GUTTER SYSTEMS.

Other than keeping gutters clean of leaves, pine needles, and debris, gutter maintenance usually involves refastening hangers that have become unfastened, repairing broken hangers, or adding new hangers where hanger spacing was excessive. Hangers come in a variety of types, including hanger and strap, hanger and bracket, spike and ferrule, and concealed (Figs. 3.12.5 to 3.12.8). The spike-and-ferrule hanger, because it fastens through the gutter, is more apt than other hanger types to pull out of the fascia due to expansion and contraction of the gutter. Concealed fasteners are becoming increasingly popular because they are easy to install, are not visible, are relatively strong, allow the gutters to expand and contract, and come in a variety of sizes and configurations with different screw and nail attachment details. Other maintenance and repair items include keeping gutter and downspout screens in place, refastening leaders, and maintaining splash blocks or connections to underground drains. Gutters deteriorate over time—wood gutters need to be oiled, metal gutters may need local repairs. At some time, usually when new roofing is required, the gutters may have weathered to the point where they are not functioning, are unsightly, or have significantly deteriorated. The choice of a replacement will be dictated by the appearance, value, physical characteristics, and age of the existing house.

ADVANTAGES: Repairs are generally easy to make and are cost-effective if the damage is localized.

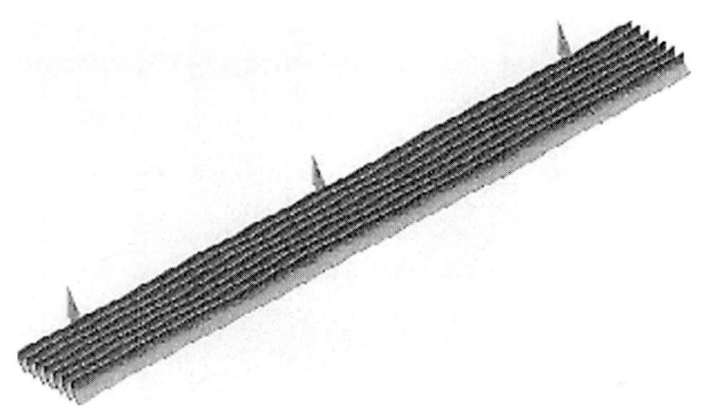

FIGURE 3.12.4 RAINHANDLER™

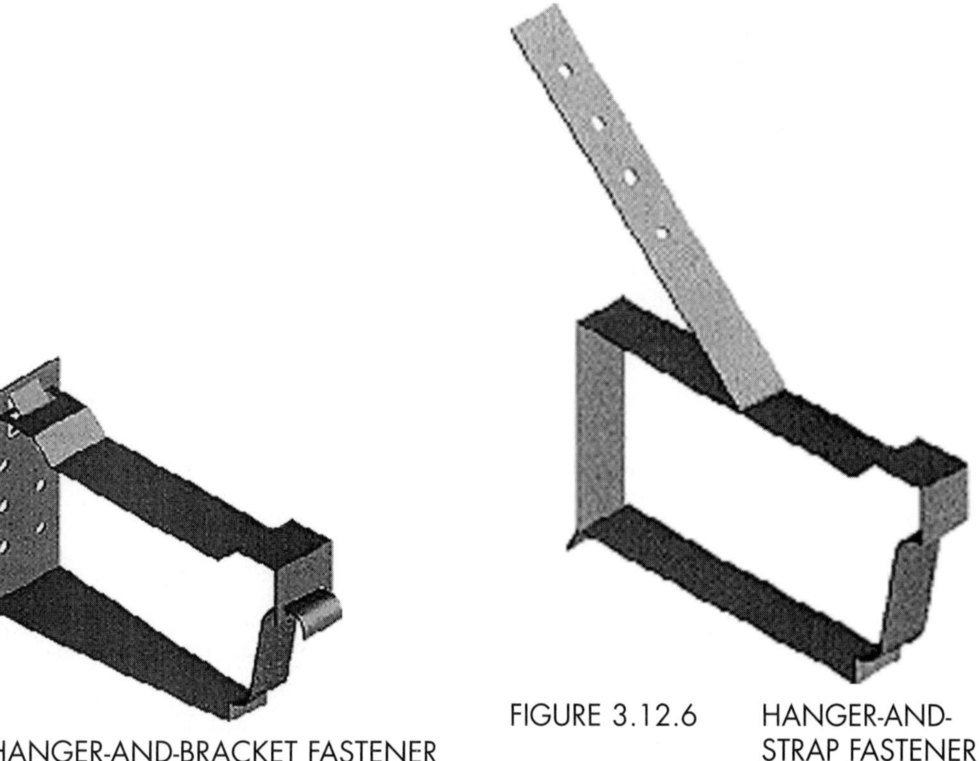

FIGURE 3.12.5 HANGER-AND-BRACKET FASTENER

FIGURE 3.12.6 HANGER-AND-STRAP FASTENER

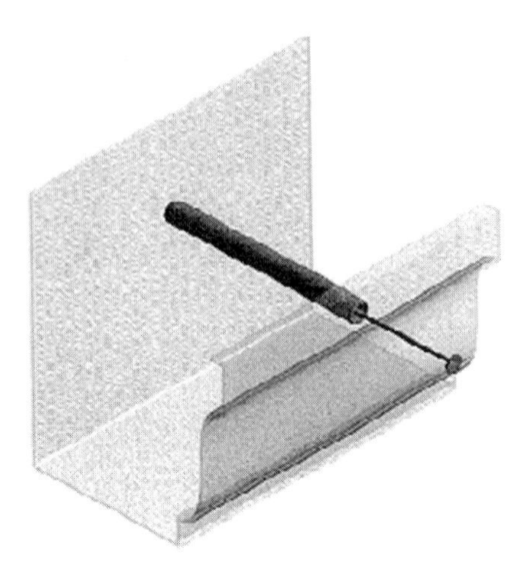

FIGURE 3.12.7 SPIKE-AND-FERRULE FASTENER

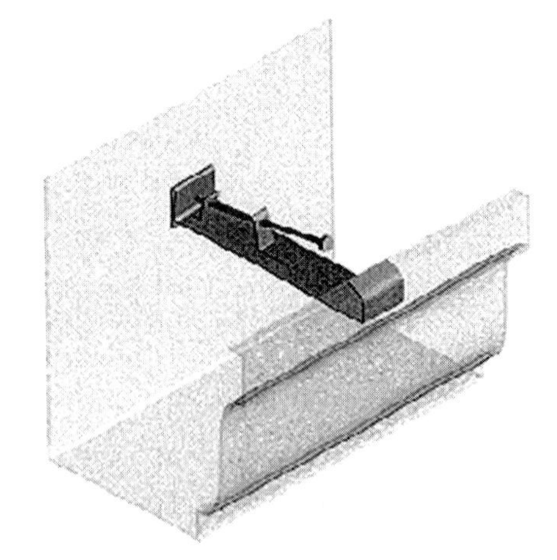

FIGURE 3.12.8 CONCEALED FASTENER

DISADVANTAGES: When gutter systems show widespread deterioration and become unsightly, they should be replaced in their entirety.

2. INSTALL NEW WOOD GUTTERS AND LEADERS.

Wood gutters are still a favorite for some traditional homes. Typically available in a 4" × 6" size in clear fir, gutters can also be milled from cedar or redwood. One supplier, Blue Ox Millwork, offers redwood gutters in any one of 16 standard patterns or can customize to suit. The company recommends a scarfed joint with both sides of the splice beveled 1/4" on the inside to allow a depression to be filled with butyl or other high-quality sealant. The sealant can be cut out and replaced when it fails (Fig. 3.12.9). The interior leg of a gutter should be 1/8" to 1/4" higher than the outside leg in order to spill water outward. Blue Ox recommends that, in lieu of lining gutters, a product called Chevron Shingle Oil be used in the gutter's interior. This product will prolong gutter life by reducing the weathering while at the same time allowing the gutter to breathe. Other products, including single-ply roofing membranes, are available that can be used to line new or deteriorated wood gutters. One of these, Deck Seal®, an SBS rubberized asphalt membrane, is offered in 1.5", 3.5", and 6" rolls (see Product Information). Leaders, which can be copper, black iron, or PVC, are attached to the gutter by means of a nipple screwed into the gutter's interior. Some traditionalists recommend boxing (enclosing) the leaders with wood to conceal the metal leader.

ADVANTAGES: Wood gutters are appropriate for more traditional housing or where the shingle style predominates. Wood is a strong, straight material that provides a crisp edge detail. Properly maintained, fir can last 50 years or more and redwood twice that. The front face can be painted to match adjacent trim.

DISADVANTAGES: Wood gutters may not be readily available in some areas. They are more costly initially than other alternatives. Redwood gutters are approximately $10 to $12/lineal ft. for 4" × 4" (used on West Coast) and $16 to $18 for 4" × 6" (plus $175 set up). Fir gutters are approximately $10/lineal ft. Gutters and leaders require maintenance to prevent wood from drying out and checking. Wood is heavier than other gutter materials and is not typically the choice for noncustom housing.

3. INSTALL NEW STEEL GUTTERS AND LEADERS.

Steel gutters and leaders are available in a variety of styles, including K-style, square (box gutters), and half-round. Available materials include electroplated and hot-dipped galvanized and Galvalume® (approximately 55% aluminum, 45% zinc by weight). Finishes also include plain galvanized, baked

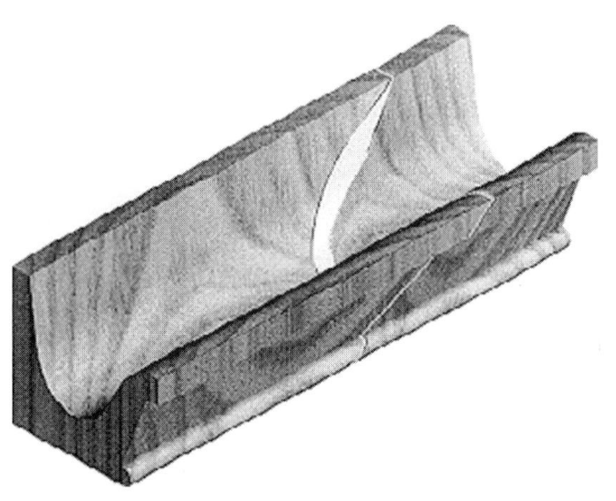

FIGURE 3.12.9 SCARFED JOINT

enamel, modified siliconized acrylic and polyester, and fluoropolymer coatings such as Kynar. Gauges run from a lightweight 28 ga to a heavier 24 ga, with 26 ga being the most typical. Gutter assemblies include "stick" systems of components sold through lumberyards or home centers (Fig. 3.12.10) with typically 10' to 20' gutter lengths, for installation by homeowners or small contractors, as well as seamless systems provided by installers with truck-mounted roll-formers. While both stick and seamless systems can work satisfactorily, seamless systems have fewer joints, are apt to leak less, and are faster to install. Stick systems are, however, apt to be less expensive on single houses or small projects.

ADVANTAGES: Steel is stronger than aluminum at equivalent thickness and contracts one-half as much. Some newer coating systems allow limited guarantees up to 50 years. Steel is popular in northern states with snow and ice conditions.

DISADVANTAGES: Some galvanized finishes have a limited life span (5 to 10 years). Hot-dipped galvanized gutters are preferred over other galvanized finishes, but are not recommended for maritime environments unless the coil has been precoated. Galvanized finishes are rarely cleaned or primed properly, and when field painted, the paint can fail prematurely.

4. INSTALL NEW K-STYLE ALUMINUM GUTTERS.

Aluminum gutters and leaders are by far the most popular gutter systems. They are available in a variety of styles, including K-style, half-round, and a K-style modified to replace a fascia board (Fig. 3.12.11). Coatings include baked enamel, polyester, and acrylic. Gutter thicknesses run from a lightweight 0.019" to a heavier 0.032", often specified on higher-end housing. The most typical thickness is 0.027". Thicknesses below 0.027" are sometimes used on low-end housing and are available through price-sensitive home centers for the do-it-yourself markets. Aluminum gutters are available in stick and seamless styles, with the latter being more common, especially on larger projects. Stick systems are frequently offered in greater thicknesses than seamless (0.032" versus 0.027") and are stronger. The benefit of seamless systems is the lack of joints (which are points of potential leaks if not maintained). Concealed hangers are becoming more popular because they are invisible and allow the gutter to expand and contract. Expansion and contraction at bracket support types that surround the gutter can lead to discoloration and wear at those points. Spike and ferrule fittings are sometimes unsightly due to sloppy installation, and can pull out of the fascia from the significant movement of aluminum resulting from temperature swings.

ADVANTAGES: K-style aluminum gutters are widely available, low cost, and relatively easily maintained. They come in both stick and seamless styles in many color options. They are easily installed. Long-term warranties on coatings are common.

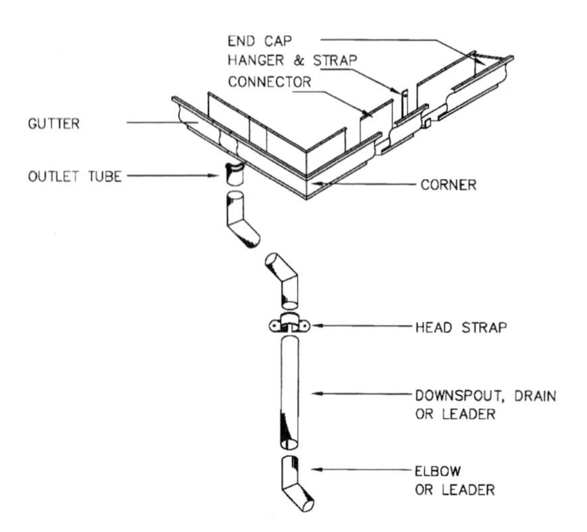

FIGURE 3.12.10 STICK SYSTEM

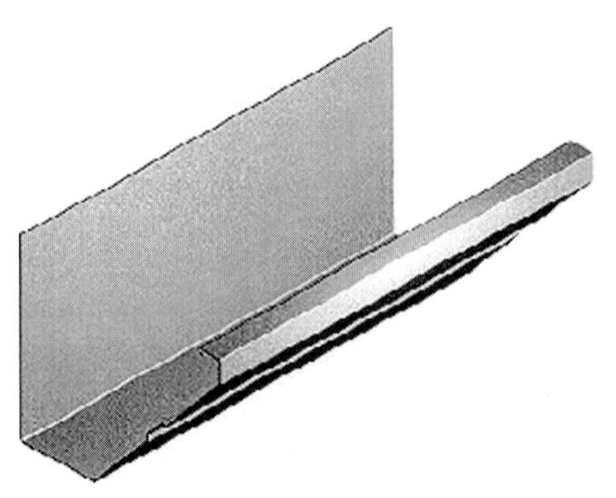

FIGURE 3.12.11 K-STYLE AND FASCIA

DISADVANTAGES: Aluminum expands twice as much as steel and 50% more than copper. It is also less strong than steel or copper and can be dented (especially in lighter thickness) more easily than other gutter materials.

5. INSTALL NEW K-STYLE COPPER GUTTERS.

Copper has been a traditional gutter material for institutional buildings and large houses and continues to be popular for high-end custom housing. Typically specified in 16 oz. or 20 oz., K-type or in half-round styles up to 8" in width, copper is available in stick as well as seamless systems. Copper has also been used as sheet flashing in boxed or built-in gutters (Fig. 3.12.12). Typically allowed to weather naturally, copper develops a blue-green coloration resulting from the formation of a protective copper oxide patina. Lead-coated copper gutters are also available from some suppliers for those preferring a gray color. According to the Copper Development Association, copper gutters are affected by the acids in red cedar shingles only to the extent that the acid may retard patination. Pitting of the gutter will not occur if recommended details are followed (see Further Reading). Copper joints and connections are usually soldered, providing continuous surfaces, which limits the potential for leaks.

ADVANTAGES: Copper gutters are considered to be a quality, premium product and made of an appropriate material for historic preservation projects. Copper patina blends well with many roofing products. Copper gutters are low maintenance, do not require paint, will not rust, and are well suited for maritime environments.

DISADVANTAGES: Copper is significantly more costly than alternative materials. Skilled installers for cutting and soldering of joints and transitions are required. Copper gutters are not a normal choice for other than custom applications.

6. INSTALL NEW HALF-ROUND COPPER AND ALUMINUM GUTTERS.

Half-round gutters were the traditional gutter style on homes built before 1950 and remain a popular choice on historic renovation projects, traditional renovations, and new upscale custom housing. The simple lines of half-round gutters complement heavily textured materials such as slate, shakes, and tiles. Where crown moldings exist in lieu of fascias, half-round gutters are hung from the roof. Where fascias exist, fascia brackets are used to attach gutters. One company, Classic Gutter Systems, sells specially designed oversized 5" or 6" half-round gutters made of heavyweight 20-oz. copper and heavyweight 0.032 aluminum. The company supplies an extremely durable hanging system as well as heavyweight cast brass and aluminum functional brackets in decorative and plain styles (Fig. 3.12.13). These are lag bolted with stainless-steel lags into the fascia board. A full line of accessories, decorative components, and screens is also available.

ADVANTAGES: Half-round gutters are the most efficient gutter shape and least affected by ice and snow. They are appropriate for historic rehabilitation and new construction, both contemporary and traditional. They are attractive and durable.

DISADVANTAGES: Half-round gutters have a higher initial material cost than some other gutter products, approximately $2.25 to $2.75/lineal foot for aluminum and $5.00 to $5.50/lineal foot for

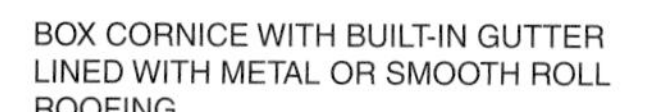

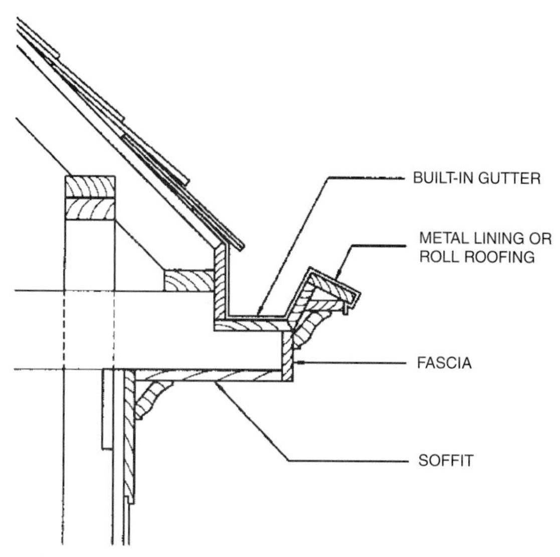

FIGURE 3.12.12 BOX CORNICE WITH BUILT-IN COVER

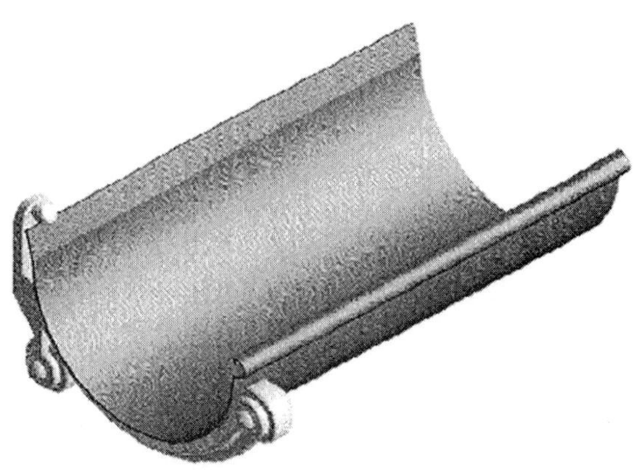

FIGURE 3.12.13 HANGING SYSTEM

copper (not including hardware — direct from Classic Gutter Systems). They are not generally used on noncustom housing.

7. INSTALL NEW VINYL (PVC) GUTTERS.

Originating in Europe, vinyl gutter systems have a small but growing following, largely among do-it-yourselfers. One U.S. company, Plastmo, has been the leading supplier of vinyl gutters in Scandinavia since 1959. Gutters can run the gamut from flexible to quite strong depending on the material's thickness, which runs from 0.062" to 0.089". Its great appeal is its snap-together simplicity and the limited number of tools required to cut and assemble it. Vinyl gutters are available in K-style, contemporary (U-shapes), and half-round styles. One small manufacturer has a half-round gutter system that can be emptied from the ground by means of a hooking device that engages a gutter that rotates on its support brackets allowing leaves and debris to fall out (see Product Information).

ADVANTAGES: Vinyl gutter systems are strong and resist dents. They will not rust, chip, or peel and are lightweight and low maintenance. They have been in use for over 35 years and are available direct from some manufacturers and at most home centers. They are effective in corrosive environments, are low cost, are easy to assemble, and can be painted if necessary.

DISADVANTAGES: Gutter stock is only available in 10-ft lengths. They are visually different from more traditional gutter systems. This is basically a do-it-yourself technology, with most gutter contractors committed to aluminum and/or steel because of their investment in those systems. They are perceived by some to be a lower-end product and come in a limited number of colors.

8. REPLACE OR PROVIDE GUTTER SCREENS OR GUARDS.

A variety of products are available to retard the buildup of leaves, twigs, dirt, and asphalt roofing granules in gutters. Historically these have been galvanized or vinyl-coated metal or plastic screening material, but recently a host of "gutter guard" products has emerged. These products range from slotted or perforated vinyl or metal extrusion to nylon mat filters. Additionally, some manufacturers and distributors of roll-forming equipment and metal coils have produced gutters that include protective hoods. One manufacturer, Englert, Inc., fabricates and sells through franchised dealers a 0.032" aluminum gutter called Leaf Guard™ that incorporates a hood (Fig. 3.12.14). The Leaf Guard™ system is guaranteed against the need to be cleaned for 20 years (or the dealer clears it). The system is claimed to be strong and to resist snow and ice damage. The cost is over three times that of conventional gutter systems. Another new system developed by Knudson Manufacturing, Inc., called K-Guard™, is made from 0.027" aluminum, has a similar profile to Leaf Guard®, but has a snap-in top that can be opened for cleaning (Fig. 3.12.15). The other gutter guard systems are typ-

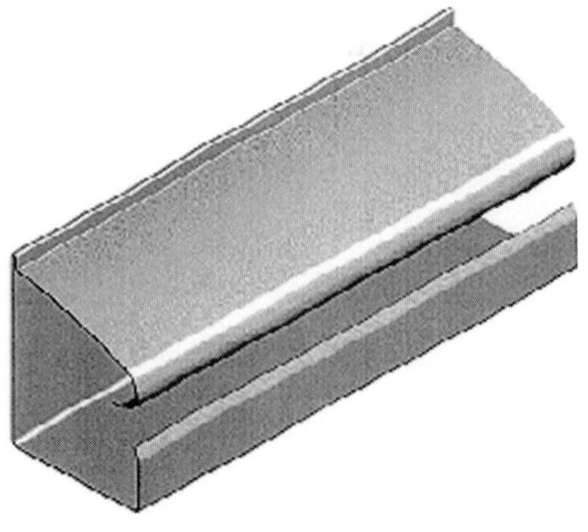

FIGURE 3.12.14 LEAF GUARD™

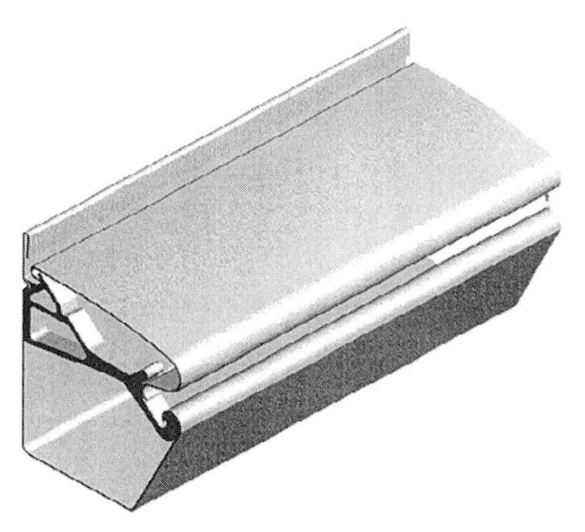

FIGURE 3.12.15 K–GUARD™

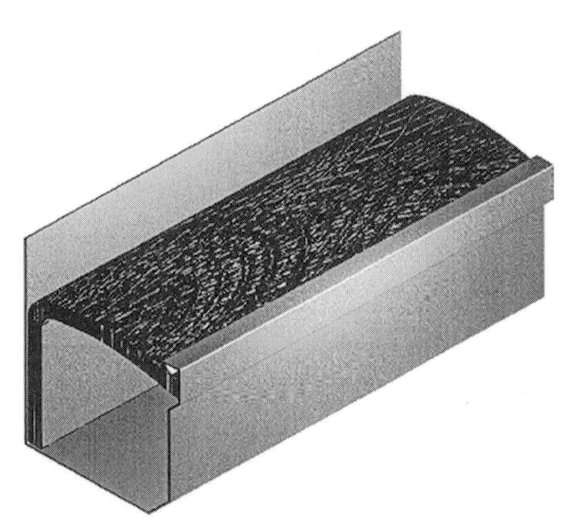

FIGURE 3.12.16 GUTTER GUARD

ically aluminum "hats" that cover the gutter and allow water to run down through slots at the gutter's front edge. One manufacturer, DCI Products and Services, produces a nylon mat that covers the top of the gutter trough but allows water through (Fig. 3.12.16).

ADVANTAGES: Gutter screens and guards reduce the need and danger of inspecting gutters. They generally keep leaves, twigs, and other debris out of gutters. They perform satisfactorily in light to moderately heavy rains. Some products are guaranteed for 20 years against debris buildup. They are designed to be self-flushing with 3" × 4" leaders.

DISADVANTAGES: Gutter screens and guards do not keep out all small twigs, dirt, asphalt roofing granules, and pine needles. They perform less well in very heavy rains or on steep roofs. They may not be considered visually appealing, and it can be difficult to remove gutter guards to access gutters.

WINDOWS AND DOORS

Chapter 4
WINDOWS AND DOORS

4.1. EXISTING WINDOW AND DOOR OVERVIEW
4.2. GLAZING
4.3. WINDOW FRAMES AND REPLACEMENT UNITS
4.4. STORM WINDOWS AND SCREENS
4.5. SKYLIGHTS
4.6. DOORS AND FRAMES
4.7. CASING AND TRIM
4.8. HARDWARE
4.9. FLASHING
4.10. CAULKING AND WEATHER STRIPPING
4.11. SHUTTERS AND AWNINGS

4.1 EXISTING WINDOW AND DOOR OVERVIEW

Windows and doors are openings in the building envelope that often have multiple and contradictory performance requirements. Windows and doors serve not only as barriers but as a mediator, providing access but preventing entry of the elements, allowing views and ventilation but also protection from the weather. Door and window selection is a balancing of desired objectives including performance, function, appearance, and cost.

Window and door performance, particularly in terms of energy conservation, has progressed significantly in the last two decades as new, better performing materials have been introduced. Issues regarding materials, manufacture, finishes, and performance of doors and windows are similar, and the discussion of such issues within this guide is usually applicable to both.

EVALUATION OF EXISTING CONDITIONS, OPTIONS, AND SELECTION

The focus of this chapter is to restore windows and doors and improve performance by means of new and innovative products. The repair of existing windows and doors, in combination with improvements such as adding a storm or screen unit and insulating the perimeter of the opening, often proves to be the most cost-effective solution. The rehabilitation of a door may simply require the adjustment of a loose hinge or strike plate and the replacement of worn weatherstripping. However, the cost of skilled labor and the conveniences provided by new technologies may justify the use of an entirely new unit or some combination of repair and new components.

New window technology has resulted in dramatic savings and increased comfort for the homeowner with the use of new or reformulated frame materials and glazing products while requiring significantly less maintenance. However, a recent study conducted by the state of Vermont concluded the energy savings realized between a renovated window with a storm unit and a replacement unit, without the benefit of high-performance glass products (such as low-e or spectrally selective) were very similar. Infiltration rates (the exchange of air) between the renovated and replacement units were also comparable. The benefits of replacement will not necessarily be energy savings, but the opportunity to provide a more comfortable, durable window or door with ease of operation and the elimination of a lead paint hazard.

Initially an evaluation should be made as to the extent of repair or replacement required. There are essentially three progressive options in addition to repair and adding storm units, which are discussed separately: a replacement window sash or door, a secondary preassembled unit, and a complete unit replacement. The first two partial replacement methods provide many of the benefits of a new window or door without disturbing the existing frame, trim, or the surrounding surfaces but do not address infiltration (leaks) at the perimeter, often a major source of energy loss and discomfort. A replacement window sash or door requires the existing frame to be in good condition and relatively square. If either of these conditions does not exist, a new sash or door will not operate properly. A secondary frame is suitable for openings that are not square, but these reduce the opening size, and the existing frame opening must be in good condition. Complete replacement provides the opportunity to improve the perimeter insulation as well as to inspect the existing construction for damage. Partial and total replacement units are available in custom sizes. Replacement units are also available in incremental stock sizes, which reduces lead time but often may require in-fill trim to enclose the existing opening.

The selection of a window should include consideration of the appearance, building type, climate, durability, orientation, expected use, and all applicable codes. A single window type may not be applicable to the entire house. Manufacturers have begun to label their windows as orientation and climate specific to achieve optimal performance. Windows that face east or north or sources of noise (such as traffic) generally should have higher insulative values. Windows in either coastal areas or high altitudes must resist higher wind loads, differential pressures, and corrosive elements. A multistory building such as a townhouse or apartment building will require low maintenance, ease of cleaning, and resistance to higher wind pressures. Historic buildings will require a matching appearance with existing materials and profiles.

Codes may require replacement window units to have minimum energy performance, safety and egress requirements, and the ability to withstand natural hazards such as wind. Building codes require safety glass, either tempered or laminated, to be installed where there is a potential for human impact. A replacement window must also comply with egress requirements in the size of the clear opening and the sill height above the floor for sleeping rooms in a home of three stories or less. Utilities, insurance companies, and financial institutions may also provide incentives to homeowners to choose units with better performance or safety in the form of premium savings or reduced rate mortgages. In addition to these factors, door selection should also consider fire-resistance requirements as prescribed by the building code.

Although not generally required of single-family homes, accessibility for the disabled may often be readily provided for doors and windows. Various accessibility regulations govern the design and construction of residential, typically multifamily, buildings. There are several prescriptive requirements, but (ANSI A117.1) *Accessible and Usable Buildings and Facilities,* is the most prevalent. Where local codes differ from the national specification the more stringent requirements should be utilized. There are numerous products available for rehab to make homes accessible (see Further Reading for sources).

WINDOW AND DOOR TYPES

Window types may generally be categorized as either fixed, sliding, pivot, or hinged, with the distinction among the many varieties described by their typical application (Fig. 4.1.1). A hinged window is either a casement, awning, or hopper according to its operation. An individual window unit may have combined properties such as a single hung window that has a fixed and sliding sash, or a projecting window with a fixed and hinged sash.

Door types may be hinged, sliding, pivot, or some variation thereof (Fig. 4.1.2). A hinged door may be described as either a passage, accordion, side hinge folding, or bi-fold. An example of a pivot door, often confused with hinged units, in a residential application is a kitchen door that swings in both directions and does not require the use of a latch. Sliding doors are either bypassing, surface sliding, or pocket sliding.

Doors are also categorized by their method of construction, such as panel, batten, or flush. Traditional wood panel doors are made of horizontal rails and vertical stiles that frame one or more

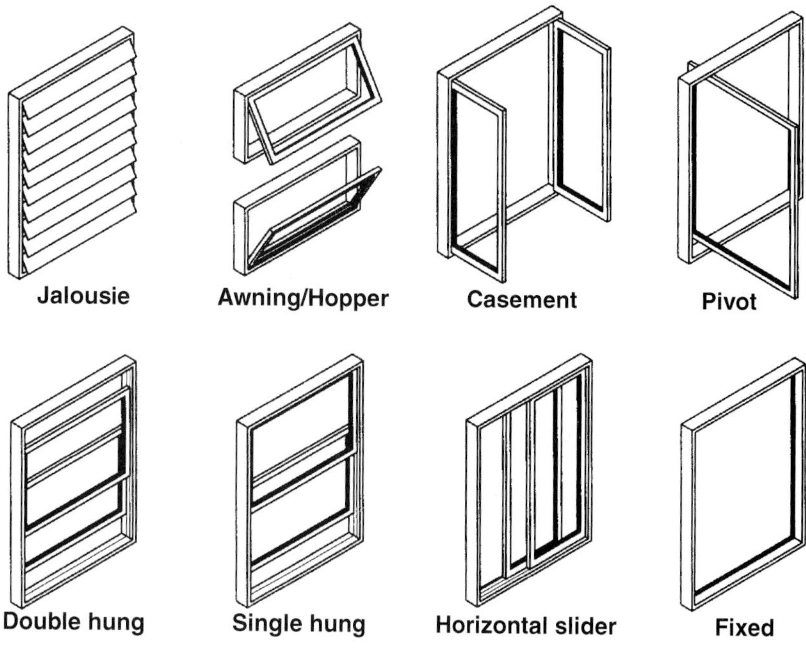

FIGURE 4.1.1 WINDOW TYPES

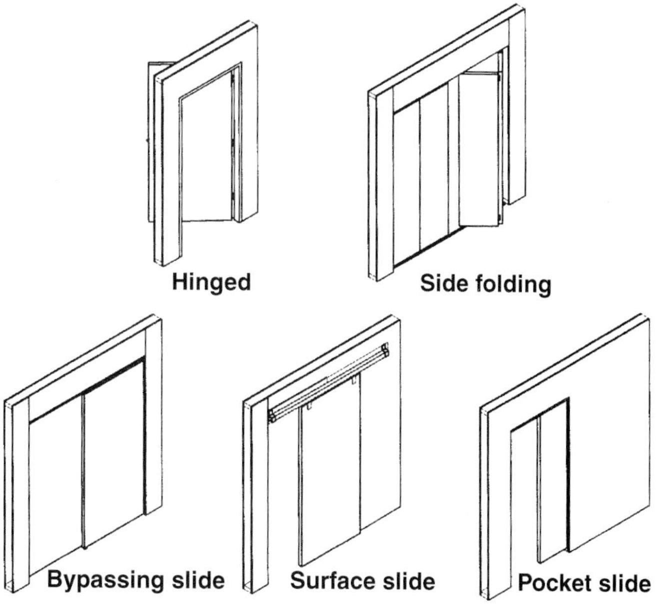

FIGURE 4.1.2 DOOR TYPES

panels. Batten doors are usually constructed of solid lumber in a series of planks that are secured with a board attached diagonally on the surface. Flush doors have interior structural cores covered with a thin surface material. This interior structure may be composed of either rails and stiles with hollow cavities or a solid monolithic material such as rigid insulation, particle board, or engineered (jointed) wood members. The exterior surface may be made of a variety of materials, including metal, plastic laminate, wood (veneer, hardboard, plywood), and fiberglass. The expressive form of a wood panel door can be simulated by flush doors with either formed (negative relief) panels or the addition of built-up (positive relief) materials to achieve similar profiles, textures, and the appearance of wood, often with the ability to be stained.

Door installation previously required the precision skills of a cabinet maker; the door constructed and hung on-site had to be installed absolutely square and plumb in order to operate properly with repeated use. Today, the door and frame are usually preassembled by either the factory, distributor, or lumberyard. Such prehung doors, which may also be predrilled for hardware, have greatly simplified door installation. Prehung doors are also available with two-piece (split) or knock-down frames, with attached trim and adjustable thresholds, for rapid installation in an out-of-square opening. A prehung insert door, similar to a secondary window frame, may be inserted within the frame of an existing door. However, the opening will be smaller, which may not be desirable or code compliant.

Hinged windows and doors, in comparison with sliding units, generally provide a tighter air seal and have less air and water leakage rates because of locking mechanisms, compression gaskets, and stronger frames. They also permit a full opening for egress and ventilation. However, the swing of hinged windows and doors may be an obstruction. Alternatively, bi-fold or accordion-style doors provide access with a minimum of space consumed. Sliding doors consume minimum space but allow only half the width of the opening for clearance. Sliding units typically use a brush-type weather seal subject to wear and tear and a shorter service life. Pocket doors permit the full width of an opening to be used, but may prove inconvenient to operate and are best suited for locations where they are infrequently operated. Ultimately, the performance of a particular unit is determined by the quality of design, construction, and materials—all of which are often difficult to evaluate by visual inspection alone.

Advances in window and door materials over time have balanced the unique qualities of materials for optimum performance. Wood, the oldest window frame material, requires more maintenance than others but is undergoing a transformation with reformulation as a composite and is being used with the protective cladding of aluminum, vinyl, and fiberglass. Wood windows and doors have remained popular because this material is easy to modify in the field for the installation of hardware, or for future adjustment. Steel doors can be coated with vinyl films to give them the appearance of wood grains and to accept stain. Aluminum, often selected for its strength and ease of manufacture and maintenance, is a poor choice in colder climates because of its conductive qualities, which transmit cold outdoor temperatures through the frame (Fig. 4.1.3). Window manufacturers responded to this deficit by combining aluminum with less conductive materials such as plastic to provide a "thermal break." Vinyl windows have enjoyed larger market acceptance in recent years for both new and rehab applications.

RATINGS AND STANDARDS

The selection of windows and doors based on energy performance criteria has been simplified with the establishment of uniform rating procedures by the National Fenestration Rating Council (NFRC). The NFRC is a nonprofit public-private organization, comprised of a diversified group which includes manufacturers, utilities, and code authorities, and sanctioned by the Energy Policy Act of 1992.

Through the NFRC certification program, participating manufacturers obtain certification authorization for total product energy ratings such as the U-value, solar heat gain coefficient, and visible transmittance. Door labels provide the U-value only. It is anticipated that in late 1998 the NFRC will also have certified ratings for both heating and cooling. The NFRC is presently working on future rating systems for air infiltration, condensation resistance, and long-term energy performance. The values are determined by licensed independent laboratories accredited by the NFRC correlating the results of computer simulations and actual physical testing for two different prescribed sizes. The performance data for both sizes, which vary by type, are designated as AA—Residential and BB—Non residential and are identified on the label.

The ratings for each individual product can be found on the product itself in the form of an NFRC temporary label (Fig. 4.1.4), and also with a permanent marking somewhere on the unit. The energy

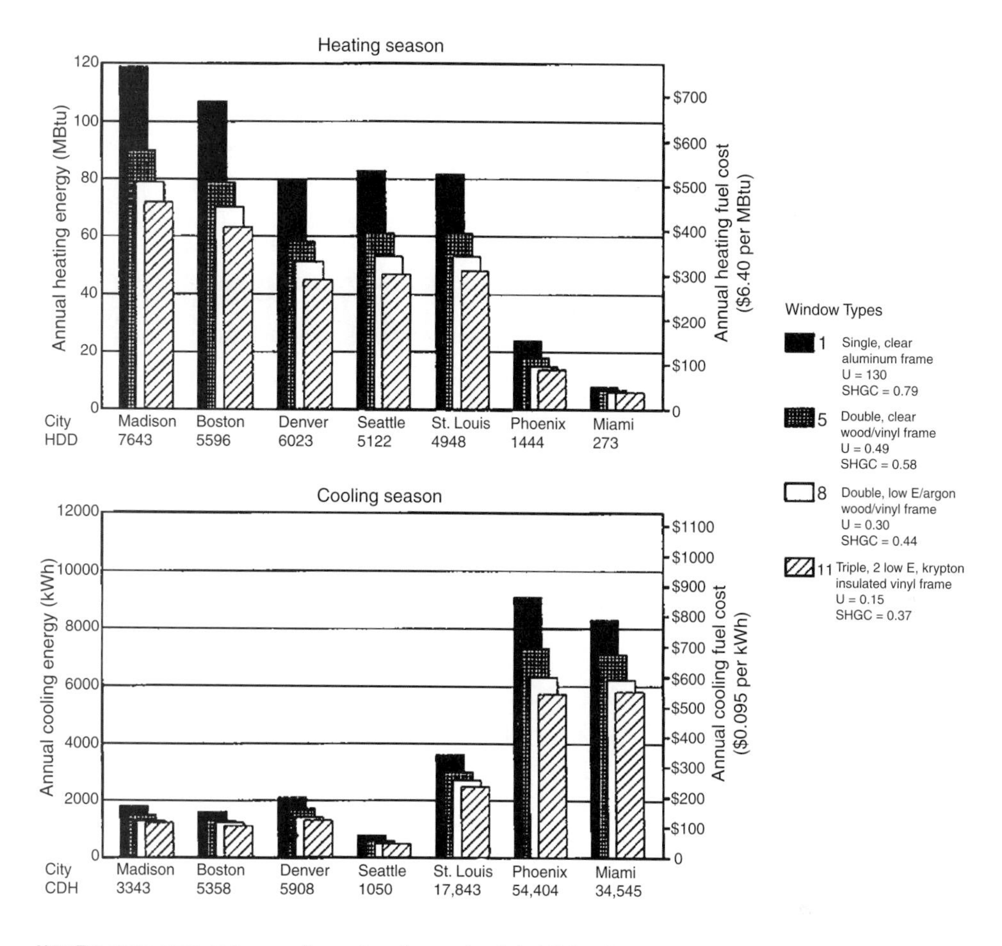

FIGURE 4.1.3 COMPARISON OF SEASONAL HEATING AND COOLING COST FOR FOUR WINDOW TYPES

performance ratings of manufacturers participating in the NFRC certification program can also be found in the NFRC Certified Products Directory, which is published annually.

Presently eight states—Alaska, California, Florida, Massachusetts, Minnesota, Oregon, Washington, and Wisconsin—and some local jurisdictions, as well as the 1995 Model Energy Code, require NFRC ratings for new windows (including those used in rehab).

Windows or doors should also be selected by their respective performance class designation, which is their ability to resist wind pressure, water, and air infiltration and their resistance to forced entry. The Window and Door Manufacturers Association (WDMA) and the American Architectural Manufacturer's Association (AAMA) have developed a voluntary standard for aluminum, vinyl, and wood windows and glass doors (AAMA/NWWDA 101/I.S.2-97). The standard combines the two national window performance standards ANSI/AAMA 101.93 and NWWDA I.S. 2.93 and is applicable to new materials such as composites and plastics.

This standard uses a design pressure designation in lieu of the former structural test pressure value such as grade 20, 40, and 60. Minimum criteria have been established for five performance classes, the lowest of which is Residential. A designation code identifies the product type, performance class, performance grade, and maximum size unit tested. Additional voluntary standards, such as acoustical performance, thermal resistance, and condensation resistance may also be evaluated by this standard. This standard, commonly employed in the selection of commercial window and door units, allows for the selection of products for specific applications.

FIGURE 4.1.4 NFRC TEMPORARY LABEL

INSTALLATION

The selection of a suitable window or door does not ensure performance; a unit is only as good as its installation. Improper or inadequate anchorage of the unit will defeat the wind and weather resistance of the best-performing window or door. Installation of a replacement window or door may require modification of an existing opening for squareness, caulking, fastening securely to structural members, and insulation of gaps with a product prescribed by the manufacturer.

A common problem encountered is rough openings not large enough to allow for expansion of unit and structural movement, in particular the space required to accommodate header deflection. Out-of-square installation will also impair the proper functioning of the unit and result in a poor weather seal. A window or door that is not functioning properly or has deteriorated may be an indication of damage elsewhere. Windows or doors that are difficult to operate may be swollen due to the effects of moisture that has leaked from a remote area such as the roof and has traveled through the wall cavity only to be discovered at the opening. Often such leaks are present for some time and rot may extend throughout the wall framing. Replacement of water-damaged windows offers an opportunity to explore the cause and to examine framing members, which may be structurally compromised.

A window or door without proper structural support will not operate properly and will be subject to infiltration around the perimeter. Insulating between the unit and the rough opening is critical to ultimate performance. Batt insulation or injectable nonexpansive foam are the two most popular means of filling this gap. Care must be exercised during installation of batt insulation that it is not too loose (permitting airflow) or too tight (reducing the thermal resistance). Foam products must be installed so as not to apply pressure to the unit itself (which can distort the frame) or chemically interact with the frame material.

Installation of new windows and doors also presents the opportunity to eliminate airborne lead and peeling paint from lead-based paints, typically present in homes built before 1978. Moving sashes and doors grind the paint into dust, which is easily transported by air movement. Typically, a window requiring lead abatement will cost about the same as a new sash replacement unit. However, replacement of the entire unit will require removal of trim, which is another potential source of lead contamination. The removal of lead-based paint requires precautions, further discussion of which is beyond the scope of this handbook (see Further Reading for more information on lead-based paints and mitigation methods).

The American Society for Testing and Materials (ASTM) is presently developing a window and door installation standard that will provide a consensus document for the installation of windows, doors, and skylights. This standard will likely serve as the comprehensive reference for both specification and instructional purposes.

COSTS AND BENEFITS

The repair and/or replacement of windows and doors can pay for itself through improved energy performance and can provide increased comfort. Reduced drafts and warmer surfaces will permit lower temperature settings. However, beware of the enthusiastic promises of manufacturers who may overstate the value of a product. All improvements may be evaluated on a payback basis of their potential savings relative to cost. Potential savings may also include incentives provided by local government or utilities and special financing referred to as an energy efficient mortgage (EEM) that enables the homeowner to finance the cost of the improvement at no additional net operational cost (mortgage plus utilities). The NFRC certified ratings for the U-factor and solar heat gain coefficient may be utilized when using computer simulation programs to assess the potential economic benefit new windows and doors will have on the energy performance of a home. RESFEN, a computer program for the purpose of calculating the annual heating and cooling energy use and cost due to window selection, is available from the Lawrence Berkeley National Laboratory. Fax a request to RESFEN Request at 510-486-4089, or e-mail your request to plross@lbl.gov. The Department of Energy (DOE) intends to provide an interactive version of this software program in the near future at the Efficient Windows Collaborative Web site (www.efficientwindows.org).

4.2 GLAZING

ESSENTIAL KNOWLEDGE

The performance of windows has increased significantly since the energy crisis of the 1970s with the use of multiple glazing layers and new glazing technology. Such insulating glass provides air space(s) between the layers to reduce the transmission of heat, cold, and sound through the window assembly. This space is typically filled with air or inert gases such as argon or krypton to further reduce the heat transmission. New glazing coatings and films enable windows to selectively transmit the majority of visible light but reflect heat and UV radiation. Such glazing can reduce heating and/or cooling costs (total solar transmittance), lighting costs (visible light), and damage to contents subject to UV degradation, and can improve comfort.

Single-glazed windows with storms and screens have steadily lost market share since 1965 to insulating glass, which now represents approximately 90% of all new window glazing in the United States A window with a low U-value (such as 0.33) or, inversely, a high R-value (such as 3.0) will reduce the transmission of heat and save energy. In addition to reducing the heating and cooling demand, a well-insulated window will minimize the effects of thermal air movement associated with drafts and greatly reduce the potential for condensation on the window or door, which can progressively deteriorate some frame materials, as well as surrounding wall finishes and furnishings. Insulating glass units have a limited lifetime due to the vulnerability of the seals and are typically warranted for 10 to 20 years. A better-performing window will not only often pay for itself but will provide increased comfort to the inhabitant.

Glazing also has a visible transmittance value—a measure of how much of the light is perceived to pass through it. Glazing with a high visible transmittance is generally desirable in most areas because it allows a clear view through the window with minimal distortion or tint and reduces the need for lighting. There are many types of low-emissivity (low-e) coatings. All low-e coatings improve the U-value, but they can be formulated to have varying solar heat gain coefficients (SHGC). Selecting the best type of low-e coating for your condition represents a tradeoff between the ability to reject unwanted heat from the sun during cooling periods and accepting heat during heating periods. Selection must be made for geographic region and window orientation. A low-e coating is typically placed on the interior surface of glazing for protection and attached to the surface closest to the source of heat to be reflected, either surface 2 for cooling climates or surface 3 for heating climates (Fig. 4.2.1). Some very high performance window units use multiple coatings on either additional glazing layers or intermediary plastic films to achieve a cumulative effect. As of 1996, low-e coatings were used on 40% of all new windows. This percentage has risen now that the industry has developed the capacity to produce such coatings at a very small cost premium.

There are three primary types of low-e coatings: (1) high-transmission, low-e coatings are typically most appropriate for heating climates allowing the most solar heat to enter the building, (2) selective low-e coatings are suitable where heating and cooling are required because they allow the greatest amount of daylight and moderate heat transmission, (3) low-transmission, low-e coatings are suitable for cooling climates, allowing the least amount of heat to enter the building and reducing the amount of visible light transmission to control glare. The NFRC rating is the simplest means of assessing the performance of each type of coating, which is often not readily identified on the unit itself.

Within the same climate different low-e glazing materials may be desirable. In a cold climate, a home may use a window with a low SHGC on elevations other than south-facing to preserve as much heat as possible where there is little opportunity for the sun's penetration, but may utilize a window with a high SHGC at the south facade to benefit as much as possible from solar gains where exposed to the direct sunlight (Fig. 4.2.2).

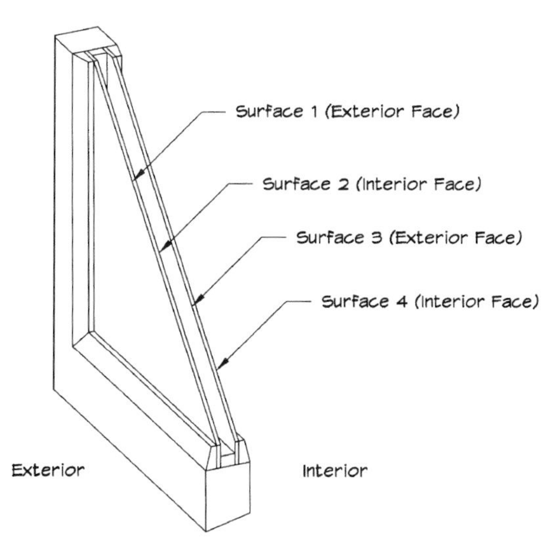

FIGURE 4.2.1 GLAZING NUMBERING SYSTEM

In addition to energy performance, glazing must also meet minimum code requirements for impact, wind, and fire resistance. The applicable codes vary by locality and should be referenced in the selection of new glazing for all types of fenestration. Numerous products and standards have been developed recently to address concerns such as high wind and theft resistance. These new products are made of acrylic and polycarbonate in corrugated and cellular forms to increase strength and insulation and reduce weight. Plastics, either by themselves or in combination with glass, improve impact resistance. New plastic materials have also been developed that provide increased fire resistance greater than glass but must be incorporated in an equally resistant frame.

Various specialty glazing products suitable for matching existing glazing materials, such as stained, textured, and salvaged glass, are available from sources identified in *Traditional Building* magazine and *Old House Journal*. An example of textured glass that is decorative and functional is prismatic glass. Prismatic glazing, characterized by a surface of textured small tile shapes assembled in a structural grid, projects light in a specific direction to illuminate areas deep in a space. Primarily used in commercial applications or street pavement lights, this material is also found in older homes. The material was employed in canopy applications appropriate for windows that are in close proximity to adjacent structures. However, the concentrating effect increases heat and glare. Prismatic glass is no longer manufactured due to the introduction of electricity and difficulty of cleaning, but it can be obtained through salvage sources.

Glass block is enjoying a revival and has the advantages of security, noise absorption, and light diffusion. The material is available with many of the same efficiency strategies as windows, employing reflective glass and fibrous inserts, with R-values equaling and exceeding some insulating glass. The mortar layers or frame material between units also serve as a shading device. Traditional glass block has been adapted for mortarless application with the use of a rigid frame material and silicone sealants. This new generation of glass block has been adapted to provide up to 90-minute fire ratings. Available as an acrylic product, the reduced weight has significantly simplified the installation process (Fig. 4.2.3). This reduced weight also allows it to be installed as an interior partition without structural modification and can be used in operable casement window units.

TECHNIQUES, MATERIALS, TOOLS

1. CLEAN AND POLISH EXISTING DAMAGED GLAZING.

Damage can occur from either physical abuse or caustic chemicals resulting in stains, etching, or discoloration. Physical abuse might include a branch brushing repeatedly against a window or impacts of

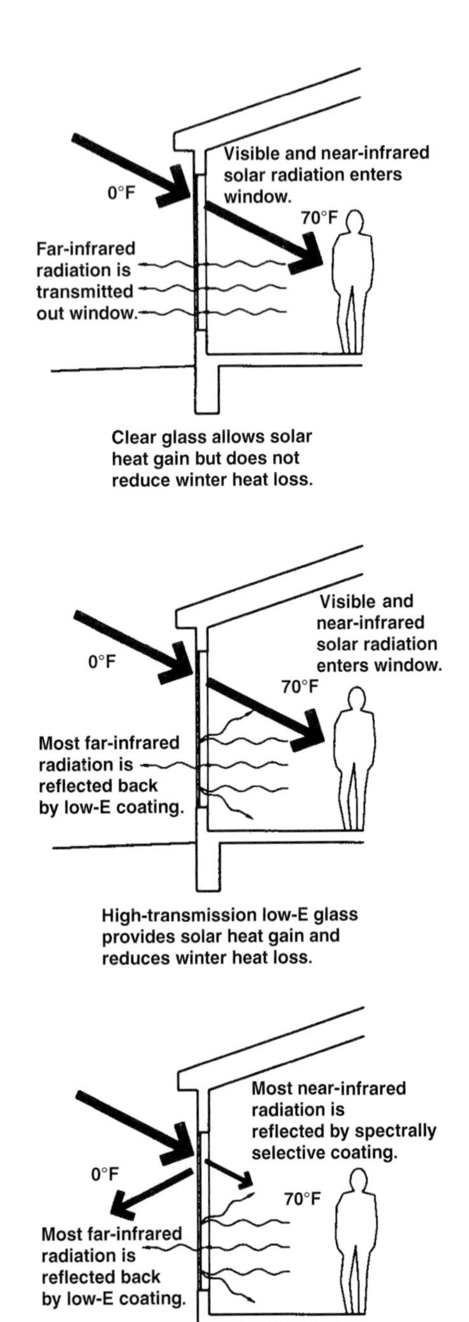

FIGURE 4.2.2 LOW-EMISSIVITY COATING HEAT TRANSFER PROPERTIES

airborne debris such as fine sand. Acid rain, alkyds from cleaning solutions, or exposure to water or rain runoff from concrete and masonry building materials are typical caustic assaults. Minimally damaged glazing often can be restored with a simple cleaning solution such as an ammonia-based solution, xylene, toluene, or trisodium phosphate. Damaged glass not readily replaceable can be repaired by a mechanical polishing method with the use of compounds such as cerium oxide paste. In all instances the source of damage should be identified and corrected before rehabilitation.

ADVANTAGES: This method preserves the original, perhaps unavailable materials and may be the most cost-effective.

FIGURE 4.2.3 INSTALLATION OF ACRYLIC BLOCK WINDOW UNIT AS MANUFACTURED BY HY-LITE PRODUCTS, INC.

DISADVANTAGES: Often other components, such as sealants, have also deteriorated and favor total replacement. A high degree of skill and experience is required to properly polish glass.

2. APPLY TINTED OR REFLECTIVE FILM TO EXISTING WINDOW GLAZING.

Numerous film products are available from manufacturers such as 3M and Courtaulds to control the amount of daylight and solar gain transmitted through existing glazing. Tinted products can significantly reduce the full spectrum of light, reducing solar heat gain and glare. Reflective products selectively reduce portions of the light spectrum to either accept or reject heat and visible light. Each product is formulated for a specific climate, and the selection of film and its specific application will determine the ultimate benefit. All of these products block the majority of UV rays to reduce fading of interior furnishings and provide limited protection from shattered glass.

ADVANTAGES: Application of film is a relatively simple, low-cost means of improving performance relative to replacement. Reduced heating and cooling loads not only provide operational savings and increased comfort but may permit smaller mechanical equipment. Appropriate films may eliminate the need for additional sun control devices, such as blinds and curtains, providing a uniform appearance to the exterior.

DISADVANTAGES: Retrofit film application will not equal the performance of factory-applied materials nor will its exposed surface perform as well over time. Low-e coatings are subject to physical and caustic damage from cleaning with ammonia-based products. Replacement of the film is a difficult, labor-intensive process. Application of film to an insulated window unit may aggravate the expansion and contraction of one side of the glazing, increasing stress on the glass and seal. Reduction of transmitted light and heat may be offset by additional heating and lighting requirements if not carefully selected.

3. APPLY SAFETY FILM TO EXISTING WINDOW GLAZING.

All film materials provide limited resistance to shattering—some more than others. Safety film products are manufactured primarily for this purpose and may be combined with sun control properties.

The increased resistance provided by these films, similar in principle to laminated glass, may reduce the risk of airborne glass and exposed shards upon impact. Films will provide resistance to accidental impact and storm debris. The resistance of the material is independent of the resistance provided by the window frame, but the preservation of the building envelope during a storm often will prevent costly water damage. The material may also prove to be a deterrence to vandalism and theft.

ADVANTAGES: Safety films may preserve the building envelope during storms. Safety films may provide a cost-effective means of increasing resistance of existing glazing. Areas subject to high winds may also benefit from films, which may preserve the building envelope in moderate storms. In addition to providing safety they may also provide comfort and savings associated with the reduction of sunlight and UV protection of furnishings.

DISADVANTAGES: All components of a building envelope must be of equal resistance in preventing either the entry of a storm or intruder. Safety glazing that is not secured in the frame itself will only maintain integrity with itself and not the window unit. Film will to some extent reduce visible light transmittance and may slightly obscure views.

4. INSTALL INSULATED GLASS INSERTS.

Existing single-layer glazing is replaced with new insulated units with the modification of the existing wood sash and a new track mechanism. Presently, only a Bi-Glass Systems licensed contractor performs this patented process on-site in a mobile workshop (Fig. 4.2.4). The exterior face of the muntin is removed and the new unit is inserted with a siliconized latex sealant. A new muntin face is applied to match the existing or may be left as a flush surface to provide for easier maintenance. Historic metal windows with deep profiles often can readily accept insulated glazed units, but due to the poor thermally conductive properties of the frame do not provide significant performance gains and are ultimately very expensive relative to the benefit. The effect of heavier insulated units on operable windows should be considered.

ADVANTAGES: Glass inserts provide improved performance while preserving the original appearance of the window. Replacement of glazing is an opportunity to choose from new technologies, such as low-e glass. The new track mechanism may also provide for ease of operation and cleaning with the addition of a tilt and turn feature. The relative ease of installation is achieved without disturbing surrounding materials and minimizing debris and use of new resources. The cost of such replacement may potentially be lower than replacement of the entire unit.

DISADVANTAGES: Bi-Glass system is presently not available nationwide and is only suitable for existing wood windows. The existing window frame and sash must be in good condition. The system will not correct for perimeter sash leakage, which may account for up to 40% of a home's air leakage.

5. REPLACE EXISTING GLAZING WITH BETTER-PERFORMING GLAZING.

This method, although rarely used, provides the opportunity to employ new glazing materials in the repair of existing windows. Repair of windows preserves the original appearance and is often the most cost-effective strategy, as new glazing technologies may be employed for little additional cost. New technologies provide increased impact, fire, and thermal resistance and may be applied to previously unavailable curved glazing and decorative glass block.

ADVANTAGES: This method is particularly cost-effective when trying to address a single attribute such as impact resistance or glare.

DISADVANTAGES: Replacement of glazing itself may be compromised by the existing frame and sash materials. Cost of repair will vary dramatically depending on the region and should be part of a comprehensive repair effort.

EXISTING WINDOW MEASURED AND GLASS UNITS FABRICATED

EXISTING SASH REMOVED, CAVITIES INSULATED AND NEW TRACKS INSTALLED

WINDOW SASH IS REMOVED AND MODIFIED TO RECEIVE INSULATING GLASS UNIT, EXISTING MUNTINS APPLIED TO SURFACE

SASH IS INSTALLED WITH NEW HARDWARE PROVIDING FOR TILT-IN OPERATION

FIGURE 4.2.4 INSTALLATION OF INSULATED GLASS INSERT

4.3 WINDOW FRAMES AND REPLACEMENT UNITS

ESSENTIAL KNOWLEDGE

Wood has historically been the primary window sash and frame material due to ease of manufacture, availability, and good thermal performance. The major disadvantage of wood has been its need for regular maintenance. Ease of maintenance has spurred competing materials such as steel in the early half of the twentieth century and, more recently, aluminum and vinyl, which have now surpassed wood in combined sales. Recent improvements in glazing technology have made the thermal performance of aluminum and vinyl frames and the method of assembly significant issues. New frame technology employs thermal breaks in metal materials and insulated cavities in vinyl frames. New materials are being developed with greater dimensional stability, some of which are less dependent on natural resources.

This new generation of materials—wood composite, fiberglass, and reinforced or reformulated vinyl (CPVC)—are stronger, more durable, and have insulative values equal to or higher than wood. The increased strength of frames also allows for narrower profiles, resulting in more glazing area relative to the size of the unit. A manufacturer of wood composite frames claims this may result in a 20% increase in the glazing area. These new materials are generally impervious to water and therefore will not swell or distort when exposed. New engineered wood materials, either composite or laminated, do not have a tendency to twist along the grain, like wood. The thermal expansion of materials such as fiberglass, composed primarily of glass fibers, closely corresponds to that of glazing material, reducing stress to seals and frame as the materials move in tandem, preserving infiltration performance. The increased dimensional stability and resistance of fiberglass, long used as a marine product, provides for consistent operation under varying climatic conditions and corrosive exposures. Fiberglass, unlike conventional vinyl material, is available in dark colors that are not subject to fading. The solid or wide extrusion profiles of either a high-density or low-density cellular vinyl (CPVC) product provide a greater surface area for the chemical welding of corners, resulting in a stronger connection. Fiberglass—with strength similar to that of aluminum—is now used as reinforcement for windows and doors with wood veneer to achieve a traditional appearance.

New materials not only provide benefits relative to wood in terms of maintenance but can surpass its thermal resistance by up to a factor of 2. Low-density cellular vinyl incorporates a large proportion of air, as does wood, and achieves comparable R-values. Engineered wood products have solid profiles with essentially the same value as conventional wood frames. Fiberglass and high-density cellular vinyl products that have cavities filled with foam insulation have the highest R-values available. These materials are not without their drawbacks, which may include higher production costs, increased weight, and UV degradation.

In addition to providing a higher overall R-value for a window assembly, frames that minimize heat loss at the glazing edge are more resistant to condensation. A center-of-glass R-value for insulated or low-e glazing is typically significantly reduced adjacent to the frame. This is because in conventional windows, the spacer between glazing layers is highly conductive. Thus, condensation on residential windows often first appears at the pane edge. New low-conductivity (warm edge) spacers significantly reduce these losses, resulting in higher overall U-values. When selecting a window, it is important to note the overall window U-value (which includes the frame) as indicated by the NFRC label. The window frame alone may account for up to a third of the window area; thus the selection of frame material has an important influence on the overall window performance.

The method of assembly is also an important consideration in overall window performance. Historically, glass was only available in small panes and a window frame joined several panes to form the desired opening. This grid of frame material is referred to as a muntin. As glass technology evolved, the size of the grid increased, ultimately only being restricted by the operable weight of the sash. With no structural requirement for muntins and the advent of insulating glass, the grid was preserved solely for aesthetic reasons (sense of enclosure, ability to define expanse of view, existing architecture). Today there are several options available dependent upon the degree of authenticity desired relative to convenience or performance (Fig. 4.3.1). A window sash with individual lites (panes) is inherently more expensive and less efficient because of the increased length of the glazing perimeter. These window muntins are described as true or authentic divided lites as they serve their original structural purpose. The width of these true muntins, however, is typically larger to accommodate the insulating glass. Multiple individually sealed glazing units also increase the likelihood of seam failure.

Alternate methods have been developed to simulate the appearance of true divided lites and provide the ease of cleaning a single pane and/or to minimize the cost of fabrication. Simulated divided lites are muntins (grilles) adhered to the surface of the glazing and are often available with an optional air space grille (spacer bar). They simulate the appearance of true divided lites at lower cost and without significantly compromising energy performance. Single-lite insulating glass reduces the likelihood of seam failure. Another type, referred to by some manufacturers as "snap-in" grilles, can be removed for cleaning but are considered to be the least convincing because of their thin profiles. They are also subject to damage with repeated use. A third option offers a combined strategy: a true divided lite sash with a single full-size panel, which is set into the sash like an integral storm unit. This provides a single surface to clean from the interior while maintaining a historically authentic appearance with good thermal performance.

Wood window frames that are not properly protected are subject to swelling and rot. The repair of these windows is often possible. However, the source of the problem should first be identified and corrected. Otherwise window repairs may be temporary at best and could potentially ruin the window if it is adjusted to accommodate for these abnormal conditions. A swollen window that has been planed to accommodate swelling will provide an insufficient barrier to infiltration when the sash

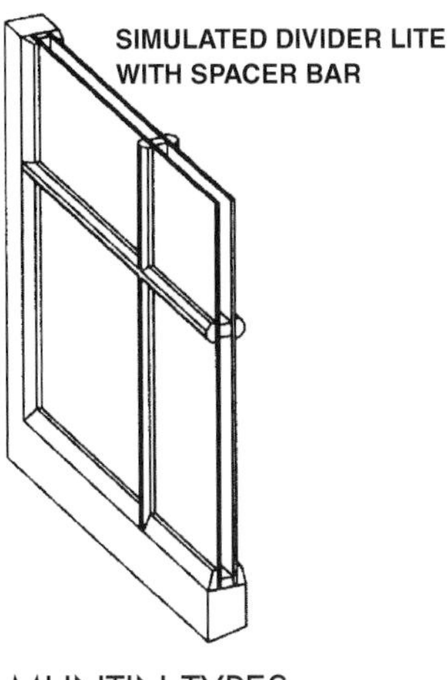

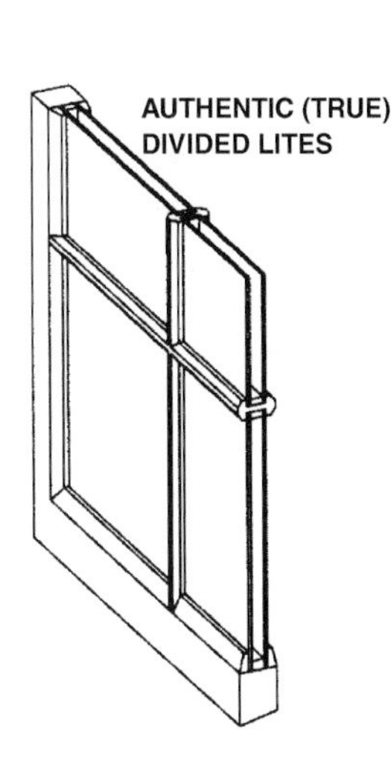

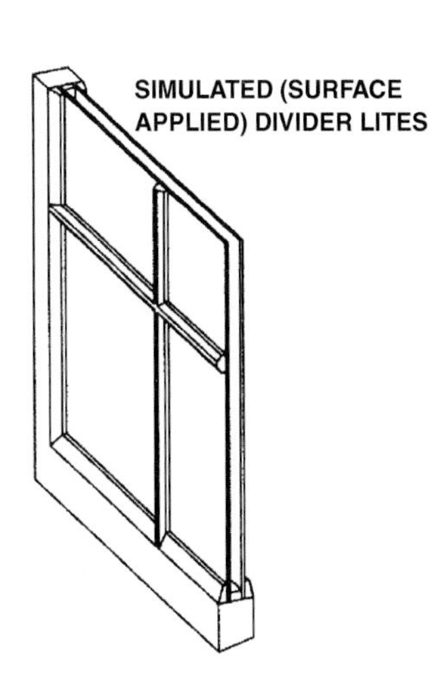

FIGURE 4.3.1 MUNTIN TYPES

shrinks. Window sashes that have been fixed in place address infiltration but compromise ventilation and safety.

The benefits provided by new materials may be retained by the use of a variety of partial replacement methods (Fig. 4.3.2). The primary benefit of partial window replacement is the preservation of the original materials including surrounding trim and surfaces, which can save labor and material costs. The preserved material's potential useful life and ability to be replicated are often the driving criteria.

TECHNIQUES, MATERIALS, TOOLS

1. REPLACE EXISTING WINDOW UNITS.

Replacement of existing window units with an entirely new unit will provide the best available performance and the opportunity to assure proper installation of the unit. Sources of damage to the unit being replaced may be corrected with the removal of the original window.

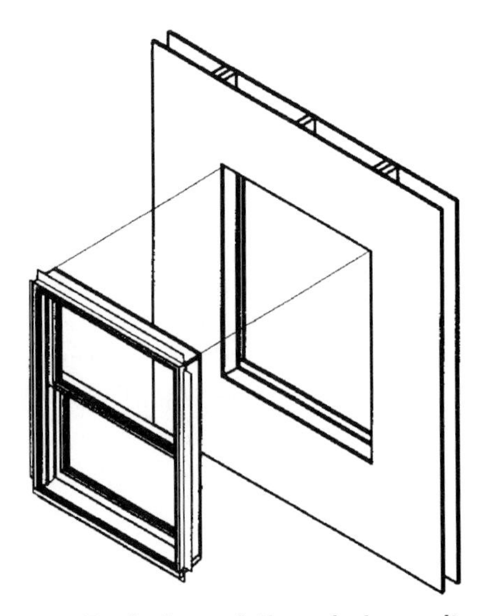

Replacing existing window unit

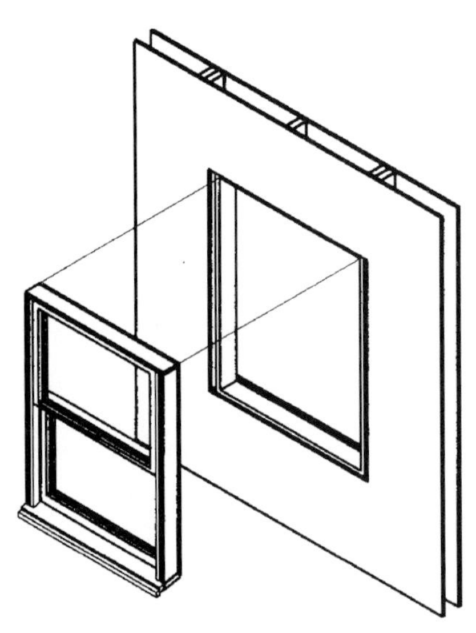

Secondary frame replacement unit

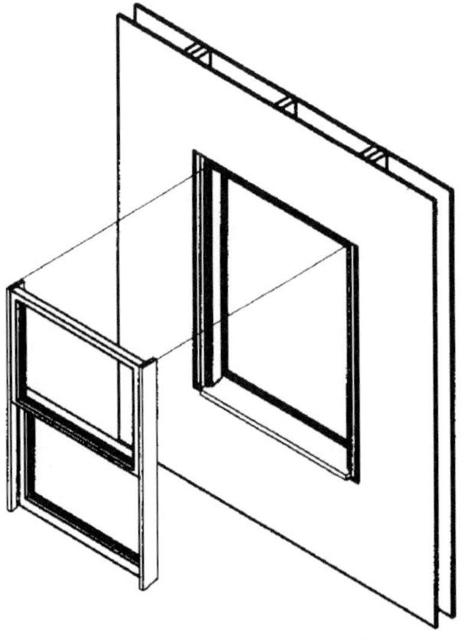

Sash replacement unit

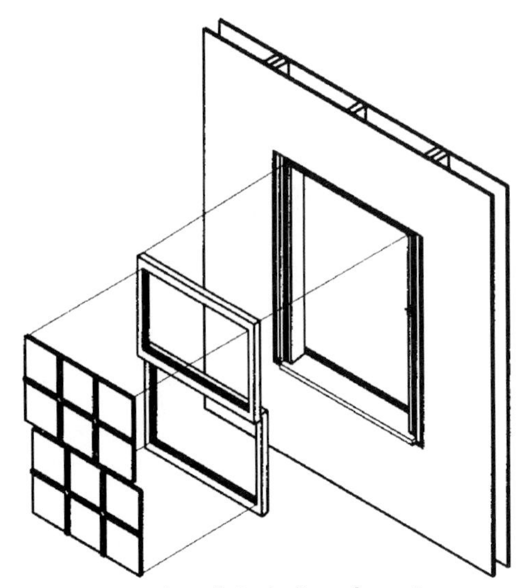

Insulated glass insert

FIGURE 4.3.2 WINDOW REPLACEMENT UNIT TYPES

ADVANTAGES: A new unit provides the longest useful lifespan and significantly improves thermal performance.

DISADVANTAGES: This is usually the most costly alternative, with the existing window discarded as waste.

2. REPLACE EXISTING WINDOW SASH AND TRACK.

Replacement sash units are a very popular choice for partial window replacement (Fig. 4.3.3) and are now produced by the majority of window manufacturers, including Caradco, Marvin, and Weather Shield. These inserts come in a kit, which includes the sash and track (jamb liners) with a counterbalance mechanism and hardware. The units are available in a wide variety of stock sizes or can be custom-fabricated with a choice of glazing.

ADVANTAGES: A low degree of effort and skill is required for installation. Adjacent surfaces and trim are preserved. Elements replaced are those subject to the greatest wear, preserving as many components as possible and thus reducing waste.

DISADVANTAGES: Existing wood frame must be in good condition with no rot and relatively square with parallel jambs. Partial replacement does not address air infiltration at the perimeter of the existing frame or causes of damage that may be found within the wall cavity.

3. INSTALL NEW (SECONDARY) WINDOW UNIT WITHIN EXISTING WINDOW FRAME.

The most popular form of replacement windows is vinyl units that fit within the existing window frame, although wood window units for this purpose are also available from Pella, and others. The secondary window unit is perceived as a unit within a unit, providing the sash and track with a preassembled narrow frame. These units are available in a wide variety of stock sizes or may be custom-fabricated with a choice of glazing.

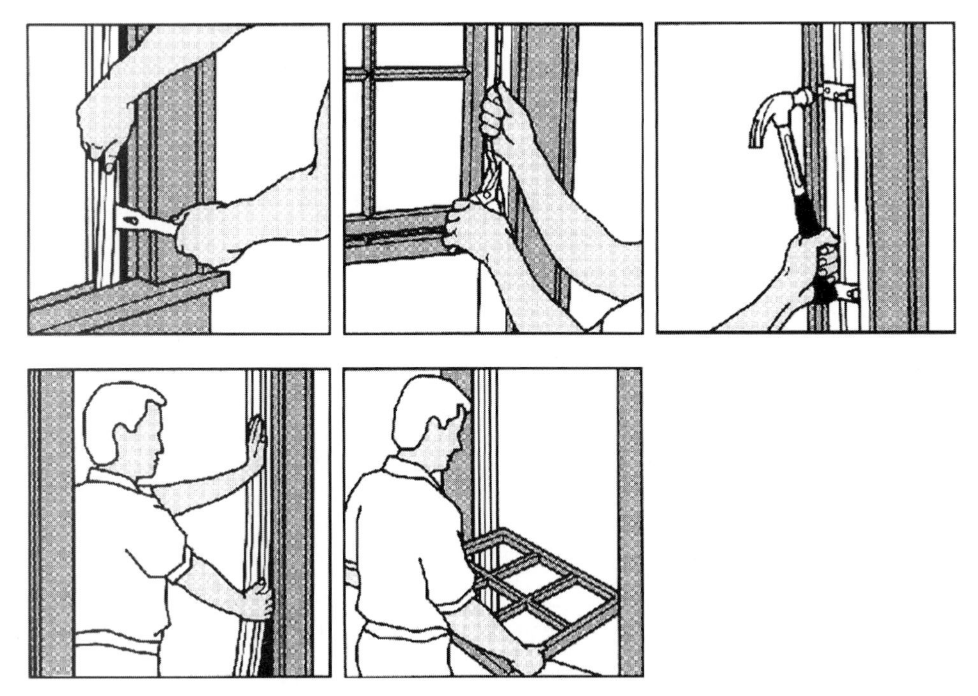

FIGURE 4.3.3 TYPICAL INSTALLATION OF A SASH REPLACEMENT WINDOW UNIT

ADVANTAGES: The secondary frame is similar in concept to the replacement of sash and track in that only those parts subject to wear are replaced while providing benefits of new technology. The secondary frame may accommodate slightly out-of-square conditions.

DISADVANTAGES: Secondary frames reduce the amount of egress and glazing area, which may be in conflict with applicable building code egress requirements. As with other partial replacement methods, this method does not address air infiltration at the existing frame's perimeter.

4. INSTALL REPLACEMENT SILLS.

A wood window sill is often the most vulnerable component of a window frame because it is possible for standing water to accumulate on this surface and form rot. This common condition may ultimately compromise the entire wall assembly by providing the means of entry for water. Replacement sills have been developed expressly for this purpose. There are essentially two means of correcting this condition. The first is to use a sheet-metal material as a cap over the existing seal to function as flashing. The second is a new generation of replacement sills made from such materials as wood composites and vinyl, as manufactured by Wenco, which are solid in profile and may be worked with conventional carpentry tools.

ADVANTAGES: Replacement sills are a necessity for preserving the weather-tightness of the building envelope and are the most economical means of addressing this common problem.

DISADVANTAGES: Both methods may serve to hide a more critical problem in which rot goes unaddressed and continues to erode surrounding materials. The metal flashing caps are considered unsightly by some and are subject to denting.

5. REPLACE EXISTING DAMAGED WOOD WITH EPOXY CONSOLIDANTS AND FILLERS.

Rotted or damaged wood frames can often be repaired with epoxy products. There are essentially two types of epoxy repairs: consolidants for use where the wood is intact, and a puttylike filler material for areas that are missing or require removal because they are beyond repair (Fig. 4.3.4). Epoxy consolidants will penetrate and bind with the wood fibers while preventing further deterioration. Consolidants, which are either poured or brushed on the surface in liquid form, bond with the wood fibers to create a surface with greater strength than wood and are water resistant. The material cures in a matter of minutes or hours (depending on the amount used) and may then be worked as wood. Consolidants may be used as a primer for the application of an epoxy filler material, to fill voids, or to achieve intricate profiles that would otherwise be difficult and expensive to replicate in wood on a small scale. The increased strength of the epoxy material is suitable for structural elements such as an operable window frame when applied as per manufacturer's instructions. Consolidants are available from numerous sources (manufacturers and distributors) including Abatron, Conservation Services, Gougeon Bros., Inc., and Repair Care Systems USA.

ADVANTAGES: Epoxy filler provides an alternative to solvent-based wood fillers, which may shrink as they cure or work themselves loose as materials expand and contract at different rates. Repair of existing wood members is often the most cost-effective solution with the least disruption. Epoxy can be worked like wood, maintaining the original appearance.

DISADVANTAGES: A degree of skill is required for proper application. The repair of damaged wood will not address the cause or progressive deterioration of adjacent materials.

6. ADJUST WINDOW OR DOOR FRAME WITH SHIM SCREWS.

Shim screws, available from GRK Canada Ltd. and Resource Conservation Technology, Inc., are often an effective means of correcting an out-of-square condition. The screw functions as two different

ROTTED WINDOW FRAME CLEANED

EPOXY FILLER APPLIED AND SANDED

EPOXY FILLER PAINTED

FIGURE 4.3.4 EPOXY FILLER

screws attached by a single length so as to allow sufficient anchorage while being able to fine-tune the position of the frame without having to use shims.

ADVANTAGES: This method requires minimal effort to correct an out-of-square condition at minimal cost.

DISADVANTAGES: An out-of-square condition may be the result of a much more serious condition such as a rough opening which is too small to allow for the deflection of a header or an improperly sized or deteriorated structural member.

4.4 STORM WINDOWS AND SCREENS

ESSENTIAL KNOWLEDGE

The storm window has traditionally been a product for cold climate regions, with small manufacturers providing custom unit sizes. Adding a storm window unit to an existing window provides several improvements. A storm window will dramatically reduce air infiltration and significantly increase the thermal performance of a single-pane window while reducing the impact of weather on the prime (original) unit. Available options such as low-e glazing will further reduce energy consumption while available tilt-in sashes allow for ease of maintenance. High-performance storm windows are also suitable for noise reduction.

The material of choice is aluminum, which provides high strength with a narrow profile. The poor conductive properties of aluminum may be mitigated by existing wood windows which serve as a thermal break. Conventional storm windows are typically not suitable for installation on vinyl prime units because the elevated temperatures between the units escalate the expansion and contraction of the frames. Storm windows may have a similar effect on windows joined with lead caming. Aluminum windows benefit from the addition of storm windows, but the differential movement between the windows must be accommodated with material such as a double-sided adhesive cork tape.

Storm windows may be installed either on the interior or the exterior of the prime unit and are typically available as sliding units operating vertically or horizontally, or as a fixed unit suitable for removal. Units are available in sizes large enough for sliding glass doors. Operable units available in double- and triple-track configurations provide for air circulation and self-storage of a screen. Fixed units are suitable for picture windows. However, they require seasonal maintenance when used in conjunction with operable prime units.

The performance of a storm unit should be evaluated as a complete assembly. The design and material of the frame, its assembly, the installation, and (perhaps most importantly) the weatherstripping details all contribute to performance, which can vary dramatically even among models from the same manufacturer. An AAMA-certified manufacturer or NFRC program participant should be able to provide the performance rating for the full product line.

Steel windows are particularly good candidates for storm windows, which will compensate for the high level of conductive heat loss of the steel window. A storm unit will reduce transfer through the individual lites typical of a steel window sash. A storm window may be applied to the existing steel sash frame with fasteners, magnetic trim, or adhesive tape, or can be affixed to material adjacent to the steel frame.

Storm windows are commonly available with screen units. The traditional aluminum screening material provides strength but is subject to denting and corrosion. Fiberglass screening, significantly less expensive than aluminum, will not dent but can stretch. A screen's primary purpose is as a barrier to insects, although some new screening materials have been developed that improve energy performance. Various fiberglass products are now available that reject unwanted heat gain in warm climates, but reduce ventilation and natural daylight (Fig. 4.4.1). The energy savings of such screening can be significant, and some utilities subsidize their cost in hot climates. These screens are particularly effective on east and west elevations where the sun is at a low inclination, and passive solar control strategies such as awnings and overhangs are not generally effective. Fine aluminum louver

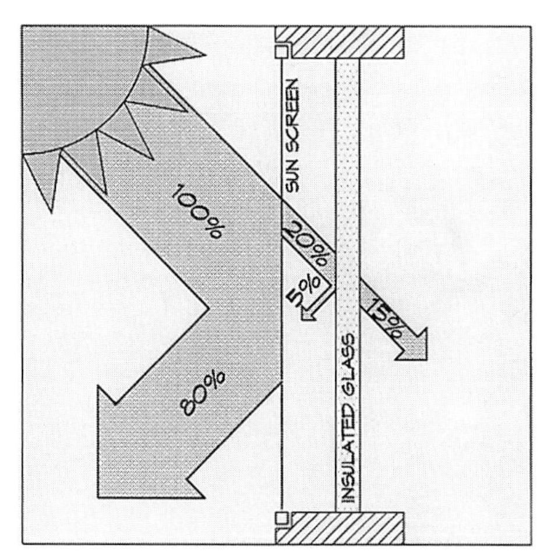

FIGURE 4.4.1 SOLAR SCREEN REFLECTION OF HEAT GAIN

shades can also reject the vast majority of the summer sun similar to venetian blinds. Both products reject heat and damaging UV light before it reaches the prime window, thus protecting the window itself and reducing heat gains.

On homes in coastal regions, copper, bronze, or stainless-steel screening is a good choice because it resists corrosion better than aluminum. Such screening should not be combined with aluminum windows, as corrosive galvanic action may occur. Aluminum windows protected by paint or vinyl finishes are satisfactory for coastal regions.

TECHNIQUES, MATERIALS, TOOLS

1. INSTALL INTERIOR (FIXED OR REMOVABLE) STORM WINDOWS.

Interior storm windows are typically secondary window units attached to the frame of an existing prime unit by a variety of means so as to provide for ease of removal (Fig. 4.4.2). With a storm window there are two separate frames, minimizing conductive transfer and air infiltration while preserving the exterior window appearance. Units are available either custom-fit or in do-it-yourself kit form, with a variety of glazing products. Acrylic glazing products are a popular choice for this application because they are lighter weight, easy to cut to size, and more damage resistant. Such units are available from national and regional manufacturers, including Allied Window, Alternative Window Company, Magnetite, Thermo-Press, Petit, and Window Saver Company.

ADVANTAGES: Manufacturers claim a reduction in air infiltration of 75% or more and almost a 100% improvement in the R-value, frequently better than new prime units. The removable fixed unit is less costly than operable units and suitable for windows of all types without changing the appearance of the exterior. The addition of another glazing layer will reduce noise transmission, the likelihood of condensation, and will provide the opportunity to utilize improved glazing (such as low-e).

DISADVANTAGES: Interior storm units may promote condensation along the cold surface of the prime unit and in some instances cause damage to wood frames. These units are typically not operable and must be removed for ventilation. Acrylic products can discolor over time and cannot be cleaned with ammonia-based products.

2. INSTALL EXTERIOR (OPERABLE) STORM WINDOWS.

Exterior storm windows, in addition to providing a second unit, may also serve to protect the prime unit. Most new operable storm units employ several tracks, sashes, and screens that are self-storing. These units are significantly stronger and more costly because they must often serve the same functions

FIGURE 4.4.2 INSTALLATION OF AN INTERIOR STORM WINDOW UNIT AS MANUFACTURED BY THE ALTERNATE WINDOW COMPANY. THE UNIQUE SPRING-TENSION FRAME COMPENSATES FOR MOST OUT-OF-SQUARE APPLICATIONS AND DOES NOT REQUIRE A FULL-PERIMETER TRACK.

as a primary unit. The criteria for selection of exterior storm windows are the same as for prime units. However, thermal conductivity is not a critical issue when attaching exterior storm windows to a wood frame. Consideration should be given to elevated temperatures and humidity that might affect the prime unit. Storm windows are available from Allied Window, Harvey Industries, Keep In Touch Restoration Products, and Larson Mfg. Co., among others.

ADVANTAGES: Exterior storm windows often provide the most cost-effective solution for poorly performing windows. Exterior storm units provide a barrier against weather and damage for existing prime windows.

DISADVANTAGES: Most units are visible from the exterior and create a flat appearance to windows. The addition of a storm unit will generally increase humidity and temperatures between units, which may damage either vinyl or wood window frames.

3. REPAIR OR REPLACE SCREEN MATERIAL.

New materials, such as fiberglass, are easier to install because of their flexibility and ease of cutting and resistance to denting, but may stretch or sag over time. Aluminum, the metal of choice, admits more light but is about twice as costly as fiberglass and can produce glare. Some aluminum products are painted gray or black to reduce glare. Other metals, such as bronze or copper, are suitable for corrosive environments but may not be used with common aluminum frames and are considerably more costly. These materials are commonly available through local building supply companies, or by mail from the McNichols Company, among others.

ADVANTAGES: The replacement of screens with the introduction of new materials allows for a simple repair.

DISADVANTAGES: Careful selection of material is necessary so as to assure compatibility with adjacent materials and environment.

4. INSTALL EXTERIOR SUN SCREENING DEVICES.

Exterior sun screening devices are effective in blocking the majority of sunlight before it reaches the prime unit, but may trap hot air between the two layers. Reduced daylight and visibility may not be significant in a cooling-dominated climate. The units require regular seasonal maintenance for optimum performance. These products are available in sunny climates throughout the nation from regional distributors. National manufacturers include Phifer Wire.

ADVANTAGES: Sun shading screens quickly pay for themselves in homes with electric central air conditioning in warm climates. Unlike new spectrally selective glazing products, the full spectrum may be recovered by removing the screen.

DISADVANTAGES: Shading devices obstruct views and daylight and require seasonal maintenance.

4.5 SKYLIGHTS

ESSENTIAL KNOWLEDGE

Skylights brought light and ventilation into buildings before the advent of artificial lighting. Skylights have developed in tandem with windows and have become similarly sophisticated and high performing. Most glazing options available on windows today are also available on skylights. Long plagued by leaks, skylights now incorporate new flashing techniques to address virtually all variations of roofs, and numerous options have been developed for these increasingly popular units. Early skylight units also lacked effective means of shading. Shades and screens are now available in a wide variety from several manufacturers, including pleated and roller shades or aluminum slat blinds. These devices can also be operated remotely. Motorized skylights and shading devices may be controlled by a single device that can be programmed to respond to rain and temperature.

Older, mass-produced skylights were typically made of a steel frame and wire glass. As artificial lighting and air conditioning became commonplace, these units have been neglected and/or painted over. As integral parts of the roofing system, however, they require regular maintenance. The conventional single layer of glazing is subject to condensation, which is collected by an integral gutter at the interior, and directed out through a weep hole. This hole, however, often becomes blocked or sealed, leading to what are perceived as leaks but what is, in fact, condensation. A well-maintained skylight of this vintage, if regularly inspected, cleaned, and painted, will last many years. The thermal performance of these units, however, may be addressed by a strategy similar to storm windows to prevent heat from escaping in a chimney fashion. Most local iron shops can repair existing units with conventional methods.

New skylights may be employed in rehabilitation to provide light and ventilation in homes on small lots while providing privacy. The recent development of new skylights that are designed to fit between conventional framing spacing and around obstacles provides new opportunities to introduce lighting and ventilation without a significant amount of structural modification. There are essentially two types that do not require modification of framing: new narrow conventional units, and units generally described as tubular (Fig. 4.5.1). Tubular skylights (also referred to as light tubes or pipes) concentrate light through a dome on the roof and direct it through a tube to a diffuser at ceiling height. These tubes do not have to be routed in a straight line, although it is preferable to efficiently distribute daylight.

Skylights can be compared by the NFRC rating system and are rated with the units in a vertical application. Sloped applications reduce the thermal resistance of the unit by inducing convective loops. Given this concern, the NFRC is reevaluating skylight performance in sloped applications.

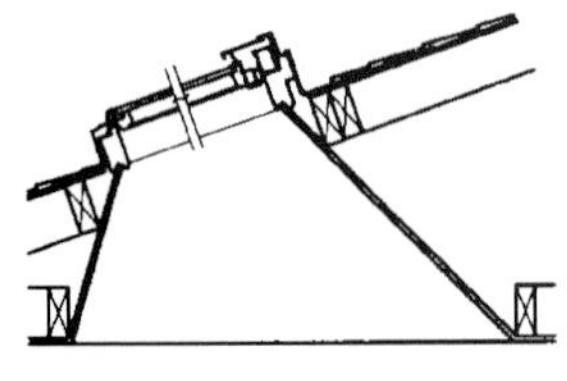

Conventional skylight   Rigid tubular skylight   Flexible tubular skylight

FIGURE 4.5.1 SKYLIGHT TYPES

TECHNIQUES, MATERIALS, TOOLS

1. REPAIR OR REPLACE EXISTING METAL SKYLIGHTS.

The repair of existing metal skylights is generally achieved by means of traditional metalwork. Architectural metal fabricators and roofing contractors often can replicate or repair existing units. A few producers of traditional metal skylights still exist, including J.S. Wagner Company and Fisher Skylights.

ADVANTAGES: Skylights often play a significant role in defining architectural spaces within older buildings. A properly functioning skylight can provide effective ventilation and lighting during much of the year. Older units are capable of long service if properly maintained.

DISADVANTAGES: Original skylights are generally subject to high energy losses and condensation. They can also siphon heat in cold climates in a chimney effect. Roof openings should be minimized, as they are a source of callbacks unassociated with work quality.

2. REPLACE EXISTING SKYLIGHTS WITH NEW UNITS.

New skylight units have been designed to minimize labor associated with installation. Innovations include specially designed flashing materials and narrower units that do not require framing modification. The Wasco E-Class skylight requires no mastic or step flashing due to its continuous flexible flange, which provides a tighter seal against suction when subjected to high winds. The gasket attaches directly to the deck and eliminates the need for a curb, thus increasing daylight admitted. The integral vinyl curb provides good thermal conductive qualities and is resistant to decay caused by condensation. Narrower units available from Roto Frank of America fit within 16" and 24" on-center framing spacing so that no structural modifications are necessary.

ADVANTAGES: New skylight products provide the opportunity to introduce ventilation and daylight in formerly inaccessible areas with minimal alteration.

DISADVANTAGES: Skylights in general are often the source of unwanted heat gain or loss, which is difficult to control. The shaft used to introduce light and ventilation to the conditioned spaces also increases the volume requiring conditioning and the amount of thermal stratification within the space, resulting in poor thermal performance relative to benefit.

3. INSTALL A TUBULAR SKYLIGHT.

Tubular skylights can be installed without modifying roof framing and can be configured around obstructions such as plumbing or ductwork. Some units are available with supplementary artificial light sources for use at night. These units are ideally suited for interior spaces, such as bathrooms, hallways, and closets. Flexible tubing is also available for ease of installation but is generally less effective at transmitting light. Tubular skylights are available from several national manufacturers, including Solatube, Sun Light Systems, and Sun Tunnel among others.

ADVANTAGES: No structural modification is required for installation. Light can be delivered to interior areas. A diffuser prevents conditioned air from traveling upward.

DISADVANTAGES: The value of this daylighting has yet to be confirmed as being a cost-effective replacement for artificial lighting, although the quality of daylight and the connection to the exterior it provides has intangible benefits. The uninsulated or poorly insulated tubes are subject to the formation of condensation. Acrylic domes and diffusers are also subject to discoloration over time.

4.6 DOORS AND FRAMES

PRIMARY ENTRY DOORS

ESSENTIAL KNOWLEDGE

Door technology has evolved with window technology, making similar improvements. Thermal performance had long been unexamined because it was of little significance relative to the entire house envelope. Solid wood panel doors, although opaque, typically have an R-value equal to an insulated window, but appearance, durability, and infiltration remain the driving forces in selection. The technology of repairing existing doors has not evolved dramatically. A wood door is repaired as per traditional wood working methods. Doors made of materials such as steel and fiberglass can be repaired with proven methods practiced by other industries such as auto and marine repair.

The most cost-effective means of repair is often the replacement of damaged components. The difficulty of modifying doors favors replacement with a prehung unit. The benefits of a new or replacement unit include lower maintenance and durability. New materials and assemblies also provide the opportunity to increase comfort and security. A restored conventional solid wood panel door will not add comfort because of its poor thermal performance, and the panels may easily be breached to gain access. As discussed below, there are several options available to repair existing exterior doors or to take advantage of new technologies in replacement units. Weather stripping and security hardware are discussed in later sections.

The introduction of steel, fiberglass, and, most recently, carbon, has raised the standard for all doors. Manufacturers such as Marvin now produce doors with large areas of glazing, such as patio and French doors. There is a convergence of window and door technologies, where new materials or assemblies are used in each. Door security requirements and the dimensional stability of doors, however, mean that such materials as vinyl are typically used in combination with other materials. Fiberglass, which is equivalent in strength to aluminum, is now used by at least one manufacturer to produce structural rails and stiles that are then clad with wood veneer. Manufacturers are readily adopting multiple materials to improve performance. For example, at their perimeter, steel doors might use wood because it is easy to modify. Or steel might have a stainable vinyl film with wood-grain texture on top of its skin, laminated over a rigid foam core.

Manufacturers make an effort to simulate wood and traditional door styles with new materials and assemblies. Wood doors have also advanced technologically with such features as fiberboard cores, or finger-jointed stock to provide solid sections with dimensional stability. Such new assemblies, in combination with new protective finishes, have greatly improved durability and performance. Wood doors now also use rigid insulated cores of either expanded polystyrene or polyurethane. The majority of these cores is manufactured of polyurethane, which has an initial R-value almost twice that of expanded polystyrene. However, some manufacturing processes emit ozone-depleting chlorofluorocarbons (CFCs), and R-values deteriorate over time. The frames of steel exterior doors typically employ thermal breaks to reduce conduction and minimize condensation at the perimeter. NFRC U-factor ratings are a means of providing a comparable standard for evaluating the overall energy performance ratings among manufacturers participating in the NFRC certification program.

Exterior doors with large glazing areas have also improved performance. Developments in window technology have made this door type possible even in cold climates, as these doors utilize many of the same technologies used in high-performance windows. However, fully glazed doors also require the security available in conventional doors. Manufacturers have begun to address these concerns with such features as multipoint locking systems, reinforced frames and stiles, and new track designs.

There are essentially three types of doors suitable for replacement (other than an entirely new unit): a knockdown frame, a prehung insert, and a split jamb. Each offers ease of installation with the benefits of a new door. The determining factor is often whether these replacement units are available in the style of door desired. A knockdown door, also known as a prefit, is delivered as separate jamb and head pieces with attached casings that interlock. An insert door is prehung within its own frame and can be inserted into an existing door frame. The profile of the secondary frame (often of steel) is narrow; however, a door often has little width to spare, especially if the existing opening is out-of-square and requires shimming. A split-jamb door is prehung and available with attached trim. The door is inserted in the existing rough opening and joined along the length of the jamb at the stop. This method allows the preservation of the full door width but requires removal of the entire unit. Finally, a new door unit provides the greatest variety of options.

TECHNIQUES, MATERIALS, TOOLS

1. REPAIR EXISTING DOOR WITH TRADITIONAL MATERIALS.

The repair of existing doors may be achieved as would the repair of a window frame. See discussion of window frame repair and products in Section 4.3.

ADVANTAGES: This is the most economical solution.

DISADVANTAGES: The end result of repair is often a poorly performing door. The proper repair of a door will generally require a high level of skill, which may be costly relative to the option of replacement, which will provide better performance.

2. REPLACE EXISTING DOOR WITH NEW DOOR SLAB.

The replacement of the door slab itself is possible if the frame is in good condition and square.

ADVANTAGES: Replacement of the slab will provide the opportunity to select from a wide variety of options. The replacement of slab and weather stripping may address a lead hazard or be desirable to change the appearance.

DISADVANTAGES: A door is often only as good as its frame. This is particularly true for security and fire resistance. Entrance doors typically come preassembled with a frame at relatively little additional cost. The perimeter condition goes unexamined and installation of hardware and locks becomes complicated when adjusting for existing conditions. Weather stripping, critical to ultimate performance, must be field applied.

3. REPLACE EXISTING DOOR WITH SECONDARY FRAME DOOR.

Steel frames, due to their inherent strength, permit thin jamb profiles, which minimize narrowing of the opening and provide the opportunity to use a steel door with higher insulative and security properties. The Benchmark Adjusta-Trim® product with integral trim is designed to encase the existing door frame and any associated lead paint (Fig. 4.6.1).

ADVANTAGES: A secondary frame allows simple installation and preservation of interior casings (Fig. 4.6.2). Replacement of the frame provides an opportunity to improve the whole unit performance.

DISADVANTAGES: Installation of a secondary frame door will reduce the opening size and may not be allowed by code. Replacement requires sound condition of adjacent framing and secure attachment. This method will not improve perimeter infiltration or allow examination of existing conditions, which may also require rehab.

FIGURE 4.6.1 SPLIT-JAMB STEEL DOOR FRAME AVAILABLE WITH PREFORMED TRIM PROFILE OF FLUSH AS SHOWN. BENCHMARK® AS MANUFACTURED BY GENERAL PRODUCTS CO., INC.

FIGURE 4.6.2 INSTALLATION OF A SPLIT-JAMB DOOR UNIT

4. REPLACE EXISTING DOOR WITH A NEW PREHUNG DOOR.

Exterior doors are commonly available prehung, with ancillary components such as sidelights. A prehung door is no assurance that all the components have been manufactured or are warranted by the same manufacturer. It is important to determine specifically what components will be used in the assembly of the entire door unit.

ADVANTAGES: Replacement with a prehung unit provides the opportunity to assure that the existing conditions and all components are designed to provide optimum performance. This and convenience of installation is available for little to no additional cost when skilled labor is employed.

DISADVANTAGES: A prehung unit may use undesirable components not recommended by the manufacturer. Some new doors utilize construction methods, such as integrally glazed lites, that do not lend themselves readily to repair.

GARAGE AND BULKHEAD DOORS

ESSENTIAL KNOWLEDGE

Traditional garage doors, which either swung or slide open, have evolved into a unique door type now utilizing many of the same materials and assemblies found in entrance doors. Modern garage doors operate as a series of track-mounted panel sections or a single panel that pivots, with both stowed overhead. Improvements to these mechanisms have evolved to address safety and security. Sectional doors have been designed to prevent trapping one's fingers in the closing panels. Spring mechanisms have been redesigned to provide for easier tensioning and have an integral cable that prevents broken springs from taking flight. The failure of garage doors in high-wind events exposes the building to a large breach, which can set off a chain reaction of envelope failures. Manufacturers have responded to these concerns with reinforced steel tracks and panel girders.

Of all the doors in a home, garage doors are perhaps the most vulnerable to security breaches due to their electronic control devices, which are subject to decryption by electronic scanners. New motorized door devices use a different code each time they operate to foil thieves. Because of their size and weight, automatic doors also pose a safety concern for children. Automatic door controls are required to have a reversing mechanism to detect objects in the door path.

Bulkhead doors, also referred to as basement doors, are a common feature (Fig. 4.6.3). Previously made of wood, which was subject to rot and abuse, bulkhead doors are now available made of steel, manufactured by Bilco and others. Such doors are subject to rust, as condensation often forms between cool basements and a warm exterior. Bulkhead doors made of fiberglass with spring-assisted hinges have recently been introduced to address these problems.

TECHNIQUES, MATERIALS, TOOLS

1. REPAIR EXISTING BULKHEAD OR GARAGE DOOR.

Conventional doors may be repaired by traditional methods as discussed in the Section 4.3. Any cause of damage should be corrected prior to repair or replacement.

ADVANTAGES: Typically the most economical solution, depending on the level of deterioration. The original appearance of the door is maintained.

REMOVE OLD DOORS

INSTALL SILL

CHECK SQUARENESS

COMPLETE INSTALLATION

FIGURE 4.6.3 REPLACEMENT BULKHEAD DOORS

DISADVANTAGES: Traditional materials such as wood and steel do not fare well in low-slope applications such as bulkhead doors or where cold air from the basement causes condensation on the door surface.

2. REPLACE BULKHEAD OR GARAGE DOOR.

New door products offer greater convenience, security, and durability.

ADVANTAGES: New materials and hardware provide for lighter, stronger, and rot-resistant doors. These doors have improved functions and are available with storm-resistant construction and safety devices.

DISADVANTAGES: New materials are not available in all styles or sizes.

STORM AND SCREEN DOORS

ESSENTIAL KNOWLEDGE

A storm door is generally the most cost-effective solution to a poorly performing primary entry door. The repetitive use and abuse of storm doors requires a product of sufficient strength and durability. Although available in wood and vinyl, storm doors made of aluminum or aluminum-clad materials are popular for these reasons. An insulating air space between the storm door and the primary door offsets the poor thermal quality of aluminum. The storm door also protects the primary door as a first barrier from weather. A storm door may trap moisture and heat within the intermediary space, which is detrimental to the primary door and weather stripping. The finish and material of a primary door may not tolerate these elevated temperature conditions. Some primary door manufacturers require the use of a ventilated storm door to preserve their warranty. Combination storm doors incorporate a glass and screen panel that may be interchanged on a seasonal basis. Screen inserts are also available as an option for most storm door units (see Section 4.4 for discussion of screen materials). A new product, by the name of Hid-N-Screen, employs a conventional roll screen mechanism as a door where the door swing or obscured view is otherwise undesirable. Existing storm doors may also be repaired, depending on age and condition.

TECHNIQUES, MATERIALS, TOOLS

1. REPAIR EXISTING STORM DOOR UNIT.

Storm doors are typically subject to significant abuse in a home and correspondingly often require regular maintenance and/or repair. Fortunately, many components of a storm door are modular in nature and may simply be replaced with little effort. Common problems include malfunctioning latches and pneumatic closers, and broken screen or glazing panels (see respective Sections 4.2 and 4.3 for discussion of these materials). Common maintenance includes replacement of weather stripping (see Section 4.10) and adjustment of the latch and closer. The frame itself may be repaired as per methods described in Section 4.3.

ADVANTAGES: New units are very durable and resistant to abuse, and the small effort required for maintenance or repair will achieve a long, useful lifespan for the storm door.

DISADVANTAGES: An existing door that does not accommodate ventilation may contribute to primary door damage as well as increased thermal gains to the home in hot weather. Newer materials, such as vinyl and aluminum, are not subject to warping (unlike wood storm doors) and should be considered for humid or exposed areas.

2. REPLACE EXISTING STORM DOOR UNIT.

The traditional stock storm door has changed dramatically to provide greater durability with new materials as well as a wide variety of styles to match most homes. Storm doors can also provide some degree of resistance to forced entry.

ADVANTAGES: New storm doors provide a wide variety of options to suit most homes and can dramatically improve the energy performance of the primary door. New doors are now more durable than ever, requiring a minimum of repair and maintenance.

DISADVANTAGES: Careful selection of a door is required to assure it will perform well in relation to the primary door. Older homes might have custom-built wood storm doors, which are costly to replicate with a degree of authenticity.

INTERIOR DOORS

ESSENTIAL KNOWLEDGE

Interior doors help define spaces in a house and provide visual and acoustical privacy. A door between a common space and a hall may have translucent glazing to allow light from perimeter rooms to penetrate interior spaces and also permit the return of conditioned air by means of an undercut door or transfer grille. Closet doors are typically opaque for visual screening, but louvers are desirable for air circulation. Interior doors are often distinguished by their operation or function: sliding, pocket,

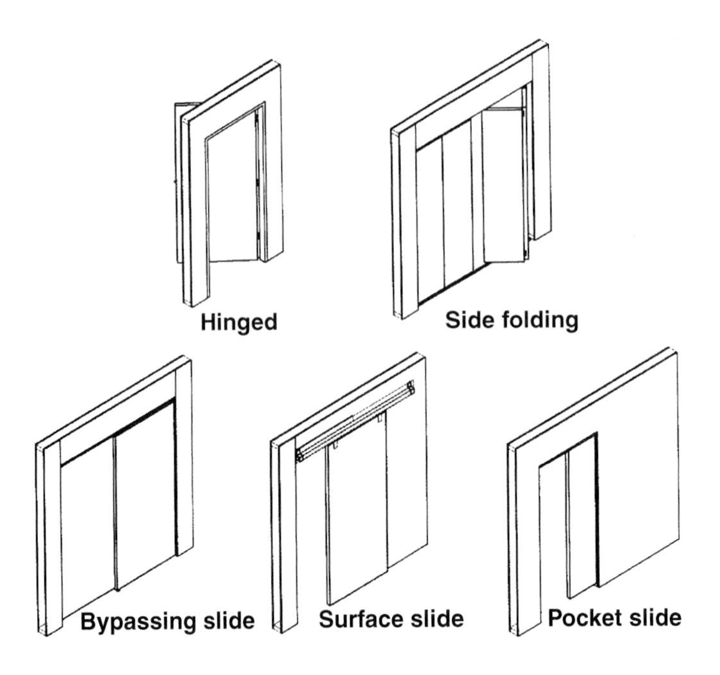

FIGURE 4.6.4 DOOR TYPES

bypass, privacy, passage, or closet (Fig. 4.6.4). Interior doors do not have sills and rarely have thresholds, except where floor levels or materials change. Pocket doors are conventional door slabs that slide on a track and do not seal tightly. Bypass doors have a double track with a guide attached to the floor. Bifold doors operate on a top track.

Common interior door slabs are also distinguished by their method of construction: flush (either hollow or solid), panel, sash, and louvered. Hollow-core flush panel doors are the most popular for interior applications. Their low cost is due to their ease of construction and engineered materials. The least-expensive doors are composed of either particleboard or jointed wood rail and stiles with a honeycomb core of cardboard and a hardboard skin material. Popular variations of panel doors include wood veneers and formed hardboard panels that simulate raised panel construction. These panels are available prefinished with simulated wood grain. Hollow-core doors are poor sound insulators, as the core air space functions like a soundboard. Solid-core doors, using a variety of core materials, have better acoustical properties. Core materials are often particleboard, solid jointed wood material (known as a stave core), or mineral cores.

TECHNIQUES, MATERIALS, TOOLS

1. REPAIR EXISTING INTERIOR DOORS.

The repair of existing interior doors, almost exclusively constructed of wood, may be achieved by conventional methods as discussed in Section 4.3.

ADVANTAGES: Repairs of cosmetic blemishes may be made with minimal effort. Doors may be stripped of lead-based paints at the time of repair to reduce potential contamination.

DISADVANTAGES: Repair of out-of-alignment doors, although typically utilizing traditional skills, may require a degree of sophistication.

2. REPLACE EXISTING INTERIOR DOORS.

Interior doors rarely require replacement but this may be desirable to redefine the relationship between rooms, by introducing either light, ventilation, or privacy. New prehung, secondary frame, and split-jamb doors require significantly less effort and skill than traditional door hanging methods.

ADVANTAGES: A change of interior doors may dramatically improve a space. New door materials such as engineered wood products and fiberglass are not as vulnerable to swelling and can prove more resistant to damage. A new unit will assure removal of a source of airborne lead contamination.

DISADVANTAGES: Doors in older homes were often custom fabricated and are difficult if not very costly to replicate. Door replacement requires disturbing existing trim, significantly increasing the degree of effort and can be a source of lead contamination.

4.7 CASING AND TRIM

ESSENTIAL KNOWLEDGE

Trim is used to conceal construction, to provide a finished appearance at gaps and joints between materials, and to accommodate slight variations between surfaces. Casing covers the gap between the window or door unit and the rough opening, and has traditionally been made of the same material as the frame (wood) and assembled on-site. With the advent of new fenestration frame materials such as aluminum and vinyl, the difficulty of joining materials requires manufacturers to supply some form of casing as an integral part of the unit. The availability and economy of installing a new window or door unit with integral casing has made the use of exterior trim primarily decorative.

A window or door installation traditionally required several trim components (Figs. 4.7.1 and 4.7.2). Sills and drip caps direct water away from the opening. A head casing, side casing, or aprons cover construction gaps. Components that extend the width of the frame are referred to as stools or jamb extensions. Standardized components have allowed manufacturers to provide units with either integral components or available options such as preassembled casing and jamb extensions. A dou-

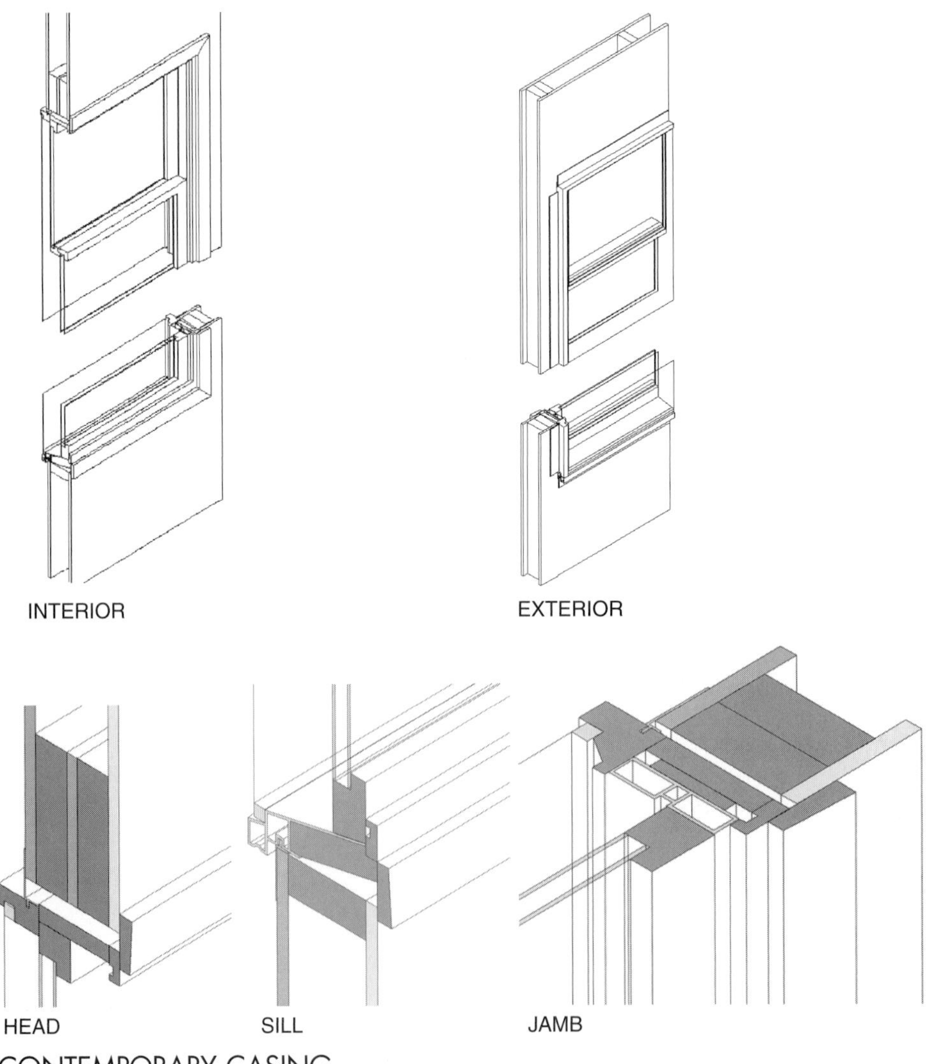

FIGURE 4.7.1 CONTEMPORARY CASING

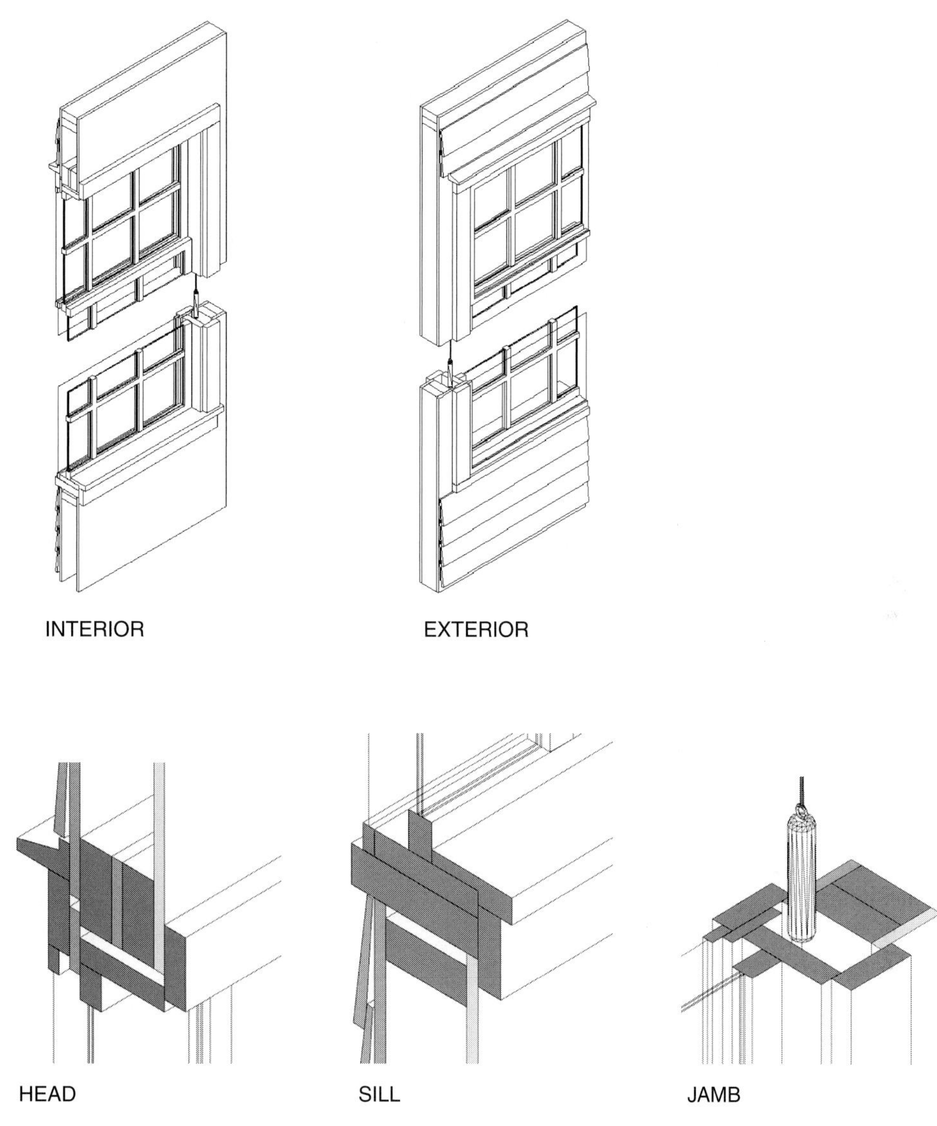

FIGURE 4.7.2 TRADITIONAL CASING

ble-hung window was traditionally trimmed differently than a casement. With today's stock units exterior casing is not required and interior casing is typically uniform on all sides. Prehung doors are available with interior and exterior trim, including thresholds.

Wood trim may be repaired by means similar to those used for window and door frames (see Section 4.3 for further discussion). Diminishing natural resources and the escalating cost of wood, in combination with the lower cost of prefabricated products and the desire for low maintenance, have resulted in a wide variety of new replacement trim products for interior and exterior. Traditional fabrication of trim by small local millwork shops has evolved into standardized, prefabricated, prefinished products that utilize a variety of new materials—either recycled, engineered, or new plastic materials—with the appearance of wood but at lower cost and reduced maintenance, uniform qualities (dimensional stability), and ease of installation. The dimensional consistency of these new products allows less-skilled workers to achieve the appearance of fine craftsmanship in areas such as miters and coping. New engineered wood exterior trim materials include finger-jointed stock, laminated veneer lumber, hardboard, and fiber cement. All of these are modified wood products that are more durable, dimensionally stable, more flexibile, and lower cost than premium solid-wood species that are increasingly scarce.

Finger-jointed stock is assembled from smaller, less-desirable pieces of premium solid woods. Laminated veneer lumber is composed of multiple layers similar to plywood products. Hardboard is made of sawdust pressed into uniform materials, some of which have properties similar to disease-resistant species such as cedar and redwood. Fiber cement, composed of a small percentage of wood fiber, is fire resistant and impervious to water and insects. All of these products are available in lengths up to 24', with uniform characteristics. These products are as little as a third of the cost of equivalent clear solid wood species. Trim for historic building styles, which is cost-prohibitive to reproduce in wood, is now available from window manufacturers in new materials that easily accommodate a wide variety of shapes. Ornate carved patterns are now achieved with new molded wood formulations or new sophisticated tooling machinery that replicates original carvings. Wood veneered trim has become a cost-effective solution. Manufacturers of these new trim materials may also provide matching jambs, sashes, and frames to assure consistency of appearance.

Factory-finished products assure a durable, consistent finish achievable only under controlled conditions and reduce on-site labor. Off-site finishing also removes flammable materials from the construction site. The traditional application of wood trim with finish nails has also changed. New materials utilize adhesives to weld plastic together on-site or use interlocking screws to hide fasteners. Corner trim is now available preassembled and butt jointed. New application methods do not necessarily reduce cost but may provide for greater consistency of work quality and reduced installation time.

TECHNIQUES, MATERIALS, TOOLS

1. REPAIR EXISTING WOOD TRIM.

Existing trim should be assessed to determine the merit of repair versus replacement premised on the existing condition, availability of matching trim, and potential for contamination from lead-based paints. Before any work proceeds, repair of existing causes of deterioration should be corrected. Almost any trim may be either repaired or replicated with the new epoxy and consolidant products, but this is a labor-intensive effort that may only be justified if the building is either historic or the amount of damage is limited.

ADVANTAGES: Repair of limited areas of damage may prove to be the most cost-effective and least-disruptive method while preserving the original appearance of the building. New wood repair products typically are more resistant to decay than wood.

DISADVANTAGES: Careful application of product and removal of existing damaged portions is required. Improper application may cause the filler to come loose over time if not properly bonded with sound material.

2. INSTALL SOLID-WOOD OR ENGINEERED-WOOD TRIM.

Traditional solid wood trim and new engineered products are popular forms of trim because they lend themselves easily to modification during installation. Trim that is to be stained or otherwise left visible typically still requires solid virgin material, but recent innovations such as veneer applied trim provide for more uniform dimensional properties. Both commonly represent the most expensive option for trim material. Wood trim is also available prefinished on all four sides, which frequently represents significant labor savings. Trim to be painted provides the opportunity to select from numerous new engineered-wood materials, which are less costly. These new materials, in addition to being dimensionally more uniform, are available in longer lengths than traditional trimboards and are free of imperfections.

ADVANTAGES: Trim may be readily modified at the time of installation. Installation of trim typically provides some form of labor savings attributable to more consistent quality of factory finish.

DISADVANTAGES: Engineered products may be more costly than wood to achieve the same appearance, depending upon the value of labor savings. Prefinished materials may prove difficult to match at the time of installation. Some new materials do not have a long performance track record.

3. INSTALL FIBER CEMENT, PLASTIC, OR POLYMER TRIM.

These materials are often only limited in length by transportation means. Longer lengths require fewer joints and associated labor, and the consistent profiles allow for ease of assembly. These materials are typically impervious to moisture and will not warp, twist, or degrade over time as readily as solid wood. The materials are often backed by exceptionally long warranties of up to 50 years. Some of these materials are formed into complex shapes and profiles and/or are flexible to accommodate irregular shapes and surfaces. Polymer trim, which has a lower density than the other materials, can be used to replicate existing ornate trim. These materials are affordable and are popular choices for replicating historical elements. Products such as Perma-Trim are designed with thin profiles specifically to encase or clad existing trim materials.

ADVANTAGES: Installation labor and maintenance costs are reduced. Materials are moisture and insect resistant and may be painted. Some products are manufactured with either waste or recycled content.

DISADVANTAGES: These materials are typically more costly than traditional materials, and the selection of sizes and shapes is limited. They are generally not able to be modified.

4. INSTALL MODULAR, PREASSEMBLED TRIM.

The most difficult element of trim is its installation. Several new products have been developed to simplify the means of joining lengths of trim with a simple butt joint, yet preserving the appearance of either a mitered or coped corner. These new products have prefabricated corners or built-up elements that snap together or are adhered in place, hiding the means of attachment. Modular units allow for low-skilled labor to replicate the appearance of craftsmanship from another era.

ADVANTAGES: Higher skill levels are not required for assembly. Typically no finishing is required, so the job is complete after installation.

DISADVANTAGES: The variety of products is limited, and the cost is significantly higher than for ordinary trim materials. These products do not accommodate irregular conditions, such as out-of-square openings, and there is no means to modify the product for such conditions.

4.8 HARDWARE

ESSENTIAL KNOWLEDGE

Hardware is often used to describe the operation of doors and windows. For example, a door slab may be either a swinging, sliding, or pocket door (depending on how the hardware operates) and either passage, privacy, or entrance (depending on the locking mechanism). Often the greatest difficulty in the repair or replacement of hardware is identifying a source of suitable components. Several large distributors and services now exist that provide exhaustive catalogs with hardware very close in appearance and function to the original, if not an exact match.

New hardware has been developed to ease the installation of new preassembled fenestration products, such as prehung and prebored doors. Installation formerly performed in the field to assure proper alignment now can be performed by the supplier, who assembles the entire unit in the shop to be installed as a modular unit. The options for repair are most often replacement. New products and tools are designed to require minimal carpentry skills.

Fenestration hardware can be classified in three primary groups: (1) hinges, tracks, guides, or a closer device for the purpose of determining movement; (2) lockset, stop, or catch as a means of securing operable parts; and (3) doorknobs, lever handles, pull handles, or push plates as operating mechanisms. Other items to supplement these include thresholds and weather stripping, discussed in Section 4.10. Most common residential hinges are a variation of the butt hinge—two plates or "leafs" secured to opposing sides that pivot around a pin connection created by alternating knuckles forming a barrel. Hinges may be mounted either on the surface, with their leafs concealed and only the barrel exposed, or entirely concealed (Fig. 4.8.1). Doors described as gliding, sliding, bypass, or pocket employ some form of a track or guide mechanism. The track or guide directs the movement of a series of rollers that are either top- or bottom-mounted. Guides are typically used where infiltration is not an issue, such as closet doors. Tracks provide a continuous barrier that can be weather-stripped, such as for a sliding patio door. Windows such as gliders also employ tracks with rollers for horizontal operation, and double-hung windows use tracks with a friction fit. Closers are hydraulic (liquid filled) or pneumatic (gas filled) mechanisms that regulate the closing speed of a door. Closers can be adjusted to allow time to travel through a doorway while reducing the impact load of a heavy door on the frame. Spring hinges are generally a less costly means of assuring closure but cannot be regulated for speed or force. Windows and doors use such counterbalances as springs, weights, and screw devices for ease of operation. Other means of controlling the swing radius of a door or sash include either wall, floor, or hinge-mounted bumpers that resist the force or hold the door in an open position.

The simplest means of securing a door or window is a catch. Magnets, friction, and spring-tension devices provide just enough resistance so as not to release the unit. Safety latches can prevent access by children to hazardous areas and typically require a combination of actions to open.

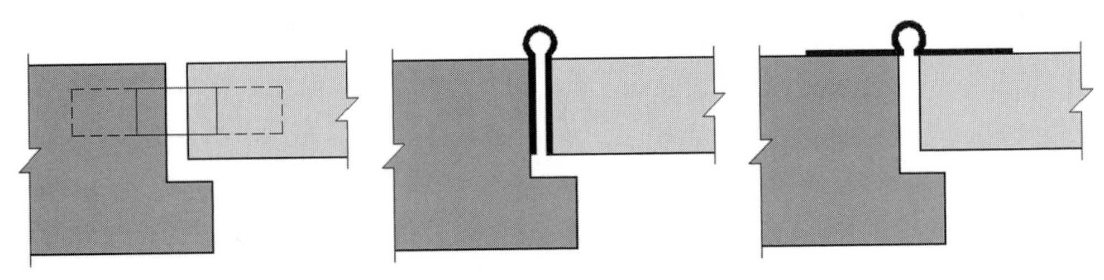

FIGURE 4.8.1 INVISIBLE, CONCEALED, AND EXPOSED HINGES

Interlocking catches provide a degree of security and a tight seal (tension). Windows may use latches in combination with sliding rods or crank mechanisms to regulate openings. Crank mechanisms have been the subject of such improvements as automated remote windows and skylights, or scissorlike hinge mechanisms to accommodate heavier sashes. Guards secure fenestration while providing a view to the exterior. Guards are simple devices that limit the swing of a door when it is open with a chain or hinged bar on the strike side that is detachable when closed.

Locksets are commonly used on doors. Three common lock types are surface-mounted (or rim lock), mortised units, and bored units (Fig. 4.8.2). Surface-mounted, or rim locks are most often associated with deadbolts and deploy a bolt through an independent strike with a lock cylinder. Increased security is provided by longer bolts that are not spring activated. Mortised locks have been considered to be more durable but difficult to install in wood doors, and are typically found in older houses. Bored locks are inserted through holes bored in the door slab and can be described as either cylindrical, tubular, or interconnected. Tubular locks, the most common, prevent the knob from turning when locked. Interconnected locks use two mechanisms operated with a single key or with the use of a single knob. All locks deploy one or more bolts that secure the operable door panel to a fixed surface (strike plate). The bolts are described as either spring latch (deploy in one direction without action, but more vulnerable to tampering) or deadbolt (requires action from both sides).

Hardware for door and window operation includes knobs and pulls (Fig. 4.8.3). Handles that lay flush are composed of bails and escutcheon plates. Operating hardware in combination with latches or locks can be either passage, privacy, or entrance devices. Passage devices provide access by means of releasing a spring-loaded latch. Privacy devices, commonly used on bathroom and bedroom doors, have a thumb turn to restrict operation from the exterior. Entrance locks preserve security, with the operating hardware usually an integral part of the mechanism.

Hardware exposed to either climatic or harsh conditions is usually made of nonferrous metals such as brass, bronze, stainless steel, aluminum, or plated steel. Recently new materials have been developed that provide similar qualities but are less expensive or easier to form into complex shapes.

When selecting hardware, consider ease of operation, installation requirements, durability, aesthetics, and security. For example, when selecting a hinge, take into account the following: frame and sash material of the door or window; number required; mounting method; the size, thickness, and weight of the door or window; clearance of the hinge to avoid casing; frequency of use; the threat of intrusion; and applicable building codes. If a change is made to the original hardware, there are other code issues that may apply, such as accessibility standards, fire codes, and egress requirements. NFPA 80, *The National Fire Protection Association Standard for Fire Doors and Fire Windows,* provides a useful guide for the selection of rated and nonrated doors and is often required by manufacturers for compliance with their warranties. Locks may be in accordance with an ANSI standard that uses such criteria as resistance to forced entry and picking. There are three security grades: Grade 1 (high security), Grade 2 (light com-

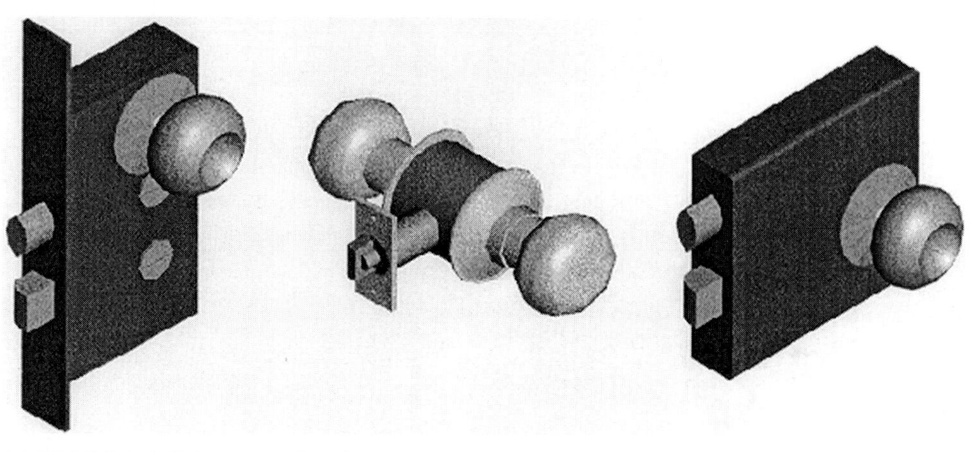

FIGURE 4.8.2 MORTISE LOCK CYLINDER BORED LOCK RIM LOCK

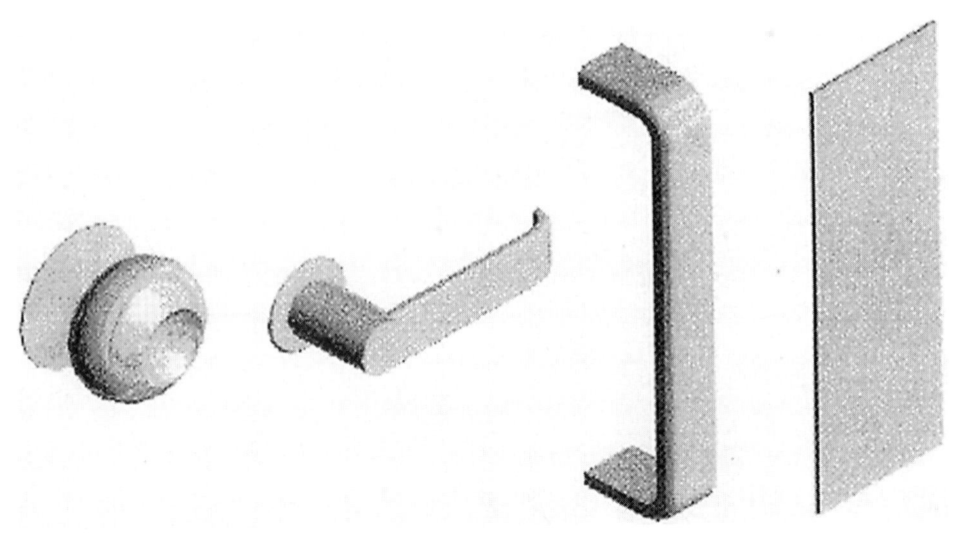

FIGURE 4.8.3 DOOR KNOB, LEVER HANDLE, PULL HANDLE, PUSH PLATE

mercial and exterior home entrances), and Grade 3 (interior applications such as bedrooms or bathrooms). Accessibility regulations govern the design and construction of multifamily residential buildings but not single-family private residences. ANSI A117.1 is the most common standard for accessibility requirements (including the Accessibility Guidelines of the Fair Housing Act), establishing minimum requirements for the location of hardware and ease of operation. Manufacturers now typically label products for compliance with the ANSI standard. There are numerous products that have been developed for retrofitting existing homes for accessibility (see Further Reading).

TECHNIQUES, MATERIALS, TOOLS

1. REPAIR OR REPLACE EXISTING HARDWARE WITH ORIGINAL COMPONENTS.

The repair of existing hardware components is often difficult at best, requiring specialized skills. Hardware is subject to repeated stress, and components that fail or have become worn should be replaced with materials of similar or equal strength. Few, if any, replacement parts are readily available.

ADVANTAGES: The primary advantage to repair is the ability to preserve the existing appearance.

DISADVANTAGES: The effort and cost associated with repair may easily exceed the cost of a complete replacement unit.

2. REPLACE EXISTING HARDWARE WITH NEW UNITS.

Hardware is often a complex mechanical apparatus, subject to both fatigue and wear. The replacement of the entire unit, available in many variations, assures all parts will have a similar life span. New materials, such as high-density polyethylene washers, provide for years of operation without the necessity for regular maintenance. Reproduction and salvaged hardware is available from national distributors who specialize in historic hardware. These distributors will often provide assistance in matching existing hardware with salvaged or new materials.

ADVANTAGES: This is often the lowest-cost alternative, with materials widely available.

DISADVANTAGES: No precise match may be available for existing hardware to match other components of a door or window.

4.9 FLASHING

ESSENTIAL KNOWLEDGE

Fenestration is often blamed for water infiltration into the building envelope because these openings interrupt the path of water traveling within the building envelope (exterior waterproofing). A building must either provide a continuous impenetrable barrier or deliberately direct the flow of water with a series of lapped materials, such as flashing. Flashing is one of the longest-lasting components of a building system. However, the proper installation of these materials requires experience and is time-consuming. New fenestration products with integral nailing fins and the use of new exterior air and moisture barrier materials, caulks, and sealants have required reconsideration of the methods and materials used to join these products (see Section 4.10 and Chap. 2 for further information).

Flashing must be durable, weather resistant, able to accommodate movement, and compatible with adjacent materials. The traditional overlapping assembly composed of multiple layers of flashing adjusts to movement like the scales of a fish and provides repetitive layers of resistant materials while covering the fasteners with each lap. Overlapping the material below prevents water from migrating in opposition to the forces of gravity when an unequal pressure condition exists, as in high-wind storms. The longer the lap the greater the force required to draw the water upward.

Noncorrosive metals such as copper, aluminum, and lead are popular flashing materials because of their durability, malleability, and impervious nature. The proper flashing system varies among window and door types, as does the method of providing an air and moisture barrier. The introduction of integral nailing flanges and large sheets of air-permeable moisture barriers has changed the common methods of providing a water barrier (Fig. 4.9.1).

Exterior doors have progressed similarly to windows and are now available prehung with casings and integral nailing fins. The traditional wood sill is a thicker profile that requires the subflooring and/or framing to be notched to provide a level entrance and flashing to protect the framing. A soldered metal door sill pan is installed below the sill to provide a protective barrier for the framing. Most doors are now produced with either extruded metal or polycarbonate sills that resist weather and have a low profile that can be attached flush to the subflooring and thus do not require flashing.

Repair of existing windows and doors often presents the opportunity to examine and repair existing flashing. The removal of a window or door may result in the penetration of the moisture barrier,

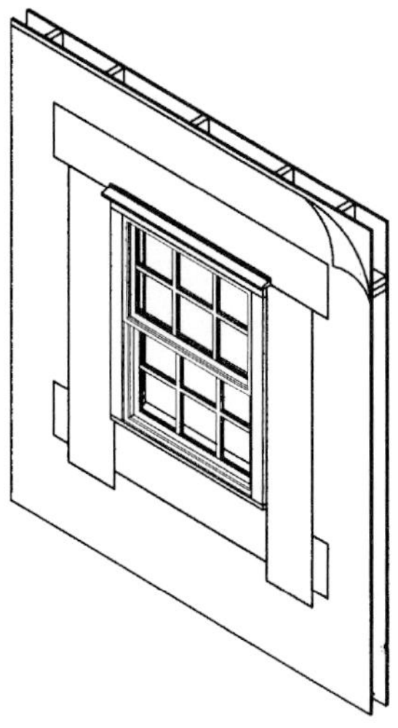

FIGURE 4.9.1 FLASHING OF INTEGRAL NAIL FLANGE WINDOW UNIT

which can be repaired with additional flashing or sealant depending on the size of the cutout. Compatibility of materials should be considered when combining new materials, caulking, or sealants and window units, which may produce an adverse chemical or electrolysis reaction. Materials that are separated may still create an electrolytic reaction if water is capable of bridging these two incompatible materials.

Flashing can be classified into three primary groups: sheet metal, vinyl products, and membrane and tape. Roofing underlayment, roll roofing materials, and tape products are examples of membranes. Conventional sheet-metal materials include aluminum, copper, zinc, and galvanized steel. Vinyl products, relatively new, are preformed in shapes suitable for particular applications. Sheet-metal and vinyl flashing are appropriate for traditional assemblies of lapped materials. New continuous drainage barriers employ self-adhering membrane and tape materials to work in conjunction with doors and windows with integral nailing fins.

TECHNIQUES, MATERIALS, TOOLS

1. INSTALL SHEET-METAL FLASHING.

Sheet-metal flashing has proven to be one of the most durable building materials, typically capable of outlasting most other components of the envelope. The common materials, in order of durability from the most to least, are as follows: stainless steel, lead, terne-coated copper, copper, Galvalume, and galvanized steel. These materials are commonly available in either sheet or preformed profiles. It is important to note, however, that special consideration must be given to the selection of materials and sealants in relation to each other so as to prevent either adverse electrolytic or chemical reactions.

ADVANTAGES: The variety of sheet-metal materials, able to accommodate most field conditions, provides a wide choice of finish appearance, durability, and costs. The installation of these materials is widely practiced and typically requires only a moderate level of skill.

DISADVANTAGES: Careful selection of flashing material and fabrication methods is necessary to avoid staining, electrolysis, or corrosion. Some materials such as stainless steel and galvanized steel are difficult to work with on-site. The costs of these materials vary dramatically but are in relation to their anticipated life span.

2. INSTALL VINYL (PVC) FLASHING.

Vinyl flashing is a relatively new material that is substituted for conventional preformed sheet-metal products. Being materially consistent with vinyl products for windows and doors, as well as with siding products, vinyl flashing has compatability with these products.

ADVANTAGES: Vinyl flashing provides for a simple, inexpensive means of flashing. The material is flexible and easily cut to conform with irregular shapes. The plastic material is not subject to galvanic action with other flashing or fasteners.

DISADVANTAGES: The variety of profiles is currently limited. Vinyl lacks tensile strength and may become brittle during cold weather. Like vinyl siding the material is subject to UV degradation and chemical incompatibility.

3. INSTALL SELF-ADHERING MEMBRANE OR TAPE FLASHING.

The increasing popularity of integral nailing fins and housewrap products has led to a variety of new methods for installing window and door units. While the methods of installation are beyond the scope of this guide, new materials, such as self-adhering tape and membranes, are commonly employed as a means of joining materials as a continuous barrier or a replacement for conventional spline and sill materials. The popularity of housewrap products is primarily attributable to the fact that they may be applied in large sheets or continuous rolls, minimizing the number of seams and reducing air infiltration. The benefits associated with reduced air infiltration may also be improved with foam gasket materials that are attached to the nailing fin either on the job site or by the manufacturer. Materials are typically available from the producers of roofing materials and housewrap products.

ADVANTAGES: The materials are easy to work with even for inexperienced users. Manufacturers claim that the membrane products self-seal around penetrations of nails and fasteners. The cost relative to other products is moderate.

DISADVANTAGES: These materials are relatively new and reliant on the adhesive bond between a wide variety of materials. Common methods of installation with these materials provide little to no redundancy for failure. Some materials when exposed to UV light, sealant materials, and excessive heat may degrade.

4.10 CAULKING AND WEATHER STRIPPING

CAULKS AND SEALANTS

ESSENTIAL KNOWLEDGE

Caulks and sealants are barriers to moisture and air infiltration with the ability to accommodate movement. However, not all joints are meant to be caulked; some provide an exit for air or moisture trapped within the wall assembly. The distinction between caulks and sealants is essentially the degree of permeability (adsorption) and the ability to conform to movement. Caulk typically provides for less movement but is easier to work and is used for interior applications, while sealants are used for exterior purposes. Here, both are referred to as sealants. The selection of sealants is made more difficult by the numerous claims and confusing terminology used by the industry to distinguish products among numerous manufacturers. There is no single product that is suitable for all uses or that provides optimal properties for a specific use. Manufacturers' instructions and technical assistance should be closely followed. The selection of caulk should be guided by knowledge of the materials that are to be adhered and the material properties that are most critical, such as durability or ease of installation. The degree of durability required and the anticipated degree of movement should also be established. Movement is expressed as a percentage of the joint's width, either positive (+) for expansion or negative (−) for contraction. The temperature at the time of installation is critical so as to utilize the sealant's full range of flexibility to expand and contract. The width of the joint should be determined during mean temperatures. Small joints prove to be the most difficult to seal because the smallest movement can represent a significant percentage of expansion. Interior applications typically do not require the same degree of elasticity because the temperature is maintained within a narrow range.

Sealants have evolved with new formulations for lower cost, ease of installation, flexibility, and durability, but no one product is ideal in all these respects (Fig. 4.10.1). Here are some of the most common sealant types.

- Latex and oil-based sealants, generally referred to as caulks with low flexibility and relatively poor durability, are low cost, easy to work, and suitable for interior applications not exposed to prolonged moisture.

- Acrylic latex, sometimes referred to as rubberized latex, is a more durable and elastic variation suitable for interior and exterior applications.

- Butyl rubber is commonly employed in insulated window assemblies because of its adhesion qualities and ability to resist water and temperature extremes. It has only moderate flexibility and is difficult to install.

TYPES OF CAULKING MATERIALS

What's Inside The Tube? Most caulks have a petrochemical base such as oil, resin, butyl rubber, vinyl acrylic, acrylic, polyethylene, polyurethane, polyvinyl acetate, etc. Additives such as stabilizers, preservatives and plasticizers give the caulks their final properties. The chart below lists the major types of caulking compounds and suggested applications.

Base Type	Retail $ (per 10 oz)	Est. Life (years)	Uses	Clean Up
Oil	1.00–2.00	1–3	Most dry surfaces*	Paint thinner
Polyvinyl acetate	1.50–2.00	1–3	Indoor surfaces only*	Water
Styrene rubber	2.00–2.50	3–10	Most dry surfaces*	Paint thinner
Butyl	2.50–3.00	4–10	Masonry and metal**	Paint thinner
Acrylic latex	2.00–4.00	5–20+	Most dry surfaces*	Water
Kraton	5.00–7.50	10–15	Most dry surfaces*	Paint thinner
Polyurethane	4.50–10.00	15–20+	Masonry**	Acetone, MEK
Silicone	4.00–7.00	20+	Glass, aluminum* (not for masonry)	Paint thinner, naphtha, toluene

** Wood, drywall, aluminum, e.g., for gaps in wood frames around perimeter of house, plumbing penetrations, gaps in rough openings around windows and doors, boots around supply and return HVAC grills, seal between bottom plates and subfloor.*

*** Gaps in masonry construction.*

The joint between a frame's seal plate and a masonary foundation has historically been sealed using caulk or adhesive, and can be a labor-intensive process. A new alternative for sealing this and other long linear joints is foam tape. This product can be faster to apply and provides a good seal by expanding to fill gaps.

FIGURE 4.10.1 TYPES OF CAULKING MATERIALS

■ Silicone, among the most flexible, is not generally paintable, is difficult to remove, and is not suitable for porous materials.

■ Polyurethanes have excellent movement and durability characteristics, but the flexibility degrades over time and they are difficult to apply and clean up.

All sealants require surface preparation and appropriate primers as directed by the manufacturer. Sealants are only able to provide for movement in two directions; if the sealant contacts a third surface, it will detach from the surface with the least adhesion. Sealants typically are applied with half the width adhered to either side of the opening in an hourglass shape (Fig. 4.10.2). The width of the opening is exposed on one side and must be prevented from adhering to materials along its other side with a nonadhering surface referred to as a bond breaker or backer material. The bond breaker material also serves to shape and support the profile of the sealant and serve as a secondary barrier. The breaker must be compatible with the sealant material and durable. Expansion (open) joints commonly use a compressible tube or rod-shaped material held in place by friction. Control (closed) joints and lap joints use a tapelike material with the sealant bridging from either side like a bandage. Although other materials are used as breakers, polyethylene and polyurethane are the most common.

TECHNIQUES, MATERIALS, TOOLS

INSTALL SEALANT MATERIALS.

Proper installation of sealant is absolutely critical to performance. The methods of installation vary among manufacturers, and it is best to consult their literature for instructions. All sealants require

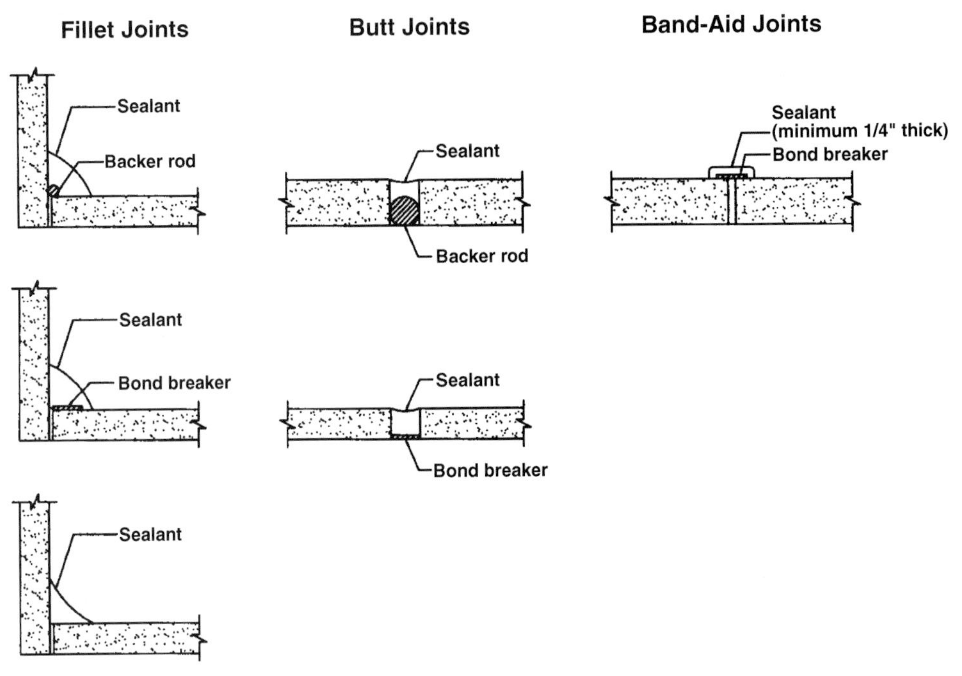

FIGURE 4.10.2 TYPICAL SEALANT JOINTS

proper preparation but the methods are beyond the scope of this book and the manufacturer's literature should be consulted. Sealants are available in essentially four types: preformed, tube, cartridge, bulk. Typically, the easier the installation method, the lower the anticipated performance. For this reason sealants available in tubes tend to be water-soluble caulk materials suitable for interior repairs. Products designed for exterior purposes, specifically windows and doors, require higher-performing materials and larger quantities. These products generally require some form of mechanical means for their application. The traditional hand-operated gun provides a relatively simple device with convenient cartridges. These guns are available in several variations to provide more control over the application. Larger bulk-loading guns are also available to provide for economy in the packaging of materials or when two-part sealants are to be combined on-site. Applications of large amounts of sealant are ideally performed with power-assisted equipment. Traditionally this was pneumatically driven, requiring an air compressor. New equipment developed specifically for this purpose is either electrically or battery driven. A new product, the Prazi™ Drill Mate, is attached to a conventional drill for this purpose (Fig. 4.10.3). With the requirements of application being so specific, these power tools and preformed materials provide for consistency of application.

ADVANTAGES: New automated tools and new materials provide for greater ease of installation and performance than ever before. The wide variety of materials is suitable for virtually any condition.

DISADVANTAGES: The improper selection of sealants has the potential of damaging or discoloring adjacent materials.

WEATHER STRIPPING

ESSENTIAL KNOWLEDGE

While sealants provide an uninterrupted barrier between materials, weather stripping provides for the movement of independently operating components such as door panels and window sashes. Weather stripping resists air and water infiltration and is also effective in reducing noise transmission and as a

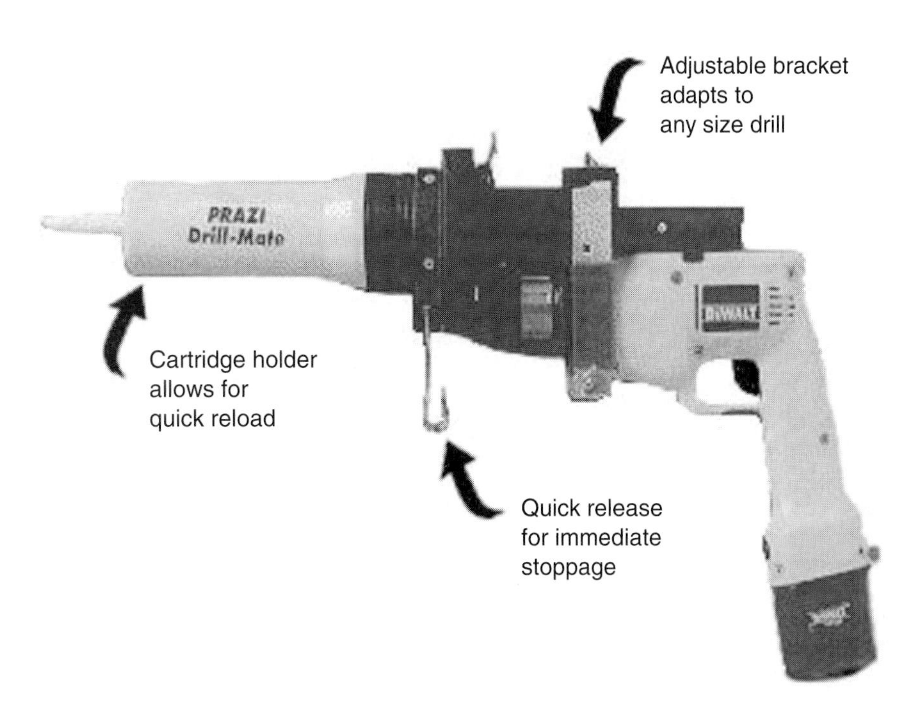

FIGURE 4.10.3 POWER CAULKING ATTACHMENT FOR CONVENTIONAL DRILL AS MANUFACTURED BY PRAZI USA

barrier to smoke and fire. Weather stripping also has value for interior applications where sound resistance is desired, presuming the door panel is not hollow core, in which case the door serves to amplify the sound.

There are three basic types of weather stripping: interlocking assembly, compression seal, and sliding seal. Interlocking seals are arguably the most durable. Constructed of corrosion-resistant metals such as aluminum and bronze, they work best on fenestration that operates consistently and does not necessarily maintain a continuous tight seal. Occupants of older homes are familiar with such materials that, when painted or bent, will prevent the door from closing. There are six common assemblies for interlocking seals in which two corresponding materials join to form a tight fit when closed.

Compression seals, particularly in combination with a locking mechanism, generally provide a tighter and more durable barrier than a sliding seal and are utilized in most new window and door products. Compression seals are described as tube seal, flipper seal, or leaf seal. The tube seal is effective and durable but is suitable for installation with only minor variation in width. Flipper and leaf seals are not as effective nor durable but can accommodate a greater variation in width. Compression seals are usually made of a material that returns to its original shape to provide a tight fit with the surface it is resisting. Such materials, which are said to have "memory," include silicone, ethylene propylene polymer (EPDM) rubber, neoprene, open and closed cell foam, vinyl, wool or synthetic pile, and spring metal. Silicone and EPDM both have good memory properties and remain flexible in cold temperatures. Neoprene is a less costly alternative but is not as durable or flexible under cold temperatures. Compression seals utilizing silicone, EPDM, and neoprene are the best materials for acoustical applications. These materials are attached by either mechanical fasteners, adhesive, or friction fit. A variation of the compression seal is a magnetic strip in combination with a flexible gasket (Fig. 4.10.4) When in combination with ferrous metal doors, windows, or a steel strip this assures a tight seal.

Sliding seals provide resistance to friction on sliding components and are described as sweep-sill or brush-seal. Sweep-sills have a bladelike profile. Sliding seals require materials resistant to friction, such as polypropylene or nylon. Face-mounted seals are most effective because there is relatively little variation in the width of the opening. Weather stripping may be applied by means of adhesive, by friction fit, or mechanically.

Typically the same material is used at the full perimeter, with the exception of a door sill. A threshold must be durable against wear of moving parts and traffic. Thresholds are typically fabricated of very durable materials such as metal in combination with inserts that function as tube seals either in compression or sliding (Fig. 4.10.5). A raised threshold provides contact with the sweep or seal only when the door is closed. In addition to interlocking and resilient types, door bottom seals are available as an automatic operable mechanism (Fig. 4.10.6). As a door closes, a button on the hinge side of the jamb compresses a seal in a downward motion against the threshold. Weather stripping may also be used in opposition with itself, two brushes, two flippers, or two compression seals.

ANSI standard A156.22, *Door Gasketing Systems*, is valuable in evaluating the performance of respective weather stripping types for appropriate selection.

TECHNIQUES, MATERIALS, TOOLS

The majority of problems associated with doors is attributable to the installation of hardware and weather stripping. Prior to any weather stripping, the window or door should be repaired or determined to be properly operating. Typically the same material is applied to all sides with the exception of the door sill. The installation of new materials is a labor-intensive process. When this work is to be performed by a professional, only the best-quality materials should be selected because they will represent a small portion of the overall cost and assure the greatest durability. Products with lower durability or effectiveness may be suitable because of the ease of installation provided to the amateur. New tools have been developed specifically for the purpose of installation, significantly reducing the skill required. If numerous windows or doors require repair, the cost of the tools may be justified.

1. INSTALL NEW OR REPLACEMENT INTERLOCKING WEATHER STRIPPING.

Installation on new interlocking weather stripping typically requires modification of the door or window and jamb to provide a pair of aligned components. Older windows and doors commonly utilize this type of weather stripping which when worn simply requires replacement. This type of weather stripping allows the door or window to operate without impedance or resistance.

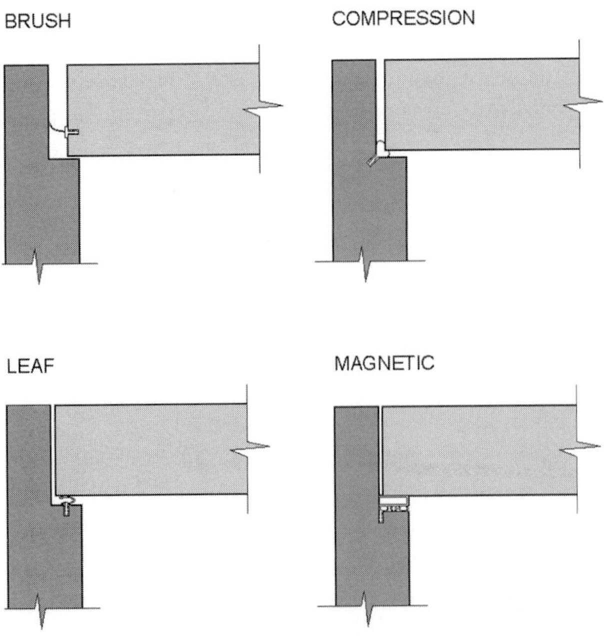

FIGURE 4.10.4 TYPES OF WEATHER STRIPPING AT JAMB

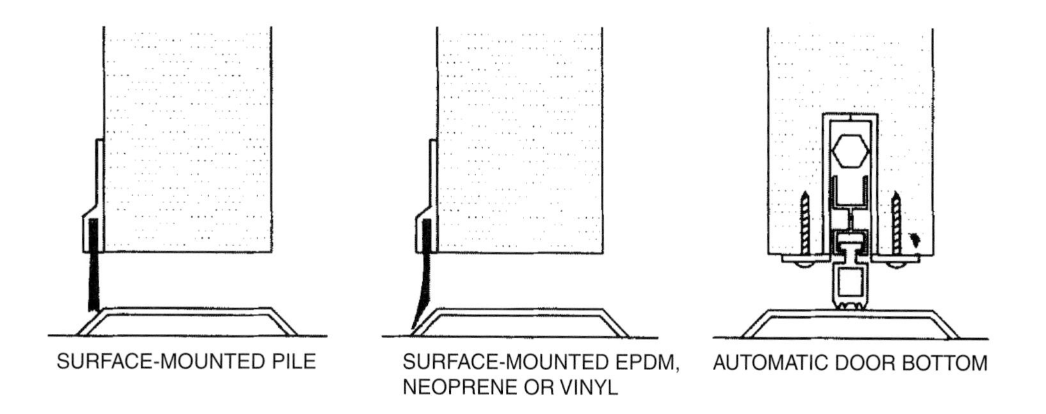

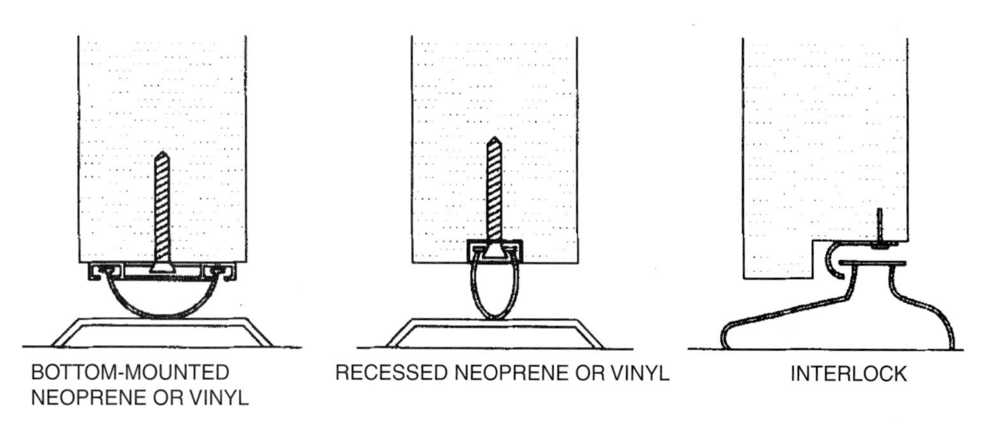

FIGURE 4.10.5 TYPES OF WEATHER STRIPPING AT THRESHOLD

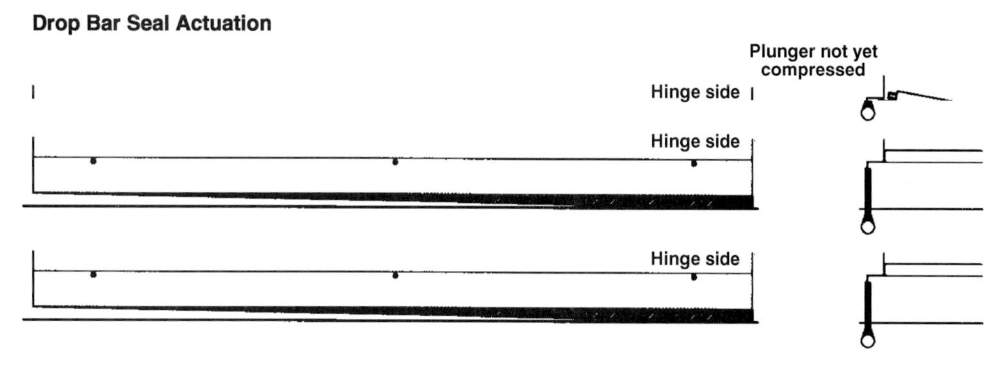

FIGURE 4.10.6 AUTOMATIC THRESHOLD SEAL

ADVANTAGES: Interlocking weather stripping provides a very durable seal that does not impede function.

DISADVANTAGES: Interlocking weather stripping does not provide for a full-contact seal and requires skilled labor for new installations.

2. INSTALL NEW OR REPLACEMENT RESILIENT (COMPRESSION OR SLIDING) WEATHER STRIPPING.

Compression and sliding types may be generally described as resilient weather stripping. The resilience of these materials provides the opportunity to accommodate irregular surfaces and those with a tight seal achieved with full contact. Numerous products are available for this purpose with a wide variety of performance. Compression does not necessarily ensure a tighter seal; manufacturer's recommendations for sizing should be carefully observed. The recently developed ANSI standard not only establishes compliance but provides guidance in the selection of appropriate materials.

ADVANTAGES: Compression materials typically provide the tightest seal and are generally very durable. Installation can usually be accomplished with a low degree of skill. New materials employed in sliding seals have significantly improved their durability while providing a tight seal. These materials are often applied to the surface of a door.

DISADVANTAGES: Resilient materials may prevent the tight closure of the door or sash, impede the function of hardware, and are generally less durable. Some materials are subject to damage under cold conditions or exposure to UV light.

3. INSTALL NEW OR REPLACEMENT DOOR THRESHOLDS.

Thresholds are available in a variety of configurations, which will determine which type of weather stripping is appropriate. Thresholds that employ a raised stop may continue the compression seal around the full perimeter. Other assemblies work in combination with either sweep or compression seals that are attached to the door unit and are compressed against the threshold. Some threshold units with integral seals are undesirable because the seals are subject to damage from foot traffic. A variation of a threshold seal is the automated door bottom that deploys a seal downward when the door is in the closed position.

ADVANTAGES: New, durable materials are designed to last many years while improving thermal performance of the assembly.

DISADVANTAGES: The replacement of the entire threshold requires a degree of skill. New, unique designs may prove to be difficult to repair in the future.

4.11 SHUTTERS AND AWNINGS

ESSENTIAL KNOWLEDGE

Shutters and awnings were the means in the past by which windows and doors were shaded or protected from storms and other intrusions. Although their original functions have been addressed by other means such as air conditioning and security systems, they remain popular decorative elements. To maintain a house's historic character, they occasionally need to be repaired or replaced. Shutters have proliferated as applied, fixed units made of contemporary, low-maintenance materials. The desire for historical authenticity and environmental concerns have also created a demand for authentic operating wood shutters, and the function of shutters and awnings is now being reevaluated.

Operable shutter louvers allow control of heat gain and glare, while providing diffuse light and promoting natural ventilation through a house and preserving some degree of privacy. Awnings provide an effective means of shade, reflecting 80% to 90% of sunlight without obstructing the view, but some awnings trap heated air against windows. Shutters also have been revived as a means of protecting windows on homes built in high-wind regions.

New innovations include automated operation and more durable materials. Traditionally made of disease-resistant wood species, shutters are now available in aluminum and vinyl. High-wind or hurricane-resistant shutters require the strength of aluminum. Awnings, once exclusively manufactured of canvas, now employ new, dimensionally stable and rot-resistant fabric materials that can withstand prolonged exposure to UV light, and significantly reduce maintenance. Similar methods and materials, which are well documented in how-to repair books, may be used for the repair of existing shutters. Awnings require regular replacement of the fabric, presenting the opportunity to employ more durable materials.

TECHNIQUES, MATERIALS, TOOLS

1. REPAIR EXISTING SHUTTERS AND AWNINGS.

Conventional shutters and awnings may be repaired by traditional methods as described in Section 4.3. Awnings are typically designed for ready replacement of the fabric material. Several sources of replacement materials are identified under product information.

ADVANTAGES: Repair of existing units with common methods will preserve authenticity and improve durability. New materials available have significantly improved the anticipated life span of these units.

DISADVANTAGES: The selection of original shutters or awnings may have been inappropriate and make them difficult to maintain.

2. REPLACEMENT OF SHUTTERS AND AWNINGS.

New shutters and awning products provide greater convenience, security, and durability.

ADVANTAGES: New products often provide distinct advantages previously unavailable. Apart from the convenience of operation or increased durability, new storm shutters and shading devices provide tangible benefits in lower energy consumption and possibly reduced insurance premiums.

DISADVANTAGES: Authentic shutters and awnings are typically expensive to replicate.

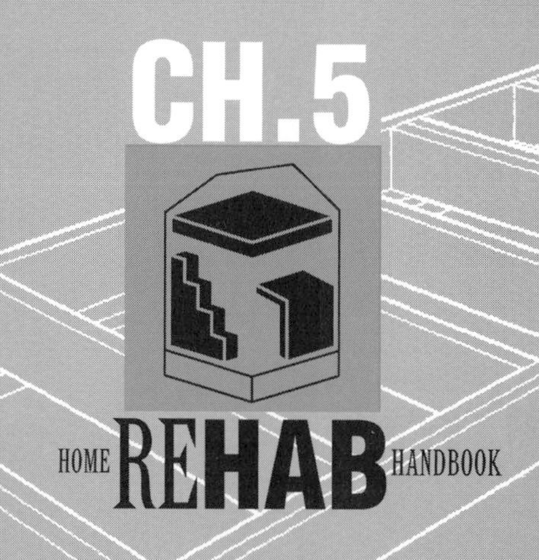

PARTITIONS, CEILINGS, FLOORS, AND STAIRS

Chapter 5
PARTITIONS, CEILINGS, FLOORS, AND STAIRS

5.1. FLOOR AND CEILING STRUCTURE
5.2. SUBFLOORING AND UNDERLAYMENTS
5.3. FINISH FLOORING
5.4. FINISH WALLS AND CEILINGS
5.5. STAIRS

5.1 FLOOR AND CEILING STRUCTURE

OVERVIEW

If a building's foundations, exterior walls, and roofs have been well maintained, and the building has been free of plumbing leaks, general abuse, overloading from heavy objects, and damage from vermin and insects, then chances are that its floor and ceiling structural assemblies will be in reasonably good shape. Exceptions to this include problems caused by deficient design, engineering, or construction, or inappropriate material selection or detailing. If, however, the building's roofs, exterior walls, and foundations have not been adequately maintained, the floor and ceiling structural assemblies (and the partitions and stairs that attach to them) will likely have suffered from the settlement, and insect- and moisture-related deterioration that affects the rest of the structure. This section addresses common structural deficiencies of floor/ceiling assemblies. Moisture-, insect- and fire-related issues are covered more extensively in Chapters 2 and 3. Additionally, because a structural rehab may involve removing and reframing a deteriorated structure, recent innovations in joists, beams, and headers will be reviewed.

COMMON STRUCTURAL PROBLEMS

ESSENTIAL KNOWLEDGE

Most problems with floor and ceiling structural assemblies are related to excessive sagging and deflection and can be attributed to a number of deficiencies, including beam strength that has been reduced by extensive notching at joist-framing connections and joists that are excessively cut at ends (Fig. 5.1.1) where they frame into girders (typical in pre-1900 houses); joists that have been excessively cut, notched, or bored (Fig. 5.1.2) to accommodate material changes, pipes, wiring, or ducts (code agencies and manufacturers of engineered wood beams and joists stipulate limits to such modification); insufficiently sized supporting beams; inadequately sized or spaced floor joists and fasteners; excessive spacing of posts supporting the beams; rotting of posts at bearing points; and insufficient or settled footings under the posts. Problems with partitions are usually related to insufficient floor support or shrinkage of the studs, which are dealt with below.

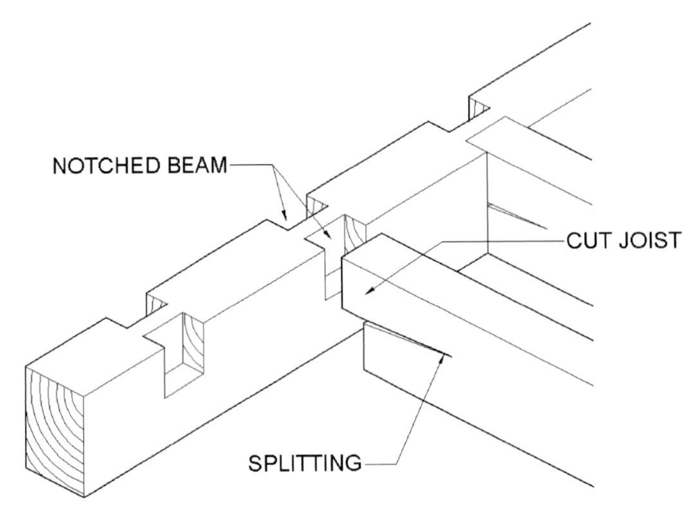

FIGURE 5.1.1 EXCESSIVE NOTCHING OR CUTTING OF BEAMS AND JOISTS

Joint size	Maximum hole	Maximum notch depth	Maximum end notch
2 × 4	None	None	None
2 × 6	1½	⅞	1⅜
2 × 8	2⅜	1¼	1⅞
2 × 10	3	1½	2⅜
2 × 12	3¾	1⅞	2⅞

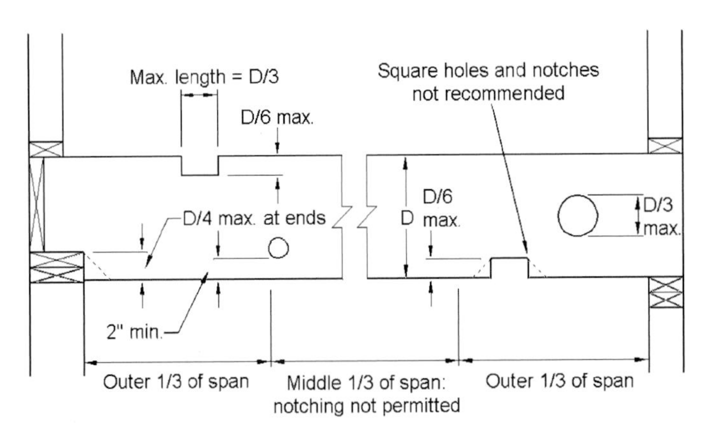

FIGURE 5.1.2 GUIDE FOR CUTTING, NOTCHING, AND BORING JOISTS

Most current codes limit deflection for floors to L/360 (length of joist/360) which is derived from longstanding standards based on the deflection at which a plaster ceiling of the space below the floor would crack. While this is generally considered adequate to control deflection, some architects, engineers, and designers believe that stiffer floors are necessary for a user's sense of well-being, and design to higher deflection limits (such as L/480) or increase the floor load requirements from 40 psf on the nonbedroom floors to as much as 100 psf in the more public spaces such as kitchens, entries, family rooms, and living rooms. Another approach is to limit deflection to a maximum, for example, ½". Many older houses built before 1920 have floor joists sized considerably below current requirements. It is not uncommon to find 3x6 and 3x8 joists in pre-1900s housing that, in some cases, have been notched where they frame into girders by as much as one-half their depth. These members may well be split, especially if they have been affected by rot or insect damage. Because of the large safety factor used in the design of newer floor systems (up to factor of four), floors will rarely fail structurally, but they may have excessive bounce and feel unsafe.

TECHNIQUES, MATERIALS, TOOLS

1. REINFORCE EXISTING MAIN BEAM BY ADDING SUPPORT.

A house's main beam that is overstressed, has deflected excessively over time, or has been affected by termites can be reinforced by adding a steel or wood column or a masonry pier to reduce the beam span (Fig. 5.1.3). It may also be possible to jack up the sagging beam to reduce or eliminate a slope in the finish floor above, although long-term settling is often difficult if not impossible to eliminate.

ADVANTAGES: This is a relatively simple and effective way to stabilize a building's major structural element.

DISADVANTAGES: This is difficult to accomplish in other than basement or crawl space areas, as columns may have to be placed in inconvenient places and will require some removal and restoration of existing finishes. Beams over crawl spaces may be difficult to access.

2. REINFORCE THE EXISTING BEAMS OR JOISTS.

Existing beams or joists can be reinforced by adding steel or wood reinforcement (sistering) along the existing members to develop additional load-carrying capacity (Fig. 5.1.4). The length and bearing of the new reinforcing beams or joists will depend on the existing conditions and should be reviewed with a structural engineer.

ADVANTAGES: Reinforcement of existing beams or joists can eliminate the use of a new column support and does not affect the space below the beam.

DISADVANTAGES: It may be difficult to insert a new support alongside the existing beam if access is a problem (such as in crawl spaces) or if joists frame directly into the beam (Fig. 5.1.5), in which case

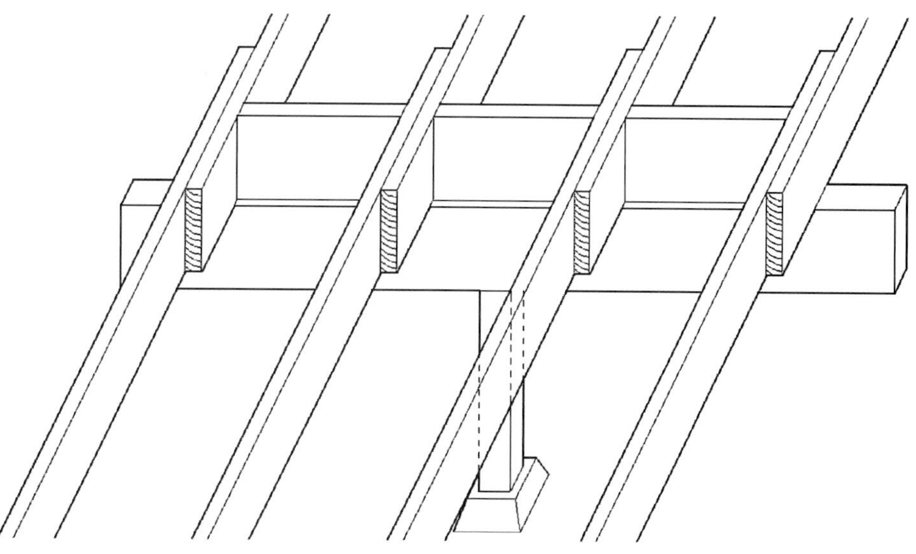

FIGURE 5.1.3 NEW COLUMN SUPPORT

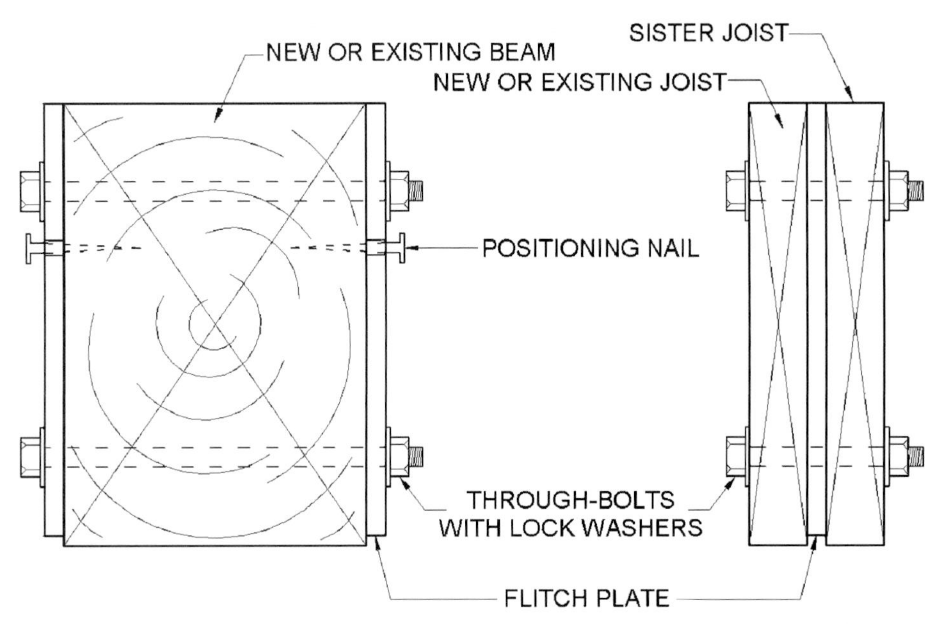

FIGURE 5.1.4 STEEL FILCH PLATE REINFORCEMENT

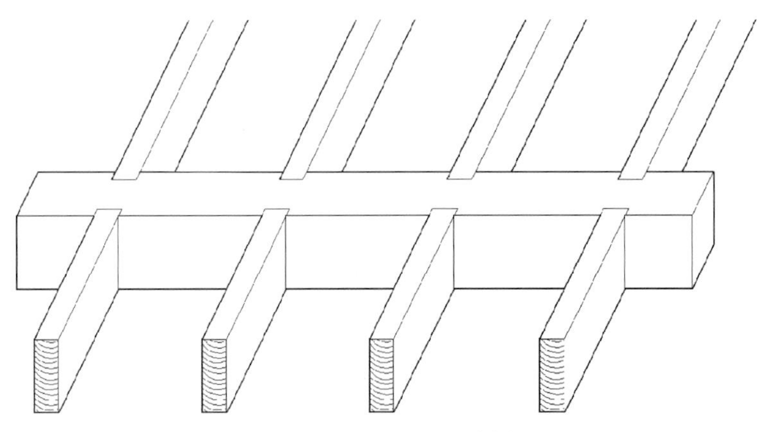

FIGURE 5.1.5 JOISTS FRAMING INTO NOTCH BEAM

the existing floor joists would have to be temporarily supported, a new beam installed, and the joists hung from the new beam with joist hangers (Fig. 5.1.6).

3. TRANSFER LOAD TO EXISTING JOISTS.

If a floor joist has been severely cut to accommodate a large pipe or wiring, it may be impossible to splice on a reinforcement member. In this instance, it may be preferable to transfer the load from that joist to adjacent joists using header joists that are end-nailed across the cut end of the interrupted joist to the adjacent trimmer joist. If the header has a span of 4' or less, a single header may be satisfactory (Fig. 5.1.7). For wider openings (up to 10') headers can be doubled up (Figs. 5.1.8 and 5.1.9). Consult with a structural engineer or architect to verify.

ADVANTAGES: This method can reinforce floors when other alternatives are not practicable.

DISADVANTAGES: This method is not possible where access is a problem.

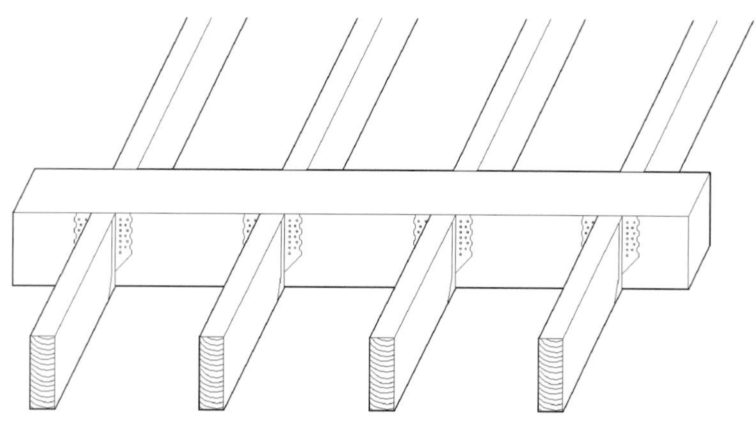

FIGURE 5.1.6 NEW BEAM WITH JOISTS ATTACHED WITH JOIST HANGERS

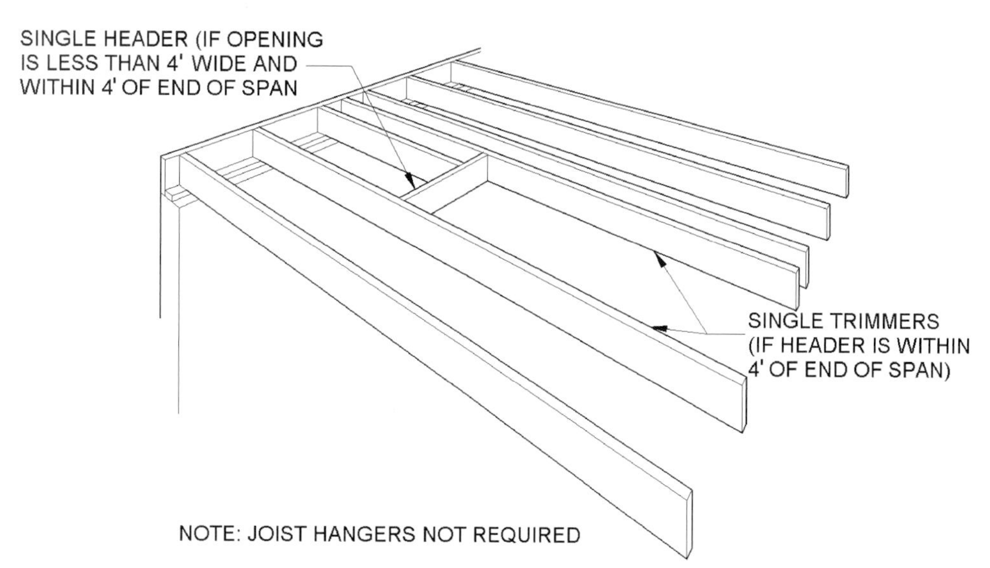

FIGURE 5.1.7 FLOOR OPENING FRAMED WITH SINGLE HEADER AND SINGLE TRIMMER JOISTS

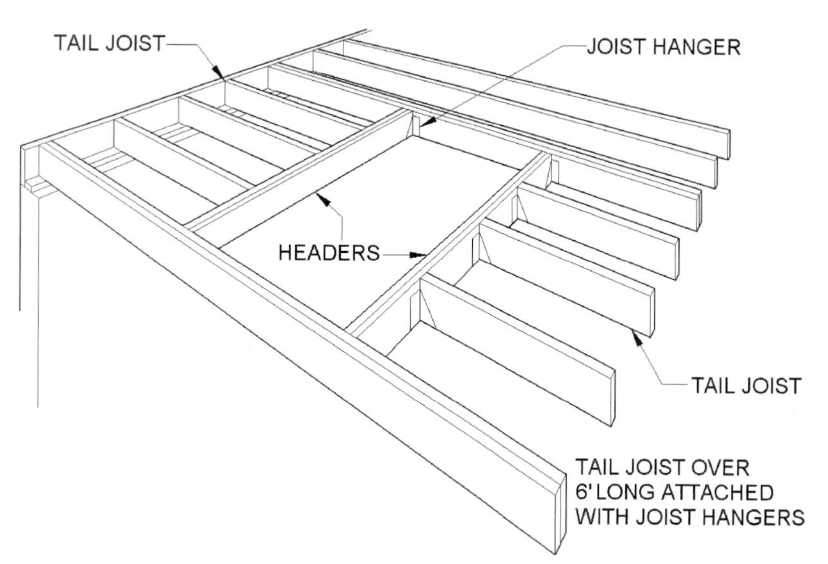

FIGURE 5.1.8 FLOOR OPENING FRAMED WITH DOUBLE HEADER AND DOUBLE TRIMMER JOISTS (JOIST HANGERS USED FOR LONGER TAIL JOIST)

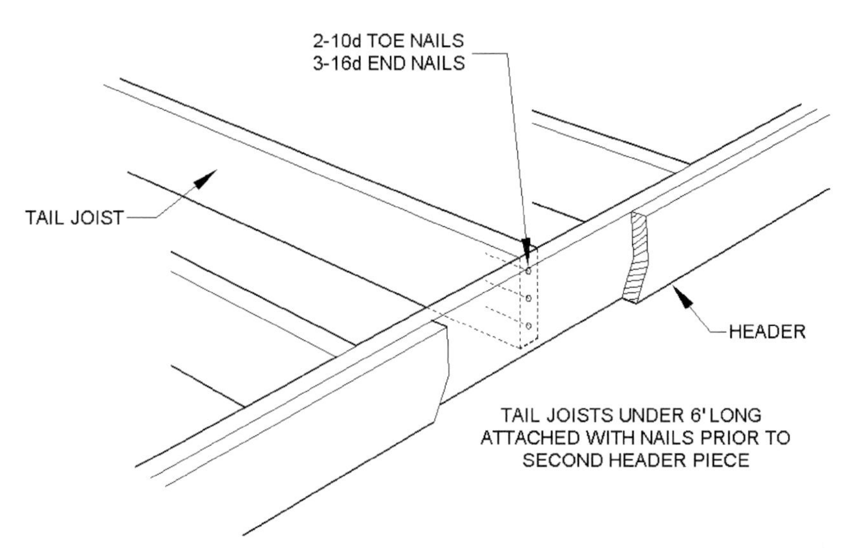

FIGURE 5.1.9 FLOOR OPENING FRAMED WITH DOUBLE HEADER AND DOUBLE TRIMMER JOISTS (NAILING TAIL JOISTS UNDER 6' LONG)

ALTERNATIVES TO SOLID LUMBER FOR FLOOR FRAMING

ESSENTIAL KNOWLEDGE

A variety of structural materials is available as an alternative to solid wood framing when existing framing needs to be reinforced or when sections of exsting houses need to be rebuilt. Light steel sections and steel flitch plates have a long history of use in wood-frame construction for carrying heavy loads over long spans. Wood floor trusses connected with metal plates and webs have been in use since the 1970s. Glue-laminated timbers (glulams), initially developed in 1890, have been used since the 1950s for both exposed beams in post-and-beam structures and concealed as headers and girders in long spans such as garage door openings. There is little possibility of repairing existing glulams if they are delaminated or overstressed, in which cases they should be replaced with new material or additional supports, respectively. Today a wide range of such products as engineered wood I-joists and structural composite lumber (SCL) has been developed and is in use. The following documents the repair of traditional products as well as the physical characteristics of some of the newer products.

TECHNIQUES, MATERIALS, TOOLS

1. REPAIR EXISTING TRUSSES.

The critical components of typical floor trusses are the metal connector plates. Damage to the plates, when it occurs, usually happens during handling, especially when trusses are lifted vertically off the ground or are raised from a horizontal to vertical position. Serious racking can cause the truss chords to break or the metal plates to disengage. Another cause of plate failure is deterioration from high-salt and moisture-laden environments. Assuming it was properly designed, if an existing truss plate shows white rust on the zinc coating or minor red rust around the edges, and the metal plate is correctly engaged, the truss is probably performing satisfactorily. If the plate exhibits significant blistering and scaling, the probable causes (typically excessive moisture) should be addressed and corrected before remedial work is undertaken. Corrective work can include the removal of scale from the plate and recoating with an appropriate rust-inhibiting paint, and the use of wood gusset plates or additional hand-driven metal plate connectors.

ADVANTAGES: This repair may allow the continued use of isolated trusses without the cost of replacement.

DISADVANTAGES: This repair is not usually possible if deterioration or damage is severe or widespread. Review by an engineer is required.

2. REPLACE EXISTING JOISTS WITH NEW METAL PLATE CONNECTED TRUSSES.

Wood floor trusses are available in a wide variety of sizes and styles both as standard commodity items and in project-specific configurations. The majority are made with 2x3, 2x4, and 2x6 lumber held together by metal plate connectors (Fig. 5.1.10). The punched connector plate acts as an array of short nails attached to a common head. Connectors are sized to ensure that the trusses can take their in-plane design load as well as out-of-plane loadings that occur during assembly, handling, transportation, storage, and erection.

ADVANTAGES: Metal plate connected trusses have a greater span capability than solid wood joists. The need for a center beam and columns is often eliminated. The open web permits running plumbing, ductwork, and wiring through trusses. These trusses are economical (they make very efficient use of small sections of dimensional lumber), are lightweight and easily handled and worked in normal lengths without heavy equipment, and are available throughout the country.

DISADVANTAGES: These trusses require more careful handling than comparable solid timbers and I-joists. Longer lengths (over 30') require lifting equipment and spreader bars. These trusses are combustible, performing less well in fire than thicker, solid framing. They are also less dimensionally stable than other framing types.

3. REPLACE EXISTING STRUCTURE WITH GLULAMS.

Glulams (Fig. 5.1.11) are typically used to replace deficient beams or headers that carry heavy loads such as ridge beams and garage door headers. They are made up of wood laminations 2½" to 10¾" wide that are bonded together to form beams of varying depths. Beams are manufactured with the highest-grade laminations on the top and bottom where the greatest compression and tension occurs. There are three appearance grades: premium, architectural, and industrial, the latter being recommended for concealed locations or where appearance is not a factor. Appearance does not affect structural characteristics.

ADVANTAGES: A glulam has greater strength and stiffness than comparable dimensional lumber, is more efficient than solid lumber for large spans and loads, and is available in wider sections than laminated

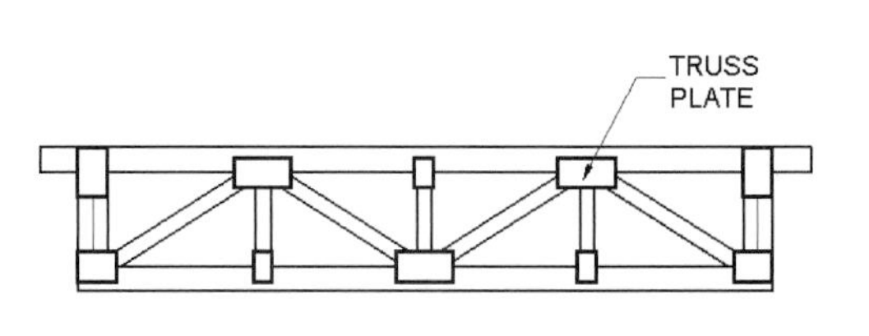

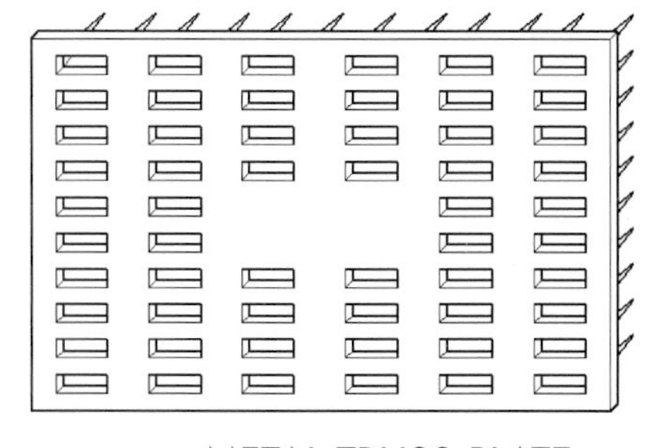

FIGURE 5.1.10 DIAGRAMMATIC FLOOR TRUSS METAL TRUSS PLATE

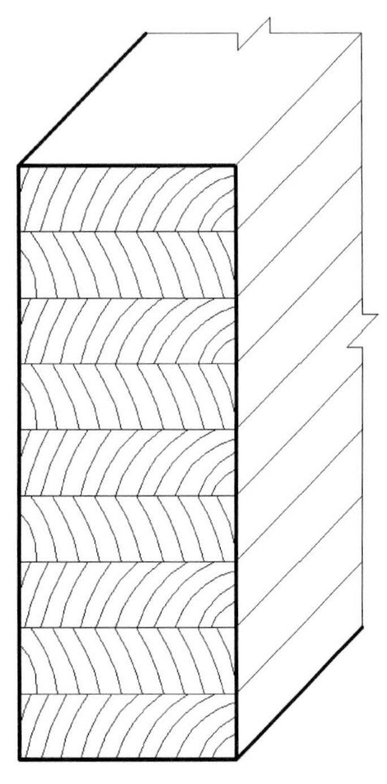

FIGURE 5.1.11 GLULAM

veneer lumbers, which have to be bolted together to carry equivalent loads. It is the only engineered wood product that can be easily cambered to reduce the visual effect of deflections or can be produced in curved shapes and arches for long spans. It has a striking appearance and can be left exposed.

DISADVANTAGES: Glulams are more costly than structural composite lumber. They cannot be nailed and require bolts and metal hangers.

4. REPLACE EXISTING STRUCTURE WITH WOOD I-JOISTS.

First marketed by Trus Joist Corporation in the late 1960s, wood I-joists (Fig. 5.1.12) are currently produced by over a dozen companies in North America. I-joists are used predominantly as an alternative to sawn lumber floor joists or parallel-chord trusses in repetitive light frame construction, although they are also used for beams and headers. The I-joist is a very efficient structural shape simulating that of a steel I-beam. I-joists typically have single webs, but at least one manufacturer, Superior Wood Products, makes a rigid-foam-insulated, double-webbed member for use as a header. Most I-joints are available with 1½" round *knockouts* spaced 12" o.c. to accommodate pipes and wiring.

ADVANTAGES: The I-joist can be used as a floor joist, header, or rafter or roof joist. I-joists are lightweight, easy to handle and install, with longer, unsupported spans and higher load-bearing capacity than solid sawn lumber of equivalent size. They are available in a wide variety of sizes and lengths; are dimensionally stable, and resist shrinking, warping, splitting, or twisting that can lead to squeaky floors. They are environmentally sound, using less wood and lower-quality trees than conventional lumber.

DISADVANTAGES: In longer lengths, I-joists require more careful handling than equivalent solid lumber members. Lack of a single industrywide performance standard makes comparison of products challenging.

5. REPLACE EXISTING STRUCTURE WITH LAMINATED VENEER LUMBER.

Laminated veneer lumber (LVL) was initially developed by Weyerhaeuser Corporation in the early 1960s, but was not produced commercially until the Trus Joist Corporation developed and began to produce Microllam® in 1968. LVL consists of layers of kiln-dried wood veneer (typically Douglas fir

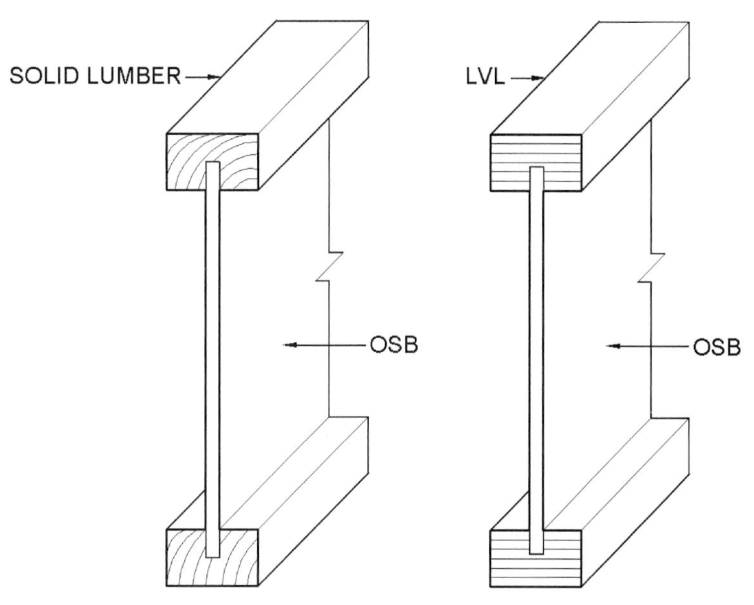

FIGURE 5.1.12 WOOD I-JOIST

and southern pine) laminated together with a structural adhesive with their veneer grains aligned with the length of the member (Fig. 5.1.13). They are typically used for girders and headers over long spans such as garage door openings, and as flange material for I-joists.

ADVANTAGES: Laminated veneer lumber combines the best qualities of natural wood with the strength and consistent performance of engineered wood products. It is easy to work with and can be cut and nailed in the field. It is dimensionally stable and resists shrinkage, warping, splitting, and checking. It is available in a wide variety of sizes from a number of manufacturers.

DISADVANTAGES: Laminated veneer lumber is relatively costly.

6. REPLACE EXISTING STRUCTURE WITH PARALLEL STRAND LUMBER.

Parallel strand lumber (PSL) was developed by MacMillan-Bloedel and introduced in 1984 under the trade name Parallam® PSL. It is currently produced exclusively by Trus Joist MacMillan. Parallam® PSL (Fig. 5.1.14) is made of strands of clear sapwood from the outer portions of Douglas fir, western hemlock, southern pine, and yellow poplar, which are not usable in the making of LVL. The strands are combined with waterproof structural adhesives, formed into a mat under heat and pressure, formed into billets, and ripped into smaller members.

ADVANTAGES: Parallel strand lumber has the highest stiffness and strength of all the composite lumber products. It is used where both strength and appearance are important and makes even greater use of wood fibers than LVL. It can be preservative and fire retardant and is available in a variety of sizes and lengths. It has thick members, which eliminate the need for multiple bolted or nailed-together LVLs or solid lumber members.

DISADVANTAGES: Parallel strand lumber is heavier than equivalent-sized sawn or glued-laminated lumber. It is more abrasive to saws and drills than LVL because of higher adhesive density. Connections must be made with metal plates and bolts rather than nails. It is relatively costly and is available only from a single manufacturer.

7. REPLACE EXISTING STRUCTURE WITH LAMINATED STRAND LUMBER.

Laminated strand lumber (LSL) was developed by MacMillan-Bloedel and has been produced since 1991 by Trus Joist MacMillan under the trade name TimberStrand® LSL and is a recent structural

FIGURE 5.1.13 LAMINATED VENEER LUMBER (LVL)

FIGURE 5.1.14 PARALLEL STRAND LUMBER (PSL)

composite lumber product (Fig. 5.1.15). LSL strands are longer (approximately 12") than those used in PSL. LSL is used primarily for beams and headers, although it is increasingly used for studs.

ADVANTAGES: Laminated strand lumber retains the advantages of both LVL and PSL products, but has the additional advantage of being able to be made from small, crooked logs of many species that are fast growing and underutilized (used for making OSB) such as aspen and yellow poplar.

DISADVANTAGES: Laminated strand lumber is not as dimensionally stable as LVL or PSL products. Connections must be made with metal plates and bolts, rather than with nails. LSL is available only from a single manufacturer.

8. REPLACE EXISTING STRUCTURE WITH STEEL FLOOR JOISTS.

Light-gauge, cold-formed steel floor framing is an innovative housing construction technology that is experiencing significant growth and acceptance in the home-building market. In general, the low

FIGURE 5.1.15 LAMINATED STRAND LUMBER (LSL)

thermal performance of steel framing, when compared to the thermal performance of wood framing, continues to be a major barrier for complete home builder and home owner acceptance. However, because steel floor framing is contained within the building envelope, the thermal concerns are minimal. As a result, steel floor framing presents a potential cost-effective alternative for home builders to explore. Cold-formed steel framing offers many benefits to the construction industry including superior structural performance, a factory-produced, quality-controlled manufacturing process; time and cost savings; price stability; dimensional uniformity; noncombustibility; termite resistance; high strength-to-weight ratios; and recyclability. Steel floor joists with preformed web openings are the most recent innovation in the steel floor framing industry. The preformed web openings provide home builders with space through which heating, ventilation, and air-conditioning (HVAC) ductwork and plumbing and electrical conduits can be installed. Typically, HVAC branch ducts are installed through the holes. However, Steel Floors, LLC, a steel floor joist manufacturer in Denver, CO, manufactures a larger-size joist (14" deep) with large web openings ($10\frac{1}{4}$" × 23") through which main duct runs can be installed, in addition to branch ducts, plumbing, and electrical systems (Fig. 5.1.16).

ADVANTAGES: Because steel floors are contained within the building envelope, heat loss is minimal. There is high structural performance when compared to wood.

DISADVANTAGES: Builders may be reluctant to use steel because workers must be trained in alternative installation techniques and materials.

MOISTURE DETERIORATION

ESSENTIAL KNOWLEDGE

Water absorbed by wood framing can raise its moisture content, reduce its compressive and tensile strength, ultimately cause rot and decay, and attract termites and other pests (see Chapter 2). Interior

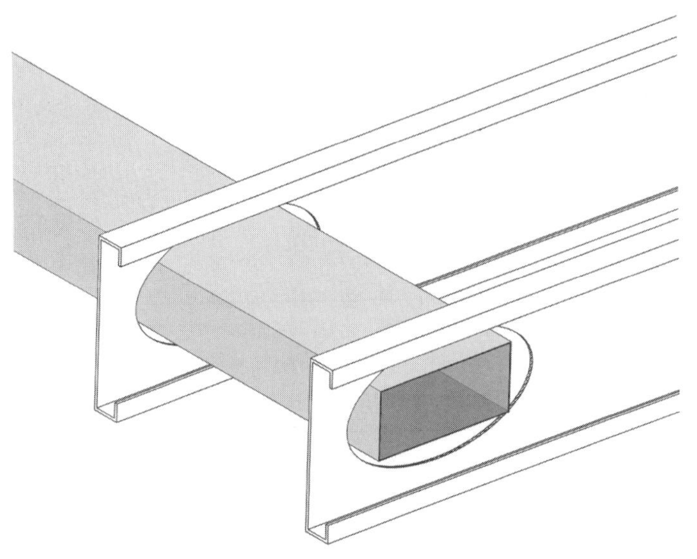

FIGURE 5.1.16 STEEL FLOOR JOIST

partitions and floors are typically not affected by moisture if the building envelope is watertight and the basement dry. Exceptions to this are problems caused by the long-term presence of moisture from sources such as faulty plumbing and interior roof drains or the generation of excess humidity as a result of cooking, clothes washing, or excessive humidification. The first floor is most typically affected by moisture because it is usually over a crawl space or basement. Fortunately, in many cases, the floor structure is exposed. The condition of the existing joist can be sampled with a sharp object such as a screwdriver or pocket knife. Sound wood will split into fibrous splinters, while decayed wood will separate into small chunks of a dark brown, black, or gray color. Decay can also be revealed by rapping on the surface of the wood member; a dull, hollow sound frequently indicates decay below the surface. Decayed lumber can be reinforced with additional structure, stabilized with structural epoxy conservation techniques, or by a combination of both.

TECHNIQUES, MATERIALS, TOOLS

1. REPAIR EXISTING JOISTS WITH EPOXY CONSOLIDANTS.

Portions of existing deteriorated joists can be repaired and reconstituted by the use of epoxy consolidants. While this technique is more typically used with supported members such as sills, it can be useful when combined with additional adjacent reinforcing members (Fig. 5.1.17).

ADVANTAGES: This repair is an effective fix for historic structures because it maintains the original structural profile.

DISADVANTAGES: This repair is expensive and not usually cost-effective when rot is widespread or in most nonhistoric situations, where the use of additional reinforcing structure is more practical.

2. REPAIR OR REPLACE EXISTING JOISTS WITH SUPPORTING STRUCTURE.

The most typical repair of a deteriorated structure is to add vertical supports or beams to shorten the spans; to add scabbed (spliced) members at deteriorated ends (Fig. 5.1.17); or to add new wood or steel reinforcement alongside the existing members (see above). Individual repairs should be supervised by a structural engineer based on site-specific conditions.

ADVANTAGES: Structural repairs are practical if the deteriorated portions of beams or joists are limited and typically near the bearing ends or where space limitations restrict the introduction and use of longer structural members.

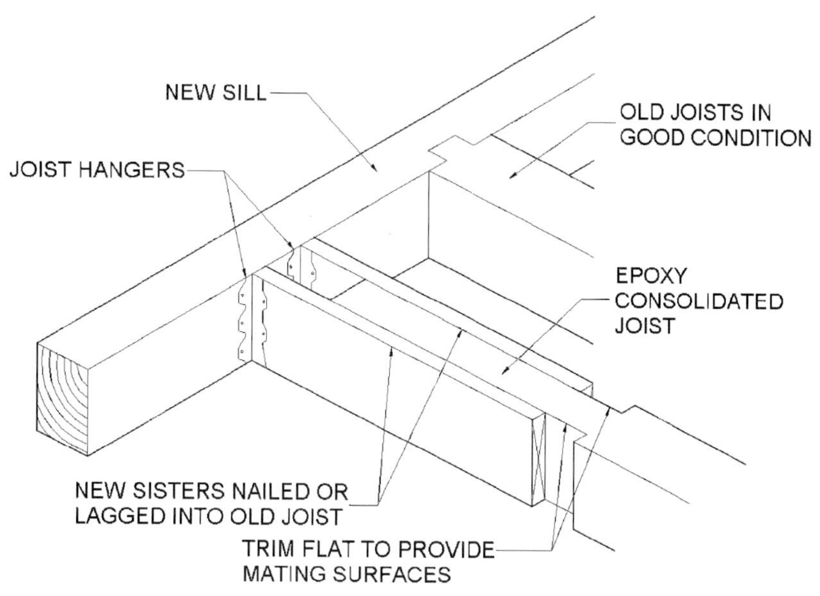

FIGURE 5.1.17 EPOXY CONSOLIDATED SISTERED JOIST

DISADVANTAGES: Reinforcing existing members is time-consuming and may not be as cost-effective as installing new adjacent members where access is feasible.

FIRE DAMAGE TO FLOOR FRAMING

ESSENTIAL KNOWLEDGE

Damage from fire can range from the total loss of a building and its contents to inconvenience from smoke odors. Unless the damage is limited, the restoration process can be complicated, involving structural, electrical, HVAC, and plumbing systems, as well as building finishes. In addition, significant health and comfort issues arise from the residual smoke, combustion gases, moisture from fire department hoses, and the existence of products containing asbestos. The selection of a restoration contractor who is experienced and knowledgeable in current techniques is critical. At least one national association, the Association of Specialists in Cleaning and Restoration (ASCR), manages training and certification programs and publishes a restoration guideline.

TECHNIQUES, MATERIALS, TOOLS

1. RESTORE FIRE-DAMAGED STRUCTURAL MEMBERS.

The first step in restoring structural members is to assess the damage to the building structure, systems, and finishes. In 2x4 construction, significantly charred members are generally removed in their entirety. Heavy timber construction can remain (according to the American Society of Civil Engineers), once the char is removed and if the remaining section is still structurally adequate (after a reduction-in-size factor of $\frac{1}{4}$" on all sides). Char is removed by scraping and abrasive blasting. It should generally be removed because it holds odors, although encapsulating coatings inhibit their transmittal. Sheathing materials (especially charred or unsound fire-retardant-treated material) should also be removed. New construction that replaces the damaged construction should meet codes for new construction. Smoke-damaged materials should be cleaned and deodorized as necessary. The use of ozone generators, to remove odors and contaminants, is controversial and considered by a number of specialists to be ineffective and potentially dangerous (see Further Reading). Water-damaged materials, such as insulation, should be replaced when the damage is irreversible.

ADVANTAGES: A large portion of the building can be salvaged.

DISADVANTAGES: Odors from fire damage may linger; the repairs can be costly.

2. SALVAGE FIRE-DAMAGED GLULAMS.

Glulams perform relatively well in fires (wood chars at the rate of $\frac{1}{40}$" per minute) and may be able to be salvaged and reused once the char is removed, if the remaining section is structurally adequate. This determination should be made by a structural engineer, based on job-site conditions.

ADVANTAGES: Fire-damaged glulams can be reused in some instances.

DISADVANTAGES: Fire-damaged glulams may retain smoke odors.

SOUND CONTROL

ESSENTIAL KNOWLEDGE

Sound travels through a structure itself in the form of vibrations caused by direct mechanical contact with a source such as a washing machine, an air-distribution motor, footsteps, a dropped object, and other impacts. The control of structureborne sound is generally treated as a floor problem, since this is where the majority of complaints originate. An excellent resource on residential acoustical theory and practice (and the source of much of the data in the following section) is *Noise Control Manual for Residential Building* by David A. Harris (see Further Reading).

TECHNIQUES, MATERIALS, TOOLS

REDUCE AIRBORNE AND IMPACT SOUNDS IN FLOOR-CEILING SYSTEMS.

Sound borne through floor-ceiling systems can be treated in much the same way as in walls, by increasing the mass, providing resilient connections, and adding sound-absorbing material to the floor surface or the floor-ceiling cavity. Fortunately, most floor systems have more mass and thickness than walls, making it easier to achieve sound-deadening properties. Impact sounds, however, are more difficult to control because the stiff, light assembly acts as a drum under such impacts as footsteps and falling or moving objects. Reducing impact sounds is best accomplished by treating the floor with cushioning material, such as carpets and underlayments. Unfortunately, there are some areas where carpet is not practical, such as in kitchens and bathrooms, or not desired. In these cases, it is possible to use such materials as sound-deadening board, cork, sound-control matting, a foam gasket between the floor joist and sheathing (Fig. 5.1.18), or sound-isolating pads or mats under the finish flooring to provide a "floating floor." Other, more complicated and costly techniques include resistant ceiling channels to "decouple" ceiling materials so they do not transmit noise; sound-absorbant blankets in the floor assembly; spray-on cellulose or urethane foam; and the addition of mass, such as lightweight concrete or gypsum over a plywood floor. This last option requires extensive door and trim adjustments, and the existing floor structure may not be able to support the additional weight.

FIGURE 5.1.18 INSTALLING FOAM GASKET

ADVANTAGES: A variety of materials and techniques is available that, individually or in combination, can substantially reduce sound transmission.

DISADVANTAGES: Remedial techniques often involve the addition of new materials and can be costly; reduction of impact sounds in lightweight wood-frame buildings is very difficult.

5.2 SUBFLOORING AND UNDERLAYMENTS

DAMAGED SUBFLOORING

ESSENTIAL KNOWLEDGE

Until the late 1930s and early 1940s the typical residential subflooring was 1x6 and larger, straight-edged or tongue-and-grooved wood boards. These boards, laid perpendicular or diagonally to the floor joists, provided a level surface to which the finished floor was attached, and also acted as a diaphragm stiffening the floor structure and transferring lateral wind loads through the floor to the building's foundation. Occasionally, in rural areas, the subflooring became the finished floor as well, although shrinkage of the boards meant that, unless they were tongue-and-grooved, cracks developed between boards. By the early 1940s, with the introduction of waterproof adhesives, plywood subfloors began to replace board sheathing. By the 1960s a variety of other subfloor sheathing products came into use, including particleboard and waferboard. Early particleboard and waferboard did not perform well in high-moisture situations and were, in turn, replaced with the stronger and more durable oriented strand board (OSB) which, because of its lower cost, has replaced plywood in many areas.

TECHNIQUES, MATERIALS, TOOLS

REPAIR DETERIORATED OR DAMAGED SUBFLOORING.

Deterioration of the subfloor sheathing, when it occurs, is often moisture-related, unless it is caused by foundation settlement, insect damage, earthquakes, or other natural disasters. It may be due to edge swelling, surface delamination, or buckling at panel joints from inadequate moisture protection during storage or construction, or it may be due to rot from roof or plumbing leaks. Corrective work should not be initiated until the cause of the deterioration is understood and corrected. Isolated repairs, typically in bathroom or other wet areas, usually entail cutting out and removing the finished floor and the damaged sheathing. If the existing subfloor shows minor damage, such as roughness, but is not decaying, it can be sanded smooth. The APA-Engineered Wood Association recommends a minimum of 11/32" undelayment be placed over lumber subflooring or uneven surfaces to provide a uniform surface. If sheathing is damaged and should be replaced, new subflooring should have a span rating that is appropriate to the framing spacing below.

ADVANTAGES: Localized repairs are cost-effective and relatively easy to make.

DISADVANTAGES: This repair requires the removal of finished flooring and often the existing sheathing and is costly over large areas.

FLOOR SQUEAKS

ESSENTIAL KNOWLEDGE

Floor squeaks can originate in a variety of places, including between finished floor and sheathing, underlayment and sheathing, sheathing and joist, and joist and bridging. Floor squeaks often result from the rubbing of the sheathing against the shank of a nail that protrudes above the surface of the sheathing. This is often due to nail "popping" or "backout" from joist shrinkage or a gap created between the sheathing and the joist from the lack of a space between panels (the APA recommends a minimum of ⅛" gap on all panel edges) and subsequent buckling from expansion due to moisture (Fig. 5.2.1). Refastening the sheathing to the joist, after identifying the probable location of the squeak, can help eliminate it. There are several ways by which this can be accomplished.

TECHNIQUES, MATERIALS, TOOLS

1. APPLY ADDITIONAL SURFACE FASTENERS.

Additional nails (preferably deformed-shank nails, such as ring or screw shank) or screws can be applied to the top of the sheathing or finished floor. Screws can be countersunk and concealed with wood plugs. Alternatively, an innovative tool, which includes an alignment device, for installing scored screws designed to break off below the surface of carpeted floors is available from O'Berry Enterprises (Fig. 5.2.2). A less costly version is available for finished wood flooring.

ADVANTAGES: This is the least-expensive and most direct approach. Underfloor access is not required. This repair stiffens floors that have been fastened with undersized fasteners, too few fasteners, or fasteners that have missed the joist below (relatively common with power-nailing since the operator has limited knowledge as to where the joists are).

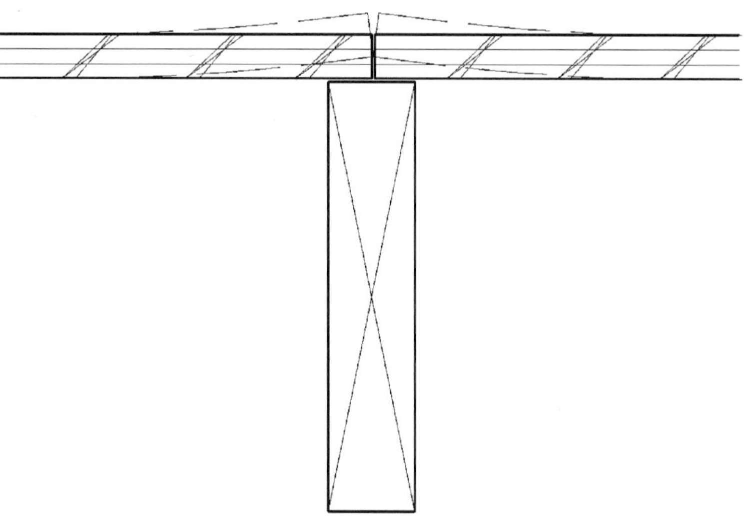

FIGURE 5.2.1 SHEATHING PANEL BUCKLING

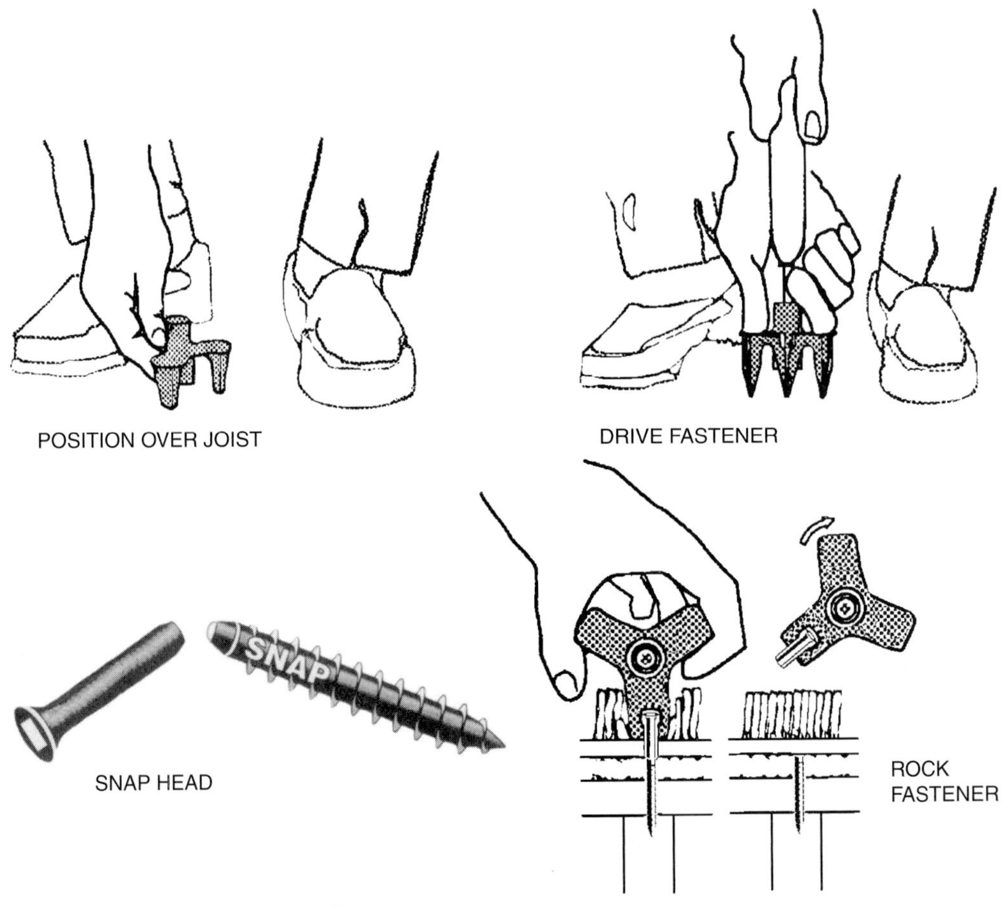

FIGURE 5.2.2 SQUEEEEEEK NO MORE® SQUEAK ELIMINATION KIT

DISADVANTAGES: This repair can affect the appearance of finished floors. The use of conventional screws can require removal of or cutting access holes through wall-to-wall carpet. This method is impractical for ceramic tile floors.

2. REFASTEN FLOOR SHEATHING TO JOIST FROM BELOW WITH LUMBER STRIP AND SCREWS.

This solution, recommended by the APA (Fig. 5.2.3), involves the use of wood blocking and a construction adhesive (where a gap exists) and floor loading to compress the adhesive. The APA recommends against the use of a shingle wedge to fill the gap between sheathing and joist as it may tend to lift the sheathing away from the joist on either side of the wedge and squeaks might result.

ADVANTAGES: Refastening does not need to be done through the top of sheathing or finished flooring.

DISADVANTAGES: Access from the underside of the floor is required.

3. REFASTEN FLOOR SHEATHING TO JOIST WITH SPECIALTY FASTENERS.

A line of specialty fasteners from E&E Consumer Products marketed as Squeak Ender™ and Seam Ender™ are specifically designed for underfloor application (Fig. 5.2.4) (see Product Information).

ADVANTAGES: These products are readily available at home centers. The application is relatively simple and mechanically draws the sheathing to the joist.

DISADVANTAGES: Access to the underside of the floor joist and sheathing is required.

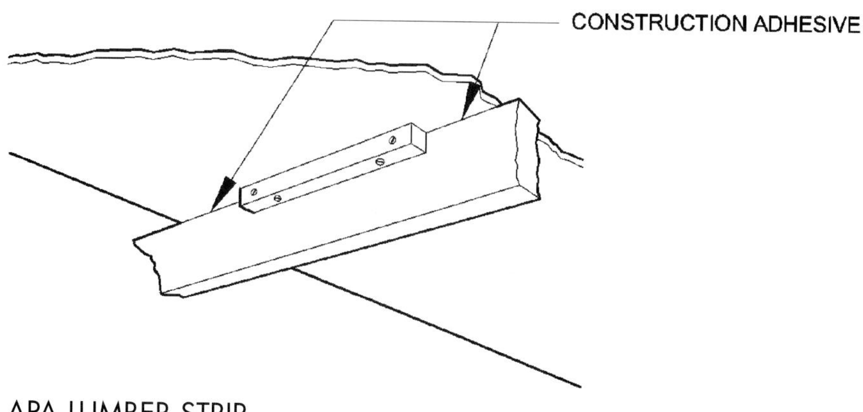

FIGURE 5.2.3 APA LUMBER STRIP

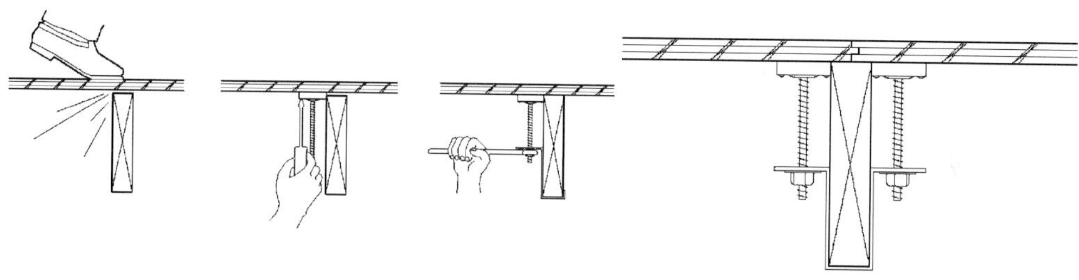

FIGURE 5.2.4 SQUEAK ENDER™ SEAM ENDER™

4. RENAIL EXISTING BRIDGING.

Occasionally, either the bottom of blocking or diagonal bridging has not been nailed, or the nails have partially pulled out. As a result, the blocking or bridging, instead of transferring load to adjacent joists, tends to rotate and slide on the fastener, resulting in squeaks. Additionally, squeaks can occur under loading from foot traffic when adjacent pairs of bridging come into contact and rub together where they cross. These problems can be controlled by renailing existing fasteners, adding fasteners, and preventing bridging members from rubbing together.

ADVANTAGES: This is a simple fix if bridging is exposed.

DISADVANTAGES: Removal of the ceiling to access bridging may be required. The repair may not be possible if access is limited, such as in shallow crawl spaces.

5. ADD NEW BRIDGING.

Solid blocking or diagonal bridging can help prevent the joists from rotating when the floor above is loaded and can be effective in reducing squeaks in existing floors. Bridging or blocking can also assist in stiffening the floor. This is useful, particularly when the sheathing is too thin (which can cause it to deflect) or when the fastening of sheathing to joists is inadequate. A manufactured product (Sag-Ender™) is available to reduce sheathing deflection (Figs. 5.2.5 and 5.2.6) (see Product Information).

ADVANTAGES: This is a relatively simple fix.

DISADVANTAGES: Access to the underside of the floor is required. There is no guarantee that the squeak will be eliminated.

6. ADJUST JOIST HANGERS.

When beams or joists are not properly fastened to their hangers or when the hangers are of the wrong width or height, the joist or beam (especially if it has shrunk) may be able to move freely in the hanger and can cause a squeak directly or indirectly by allowing the sheathing above to move. The beam or

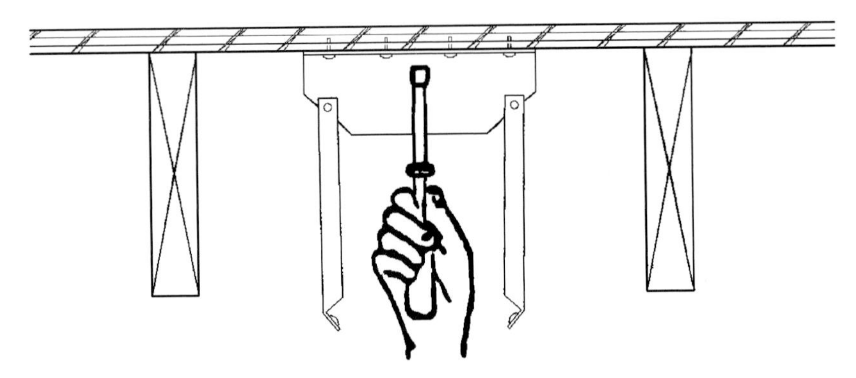

FIGURE 5.2.5 INSTALLATION OF SAG ENDER™

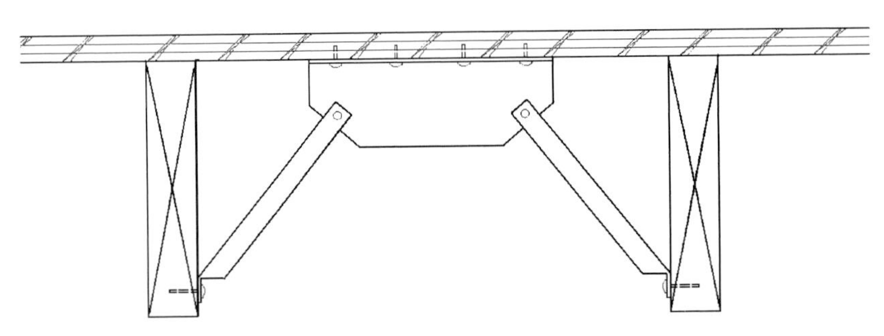

FIGURE 5.2.6 SAG ENDER™ FULLY INSTALLED

joist should be shimmed to fit tightly into the hanger and refastened (Fig. 5.2.7). If the hanger is the wrong size, it should be replaced with a properly sized one.

ADVANTAGES: This is a simple but effective fix.

DISADVANTAGES: Access to the underside of the floor is required.

7. ADJUST DUCTWORK WITHIN FLOOR CAVITY.
Heating and cooling ducts frequently run beneath the floor where they connect to floor registers. Floor squeaks can result when the hole cut in the floor for the register is too tight. A deflection of the floor near the register, from foot traffic, can cause a squeak. The heating ducts themselves may cause noise due to movement from expansion and contraction if they are not supported to allow for movement without rubbing against adjacent structures.

ADVANTAGES: This is a relatively simple fix.

DISADVANTAGES: Access to the underside of the floor is required.

UNDERLAYMENTS

ESSENTIAL KNOWLEDGE

Underlayments, typically plywood or cement board, and in some instances particleboard or hardboard, are used on top of floor sheathing and under vinyl tile, sheet vinyl, linoleum, and other materials that

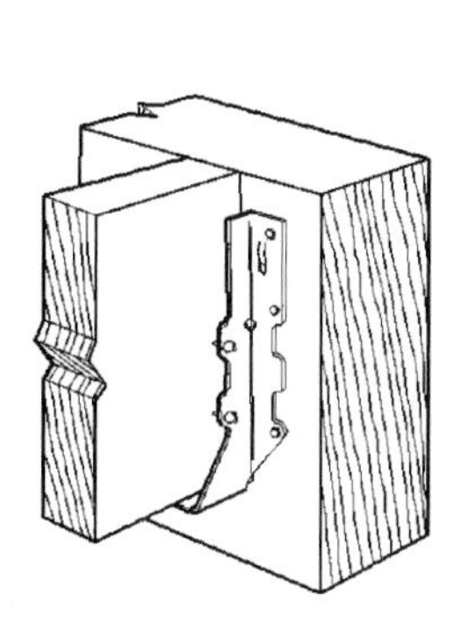

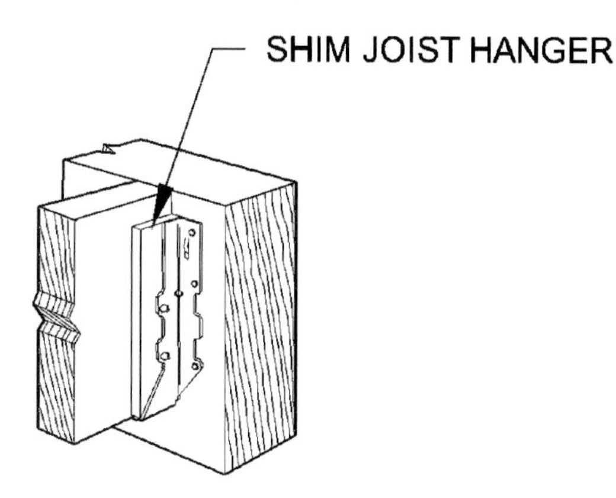

FIGURE 5.2.7 SHIM JOIST HANGER

"telegraph" imperfections in rough, delaminated, or weathered floor sheathing (subflooring). They are also used under ceramic tile, especially thin-set tile. Underlayments are typically installed when a home is fully enclosed and protected from the weather. Existing underlayments may have been damaged by excessive wear or abuse due to deteriorating finishes above them, by moisture from excessive humidity, or by exterior or interior plumbing leaks. Flooring manufacturers and their industry associations make recommendations as to appropriate underlayments. These recommendations should be followed to assure that product warranties remain in force.

TECHNIQUES, MATERIALS, TOOLS

1. REPAIR EXISTING UNDERLAYMENT.

There is very little possibility of salvaging existing deteriorated, delaminated, or otherwise damaged underlayment. Localized deficient areas should be cut out, removed, and replaced with an appropriate material of equal thickness.

ADVANTAGES: Localized patching repairs are practical if the affected area is small.

DISADVANTAGES: Repair of underlayment necessitates the removal of the finished floor, which may be (as in the case of ceramic tile) very costly.

2. REPLACE EXISTING UNDERLAYMENT.

New underlayment should be used to replace deteriorated underlayment or used over existing underlayment, hardwood flooring, and resilient flooring, if those materials are basically sound. It should not be laid directly over existing ceramic tile floors without the addition of an appropriate sheathing below. Typically, underlayment material for hardwood or resilient flooring is plywood. Cement board is frequently recommended for ceramic tiles. The APA-Engineered Wood Association recommends a minimum of ¼" plywood over smooth subfloors and $^{11}/_{32}$" over lumber subfloors or uneven surfaces. Recommended grades for use under adhered carpet, resilient sheet goods and tile, and ceramic tile include underlayment of C-C Plugged with sanded face; plywood or Com-Ply® Sturd-I-Floor ($^{19}/_{32}$" or thicker) with sanded face; Underlayment A-C; Underlayment B-C; Marine EXT; or sanded plywood marked Plugged Crossbands Under Face, Plugged Crossbands (or Core), Plugged Inner Plies, or Meets Underlayment Requirements. Other non-APA-rated plywood panels are frequently sold through supply outlets. These materials, including Luan plywood or other species, may not have plugged lamination or exterior glue and may be susceptible to deterioration from moisture when used in wet environments. Hardboard and particleboard are sometimes used, but are often not recommended because they can be susceptible to swelling when wet. Specific materials and installation procedures should be reviewed with manufacturers and industry associations, such as the Tile Council of America.

ADVANTAGES: Appropriate new underlayment, properly applied, can provide a smooth base for finished flooring materials.

DISADVANTAGES: Appropriate underlayment can be relatively costly; may add height that causes difficulties with door clearance, built-ins, etc.; and may require reducing strips between new and existing material.

5.3 FINISH FLOORING

WOOD FLOORING

ESSENTIAL KNOWLEDGE

Wood flooring has been a traditional residential flooring material for hundreds of years. Its use diminished after World War II, however, with the introduction of plywood subflooring and underlayments, and the increased popularity of such less costly alternatives as wall-to-wall carpeting and resilient flooring. In the last two decades wood floors have gained substantial new popularity as a "natural," low-maintenance, attractive, quality floor covering. As a result, there has been a recent proliferation of new manufacturers and products.

Types of wood flooring include solid-wood flooring, both hardwood (such as oak, maple, hickory, ash, cherry, and other domestic and imported hardwoods) and softwoods (including eastern white, southern yellow, and ponderosa pine; Douglas fir; and hemlock); laminated wood flooring (comprising three to five veneers of wood and engineered wood products); and acrylic impregnated flooring (a very durable flooring available in both solid and laminated forms).

Styles of wood flooring include strip, plank, and parquet. Strip flooring is the most common, least costly flooring, typically 1½", 2¼", or 3¼" wide oak. Plank flooring is common in hardwood in widths of 3", 4", 5", and 6" with softwood flooring available up to 18" and sometimes wider. Plank flooring can be made straight-edged, tongue-and-grooved, or shiplapped. Wider sections are typically face-screwed and plugged, or with softwoods, face-nailed with cut nails to restrict cupping and provide a traditional appearance. Parquet flooring is made up of small sections of solid or laminated wood that are combined to create geometric designs and patterns.

Aside from structural problems with the building's frame, floor abuse, and overloading from furniture or occupants, most of the problems with wood floors relate to a lack of required maintenance or to the effects of high levels or changes in the moisture content of the flooring, sheathing, and underlayments. Changes in moisture content may be due to a variety of factors including excessive moisture conditions in basements, crawl spaces, and under on-grade slabs (see Chapter 1); inadequate drying out of framing and finishes in new construction prior to the installation of flooring; inadequate or missing moisture retarders under slabs, in crawl spaces, or between sheathing and flooring; inadequate protection of flooring material; and leaks in the building's envelope, plumbing, or mechanical systems. The best way to assure correct floor installations is to measure the moisture level of adjacent surfaces with a moisture meter. For radiant heat installations, the only sure way to dry the slab and subfloor is to turn on the radiant heating systems before installing the wood flooring. When wood floors are not maintained properly, they can be refinished. Severe moisture changes, however, can lead to significant cracks, movement, cupping, and/or buckling of floors.

TECHNOLOGY, MATERIALS, TOOLS

1. MAINTAIN WOOD FLOORING.

Weekly (or as necessary) vacuuming and/or dry dust mopping will remove grit and dirt that can scratch and erode floor finishes. If the finish is properly maintained, flooring will require only periodic refinishing, and will last almost indefinitely. A damp mop, cloth, or sponge can be used for spills and cleanup on non-waxed polyurethane or similar surface finishes, although care should be taken to keep the floor from becoming too wet, as excessive water can seep between boards and into small scratches, causing deterioration of finishes. Small dents, where the wood fibers are not broken, can sometimes be removed by covering with a damp cloth and pressing with an iron to draw fibers up. Dark stains can sometimes be removed by lightly abrading the floor surface with fine sand paper and covering with a damp cloth containing 50/50 proportions of water and household bleach for 30 minutes. Let dry, recolor, and refinish if necessary. Waxed floors can be cleaned with paint thinner. Many manufacturers, particularly those that make laminated flooring, supply specially formulated cleaning products. Waxed finishes should not be cleaned with water in any form, as it can leave spots. Waxed floors should be buffed occasionally to redistribute the wax. Scatter rugs can be placed at entrances and areas of high traffic to reduce wear at those locations.

ADVANTAGES: Maintaining wood flooring properly is relatively simple and will reduce the need for refinishing and extend its life.

DISADVANTAGES: Even with continuing maintenance, wood floors will require periodic restoring or refinishing.

2. REPAIR DAMAGED WOOD FLOORS.

Floors that have rot; deep stains; bad gouges; broken wood; or permanently cupped, warped, or crowned boards may not be able to be restored by normal sanding and refinishing. A common repair technique is to cut out the defective flooring using a rotary saw and carbide-tipped blade. Clean cuts can be more easily made by nailing down a board temporarily as a straight edge (Fig. 5.3.1). After being cut, the defective section can be removed with a pry bar (and chisel if necessary), the subfloor repaired or replaced, and new flooring installed. A repair that can be effective on wide bulging boards that are to be painted or covered with carpet includes the application of a saw kirf/relief joint down the center of the bulge and subsequent screwing down of the sections of board adjacent to the saw cut.

ADVANTAGES: Repairs of small sections of flooring are cost-effective and relatively simple.

DISADVANTAGES: This repair will not be effective if large sections of the floor have been affected.

3. STABILIZE AND REPAIR MOISTURE-DAMAGED WOOD FLOORS.

Solid-wood flooring will contract during periods of low humidity (usually during the heating seasons) leaving cracks between flooring boards (a 2¼"-wide-strip oak floor can easily develop a crack the thickness of a dime, and wider boards will develop correspondingly wider cracks). Alternatively, flooring will expand during periods of high humidity. These cracks, more noticeable on bleached or white floors, are normal and usually not objectionable, and can be controlled by stabilizing the environment of the build-

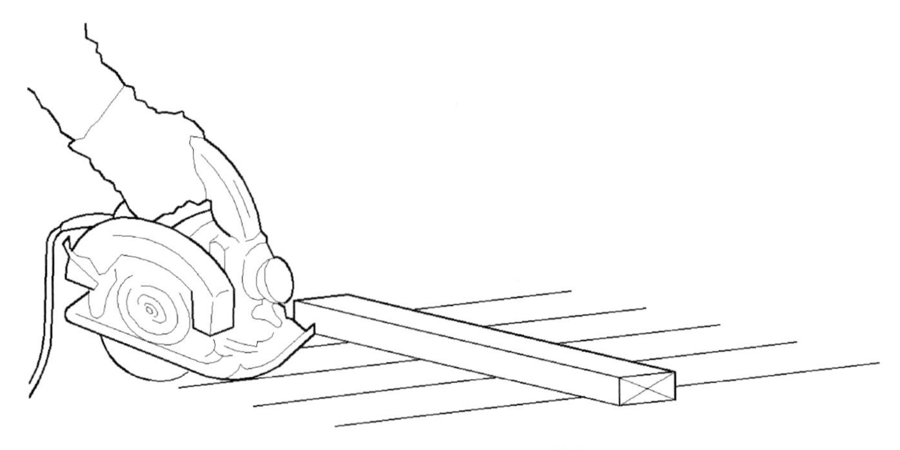

FIGURE 5.3.1 CLEAN CUT WITH NAILED STRAIGHT EDGE

ing through temperature and humidity control. Excessive moisture changes, however, can cause severe opening or expansion of flooring, which can lead to cupping, crowning, or buckling. Before strip flooring is installed, its moisture content should be within 4% of the subfloor below, and plank flooring should be within 2%, or excessive expansion or contraction can occur. For example, if buckling floors are treated early, several boards may be able to be removed, allowing air to circulate under the boards, particularly if they are on sleepers. Once the floors have dried to a more stable condition, repairs can usually be made. In some cases, however, the flooring may have to be removed. Given normal conditions (70°F interior temperature and 40% relative humidity), a 5" oak board has a moisture content of 7.7%. If the relative humidity falls to 20%, the moisture content of the board will fall to 4.5% and the board will shrink by 0.059". Across 10' of flooring, this could amount to as much as 1.4". If the humidity were to rise to 65%, the board's moisture content would be 12% and the board would expand by 0.079". Across 10' of flooring, this could translate to 1.9". A burst pipe could cause buckling and render a floor unusable. Cupping and crowning are common problems that develop with high humidity (Fig. 5.3.2). Once the source of moisture is controlled, fans can be employed to assist in drying out the floor. Removing a strip of flooring against a wall may reduce pressure and assist in drying flooring. When this is accomplished, the floor can be left as is or sanded and refinished if necessary.

ADVANTAGES: Minor moisture-related damage may correct itself when the source is eliminated and the floor dries out.

DISADVANTAGES: Extensive moisture damage may involve sanding, refinishing, or removal and replacement.

4. SAND WOOD FLOORS.

When the surface of wood floors becomes scratched, worn, gouged, discolored, or distorted in profile, or if a new finish is desired, the appropriate step is to sand prior to refinishing. For hardwood floors, the National Oak Flooring Manufacturers Association estimates that a typical sanding will remove between 1/64" and 1/32" of wood. Tongue-and-groove, 3/4" oak flooring has 19/64" above the tongue. Therefore, a floor could theoretically be sanded and finished 6 to 10 times or more before the top of the groove is weakened. Under normal conditions, refinishing occurs at approximately 15-year intervals. This suggests that hardwood flooring, if not abused, can last as long as the structure. Softwood flooring will be affected to a greater degree by sanding, depending on the hardness of the wood. Great care should be taken to assure that the sander is not left stationary while it is on, as it will create ridges in flooring in a very short period of time. Recommended sanding and refinishing techniques are discussed in detail in the National Oak Flooring Manufacturers Association pamphlet entitled *Finishing Hardware Flooring* and other publications referenced in Further Reading.

ADVANTAGES: Spot or more extensive sanding can provide the opportunity to remove surface flaws without destroying the long-term performance of wood floors, especially hardwood floors.

DISADVANTAGES: This is an inherently messy operation requiring removal of all furnishings and protection of adjacent spaces. Ridges and marks can be left unless sanding is done carefully. Softwood floors can be sanded a very limited number of times.

5. REFINISH WOOD FLOORING.

Wood flooring that has been excessively worn, overloaded, abused, or subject to water damage will have to be restored (sanded) and refinished. Wood flooring is available both prefinished, including

FIGURE 5.3.2 CROWNED FLOOR CUPPED FLOOR

acrylic impregnated, and site-finished. Recommendations for refinishing prefinished as well as site-finished flooring can be obtained from the manufacturer. Each floor finish has its advantages and disadvantages. Some of these are subjective, and there is no consensus on the best finish, especially since job conditions vary considerably. However, general attributes for the most popular finishes follow.

5.1 WATER-BASED URETHANES.

Water-based urethanes are usually combinations of urethanes and acrylics with a catalyst mixed prior to application. In general, the higher the percentage of urethane, the more durable and expensive.

ADVANTAGES: Water-based urethanes contain fewer volatile organic compounds (VOCs); are less noxious than other finishes; are clearer, less yellowing than other finishes; have good durability; are fast drying and nonflammable and are becoming increasingly popular.

DISADVANTAGES: Water-based urethanes are somewhat less durable than other urethane finishes and require more coats than solvent-based urethanes to achieve comparable film thickness (up to four coats). New coats may not adhere well to old coats.

5.2 OIL-MODIFIED URETHANES.

Oil-modified urethanes are technically oil-based; examples include linseed and tung oil.

ADVANTAGES: Until recently, oil-modified urethanes were the most popular urethane finish and are still favored by many users. Fewer coats than water-based methanes are required. Oil-modified urethanes are very durable, commonly available and easy to recoat.

DISADVANTAGES: Oil-modified urethanes impart a yellower cast than other urethanes, are slow to cure, and may require sanding between coats. They have a high-VOC content and proper lung, eye, and skin protection is required. They are combustible.

5.3 MOISTURE-CURED URETHANES.

Moisture-cured urethanes react with the humidity in the air to dry.

ADVANTAGES: Moisture-cured urethanes have excellent durability, provide the hardest wearing surface, dry rapidly in moist environments and are recoatable.

DISADVANTAGES: Moisture-cured urethanes are difficult to apply and thus application should be left to professional finishers. They are available only in glossy finishes. They have high levels of VOCs and careful lung, eye, and skin protection is required. They are extremely flammable. Significant changes in humidity can lead to blistering or other defects.

5.4 "SWEDISH" FINISHES.

This is a type of finish, typically acid-cured (containing formaldehyde) and sometimes water-based, that is high-performing but expensive.

ADVANTAGES: This finish has excellent durability, transparency, and elasticity. It is a popular product among professionals. It is recoatable and fast drying.

DISADVANTAGES: This finish has a high-VOC content, and the presence of formaldehyde in acid-cured formulas restricts application to professionals. It is difficult to apply, requires a carefully sanded floor, and is combustible.

5.5 OIL FINISHES.

Most penetrating oil sealers and finishes are combinations of highly modified natural oil, such as linseed or tung oil, with additives to improve hardness and drying. Adding wax to an oil-finished floor will afford protection against spills and abrasion, although the manufacturers of some finishes such as Velvit™ oil maintain that their products do not require wax.

ADVANTAGES: Oil finishes are easy to apply and repair (just brush or rub on another coat); have good durability; will not crack, craze, or peel; low luster finishes are popular with installers and users of traditional softwood flooring.

DISADVANTAGES: Oil finishes are not as durable as other finishes and can take a long time to completely cure. The surface may collect dust or can water-spot. Some finishes require waxing. Oil finishes have a strong initial odor and are combustible.

5.6 WAXES.

With the increased use of urethane finishes, waxes (typically paste waxes) are not as common as they once were. Most manufacturers of urethane finishes do not recommend the use of waxes over urethanes because of added maintenance. Waxed surfaces require a stain or grain sealer prior to waxing.

ADVANTAGES: Waxes protect and extend the life of oil finishes. They are easy to apply, surprisingly durable, and fast drying.

DISADVANTAGES: Waxes require maintenance (touch-up, buffing, and periodic removal with wax removers or sanding, and rewaxing). Waxes become brittle and can yellow flooring. They can be slippery when wet and are not suggested for use in kitchens, entryways, or bathrooms or powder rooms. Waxes contain VOCs, have a strong initial odor, and can water-spot.

6. REPLACE OR RECOVER WOOD FLOORS.

Flooring may have cupped, split, buckled, or deteriorated to the point where it has to be replaced. The type of replacement flooring can be similar to that removed or may be another material, such as laminated flooring, which is less susceptible to moisture, or another wood species considered more visually appropriate. The wide range of choices can be clarified by reading reference material or contacting industry associations and individual manufacturers (see Further Reading).

ADVANTAGES: Replacing or recovering flooring may provide a chance to install more appropriate material.

DISADVANTAGES: This method is disruptive and expensive.

VINYL SHEET FLOORING AND TILE

ESSENTIAL KNOWLEDGE

Vinyl sheet flooring and tile are probably the most popular choices for kitchen and bathroom floor coverings in homes across the United States. Although the sheet vinyl material available today is sometimes referred to as linoleum, it differs from that material. Linoleum contains cork, wood products, and oleoresins, while resilient sheet material contains vinyl resins with a fiber back. Today's sheet vinyl material provides a soft cushioned walking surface that is not associated with linoleum-covered surfaces. Resilient sheet vinyl flooring and vinyl tile are a low-cost, easy-to-install flooring option. They are available in styles and patterns that can satisfy any taste and blend well with any home's decor. Although usually found in kitchens, laundry rooms, and bathrooms, they can be installed anywhere in the home. Manufacturers have made the job of tile installation easy with adhesive-backed tile that does not incur the mess, expense, and extra time associated with adhesive applications.

Resilient vinyl flooring and linoleum installed in homes from the 1920s through the 1980s was manufactured with asbestos. Even though some flooring still contains small amounts of asbestos, most manufacturers have eliminated it from their manufacturing process. As some resilient vinyl flooring or linoleum wears, asbestos-containing fibers can be released into the air causing a potential health risk. Sheet goods and resilient tile contain a polymerlike top layer and a second, fibrous layer that may contain up to 40% chrysotile asbestos. This asbestos is released into the air only when the fibrous layer becomes agitated. Since the asbestos is very tightly bonded to the polymerlike top layer, there is minimal potential health hazard if the fibrous layer is not damaged. Extreme caution and professional help should be used when removing asbestos-containing tiles. Although this section focuses on rehabilitating vinyl sheet flooring and tile, the same techniques can be applied to rehabilitating linoleum floor coverings.

TECHNIQUES, MATERIALS, TOOLS

1. REMOVE STAINS FROM SHEET VINYL AND VINYL TILE FLOORING.

If vinyl floor surfaces have yellowed or become stained with age, contact a professional floor-refinishing contractor or the manufacturer for cleaning recommendations. Using household bleaches or dyes to clean stains on vinyl materials may dull or damage surfaces. Sometimes resilient flooring may be installed incorrectly, leaving gaps between tiles or sheet-flooring seams, or gaps between tiles or sheets and appliances or cabinets that may be too wide. Dust and dirt may settle into these gaps. If grime has accumulated in these gaps, a soft tooth brush can be used to gently loosen dirt particles. If the dirt will not loosen, a pin can be used to clear dirt from gaps or seams.

ADVANTAGES: Because resilient sheet goods and tile are moisture resistant and nonporous, they are an easily maintained floor-covering option.

DISADVANTAGES: Improper cleaning methods and agents can mar vinyl surfaces beyond repair.

2. REPAIR VINYL FLOOR COVERING.

Resilient vinyl sheet flooring and vinyl tiles may tear or puncture if proper care is not taken to protect floor surfaces. Moving appliances over an unprotected vinyl floor surface is one of the most common causes of surface damage. Any type of rip or tear should be repaired to prevent further damage. The best way to repair ripped or torn resilient flooring is by cutting out the damaged section and replacing it with a matching patch (Fig. 5.3.3). This can be accomplished by cutting around the damage along the pattern lines. Using the pattern lines as a guide for cuts will conceal the repair. Remove a matching patch from an inconspicuous area of the room—under cabinets or appliances is usually a good place. Repair damage to a nonpatterned floor using the same technique. Because the seams may appear more obvious when patching nonpatterned flooring, be sure to keep the cutout as small as possible. Once the damaged piece is removed, use mineral spirits to remove the old adhesive from the matching patch. Apply the new adhesive to the patch and fit it into place. Use a roller or rub with a cloth to ensure proper adhesion. Place a heavy object over the newly patched area until the adhesive is set.

ADVANTAGES: Repairs to vinyl floor coverings are relatively quick and easy using a matching patch cut from a hidden area of the room.

DISADVANTAGES: Cutting out the damage to nonpatterned floor coverings and replacing it with a matching patch may reveal seams. Matching patches may not be available.

3. REPLACE VINYL FLOOR TILE.

Cracked, torn, or badly stained vinyl floor tile should be replaced. If a matching tile is not available, remove a tile from an inconspicuous area, under kitchen appliances or inside pantry closets, for

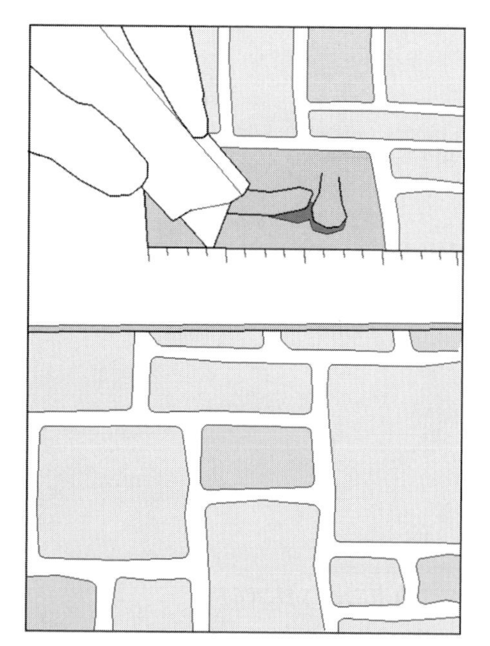

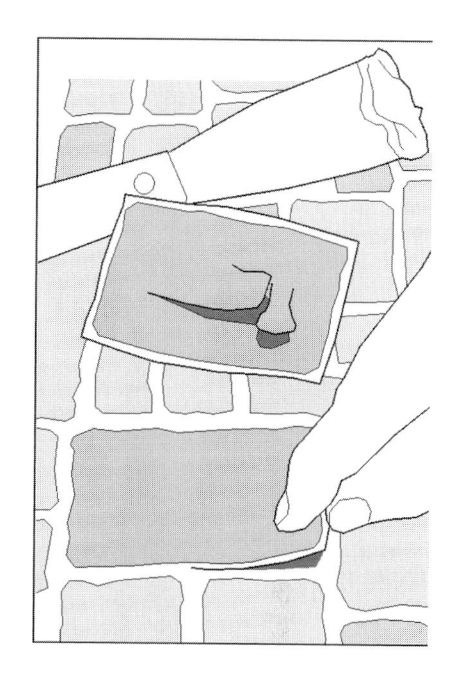

FIGURE 5.3.3 REPAIRING VINYL FLOORING

example. A dry removal of tile, which involves cracking the material and lifting the pieces off the floor with a scraper, may release asbestos fibers into the air, so proper precaution must be taken. Removing tiles in one piece using solvents to loosen any adhesive will not break the asbestos-containing fibrous layer and greatly reduces any health risk. If it is uncertain whether or not the tile contains asbestos, consult a flooring professional for advice on proper removal and repair.

Use a heat gun or hot iron over the damaged tile to loosen the adhesive, and pry up the tile with a putty knife (Fig. 5.3.4). If it does not lift off easily, continue to apply heat and pry again. Another technique is to place a pan of ice cubes over the tile until the adhesive becomes brittle and the tile pops out. However, this process may cause some tiles to crack and release asbestos-containing fibers into the air. Remove the adhesive from the underlayment where the damaged tile was removed and from the underside of the matching patch tile. Mineral spirits and a scraper will remove all adhesive. Apply new adhesive to the underlayment and fit the replacement tile in place. Use a roller or rub with a cloth to ensure proper adhesion. Place a heavy object over the patched area until the adhesive is set. Clean excess adhesive with a damp cloth.

ADVANTAGES: Damaged resilient floor tiles can be easily and inexpensively replaced.

DISADVANTAGES: Matching tile may be difficult to locate. Asbestos particles can be released into the air if proper precautions are not taken.

4. REMOVE RESILIENT TILE AND SHEET GOODS.

Resilient vinyl tile and sheet goods can be removed if damaged beyond repair or not properly bonded to the subfloor. Vinyl wall base is removed by using a heat gun and a putty knife and gently prying it from the wall. For tile, a heat gun can be used to direct heat over the material to soften old adhesive and pry tiles from the subfloor using a putty or spackle knife. The old adhesive should be scraped from the subfloor. Floor scrapers with long handles are available that allow the old adhesive to be scraped from the subfloor from a standing position. To remove linoleum or sheet goods begin by using a utility knife to slice the flooring in strips 10" to 12" wide, which can be peeled away. A paint-roller tube is a handy tool to wrap the strips around as they are peeled (Fig. 5.3.5). The top layer of some cushioned sheet goods may separate from the backing during removal, leaving it adhered to the subfloor. This backing can be loosened from the subfloor with a soap-and-water solution and then removed by scraping.

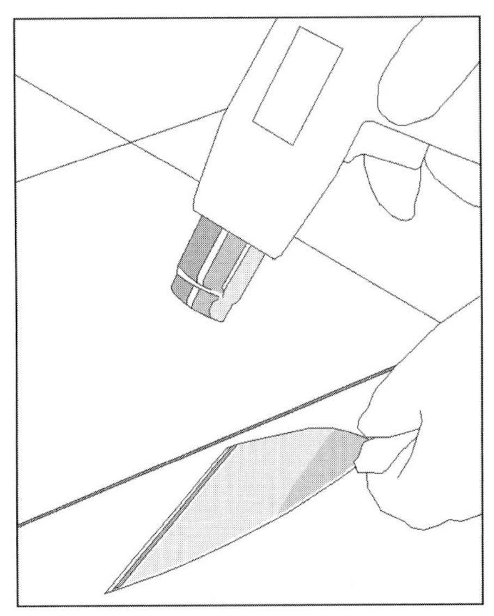

FIGURE 5.3.4 HEAT GUN AND BLADE TO REMOVE TILE

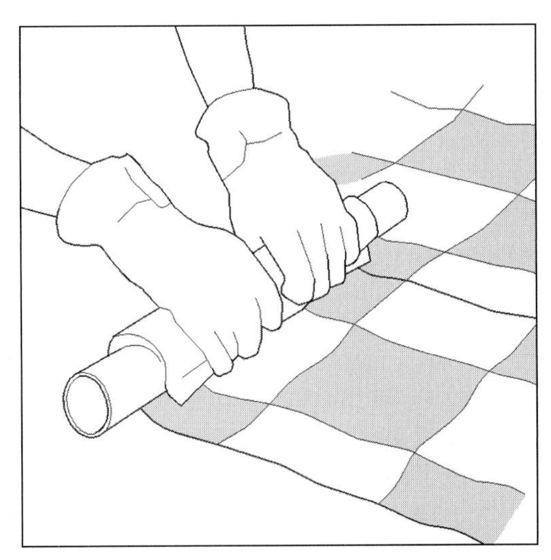

FIGURE 5.3.5 ROLLING UP SHEET FLOORING

ADVANTAGES: Removing resilient flooring will ensure proper adhesion of the new floor.

DISADVANTAGES: This can be a time-consuming job. Old vinyl-asbestos tile or linoleum may break in the process, causing asbestos fibers to be released into the air.

5. INSTALL RESILIENT VINYL TILE.

Resilient vinyl tile is either solid vinyl or through-pattern vinyl. *Solid vinyl* tile is made of vinyl resins that make up approximately 60% of the material's weight. *Through-pattern vinyl* tile is made of thermoplastic binders, fillers, and pigments and is less costly than solid vinyl. This tile is commonly available in 9" or 12" squares, either self-sticking or adhesive-applied. Begin tile installation along perpendicular layout lines by dry-fitting the tile to make sure that the pattern is suitable. Start in the center of the room and work toward the walls leaving about a ¼" expansion gap between tile and walls.

ADVANTAGES: Installation is quick and easy.

DISADVANTAGES: An unprofessional appearance will result if tiles are not installed along straight lines.

6. INSTALL RESILIENT SHEET FLOORING.
Sheet material is usually available in 6'-, 9'-, 12'-, and 23'-wide rolls and is either inlaid, rotovinyl, or modified rotovinyl. Inlaid, the most costly choice, has an integral pattern or color from the surface through the backing. Rotovinyl sheets have a backing that is covered with a foam layer on which a patterned coat of vinyl is applied. A protective top layer is applied to the printed layer. It is the thickness of this top layer that determines its durability. Modified rotovinyl sheets have chips of color that are spread between the printed and top layers. Because it lacks the durability and color quality of its counterparts, it is the least costly choice. When installing sheet flooring, try to locate seams in inconspicuous areas of the floor. To minimize cutting errors, make a template of the floor to be covered using sheets of paper taped together and to the floor along the perimeter of the room. Transfer the template to the unrolled sheet flooring, trace the outline of the template, and carefully cut along the template lines. Be sure to locate seams along pattern lines. Staple the sheet flooring to the floor along one edge of the sheet. Begin applying adhesive as necessary, working from the stapled edge toward the room's interior. Use a roller to bond the sheet to the floor.

ADVANTAGES: Seams are virtually invisible, providing a smooth, uniform appearance.

DISADVANTAGES: Installation is time-consuming.

CERAMIC TILE

ESSENTIAL KNOWLEDGE

Because it is impervious to water, ceramic tile has traditionally been used as a floor finish in damp areas such as bathrooms, kitchens, basements, and entryways. Because of its relative durability, low maintenance, and decorative qualities, ceramic tile is increasingly used in other spaces as well. Typical installation methods include thick-set (the most traditional method, sometimes called mud-set), set in ¾" to 1 ¼" of portland cement paste or mortar laid over a previously set mortar bed or concrete slab; and thin-set, set in an organic or epoxy adhesive. Ceramic tile is divided into glazed and unglazed varieties. Most historic floor tiles were unglazed and were the color of the clay and added oxides or pigments from which they were made (current examples are quarry tiles). Glazed tile is colored with a variety of glossy or mat glazes applied to the tile surface. Glazed tile is subject to scratching and abrasion from extended use and is usually installed in low traffic areas or covered with rugs. Tile failures are caused by a number of factors: improper maintenance, including the use of inappropriate cleaning agents or the degenerative effects of standing water on the grout, and the erosion of grout over time from traffic and cleaning. Structural problems can also cause failure. These include cracking and loosening of tile from overloading, sudden impacts, or frequent vibrations; defective or deteriorated substrates, such as concrete floors that have cracked, heaved, or settled; wood floor substrates that deflect excessively (too springy), have buckled, swelled, or deteriorated; and concrete or wood floors that have improperly mixed or applied tile bonding materials.

TECHNIQUES, MATERIALS, TOOLS

1. MAINTAIN AND RESTORE CERAMIC TILE.
When properly specified and applied, ceramic tile requires relatively little maintenance other than damp mopping or vacuuming. Normal cleaning is done with a neutral pH cleaner designed specifically for ceramic tile. Tile and floor covering stores sell cleaning products developed specifically for cleaning tile. These specialized cleaners generally outperform products available in supermarkets or hardware stores. Most glazed tile does not require sealers, but some tile such as quarry, saltillo, and oth-

ers do. Sealers are also used to protect grout joints. Use cleaners and sealers recommended by the manufacturer or sealers designed specifically for ceramic tile. Restoring or rejuvenating existing tile may involve the removal of deep stains, grease, oil, or various coatings and sealers. The Tile Council of America (TCA) recommends nonflammable, nonacid, methylene chloride-based solvents. If the cleaning is extensive, a company experienced in this work should be employed. The TCA-recommended cleaners are used to remove stains, such as light haze of grout. Acid will not remove heavy grout stains, oil, grease spots, or paint. Sulfamic acid does not give off noxious or damaging fumes. The use of muriatic acid is not advised, as it can erode tiles and grout joints and it can cause serious skin burns and internal injuries to individuals using it. While glazed tile used for countertops or kitchen floors must be able to resist etching from vinegar and citric acid, it may not resist bottled acid containing sulfamic and phosphoric acids. Muriatic acid should not be used in any case. Acid, if it is used, should be completely removed by continuous scrubbing and rinsing with clean water. Adjacent surfaces and materials should be protected.

ADVANTAGES: Some cleaners are relatively inexpensive and easy to use.

DISADVANTAGES: The cleaning of large areas or the use of professional cleaners or acids should be undertaken by floorcare experts.

2. REPLACE LOOSE, CRACKED, OR DAMAGED CERAMIC TILE.

Assuming that the substrate is satisfactory, the loose, damaged, or cracked tiles should be removed. If necessary, the grout joint can be carefully cut out with a grout saw, carbide, or diamond-tipped saw blade. The tile can then be chiseled out to reduce the risk of damage to adjacent tiles. If the substrate is concrete, the surface of the concrete can be scarified (scratched) using a cup grinder or other appropriate tool to remove any old bonding agents. Installation of reused or replaced tiles and mortar joints should be in accordance with the TCA's *Handbook for Ceramic Tile Installation*, or individual manufacturer's recommendations (see Further Reading).

ADVANTAGES: For small areas, this is a relatively inexpensive and easy fix.

DISADVANTAGES: Careful use of cutting tools is required; materials can easily be damaged.

CARPET AND RUGS

ESSENTIAL KNOWLEDGE

For many rehab applications, carpeting presents a low-cost, easily installed solution to flooring problems, and it can cover a multitude of sins. Existing carpet may be salvageable if wear and tear is not extensive. Today, carpet is available in a variety of designs and materials that extend the life of this popular floor covering.

TECHNIQUES, MATERIALS, TOOLS

1. MAINTAIN CARPET.

Carpeting can collect dust, pesticides, pet dander, and other allergens, and can be a breeding ground for dust mites under certain temperature and humidity combinations. If a carpet becomes wet and is not dried thoroughly within 24 hours, it can support mold and mildew, resulting in unpleasant odors and possible allergic reactions from occupants. Also, among health concerns is the effect on people with allergies of high levels of volatile organic compounds (VOCs) given off by some carpet backing material, especially older types of installation adhesives. In addition, carpeting, because of its soft, fibrous nature, has the ability to absorb VOCs from the environment (such as paint, cleaning products, smoke from cigarettes, building materials, and furnishings), and slowly re-release them over time. Carpet manufacturers are increasingly conscious of these issues and most provide new carpets with low or no VOCs. A carpet should be vacuumed regularly, and to retain its luster and attractiveness, it should be cleaned every 12 to 18 months or before it shows significant soiling. Refer to the carpet manufacturer's warranty for recommendations for appropriate cleaning methods. The Carpet and Rug Institute (CRI) has recommendations for the selection of professional carpet cleaners (see Further Reading). Carpets that have sustained prolonged water damage will most likely have to be replaced in their entirety. Rugs and carpeting must be vacuumed regularly with a strong, well-functioning vacuum cleaner. Most new carpet is protected with special finishes such as 3M's Scotch Guard™ to resist soil and stains. These finishes hold the spill on the fiber's surface, allowing the liquid to be removed before penetrating the fiber. The longer the stains are left, the more difficult removal will be. Stains should be blotted with a soft, white absorbent cloth or paper towel from the spot's edges toward the center. Use clean water to remove any remaining cleaning agent and absorb remaining moisture with paper towels. A comprehensive guide to spot removal is found on CRI's Web site: www.carpet-rug.com/athome/default.htm.

ADVANTAGES: Continuous, timely maintenance will prolong the life of carpets.

DISADVANTAGES: All carpet has a limited lifetime based on use.

2. RESTORE WATER-DAMAGED CARPET.

Determine whether the water is sanitary (uncontaminated sink or toilet overflows), gray water (some degree of biopollutant contamination including punctured water beds, dishwasher overflows, contaminated sink water), or black water (water that has come into contact with the ground or that contains raw sewage). Only carpet damaged by sanitary water can be treated nonprofessionally. Cleaning professionals should handle carpet damaged by unsanitary or black water, as it may harbor disease-carrying bacteria. Carpet damaged by black water must be discarded. The CRI and the Institute of Inspection, Cleaning and Restoration Certification (IICRC) have toll-free telephone numbers and can provide additional restoration information (see Further Reading).

ADVANTAGES: Carpet subject to uncontaminated or gray water flooding may be able to be restored.

DISADVANTAGES: Restoration may not be cost-effective.

3. REPAIR STAINED CARPET.

Isolated damage to carpet, such as stains and burns, can be removed and patched. If the damage is small, locate a matching piece of carpet from an inconspicuous location (inside closets for example). Cut out the damaged area using a straight edge and a utility knife. Transfer the measurements of the cutout onto the spare piece of carpet. Cut the patch to size, coat with carpet adhesive or apply double-sided carpet tape, and position. If the damage is extensive, consider removing a large section of carpet from the nearest seam to a wall or door jamb. If a matching piece of carpet cannot be located, consider taking a swatch to a home center or a carpet manufacturer who may be able to order a matching piece. Carpet discolored from bleach or household cleaners can be restored if the carpet fibers are not damaged. Professionals can spot-dye carpets or remove them from the home and dye them in the factory, restoring them to their original color and condition.

ADVANTAGES: Patching isolated problem areas will restore the carpet's appearance.

DISADVANTAGES: A matching patch may be hard to locate. Professional carpet dying, cleaning, and restoration services may not be cost-effective.

4. RESTORE CARPET FROM SMOKE DAMAGE.

Carpet damage from smoke usually requires a more extensive treatment than that for small burns or stains. As a result of smoke, dry soot settles on the carpet and penetrates the fibers. The content and amount of the soot depends on the fire's source and burn rate. Strong odors usually accompany any smoke or fire damage. Proper ventilation will help eliminate these odors, although unless the carpet is cleaned, the odor may persist. While smoke residue can be cleaned by a professional carpet cleaning or restoration professional, any burned areas must be removed.

ADVANTAGES: Smoke-damaged carpet can be restored to its original condition if there is no damage to carpet fibers.

DISADVANTAGES: Carpet with extreme smoke damage may require removal and factory cleaning. Carpet with damaged fibers may require replacement.

5. INSTALL CARPET OVER DAMAGED FLOORS.

Installing carpet over damaged floors is a cost-effective way to enhance the floor's appearance. If the damage to the existing floor is extensive (a cracked concrete slab, for example), higher-pile carpet, such as Saxony or plush styles, usually work best. Low-pile carpet, such as Berber or level loop, are less apt to mask extensive damage. If floor damage is slight, any style of carpet will work. Be sure to check the manufacturer's recommendations for installing carpets over damaged floors. Some carpet will require underlayment, although many may be installed without it. In some cases, a latex patching compound can be used to even out floors for carpet installation.

ADVANTAGES: Installing carpet over damaged floors will enhance their appearance.

DISADVANTAGES: This may not be a cost-effective solution if underlayment is required.

5.4 FINISH WALLS AND CEILINGS

With the exception of solid-wood paneling found in high-end housing or various forms of plywood paneling, the dominant wall material has been plaster or, since the 1960s, gypsum board (typically known as drywall, wallboard, and Sheetrock, which is a trademarked name for US Gypsum's gypsum board). Damage to wall and ceiling surfaces can result from a variety of causes, including wood shrinkage, undersized structural framing, building settlement, impacts and vibrations, high winds or seismic events, moisture, fire, and insects. This section will review various wall and ceiling materials, finishes, and trim, and typical problems and corrective measures.

PLASTER AND DRYWALL

ESSENTIAL KNOWLEDGE

Assuming they have been properly applied, the most common damage to older plaster ceilings is from the breaking of the plaster keys that attach the plaster to the supporting wood lath. This is typically caused by deflection from shrunken, warped, or inadequately sized joists and rafters; heavy vibrations from foot traffic on the floors above; construction activity including the cutting in of plumbing lines; and the installation of new wiring and recessed lighting, or softening of the plaster keys from moisture infiltration. This can lead to a particularly dangerous situation, as the condition of the plaster keys is not visible, and instances of the entire ceiling separating from the ceiling lath and falling to the floor below are not uncommon. Damage to wall surfaces is less serious, and if some of the plaster keys have split, it is usually not a serious problem.

In general, the least expensive and most desirable way to repair old plaster wall and ceiling surfaces is to carefully patch and then paint or (in the case of walls) wallpaper them. Covering cracked or deteriorated walls with an additional layer of gypsum board or paneling is the next least expensive option. By far, the most expensive option is to tear out old plaster in its entirety and replace it with new drywall and trim. The Enterprise Foundation, a nonprofit organization that has extensive rehab experience, has found that it is less expensive to repair up to 50% of walls and ceilings that have otherwise sound plaster, than to remove all the existing plaster and replace it with drywall. One reason for this is that the studs in old walls are typically uneven and will require extensive shimming or furring before new drywall can be applied. In addition, the existing trim will likely have to be replaced, which is labor-intensive and expensive.

Damage to existing drywall can be caused by structural or moisture-related problems, but may also be caused by abuse or daily wear and tear. Drywall is thinner than plaster (½" or ⅝" rather than ¾") and spans greater unsupported distances, typically 16" to 24" on interior partitions. The following highlights some of the repair techniques for both materials.

TECHNIQUES, MATERIALS, TOOLS

1. REFASTEN BOWING OR DEFLECTING PLASTER.

Plaster, if it has separated from its metal, gypsum, or wood lath base, is difficult to repair, and extreme care is required, especially in ceilings, to assure that the remedial work secures the plaster adequately. Stabilization methods include the use of plaster washers, 1" galvanized metal disks that are screwed into the wood lath and preferably into the joists above at a spacing that varies with job conditions. The metal washers can be countersunk into three-coat plaster using a 1¼" spade bit or can be left surface-mounted. The metal plaster washers are tightened against the lath or joists, covered with fiberglass tape, and skim-coated with plaster or gypsum drywall joint compound (Fig. 5.4.1). A more sophisticated technique, used for reinforcing the bond between the lath and the plaster in historic restoration, or where preserving ornamental plaster detailing is important, was developed by the Society for Preservation of New England Antiquities (SPINEA) and is currently employed by a few restoration contractors. This system, which requires access from above, comprises the drilling of two or three 3⁄16" holes into each lath between joists over the ceiling surface. The holes penetrate the lath and not the plaster scratch coat. After the sawdust is vacuumed away, a prewetting solution of water, alcohol, and acrylic adhesives is injected into the holes. Additional acrylic adhesive is then injected under pressure which causes the adhesive to travel along the interface between the scratch coat and the wood lath forming a continuous bond when it hardens (Fig. 5.4.2).

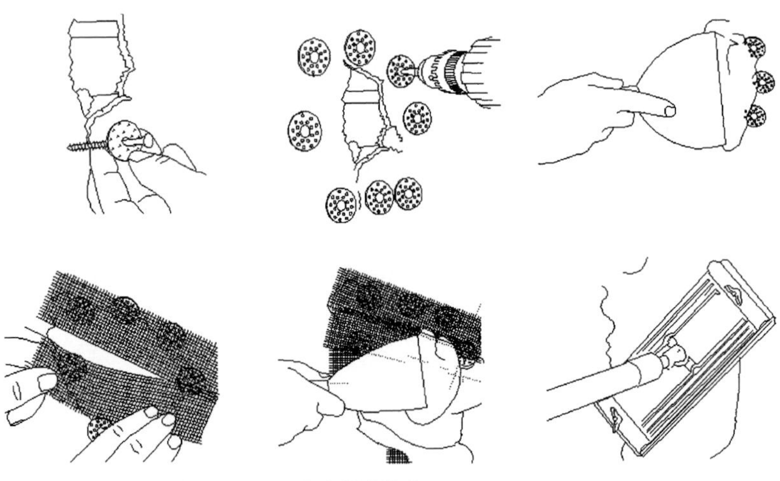

FIGURE 5.4.1 PLASTER REPAIR WITH WASHERS

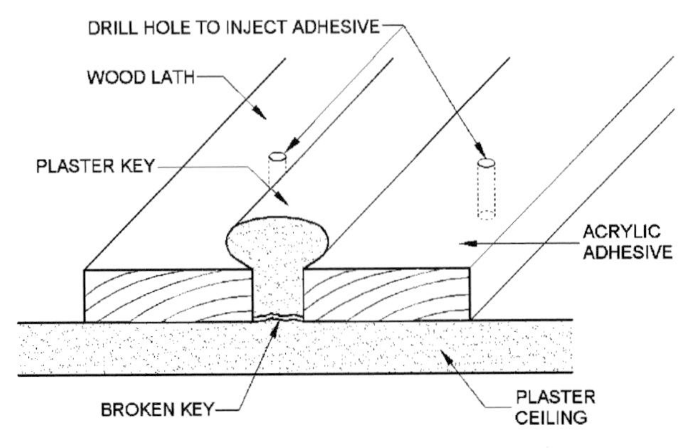

FIGURE 5.4.2 ADHESIVE INJECTION TECHNIQUE

ADVANTAGES: Plaster washers and acrylic adhesives can stabilize existing ceiling and wall plaster. This method helps preserve architectural details and saves the cost of new plaster or gypsum board.

DISADVANTAGES: This method does not eliminate underlying deflection problems. Locating structural members to screw the plaster washers into may be too difficult and time-consuming. Acrylic adhesive restoration must be undertaken by professionals, is expensive, and requires access to the wood lath.

2. REFASTEN BOWING OR DEFLECTING DRYWALL.

Drywall that has pulled away from studs or joists can typically be refastened, unless the paper face is severely damaged or the gypsum material is crumbly, in which case it should be replaced. Fasteners for wood studs and joists include nails (ring shanks hold better than plain shanks), or preferably screws, which do less damage to the paper facing and hold the gypsum board tighter against the framing. Screw heads should sit just below the surface of the paper face without tearing it. Nails should be set in a shallow dimple, but not so deep that they break the paper and damage the gypsum core (Fig 5.4.3).

ADVANTAGES: Drywall, if it is sound, can be refastened relatively easily.

DISADVANTAGES: If the surface is significantly cracked or the gypsum core has been damaged, refastening may not be adequate and that section may need to be replaced.

3. REPAIR CRACKS AND HOLES IN PLASTER.

If the cracks are minor and not the result of structural problems, which should be corrected prior to repairs, the repairs are relatively simple. Plaster cracks can be cut out, coated with a bonding agent such as polyvinyl acetate (PVA), and filled with patching plaster or drywall joint compound. Or they can be covered with self-adhering fiberglass mesh tape which does not require a bedding compound, and covered with drywall joint compound (Fig. 5.4.4). Larger holes should be cleaned of loose material, metal lath affixed to the substrate, the bonding agent applied, and the holes replastered. Some patches may require complete plaster mixes, including scratch, brown, and finish (white) coat.

ADVANTAGES: Holes and cracks in plaster are relatively easy to repair with conventional techniques and materials.

DISADVANTAGES: Large sections of walls requiring repair may need laminating with new drywall or complete removal and replacement with new materials.

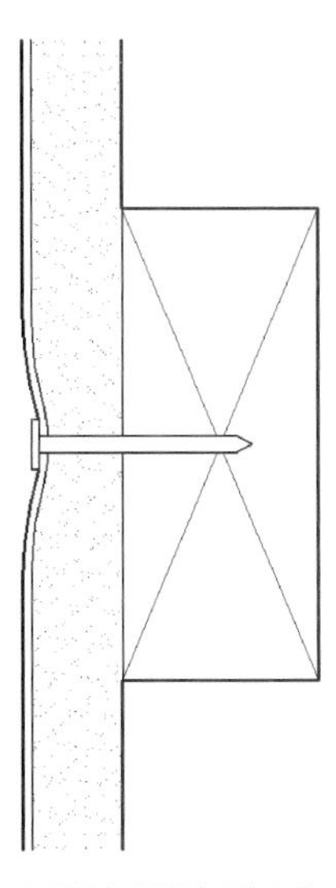

FIGURE 5.4.3 REFASTENING DRYWALL

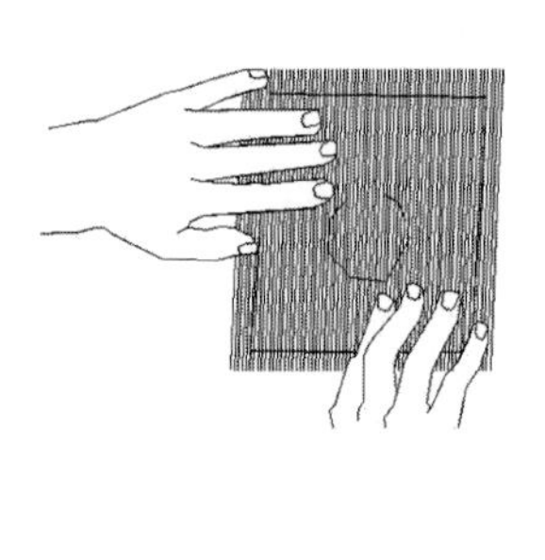

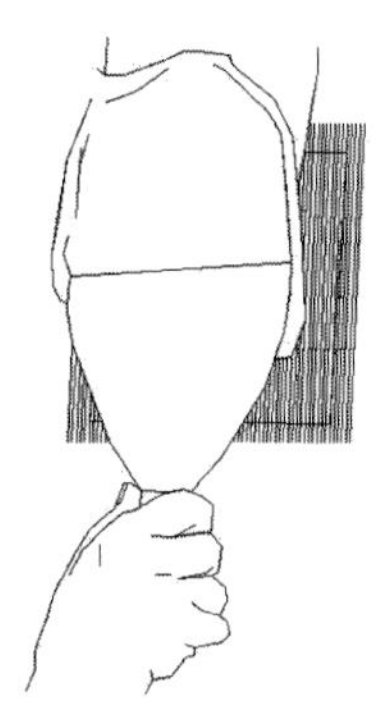

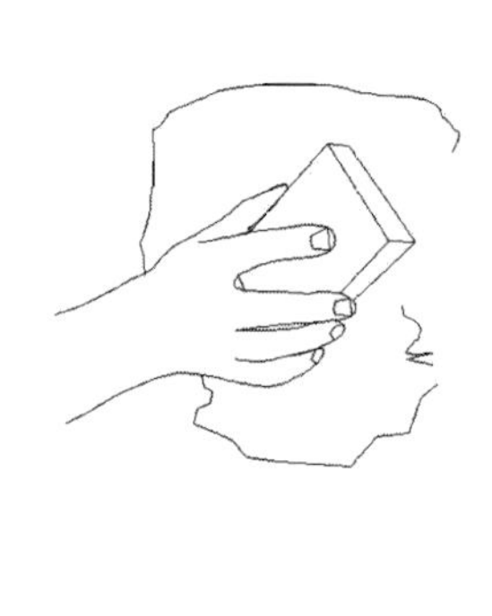

FIGURE 5.4.4 MESH TAPE AND JOINT COMPOUND

4. REPAIR SMALL HOLES, CRACKS, DENTS, AND POPPED NAILS IN DRYWALL.
Small holes and dents are typically caused by impacts from furniture or other sharp or heavy objects. Repairs to larger sections are often due to structural or moisture problems. To repair small holes, cracks, and dents, wipe the area clean, fill with joint compound using a taping knife, let harden, add a second coat if necessary, and sand and prime when dry. Nail pops occur when a stud shrinks or twists, leaving a space between the stud and the drywall. When pressure is applied against the drywall, it is forced against the stud, causing the nail head to protrude beyond the face of the drywall (Fig. 5.4.5). To repair a popped nail, drive and dimple a new nail 1 ½" from the popped nail, drive and dimple the popped nail, cover with joint compound, and sand and prime when dry.

ADVANTAGES: These are simple repairs that do not require special skills.

DISADVANTAGES: Joint compound may not adhere well; these repairs may require application of fiberglass or paper tape as well as compound.

5. REPAIR MEDIUM HOLES.
Bridge opening by criss-crossing two or three strips of self-adhering fiberglass mesh tape, fiberglass mesh repair patch, or conventional paper joint tape set in compound over the opening. Apply joint compound over the tape and around the edges of the hole with a taping knife. Wipe away excess compound and let the remaining compound harden. Apply a second coat of compound, and sand and prime when dry.

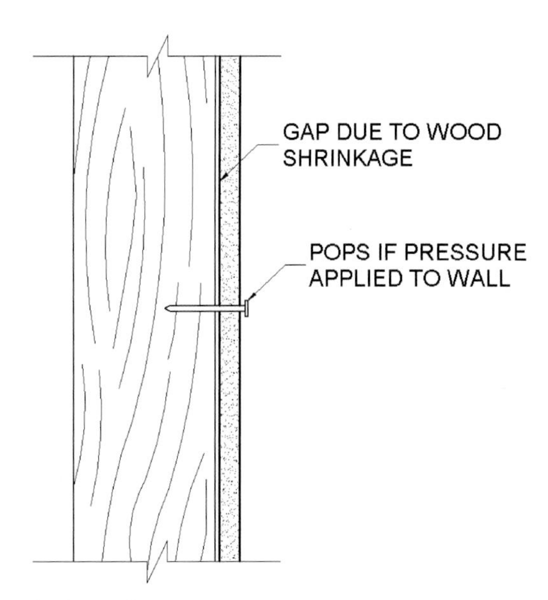

FIGURE 5.4.5 NAIL POPS

ADVANTAGES: This is a relatively simple repair.

DISADVANTAGES: Some skill is required to cover the tape and provide a smooth, uniform surface.

6. REPAIR LARGE CRACKS (⅛" TO ¾").

Bridge opening with fiberglass mesh tape. Press and smooth joint compound into the tape with a taping knife. Apply a second coat, and sand and prime when dry.

ADVANTAGES: This is a relatively simple repair.

DISADVANTAGES: Some skill is required to provide a smooth, uniform surface.

7. REPAIR TORN DRYWALL PANEL FACE PAPER.

Peel and remove loose face paper. Apply a skim coat of joint compound to the damaged area with a taping knife and feather for a smooth finish. Let harden, apply a second coat if necessary, and sand and prime when dry.

ADVANTAGES: This is a relatively simple repair.

DISADVANTAGES: This repair will not be effective if gypsum under the removed paper is crumbly, which calls for replacement.

8. REPAIR LARGE HOLES OR WATER-DAMAGED AREAS WITH US GYPSUM REPAIR KIT.

Cut out the damaged panel section using a utility knife or keyhole saw along and between studs. Remove damaged sections and old fasteners. A repair method (Fig. 5.4.6) recommended by US Gypsum Company includes slipping a drywall repair clip onto the edge of the damaged section and screwing through the wall into each drywall repair clip, positioning each screw about ¾" from panel edge and centered between tabs. This will line up screw with perforations in the clip. Measure and cut new drywall panel section to fit damaged area. Screw through new drywall into each drywall repair clip, positioning screw opposite screw holding clip to existing wall, and about ¾" from edge. Remove the tabs from each drywall clip. Apply fiberglass or paper tape and two to three coats of joint compound as needed, feathering out from previous coats. Sand and prime when dry.

ADVANTAGES: The repair kit makes patching easier.

DISADVANTAGES: Some knowledge of drywall repair is required; large patches require replacement with drywall.

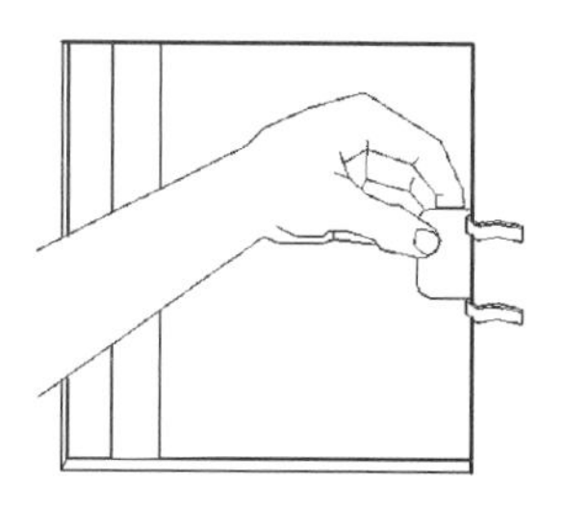

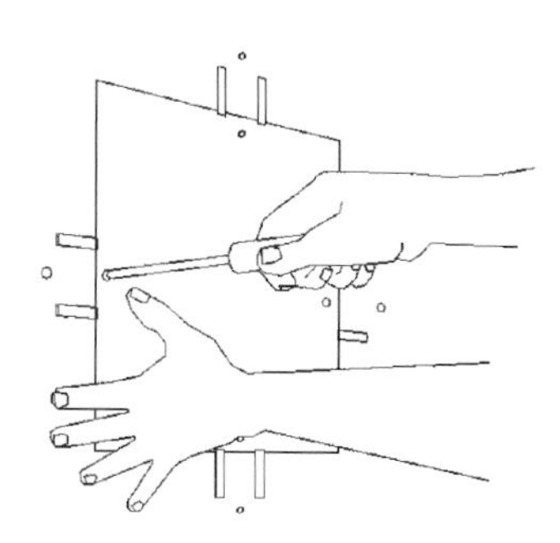

FIGURE 5.4.6 DRYWALL REPAIR CLIP TECHNIQUE

9. REPAIR LARGE SECTIONS OF DRYWALL.

Cut out the damaged area with a utility knife or keyhole saw along and between the studs. Remove damaged sections. Frame the opening with blocking to support the new drywall. Cut a new drywall panel section to fit the damaged area leaving an approximately ⅛" gap all around. Tape and apply two to three coats of drywall compound as needed, and sand and prime when dry.

ADVANTAGES: Sections of drywall are relatively simple to replace.

DISADVANTAGES: Seamless taping and compounding requires skilled applicators.

10. LAMINATE WALLS AND CEILINGS WITH NEW DRYWALL.

When more then half of a plaster wall needs repair, the most cost-effective treatment may be laminating the existing surface with new drywall. For walls, ⅜" or even ¼" thick material is usually sufficient. For ceilings ⅜"-thick material is usually adequate if fastened directly to joists, or to furring strips running perpendicular to the joists, if the spacing does not exceed 16" o.c. For wider spacing of supports, ½" will deflect less between supports and may be required. Before laminating ceilings, the capability of the existing framing should be reviewed with an engineer to determine if it can support the additional load. Preferably, the gypsum board should be adhered to the existing surface with screws and adhesives, making sure that the screws are long enough to penetrate well into the structure, not just the lath.

ADVANTAGES: This is a cost-effective repair when there are extensive problems with existing wall and ceiling materials.

DISADVANTAGES: This repair adds significant weight to the existing surfaces which may not be able to be accommodated and will cause detailing problems with the existing door and window trim. Expertise is required.

11. REPAIR CRACKED OR "ALLIGATORED" WALLS AND CEILINGS WITH FIBER GLASS MATS.

Historically, painters have used canvas applied over a wheat paste binder to cover moderately damaged, alligatored, or slightly uneven plaster surfaces prior to painting. A more recently developed ceiling material, NU-Wal®, is a fiberglass mesh fabric roll that is applied to the damaged surface over an acrylic saturant (Fig. 5.4.7). After installation, an additional coat of saturant is applied to the wet mat. When the mat has dried, the wall is ready for painting. Other manufacturers, including Permaglas™, provide self-adhesive rolls of fiberglass mesh that are designed to be skim-coated with plaster or drywall compound after application.

ADVANTAGES: These products are relatively easy to apply and coat.

DISADVANTAGES: Careful applicators are required to cover the mat with the skim coat and to smooth out the skim coat; bumps must be sanded smooth prior to installation.

12. APPLY NEW REPLACEMENT DRYWALL TO DAMAGED WALL SECTIONS.

Replacement drywall products range from conventional paper-faced products available in regular and moisture- and fire-resistant configurations; abuse-resistant products, such as National Gypsum's Hi-Abuse® board with reinforced paper facing; impact-resistant board such as National Gypsum's Hi-Impact® board with a plastic Lexan facing on the interior (stud) side; and US Gypsum's FiberRock®, a non-paper-faced board made with recycled newsprint and wood particles. Gypsum board products that are reinforced with fiberglass mats on the fronts and backs, such as Georgia-Pacific's Dens-Glass® products, can be, but are generally not, used on interior walls, because the rough surfaces cannot be painted without a skim coat of plaster, which is expensive and time-consuming to apply.

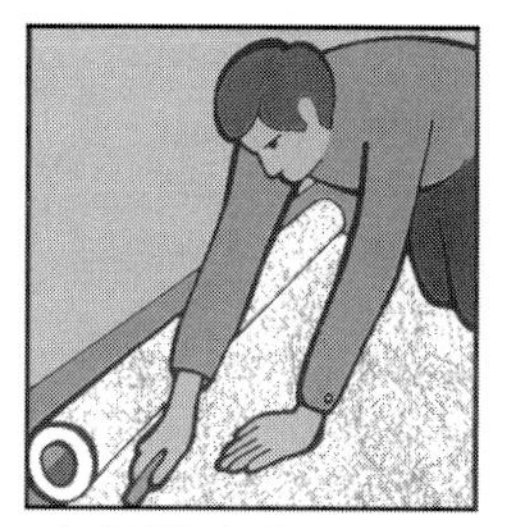
1. CUT FIBERGLASS MAT 2" LONGER THAN HEIGHT.

2. APPLY SATURANT TO AREA TO BE COVERED.

3. APPLY FIBERGLASS MAT TO WET SURFACE.

4. TRIM EXCESS MAT WHERE WALL MEETS CEILING.

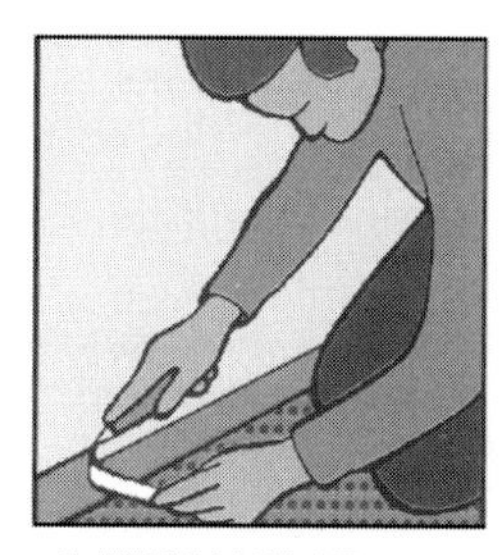
5. TRIM MAT AT BASEBOARD AND WINDOW.

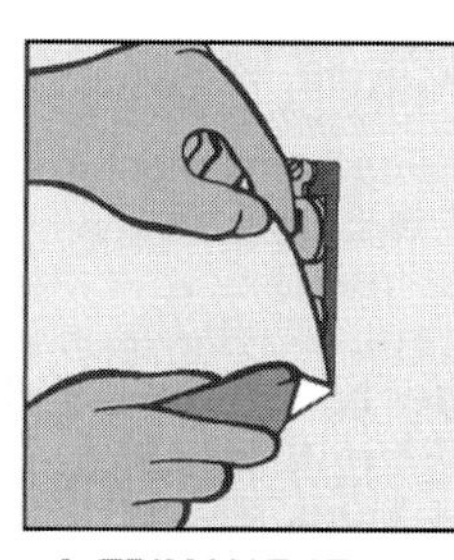
6. TRIM MAT AT OUTLETS, SWITCHES, ETC.

7. APPLY SECOND COAT OF SATURANT TO WET MAT.

8. APPLY 1ST COAT OF SATURANT TO ADJACENT AREA.

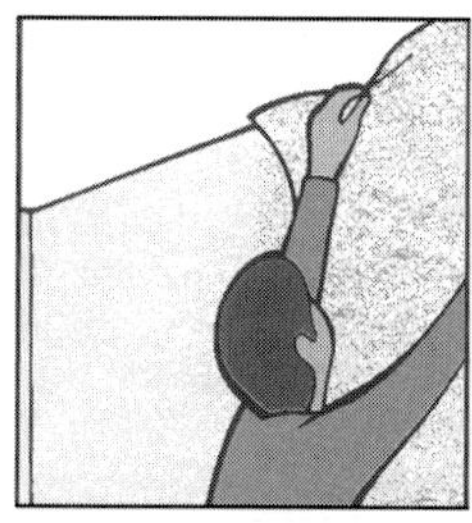
9. APPLY MAT TO 2ND AREA, OVERLAPPING BY 1".

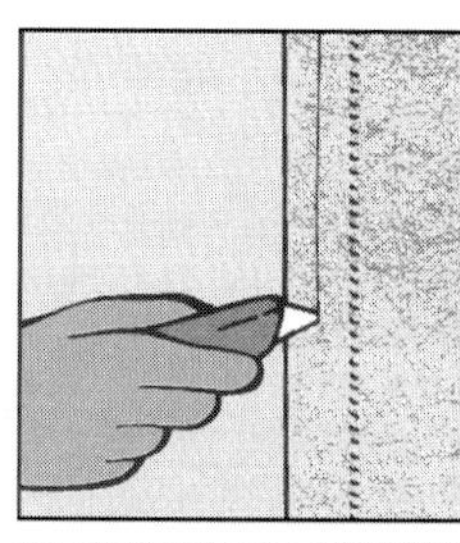
10. CUT DOWN CENTER OF OVERLAP (BOTH LAYERS).

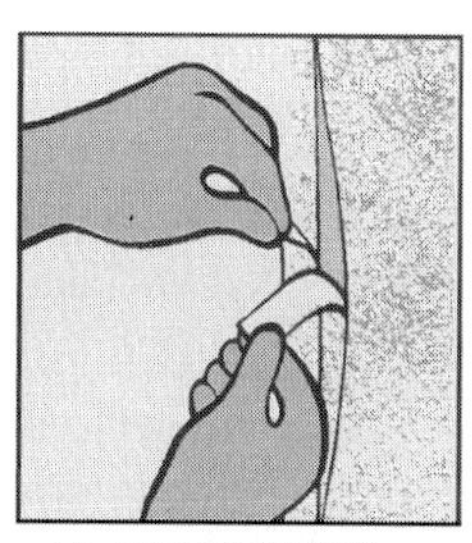
11. REMOVE MAT STRIPS ON BOTH SIDES OF CUT.

12. APPLY 2ND COAT OF SATURANT (INCLUDE SEAM)

FIGURE 5.4.7 NU-WAL RESURFACING TECHNIQUE

ADVANTAGES: Gypsum board comes in a variety of types and is the most used and cost-effective of all wall materials.

DISADVANTAGES: Requires proficient applicators and finishers as poorly installed, taped, and compounded drywall seriously distracts from appearance.

13. APPLY CORNER BEAD AND JOINT COMPOUND TO NEW DRYWALL.

A variety of new fast-setting joint compounds and corner bead systems have been developed recently that speed up the completion of projects by requiring less drywall compound than previous all-metal corner bead systems, resulting in savings of labor, material, and time. These systems include paper-faced metal drywall beads and trim (US Gypsum), rigid vinyl drywall accessories (Trim-Tex, Inc.), and plastic and paper drywall (No-Coat® and Straight Flex®).

ADVANTAGES: Paper and plastic/metal taping systems lay flatter than all-metal systems, use less compound (and therefore dry more quickly), are easier and faster to apply, and are less costly.

DISADVANTAGES: Not as strong as all-metal systems.

14. ALTERNATIVE TAPING TOOLS

Among the products on the market that combine the taping and joint compound application is the Homax Drywall Taping Tool, which accommodates 5 lb of mesh and up to 500 ft of tape.

ADVANTAGES: This too maintains a consistent amount of compound and applies tape smoothly to seams without folds or wrinkles.

DISADVANTAGES: This tool requires some skill to use and does not work on exterior and interior corners.

PAINTS AND WALL COVERINGS

ESSENTIAL KNOWLEDGE

Paint and wallpaper are the two most common protective and attractive finishes applied to walls. Paint is either oil (alkyd) or water-based (latex, vinyl, or acrylic). Oil-based paint is less permeable, shows streaks less, is more durable, and usually takes longer to dry than water-based paint. Because it does not take abrasion as well as oil-based paint, water-based paint was historically less commonly used. However, today's water-based formulations, especially the all-acrylic paints, have improved significantly, and it is the most common type of paint used because it is easy to maintain, quick drying, and does not require thinning agents for clean-up. Although today's paints are lead-free and low in volatile organic compounds, it is important to consult applicable state and federal regulations on lead paint abatement, safety, and disposal when rehabilitating an older home. Wallpapers are used to provide a quick and easy finish. Historically, they were made of colored paper and applied to walls with adhesives. Today, these coverings commonly contain vinyl and are prepasted, applied with water and a sponge. Vinyl coverings are easy to maintain and are more durable than papers.

TECHNIQUES, MATERIALS, TOOLS

1. TREAT STAINS ON PAINTED WALLS.

The most effective cleaning agent for painted walls is a mild soap-and-water solution. However, painted surfaces can become stained and may require other cleaning techniques. Stains on painted walls are usually of two types: solvent or water soluble. Water-soluble stains can usually be removed by applying a solvent-thinned primer over them. Conversely, remove solvent-soluble stains by coating them with a water-thinned primer. Shellac is another effective product to use when trying to remove water-soluble stains. However, it is ineffective against most solvent-soluble stains.

If cleaning stains on painted walls with soap and water does not remove them, they will most likely require repainting. However, before repainting, be sure to prime stains with the appropriate solution. Mold or mildew growth will stain painted surfaces and must be eliminated before repainting, or they will reappear. A strong bleach-and-water solution applied over mold and mildew will usually work. Remove any dead spores by scrubbing with a strong detergent and flushing with clear water.

Water is commonly the source of stains on painted walls. If water stains are not primed correctly, they will always show through a new coat of paint. Water stains cannot be cleaned from painted surfaces, but they can be prevented from bleeding by covering them with two or three coats of an oil-based primer.

ADVANTAGES: Cleaning stains is an alternative to repainting.

DISADVANTAGES: Removing stubborn stains may require repainting.

2. CLEAN PAPER AND VINYL WALL COVERING.

Some of the more common wall coverings are paper, vinyl-coated, vinyl acrylic, solid sheet vinyl, and laminated vinyl. Most of them are nonwashable, washable, or scrubbable. Test a small area in an inconspicuous location by wiping with a damp cloth. If the covering is paper and it does not change in color, it is most likely washable. If the wall covering is vinyl-based and does not change color when wiped with a damp cloth, it is most likely scrubbable. Before cleaning the wall covering, use a vacuum to remove dust and loose dirt. Dusting walls with a cloth tied to a broom or mop will also remove dust particles.

Nonwashable papers should be dusted frequently to avoid dirt or dust buildup. Puttylike commercial wall paper cleaners are available, and when used correctly they are effective. However, be sure to test them on an inconspicuous area first. Washable papers are commonly coated in plastic. Clean with a damp cloth and avoid wetting. After cleaning, wipe walls down with a clean dry cloth to remove any moisture. If the walls need a second cleaning, allow them to dry completely before applying the damp cloth again. Scrubbable wall coverings are commonly made of vinyl or are vinyl-impregnated. These types of coverings can be cleaned with appropriate foam cleaners or detergents. Be sure to avoid abrasive cleaners. Rinse walls with a damp cloth after using any type of cleaner. Remove any moisture with a clean dry cloth.

A variety of dirt and stains can be removed by wiping with art gum or commercially available cleaners. Try a soap-and-water solution on washable or scrubbable coverings before using commercial cleaners. Use a warm iron over white paper towels to remove grease spots. Some spot removers applied to grease stains will turn them into a powder that can be brushed away. Use a soap-and-water solution to remove any remaining grease from washable or scrubbable wall coverings.

ADVANTAGES: Most wall coverings are easy to clean.

DISADVANTAGES: Nonwashable papers are high maintenance and cannot be cleaned by common techniques.

3. REPAIR PAPER AND VINYL WALL COVERINGS.

Seams between strips of paper and vinyl wall coverings may separate from the wall and begin to peel back over time. Paper that isn't pasted correctly is usually the cause. There is commercially available seam adhesive that does an effective job in repairing these minor problems. Use a small paint brush or sponge to coat the seam that has peeled away. Press the paper back into position with a damp cloth or a seam roller. Horizontal tears can be repaired in the same way. Major rips or punctures must be cut out and repaired with a matching patch. With a sharp blade and a straight edge, remove a matching patch from inside a closet or behind a cabinet and cut out the tear (Fig. 5.4.8). Apply adhesive to the exposed wall and to the matching patch. If the paper is prepasted, soak the patch in water for about 30 seconds to activate the adhesive. Place into position and apply pressure with a damp cloth or seam roller. Tape the patch into position with masking tape to ensure proper adhesion. Blisters or air bubbles are usually caused by incorrect installation. If the blister or bubble is caused by something that has been trapped between the paper and the wall, it must be removed. Cut around the particle, remove

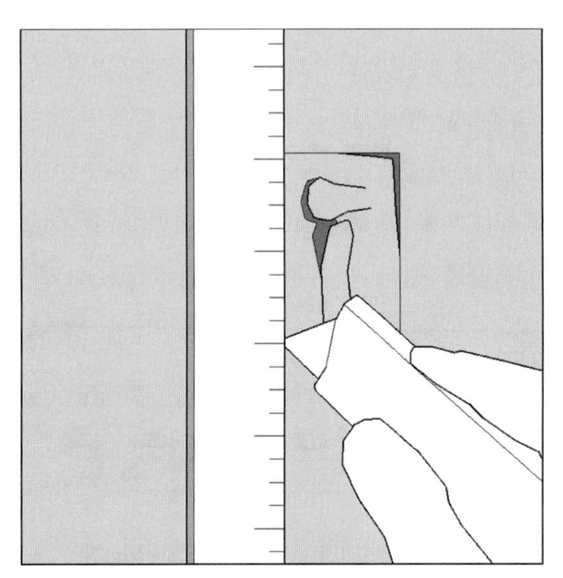

FIGURE 5.4.8 REMOVING RIPPED WALLPAPER

it, repaste, and install the patch that has been removed. If the bubble or blister is just an air pocket, inject adhesive into this pocket with a commercially available glue-injecting syringe. Flatten with a seam roller and use a damp cloth to remove any excess adhesive.

ADVANTAGES: Minor damage to wall coverings is easy to repair.

DISADVANTAGES: A matching patch of wallpaper may not be available.

4. REMOVE PAPER AND VINYL WALL COVERINGS.

If paper and vinyl wall coverings have become too worn or outdated, they may require removal. The best practice is to remove old coverings before applying new ones. The removal process may be time-consuming but is easy thanks to today's effective nondrip, nontoxic enzyme strippers. Wallpapers can be removed with hot water or commercial strippers. Because they are made of paper, they easily soak up water or strippers and can be peeled from the wall. Vinyl wall coverings are usually tear resistant and do not soak up water or strippers. The vinyl strips can usually be peeled away from the wall, but the adhesive backing will remain. Remove the adhesive backing by soaking with hot water or strippers. Several layers of paper, or paper and vinyl wall coverings that have been painted, are difficult and time-consuming to remove because they are sealed and the adhesive cannot absorb water or strippers. Commercially available scoring tools are most effective (Fig. 5.4.9). These tools can be run across the wall surfaces and create tiny punctures in the covering without damaging the wall. Strippers applied to the scored paper will penetrate the holes, loosening the adhesive from the wall.

ADVANTAGES: Commercial strippers make removal simple.

DISADVANTAGES: Removing some wall coverings, layers of coverings, and painted coverings is time- consuming.

MOLDINGS AND TRIM

ESSENTIAL KNOWLEDGE

Wall trim and molding at floor level to about 6' above the floor can become abraded from furniture pushed up against it, touching, kicking, and cleaning. Chair rails were originally intended to protect

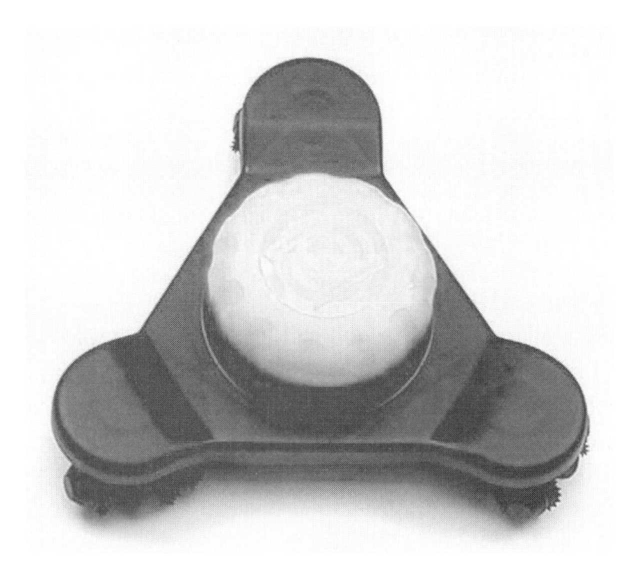

FIGURE 5.4.9 WALLPAPER SCORING TOOL

FIGURE 5.4.10 SCORE WALL TO REMOVE TRIM

plaster or wallpapered walls from damage caused by chairs pushed up against them. Older plaster trim may have been custom molded and woods custom milled. Today's moldings are usually hardwoods and are readily available at home centers. Minor abrasions to any type of molding can be easily repaired. Replacement may be required for severe damage.

TECHNIQUES, MATERIALS, TOOLS

1. REPAIR OR REPLACE DAMAGED TRIM.

Minor damage to plaster trim can be repaired with patching compound. Sand smooth, prime, and repaint. Severely damaged or missing trim pieces can only be repaired by replacing with wood plastic or foam-based trim, spackling the seams, priming, and repainting. Trim to match old plaster moldings may not be available. However, there are many commercially available plastic or foam-based moldings that resemble original styles. Even wood trim that closely resembles the plaster can be used. Damaged wood trim can be repaired in the same fashion. Fill dents and cracks, sand smooth, and repaint. Prime all surfaces before applying a fresh coat of paint. Use nontoxic strippers to remove paint layers from moldings to restore them to their natural condition.

ADVANTAGES: Minor damage to molding and trim are repaired quickly with spackle.

DISADVANTAGES: Major damage may require replacement of the molding or trim.

2. PATCH DAMAGED TRIM.

Severely damaged trim may require a replacement patch. Cut out the piece of molding that needs replacing and try to locate matching wood or plastic trim to replace it. If matching molding cannot be located, one can be custom milled using the damaged piece as a template. Cut a matching piece to length and nail in place. Prime any unpainted surfaces before painting. When removing sections of molding, score the wall along the trim before it is removed (Fig. 5.4.10). This will prevent damage to the drywall or plaster during the removal process.

ADVANTAGES: Replacement patch pieces are easily disguised when painted to match the existing molding or trim.

DISADVANTAGES: Matching profiles may not be available.

5.5 STAIRS

REPAIRING TREADS AND RISERS

ESSENTIAL KNOWLEDGE

Stairs can have either open or closed risers. When staircases are constructed of both risers and treads, they are called closed-riser stairs. When staircases are constructed without risers, they are referred to as open-riser stairs, or ladder stairs. In all cases, treads are the essential element that allows one to negotiate a stairway, moving vertically from one floor to the next.

Over time, the natural course of a home's settlement, wood shrinkage, and the constant use of stairs begins to separate treads from the supporting carriage or risers, causing squeaks. Squeaks may also be caused by a riser that rubs against a tread or the carriage. Squeaky treads and risers are the most common complaints from owners or occupants of homes with older stairs.

TECHNIQUES, MATERIALS, TOOLS

1. USE GRAPHITE POWDER TO FIX SQUEAKY TREADS AND RISERS.
In order to stop treads or risers from squeaking, they must first be located. Walk up or down the stairs to locate where the squeak occurs. A quick fix to silence squeaky treads or risers can be achieved by blowing powdered graphite into the joint between the tread and the riser. The graphite lubricates the joint and eliminates squeaks by reducing the friction that results when wood members slide against each other.

ADVANTAGES: This is an inexpensive and quick way to silence squeaks.

DISADVANTAGES: Graphite powder is a temporary fix because it will wear away eventually. As a result, the squeaks will reoccur.

2. REFASTEN TREAD FROM ABOVE.
More permanent remedies for squeaks include nailing the tread down with angled nails or trim head screws, or wedging the tread tight. If squeaks occur near the center of the tread, angle finishing nails into the riser below. If the squeak occurs at the ends of the tread, angle finishing nails into the carriage. Whenever possible, nail the tread into the carriage or stringer. For example, if a staircase has three carriages, one on each end and one in the center, and the squeak occurs near the center of the tread when it is stepped on, it is best to angle finishing nails into the center carriage. Because staircases in older homes are usually constructed of hardwood, pilot holes must be drilled before nailing or screwing to prevent splitting. Countersink nails and screws and fill with a wood filler.

ADVANTAGES: Nailing or screwing treads in place is a permanent solution to silencing squeaks. Sometimes access to the underside of a staircase may not be an option. As a result, working from above is the most convenient and least time-consuming method of repair.

DISADVANTAGES: Repairing squeaky treads or risers from above may not solve the problem. If this method is not successful, one must consider working from below. However, working from below can present a challenge because the stair structure may be hidden behind a finished surface or may be out of reach because it may connect levels of the building which are above the first floor.

3. REFASTEN TREAD FROM BELOW.

If access to the underside of stairs is possible, wood blocks can be glued at the joint between the riser and tread. If old blocks are already present, they may be removed, reglued, and refastened. Metal angle brackets can also be installed at this joint to secure the treads and risers (Fig. 5.5.1). If the squeaks are caused by a tread which has been loosened significantly from the carriage or has become too warped, it can be refastened to the carriage from below using screws instead of finishing nails. Lubricate screws with paraffin wax if they are used to fasten oak treads. The wax will allow the screw to pierce hardwood more easily. Countersink the screws and fill with a wood filler.

ADVANTAGES: Wood blocks or metal angle-brackets fastened to the underside of stairs are strong supports that are hidden from view.

DISADVANTAGES: If the underside of stairs is finished, the surface must be removed to reveal the stair structure. The underside of stairs may be out of reach, and complicated equipment may be required to gain access.

4. USE WOOD WEDGES TO FIX SQUEAKY STAIRS.

Instead of using nails or screws, inserting wooden wedges into the joint between the riser and the tread from above may be a simple and quick fix to eliminate squeaks. Many older staircases have moldings that run under nosings and at the back of treads where they join the risers. This molding can be removed and wedges can be inserted. Once the molding is removed, insert a knife or similar tool into the joint between the tread and riser. This will reveal the type of joining system used in the stair's construction. The joints used to fasten treads to risers are either butt, rabbeted, or tongue-and-groove. Coat hardwood wedges with glue, and hammer them into the joint as far as possible. Once the wedge is inserted, cut off the visible end and replace the molding.

ADVANTAGES: Using wood wedges to secure treads and risers is a quick and inexpensive method of reinforcement which is hidden from view. This method eliminates the need for nails or screws.

DISADVANTAGES: Damage to moldings may occur if they must be removed from under nosings or at the back of treads where they join the risers.

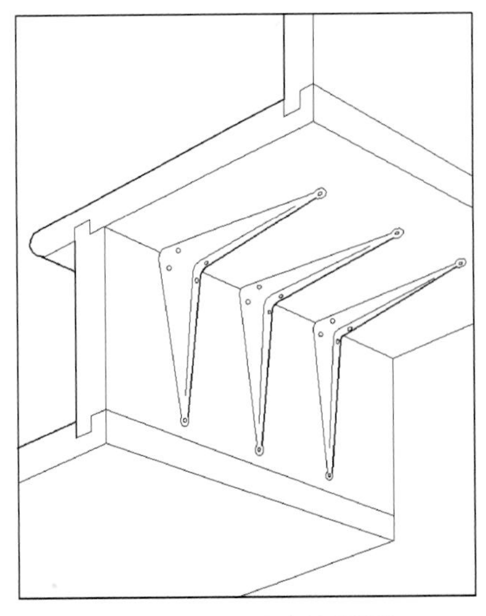

FIGURE 5.5.1 METAL ANGLE BRACKETS

REPLACING TREADS AND RISERS

ESSENTIAL KNOWLEDGE

Treads are the stair component that gets the most wear. As a result, they may become unevenly worn, crack, split, or even become so scratched that replacing them is the best option. Conversely, risers are probably the least likely to become worn; however, over time they may become subject to cracks or dents and need replacing as well. Treads and risers can be replaced with readily available stock treads and custom-cut hardwoods. Stock treads are available with integral factory-milled nosings. Whether the staircase is open on one or both sides, or located between walls, replacing treads and risers is a relatively simple procedure.

TECHNIQUES, MATERIALS, TOOLS

1. REPLACE TREADS AND RISERS.

If a rehab project involves replacing all treads and risers, perhaps the most obvious recommendation is to begin at the bottom of the staircase. Before starting on the treads and risers, all of the balusters must be removed (see Damaged or Broken Balusters later in this section). If the project requires the replacement of isolated treads or risers, the same method applies; however, only remove those balusters that attach to the damaged tread. Begin by prying the first riser and tread from the carriage or stringer. Hammering a pry bar into the joint between the first riser and the first tread will help lift the tread. If nails are exposed and prevent the tread or riser from being removed, cut them using a utility saw.

If the tread or riser cannot be lifted, try drilling holes into the tread in two places to allow access for a saw (Fig. 5.5.2). Cut into the tread from the back of the riser below to the face of the riser at the back of the tread. Chisel off the nosing attached to the tread. Begin removing the tread by prying the cut sections of the tread from the carriage using a pry bar and a hammer. After removing the first three or four treads and risers, install the new ones, being careful not to close the staircase in order to allow access for removing the rest of the treads and risers. Cut the risers so that their ends are flush with the face of the carriage. If using a stock tread with an integral

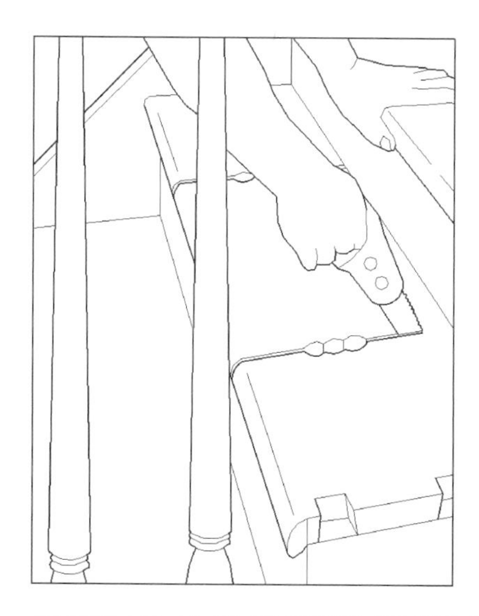

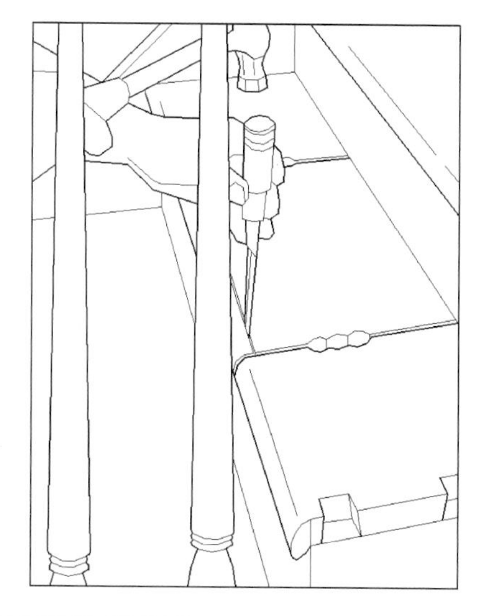

FIGURE 5.5.2 CUTTING AND CHISELING TO REMOVE TREAD

nosing, cut off the side nosing that protrudes over the carriage. Begin at the right angle at the front of the tread in the corner that meets the carriage, and make a 45° cut into the tread for the depth of the nosing. Then saw off the rest of the integral nosing along the entire side of the tread (Fig. 5.5.3).

Install the newly cut tread so that the ends are flush with the face of the end carriages and it abuts the riser above and below. Fasten it to the riser below using nails or screws and fasten it to the back riser using predrilled pilot holes and driving nails or screws on an angle from the top of the tread into the back riser. If stairs can be accessed from below, fasten risers to treads from under the stair using nails or screws. If the treads and balusters are dovetail jointed, hold the balusters against the treads and mark where the joint must be cut. Remove the tread and cut the joint that will receive the baluster. Install and fasten the treads to the risers and the carriage. Insert the balusters in place and fasten the return nosing.

ADVANTAGES: Replacing treads and risers is a permanent fix for those that have become unevenly worn or damaged. Because damaged treads may present safety concerns, replacing them may reduce hazardous conditions that risk safety.

DISADVANTAGES: Replacing treads and risers along an entire stair run may be costly and time-consuming. Removing the balusters without damaging them and reinstalling them intact is not easy. If damage to balusters occurs, they may require replacement.

SAGGING CARRIAGES

ESSENTIAL KNOWLEDGE

Over time, stair carriages may begin to sag or bow. Sagging carriages may loosen the connection between risers and treads, causing riser heights to vary along the stair run and creating a potentially hazardous condition. If this is the case, the staircase may need to be replaced entirely. However, if the sagging carriage has not jeopardized the safety of the staircase, it may be reinforced to prevent further sag.

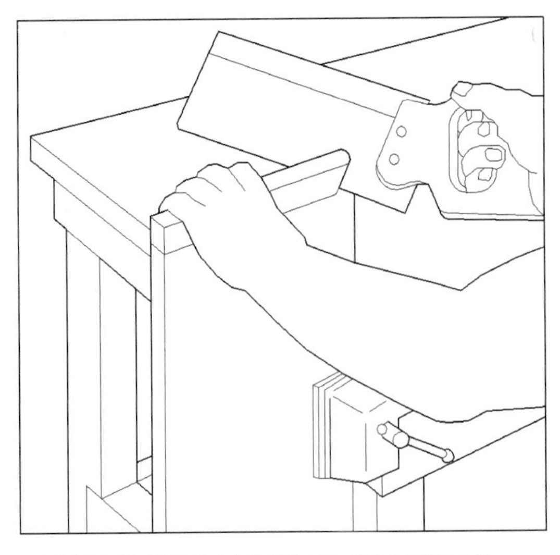

FIGURE 5.5.3 REMOVE SIDE NOSING ON TREAD

TECHNIQUES, MATERIALS, AND TOOLS

REINFORCE A SAGGING CARRIAGE.

More often then not, the underside of the sagging stair will have a finished surface. This surface must be removed to expose the stair structure. Once the structure can be accessed, it can be reinforced to prevent further sag by screwing metal angle-brackets or wood blocks through the carriage and into the supporting structure. For added support, one or more new carriages can be installed. For a staircase that is open on one side, constructing a knee wall below the stair may be a practical reinforcement solution. Because a sagging carriage may be fixed in a number of ways, a qualified structural engineer will be able to recommend the best option for repair.

ADVANTAGES: Reinforcing sagging carriages from below the stair structure is usually a simple and permanent process.

DISADVANTAGES: Reinforcing a sagging carriage may require finish surfaces under stairs to be removed to expose the stair structure. If the underside of stairs is not reachable, accessing the sagging carriage can be challenging.

DAMAGED OR BROKEN BALUSTERS

ESSENTIAL KNOWLEDGE

Balusters are the individual vertical elements that attach the railing to the staircase. The railing is attached to a newel post usually found at the top or bottom of the run. Balusters may not be present if the railing is attached directly to a wall, as is the case with a closed staircase encased between walls, or a staircase that is open on only one side. Over time, balusters can become loose or broken. It is important to remedy this situation because the balusters provide the structural support to the handrail. Repairing a loose balustrade can be a simple process. However, replacing balusters may be more involved, depending on the way they are connected to the handrail and stair. There are generally three different types of connections: filleted, doweled, and dovetailed.

A filleted baluster is usually square-topped and fits into a groove at the underside of the railing. Fillets, or small blocks of wood, are secured into this groove between balusters, which holds them in place. These types of balusters may also be inserted into the groove of a lower rail that is attached to the stringer. Balusters can also be doweled at the bottom or the top, or doweled at the top and dovetail jointed to the tread at its base. The exact profile of balusters in need of replacement may not be available. Using the broken baluster as a template, a skilled millworker can turn an exact replica.

TECHNIQUES, MATERIALS, TOOLS

1. STRENGTHEN EXISTING BALUSTERS.

Loose balusters can be secured by reattaching them to the railing with nails or screws. Pilot holes can be bored on an angle at the top of the balustrade and into the railing. Drive the nails or screws into the pilot holes through the baluster and into the railing. The same technique may be used to fasten the bottom of the baluster to the stair.

ADVANTAGES: Because this type of repair requires fastening from the underside of the railing, it usually goes unnoticed if nails and screws are countersunk and filled.

DISADVANTAGES: Loose balusters may be caused by damage to the baluster itself. If this is the case, securing the baluster to the rail or to the tread below will not solve the problem. If damage to the baluster occurs, it will have to be repaired in place or removed and replaced.

2. REPLACE FILLETED BALUSTERS.

To remove a filleted baluster, chisel out the fillets at the railing and at the base. The baluster may be easily hammered out and removed. Be careful to remove any old glue from the grooves into which the new baluster will be inserted. The proper angle for the new balusters can be obtained by holding the old one at its side and marking the angle on the new one. Insert the new balusters into the grooves at the rail and base and fasten by toe-nailing. Measure and cut new fillets using the old baluster to determine the correct angle. Coat with glue and insert into the grooves between balusters.

ADVANTAGES: Because the actual joint between the balusters and the rail or base is hidden by the fillets, slight inaccuracies in cutting or fastening will go unnoticed. Removing the return nosing on treads is not required when installing filleted balusters.

DISADVANTAGES: Replacing filleted balusters along an entire stair run may be time-consuming because it requires measuring and cutting blocks of wood into many small lengths.

3. REPLACE DOWELED BALUSTERS.

A broken doweled baluster may be removed by sawing it in half and prying it loose from the glue joint at the tread and from the underside of the railing (Fig. 5.5.4). Breaking the glue joint is not always possible. If this is the case, saw the baluster flush with the tread or underside of the railing and use a drill with a bit that is the same size as the dowel to bore a new dowel hole. Coat the dowels on the new baluster and the dowel holes in the railing and tread with glue. Angle the baluster and insert the upper doweled end into the hole in the railing first. Lifting the railing, drag the bottom end of the baluster across the tread and insert it into the hole. If the railing will not lift, trim the dowel ends to shorten the baluster.

ADVANTAGES: Doweled balusters are easy to remove and do not require the time-consuming process of prying them from the rail or base or measuring and cutting fillets. Boring a new dowel hole is also a simple procedure. Removing the return nosing on treads is not required when installing a doweled baluster. Nails or screws are not required for fastening.

FIGURE 5.5.4 BREAK GLUE JOINT WITH WRENCH

DISADVANTAGES: Once the new doweled baluster is inserted into the rail, it may not be easy to lift the rail to insert the base of the baluster. As a result, if precautions aren't taken to protect the tread, dragging the baluster along the tread may damage its surface.

4. REPLACE DOVETAILED BALUSTERS.

In order to remove balusters that are fastened to the tread with a dovetail joint, first remove the cap molding or return nosing that attaches to the side of the tread and covers the dovetail joint. This nosing can be pried away from the tread using a pry bar. The joint between the tread and the nosing may need to be cracked before it can be pried away. The baluster may be hammered out once the dovetail is revealed. To replace a dovetail baluster, insert its top into the underside of the railing first and then insert the dovetail into the tread. Drill pilot holes through the dovetail and into the tread and secure with nails or screws. Replace and secure the return nosing, being careful to countersink any nails or screws and fill with putty.

ADVANTAGES: A dovetail joint is the strongest type of joint and provides the most stability. As a result, a dovetail baluster is less likely than balusters connected to rails or treads by other types of joints to become loose or need replacing.

DISADVANTAGES: Removing dovetailed balusters from the tread is a time-consuming process because the joint may be difficult to separate. The return nosings on treads must be removed before dovetailed balusters can be installed, and nails or screws are required for fastening.

PREFABRICATED STAIRS

ESSENTIAL KNOWLEDGE

If rehabilitating an existing staircase is not possible, or if constructing a site-built staircase to replace an older one is too costly, factory-built staircases are an economical solution to stair replacement. There is a variety of styles for both closed- and open-riser stairs manufactured in a variety of materials and finishes to fit almost any taste. Whether installing a prefabricated staircase housed between walls or open on one or both sides, the technique is the same.

If a prefabricated staircase is to be installed between walls, handrails must simply be fastened to one or both walls, depending on code requirements. If a staircase is open on one or both sides, then a balustrade must be installed. This is the most complex and time-consuming part of installing a prefabricated staircase. If you are installing a prefabricated staircase and have removed the preexisting staircase structure, chances are that access to the stair from below is possible. However, if you are installing a stair housed between walls and access from below is not available, remove the wall's surface material to allow for easy access.

When replacing an existing staircase with a prefabricated staircase, be cautious with the demolition work. Older stairs are sometimes part of the home's structure and may be difficult to remove. Before ordering the prefabricated staircase, be sure to check with the local building department for code requirements.

TECHNIQUES, MATERIALS, TOOLS

1. INSTALL A PREFABRICATED STAIRCASE.

Once the prefabricated staircase is delivered, it is ready for installation. First drill pilot holes in the top riser so that it can be fastened to the header in the staircase opening. With someone standing at the opening above, position the staircase so that the top riser rests against the header in the staircase opening and is flush with the subfloor. Shim under the stair which rests on the floor below to ensure that the treads are exactly horizontal. Secure the top riser to the header using finishing screws. With the stair in position, locate studs on the walls to which stringers will be attached. Fasten the top plate to the underside of the staircase about 1½" in from the open end. Hammer nails through the top plate up through the risers. Fasten a bottom plate to the floor below and toe-nail studs in place. A prefabricated staircase can be ordered with a balustrade that is precut and ready for installation after the staircase is secured in position. Since this may be the most difficult and time-consuming part of the prefabricated staircase installation, it is important to carefully follow the manufacturer's installation recommendations.

ADVANTAGES: This is an economical solution to replacing or adding a stairway. Many styles are available from many different prefabricated staircase manufacturers. Manufacturers offer step-by-step instructions to ensure that the staircase is assembled correctly.

DISADVANTAGES: Custom options are usually not available. Make sure that measurements are accurate to ensure that the staircase fits properly.

2. USE A PREFABRICATED STRINGER FOR STAIR CONSTRUCTION.

An innovative prefabricated stair system called the Easy Riser provides an alternative to traditional notched-stringer stair construction. The Easy Riser stair is a two-stringer, engineered wood system that uses prefabricated individual components which, when fastened to a 2x6 or 2x8, create the effect of a notched stringer onto which the risers and treads are fastened (Fig. 5.5.5).

ADVANTAGES: This system significantly reduces the time and labor involved in constructing traditional notched-stringer stairs

DISADVANTAGES: The cost for individual tread and riser components may be more than the labor and time savings.

ATTIC LADDERS

ESSENTIAL KNOWLEDGE

Providing access to attics is a practical solution to the lack of available storage space common to many homes. However, accessing the attic may not have been part of the original design of the home. As a result, floor space may not be available to allow for the construction of a new staircase to the attic.

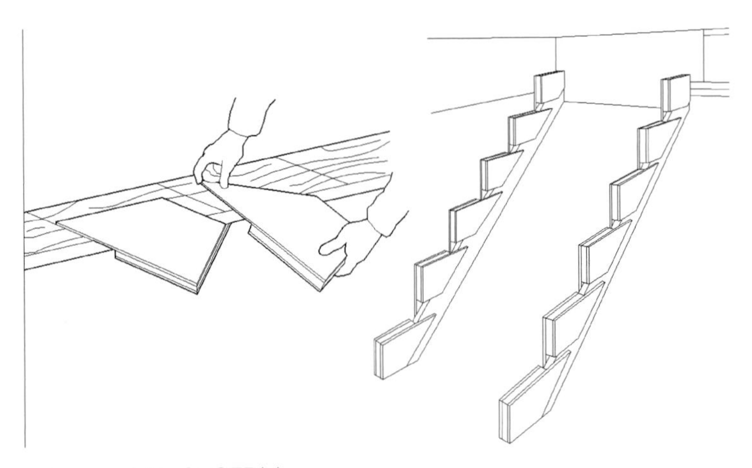

FIGURE 5.5.5 EASY RISER SYSTEM

Installing an attic ladder, or disappearing stair, may be a simple and practical way to gain access to the attic. Factory-built attic ladders are available in a number of standard sizes. Because different attic ladders have different clearance and headroom requirements, it is important to check these measurements. Before making any cuts in the ceiling or attic floor be sure to install shoring or supports. This is especially important if the opening will be cut so that the long side will run perpendicular to the joists. A staircase opening of this type will require approximately six joists to be cut. To minimize the number of joists to be cut, locate the staircase opening so that the long side runs parallel to the ceiling joists.

TECHNIQUES, MATERIALS, TOOLS

INSTALL AN ATTIC LADDER.

The factory-built attic stair (Fig. 5.5.6) will require that the hole to receive the stair be cut to specific dimensions. Mark those dimensions on the ceiling, and drill holes into the ceiling at the four corners. Snap chalk lines that connect these drill holes, and then snap chalk lines that extend 3" outside the perimeter of the original markings. Saw the attic floorboards along the perimeter of the outside chalk lines. Remove the attic floorboards and subfloor. From below, remove the ceiling section which covers the opening in the same way the attic floor was removed. The exposed joists should be sawed so that they are flush with the opening. Install trimmer joists along the long sides of the opening. Using joist hangers, install double headers against the cut ends of the tail joists. Once the opening is cut according to the manufacturer's dimensions, the attic staircase is ready to be installed. Carefully follow the manufacturer's recommendations for installation. If possible, mount insulation material on the top side of the access hatch so ceiling insulation performance is not degraded.

ADVANTAGES: The attic ladder is a simple solution that provides access to attics and can be hidden from view when not in use. It is very economical, arrives at the job site fully assembled, requires minimal adjustments, and can usually be installed in a few hours.

DISADVANTAGES: An opening must be cut in the ceiling for its installation. To avoid cutting too many joists and to ensure that the structural integrity of the ceiling is not compromised, openings should be located so that the long side runs parallel to the joists. If this cannot be accomplished, a structural engineer should be consulted for advice on proper reinforcement.

FIGURE 5.5.6 PREFABRICATED ATTIC LADDER

CH.6

HOME REHAB HANDBOOK

KITCHENS AND BATHS

Chapter 6 KITCHENS AND BATHS

6.1. KITCHENS AND BATHS OVERVIEW
6.2. CABINETS
6.3. COUNTERTOPS
6.4. APPLIANCES
6.5. SINKS AND LAVATORIES
6.6. TUBS AND SHOWERS
6.7. TOILETS AND BIDETS

6.1 KITCHENS AND BATHS OVERVIEW

The rehab specialist might have to rethink the kitchen and bath when called on to restore these areas. These two rooms are used by every member of the household every day, and they have systems with component parts that must function in harmony for the space to work. The National Kitchen and Bath Association (NKBA) has developed two books, *41 Guidelines of Bathroom Planning* and *40 Guidelines of Kitchen Planning* to address function, storage, layout, safety, access, design assistance, and product selection.

Being the most-used rooms in any home, the kitchen and bath will likely be ripe for rehab work. Appliances and other equipment and fixtures become worn, obsolete, or energy hogs as they age. The preparation of food in the kitchen, and high-moisture levels in both the kitchen and bathroom take their toll on finishes, particularly countertops, cabinets, and floors.

Traditionally, the kitchen *work triangle* is comprised of the sink, range, and refrigerator (Fig. 6.1.1). According to the National Kitchen and Bath Association, each leg of the triangle should be between 4' and 9' long to reduce the walking distance between the sink, range, and refrigerator. The work triangle should be outside of traffic patterns through the house so preparation is uninterrupted. In older homes, the back door and stairs to the basement are often off of the kitchen, and the only route is through the center of the kitchen and usually through the work triangle. Improving the function of the kitchen might involve changing the shape of the space by adding a peninsula or island. Although the parameters of the room dictate the size of the kitchen, its shape depends on the layout and combination of the work triangle and activity centers. Common shapes include the one-wall kitchen, corridor-shaped kitchen, L-shaped kitchen, and U-shaped kitchen.

When possible, a rehab project should incorporate these work triangle guidelines into the layout of the kitchen to enhance its flow and efficiency. Lifestyle trends have changed the way these spaces are used. The kitchen has gone from being a preparation area where one person cooks and serves meals, to the central living center where food preparation, household tasks, and socializing are combined. To support this centralized-family concept, the kitchen may also include centers of activity such as the secondary sink center, an additional microwave oven center, serving center, home office center, media center, and socializing center.

Time constraints and stress also affect today's kitchen and bathroom. With most homeowners' time being at a premium, efficiency in the kitchen and bathroom is critical, and technology is having an impact on the way the kitchen and bathroom are used, as well as on the products available. Today the height of the sink or work surface can be adjusted with the touch of a button, or a chicken can be roasted in a fraction of the time it used to take. In the bathroom, a combination of rehabed fixtures and finishes with new technologies will enhance efficiency. When space is minimal, the bathtub, if not the only one in the home, is often exchanged for an oversized shower, often with a seat, grab bars, dual shower heads, body sprays, or steam.

The most essential element in rehabing a kitchen or bathroom is the need to incorporate universal design. With household members of every age, stature, and level of physical ability, the issues of safety and access within the existing space should be addressed. The Center for Universal Design has developed a set of principles that can serve as a guide to the rehab professional.

- *Equitable use.* The design of products and environments is usable by all people to the greatest extent possible, without the need for adaption or specialized design.

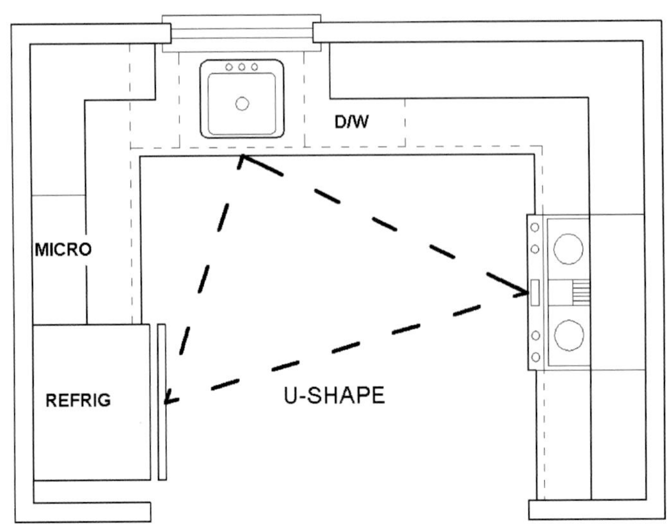

FIGURE 6.1.1A U-SHAPED KITCHEN

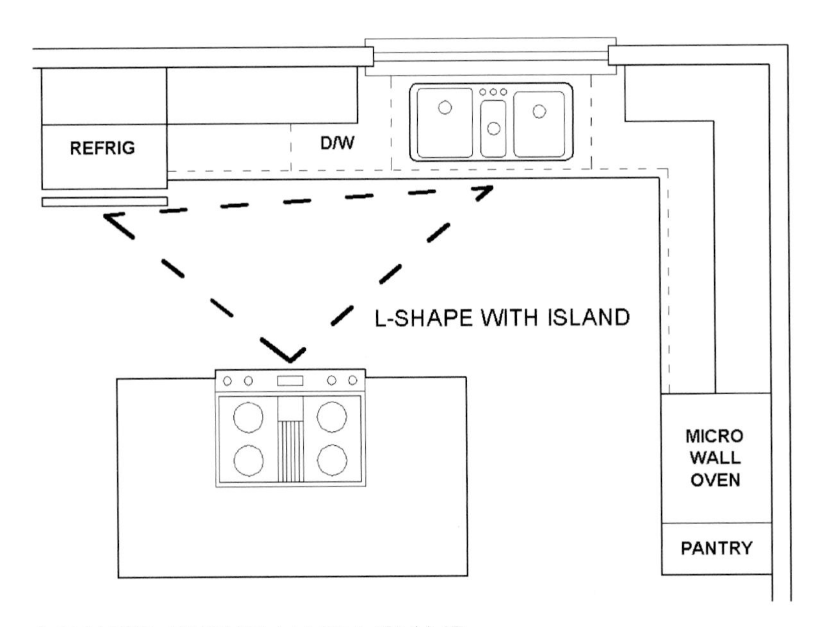

FIGURE 6.1.1B L-SHAPED KITCHEN WITH ISLAND

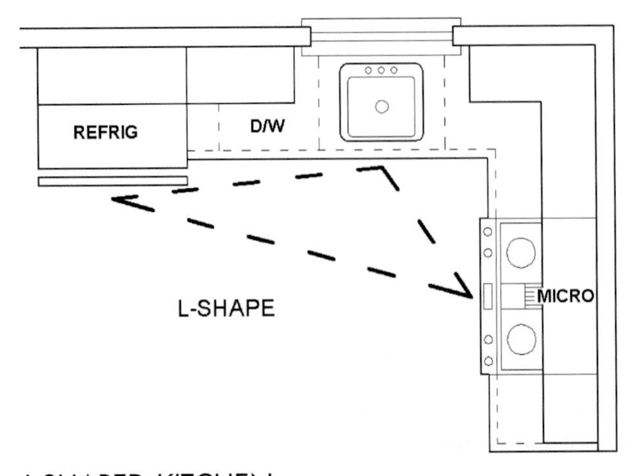

FIGURE 6.1.1C L-SHAPED KITCHEN

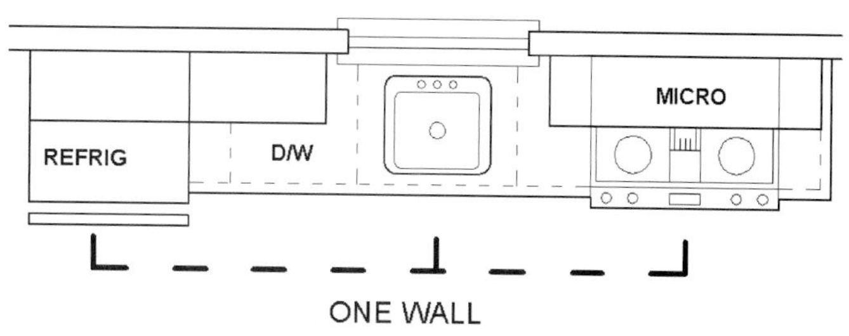

FIGURE 6.1.1D ONE-WALL KITCHEN

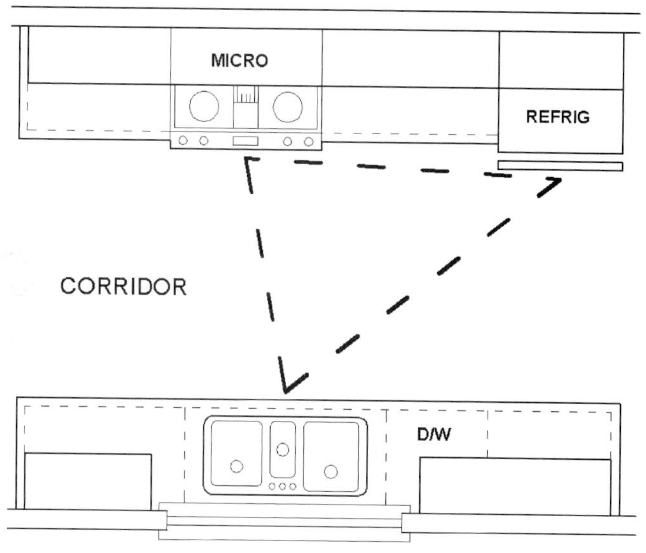

FIGURE 6.1.1E CORRIDOR-SHAPED KITCHEN

- *Flexibility in use.* The design accommodates a wide range of individual preferences and abilities.

- *Simple and intuitive use.* The design is easy to understand, regardless of the user's experience, knowledge, language skills, or current concentration level.

- *Perceptible information.* The design communicates necessary information effectively to the user, regardless of ambient conditions or the user's sensory abilities.

- *Tolerance for error.* The design minimizes hazards and the adverse consequences of accidental or unintended actions.

- *Low physical effort.* The design can be used efficiently and comfortably and with a minimum of fatigue.

- *Size and space for approach and use.* Appropriate size and space is provided for approach, reach, manipulation, and use regardless of user's body size, posture, or mobility.

In addition, references to the clear floor space throughout the Kitchen & Bath volume are minimums for people who use stationary seats, walkers, crutches, or wheelchairs while using or working at a fixture, appliance, or piece of equipment. Because of the way the kitchen and bathroom are used today, awareness of the need for concepts and products that adapt to the user rather than the reverse, has increased; responsible rehab specialists will improve kitchens or bathrooms by attending to these details.

6.2 CABINETS

SURFACE MAINTENANCE AND REPAIRS

ESSENTIAL KNOWLEDGE

Cabinetry is most commonly made from a combination of plywood or fiberboard with laminate, veneer, or a vinyl or foil wrap as the finish material. Repeated abrasion and standing moisture are the two major sources of damage and deterioration to the finish on kitchen and bathroom cabinetry. Once the finish is disturbed, cabinet surfaces, particularly those made of fiberboard, can rapidly deteriorate, absorbing moisture and degrading in appearance. This deterioration can also affect the function of the doors and drawers. Successful resolution requires refinishing, repair, or replacement of cabinetry fronts or entire cabinets.

TECHNIQUES, MATERIALS, TOOLS

Options for restoring the exterior of cabinetry include:

1. TOUCH-UP OR PATCH SURFACES.
Cracks or gaps in seams or on the wood surface can be filled with wood or seam fill products such as Kampel's Woodfil or Seamfil. In the case of less-expensive cabinets, edges or interiors may never have been sealed. Multiple coats of a sealer help cut down on volatile organic compounds (VOCs) outgassing in the home.

ADVANTAGES: This repair discourages or stops further moisture absorption.

DISADVANTAGES: There may be a noticeable difference in the appearance between the filler and the original material, particularly in wrapped, veneered, or laminate finishes. Although the coats of sealer reduce emissions of VOCs, they do not completely stop them. Some of the products have an odor during application.

2. REFINISH SURFACES.
On the solid-wood portions of the cabinetry, existing stains can be stripped. Ideally, this will include the face frame, doors and drawer fronts, and exposed ends and trim. A diluted solution of trisodium phosphate and water sponged on the surface will remove grime, cooking grease, and layers of wax. Lightly sand the surface with the grain to smooth. The stripping should be done outside or in a well-ventilated area. A new stain or paint can be applied and then a sealer applied to complete the process. Greater success may result in subcontracting to a specialist to do the finishing, particularly if a faux finish or unusual paint finish is desired.

ADVANTAGES: Refinishing can dramatically change the appearance of and rejuvenate wood. The material cost is relatively low. A tremendous variety of finishes are available including faux finishes, paints, and glazes.

DISADVANTAGES: Refinishing is time and labor intensive. These finishes will not usually be as durable as the manufacturer's. Aside from limited success with painting, this process does not work for veneer, laminate, or wrapped finishes.

3. REFACE FRONTS AND SIDES.

Cabinetry made from the most common materials—wood or fiberboard with veneer, laminate, or vinyl wrap—can be refaced, provided the surfaces are uniform and devoid of significant ornamentation. Laminate or veneer is applied to exposed or visible surfaces using an adhesive backing or less-toxic contact cement such as Elmer's Safe-T to suit the material.

ADVANTAGES: Refacing is somewhat less costly than replacing cabinetry.

DISADVANTAGES: Refacing will not work on doors or drawer fronts with multiple levels; frayed or gouged edges or surfaces will not be restored.

4. REPLACE CABINET DOORS AND DRAWER FRONTS.

Usually limited to framed cabinetry, the face frame and case can be maintained in the original finish, and doors and drawer fronts can be replaced in a complementary finish, providing a pleasing contrast (Fig. 6.2.1). The contrast can be extended to toekicks, counter trim, light valances, or other moldings.

ADVANTAGES: Replacement of cabinet doors and drawer fronts cuts down on labor, eliminates the need to strip or match existing finishes of damaged door and drawer fronts, and provides for the addition of new cabinetry that might improve the function of the space.

DISADVANTAGES: Damage in the case or face frame is not resolved. Strong cases and joints are required.

HARDWARE TO MAXIMIZE ACCESS AND FUNCTION

ESSENTIAL KNOWLEDGE

Hinges, pulls, or drawer glides that are worn function poorly, if at all. Cabinetry that does not maximize storage or is not accessible to a variety of people can be modified for easy access. Concerns include not only inconvenience, but also safety hazards. Short of a total "gut and renovation," there are solutions that involve replacing hardware or adding accessories.

TECHNIQUES, MATERIALS, TOOLS

1. REPLACE PULLS OR KNOBS.

Open pull or D-ring handles on doors and drawers are recommended because they allow people of all ages and abilities to easily access the cabinet. The smooth D-ring is preferred over handles with square corners or pointed ends because minor accidents such as bumping the hip or head can be avoided. If existing drilled holes do not have the same center dimension as the new hardware, cover-plates can be used between the pull and the door or drawer front (Fig. 6.2.2).

ADVANTAGES: Adding new hardware will update the existing cabinetry and allow even a weak hand to easily access the cabinet door and drawer.

DISADVANTAGES: Cover-plates may discolor or age at a different rate than exposed cabinetry.

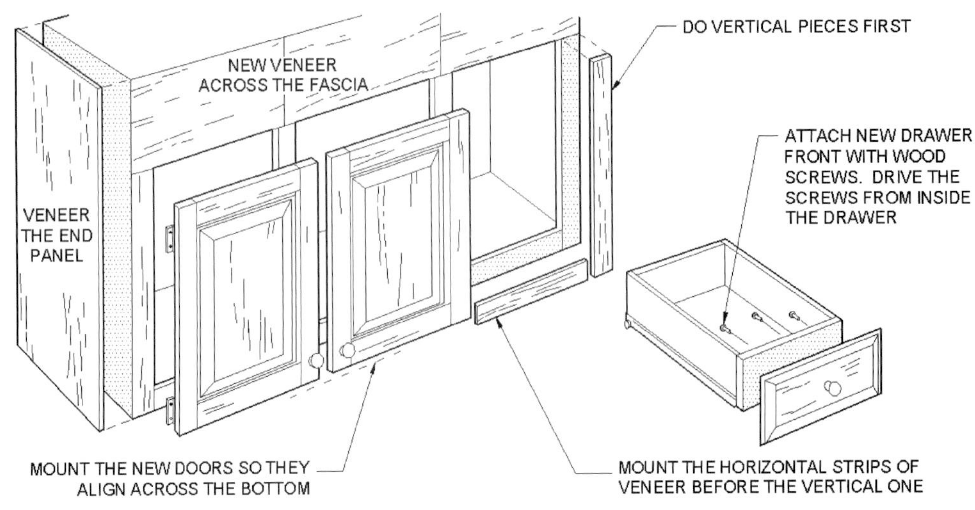

FIGURE 6.2.1 CABINET REFURBISHMENT

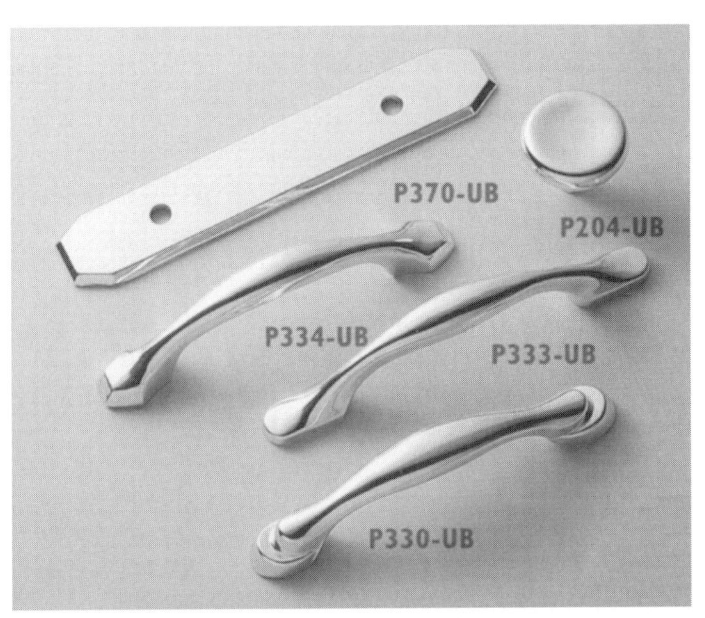

FIGURE 6.2.2 COVER-PLATE TO COVER EXISTING DRILLED HOLES

2. REPLACE HINGES.

Depending on the quality of original door hinges and the level of wear and tear, the existing condition may include doors that will not stay closed, won't close at all, or do not align. Some hinges will allow adjustment to improve the situation; replacing the hinges will otherwise eliminate the problem. In order to replace hinges, it is first necessary to determine the door and case style. Cabinet cases are typically framed or frameless, and typical door styles are inset, standard offset, or marginal overlay and full overlay (see Replace or Add Cabinetry later in this section for typical cabinet case and door types). Based on an examination of the existing hinge and its installation and condition, one can choose knife, concealed, or various barrel hinges (Fig. 6.2.3).

ADVANTAGES: Variations in quality and style are available in hardware today; the improvement can be significant.

DISADVANTAGES: The new hinges must be compatible with the type and installation of the hinges being replaced.

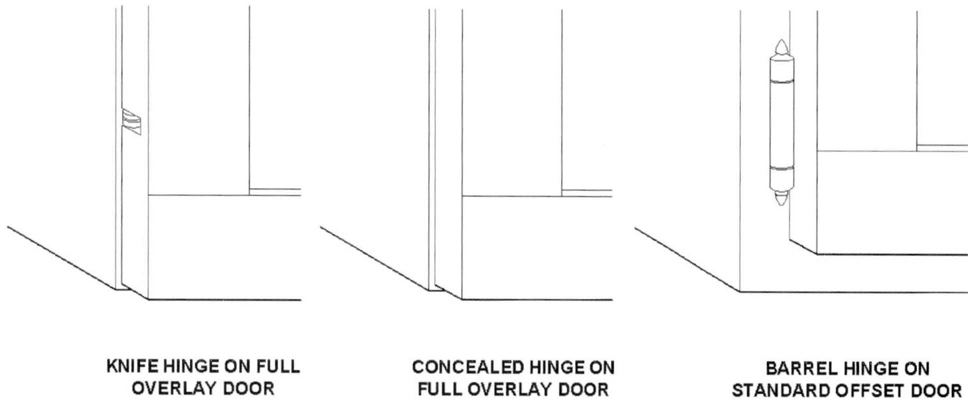

FIGURE 6.2.3 KNIFE, CONCEALED, AND BARREL HINGES

3. UPGRADE TO SPECIALTY BIFOLD HINGES TO IMPROVE ACCESS.

Where space is tight or mobility and endurance are issues, converting existing hinges to a bifold system improves access, reducing the space required for the door swing by half. These hinges allow the door, which is cut lengthwise into two doors half the size of the opening, to be moved to the right or left of the cabinet. An additional piece of hardware prevents the door from swinging wide.

ADVANTAGES: Bifold hinges can be used in corner cabinets to improve access or under sinks or cooktops to create a storage space or knee space for a seated cook. They allow doors to open in approximately half the traditional space, increasing clear floor space for access (Fig. 6.2.4A and B).

DISADVANTAGES: Adjacent obstructions, such as dishwashers or hoods, inhibit the door and hinge from swinging open fully.

4. UPGRADE TO SPECIALTY RETRACTABLE DOOR HINGES.

These hinges allow the doors to open to 90° and then slide back just inside either side of the cabinet (Fig. 6.2.5).

ADVANTAGES: Cabinet doors can stay open with no obstruction in front of the unit. These hinges are commonly used with television cabinets, but can also be used to house a waste receptacle that can be left exposed when in use. A knee space is created in a base cabinet for a seated user.

DISADVANTAGES: The hardware and retracted doors will consume interior cabinet space, as much as 3" per side, which can be critical if the space is being used as a knee space. Retracting hinges less than top quality require ongoing adjustment.

5. INSTALL UP-SWINGING HINGES.

These hinges carry the door up and sometimes over the cabinet (Fig. 6.2.6).

ADVANTAGES: These hinges allow the wall cabinet door to move out of the path of cook.

DISADVANTAGES: These hinges can only be applied to the wall or upper sections of cabinetry and can move the door out of the reach of shorter or seated cooks.

6. UPGRADE TO DROP-DOWN DOOR HINGE.

These hinges allow a door to hinge at its bottom with locking side bars to control how far the door opens (Fig 6.2.6).

ADVANTAGES: Access for bins or hampers used for storing bulk items (dog food, recycle bin, laundry hampers, or access under a seating area) is improved.

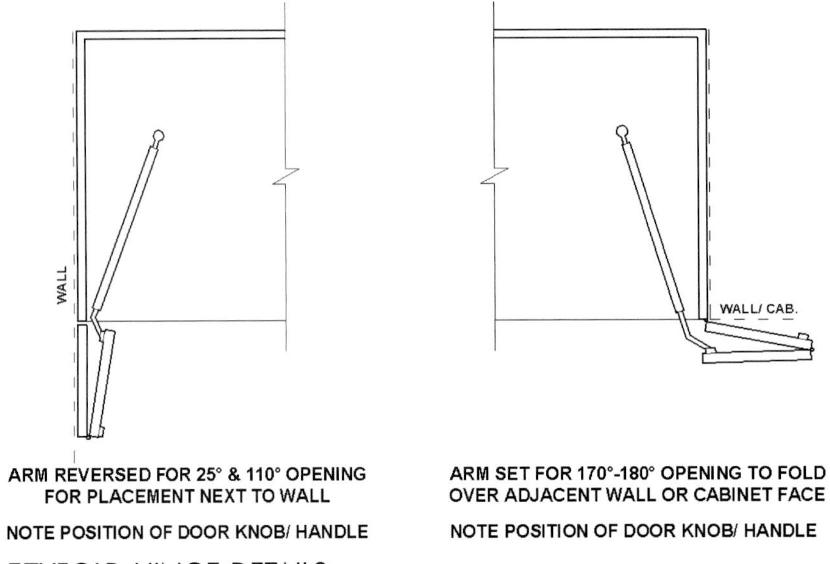

FIGURE 6.2.4A EZYFOLD HINGE DETAILS

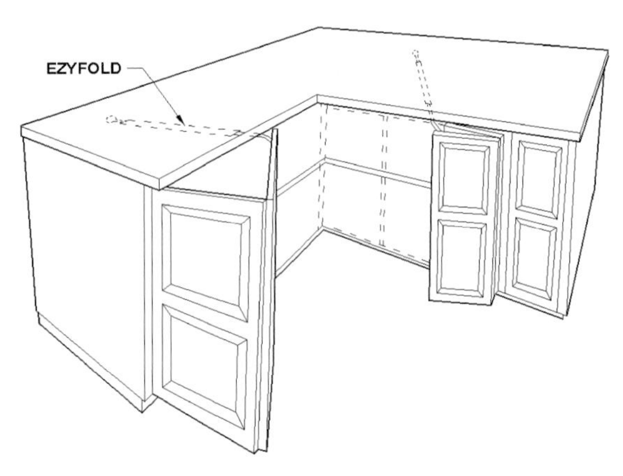

FIGURE 6.2.4B EZYFOLD HINGE

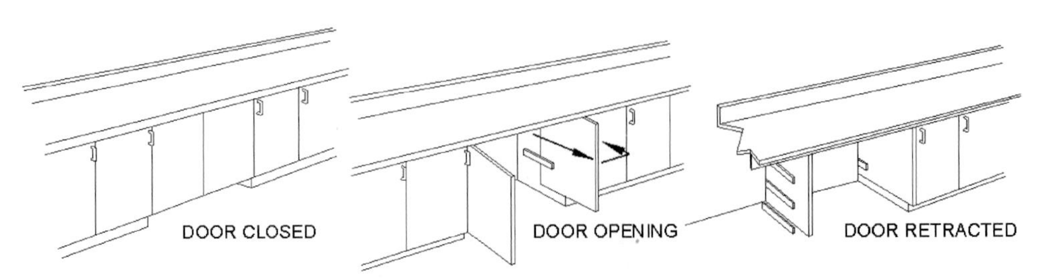

FIGURE 6.2.5 RETRACTABLE DOOR HINGE IN BASE CABINET

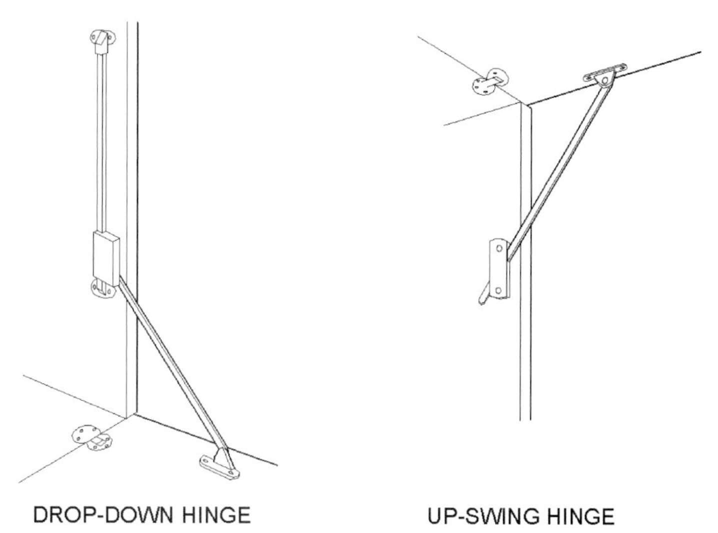

FIGURE 6.2.6 DROP-DOWN AND UP-SWING HINGES

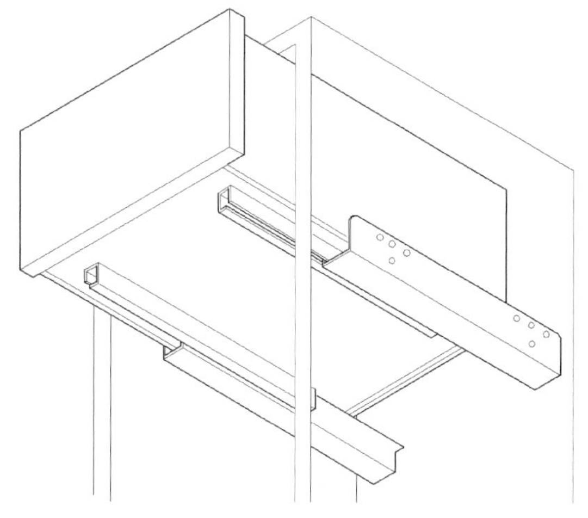

FIGURE 6.2.7 BOTTOM-MOUNT DRAWER GLIDE

DISADVANTAGES: There is limited application for these hinges.

7. REPLACE OR UPGRADE TO BOTTOM-MOUNT DRAWER GLIDES.

The most popular form of drawer hardware, this type of system is installed on the bottom of the drawer (Fig. 6.2.7). It offers smooth and quiet operation and allows for construction of a larger drawer box.

ADVANTAGES: The bottom location of the glide allows for a cleaner appearance. Glides can be adjusted over time as use requires.

DISADVANTAGES: Glides will not easily fit onto existing drawers that have previously had traditional side-mounted glides. Glides are not commonly able to feature full extension.

8. REPLACE OR UPGRADE TO SIDE-MOUNT DRAWER GLIDES.

These glides are available in a variety of types, from inexpensive and no frills, to more high-end with bearing movement and integral adjustment (Fig. 6.2.8).

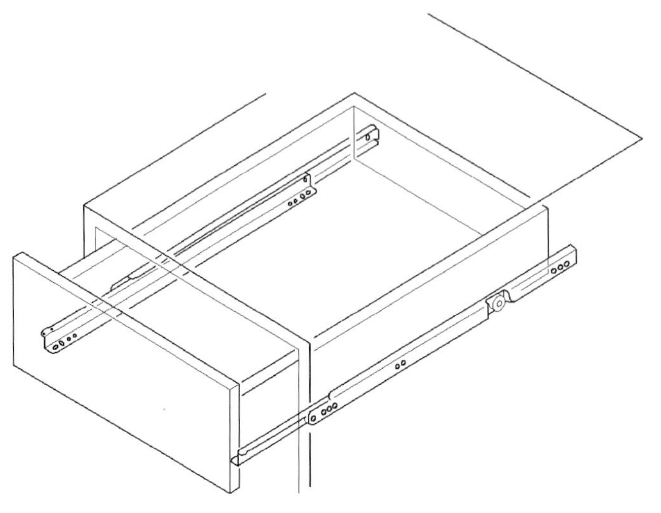

FIGURE 6.2.8 SIDE-MOUNT DRAWER GLIDE

ADVANTAGES: Side-mount drawer glides can be inexpensive, require the least labor to install. They can include a self-closing feature.

DISADVANTAGES: These glides may have no provision for adjustment, so maintenance can be more difficult. Lower-priced versions operate less smoothly and with more noise and effort.

9. REPLACE OR UPGRADE TO FULL-EXTENSION DRAWER GLIDES.

Typically side-mounted, these glides provide easy access to the entire drawer and often support heavier storage (Fig. 6.2.9).

ADVANTAGES: Full-extension drawer glides allow for complete access to drawer contents without bending or reaching. These glides are typically rated to support heavier weights.

DISADVANTAGES: These glides are usually bulkier, less smooth, and more difficult to operate.

ACCESSORIES TO MAXIMIZE ACCESS AND IMPROVE STORAGE

ESSENTIAL KNOWLEDGE

Acknowledgment of the human aging process and an increasing appreciation for the diversity of physical characteristics in household members have changed the approach to access and storage in both the kitchen and the bathroom. In general, rehab work should take into account that access and storage can be improved by following basic universal design principles and practices (see Section 6.1). These guidelines establish flexibility and improve access and support to make the space usable by more types of people most of the time. Critical to access is creating storage and work centers within the comfort zone of most people. Storage should be concentrated in the universal reach range of 15" to 48" above finished floor height, and near its point of use. Work surface heights should vary to accommodate standing or seated people of varying heights. While the kitchen or bathroom cannot be totally

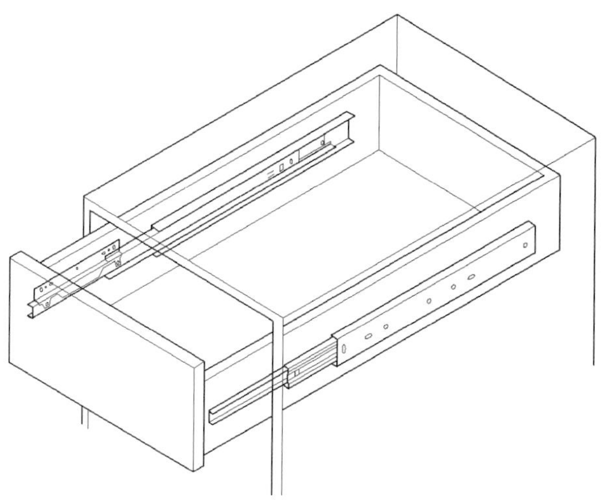

FIGURE 6.2.9 FULL-EXTENSION DRAWER GLIDE

redesigned when keeping existing cabinetry in rehab work, much can be done to enhance the storage and other functional aspects of the space.

TECHNIQUES, MATERIALS, TOOLS

The following are examples of accessories available in a range of costs with a variety of benefits to improve existing cabinetry:

1. INSTALL A STEP STOOL.

Step stools can easily be installed on the inside of a base or tall cabinet door for use in a single location, or they may be stored in either of these spaces or in the toekick of a cabinet, to be removed and used where desired (Fig. 6.2.10). Typical units store in 4" of space and unfold to a height of 15" as single-step units or higher in the two-step version. Both Hafele and Rev-A-Shelf offer step stool accessories.

ADVANTAGES: Step stools provide safer access to storage above the universal reach range. Built-in units provide storage at the point of use. Installation is easy to accomplish in a rehab project.

DISADVANTAGES: Deeper units may absorb some base cabinet storage. Built-in units can only be used where installed.

2. IMPROVE CORNER STORAGE.

A corner revolving shelf or corner swing-out shelf will improve access to previously blind corners (Fig. 6.2.11). Recent needs for recycling have brought responsive design in the form of rotating multiple bins that make good use of the otherwise poor storage in the corners.

ADVANTAGES: Access is improved.

DISADVANTAGES: To fit within the cabinet, the movable storage is often smaller than the overall space available. Successful installation in rehab requires precise dimensioning of available interior space and the opening.

3. REPLACE FIXED SHELVES WITH ADJUSTABLE OR ROLL-OUT SHELVES.

In wall, tall, or base cabinets, converting to adjustable shelves allows the homeowner to maximize storage (Fig. 6.2.12). In base or tall cabinets with a typical depth of ±24", roll-out shelves and accessories

FIGURE 6.2.10 REV-A-SHELF STEP STOOL INSTALLED ON DOOR TO IMPROVE ACCESS

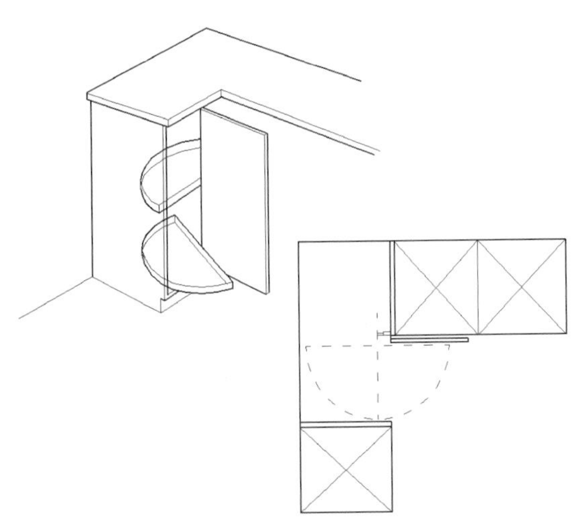

FIGURE 6.2.11 SWING-OUT SHELF TO IMPROVE ACCESS TO CORNER STORAGE

further increase access to storage. Note that when cabinet depth is less than 18" roll-outs are usually not necessary.

ADVANTAGES: Items can be easily stored and retrieved without reaching into cabinets.

DISADVANTAGES: The weight of items to be stored must be considered for proper function of the roll-out shelf (hardware and shelf ratings for maximum load are usually available).

4. INSTALL SPECIFIC PURPOSE ACCESSORIES.

Accessories are available to support recycling, tray storage, spice storage, appliance storage, and more (Figs. 6.2.13 to 6.2.15). Typically, these accessories are designed to be removable to allow for flexibility in storage.

ADVANTAGES: Organization and efficiency are improved.

DISADVANTAGES: Permanently installed items eliminate flexibility. After-market accessories often use less than the full cabinet interior.

FIGURE 6.2.12 ROLL-OUT SHELVES IN BASE CABINET

FIGURE 6.2.13 SPICE RACK IN DRAWER. (*Feeny Photograph.*)

5. INSTALL BACKSPLASH STORAGE ACCESSORIES.

Easy-to-install "appliance garages" and railing systems supporting a variety of storage options can be added to an existing backsplash area to increase and improve flexible and accessible storage. Railing systems, manufactured by Hafele and Rev-A-Shelf, are growing in popularity. Storage provisions include knives and utensils, wraps, cutting boards, condiments and spices, paper towels, cookbooks, dish draining racks (Fig. 6.2.16). Tambour, sliding, or (occasionally) hinged doors are installed in the backsplash below the wall cabinets to conceal stored appliances. Including outlets provides true storage at the point of use.

ADVANTAGES: Backsplash storage accessories can increase and improve storage and are easily installed in a rehab situation.

DISADVANTAGES: These accessories can be costly and can interfere with countertop use. In the case of appliance garages, appliances to be stored must be measured for a good fit.

6. INSTALL HEIGHT-ADJUSTABLE STORAGE.

In existing cabinetry and design, wall cabinet storage can be made height-adjustable via an Accessible Design Adjustable Systems (ADAS) motorized unit or a Hafele mechanical system (Fig. 6.2.17).

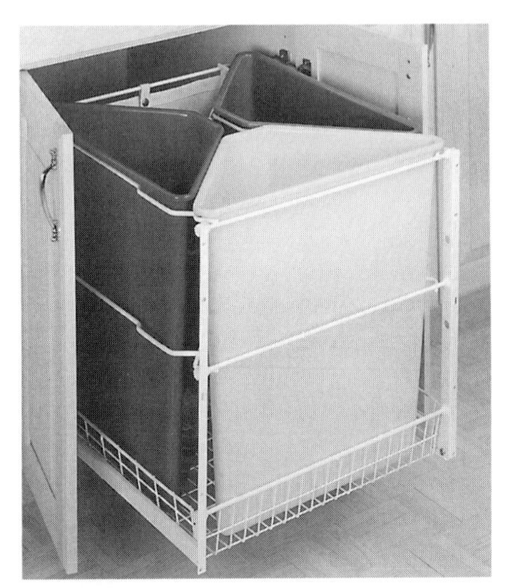

FIGURE 6.2.14 BASE CABINET RECYCLING BINS. (*Rev-a-Shelf Photograph.*)

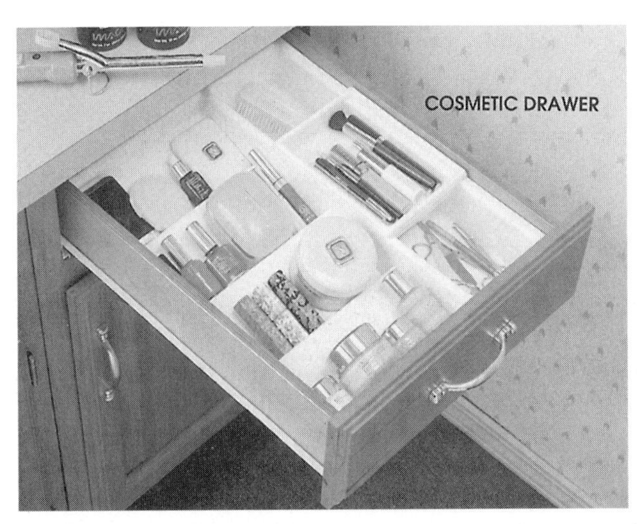

FIGURE 6.2.15 COSMETIC DRAWER DIVIDER. (*Feeny Photograph.*)

ADVANTAGES*:* These systems bring wall cabinet storage within the reach of most people. Their growing use has brought the cost down.

DISADVANTAGES*:* These systems are difficult to install in an existing location and require approximately 3" in depth, either in addition to the cabinet depth or absorbed from the cabinet depth. When used, the cabinet is in a lowered position and can interfere with the counter.

REPLACE OR ADD CABINETRY

ESSENTIAL KNOWLEDGE

In rehab projects where the condition, appearance, and layout of the existing cabinetry is deemed beyond repair, some or all of the cabinetry can be replaced. This may be the eventual outcome of a plan to replace other components of the space or a desire to change the look and function of the space.

FIGURE 6.2.16 HAFELE BACKSPLASH RAILING SYSTEM

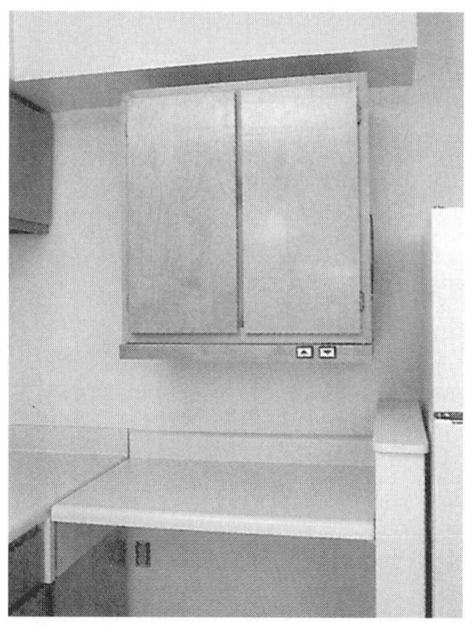

FIGURE 6.2.17 ADAS ADJUSTABLE HEIGHT CABINET

It is important to recognize that this option allows for a change in the overall layout of the space, as none of the previous options do.

TECHNIQUES, MATERIALS, TOOLS

The most successful techniques for replacing cabinetry include:

1. DETERMINE CABINET CONDITION.

Before cosmetic rehabilitation is attempted, the cabinet's structure and components should be examined. If the horizontal rails, vertical stiles, or side and back panels are warped and damaged, the cabinet may not be level or plumb. This will affect the fit of new cabinetry installed next to it.

ADVANTAGES: Time and money of rehab will be saved if the cabinet can be salvaged.

DISADVANTAGES: Mixing and matching cabinet pieces may result in a weak cabinet.

2. SUPPLEMENT EXISTING CABINETRY.

Before adding cabinetry to an existing or improved kitchen layout, the occupants' dining needs and cooking and study habits should be determined (if possible). Refer to NKBA Guidelines for clearance dimensions required around an island, between opposing work counters, and relating to kitchen work centers. Tall units, such as pantries, should be placed at the end of a run to maximize continuous counter space and maintain an uninterrupted work triangle. Storage for canned goods is best at 12" deep, but 24"-deep cabinetry may work better for appliances and can be accessories for storage of canned goods. Shorter vanity cabinets add interest and function to an existing kitchen because a 30"- to 33"-high counter is the recommended height for a baking center, desk area, or seated work area.

ADVANTAGES: Separate areas are created within the kitchen to improve function.

DISADVANTAGES: Existing and new cabinet exteriors should complement rather than match each other. Detailed specifications should be examined to determine if existing and new cabinetry have the same toekick height, drawer head height, and overall height dimensions if they are to be placed next to each other (Fig. 6.2.18).

3. SELECT TYPES OF CABINETRY.

If existing cabinetry is supplemented with new cabinetry, the new cabinetry should have the same structure as the original. The construction will either be framed or frameless, and the door type will either be standard offset, full overlay, or inset (Fig. 6.2.19). When comparing construction of different cabinet lines, the joint technique, component parts thickness, hardware quality, and core materials should be examined. If unsure about the quality of cabinets, refer to the Kitchen Cabinet Manufacturers Association, which certifies cabinets that meet its performance standards. Typically, inset doors are more expensive than full overlay, and full overlay are more expensive than standard offset. Depending on the style of the kitchen or bathroom, the door styles can complement, yet differ, from one another.

3A. INSET DOORS.

Inset doors are set into the door frame.

ADVANTAGES: Inset doors have the look of craftsperson-built or furniture quality.

DISADVANTAGES: Inset doors are the most expensive and require the most precision work to be done well. This style gives a strong traditional message and so is rarely used in contemporary kitchens.

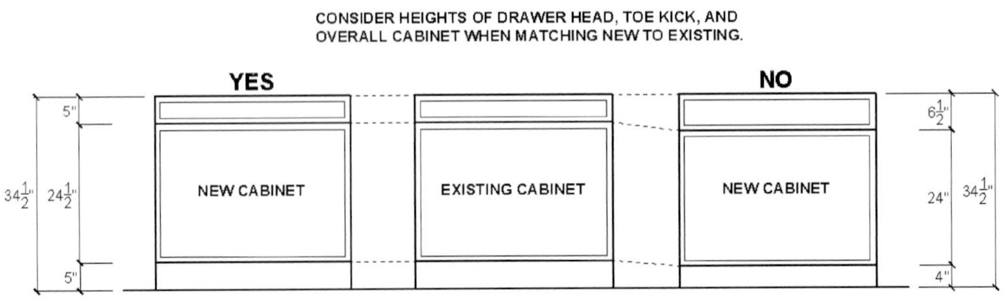

CONSIDER HEIGHTS OF DRAWER HEAD, TOE KICK, AND OVERALL CABINET WHEN MATCHING NEW TO EXISTING.

YES | NO

NEW CABINET | EXISTING CABINET | NEW CABINET

FIGURE 6.2.18 DRAWER AND TOEKICK HEIGHTS IN SUPPLEMENTARY CABINETS

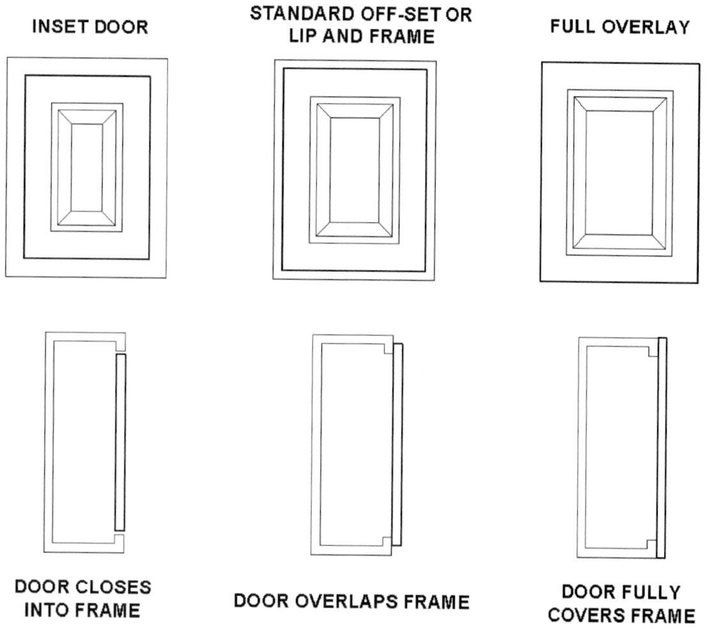

FIGURE 6.2.19 INSET, STANDARD OFFSET, AND FULL OVERLAY DOORS

3B. STANDARD OFFSET DOORS.

Standard offset doors extend slightly into the face frame.

ADVANTAGES: These doors are the most reasonably priced and most forgiving of poor craftmanship. The door laps the frame with additional space surrounding the door so there is room for adjustments and it is easier to conceal slight inaccuracies in door position. These doors can appear traditional or as a transition to contemporary and can be used with concealed or visible hinges.

DISADVANTAGES: Edges of doors are exposed to damage.

3C. FULL OVERLAY.

Full overlay doors completely overlay the front face frame.

ADVANTAGES: These doors have clean lines, are easy to clean, and use concealed hinges.

DISADVANTAGES: This type of door must be hung and adjusted accurately because there is often no frame behind it, and the look emphasizes precision.

4. IMPROVE INDOOR AIR QUALITY WITH CABINETRY.

A growing concern for natural resources and the environment makes it critical to address resource efficiency and air quality. Neil Kelly Signature Cabinets addresses resource efficiency and indoor air quality concerns in its Naturals Collection. Certified wood components and environmentally friendly finishes are standard, and add-ons include a natural oil/wax finish and nonformaldehyde Medite II medium-density fiberboard (MDF). Neff Cabinets offer 98% reduction in formaldehyde content. Metal cabinetry provides an attractive alternative to nonformaldehyde cabinet construction. Ampco, Arctic Metal Products, Dwyer Products, Cervitor, Heritage, and St. Charles distribute metal cabinetry with different options, styles, colors, and add-ons.

ADVANTAGES: The costs for this type of cabinetry are comparable to middle to high-end semi-custom cabinetry. Natural resources are conserved.

DISADVANTAGES: Limited door styles and finishes are available.

6.3 COUNTERTOPS

SURFACE MAINTENANCE AND REPAIRS

ESSENTIAL KNOWLEDGE

The most common materials used for kitchen and bathroom counter surfaces include laminate, solid surface, solid surface veneer, cultured stone, stone, tile, wood or butcher block, and (more recently) concrete and aggregate stone. None of these surfaces is indestructible, and a combination of knife scratches, chips, burns, or imperfect seams may occur over time. Wherever these blemishes are present, maintenance, bacteria, and further degradation of the surface increase. Prompt repair reduces health risks and further damage which is more difficult to repair.

TECHNIQUES, MATERIALS, TOOLS

Options for repair of counter surface blemishes relate directly to the surface material:

1. REPLACE LAMINATE SURFACES.
Laminate cannot be repaired, but a laminate counter can sometimes be salvaged if the substrate has not absorbed moisture. The damaged section can be cut out and an inset of a contrasting material such as tile, stone, or butcher block can be installed.

ADVANTAGES: The inset can enhance appearance and function of the top; tile or stone will be heat resistant, and butcher block will provide a built-in cutting surface.

DISADVANTAGES: Repair of a laminate counter in place is labor intensive and the cost of repair may come close to the cost of replacing the top, as laminate is a moderate-priced material.

2. REPAIR CULTURED STONE.
Made from chips of natural stone embedded in a polyester resin, cultured stone is durable and easy to maintain. Cultured onyx consists of polyester resin and alumina trihydrate, which is combined to reproduce the veined appearance of natural onyx or alabaster. Cultured stone is also called *cast polymer*. Minor scratches can be buffed out using a polishing pad, and deep scratches or chips should be repaired by a porcelain repair company.

ADVANTAGES: Cultured stone is available in many colors and designs and can be made in large sizes so there are no seams.

DISADVANTAGES: Cultured stone scratches easily and may need waxing to maintain its finish.

3. REPAIR AGGREGATE STONE.
New on the market, aggregate stone is made of natural stone particles in a polyester resin and has the characteristics of natural stone at a lower cost. Silestone from Cosentino USA is an engineered stone

that resembles granite. Aggregate stone can be repaired using the same techniques used for natural stone (see option 6, Repair or Replace Stone).

ADVANTAGES: Aggregate stone has the look of natural stone without the cost.

DISADVANTAGES: Aggregate stone is available in limited colors and product types.

4. REPAIR SOLID SURFACE.

Solid surface countertops, such as Corian, are homogeneous materials such as polyester, acrylic, or a polyester and acrylic blend. Every manufacturer has developed a unique composition, some including mineral fillers to improve durability or fire-retardant properties. Because of the homogeneous make-up, scratches or burns in the solid surface can be removed with a scouring pad, 320-400 grit sandpaper, or steel wool. The manufacturer can refer a certified installer to repair more serious damage.

ADVANTAGES: Stains will not penetrate the nonporous surface, and minor scratches and burns can be buffed out.

DISADVANTAGES: The solid surface material will expand when hot and will crack if a hot pot is placed on the surface. Deep cuts that cannot be buffed may need to be filled with the manufacturer's filler.

5. REPAIR OR REPLACE CERAMIC TILE AND GROUT.

Ceramic tile is heat resistant but varies in its resistance to scratches and chips. Tile countertops are susceptible to water penetration, and if the grout between the tiles gets wet, water can seep behind the tile and damage the substrate. An individual tile or section of tile can be removed and replaced if a tile is chipped, cracked, or loses adhesiveness. Using a cold chisel, break the damaged tile into several pieces and pry the small sections of tile from the backer board. A chisel (rather than a hammer) will not cause damage to the substrate. Grout is porous and susceptible to cracks and stains, and repairs to the counter surface often involve removing the existing grout with a grout saw and regrouting between the tiles. Grout is available as cement, vinyl, or epoxy-based. Epoxy-based grout is stronger and more impervious to water, mildew, and stains, but more care is required during installation. Penetrating or surface silicone cleaner applied to grout increases its stain resistence. Household cleaners containing acid, bleach, or vinegar should be avoided because they will etch the tile and grout over time.

ADVANTAGES: Though ceramic tile rarely chips, bacteria can collect in the damaged area. If extra tiles are available, this repair is relatively easy.

DISADVANTAGES: Care must be taken to remove tile without damaging surrounding tiles; a hand-held grout saw or diamond blade grinder should be used. It is sometimes difficult to match grout colors. Epoxy-based grouts tend to be more expensive than others.

6. REPAIR OR REPLACE STONE.

When polished and sealed, slate and granite are the most appropriate types of stone used for countertop application because of their durability. Although a fragile ¾" thickness is available, a 1 ¼" thickness can support a 12" overhang, and it is stronger and thus easier to transport and install. Stone counters are available in slabs up to 4' wide and 9' long. If there are minor cracks or chips, the countertop can be repaired using an overfill and grind technique. The chipped or damaged area is cleaned and dried before it is filled with an epoxy or polyester matrix combined with a ground stone or sand aggregate to produce a mortarlike mixture. After the mixture cures, the fill is ground and polished. These repairs should only be done by an experienced professional. Any cracks in the surface can be repaired with the same mixture described above. Cracks repaired using the overfill and grind technique will be interpreted as a vein in the stone. To eliminate an obvious straight-line repair, the edges of the crack can be chipped and widened, replicating the existing veins.

ADVANTAGES: The aggregate repair will improve the similarity between stone and fill.

DISADVANTAGES: Epoxies are structurally superior to polyester, but epoxy may yellow.

7. REPAIR OR REPLACE WOOD OR BUTCHER BLOCK.

Laminated maple countertops can be used throughout the kitchen or as a butcher block insert. The unfinished wood will need to be oiled weekly with a mineral oil, and prefinished wood must be finished with a penetrating sealer and a nontoxic lacquer finish. Urethane varnish will protect a dining counter wood top from moisture and liquids, but it should not come in contact with foods, and should not be used as a chopping surface. Burns or deep scratches can be sanded and retreated. Balley Block, Block-Tops, John Boos, and Taylor Wood-Craft all manufacture butcher block countertops.

ADVANTAGES: Natural wood butcher block may be more sanitary than synthetic surfaces.

DISADVANTAGES: Maintenance is critical to ensure the wood is sealed.

8. REPLACE WITH STAINLESS STEEL.

Stainless steel is heatproof and although not intended as a cutting surface, it will continue to perform with the scratches that occur with use. An 18-gauge stainless steel with either 8% or 10% nickel should be used to prevent staining, scratching, and corrosion. Counters are custom fabricated, and there are many design options such as integral sinks, drainboards, and backsplashes. Custom Copper and Brass, along with commercial kitchen product manufacturers, offer stainless-steel countertops.

ADVANTAGES: Joints and welds are polished out for a seamless appearance.

DISADVANTAGES: Custom fabrication adds to the cost of the counter.

9. REPAIR OR REPLACE CONCRETE SURFACES.

Concrete countertops are an emerging technology in the kitchen. Hot pots can occasionally be placed on the surface, but it is not recommended as the concrete may crack. Silicone sealers, acrylic topcoats, or butcher's wax will protect the porous surface from moisture and staining. Although small hairline cracks are characteristic of concrete countertops, scratches and chips are not, and are a breeding ground for bacteria. Get Real Surfaces and Counter:culture from Soupcan fabricate concrete counters. Chips can be filled with concrete filler supplied by the fabricator.

ADVANTAGES: Concrete countertops can be tinted, and any fillers needed can also be tinted to match the existing concrete color.

DISADVANTAGES: Sealant must be reapplied if the concrete countertop is cut or scraped, and cutting food on the surface is not recommended. Sealant must also be applied after repeated exposure to water, as is typical in a kitchen or bathroom.

IMPROVE INDOOR AIR QUALITY

ESSENTIAL KNOWLEDGE

When particleboard is used as a substrate for laminate and solid surface counters, there is a threat of harmful emissions from the glue that binds the wood that forms the medium density fiberboard (MDF). Formaldehyde emissions are a hazard to indoor air quality, a suspected carcinogen, and the emissions cause skin irritations and asthmalike conditions in those with no previous symptoms.

TECHNIQUES, MATERIALS, TOOLS

The following are suggestions to reduce formaldehyde emissions:

1. REDUCE PARTICLEBOARD OUTGASSING.

Store the particleboard outdoors or in an unoccupied storage building. Three months of outgasing is recommended before use. Good ventilation will reduce outgasing in installed cabinetry.

ADVANTAGES: There will be decreased outgassing once the cabinetry is installed.

DISADVANTAGES: Fabrication time, storage needs, and expenses are increased.

2. COVER ALL EDGES OF SUBSTRATE.

When storage of particleboard is not possible to outgas formaldehyde, a urethane, polyurethane sealant, or melamine laminate applied to the bottom, top, and edges of the particleboard will act as a barrier and reduce emissions. Seams should be sealed in existing countertops.

ADVANTAGES: Sealed surfaces drastically limit emissions.

DISADVANTAGES: The material and labor expense for fabrication are increased.

3. USE LOW OR NONFORMALDEHYDE SUBSTRATE.

Most particleboard countertop substrates use urea-formaldehyde. A healthier choice is a phenol formaldehyde-based particleboard. Another option is Medex, a formaldehyde-free particleboard, which uses a polyurea resin matrix to bond the MDF. Compared to standard MDF, Medex results in virtually no outgasing.

ADVANTAGES: Outgassing is low or nonexistent; particleboard is machinable.

DISADVANTAGES: This is a high-cost product.

IMPROVE BACKSPLASH AND COUNTERTOP SEAMS

ESSENTIAL KNOWLEDGE

The backsplash, whether 4" high or full height between the countertop and the underside of the cabinet, is a shield to protect the walls surrounding sinks, cooktops, and work areas. However, seams between the backsplash and the countertop are susceptible to water, bacteria, and dirt buildup. Maintenance is increased as a result. In climates with large swings in humidity levels, the situation is complicated by the seasonal swelling and shrinking of the material. Removing or covering these seams will eliminate areas for dirt to accumulate and bacteria to thrive.

TECHNIQUES, MATERIALS, TOOLS

Suggestions to improve or eliminate seams include:

1. FILL SEAMS IN LAMINATE COUNTERTOPS.

Products such as Kampel's SeamFil can be used to fill in gaps in laminate countertops. They bond to the substrate, and when used with a retardant may be mixed to match the surface color.

ADVANTAGES: This method improves the appearance of any visible seams between sections of laminate, whether field joints or laminate seams, and helps to preserve the countertop and reduce maintenance.

DISADVANTAGES: There may be a noticeable difference between the filler and surface color or texture. Because these products are fairly rigid, they cannot be used where the decks meet the backsplash.

2. FILL SEAM BETWEEN COUNTERTOP AND BACKSPLASH.

Although there are many caulks available, a 100% silicone caulk can be used to fill in the seams and resist water.

ADVANTAGES: Silicone caulk is durable and somewhat flexible, doesn't need replacement for long periods, and minimizes exposure to toxins.

DISADVANTAGES: Caulk cannot be painted; colors are usually limited to clear, white, brown, and black.

3. INTEGRATE BACKSPLASH WITH COUNTERTOP.

Integral countertop and backsplash combinations available in solid surfaces and formed laminate have either an invisible bonded seam, where pieces of like material are chemically fused to make the seam nearly invisible, or no seam (Fig. 6.3.1).

ADVANTAGES: The combination decreases maintenance, eliminates the critical joint where the counter meets the wall, and improves sanitary conditions.

DISADVANTAGES: The possibilities for the use of any contrasting materials on the backsplash is reduced, but this can still be accomplished if the splash is not full height.

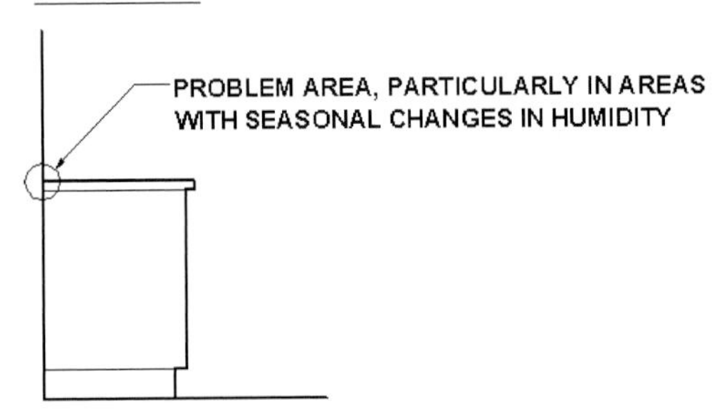

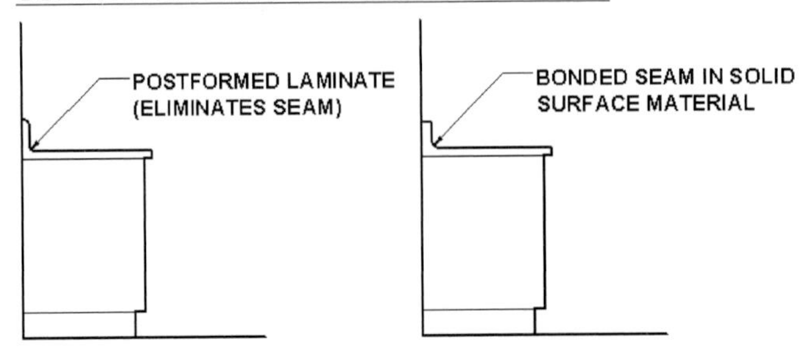

FIGURE 6.3.1 COUNTERTOP-BACKSPLASH SEAM DETAILS

MAXIMIZE ACCESS AND FUNCTION

ESSENTIAL KNOWLEDGE

Because kitchen and bathroom counters and work surfaces are used for a variety of tasks by a variety of household members, universal design principles should be considered. Rehab is an opportunity to create flexibility in these areas so that different tasks can be performed safely and comfortably by users of differing age, stature, and abilities. To provide for this improved access, countertop space needs to be planned at a variety of heights within the recommended 30" to 45".

TECHNIQUES, MATERIALS, TOOLS

A few options available to accommodate all users are given here:

1. INSTALL COUNTERS WITH EASED OR BEVELED CORNERS.

Countertop corners should be rounded or beveled to eliminate a sharp 90° corner. When designing a beveled corner, the overhang must be sufficient to allow for the bevel (Fig. 6.3.2).

ADVANTAGES: These corners help prevent minor injuries such as bumping one's head or hip. and may prolong the useful life of the countertop.

DISADVANTAGES: Equal overhang dimensions look best, but may not be available.

2. ADD DIFFERENT COUNTER HEIGHTS.

The shorter 30"-high base cabinet traditionally used for bathroom vanities is being replaced by a 34 ½" cabinet, which is more practical and comfortable for face washing or shaving. The 30"-high cabinets can be used in the kitchen in an area dedicated to baking and preparation because the mixing action is more comfortable and efficient when the arms are extended. This height also suits a seated

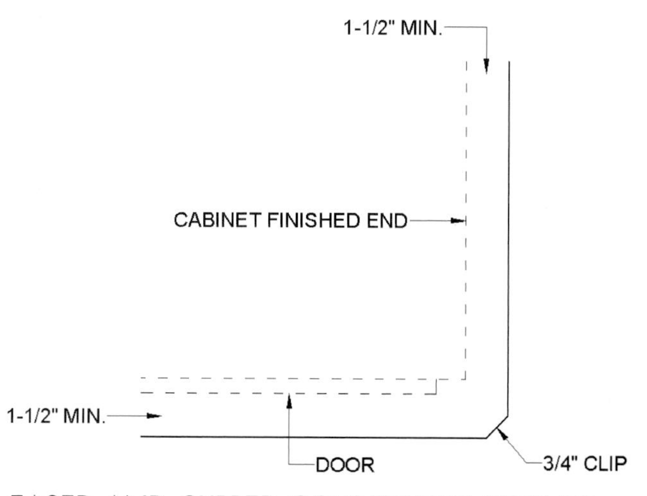

FIGURE 6.3.2 EASED AND CLIPPED COUNTERTOP CORNER

user. In addition to the traditional 40½" stub wall constructed to support a dining counter, a 42"-high counter can be supported by raising the dishwasher, single oven, or microwave.

ADVANTAGES: Various heights allow for different work areas throughout the kitchen.

DISADVANTAGES: Some cabinetry may have to be replaced and half walls may have to be constructed to vary the countertop heights. In a kitchen with minimal counter surface, this reduced continuous work surface may be inconvenient.

3. INSTALL ADJUSTABLE HEIGHT COUNTER.

A 36" to 48"-wide counter can be adjusted 8" vertically with the use of a motorized system, such as those available from Accessible Designs Adjustable Systems or Auton, to allow height adjustment at the touch of a button (Fig. 6.3.3). Another system from Hafele provides the same adjustment in height via a crank system. Specifically for the bath, systems are available that allow the sink to adjust in height.

ADVANTAGES: The unit plugs into a ground fault circuit interrupter (GFCI) outlet, and the countertop adjusts to a height comfortable for a baking center, desk area, or for users of different heights.

DISADVANTAGES: No cabinet space is available below the countertop.

IMPROVING FUNCTIONAL COUNTERTOP SPACE

ESSENTIAL KNOWLEDGE

In most kitchens, countertop space is minimal. It is sometimes believed that a work surface on either side of the sink is sufficient. However, counter surfaces should be available for use as a landing for items removed from the cooktop, oven, microwave, refrigerator, and dishwasher. By planning additional and flexible counter space, food preparation, serving, dining, and dish cleaning areas can be created in a rehabbed kitchen without adding space, and the function of the space is improved.

FIGURE 6.3.3 ADAS ADJUSTABLE HEIGHT COUNTER

TECHNIQUES, MATERIALS, TOOLS

Options to make more functional countertop space include:

1. INSTALL SMALL APPLIANCES UNDER WALL CABINETS.

Small appliances clutter the counter and minimize the work surface. Installing undercabinet appliances at their point of use will also improve function. For example, undercabinet microwaves are available with a hanging kit to suspend the oven from the bottom of a wall cabinet. A metal bracket is screwed to the bottom of the wall cabinet, and the oven is bolted to the bracket in predrilled locations.

ADVANTAGES: By removing the appliances from the countertop, a full-depth, unobstructed, and continuous work space is created.

DISADVANTAGES: Appliances may be at a new height that is not accessible to all users.

2. INSTALL A PULL-OUT WORK SURFACE.

A standard drawer can be replaced with a pull-out table. Drawer fronts are screwed directly to the table trim panel and accompanying hardware. The unit is not visible when installed. The Hafele pull-out table unit (Fig. 6.3.4) is self-supporting and has a load capacity of 220 lb.

ADVANTAGES: Additional work space can be added when needed, and the table can be closed when not in use.

DISADVANTAGES: Drawer space is eliminated, and a minimum drawer width of 23⅝" is required.

3. INSTALL A HEAT-RESISTANT COUNTER MATERIAL

Small sections of heat-resistant surfaces such as stainless steel, ceramic tile, and stone can be installed adjacent to the cooking or dining area.

ADVANTAGES: The surface can be used as a landing space for hot items and adds a contemporary look to the countertop.

DISADVANTAGES: There will be a seam where two different surfaces join.

FIGURE 6.3.4 HAFELE PULL-OUT WORK SURFACE

4. INSTALL A ROLLING TABLE OR CART

Castors and a countertop can be attached to any existing shelf unit. A cart can be constructed with a base cabinet door and drawer front so it will fit in the base space when not in use. The table will act as a mobile island with additional countertop space and storage below. The table can be placed below the oven to provide a close landing surface for hot food.

ADVANTAGES: A rolling table improves flexibility and increases counter space.

DISADVANTAGES: A rolling table compromises or limits drawer and base cabinet storage.

ENHANCE BACKSPLASH WITH SURFACE MATERIALS

ESSENTIAL KNOWLEDGE

Sometimes a rehab kitchen or bathroom project can be greatly improved in appearance with the addition of countertop and backsplash surface materials. Surfaces not advised for countertop use (such as metallic laminate, marble, or decorative tiles), or too costly to use throughout the entire countertop and backsplash (such as stainless steel, wood, or granite), can be used in the backsplash only. Because of the vertical plane of the backsplash, surface materials on the backsplash are more visible than the horizontal countertop plane and can easily become a focal point. To further maximize visibility, the backsplash can extend to the underside of the cabinet above it, rather than the typical 4"-high backsplash.

TECHNIQUES, MATERIALS, TOOLS

Options to visually improve the backsplash:

1. INSTALL METALLIC LAMINATE BACKSPLASH.

Metallic laminates are applied with adhesive and heat in the same fashion as plastic laminate. Applying the laminate to a ¼" or ½" substrate will decrease any risk of problems at the seam where the counter meets the splash. Abet Laminati, Lamin-Art, WilsonArt's Artisan Group, and Formica's DecoMetal collections offer a variety of colors and textures to choose from.

ADVANTAGES: Metallic laminates are less expensive than metallic tiles or solid metal sheets.

DISADVANTAGES: Metallic laminates are available in limited colors and textures and have a higher cost than plastic laminates.

2. INSTALL DECORATIVE CERAMIC TILE BACKSPLASH.

Decorative wall tiles come in a wide variety of colors, textures, sizes, and shapes, and can be used in combination with the field tile to distinguish and outline separate work areas along the backsplash. Decorative tile can be used to create a motif or accent a space. Crossville manufacturers a variety of metallic tiles.

ADVANTAGES: White or light-colored grout can be used on the backsplash because it will not be stained, unlike grout used on floors and countertops.

DISADVANTAGES: Large tiles may require additional time to install, and grout may need sealing and minor maintenance during the life of the tile.

3. INSTALL STONE BACKSPLASH.

Granite, soapstone, slate, and even marble can be used in the backsplash. Because marble is soft and porous, it is not recommended for heavy countertop use, but it can be used on the backsplash. While some stone is available only in slab, most marble is also available in tiles. For optimum results, outlets and switches should be moved (unless this results in impractical access) to provide an uninterrupted backsplash. Typically, plug molding is installed at the top of the splash or receptacles are installed at drawer height under the counter.

ADVANTAGES: This type of backsplash adds the visual impact of stone without the high cost of solid stone counters.

DISADVANTAGES: Stone splash material is higher in cost than most other options, and it will require sealing and maintenance similar to that of tile.

4. INSTALL OTHER MATERIALS.

A wood or mirrored surface can be used on the backsplash for visual impact. A beadboard panel complements the existing cabinetry and can be purchased locally. A mirror can be cut and installed by a glass supplier.

ADVANTAGES: These materials can brighten, enlarge, and otherwise enhance the existing space.

DISADVANTAGES: It can be difficult to remove water spots from these materials, and some high-end materials can be costly.

6.4 APPLIANCES

REPLACE OUTDATED OR NONFUNCTIONING APPLIANCES

ESSENTIAL KNOWLEDGE

When a kitchen appliance is nonfunctioning, the solution may be as simple as repair by an authorized service person, and refinishing to update the appearance. More often, the problem will be better solved by replacing the appliance because it no longer meets the family's needs or its performance and energy use are inferior to those of newer models. ENERGY STAR ratings from the U.S. Department of Energy and the Environmental Protection Agency list, by brand, the most energy-efficient appliances. Impact on the overall kitchen space must be determined by examining the existing appliances and the space in which they had been installed, and whether or not they are structurally connected to the surrounding space. Once the space available for the new appliance has been determined, a new model can be selected. This process can be facilitated by information available on the Web sites of most of the major manufacturers. This is also an opportunity to improve access to the appliance and surrounding work areas and to change the layout of the entire space.

TECHNIQUES, MATERIALS, TOOLS

1. REPLACE REFRIGERATOR.

Refrigerators with inaccurate temperature control, broken bins and racks, and irreparable damage to the door gaskets should be replaced. In the recent past, refrigerators have undergone major changes in design and technology to improve function and save energy. Current models include freestanding 30"-deep cases, built-in style freestanding 24"-deep cases, and fully integrated styles that must be built into walls or cabinetry. A new refrigerator should last about 15 years.

ADVANTAGES: Current models can result in energy savings that over time will recover the added cost of the appliance. Convenience can be greatly enhanced.

DISADVANTAGES: Changes in width, height, and style of the refrigerator can force adjustments in surrounding cabinetry, counters, and walls. Changes in depth can require rehabbing the flooring. Energy Star appliances may be higher in cost.

2. REPLACE RANGE.

Ranges with burners that won't heat, ignition malfunction, lack of temperature control, or nonoperating self-cleaning features should be replaced. Older gas ranges with pilot lights waste energy and degrade indoor air quality. Although there are various sizes, most models are available in the standard 30" width. A review of desired performance will assist in selection. The desired power source, whether gas (liquid propane or natural) or electric, must be accommodated and the conversion made if necessary. There are three main types of ranges available: freestanding, slide-in, and drop-in. If the existing range is freestanding, replacement is straightforward. Slide-in or drop-in ranges will require a close examination of the new range for proper fit. Commercial style freestanding ranges have increased power and heat and require attention to ventilation requirements. Consult the manufacturer for exact specifications. A new range should last about 15 years.

ADVANTAGES: A new range will provide improved performance, added safety features, and aesthetics.

DISADVANTAGES: Replacing the range may require that adjustments be made to surrounding cabinetry, counters, and ventilation.

3. REPLACE COOKTOP.

Cooktops with burners that won't heat, ignition malfunction, downdraft ventilation mechanical difficulties, or a damaged surface should be replaced. Similar to a range, the power source must be determined before a new cooktop can be selected. Three general types of electric cooktops available are conventional coil elements, cast-iron solid disk elements, and glass ceramic cooktops, with induction and halogen technologies used with glass ceramic cooking surfaces. Cooktop sizes and cutouts vary tremendously and must be determined for proper fit in the existing counter. Consult the manufacturer for exact specifications. A new cooktop should last about 15 years.

ADVANTAGES: A variety of styles and features are available; aesthetics are improved. This may be an opportunity to improve access to the cooking area.

DISADVANTAGES: Installation may require reworking or replacing counter and cabinet.

4. REPLACE OR ADD A RANGE VENTILATION SYSTEM.

The kitchen produces significant indoor air pollution with water vapor, grease, smoke, and odors, and a ventilation system that does not remove pollution from the kitchen should be replaced. The exhaust rate of the fan [cubic feet per minute (cfm)] must be adequate to accommodate today's cooking systems. As a minimum, a 0.1 multiplier of the range's BTU rating should be used to determine the cfm. For example, a professional gas cooktop with a 900 BTU rating should have a hood exhaust system equal to, or greater than, 90 cfm. In all cases, the manufacturer's specifications regarding the exhaust rate should be met as a minimum. A recirculating vent with a carbon filter only filters the pollution and reintroduces it back into the room. Such systems should be replaced with a ventilation system that exhausts to the outside. Hood systems, microwaves with built-in ventilation, and proximity or downdraft systems are ventilation options; the system selected depends on the kitchen arrangement, fan noise, and cooking load. Microwaves with built-in ventilation placed above the cooking surface are not accessible or safe for all users and do not comply with universal design guidelines.

ADVANTAGES: Grease and smoke particles will not accumulate on finishes when adequate ventilation is used.

DISADVANTAGES: Make-up air may be needed to avoid negative air pressure.

5. REPLACE A WALL OVEN.

Though energy usage is not affected when a separate oven and cooktop are used, individual units are usually more costly. Similar to the cooktop, the energy source and cutout size of a wall oven are critical to successful replacement; both gas and electric ovens have grown in size from 24" to 27" and recently, 30" widths. If a change of energy source is desired, a rehab project is the time to switch from either gas or electric because gas pipes, vents, and electrical connections may restrict the design. Gas ovens are heated by a burner below the oven floor, and utilize convection and radiation heating methods. For energy savings, gas ovens with an electronic ignition should replace those with a continuously burning pilot light. Placement of the heating units within electric ovens is usually on the floor and ceiling of the oven chamber. The use of convection heat in electric ovens is preferable because convection ovens offer even baking in less time and with less energy. Convection heat is created when an element heats the air and a fan circulates the air evenly throughout the oven. For energy source and cutout size information, consult the manufacturer's specifications.

ADVANTAGES: A new oven will provide improved performance; a safety lockout is available.

DISADVANTAGES: New ovens rarely fit the existing cutout and require some amount of redesign of the space.

6. REPLACE MICROWAVE.

Microwave ovens that spark or cycle while the door is open should be replaced. If a microwave is built in or combined with a convection oven in a single unit, replacement can be treated as a wall oven (see option 5, Replace a Wall Oven). If the microwave is freestanding, replacement is straightforward. Consider the option of selecting a microwave that can be hung from a wall cabinet to clear counter space.

ADVANTAGES: Newer models are smaller and have added features.

DISADVANTAGES: Built-in microwaves will require coordination with the surrounding cabinetry or walls.

7. REPLACE DISHWASHER.

Dishwashers with motor failure and excessive water leakage, noise, and water usage should be replaced. American manufactured dishwashers are typically designed to fit into a space 24" wide, so replacement is fairly straightforward. If the finish floor has been installed after the dishwasher and the floor height is greater, it may be difficult to remove the old dishwasher and install the new one. European dishwashers are gaining in popularity because of their water-saving features, and they are often slightly narrower than the American models. In either case, front panels to match existing cabinetry vary in size. Again, check the manufacturers' specifications. When redesign is necessary or possible, consider raising the dishwasher to a 42" height to improve access. A new dishwasher should last about 12 years.

ADVANTAGES: A new dishwasher will provide improved energy and water conservation, improved performance, reduced noise, and easier access.

DISADVANTAGES: A new dishwasher may create problems with flooring, custom front panels, and adjacent cabinetry. Energy Star appliances may be higher in cost.

8. REPLACE FOOD WASTE DISPOSER.

Feeding garbage too quickly into a waste disposer or failing to run enough cold water to completely flush drainpipes during processing can result in a clogged system. Old disposers are subject to rusting and motor burnout. Motor size, insulation, and antijam features differ from model to model. Continuous feed and batch feed options are available. A continuous feed disposer operates continuously from a switch, typically a toggle switch located on the wall, and a batch feed disposal operates when the lid is turned. New to the market, In-Sink-Erator offers a disposer designed for use with a septic system.

ADVANTAGES: New waste disposers include features that meet users' needs and practices.

DISADVANTAGES: If a home uses a septic system, local codes may not allow a disposer; some septic systems may need to be enlarged and cleaned more often if a disposer is used.

IMPROVE ACCESS AND FUNCTION WITHIN WORK CENTERS

ESSENTIAL KNOWLEDGE

The work triangle is the main working area of the kitchen, and it is created by the work flow between the work centers of the cooktop, the sink, and the refrigerator. The National Kitchen and Bath Association guidelines suggest that the work triangle should not exceed 26 lineal feet, with no leg being less than 4' or more than 9'. Relocating the appliances within the work triangle or creating a secondary work triangle can maximize efficiency and function of the space. Food preparation begins with the refrigerator, moves to the preparation area between sink and refrigerator, and on to the mixing area between the sink and the cooktop. The last steps occur in the serving area from cooktop to the table. Cleanup occurs from the table to the dishwasher, and from there to storage. This arrangement should be considered when planning rehab work in kitchens, and new appliance layouts should be designed to better facilitate use. Today's kitchens often include more than one cook and a variety of activities at the same time. To accommodate this, duplicates of certain appliances, such as cooktop, oven, or dishwasher, can be desirable.

TECHNIQUES, MATERIALS, TOOLS

Suggested locations and placement of appliances:

1. CONSIDER REFRIGERATOR LOCATION AND RELOCATE IF NECESSARY.

The refrigerator is often placed at the end of a cabinetry run or on a separate wall because it should not interrupt a continuous work counter. However, the landing space needed in front of or adjacent to the refrigerator is often eliminated, causing inconvenience when opening the refrigerator door. A landing counter with a 15" minimum width is needed on the freezer side of a side-by-side style, on the handle side of a top-mount or bottom-mount style, or directly across from either model on an island (Fig. 6.4.1). Return walls that extend farther than the depth of the refrigerator case intrude into the required clear floor space and can limit the door swing. Maintaining return walls no greater than the depth of the refrigerator will allow a full door swing. Adding another refrigerator station such as an under-counter refrigerator or beverage chiller in the secondary work triangle allows for multiple cooks in the kitchen. Refrigerator and freezer drawers from Sub-Zero permit greater flexibility and easier access.

ADVANTAGES: A well-placed refigerator can improve function; a second appliance can increase the amount of work stations without eliminating any work counter surface.

DISADVANTAGES: The rehab required to situate the refirgerator as suggested here may be greater than is practical. The cost of a second appliance may be considerable. Some base cabinet storage sacrificed.

2. CONSIDER COOKTOP AND SEPARATE OVEN RATHER THAN RANGE.

When a kitchen is used by more than one cook, the range can become crowded and it can be difficult to transfer hot items from the oven to the nearest landing space. Separating the cooktop and oven improves flexibility in the kitchen and can improve the landing and work space surrounding each

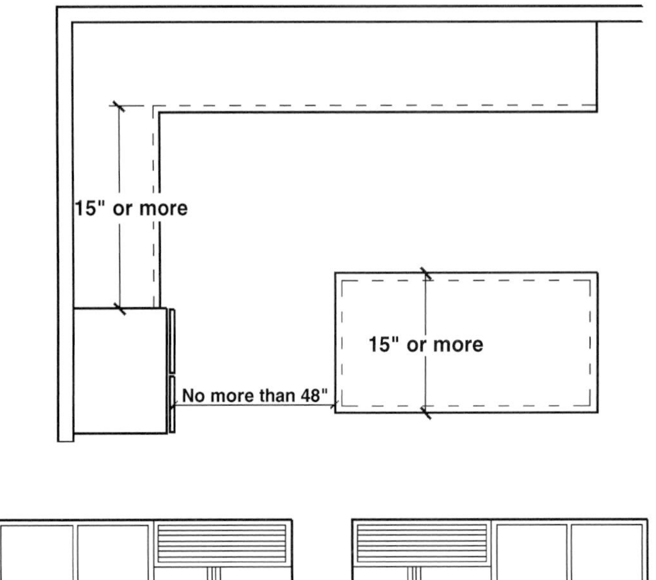

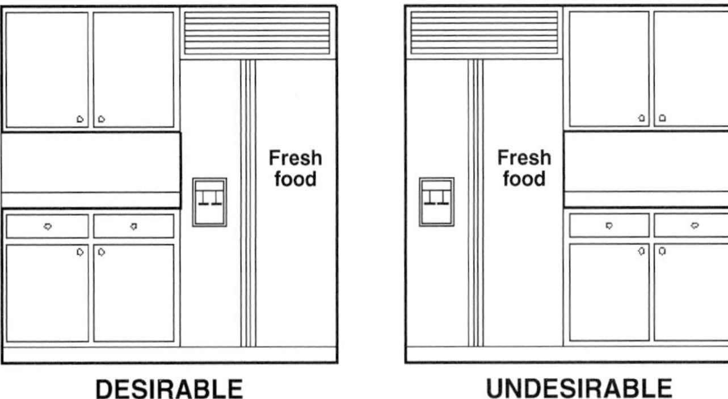

FIGURE 6.4.1 REFRIGERATOR DOOR SWING TO OPEN TOWARD COUNTER

appliance. Because the oven is the least-used appliance in the kitchen, it is acceptable to place it outside the concentrated working area of the kitchen.

ADVANTAGES: Separating these two appliances can greatly improve their use and access.

DISADVANTAGES: It is more costly to separate these two appliances than to simply replace the existing range.

3. CONSIDER COOKTOP LOCATION AND RELOCATE IF NECESSARY.

A safety concern in many existing homes is the lack of counter space surrounding the cooktop burners. A landing space of 15" is needed on at least one side of the cooktop (Fig. 6.4.2). To maintain a safe distance from the burners and pot handles, at least 9" of open counter space or 3" to a heatproof wall is needed on the other side of the landing space. Cooktops placed on islands should have 9" of clearance or a change in height on the back side of the island for safety. The cooktop should be relocated, the counter extended, or a wall constructed to maintain the required distance on either side of the cooktop. The cooktop and adjacent counter may be installed at a height other than the standard 36"; a height between 30" and 34" above the floor improves access for most cooks. Maintaining open space or flexible storage below the cooktop allows the option to sit while cooking.

ADVANTAGES: Relocating the cooktop as suggested here will greatly improve its use and access.

DISADVANTAGES: Safety considerations may eliminate the option of situating a cooktop on a small island.

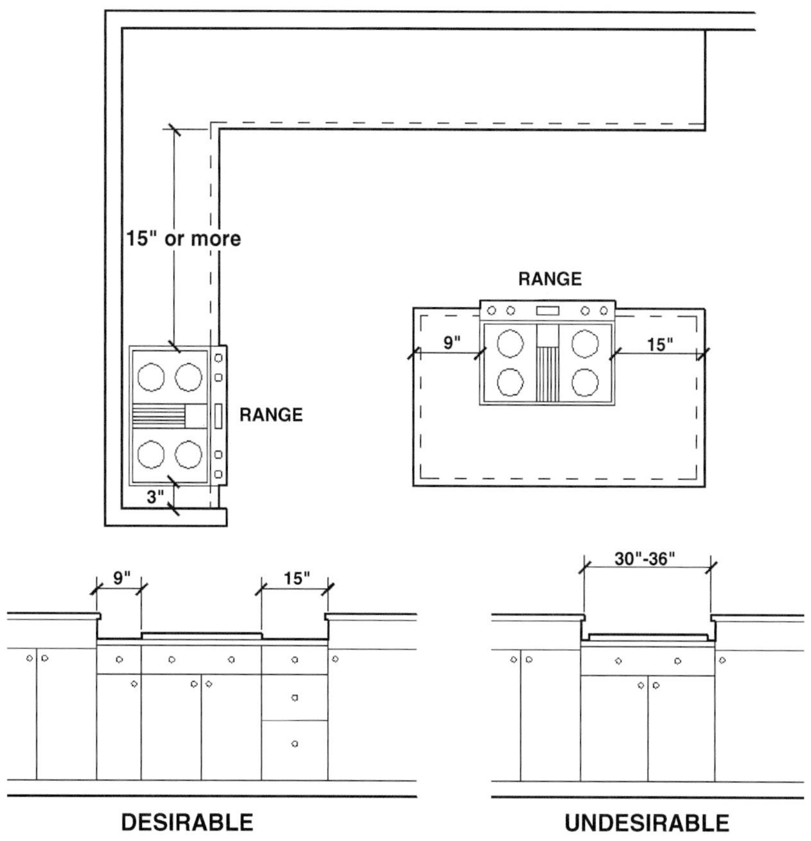

FIGURE 6.4.2 REQUIRED LANDING COUNTER CLEARANCES FOR COOKTOP AND RANGE

4. CONSIDER OPTIONAL OVEN PLACEMENT.

A tall cabinet that houses double ovens is a remedy to ease congestion in the kitchen work area. However, this results in oven placement at a difficult height. A better solution is to install one oven with the bottom ±24" above the finished floor, a comfortable distance between the universal reach range of 15" to 48". An oven with a side swing door, installed at ±30", and a pull-out landing surface directly below the oven door is another alternative. A third option is to raise the height of the base cabinet housing of the oven to 40½" above the floor. This allows for a possible drawer below and it creates a standard snack-bar height of 42".

ADVANTAGES: Oven placement as suggested here greatly improves use and access.

DISADVANTAGES: The existing kitchen may have space restraints.

5. CONSIDER MICROWAVE LOCATION AND RELOCATE IF NECESSARY.

Microwaves are often an afterthought in kitchen design, resulting in a less-than-desirable space for this popular appliance. Although the height of the standard 36"-high countertop is ideal for most sizes, ages, and abilities, locating the microwave on the counter eliminates valuable work space. If the depth of the microwave leaves a counter space of less than 16", a pull-out shelf or rollout cart directly below can be used. The ideal location for access is to mount the unit so that its bottom is 24" to 48" above the floor. Built-in styles can be installed in a base cabinet, 40½" high (under a 42" snack bar), to reduce bending. They can also be placed above a single oven in a tall cabinet or on a shelf below a wall cabinet, depending on depth. Smaller compact styles can be hung from wall cabinets, installed above a raised dishwasher or oven, or hung from a 42" dining counter above a table (Fig. 6.4.3). The best location for the microwave in the main kitchen area for food preparation tasks is between the sink and the refrigerator. However, the role of the microwave in

FIGURE 6.4.3 UNDERCABINET MICROWAVE WITH HANGING KIT

FIGURE 6.4.4 DISHWASHER RAISED TO 42″

the kitchen has expanded, and multiple users can result in waiting and congestion in front of the microwave in the main preparation area. The addition of a second microwave provides for simultaneous food preparation when one unit is placed in the work triangle and the other outside of it.

ADVANTAGES: Placement of the microwave as suggested here improves its access, efficiency, and safety. The cost of an additional microwave is reasonable.

DISADVANTAGES: The size of the microwave should be based on the intended use. Lack of space in the kitchen may limit the size of the second microwave.

6. CONSIDER DISHWASHER LOCATION AND RELOCATE IF NECESSARY.
The traditional dishwasher location is beneath the standard 36"-high counter, which requires a standing user to bend to access the bottom rack. Mounting the dishwasher 6" to 9" above the floor puts the racks within the 15" to 48" universal reach range to improve comfort (Fig. 6.4.4). The dishwasher does not need to be immediately adjacent to the sink. National Kitchen and Bath Association guidelines suggest the dishwasher be no farther than 36" from the edge of the sink. This will provide continuous work counters of 12" and 24" on either side of the sink when the dishwasher is raised to a different counter height. Often a dishwasher is placed so it can be used by only one person at a time, and there should be 21" clear floor space on either side of the dishwasher when another appliance, cabinet, or counter is at a right angle to it to allow access from either side. Adding a second dishwasher in the butler's pantry or near dining improves function. Dishwasher drawers, such as those from Fisher & Paykel, allow for maximum flexibility.

ADVANTAGES: Placement of the dishwasher as suggested here greatly improves its use and access.

DISADVANTAGES: Counter space surrounding the sink may not be sufficient to allow the dishwasher to be raised. Installing a new dishwasher may require new patterns for loading dishes.

UPGRADE APPLIANCE APPEARANCE

ESSENTIAL KNOWLEDGE

Many older appliances show their age because of an outdated color, or scratches on the visible sides and edges. In many rehab projects the appliances may function well, but should be refurbished or better coordinated to the updated kitchen if they are to remain.

TECHNIQUES, MATERIALS, TOOLS

Techniques to improve exterior of appliances:

1. PAINT APPLIANCE EXTERIOR.
Specialty-paint or automobile-paint stores may be able to recommend paint for an appliance exterior. The surface must be clean and free of oils; otherwise, the newly applied paint may fade, peel, or crack. The unit should also be empty and unplugged before it is painted, and the paint should be applied in thin coats. If a more uniform surface is desired, a professional should be contacted to apply the paint.

ADVANTAGES: Older appliances can be updated.

DISADVANTAGES: The appliance interior may not match the exterior; the appliance life will be shorter than the rest of the kitchen. This is a labor-intensive repair.

2. INSTALL PANELS.
Trim kits for many older model dishwashers and refrigerators may still be available. Call the appliance manufacturer with the model and serial numbers to obtain the trim kit model number. A custom look is achieved with wood or laminate panels that match the cabinetry and can be ordered through a cabinet supplier. Caution should be used to maintain a comfortable grip at the handle if choosing a thick raised panel.

ADVANTAGES: An older, functional appliance can be updated.

DISADVANTAGES: This method does not take advantage of improved energy efficiency available in new models. It is sometimes difficult to match existing cabinet finishes.

3. CREATE A BUILT-IN LOOK FOR THE REFRIGERATOR.

The average freestanding refrigerator case is 30" deep, which means 6" extends past the cabinetry into the work aisle (refer to user's manual to determine the coil and compressor location and air space needed for ventilation surrounding the unit). Pulling forward and blocking the surrounding base and wall cabinets 4" to 6" will create a 30"-deep countertop flush with the refrigerator case. The deep counter is an ideal work space situation for small appliance storage. End panels surrounding the refrigerator are extended to 30" deep to house the refrigerator.

ADVANTAGES: This method adds work space and improves appearance.

DISADVANTAGES: A new adjacent counter for added depth is required. Air circulation should not be blocked.

4. RECESS REFRIGERATOR.

A refrigerator can sometimes be recessed into a partition wall, a closet wall, or a garage wall behind it. The wall should not be load-bearing. The sheetrock or plaster is removed and the area surrounding the unit is framed. The depth of the recess can be determined by the allotted space needed to make the refrigerator flush with the cabinets (usually 4" to 6") and a plywood backer is then screwed to the sheetrock (Fig. 6.4.5). The recess should not be deeper than the case. Alcove walls can be built on both sides of the refrigerator with 2x4 studs and sheetrock. Mounting the electrical outlet above the back of the refrigerator eliminates space lost from a plug directly behind it. The adjacent counters should be clipped so that the refrigerator doors can swing fully open.

ADVANTAGES: The refrigerator will be nearly flush with adjacent cabinetry.

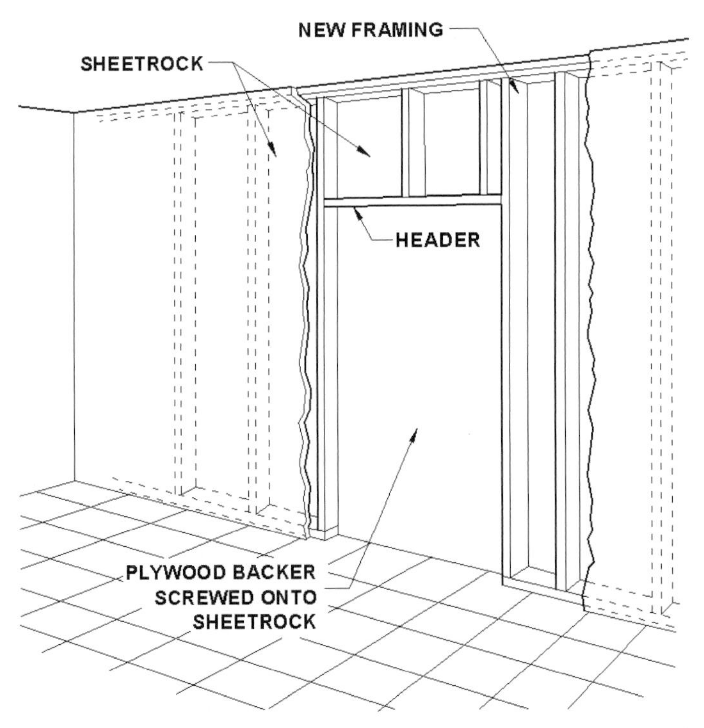

FIGURE 6.4.5 NEW FRAMING TO RECESS REFRIGERATOR INTO WALL

DISADVANTAGES: This method will increase sound penetration from the kitchen to the next room. Air circulation to the refrigerator's rear cooling coil can be reduced. If a garage wall is used, fire codes may prohibit this option.

IMPROVE RESOURCE AND ENERGY EFFICIENCY OF EXISTING APPLIANCES

ESSENTIAL KNOWLEDGE

According to *Home Energy Magazine*, the kitchen consumes 20% to 40% of a household's energy. Therefore, small changes in the location and adjacencies of existing appliances can result in reducing electric and gas bills.

TECHNIQUES, MATERIALS, TOOLS

Suggestions to improve appliance energy efficiency:

1. CONSIDER APPLIANCE ADJACENCIES AND AIR TEMPERATURE.

A refrigerator placed in direct sunlight will have to work harder to maintain its interior temperature. Higher air temperature and humidity in the room will also force the compressor to use more energy. When the refrigerator is placed next to a high-temperature appliance such as the cooktop, oven, or dishwasher, the compressor must work harder to maintain the refrigerator temperature. The layout of the kitchen can be redesigned to separate the appliances, or a barrier of 3" fiberglass or 1" foil-faced insulation between the refrigerator and the hot appliance will improve efficiency.

ADVANTAGES: Efficiency and energy performance are improved.

DISADVANTAGES: Redesign for needed additional space may be required.

2. PROVIDE VENTILATION AND CIRCULATION.

Often, appliances are "suffocated" because a built-in look is incorrectly executed. Consult manufacturers' specifications and, particularly in the case of true built-in appliances, follow them carefully. Consider options for additional circulation by creating openings behind the cabinetry above the appliance. Freestanding refrigerators require space surrounding the unit so heat from the condensing coil and compressor can escape. The coils should be cleaned twice a year to remove cooking grease, dust, and lint. To maximize efficiency, some sources recommend doubling the manufacturer's specified clearance for refrigerator installation.

ADVANTAGES: Efficiency and energy performance are improved.

DISADVANTAGES: More room dedicated to free space around the appliances will be required.

3. PROVIDE ADEQUATE VENTILATION TO EXTERIOR.

Clothes that do not dry in a timely manner may be an indication that a dryer is not running properly. When the dryer is connected to a long run of ductwork, the warm moist air is not exhausted as intended, and the dryer interior remains moist and uses more time and energy to dry clothes. Fantech offers an inline fan (Fig. 6.4.6) that compensates for the duct system length by boosting the airflow to the exterior.

ADVANTAGES: Efficiency and energy performance are improved.

DISADVANTAGES: The system must be accessible.

INSTALL RESOURCE- AND ENERGY-EFFICIENT APPLIANCES

ESSENTIAL KNOWLEDGE

Outdated appliances are often oversized for their limited technology and, compared to today's efficient models, consume more gas, electricity, or water. This combination often means that more space must be dedicated to older appliances whose performance is poor. One must consider the long-term, life-cycle cost and energy savings. The ENERGY STAR® rating and Energy Guide compare the initial cost and lifetime operating costs. The ENERGY STAR rating lists and rates only selected high-efficiency refrigerators, dishwashers, clothes washers, water heaters, and heating and cooling equipment, and the Energy Guide compares the initial cost and the lifetime operating costs of all similar appliance models.

TECHNIQUES, MATERIALS, TOOLS

Energy-saving features to look for in new appliances:

1. CONSIDER REPLACING THE REFRIGERATOR.

The refrigerator consumes more energy than any other kitchen appliance, and care should be taken to choose an energy-efficient model. The Energy Guide compares the initial cost and long-term energy savings to other similar models with the same features, style, and capacity, and when energy efficiency is a priority over all other considerations, the ENERGY STAR® rating should be used. Although larger refrigerators are more energy efficient, they still require more energy to operate and they take up more space. A top- or bottom-mount refrigerator is more efficient than a side-by-side refrigerator. Refrigerators with additional conveniences, such as ice makers, water dispensers, and automatic defrost, improve function but consume more energy. Northland manufactures a custom refrigerator with glass doors to allow the contents of the refrigerator to be viewed without opening the doors.

ADVANTAGES: Cost can be offset by local power company credits, long-term savings.

DISADVANTAGES: An energy-efficient refrigerator will have a higher initial cost. Not all models from a manufacturer will have an ENERGY STAR® rating.

2. CONSIDER REPLACING THE DISHWASHER.

The cost of running a dishwasher is incurred mostly from heating the water. When selecting a new dishwasher, consider features that will reduce the amount of hot water used. For dishwashers to clean

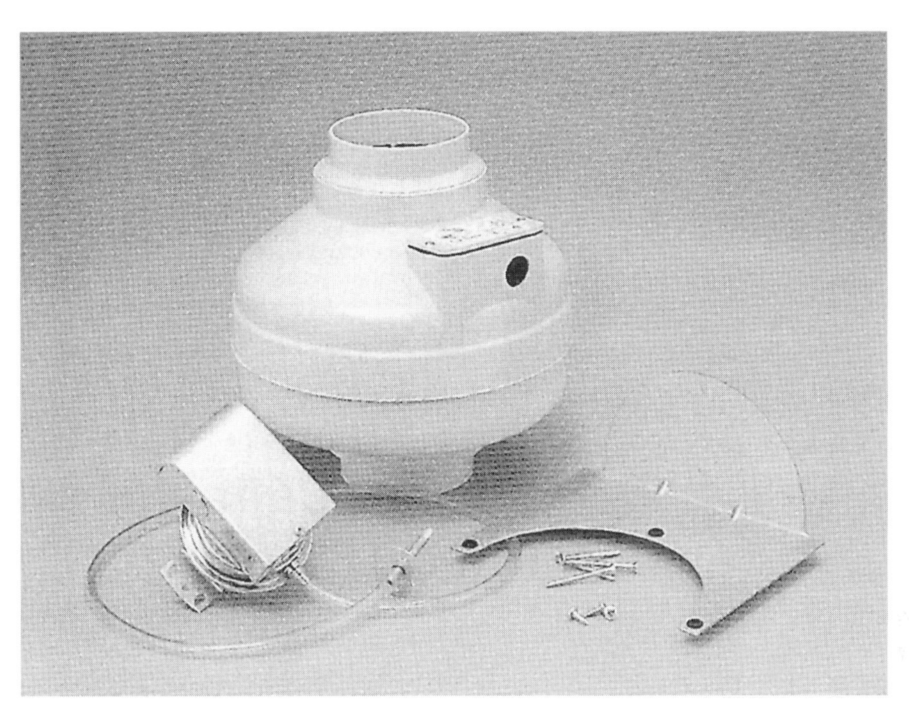

FIGURE 6.4.6 FANTECH DRYER BOOSTER

properly, 140°F is the recommended water temperature. The cost of running a dishwasher is increased because the domestic hot water heater is set to 140°F, although no other fixtures in the house require water hotter than 120°F. An integrated water booster provides the 140°F temperature needed for dishwashing without overheating the domestic water supply, and the whole-house water heater can then be set to 120°F. To keep the interior of the dishwasher clean during the wash, hot water fills the unit, increasing wash time and water usage. A dual-pump design in the dishwasher uses half of the water of a one-pump design. The primary pump circulates fresh incoming water, while the secondary pump eliminates any soil residue from previous wash cycles. Variable cycle selections or options for running smaller or less-powerful loads will save water when the unit is not as full or the contents are not so dirty. The Energy Guide compares the initial cost and long-term energy savings to other similar models with the same cycles, style, and capacity, and when energy efficiency is a priority over all other considerations, the ENERGY STAR® rating should be used.

ADVANTAGES: An energy-efficient dishwasher will save water and energy.

DISADVANTAGES: Energy-efficient dishwashers are not available from all manufacturers and have a higher initial cost.

3. CONSIDER REPLACING THE OVEN.

To improve overall efficiency, the best selection may be a self-cleaning oven. These have more insulation because of the extreme cleaning temperature, and this reduces energy loss during normal use. In addition, a convection oven uses a fan and the heating elements to cook more evenly and quickly, resulting in considerable energy savings over a conventional oven. Speed cooking is a new technology using a combination of heat from lightwave energy and microwaves to reduce the amount of cooking time without sacrificing taste. Similar to the traditional oven, food can be roasted, baked, or broiled. Speed cooking ovens are available from several manufacturers including Wolf and GE Appliances.

ADVANTAGES: An energy-efficient oven will provide energy savings. Some over-the-range speed-cook ovens will retrofit in the hood location.

FIGURE 6.4.7 FRIGIDAIRE HORIZONTAL AXIS WASHER AND DRYER

DISADVANTAGES: An energy-efficient oven will have a higher initial cost; some will require a 240-V hookup.

4. CONSIDER REPLACING THE WASHER AND DRYER.

If the home being rehabbed has a laundry room or area with equipment, these appliances may need replacement. Horizontal axis washing machines significantly reduce water consumption as a result of improved technology and the reduced water fill level. There is no agitator; rather, the clothes are forced by gravity into the water as the tub cycles. In combination with horizontal axis dryers, they can be stacked, installed below a countertop, or raised to a comfortable height. Frigidaire, GE, and Maytag are some of the manufacturers who offer horizontal axis washers and dryers (Fig. 6.4.7). The Energy Guide compares the initial cost and long-term energy savings to other similar models with the same cycles, style, and capacity, and when energy efficiency is a priority over all other considerations, the ENERGY STAR rating should be used.

ADVANTAGES: An energy-efficient washer and dryer will provide energy savings and improve access.

DISADVANTAGES: There will be a higher initial cost.

6.5 SINKS AND LAVATORIES

SURFACE MAINTENANCE AND REPAIRS

ESSENTIAL KNOWLEDGE

Sink and lavatory surfaces can be damaged through years of use. The maintenance and repair of a sink or lavatory will depend on its materials and fabrication. If the fixture is otherwise functional, rehab work might focus on refinishing or rejuvenating the fixture surface. Dirt and bacteria are often found in seams between the countertop surface and the sink or lavatory; integral or undermount sinks can reduce or eliminate such seams.

TECHNIQUES, MATERIALS, TOOLS

1. REFINISH CULTURED STONE.
Also referred to as cast polymers, cultured stone is durable and easy to maintain. Cultured stone is made from chips of natural stone embedded in a polyester resin, which consists of polyester resin and alumina trihydrate combined to reproduce the veined appearance of natural onyx or alabaster. Tiny fractures in the gel topcoat are common problems in cast polymers. Fractures usually occur around the drain which is exposed to the thermal shock of alternate hot and cold water temperatures. Tightening the lock nut at the drain only slightly more than finger tight can help eliminate the cracking and crazing that occurs over time. Minor scratches in the gel topcoat are buffed out using a polishing pad, and deep scratches or chips should be repaired by a porcelain repair company.

ADVANTAGES: Cultured stone is available in many colors and designs. Integral sinks eliminate seams.

DISADVANTAGES: Care must be taken when installing the drain to prevent tiny fractures in the surface coat.

2. REPAIR VITREOUS CHINA.
Vitreous china is a common material for lavatories, and it is a form of ceramic and porcelain that is vitrified. Patterns on china may be applied at several stages of the firing process, some more durable than others. Vitreous china is naturally stain resistant and durable, although pitting can form from use of abrasive cleaners. A ceramiclike epoxy from Abatron can be used to fill any cracks or pits on the surface. The surface must first be cleaned and free of dust and debris, and after two or three recommended coats, curing continues for up to 2 weeks. The hardening and curing process can be accelerated with temperatures around 200°F.

ADVANTAGES: The glasslike appearance is favored for its sanitary characteristics. Epoxy is effective on moist or wet surfaces and ideal for small areas.

DISADVANTAGES: Refinishing the entire fixture is labor intensive and should be left to a professional for a uniform appearance. Decals applied to the glaze will not be as durable if the fixture was not fired at a high enough temperature.

3. REPAIR ENAMELED STEEL.

Although the surface has the same properties as enameled cast iron, the steel base is not as strong or durable. Because the steel is smooth, these sinks have a greater tendency to chip and wear. Rust stains that appear as a result of chipped or worn enamel can be removed with naval jelly, muriatic acid, or diluted phosphoric acid. Touch-up paints and epoxies that resist corrosion, such as those from Abatron, are available. The cost of refinishing enameled steel does not exceed the cost of a replacement sink. If the existing sink cannot be easily removed or replaced, refinishing usually will exceed the cost of a replacement sink.

ADVANTAGES: Less costly material and touch-ups.

DISADVANTAGES: Color matching can be difficult; patched areas do not wear as well as the original surface.

4. REPAIR OR REFINISH ENAMELED CAST IRON.

Finished with a smooth enamel topcoat, cast-iron sinks are durable and will last indefinitely. Nonabrasive cleaners remove surface dirt, but a mild abrasive should be used to remove ground-in dirt. Rust stains that appear as a result of worn enamel can be removed with naval jelly, muriatic acid, or diluted phosphoric acid. The enamel can be chipped if a hard blow bends the cast-iron base. If the sink is irreplaceable or cannot be removed easily, the surface can be professionally refinished with a polyurethane-base coating. First, the surface is acid-etched, and scratches and dents are filled. The sink is then sanded and primed with a bonding coat, and glaze layers are applied. Any glaze applied to the surface without a chemical bond or acid-etching will likely peel.

ADVANTAGES: Enameled cast iron is durable and does not chip easily.

DISADVANTAGES: Often the cost of refinishing exceeds the cost of a replacement sink.

5. MAINTAIN STAINLESS STEEL.

Steel gauge, nickel content, and finishing technique determine the quality of the stainless steel. The most durable sink is 18 gauge; 20 gauge is acceptable, and 22 gauge should only be used in budget construction. Better sinks will be insulated on the underside to deaden sound. Stainless steel is stain resistant and hard-wearing. The surface should be free of dirt, stains, and fingerprints to preserve corrosion resistance. Scratches in the surface cannot be repaired, but a brushed finish will conceal any fine scratches and watermarks on the surface.

ADVANTAGES: A large number of design and style options are available.

DISADVANTAGES: Middle- and low-range products usually do not have insulation to reduce sound. The surface should be cleaned whenever dirt or stains are visible.

6. REPAIR SOLID SURFACE.

All solid surface sinks and lavatories are homogeneous, made of polyester, acrylic, or a polyester and acrylic blend. Each manufacturer has developed a unique composition, some including mineral fillers to improve durability or fire-retardant properties. Sinks can be integral with a solid surface counter or undermount. Because of the homogeneous makeup, scratches or burns in the solid surface can be removed with a scouring pad, 320 to 400 grit sandpaper, or steel wool. More serious cuts or burns can be buffed out by a solid surface technician.

ADVANTAGES: Integral solid surface sinks create an invisible seam between the counter and the sink; many colors and styles are available.

DISADVANTAGES: Repairs can be costly.

WATER PURIFICATION

ESSENTIAL KNOWLEDGE

Lead pipes and solder used in residential and municipal water systems at the beginning of the 1900s contribute to water pollution. Older homes with vintage plumbing may benefit from a purification system to ensure water quality. No filtration system can remove all pollution from water, and the correct filtration system depends on the contaminants present in the water.

TECHNIQUES, MATERIALS, TOOLS

1. INSTALL A CARBON FILTER ADAPTER OR FAUCET.

Carbon removes many organic contaminants from water that result in odor and bad taste, as well as dissolved gases (including chorine), most pesticides, many chemicals, and radon gas. Although there are many varieties, block and granulated carbon are most commonly used in filters. Granulated carbon has a pitted surface to increase its area and absorption capacity, and carbon powder in a block forms a matrix structure to absorb contaminants. In addition, carbon block eliminates heavy metals and matter particles. Moen, Price Pfister, and Ultraflo all offer integrated filter faucets. A carbon filter attachment screws onto the existing faucet, and the filter is either granulated carbon or carbon block. A faucet with an integrated filter in the neck or base of the faucet allows the user to select unfiltered cooking and cleaning water or filtered drinking water.

ADVANTAGES: No separate tap is needed; the base cabinet storage space below is maintained. An adapter is inexpensive and easy to install.

DISADVANTAGES: Filters need to be replaced every three months and can become a source of pollution if not maintained. Other fixtures in the house are not filtered.

2. INSTALL AN UNDERCOUNTER CARBON FILTRATION SYSTEM.

Undercounter carbon filtration systems often use two separate carbon block filters to improve reduction of pollution. Filters should be replaced every six months. A combination of carbon filters is available and should be selected to filter anticipated pollution.

ADVANTAGES: This system is less expensive than a reverse osmosis system; it can be installed in the basement or elsewhere in a remote but close location to the sink.

DISADVANTAGES: This system eliminates some storage space under the sink. Some farmhouse-style sinks with an integral front apron may not accommodate the system. A separate filtered water faucet is needed in addition to the primary faucet.

3. INSTALL AN UNDERCOUNTER REVERSE OSMOSIS AND CARBON COMBINATION FILTRATION SYSTEM.

Reverse osmosis (Fig. 6.5.1) is the process by which water is forced under pressure through a semipermeable membrane that separates water from dissolved solids, heavy metals, asbestos, radioactive particles, and some bacteria. When lead and nitrates are present in the drinking water, a reverse osmosis system is a better solution. A combination of reverse osmosis and carbon is needed because reverse osmosis systems cannot be used independently with chlorinated supply water. Chlorinated water first needs to be filtered through the carbon.

ADVANTAGES: The combination of carbon and reverse osmosis will block almost all contaminants. The system can be installed in the basement within close proximity to preserve base cabinet storage.

DISADVANTAGES: Reverse osmosis systems sometimes process slowly and can eject three or four gallons of water for every gallon filtered. Membranes will need to be replaced every one to three years. Some storage space under the sink is eliminated. Some farmhouse-style sinks with an integral front apron may not accommodate the system. A separate filtered water faucet is needed in addition to the primary faucet. Reverse osmosis systems can be expensive compared to other filtration systems.

WATER CONSERVATION

ESSENTIAL KNOWLEDGE

Water conservation is a critical aspect of energy and resource management, and it should be considered in any kitchen rehab project. Many of the devices discussed are available at little or no cost from utility companies.

FIGURE 6.5.1 GE SMARTWATER REVERSE OSMOSIS FILTRATION SYSTEM

TECHNIQUES, MATERIALS, TOOLS

1. INSTALL LOW-FLOW FAUCET.

Although there are no code requirements for rehab construction, a flow rate of 2.0 gallons per minute (gpm) in lavatory faucets and 2.5 gpm in kitchen faucets is recommended.

ADVANTAGES: Water consumption is reduced.

DISADVANTAGES: The pressure of water flow is reduced, but an aerator can increase the pressure of the available water as it leaves the faucet.

2. RETROFIT A FAUCET AERATOR.

Replace the faucet head screen with a faucet aerator. Air is added to water to provide good pressure and a steady stream of water, and the flow seems greater than it actually is. Neoperl and Resource Conservation offer aerator attachments. Again, a flow rate of 2.0 gpm in lavatory and 2.5 gpm in kitchen faucet aerators is recommended. Periodically, the aerator should be unscrewed from the spout and cleaned.

ADVANTAGES: A faucet aerator is highly efficient, inexpensive, and easy to install.

DISADVANTAGES: The type of spray or flow may not be desirable.

3. REPLACE FITTINGS.

Fittings for the sink can be repaired by removing the faucet and flushing out any debris. Leaking fixtures that are older may be repaired by replacing washers, but most newer fittings are washer-less and eliminate future leaks from worn washers. If this does not improve the flow and the fitting is not repairable, replacement provides the opportunity for improved access and function (refer to Maximize Access and Function later in this section).

ADVANTAGES: Replacement of fittings is inexpensive.

DISADVANTAGES: Conditions that are not obvious may exist and may need professional attention.

MAXIMIZE ACCESS AND FUNCTION

ESSENTIAL KNOWLEDGE

Kitchen access is important and bathroom access is critical to all household members. Rehabbing presents an opportunity to create flexibility at the sink so that different tasks can be performed safely and comfortably by users of differing age, stature, and ability.

TECHNIQUES, MATERIALS, TOOLS

Products to improve access at the kitchen or bathroom sink include:

1. INSTALL DIVERTER FOR SPRAY.

On a kitchen or bath faucet, a diverter permits the user to easily switch from an aerator spray to a needle spray. The diverter system is designed into a plastic integrated pull-out faucet or the traditional faucet with separate spray. Since 1994, manufacturers have fused the rubber diverter to the plastic handle. In some cases, a pull-out spray in the bath will make it easier to wash hair in the sink or to give sponge baths; in the kitchen it makes rinsing food and dishes easier.

ADVANTAGES: Flexibility is improved.

DISADVANTAGES: An additional drilled hole in the sink is needed for a separate diverter spray.

2. INSTALL SCALD PROTECTION DEVICE AT FAUCET.

Changes in water pressure in the house can cause sudden changes in water temperature at a kitchen or bathroom sink. Water temperature can be controlled by balancing the hot and cold water inlets with either temperature or pressure mixing. While presetting the water temperature can be considered a luxury, temperature control is also a safety feature. For scald protection, the water flow is controlled by a thermostat. When the temperature reaches the 120°F range, the faucet will automatically shut off. Some faucets have a restart button that purges the hot water and restarts the supply. In a retrofit, a temperature-limiting valve can be installed under the lever and escutcheon, allowing the temperature to be adjusted. Many manufacturers, including Grohe and Resource Conservation, manufacture faucets with temperature and pressure balancing protection.

ADVANTAGES: Some scald-protection devices can be retrofitted into an existing faucet. Temperature control is safe and convenient.

DISADVANTAGES: Liquid elements and metal may not respond well over time due to damaging water conditions.

3. INSTALL GOOSENECK OR HIGH ARC SPOUT.

When the kitchen sink depth is limited, a tall gooseneck spout, often used in laboratories, facilitates filling tall pots.

ADVANTAGES: Clearance for tall pots is increased.

DISADVANTAGES: The increased height of the spout can result in splashing in shallow bowls.

4. INSTALL INSTANT HOT AND INSTANT COLD DISPENSERS.

Hot water dispensers at the sink improve function in the kitchen by providing instant hot water at about 190°F for tea, soup, or coffee. Cold water dispensers provide instant cold water available for drinking or cooking. Franke offers either hot water or cold water dispensers, while Steamin' Hot from In-Sink Erator provides hot water only.

ADVANTAGES: The need to heat water is eliminated. The cold water dispenser eliminates the need for a water dispenser at the refrigerator.

DISADVANTAGES: A cold water dispenser is more costly because the dispenser and chiller are usually sold separately.

5. INSTALL POT-FILLER FAUCET.

The pot-filler faucet, which is mounted to the wall over the stove, fills large pots on the front and back burners. The jointed arm can be folded out of the way when cooking. Chicago Faucets and Franke are two manufacturers that offer different variations of pot-filler faucets.

ADVANTAGES: The need to carry a heavy pot full of water from the sink to the stove is eliminated.

DISADVANTAGES: The pot-filler faucet requires water lines and may not be cost-effective if pipes have to be installed.

6. INSTALL PEDAL VALVE CONTROLS.

Pedal Valve's Pedalwork foot controls can be easily installed to allow the user to operate the faucet by pressing foot pedals, in addition to traditional hand operation. The pedals are located in the toekick space under the base cabinet.

ADVANTAGES: Pedal valve controls are easy to install, inexpensive, and color coordinated.

DISADVANTAGES: Minor adjustments must be made to the sink cabinet.

7. CONSIDER SHAPE, SIZE, AND NUMBER OF SINKS.

In both the kitchen and the bath, adding a sink increases the available work space for multiple users. A variety of sizes and shapes is available to fit many spaces and allow for a second sink at a different height than the original to accommodate people of varying stature.

ADVANTAGES: A second sink will improve access and increase function.

DISADVANTAGES: The expense of a second sink can be considerable, and space may not be available.

8. INSTALL TOUCHLESS FAUCET CONTROL.

Touchless faucet controls turn on when they sense motion. Because water cannot be left on, touchless controls improve water conservation. Although electronic-operated models are available, a battery-operated model may work better for a rehab project. Temperature and flow rate are preset, but some models include handles for manual override. Residential bathroom and kitchen faucets are available from Aqua Touch, KWC, and Geberit.

ADVANTAGES: A touchless faucet eliminates the need to turn handles off and on when strength is minimal. Unnecessary water consumption is decreased.

DISADVANTAGES: Battery-operated faucet needs battery replacement about once a year. A touchless faucet occasionally responds to unintended motion.

6.6 TUBS AND SHOWERS

SURFACE MAINTENANCE AND REPAIRS

ESSENTIAL KNOWLEDGE

Tub and shower surfaces can become damaged through years of use. The tub and shower are often the areas in the bathroom that show the most wear. The maintenance and repair of a tub or shower will depend on its materials and fabrication. If the fixture is otherwise functional, rehab work might focus on refinishing or rejuvenating the surface.

TECHNIQUES, MATERIALS, TOOLS

1. REPAIR FIBERGLASS TUB AND SURROUND.

Fiberglass is the backing material used to reinforce the polyester-gel-coat finish surface. If the gel coat becomes damaged, a fiberglass repair expert should be contacted to repair the surface.

ADVANTAGES: A fiberglass tub has the least-expensive finish and the easiest to repair.

DISADVANTAGES: The polyester gel coat is not as durable as other finish materials.

2. REPAIR ACRYLIC TUB, BASE, AND SURROUND.

Acrylic and acrylonitrile butadiene styrene (ABS) sheets are thermoformed to fabricate the unit, and are harder finishes than polyester gel coats. The color is continuous throughout, and the material can usually be sanded out and then buffed if the surface is scratched. An automotive-type rubbing compound and a nongrain paste wax will help polish a dulled surface.

ADVANTAGES: Formed tub and surround, shower base, or separate surrounds are available.

DISADVANTAGES: Limited styles and shapes are available.

3. REPAIR CULTURED STONE BASE AND SURROUND.

Also referred to as cast polymer, cultured stone is durable and easy to maintain. It is made from chips of natural stone embedded in a polyester resin; cultured onyx consists of polyester resin and alumina trihydrate, which are combined to reproduce the veined appearance of natural onyx or alabaster. Minor scratches in the gel coat can be buffed out using a polishing pad. Tiny fractures in the gel coat are common problems in cast polymers. Damage usually occurs around the drain, which is exposed to the thermal shock of alternate hot and cold water temperatures. Deep scratches or chips should be repaired by a porcelain repair company.

ADVANTAGES: Many color and design options are available.

DISADVANTAGES: Repairs do not always match the original surface.

4. INSTALL A LINER OVER EXISTING TUB OR SHOWER BASE.

Rusted, chipped, and worn bathtubs and shower bases often cannot be rejuvenated. A liner (Fig. 6.6.1) is formed as an exact fit over the existing fixture. Manufactured from coextruded ABS and acrylic, ReBath bathtub liners, shower base liners, and wall surrounds are nonporous.

ADVANTAGES: Liners are more affordable than replacement, and there is minimal installation time.

DISADVANTAGES: The installer must be authorized by the manufacturer to perform installation. There is a limited color choice.

5. REPAIR ENAMELED STEEL TUB.

Enameled steel surfaces have the same properties as enameled cast iron, although the steel tub base is not as strong or durable. Because the steel is smooth, the fixtures have a greater tendency to chip and wear. Rust stains may appear as a result of worn enamel, and can be removed with naval jelly, muriatic acid, or diluted phosphoric acid. Touch-up paints and epoxies that resist corrosion, such as those from Abatron, are also available. Enameled steel can be refinished using a similar process to enameled cast iron, but often the cost of repair exceeds the cost of a replacement tub.

ADVANTAGES: The materials are less costly.

DISADVANTAGES: Enameled steel tends to chip more easily than enameled cast iron. Color matching can be difficult, and patched areas do not wear as well as the original.

6. REPAIR OR REFINISH ENAMELED CAST-IRON TUB.

Finished with a smooth enamel topcoat, cast-iron tubs are heavy, durable, and quiet. They will last indefinitely, although the enamel can be chipped if a hard blow bends the cast-iron base. Nonabrasive cleaners should be used to remove surface dirt, but a mild abrasive is the best choice to remove ground-in dirt. If damage to the enamel is significant, or the tub cannot be replaced, the surface can

FIGURE 6.6.1 INSTALLATION OF REBATH BATHTUB LINER

be professionally refinished with a polyurethane-base coating. The surface is first acid-etched, and scratches and dents are filled. The entire tub is then sanded and primed with a bonding coat, and glaze layers are applied. Any refinishing product applied to the surface without a chemical bond or acid-etching will likely peel. If an existing cast-iron tub is beyond repair, it can be demolished with a sledge hammer and removed.

ADVANTAGES: Enameled cast iron does not chip easily.

DISADVANTAGES: Cast iron is a good conductor of heat, and will cool bath water more quickly than a plastic-based material. Refinishing is costly.

7. REPAIR SOLID SURFACE BASE AND SURROUND.

All solid surface shower bases and surrounds are homogeneous polyester, acrylic, or a polyester and acrylic blend. But each manufacturer has developed a unique composition, some including mineral fillers to improve durability properties. Because of the homogeneous makeup, scratches or burns in the solid surface can be removed with a scouring pad, 320 to 400 grit sandpaper, or steel wool. More serious cuts or burns can be buffed out by a solid surface technician. The invisible seam between the base and the surround eliminates water seepage through the seam into the floor and wall. Installation and major repairs should be made by a technician authorized by the manufacturer.

ADVANTAGES: There are invisible seams between the base and surround; many colors, sizes, and custom shapes are available.

DISADVANTAGES: Solid surface bases and surrounds are more costly than other plastic bases and surrounds.

8. REPAIR OR REPLACE CERAMIC TILE AND GROUT.

Though ceramic tile rarely chips, bacteria can collect in the damaged area. Household cleaners containing acid, bleach, or vinegar should be avoided for daily cleaning because they will etch the tile and grout in time. Mold in the grout lines can be killed with a 1-to-3 ratio of bleach to hot water. Tile is susceptible to water penetration at the joints; if the grout becomes wet, water can seep behind the wall tile and damage the substrate (refer to Moisture Control later in this section for installation techniques). If the tile is broken, chipped, cracked, or loses adhesiveness, an individual tile or a section of tile can be removed and replaced. Care must be taken to remove tile without damaging surrounding tiles. Remove the surrounding grout with a hand-held grout saw or diamond blade grinder. Then break the damaged tile into several pieces, and pry the small sections of tile from the substrate. A chisel should be used to remove the tile from the backer board without damaging the substrate. Insert new tile with adhesive and level to adjacent existing tiles. On walls, tape the tile into place until dry. Any loose grout should be picked off. Grout is available as cement, vinyl, or epoxy-based. Epoxy-based grout is stronger and more impervious to water, mildew, and stains, but more care is required during installation. Penetrating or surface silicone cleaner applied to grout increases its stain resistance.

ADVANTAGES: If extra tiles are available, this repair is relatively easy.

DISADVANTAGES: It is sometimes difficult to match grout colors. Epoxy-based grouts tend to be more costly than others.

9. REPAIR STONE BASE AND SURROUND.

If there are minor cracks, stone can be repaired using an overfill and grind technique. The damaged area is cleaned and dried before it is filled with an epoxy or polyester matrix combined with a ground stone or sand aggregate to produce a mortarlike mixture. To eliminate an obvious straight-line repair, the edges of the crack can be chipped and widened, replicating the existing veins. After the mixture cures, the fill is ground and polished. A professional stone fabricator for the repairs can ensure a good bond and lasting results. Marble should not be used in steam showers because the marble can pit, delaminate from the substrate, or deteriorate around steam vents. Some types of marble will develop rust spots or discolor. Sealers applied to the stone surface will not prevent such problems.

ADVANTAGES: The aggregate repair will improve the similarity between stone and fill; epoxy is effective on wet surfaces.

DISADVANTAGES: Epoxies may yellow.

MOISTURE CONTROL

ESSENTIAL KNOWLEDGE

Moisture control can be an issue in the kitchen and is a major concern in the bathroom. Lack of proper ventilation and improper installation and sealing of fixtures at walls can lead to excessive moisture, which can result in damage if not quickly removed. Indicators of moisture buildup in the bathroom include mold, mildew, failing grout, dislodged tiles, water stains, or rotted walls. Moisture can permeate wallboard, moving from the warm side of the wall to the cool side, which varies according to climate. Steam can condense as it hits a cool surface and pool at the base of that surface. Any damp area must completely dry out. Moisture damage will occur if there is not proper sealing or ventilation. In a rehab project, both the visible problem and its source must be repaired and corrected.

TECHNIQUES, MATERIALS, TOOLS

A number of approaches to reducing the effect of moisture problems in a rehab bathroom should be considered:

1. IMPROVE VENTILATION.

A good exhaust system is critical to moisture control in bathrooms because it will help prevent water from condensing and building up in the walls. Determine the proper cubic feet per minute (cfm) rating for the bathroom with the following formula:

Cubic feet of room (L x W x H) x 8 air changes per hour x 60 minutes = required fan cfm rating

Controls should be placed on a timer so that the fan is on for 15 to 20 minutes after someone has showered. A combination low-voltage fan and light, the Vent-Axia Fan n' Light from Coast Products, is a ceiling-mount fixture that can be used in a tub or shower to help reduce moisture in the rest of the bathroom. The duct system should take the shortest and most direct route to the outside. It is important to plan for return or makeup air. In hot and humid climates, a powered system that cools and dehumidifies the fresh air is recommended. For more detail on exhaust systems, see Chapter 8.

ADVANTAGES: Exhaust systems remove some moisture and odors.

DISADVANTAGES: Makeup air must be considered. Long lengths and bends in the ductwork reduce efficiency.

2. REDUCE MOISTURE AT RECESSED LIGHTS.

Moisture can get into attic or other overhead spaces through recessed lighting (Fig. 6.6.2). Even when an "airtight" recessed light is used in the bathroom, some moisture will penetrate into the structure above. The warm, moist air will condense or freeze on contact with a cold surface. Either an airtight

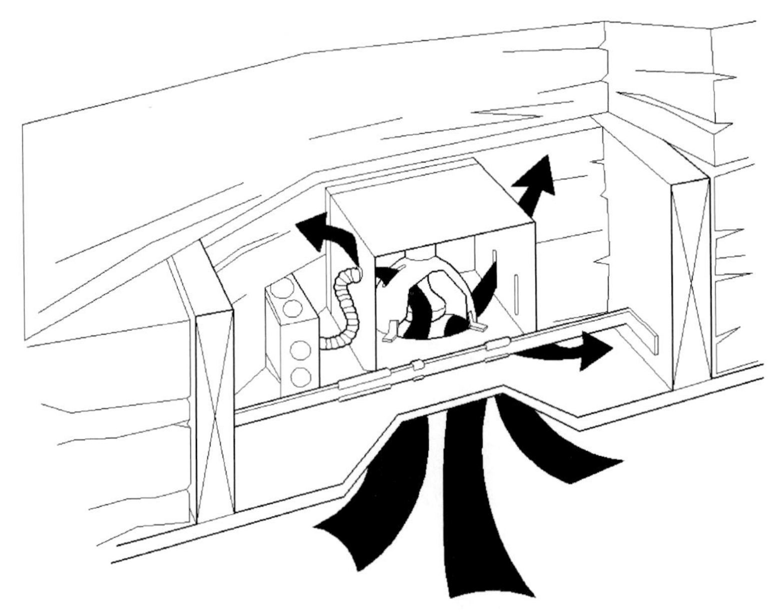

FIGURE 6.6.2 MOISTURE PENETRATING THROUGH RECESSED LIGHT FIXTURE INTO OVERHEAD SPACE

box built around the fixture with insulation or a dropped soffit with an air barrier at the ceiling plane will reduce or eliminate moisture penetration through a recessed light.

ADVANTAGES: The amount of moisture trapped in attic or proximate walls is reduced.

DISADVANTAGES: Thorough sealing and double sealing at and above recessed lighting adds labor and costs.

3. MANAGE WATER PENETRATION AT TILE

Small amounts of water penetrating the tile in showers or tub surrounds will be absorbed by cement backer board, and by the framing if the water passes beyond the backer board. To avoid the formation of rot and mildew, these materials must have a chance to dry out before they become saturated. Interior walls and exterior walls or roofs that receive direct sunlight will usually dry satisfactorily, but shaded exterior walls or roofs require special attention. Insulate the exterior wall or roof and apply a good, tight air barrier to the inside (4- or 6-mil poly, Tyvek, gypsum board, or paper-based wall sheathing; asphalt-impregnated felt should not be used inside the home). Then furr the wall or roof with flat 2x's running horizontally, and apply the backer board and tile. Gypsum board used behind tile (even water-resistant "green board") will not dry out as quickly as cement-based backer board. Behind leaking tile grout and cracks, the accumulated water will soon reemulsify the adhesive and the tile will fall off. Also, mildew will form, fed by the cellulose in the paper face on the gypsum board. For rehab work, use cement-based backer board.

ADVANTAGES: These methods recognize that tile work cannot be made permanently waterproof and help prevent mold growth in walls and possible structural failure of tile.

DISADVANTAGES: Cement backer board is more difficult to cut and attach than gypsum board.

4. CHANGE INSTALLATION OF TILE SUBSTRATE AT TUB OR SHOWER SEAM.

When tile overlaps the seam created by the green board and the shower base, the tile acts like a dam and a water reservoir is created. Problems often occur when the cut, unprotected end of the green board is installed at the seam of the shower base, exposing the capillary tubes that run from the reservoir up into the green board. Once the capillary tubes fill, the saturated substrate will eventually deteriorate. One way to solve this problem is to seal the cut edge of the green board with tile mastic in the field. The substrate and the tile should both be held off the shower pan or tub by at least ¼". The caulk will fill this gap between the tile and the surround, and a 1" weep hole in the caulk will allow water in the reservoir behind the tile to freely exit (Fig. 6.6.3).

ADVANTAGES: The structure of the substrate is preserved.

DISADVANTAGES: Installation is not widely practiced; consult Further Reading sources for further details and clarification.

MAXIMIZE ACCESS AND FUNCTION

ESSENTIAL KNOWLEDGE

A rehab project that includes the tub, shower base, or surround is an opportunity to incorporate universal design principles to safely and comfortably accommodate users of differing age, stature, and ability.

TECHNIQUES, MATERIALS, TOOLS

Suggestions to improve safety and access can be applied to faucet fittings and fixtures.

1. INSTALL SCALD PROTECTION DEVICES.

Although many devices regulate water temperature, devices that stop the flow of water, such as ScaldSafe from Resource Conservation, are recommended to provide scald protection. A temperature-limiting

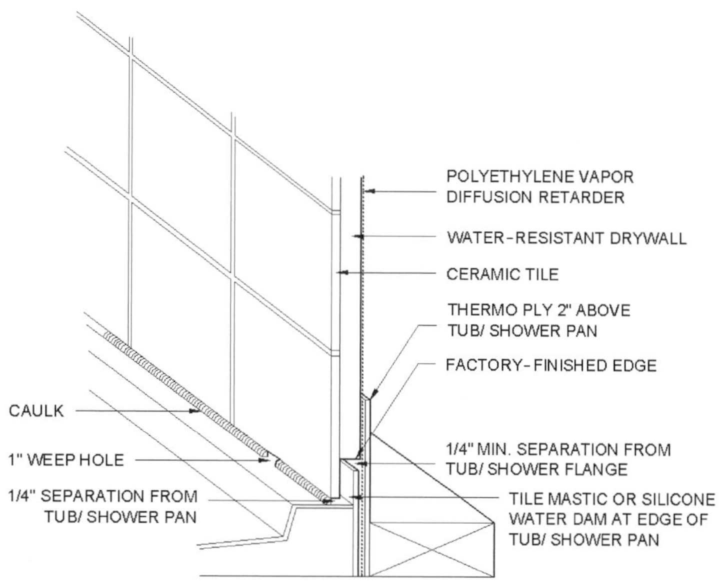

FIGURE 6.6.3 ONE-INCH WEEP HOLE IN CAULK WITH SUBSTRATE HELD OFF TUB OR SHOWER FLANGE

valve is easily installed under the lever and escutcheon, and the limit can be set and later adjusted. Liquid and metal thermostatic systems expand and contract with temperature. When the temperature reaches the set limit, the faucet will shut off. Some faucets have a restart button that will purge out the hot water and restart the actuator. Pressure mixing and balancing systems adjust the flow of hot and cold water to protect against dramatic temperature changes due to sudden changes in water pressure.

ADVANTAGES: A scald protection device can be retrofitted into an existing faucet.

DISADVANTAGES: Fluid elements and metal may not respond well over time due to damaging water conditions.

2. RELOCATE CONTROLS.

If the tub or shower surround is to be replaced, the existing control and diverter can be offset 6" toward the outside of the tub. The offset will allow one to turn on and test the water temperature and flow before entering the tub or shower. In the bathtub, the controls should be placed 17" to 30" above the floor; up to 48" in the shower. In a custom walk-in shower the controls should be placed at the point of entry (Fig. 6.6.4).

ADVANTAGES: Access is improved.

DISADVANTAGES: Replacement or repair of the tub or shower wall is required.

3. INSTALL BUILT-IN SEAT TO IMPROVE TRANSFER.

A built-in seat framed at the head of the tub or in a custom shower will provide a surface for transfer into the tub or shower. A depth of 15" is recommended with a preferred height of 18", and it should gently slope toward the tub or shower base at ¼" per 12" to avoid standing water (Fig. 6.6.5). The transfer surface needs to withstand a minimum load of 300 lb. The ceramic tile deck overlaps the tub flange to eliminate a permeable seam, and the tile should continue down the front of the seat. If glass doors are used, they must open up to the seat and a custom size may be needed. A shower curtain rod must extend over the seat, and an extra-wide curtain or two curtains must be used.

ADVANTAGES: Access is improved.

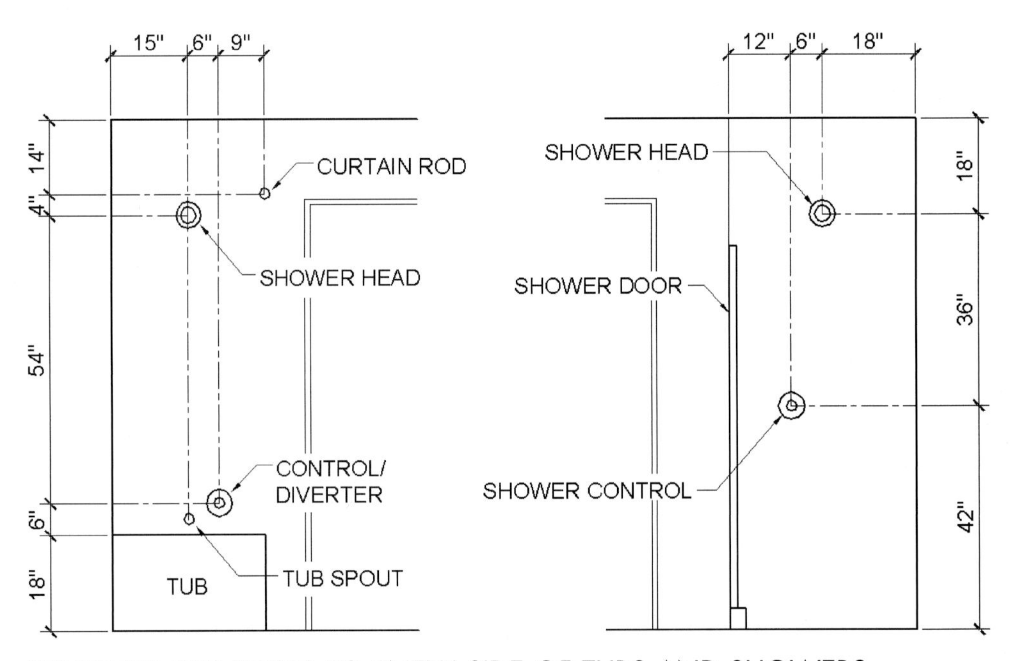

FIGURE 6.6.4 RELOCATE CONTROLS TO ENTRY SIDE OF TUBS AND SHOWERS

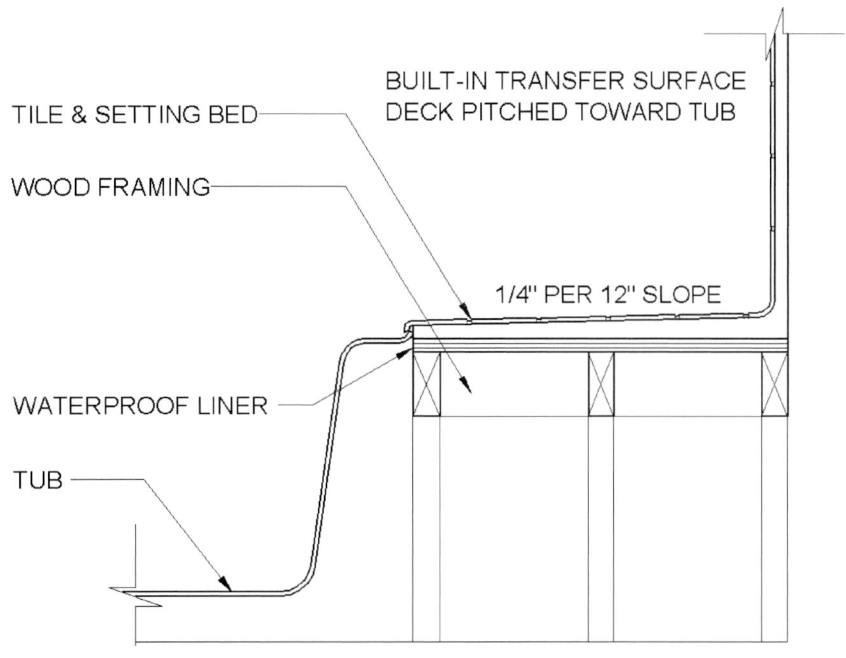

FIGURE 6.6.5 BUILT-IN TRANSFER SEAT DETAIL

DISADVANTAGES: A built-in seat remains at a fixed height and cannot be adjusted for changes in need. The recommended height of 18" is often higher than the top of the tub.

4. INSTALL BLOCKING FOR GRAB BARS.

Grab bars improve safety and stability, and blocking should be installed throughout the bathtub area so that grab bars can be installed when and where they will be most effective. Grab bars should be placed at the point of entry and where a person shifts positions. Horizontal grab bars should be placed on all three walls at a height 33" to 36" above the floor. Vertical grab bars may be placed on the sidewalls for additional safety. In a rehab situation, with finished walls in place, grab bars can be added only where there are studs or if there is known blocking. When these options are not clear or not in the right location, a solid piece of wood (2x4 or 2x6) can be attached to the studs, and the grab bars attached to the solid wood. The top edge should be beveled so that it slopes slightly, and the entire surface should be covered with a waterproof material. Another solution, relatively new to the market, is the WingIt Grab Bar Fastening System from Pinnacle (Fig. 6.6.6), which can be installed into substrates such as drywall to fasten grab bars to the wall. Grab bars are available from many manufacturers including Hewi and Otto Bock.

ADVANTAGES: Access is improved when needed. The risk of slipping and falling is reduced.

DISADVANTAGES: Future needs may change.

5. USE A NO-THRESHOLD SHOWER BASE.

When possible, the curb entering the shower should be eliminated for safety and access. This works best in an oversized shower with the flow of water from the showerhead directed away from the entrance. When this is done, the floors should be set in mud, rather than thinset and grout, and sloped gently (¼" per foot maximum) toward the drain. A waterproof membrane below the shower area should be extended into the room to reduce the possibility of leaks from standing water. During framing, the floor should be planned to accommodate the thickness of the mud underlayment required to create the slope in the tile. This can also be done using a solid surface custom-shaped base or manufactured curbless shower base.

ADVANTAGES: This type of shower base improves safety and access and is easier to maintain.

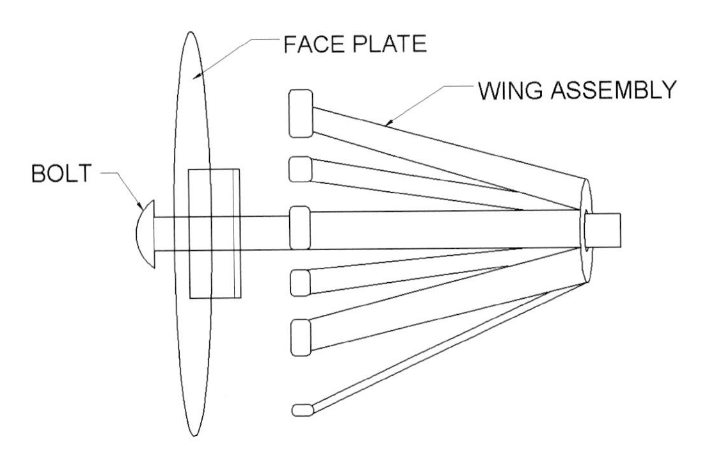

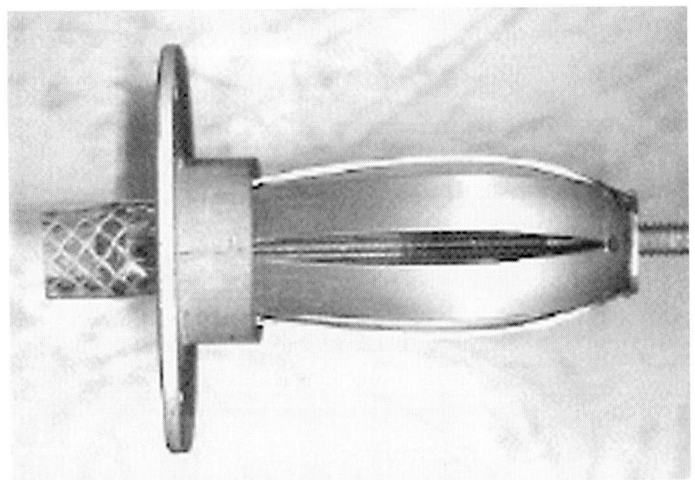

FIGURE 6.6.6 WINGIT FASTENER IN OPEN AND CLOSED POSITIONS

DISADVANTAGES: Attention to water flow and standing water is required to reduce the risk of leaking.

WATER CONSERVATION

ESSENTIAL KNOWLEDGE

It is an opportune time when rehabbing a bathroom to consider strategies for water conservation. Standard showerheads typically provide 2.5 to 3 gpm. Showerheads are available with a flow restriction of 1.7 gpm, which reduces water use. Products that are a substantial upgrade over the existing fixture should be selected to achieve maximum long-term water conservation.

TECHNIQUES, MATERIALS, TOOLS

Suggestions to limit the flow of water without reducing function:

1. REPLACE FITTINGS.
Tub and shower controls may be difficult to operate due to years of use. Controls can be repaired by removing the nozzle or spray head and flushing out debris. Leaking fixtures can be repaired by replac-

ing washers, but most newer fittings are washerless and eliminate future leaks from worn washers. If this does not improve the flow and the fitting is not repairable, replacement provides the opportunity for improved access and function (refer to Maximize Access and Function earlier in this section).

ADVANTAGES: Replacement of fittings is inexpensive.

DISADVANTAGES: Some repairs may require professional service.

2. INSTALL HIGH-EFFICIENCY, LOW-FLOW SHOWERHEAD.

Limiting devices can be installed behind an existing showerhead. Hansgrohe has a three-part device that reduces the quantity of water supplied by 20%, 30%, or 40%, depending on the configuration of the device's three parts. Showerheads that limit the outflow of water to 2.5 gpm or less need to compensate for pressure. Showerheads should have a water pressure of 60 psi.

ADVANTAGES: Fittings can be retrofit in existing tub or shower.

DISADVANTAGES: There is a greater potential for scalding in low-flow showerheads with 2 gpm than the standard 3 gpm.

6.7 TOILETS AND BIDETS

MAINTENANCE AND REPAIRS

ESSENTIAL KNOWLEDGE

Over time, toilets and bidets may be subject to cracks or chips in the surface of the fixture, or wear and tear of the mechanical devices that allow for smooth functioning. Disabled tank floats, stuck valves, and leaky connections are some of the most common maladies that require attention. Examining the existing condition of these fixtures is the first step in determining the rehab measures necessary to return them to good working order. Dirt and bacteria are often found in seams between the fixture and the floor and should be scoured. If fixtures need replacement, wall-hung fixtures are one way to eliminate such seams.

TECHNIQUES, MATERIALS, TOOLS

1. REPAIR VITREOUS CHINA SURFACE.
The most common surface material for toilets, bidets, and urinals is vitreous china—a high-fired, nonporous, ceramic material with a high-gloss glaze. A ceramiclike epoxy from Abatron can be used to fill cracks or pits on the surface. The surface must first be cleaned and free of dust and debris. After an application of two or three coats, curing continues for up to two weeks. The hardening and curing process can be accelerated by applying heat around 200°F.

ADVANTAGES: The glasslike appearance is favored for its sanitary characteristics. Epoxy is effective on moist or wet surfaces, and ideal for small areas.

DISADVANTAGES: Refinishing the entire fixture is labor intensive and should be completed by a professional for a uniform appearance. Decals applied to the glaze will not be as durable if the fixture was not fired at a high enough temperature.

2. REPAIR WATER SEEPAGE.
Water seepage at the base of the toilet is likely a sign of a damaged wax ring gasket seal. Water under the base of the toilet can result in damage to floor finishes and the subfloor. Remove as much water from the bowl as possible, disconnect the riser tube, and remove the reservoir tank from the toilet bowl. Brittle or cracked rubber washers and gaskets should be replaced. Unbolting the bowl from the floor will reveal the wax ring, which should be removed. Clean surfaces and insert a new ring.

ADVANTAGES: The floor and structure are saved and water is conserved.

DISADVANTAGES: The repair can be labor intensive.

3. REPAIR RUN-ON TOILET.

A run-on toilet can be caused by a faulty tankball, inlet valve, or float (Fig. 6.7.1). A rubber tankball controls water flow into the reservoir tank. The float (usually a large ball) rises and falls with the water level in the reservoir tank to control the inlet valve. The inlet valve opens as the float moves down. The float ball should resist any pressure to push it under water; if it remains partially submerged, it may be taking on water and should be replaced. If the water continues to run, the float arm or tankball rod can be adjusted by bending it down until the water shuts off. Leak savers are available from Resource Conservation that lock the float ball in the off position after every flush to stop the toilet from cycling on and off.

ADVANTAGES: This repair conserves water and can be inexpensive.

DISADVANTAGES: Conditions that are not obvious may require repair by a professional.

WATER CONSERVATION

ESSENTIAL KNOWLEDGE

Toilets more than 20 years old use between 5 and 7.5 gallons of water per flush (gpf); toilets manufactured after 1980 use 3.5 gpf. According to the Federal Energy Act of 1994, toilets manufactured today must use no more than 1.6 gpf and urinals no more than 1.0 gpf. Toilets are either gravity-assisted (relying on gravity to flush the bowl) or pressure-assisted (relying on gravity and a pressurized chamber to flush the bowl).

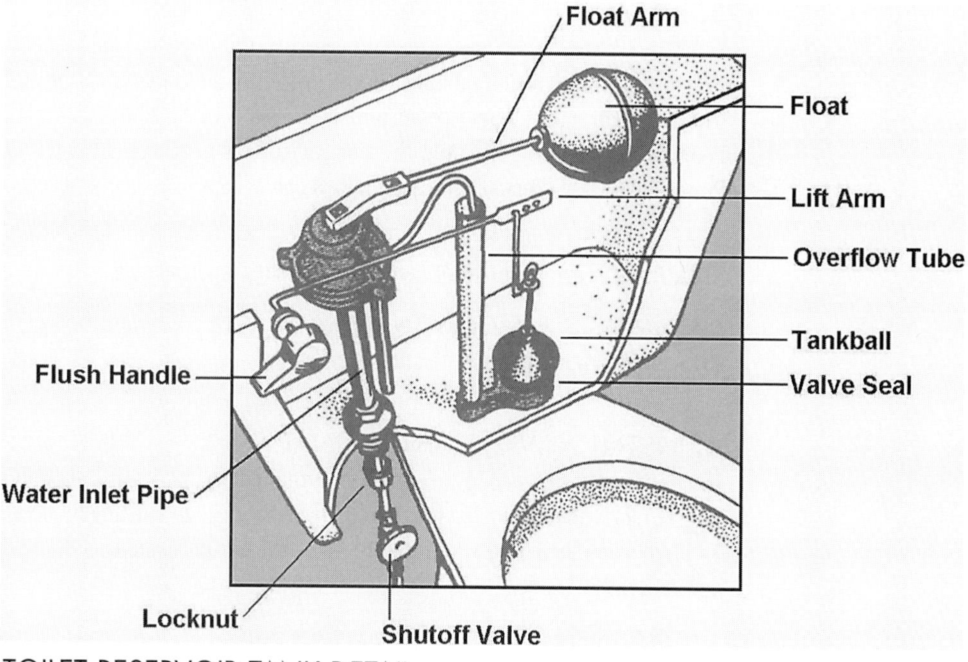

FIGURE 6.7.1 TOILET RESERVOIR TANK DETAIL

TECHNIQUES, MATERIALS, TOOLS

1. INSTALL WATER-LIMITING DEVICES.

A variety of devices can be installed in existing toilets that use 3.5 gpf to conserve water. Resource Conservation offers stainless-steel and vinyl panels that are placed in the tank to dam off a portion to limit the flow of water into the tank; the toilet flushes normally with less water. When the toilet is flushed, water directed into the overflow pipe refills the toilet bowl. In many toilets, water will continue to run into the bowl while the tank fills, although the bowl is full. Water then drains out of the bowl. The AquaSaver from The Fuller Group, and a similar device from Niagra Conservation Group, fills the tank faster, and the bowl slower, to reduce water drainage through a full bowl.

ADVANTAGES: These devices are inexpensive, easy to install, and a short-term alternative to low-flow toilets.

DISADVANTAGES: These devices cannot be used in all toilets and do not compare to the long-term water savings of low-flow toilets.

2. INSTALL A GRAVITY-ASSISTED TOILET.

In gravity-assisted toilets, the pressure of the water rushing down the drain creates a vacuum or siphon effect that draws waste down with it. Clogs and backups can occur when there is not enough force to push waste through the 4"-to-6" diameter pipe found in many older homes. The interior surface of cast-iron pipes, found in many older homes, is rougher than plastic pipe. The vent, pressure level, piping size, bacteria buildup, or reverse pitches in the drain should be evaluated before installation. Vacuum-assisted technology was developed by Fluidmaster and is used in the Vacuity model manufactured by Briggs Industries. Two plastic tanks within the toilet tank hold 1.6 gallons; a vacuum is created when the tank is flushed that forces water into the bowl.

ADVANTAGES: Low-end to high-end models are available, and there is a wide choice of designs.

DISADVANTAGES: There is a frequent need to double flush; streaks on the bowl and clogs are common.

3. INSTALL A PRESSURE-ASSISTED TOILET.

Pressure-assisted toilets use waterline pressure to increase flush velocity. Water is stored in a small pressure chamber inside the tank, and a pocket of air releases water at 25 psi into the bowl. Toilets with a Sloan Flushmate pressure-assisted tank include the American Standard Cadet, Crane Economiser, Eljer Aqua-Saver, Gerber Ultra Flush, Mansfield Quantum, and Universal Rundle Powerflush. Only specially designed toilets can accept pressure-assisted units. The technology for an adapter unit to convert gravity-assisted toilets to pressure-assisted fixtures has yet to be developed.

ADVANTAGES: The problem of incomplete flushes is eliminated.

DISADVANTAGES: Pressure-assisted toilets are more costly and noisy. They are available in limited styles and from limited manufacturers.

4. INSTALL A WALL-HUNG URINAL.

To supplement an existing toilet, a wall-mounted urinal, such as a compact unit from MisterMiser (Fig. 6.7.2), can be installed between the wall studs in an existing bathroom. The urinal is rinsed with 10 oz of water, activated when the lid is closed, which is far less than the 1.6 gpf needed to flush a new toilet.

ADVANTAGES: Cleanup, water, and space are saved.

DISADVANTAGES: The cost is comparable with that of a high-end toilet.

FIGURE 6.7.2 WALL-HUNG URINAL RETROFIT INTO WALL IN CLOSED AND OPEN POSITION

MAXIMIZE ACCESS AND FUNCTION

ESSENTIAL KNOWLEDGE

A clear floor space of 48" x 48" is recommended in front of toilets and bidets, and a 15" minimum centerline clearance to the toilet or other fixtures should be maintained. From the fixture to a sidewall, at least 18" should be maintained. A rehab project is an opportunity to apply universal design principles to improve access and flexibility.

TECHNIQUES, MATERIALS, TOOLS

Ideas to improve access to the toilet or bidet area:

1. INSTALL SPECIAL SEAT TO INCREASE SEAT HEIGHT.

Standard toilets have a height of 15" to the top of the seat; standing from a low seated position can be difficult for people with mobility impairments, back restraints, reduced strength, or joint conditions. Increasing the toilet seat height with special, thick seats or spacers that fit between the rim of the bowl and the seat can solve this problem.

ADVANTAGES: These seats provide flexibility as needs change.

DISADVANTAGES: Special seats have an institutional look.

2. INCREASE HEIGHT OF TOILET SEAT.

A raised base will elevate the toilet seat to a desirable 18". The base can be constructed of 2x4s. The existing drain line must be extended up to meet the fixture, and existing flooring material can be matched to cover the base (Fig. 6.7.3).

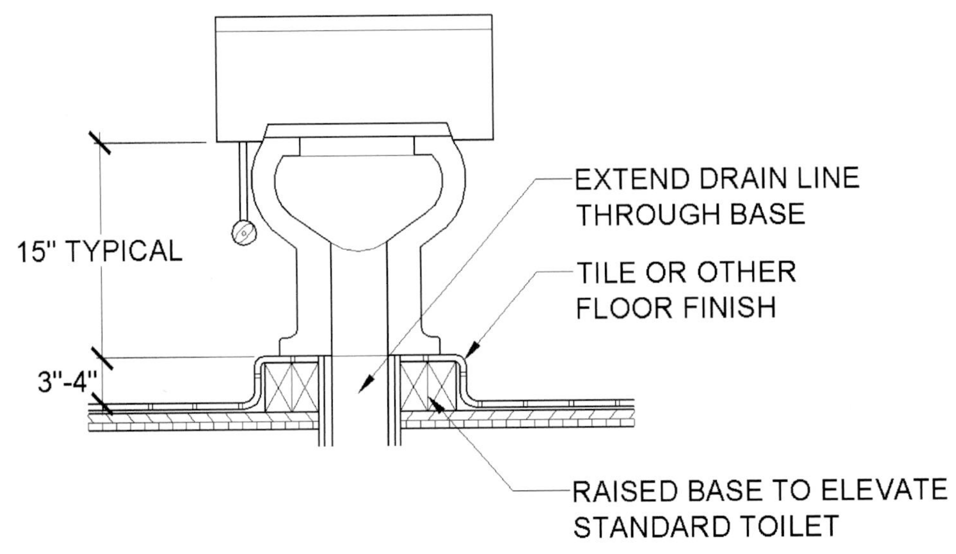

FIGURE 6.7.3 ELEVATED TOILET TO IMPROVE ACCESS

ADVANTAGES: Comfort and access are improved.

DISADVANTAGES: The elevated toilet seat may not be easy to access by small children.

3. INSTALL BLOCKING FOR GRAB BAR.

Grab bars improve safety and stability; blocking should be installed on the walls surrounding the toilet and bidet so grab bars can be positioned where they will be most effective (Fig. 6.7.4). A horizontal grab bar at least 42" long should be placed on at least one sidewall at a height 33" to 36" above the floor. An optional grab bar may be located on the rear wall. In rehab work, with finished walls in place, grab bars can be added only where there are studs or if there is known blocking. When these options are not clear or not in the right location, a solid wood member (2x4 or 2x6) can be attached to the studs and the grab bars attached to it. Another solution, relatively new to the market, is the WingIt Grab Bar Fastening System from Pinnacle, which can be installed into substrates such as drywall to fasten grab bars to the wall (Fig. 6.6.6).

ADVANTAGES: Access is improved when needed. The risk of slips and falling is reduced.

DISADVANTAGES: Future needs may change.

4. INSTALL WALL-HUNG TOILETS AT DESIRED HEIGHT.

Concealed tank wall-hung toilets are mounted on the wall with the tank concealed in the wall. Pipes are not exposed, allowing for fast and easy cleaning. Wall-hung units can be installed on an existing wall, which must be at least 6" in depth. The flush actuator and access panel are horizontally mounted on a 6"-wide ledge built in front of the existing wall. The height of the seat can be determined at installation.

ADVANTAGES: Insulation in the wall around the tank results in a quieter flush.

DISADVANTAGES: Access to the tank parts through a narrow access panel may be difficult.

5. INSTALL TOILET-BIDET COMBINATION.

While considerable space, materials, and labor may be required to add a bidet to an existing bathroom, it is possible to add a personal hygiene system within the existing toilet. Because a common

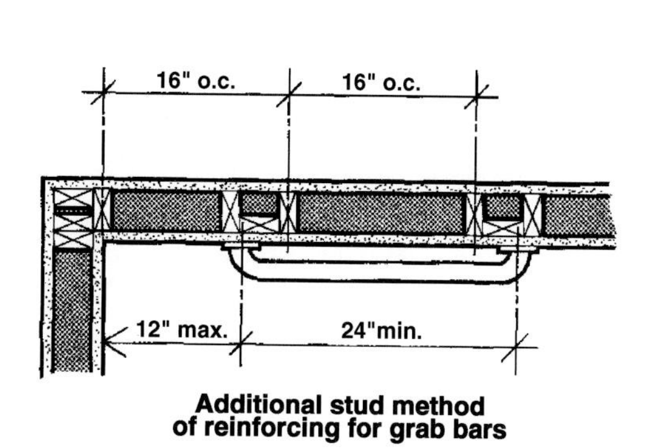

Additional stud method of reinforcing for grab bars

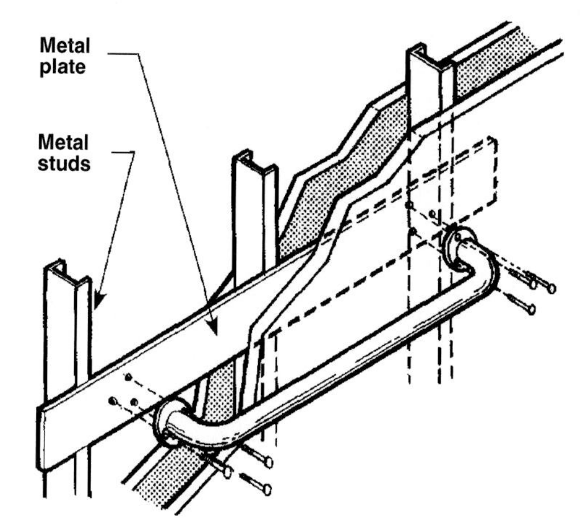

Reinforcing at metal studs

FIGURE 6.7.4 GRAB BAR AND BLOCKING DETAILS

challenge to an aging person is personal hygiene, this can be a valuable addition. Whether a separate fixture or an addition to the toilet, this system should be considered in every rehab bathroom project. The Geberit and Toto combination toilet and bidet converts to a bidet when a spray button is pressed. The rinsing spray arm retracts into a protected sleeve in the bowl after the spray button is released. Argenta and Lubidet offer an add-on bidet system for existing toilets.

ADVANTAGES: This one fixture provides the functions of toilet and a bidet.

DISADVANTAGES: The air purifier and water heater in this fixture add to the energy use.

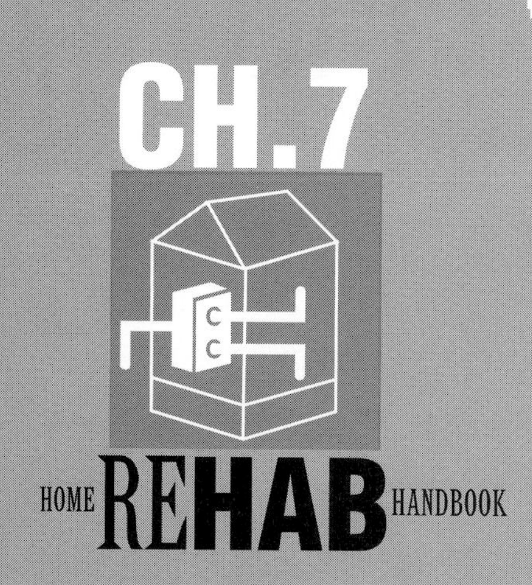

1 IN.

ELECTRICAL AND ELECTRONIC SYSTEMS

Chapter 7
ELECTRICAL AND ELECTRONIC SYSTEMS

7.1. ELECTRICAL AND ELECTRONIC SYSTEMS OVERVIEW
7.2. SERVICE PANELS
7.3. WIRING AND RECEPTACLES
7.4. LIGHTING AND CONTROLS
7.5. ELECTRIC BASEBOARD HEATING
7.6. PHONE, COMPUTER, AND TV CABLING
7.7. SECURITY SYSTEMS

7.1 ELECTRICAL AND ELECTRONIC SYSTEMS OVERVIEW

ESSENTIAL KNOWLEDGE

Electrical systems that distribute power within a house must be safe, reliable, and efficient in the utilization of power. Defects in the performance of the system can result in damage to sensitive electronic equipment, electrical shocks, or fire. Recent studies by the Consumer Product Safety Commission reveal that fires of an electrical origin damage more than 400 homes a day, sometimes causing injury or death. Electronic detection devices such as carbon monoxide and smoke detectors are also important parts of the electrical system and must be reliable to warn house occupants of hazardous conditions.

This guide will provide an overview and a reference resource for information about electrical and electronic systems, subsystems, and materials; a review of current theory in terms of performance of these systems; a discussion of new materials, techniques, or components that have recently been improved or that represent totally new product lines; and state-of-the-art practice and new standards in cabling for telephones, television, and computers. Because it is primarily about rehabilitation, this chapter will cover retrofitting electrical systems to protect against hazardous situations. It will address problems with receptacles and controls to wiring and light fixtures to security and detection devices.

In comparison with other house systems, such as framing or plumbing, the electrical system is relatively new. From their infancy in homes a little over 100 years ago, electrical systems and materials have gone through significant transformations, all of which have contributed to improved safety. Most new product developments today are in low-voltage wiring for electronics, communications, and control systems.

Because the electrical system in a home is old does not mean it needs to be replaced. Most older electrical systems or components of the system are "grandfathered" in, permitted to remain in place and continue to operate, unless the building inspector believes them to be unsafe. In remodeling, the general rule of thumb is: If it is touched, it must be brought up to code.

Before beginning any electrical work, consult the *National Electric Code* (NEC), local building codes, and the local building code official. The NEC, first developed in 1897, is a model set of electrical safety requirements published by the National Fire Protection Association (NFPA) for building and insurance inspectors and electrical contractors. It has no legal standing of its own. Most towns or jurisdictions simply adopt it, and are free to interpret or enforce it as they see fit. Some localities also add their own electrical code requirements.

One of the requirements often found in the NEC is that appliances or equipment on an electrical circuit be *listed.* This means that the device has been certified by an independent testing laboratory. The function of an independent testing laboratory, such as Underwriters Laboratory (UL), is to perform tests on a product to make sure that it fulfills the manufacturer's claims and that it is safe. Without certification, an installation may not pass the electrical inspection. In this case the field inspection service of a Nationally Recognized Testing Laboratory (NRTL) can perform a single unit investigation and certify the product.

7.2 SERVICE PANELS

ESSENTIAL KNOWLEDGE

A residential service panel serves two functions. It is a master switch that can cut off all of the power in the house, and it divides utility-supplied electrical power into branch circuits, which safely distribute power throughout a house. The wiring of each branch circuit is protected by a fuse or circuit breaker, which cuts off the power when a circuit is overloaded. Without adequate protection, overloaded wires heat up. This damages the wire's insulation and may eventually lead to a fire. The NEC lists the specific fuse or breaker size and the wire gauge that it is meant to protect. For example, 12- and 14-gauge wire, which are commonly found in residences, require overcurrent protection of 20 and 15 amperes (A), respectively.

There are several causes of circuit overloads. They may result from short circuits, ground faults, or appliances drawing more current than the rating of the fuse or circuit breaker (Figs.7.2.1 and 7.2.2). A direct short circuit occurs when the hot and neutral wires are either directly or indirectly touching. A nail driven into the wires would cause this type of short circuit. A ground fault occurs when a hot wire touches a ground wire. An example is if the hot wire accidentally touches the grounded frame of a tool or appliance.

A fuse uses a fusible link to protect the circuit. When overloaded, the fusible link melts, opening up the circuit. There are two types of fuses: cartridge fuses and plug fuses. Cartridge fuses, which are no longer common in residences, can still be found in older homes. Circuit breakers use a two-part system to protect the circuit. For mild overloads, a bimetal strip heats up, bends backwards, and eventually trips the breaker. For severe short circuits an electromagnet helps bend the strip faster, providing an almost instantaneous response to the open circuit. In general, the higher the current, the faster the breaker trips.

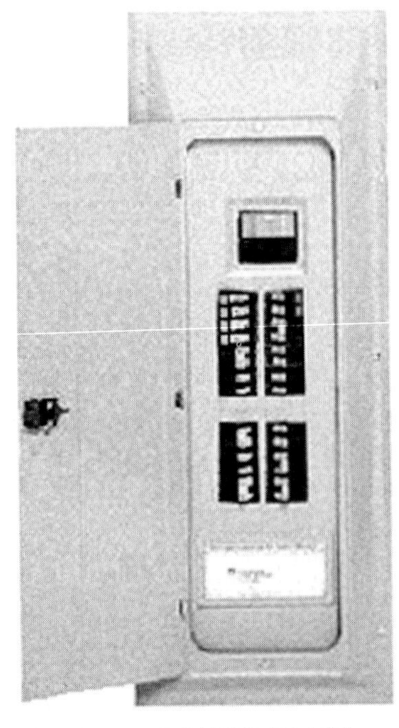

FIGURE 7.2.1 CIRCUIT BREAKER PANEL

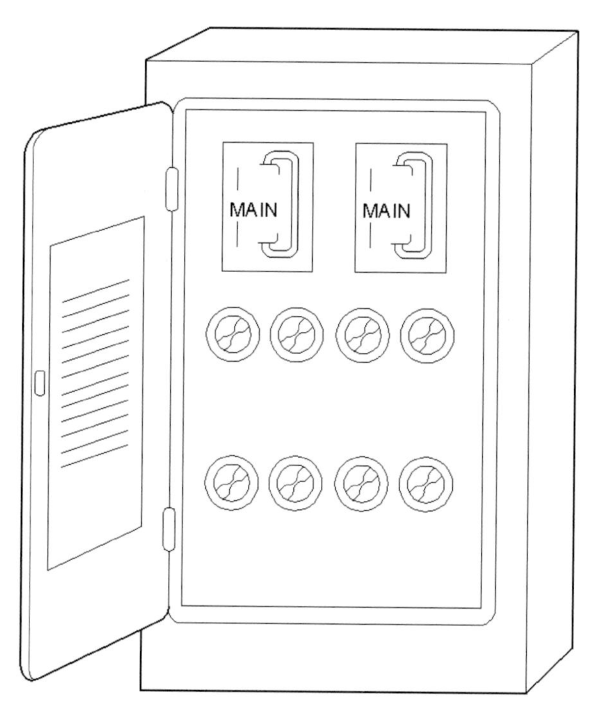

FIGURE 7.2.2 FUSE BOX

Whereas standard circuit breakers and fuses are meant to protect the wiring of a house, the ground fault circuit interrupter (GFCI) is meant to protect life. There are two types of GFCIs available: circuit breakers and receptacles. The circuit breaker type (Fig. 7.2.3) was first introduced around 1968, which coincides with the year they were required by the NEC for underwater pool lights. Since then, the NEC has gradually required their use in more and more locations. Today, they are generally required in kitchens, bathrooms, garages, outdoors, and unfinished basements or crawl spaces (Fig. 7.2.4). A GFCI works by monitoring the current going to the load and comparing it with the current returning. If there is a difference between the two (up to four to six thousandths of an amp), current must be leaking out and the GFCI will open the circuit. Therefore, if this leak is to the ground through a person holding a tool or appliance, the GFCI will open the circuit in between $\frac{1}{25}$ and $\frac{1}{30}$ of a second. A person may still receive a shock, but it will last less than $\frac{1}{30}$ of a second.

Until the late-1950s, fuse boxes dominated the residential service panel market. Circuit breaker panels were introduced around 1951 and became more popular, and by the 1960s circuit breaker panels surpassed fuse boxes in market share. Today, new residential fuse panels are difficult if not impossible to find.

Service panels of older homes are often overloaded and require a service upgrade. Because of the increasing number of electrical devices used in today's households, electrical demands have steadily increased. In the early 1900s houses were usually equipped with 15-A and 20-A services. As electrical appliances became household necessities in the 1930s and 1940s, 40-A and 60-A service became common. Today 100-A service is the minimum required by code for new construction. Generally, if the existing service in a home is below 100 A and additional circuits are required, the service will have to be upgraded. This is especially true if the service to the house and panel is 120 volts (V) which cannot support an appliance requiring a 240-V line, such as a clothes dryer.

The service disconnect indicates the size of the existing service. If there is no service disconnect, a licensed electrician will have to determine the service by verifying the size of the service entrance wire and the rating of the panel itself. Signs that may indicate an overloaded service panel include a frequently failing main fuse or circuit breaker, no room left in the service panel for additional circuits, dual or half-size breakers installed where they shouldn't be, and two or more hot wires (each representing a

FIGURE 7.2.3 GFCI CIRCUIT BREAKER

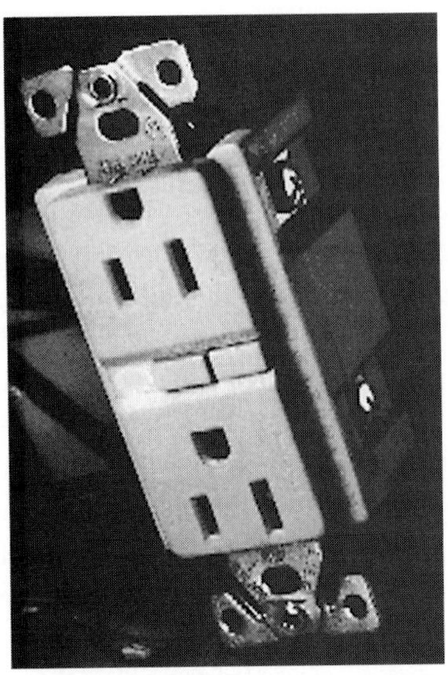

FIGURE 7.2.4 GFCI RECEPTICLE

circuit) connected to one fuse or circuit breaker. Fuses that exceed the maximum amperage allowed (15-A fuses replaced with 30-A fuses) or lights dimming when such equipment as the furnace motor or refrigerator compressor turns on may also indicate an overloaded service panel or circuit. However, the best way to determine whether the existing service panel or circuit is overloaded is to calculate the electrical loads according to NEC guidelines.

Loose connections and corrosion are other common problems found in service panels that may require panel replacement. Corrosion can be a result of rainwater entering the service panel by traveling along the service entrance cable (Fig. 7.2.5), or from the service panel being in a damp location. A drip loop or conduit weatherhead, along with proper sealants, will stop the former problem. The latter may require eliminating the source of moisture or relocating the service panel. Corrosion negatively affects the performance of fuses and circuit breakers. Corroded fuse connections allow only partial voltage to be available to a circuit. Corroded circuit breakers create a fire

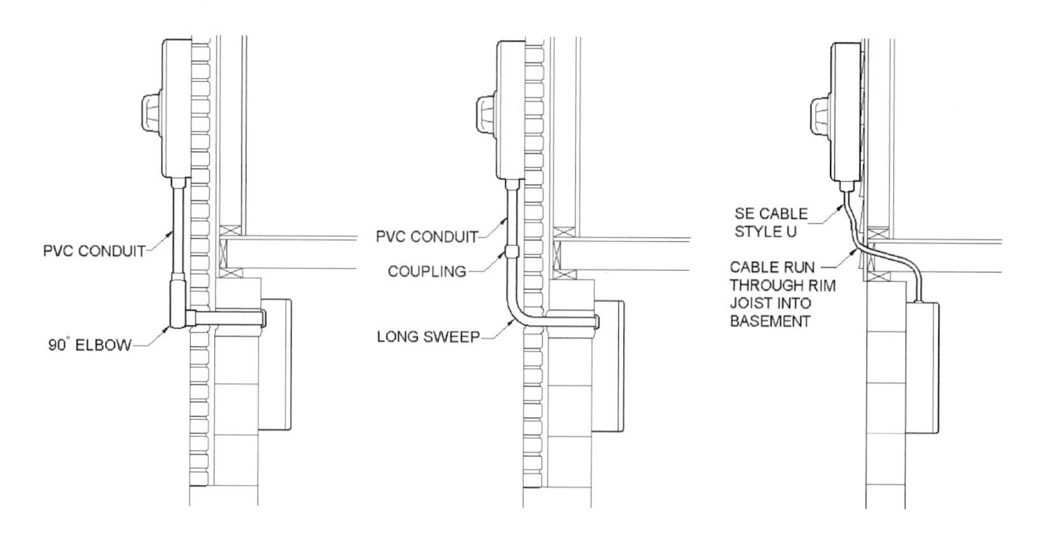

FIGURE 7.2.5 SERVICE ENTRANCE CONFIGURATIONS

hazard by increasing the amount of current required to trip a breaker. Loose fuse and circuit breaker connections also create a fire hazard by causing arcing, burning, and overheating of the overcurrent protection device, and sometimes the entire service panel. The smell of burnt insulation, a deformed bus bar, or the presence of heat can alert you to the presence of loose connections in the service panel.

TECHNIQUES, MATERIALS, TOOLS

1. INSTALL NEW PANEL WITH GREATER CAPACITY.

Installing a new service panel of greater amperage is the best, if not the only, practical solution available when the existing service panel is overloaded and the service needs to be increased. Where the existing panel is not overloaded but needs replacement, a larger panel should be considered, especially if future plans may require a service upgrade. To upgrade the power, an electrical construction permit will be required and the power company will have to be notified. The power company will remove the old meter before the old distribution panel is taken out and will restore power after the new panel is installed and has been inspected. In addition to a new service panel, the electric meter, weatherhead or underground connection, and the cable, conduit, and wire that link these service components will also have to be upgraded.

ADVANTAGES: Ample power for present and future needs will be provided.

DISADVANTAGES: This is a costly solution because of a combination of labor, materials, and permitting fees.

2. INSTALL A HOME AUTOMATION-READY SERVICE PANEL.

When a service panel needs replacement, a panel with provisions for a home automation system (Fig. 7.2.6) can be installed. Such systems centralize the control of many electrical elements, including security, lighting, and communication. They also allow the homeowner to take advantage of incentives provided by the energy supplier by monitoring and controlling energy use. For example, electric hot water heaters and air conditioners can be shut off when no one is home. Home automation systems start at around $2,000. For more information contact Cutler-Hammer about its Advanced Power Center™ Systems.

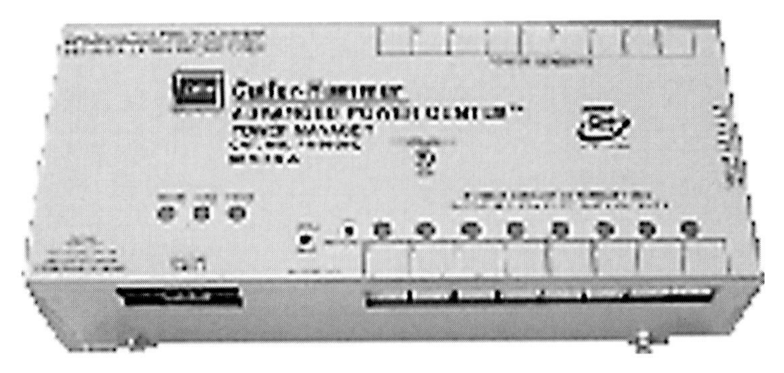

FIGURE 7.2.6 HOME AUTOMATION PANEL

ADVANTAGES: A home automation-ready service panel provides future flexibility at only a slightly higher cost than a standard load center.

DISADVANTAGES: It is questionable whether real economic benefits now exist for installing a home automation system.

3. INSTALL A NEW PANEL WITH SAME CAPACITY BUT MORE CIRCUITS.

Generally speaking, residential service panels are available with anywhere from 8 to 40 circuits. Often, service panels will seem fully loaded because all the circuits are full. This is not always the case. If, after calculating the house loads, there is still electrical capacity available, it may be possible to replace the existing service panel with a new panel containing more circuits but with the same rating.

ADVANTAGES: Installing this type of panel is less costly than installing a higher-rated service panel because less labor and fewer materials are required. A construction permit may not be required because the work is considered a repair.

DISADVANTAGES: This is an option only if the existing service panel has additional capacity. This solution may not provide adequate capacity for increased electrical demands in the future.

4. INSTALL A SUBPANEL.

This is usually used when bringing electricity to a remote location on a house or property. It can also be used to add more circuits to a service panel (Fig. 7.2.7). For instance, because fuse panels rarely have room for additional circuits, more circuits can be added by placing a subpanel next to the main fuse box. Before adding a subpanel, loads must be calculated to make sure that the main panel will not be overloaded by the additional circuits added. A typical subpanel is fed from the main panel with a service entrance style "R" (SER) or round cable and contains breaker positions for two to six circuits. The installation of a subpanel may require approval by an electrical inspector prior to the feeder being connected to the main panel.

ADVANTAGES: For remote wiring applications, it is easier to connect new circuits to a subpanel than to run them back to the main distribution panel. When additional circuits are required, it is also easier to install a subpanel next to the main panel than to install a new service panel with more circuits, but the same rating.

DISADVANTAGES: SER cable is costly and may outweigh the cost of labor to run multiple circuit branch wires to the main service panel. Additionally, SER cable may be hard to find. This option is not available if the service panel is already at full electrical capacity.

5. INSTALL A DUAL OR HALF-SIZED BREAKER.

In older homes, it is often necessary to add additional outlets in rooms such as the kitchen for convenience and safety. As a result an additional circuit may be needed to accommodate these new outlets.

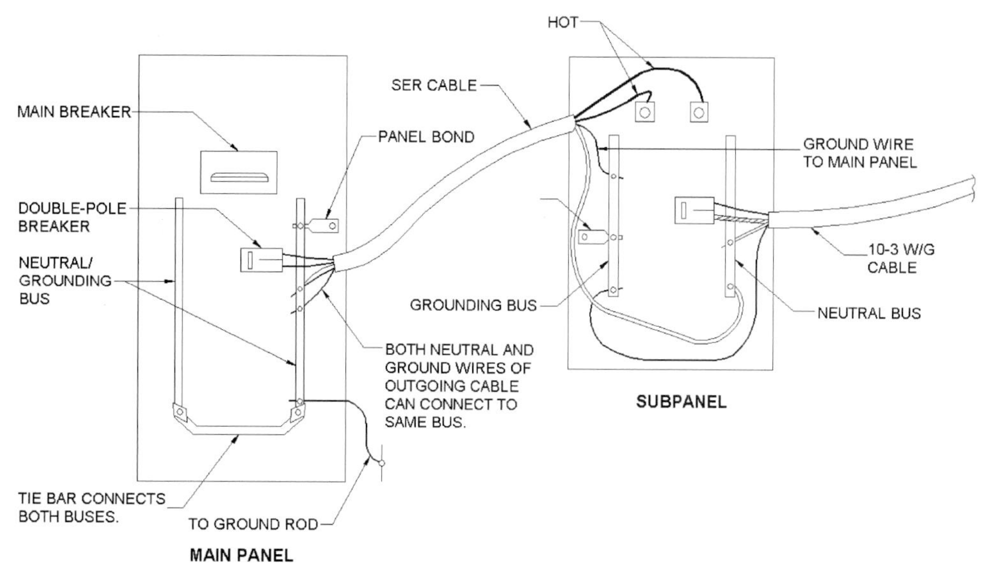

FIGURE 7.2.7 INSTALLING A SUBPANEL

If there is no room in the service panel for additional circuits, a dual or two half-sized breakers may be used to replace a full-sized single-pole breaker (Fig. 7.2.8). This is only possible if the existing service panel is designed to accommodate dual or half-sized breakers. Altering the panel to accommodate dual and half-sized breakers where they are not supposed to be is not only a code violation, but also a fire hazard.

ADVANTAGES: This is the easiest and least costly way to add additional circuits to a service panel.

DISADVANTAGES: Not all service panels are designed to accommodate dual and half-sized breakers. Because of their small size these breakers can be difficult to install. This option is not available if the service panel is already at full electrical capacity.

6. INSTALL TYPE S ADAPTERS IN EXISTING FUSE SOCKETS.

Although the Code does not require Type S fuses to be used for existing fuse boxes, they should be considered. All Edison-base fuses up to 30 A are interchangeable. Therefore, nothing prevents a homeowner from substituting a 30-A fuse for the 15-A fuse that should be used. This defeats the purpose of the fuse and presents a severe fire hazard. To prevent overfusing, Type S fuses were developed. These fuses can only be used with adapters that are screwed into an ordinary Edison-base fuse holder. The 15-A adapter accepts only a 15-A fuse or smaller. Once these adapters are installed, they cannot be removed without damaging the fuse holder.

ADVANTAGES: The use of Type S fuse adapters makes it nearly impossible to overfuse a circuit, and thus significantly reduces the risk of fire.

DISADVANTAGES: Proper installation of Type S fuses requires that they be turned firmly to flatten out the spring under the shoulder of the fuses to make proper contact.

7. INSTALL GROUND FAULT CIRCUIT BREAKER.

Replacing circuit breakers in good working order with GFCI circuit breakers is not required by most codes. However, because of the protection they provide, they should be installed in locations specified by the NEC. There are electrical devices that should not be protected by GFCIs, such as lighting, because a tripped circuit would leave a room dark. Freezers, refrigerators, sump pumps, and medical equipment are others. Even though the cost of a GFCI circuit breaker is about seven times the cost of a standard breaker, most people would agree that this is a small price to pay for a device that may save a life.

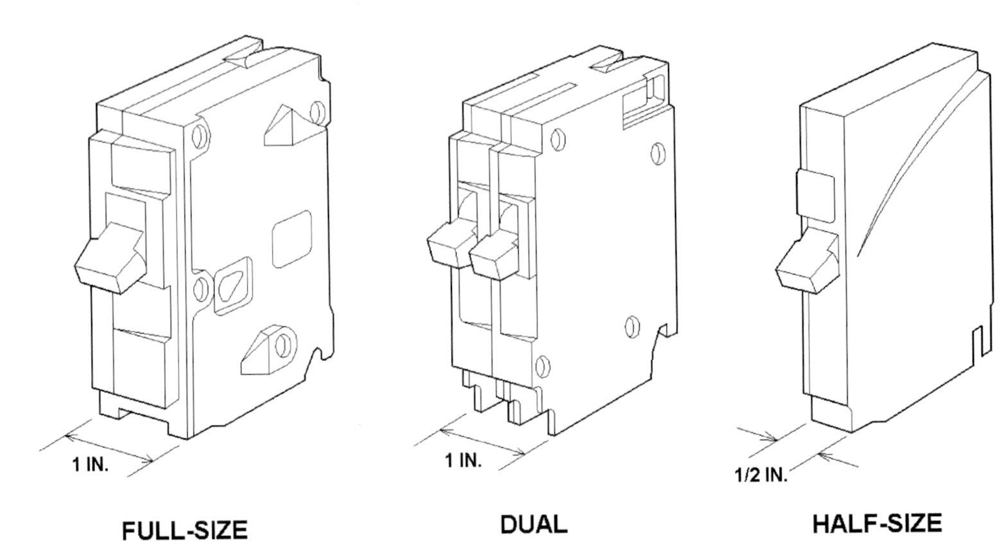

FIGURE 7.2.8 SINGLE-POLE CIRCUIT BREAKERS

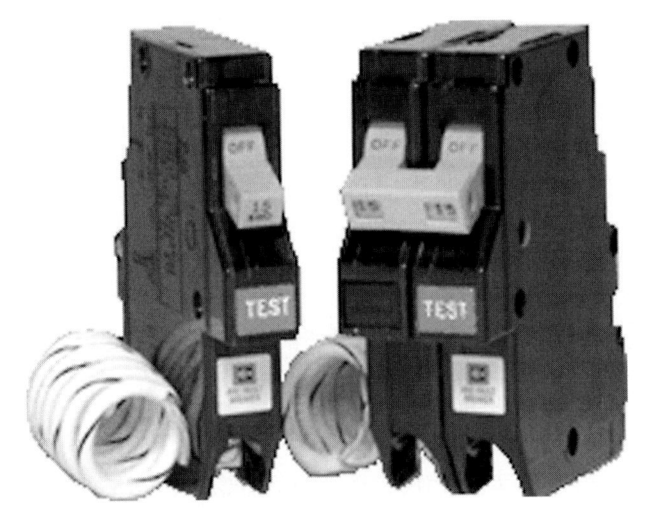

FIGURE 7.2.9 AFCI CIRCUIT BREAKER

ADVANTAGES: A GFCI circuit breaker will protect all receptacles on a circuit. For outdoor circuits, a circuit breaker–type GFCI will outlast a receptacle-type, which is affected by moisture.

DISADVANTAGES: Circuit breaker–type GFCIs are more costly than receptacle-type GFCIs. A circuit breaker–type GFCI is less convenient than a receptacle-type because one has to go back to the main panel to reset the breaker, whereas the receptacle-type is reset at the receptacle itself. The circuit breaker-type GFCI cannot be used in old wiring systems that use shared neutrals; a receptacle-type GFCI should be used.

8. INSTALL AN ARC FAULT CIRCUIT INTERRUPTER (AFCI).

An AFCI is a newly introduced circuit breaker (Fig. 7.2.9) that detects electrical arcing by monitoring the waveform of the voltage. Arcing generates high-intensity heat and expels burning particles that can easily ignite combustible materials. Arc faults occur when electrical products or wires are damaged, aged, or improperly used. An example is an extension cord that is repeatedly closed in a door, or a wire punctured by a nail or screw. Beginning in 2001, the NEC required that all bedrooms be protected with AFCIs.

ADVANTAGES: AFCIs provide additional protection against the risk of fire. They provide great protection in old house wiring where the condition of the wiring insulation is uncertain.

DISADVANTAGES: An AFCI is approximately eight times the cost of a standard breaker. It cannot be used in old wiring systems that use shared neutrals (see Section 7.3.4).

7.3 WIRING AND RECEPTACLES

WIRING OVERVIEW

ESSENTIAL KNOWLEDGE

Wire provides the means by which electrical current is conveyed from the point of generation to the point of use. Wire types include transmission wires, service wires, branch circuit wiring, extension cords, and appliance power cords. This chapter focuses on the rehabilitation of residential branch circuit wiring.

Wire is sized according to the amperage it is designed to carry. Codes specify the maximum current-carrying capacity that is safe for different size wires. Wire diameter sizes are measured using the American Wire Gauge (AWG) system. In this system the bigger the number, the smaller the wire diameter. For example, No. 14 wire, which is commonly used for general service wiring in a house, is larger than No. 16. A problem in older houses is determining the size of existing wiring. Although the Code requires that wire sizes be continually marked on the wire insulation, it is often illegible. If this is the case, a wire gauge or wire samples should be used to determine the wire size.

Both sizes and composition of wiring systems have changed over the years. Knob-and-tube was the first widely used electrical system, prevalent from the 1890s to 1920. Knob-and-tube is a two-wire system in which nonconducting porcelain knobs hold wire 1" off the surface of studs and joists; porcelain tubes protect wire where it penetrates framing or crosses other wire. Electrical connections were not required to be made in an electrical enclosure. One wire was wrapped around the other and then the joint was soldered and taped. Wires that terminated in outlets and switches were protected with a tube of woven fabric called loom. This system contained no ground, and the outlets were not polarized.

Less expensive than knob-and-tube, raceways made of wooden molding were used from about 1900 to the 1930s when they became illegal. In this system, wood moldings carried two or three wires in a grooved strip that was attached to the wall, and covered with a cap. This system had two major flaws: Wood is quite combustible and homeowners mistaking the molding for picture molding would drive nails through it and the wires. If wooden molding raceways are encountered in rehab work, they should be removed. Metal raceways, once considered to be moldings, were first recognized by the NEC in 1907 and are still used today.

The development of Greenfield—flexible-steel conduit for holding wires—in the late 1890s made it easier to snake conduit into walls and floors of houses. The flexibility was accomplished by wrapping galvanized steel strips into a tube (Fig. 7.3.1). The next development was armored cable or Type AC (also called by its tradename *BX*) where wires were bound in a continuous spiral of galvanized metal. Although it was recognized by the NEC in 1899, Type AC use did not become widespread until 1920, and it did not achieve major popularity until the late 1920s. Early armored cable contained no ground; the armor itself was used as a ground. Safety was significantly improved in 1959 when the NEC required that all armored cable contain a slender aluminum bonding strip as a ground. Moisture plagued all these wiring products. In damp locations, rust would deteriorate the insulation and eventually lead to a ground fault. This problem was solved with the introduction of vinyl insulation after 1940.

Nonmetallic sheathed cable (Fig. 7.3.1) or Type NM was first recognized by the NEC in 1926. This cable consisted of two rubber insulated conductors sheathed in cloth. In 1928, the NEC required that ground conductors be installed in NM cable; however, in some jurisdictions ungrounded NM continued to be used until the mid-1960s. After World War II the conductor insulation changed from rubber to plastic; the transition from cloth to plastic sheathing was completed in the early-1960s.

The most common problems found in residential electrical systems are loose or broken connections and deteriorated insulation. Usually loose or broken connections occur in electrical enclosures and can easily be repaired. Knob-and-tube is the exception to this because wire splices occur inside walls. Deteriorated insulation, which is found in many older electrical systems, may not be as easily repaired. Deteriorating insulation has several causes, including the age and type of the insulation used, and its exposure to air, heat, or moisture. Rubber, the standard wiring insulation before 1930, has a life expectancy of about 25 years. Because rubber deteriorates rapidly in open air, most of this exposed wiring will be extremely brittle and potentially hazardous. In contrast, insulation on wiring in conduit or armored cable may only be cracking where it leaves the enclosure to make connections to switches or outlets. Exposure to the excessive heat from overloaded wires, or a wall- or ceiling-mounted light fixture, can cook wiring insulation, making it brittle. Wires that are overloaded will also stress the conductor, making it more likely to break. Old wiring insulation exposed to moisture should be replaced.

TECHNIQUES, MATERIALS, TOOLS

1. REWIRE THE ELECTRICAL SYSTEM.

Rewiring a house or specific circuit is the safest way to solve existing wiring problems such as deteriorated insulation, stressed conductors, and overloaded circuits. However, it may not always be necessary. The trick to rewiring is to get wires to where they are needed while disturbing the walls and ceilings as little as possible. In general, to get wires from one location to another, a fish tape is used to pull wires through wall and ceiling cavities. For tips on rewiring, refer to Rewiring Old Homes on the *Journal of Light Construction* Web page: www.jlconline.com/.

ADVANTAGES: Rewiring is the safest solution for wiring that is in poor condition.

DISADVANTAGES: Rewiring is labor intensive and can be very costly.

2. INSTALL WIRING RACEWAYS.

Raceways provide a simple way to rewire or add a circuit (Fig. 7.3.2). In this system, base pieces are attached to walls or floors, wires are laid in the base, protective U-shaped wire clips are set over them at

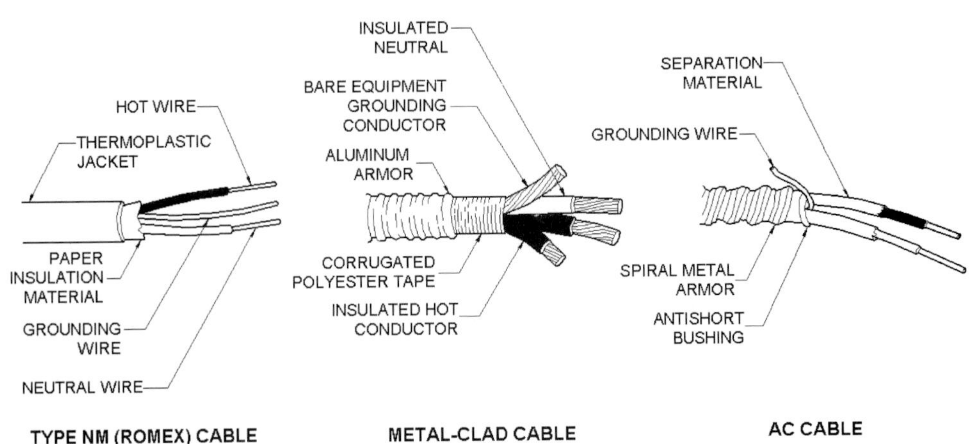

FIGURE 7.3.1 ELECTRICAL CABLE TYPES

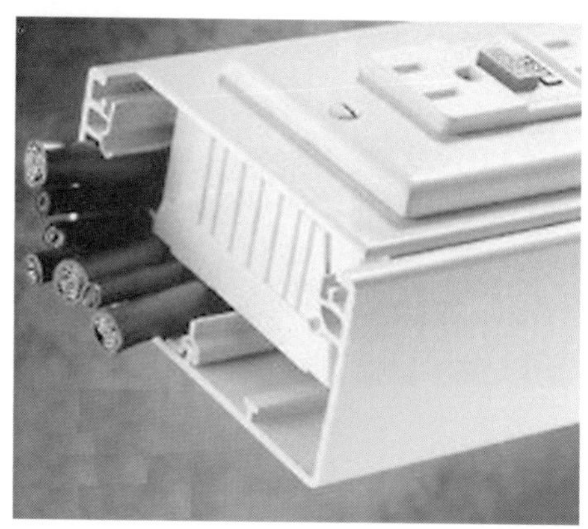

FIGURE 7.3.2 PVC MULTICHANNEL RACEWAYS

12" intervals, and the covers are snapped on (Fig. 7.3.3). They are attached to either the service panel or an outlet with the use of special adapters. Raceways are available in metal or polyvinyl chloride (PVC), the latter of which is available in white or wood laminate versions. Models are available that also encase both electrical and communication wires (Fig. 7.3.4).

ADVANTAGES: This method is significantly less costly than snaking wires through existing wall and ceiling cavities. Wiring is accessible for future changes, and locations of outlets and jacks can be easily changed. Outlets, jacks, and switches of some models are located inside the raceway channel, not in full-sized obtrusive boxes.

DISADVANTAGES: This method may not be acceptable aesthetically.

3. REPAIR INSULATION.

Crumbling insulation may often be repaired with one of several techniques: insulating varnish can be applied over old rubber insulation, heat-shrink tubing can be installed, and bare wires can be wrapped with electrical tape. Where there is enough slack in armored cable, cutting 12" off the armor to expose wire with good insulation is another solution. If there is not enough slack, the cable can be cut back 12", a junction box installed, and a new armored cable spliced in.

ADVANTAGES: Repairing insulation is significantly less costly than rewiring and can add years to the life of the existing electrical system.

DISADVANTAGES: Solutions may not be acceptable to local code officials. This is a temporary fix. Installation of a new junction box may not be aesthetically acceptable.

ALUMINUM WIRING

ESSENTIAL KNOWLEDGE

When the price of copper soared in the early 1960s, manufacturers responded by making residential electrical wires out of aluminum. Between 1962 and 1972, nearly 2 million homes were wired with aluminum, and many of these have not been upgraded. The hazard with aluminum wiring arises from two types of corrosion: (1) the connection of dissimilar metals, and (2) the oxidation of exposed aluminum. Both increase the electrical resistance of the wiring, making it hotter when in use and therefore a fire hazard. Also, because aluminum expands and contracts significantly more from changing temperatures than copper, unless all connections are made very tightly, the wiring may pull loose as a result of the heat of resistance.

FIGURE 7.3.3 COVERED RACEWAYS

FIGURE 7.3.4 ELECTRICAL AND COMMUNICATION WIRING RACEWAY

Aluminum wiring can be identified by the dull gray color of the wire or by "AL" on the sheathing. Receptacles and outlets used with aluminum wiring must be marked "OC/ALR" and are designed to prevent contact between dissimilar materials. Warning signs of problems include warm coverplates, devices that fail to work for no apparent reason, and strange odors or smoke.

TECHNIQUES, MATERIALS, TOOLS

There are three basic ways of correcting aluminum wiring:

1. REPLACE EXISTING WIRING.

The safest way to eliminate the fire hazard potential of aluminum wiring is to remove all of the aluminum wire and install new copper wire. It may be possible to use the existing wire as a guide to bring a new cable through the wall. If not, the wire will have to be snaked. This will be difficult if all wiring is enclosed in finished walls and ceilings.

ADVANTAGES: This fix eliminates the fire hazard potential of aluminum wiring connections in its entirety.

DISADVANTAGES: Rewiring an entire house is costly.

2. SPLICE WIRE USING AMP COPALUM CONNECTORS.

A more practical method for reducing the fire hazard potential of aluminum wiring connections is by splicing or "pigtailing" a short length of copper wire to each aluminum wire using an AMP COPALUM connector and heat-shrink tubing system. It is applied using special tooling and is only available with installation by specially trained electricians. This system is the only method considered by the U.S. Consumer Product Safety Commission (CPSC) to be a permanent repair. Its recommendation is based upon extensive testing.

ADVANTAGES: This repair avoids the expense of rewiring the entire house and therefore is less costly.

DISADVANTAGES: This system is not available in all parts of the country; installation requires a specially trained electrician and specialty tools.

3. SPLICE WIRE USING SCOTCHLOK TWIST-ON CONNECTORS.

Pigtailing with certain types of connectors, although they might be presently listed by UL for the application, can lead to increasing the fire hazard of the connection. Following special installation procedures, the 3M Scotchlok connectors are considered by CPSC to be the best available alternative to the COPALUM crimp. 3M Scotchlok has several features that make it safer than other connectors: a nonflammable shell, a metal shell around the spring, and a heavier spring wire. In this system, the bare aluminum wire is abraded under a coating of nonflammable oxide inhibitor. The connector spring is then filled with an oxide inhibitor before the connector is applied to the pretwisted wires. This work must be done by a qualified electrician.

ADVANTAGES: This repair avoids the expense of rewiring the entire house. Specialty tools are not required, and its availability is not limited to certain parts of the country.

DISADVANTAGES: This system is only as good as the installation technique; installation requires a qualified electrician.

RECEPTACLES

ESSENTIAL KNOWLEDGE

Receptacles (also known as outlets) supply power to portable equipment used in houses, such as floor lamps, radios, and toasters. Up until the mid-1960s, ungrounded receptacles were installed in most houses (Fig. 7.3.5). In older homes, these two-prong receptacles were not polarized. Grounding and polarization are important features of an electrical system. Equipment grounding reduces the shock hazard from electrical boxes or equipment that may become "hot" because of a conductor insulation failure or loose connection. In this system each metallic, noncurrent-carrying part of the electrical system (electrical boxes, equipment frames, appliances, motors, and conduit) is connected to the ground wires of each branch circuit, to the neutral bar of the load center, and then to the earth. If a hot wire accidentally touches the metal housing of an electric drill, the fault current will flow back to the ser-

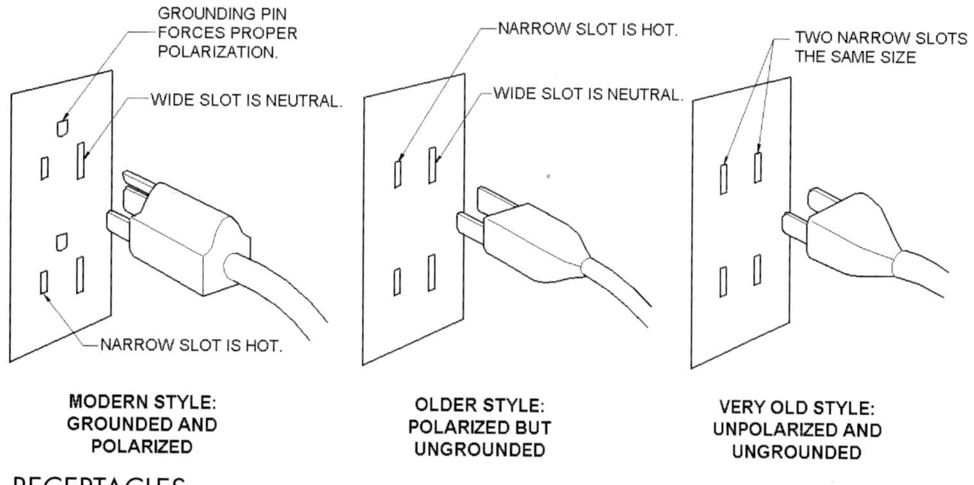

FIGURE 7.3.5 RECEPTACLES

vice panel via the equipment grounding conductor, and the circuit breaker will trip. If this system was not in place, a person could receive a serious shock from a faulty appliance if he or she were not insulated from the ground. Grounded receptacles are also required for the proper functioning of plug-in surge protectors used to protect sensitive electronic equipment. Surge protectors work by directing excess voltages and currents into the grounding system. Therefore, without a high-quality ground in place, the surge protector is useless.

In an electrical circuit, keeping the hot and neutral separate is called *polarization*. It is achieved by making the neutral blade of a plug and the neutral slot of a receptacle wider than their hot counterparts. This ensures that the hot and neutral wires of an ungrounded appliance cannot be reversed. If polarity is accidentally reversed, the exposed socket of a properly wired lamp could deliver a shock even when the switch is off. Reversed polarization does not occur in grounded appliances because the grounding pin forces proper polarization.

In most cases, nonoperable receptacles can easily and inexpensively be repaired. Loose wire connections can be tightened, and failed receptacles can be replaced. The NEC requires that all new receptacles, whether for replacements or new installations, be grounded. This does not mean that existing two-prong receptacles in a house need to be replaced, although this would improve the safety of the existing electrical system and therefore should be considered.

TECHNIQUES, MATERIALS, TOOLS

There are several methods for converting a two-prong outlet into a three-prong, grounded receptacle:

1. REPLACE EXISTING WIRING.

Running new electric wiring with a ground from the service panel to the outlet is the most certain way of providing an uninterrupted ground where the existing wiring system has none (early nonmetallic sheathed cable) or where the continuity of the ground is questionable (early armored cable with no bonding strip). That certainty may be desired on circuits where plug-in surge protectors are used to protect expensive electronic equipment. For rewiring options, refer to earlier sections. This method is only as good as the grounding of the service panel.

ADVANTAGES: This is a safe, reliable way to provide a grounded receptacle.

DISADVANTAGES: Replacing existing wiring is relatively costly. Existing wiring may be in good condition and not worth the cost of removing it to install a new grounded cable.

2. INSTALL A NEW GROUND CONDUCTOR.

Where an uninterrupted ground does not exist, installing a new, adequately protected ground conductor to a reliable ground is the next best thing to installing a new cable. To provide adequate protection, the ground conductor will most likely have to be fished through walls. It cannot be tucked under baseboards or tacked in the same fashion as telephone wires. The ground can be run either back to the service panel, or to the nearest accessible point of the grounding electrode system. This method is only as good as the grounding electrode system.

ADVANTAGES: Installating a new ground conductor is less costly than installing a new electrical cable. If the grounding electrode system is accessible, the ground conductor will not have to be run back to the service panel.

DISADVANTAGES: The ground conductor may have to be run back to the service panel because it is less likely to be disturbed there than if it is connected to another part of the grounding electrode system.

3. CONNECT RECEPTACLE TO THE GROUNDED RECEPTACLE BOX.

Where the existing receptacle box is grounded, a grounded receptacle can easily be installed by connecting a wire jumper from the receptacle's grounding terminal to the back of the box. The absence of a copper grounding conductor does not mean that the box is not grounded. Where armored cable is used, it does not have a separate ground wire because the steel jacket acts as the ground conductor. The addition of a bonding strip in the armored cable, a 1959 NEC requirement, greatly improved this system's continuity to ground. However, it cannot be assumed that the presence of armored cable means that the receptacle box is adequately grounded. Both the box and armor need to be tested with a circuit tester or another electrical device for continuity to ground.

ADVANTAGES: This can be easily and inexpensively accomplished.

DISADVANTAGES: There is a potential for the armor to rust and create a discontinuity, which would result in the false impression of the outlet being grounded when it is not.

4. INSTALL A RECEPTACLE-TYPE GROUND FAULT CIRCUIT INTERRUPTER.

The NEC forbids replacing a two-prong outlet with an ungrounded three-prong outlet unless it is a GFCI type. The reason is that although the GFCI is not grounded itself, it protects users by shutting off the current in the event of a ground fault. It also protects receptacles downstream. Therefore the code allows all downstream receptacles to be replaced with non-GFCI three-prong receptacles if they are labeled with a sticker stating "GFCI Protected." A new GFCI outlet produced by Leviton, called the Smart Lock GFCI, contains two new safety features: (1) the reset button is mechanically prevented from being reset if the GFCI is damaged after a voltage surge, and (2) a diagnostic feature prevents downstream receptacles from operating if the GFCI outlet is wired incorrectly.

ADVANTAGES: A receptacle-type GFCI can be installed easily and is inexpensive.

DISADVANTAGES: A surge protector plugged into this type of circuit is rendered useless. A ground is needed to make the surge suppressors or line filters, which are commonly used to protect computer systems, effective.

7.4 LIGHTING AND CONTROLS

INTERIOR LIGHTING

ESSENTIAL KNOWLEDGE

Lighting provides illumination for safety, security, and the performance of visual tasks. Electric lighting systems are comprised of two parts: the lamp or bulb and the light fixture. About 95% of the lamps used in older residential homes are incandescent; the rest are fluorescent.

In general, an incandescent lamp works by passing electrical current through a filament which heats up to the point where it produces light. The use of electric lighting became affordable and practical when Thomas Edison invented the incandescent lightbulb in 1879. The nature of these early lamps became a critical influence on the design of the fixtures.

For nearly 100 years, prior to Edison's invention, gas was the primary lighting system in the United States. With the advent of the incandescent lightbulb, gas began to lose its hold on the market. From the early 1880s until about 1910, electricity was not the most reliable energy source. As a result, early electric light fixtures were a combination of gas and electric: gas fixtures upgraded with one or two electric arms. Combination fixtures remained in catalogs as late as 1912, but were sharing their pages with all-electric fixtures by this time. After 1913 gas lost its dominance to electricity.

Early electrical fixtures did not have shades because the modest light output was best left unobscured (Fig. 7.4.1). As the quality improved, a new breed of all-electric lighting emerged free of accoutrements of earlier gas light fixtures. Three general illumination systems emerged: direct illumination, indirect illumination, and semidirect illumination.

Although Edison patented a fluorescent lamp in 1896, the first commercial production of fluorescents was not until 1938. In general, a fluorescent lamp works by sending an electric arc down a phosphorus-lined glass tube, energizing the phosphors causing them to emit light. This is accomplished with the help of two additional devices: a ballast and a starter. The ballast limits the current flowing through the lamp to a predetermined value, and the starter performs the preheat function used to light the fluorescent lamp. This preheat circuit is seldom used today, having been replaced by instant-start and rapid-start circuits. The major advantages of fluorescent over incandescent is that fluorescent can produce two to four times as much light, and can last up to 20 times longer. Long-tube fluorescents also have lower surface brightness, producing less reflected glare and shadow. Many improvements have been made to fluorescents since they were first introduced. Today the color rendition of fluorescents can be almost the same as incandescents, and many are dimmable.

The electrical parts of a light fixture have a finite life. As mentioned in Section 7.3, incandescent lights can heat up fixture wires, causing insulation to deteriorate. This can also happen to wires near a fluorescent light's deteriorating ballast if it is not thermally protected. Poor contact between the bulb and the fixture is another problem: Over time and through repeated lamp replacement, contacts can corrode or lose their resilience and cause the fixtures to function poorly or not at all. Fluorescent light fixtures can have malfunctioning starters and ballasts.

Beyond the functioning of fixtures, a broader issue may exist. Existing fixtures may not provide sufficient light for contemporary tastes and needs. Lamps with higher lumen output are not neces-

FIGURE 7.4.1 PAN LIGHT (1910–1930)

sarily the best solution. The fixture may not be designed to handle the additional power and the heat generated by a higher output lamp. A lamp that is too large, too bright, or too frosted will compromise the effect of a period light fixture. For instance, pan lights (flush-mounted ceiling fixtures with bare bulbs) were popular in bedrooms and less formal living rooms from about 1910 to 1930 and were designed for small round globe lights in the 15- to 40-watt (W)range. Installing a larger lamp may actually have a negative effect by creating glare.

TECHNIQUES, MATERIALS, TOOLS

1. REWIRE EXISTING FIXTURE.

If the existing fixture in a home has historical value or is otherwise unique, rewiring may be an option. As the fixture is disassembled, document how all of the parts go together. Replace a failed lamp socket by prying off the shell with a screwdriver. When the new socket is connected, make sure the wire wraps clockwise around the terminals and that there is a cardboard insulator between the socket and the shell. When removing old wiring with cracked or dry insulation, first lubricate the fixture insides with silicone spray; don't force the wiring. Use 18- or 20-gauge lamp cord for replacement.

ADVANTAGES: Repairing an original fixture has more historic value than installing a new fixture, and a matching reproduction may not be available.

DISADVANTAGES: Rewiring an existing fixture requires more labor than replacing the fixture and, therefore, depending upon the price of a new fixture, may be more costly.

2. INSTALL A NEW, PERIOD REPRODUCTION FIXTURE.

Many lighting manufacturers now offer reproductions of historic light fixtures. One resource that lists some of these companies is *Old-House Journal's Restoration Directory.*

ADVANTAGES: Installing a new fixture requires less labor than repairing the existing fixture and therefore, depending upon the fixture price, may be less costly.

DISADVANTAGES: Replacing the fixture may be costly. A new fixture has less historic value.

3. INSTALL ADDITIONAL LIGHT FIXTURES.

Where the existing light fixture does not provide adequate lighting, the number and variety of light sources may be increased. In addition to the conventional incandescent and fluorescent light fixtures, other lighting systems may be considered:

Low-voltage lighting fixtures utilize halogen lamps, which provide a more precise focus of the light, better light quality, and lower operating costs than traditional incandescent lighting. Halogen lamps are 50% brighter per watt than typical lightbulbs and last twice as long. The smaller size of some of these fixtures makes them useful in places where larger incandescent fixtures will not fit, such as under counters. A transformer is required to convert the current from 120-V alternating current (AC) to 12- V direct current (DC). The transformer should be out of view but easily accessible.

Dedicated compact fluorescent light fixtures provide three to four times more light output per watt than traditional incandescent lighting. Although fixtures can be considerably more costly than a comparable incandescent, utility companies often provide rebates offsetting this added cost.

White LED (light emitting diode) lighting consumes less than a quarter of the electricity that fluorescent lighting does, and lasts about 10 times longer. A 1.2-W white LED cluster is as bright as a 20-W incandescent lamp. Light quality is comparable to that of cool white compact fluorescent lights (CFLs). Currently considered expensive, look for their costs to come down as they become more popular and more manufacturers begin to make them. LEDs are available for direct replacement of incandescent lamps, as strip lights for installation under cabinets, and in custom arrays for custom-built down lights, sconces, and surface-mounted fixtures.

ADVANTAGES: In historical applications, wall sconces and brackets will add to the light level without detracting from a historic ceiling fixture. Additional light fixtures can increase the quality of light within a room and reduce glare.

DISADVANTAGES: Certain fixtures can be expensive.

4. INSTALL A DIMMER SWITCH.

To reduce a light fixture's glare, replace the existing light switch with a dimmer switch. A dimmer switch allows one to simulate the low light levels of a historic light fixture or run them at full output as required for contemporary needs. Dimming a light fixture saves electricity and extends the life of incandescent bulbs.

ADVANTAGES: A dimmer switch is relatively easy to install and has a reasonable cost.

DISADVANTAGES: Dimmers reduce the efficiency of incandescent lamps. They can produce an audible buzzing sound, and put radio frequency interference (RFI) into a branch circuit.

EXTERIOR LIGHTING

ESSENTIAL KNOWLEDGE

In its most practical sense, exterior lighting provides illumination for safety and security. It can illuminate a safe path to the entrance of a house, or the grounds around the house so that any prowlers are visible. Some of the same problems encountered in interior light fixtures are encountered outside (see previous section). Additionally, because of their exposure to sunlight and moisture, exterior light fixtures are susceptible to corrosion.

Exterior light levels at an existing home are often insufficient. Walking on an unlit path can be difficult or even hazardous, especially at changes of grade. Too much light can create a blinding glare, which can also make it easier for a burglar to remain in the shadows unseen, defeating the purpose of security lighting. In general, lower levels of light are preferable for both security and safety. Exterior lighting should also provide a gentle transition from darkness outside to brightness inside, and vice versa, which is better for older people who may have poor night vision and poor depth perception.

There are special requirements for running standard 120-V wiring to a new exterior light fixture, such as a lamp post. Conduit is required to protect all wiring above ground and is often necessary underground. It can be made of PVC, thin-wall metal, or heavy-wall (rigid) conduit. PVC and thin-wall metal conduit are easy to work with, but must be buried at least 12" underground. Rigid metal conduit is more costly and harder to work with, but it only needs to be buried 6" underground. If local codes permit, a heavily sheathed nonmetallic cable called UF cable can be used instead of conduit. It also needs to be buried at least 12" underground.

All exterior lamps and light fixtures should be rated for outdoor use. Lamps should have a corrosion-resistant nickel or copper-coated base and a hardened glass bulb. Fixtures should be made of corrosion-resistant materials such as copper, brass, or aluminum and have watertight housings. Plastic fixtures can also be used, but they may degrade from exposure to heat and sunlight. Fluorescent and high-intensity-discharge (HID) lamps require additional current transformers called ballasts, which are integral parts of the fixture and should also be rated for outdoor use.

TECHNIQUES, MATERIALS, TOOLS

1. REWIRE EXISTING FIXTURE.

Refer to Interior Lighting, option 1, at the beginning of this section.

2. INSTALL NEW, PERIOD REPRODUCTION FIXTURE.

Refer to Interior Lighting, option 2, at the beginning of this section.

3. INSTALL A LOW-VOLTAGE LIGHTING SYSTEM.

Where additional lighting is required, low-voltage lighting may be adequate. Low-voltage lighting fixtures are powered by a transformer that lowers the voltage and converts the current from 120-V AC to 12-V DC. Lowering the voltage eliminates the risk of electrical shock, and therefore this system's wiring does not have to be buried or protected by a conduit. These systems are readily available in kits.

ADVANTAGES: This system is easier and less costly to install than standard voltage fixtures; it is easily relocated.

DISADVANTAGES: These systems are limited to runs of about 100', after which the voltage drop will start to affect the light intensity.

4. INSTALL PHOTOVOLTAIC LIGHTING.

Where new lighting needs to be added, a photovoltaic (PV) lighting system can be installed. In this system, solar panels convert sunlight to electricity that is stored in compact, highly efficient batteries for use at night (Fig. 7.4.2). These systems can be used for decorative lighting, path lighting, patio, and security lighting. They are available as self-contained units or as lights grouped together with a wire. The self-contained units must be installed in a sunny location. The wiring for the latter system is low voltage, and therefore does not have to be protected by conduit or direct burial.

ADVANTAGES: PV systems are easy to install and can be inexpensive. No additional power from the service panel is required because PV systems are self-contained.

DISADVANTAGES: The number of hours per night that the lighting will operate is dependent upon the amount of sunlight the system receives, which is a function of geographical location, weather conditions, nearby objects, and time of year.

FIGURE 7.4.2 PHOTOVOLTAIC FLOOD LIGHT

5. INSTALL A FULL CUT-OFF FIXTURE.
This type of fixture directs light rays below the horizon of the fixture; down and out instead of up and sideways. This results in a fixture with less glare and more light on the walking surface.

ADVANTAGES: Full cut-off fixtures reduce light trespass (light shining into a neighboring property); save energy by focusing light only where it is intended, and reduce glare and light pollution.

DISADVANTAGES: There is a higher initial cost.

6. INSTALL WHITE LED LIGHTING.
If an exterior fixture needs to be replaced, consider LED lighting (Fig. 7.4.3). Currently, manufacturers consider applications such as gardens, walkways, and decorative fixtures outside garage doors to be the most cost-efficient use of LEDs. For a further description refer to Interior Lighting option 3, at the beginning of this section.

ADVANTAGES: White LED lighting consumes less than a quarter of the electricity used by fluorescent lighting and lasts about 10 times longer.

DISADVANTAGES: Because white LED lighting is relatively new to the marketplace, few manufacturers offer it; it can be costly.

CONTROLS

ESSENTIAL KNOWLEDGE

Switches control the power used in an electrical circuit. A standard switch has two settings, on and off. When on, electricity flows through the circuit from its source to a point of use. When off, the circuit is opened, interrupting the flow of electricity. Most light fixtures and many receptacles are controlled by at least one switch. A dimmer has more than two settings. It is able to control the level of light from very dim to bright by controlling the voltage of the electrical current reaching a light fixture.

A key-type switch built into the fixture operated early wall, ceiling, and table fixtures. In 1890, the push-button switch was introduced and remained the most popular switch used for the next 40

FIGURE 7.4.3 LED BULBS

to 50 years. In order to minimize arcing, it was designed with springs that would "snap" the blades away from the contacts quickly. This resulted in the switch's distinctive snapping sound. For a quiet switch, mercury switches were developed, with contacts made and broken at the boundary of a small pool of mercury. There were no springs to fatigue or metal contact surfaces to fail. Mercury switches are still made but are difficult to find. "Tumble" switches were available around 1898, but did not become popular until after alternating current became the standard.

Wall switches are one of the most reliable devices in the house. A good-quality switch, under normal use, will last 20 years or more, but at some point its springs break or its contacts pit. The slight mechanical movement caused by flipping the switch on and off can also loosen wiring connections. Because replacing a switch is easy and inexpensive, the homeowner may want to replace the switch long before it fails.

There are a few problems associated with dimmers: They reduce the efficiency of incandescent lamps, they often produce an audible buzz, and they may put radio frequency interference (RFI) into a branch circuit. Any load on this branch circuit such as an AM radio, TV, or cordless telephone will receive this RFI noise. This is not a problem where lighting and receptacle circuits are separate. Standard dimmers should not be used to control fans or any motorized device.

TECHNIQUES, MATERIALS, TOOLS

1. INSTALL A NEW, PERIOD REPRODUCTION SWITCH.

With the restoration boom of the late 1970s a new interest was generated in the push-button switch. In 1985, Peter Brevoort of Michigan began manufacturing new push-button switches that he and an electrical engineer had redesigned to meet current codes. It retained the outward appearance and the mandatory snap of the originals. Push-button switches are available from a number of companies that specialize in historic reproductions.

ADVANTAGES: The historical accuracy of the home is preserved.

DISADVANTAGES: The cost is more than for a standard switch.

2. INSTALL A DIMMER WITH A BUILT-IN NOISE FILTER.

To eliminate radio frequency interference associated with the use of dimmers, install a dimmer with a built-in noise filter.

ADVANTAGES: A separate device is not needed to control RFI, and therefore another junction box is not required.

DISADVANTAGES: The cost is more than for a standard dimmer switch.

3. INSTALL A LAMP DEBUZZING COIL (LDC).

To eliminate dimmer-produced radio frequency interference, an LDC can be installed at the dimmer. LDCs fit into a 4" x 4" junction box that for best results should be located as closely as possible to the dimmer.

ADVANTAGES: This is the most effective way to reduce RFI.

DISADVANTAGES: LDCs have their own audible buzz. Another junction box may not be aesthetically acceptable.

4. INSTALL A SELF-CONTAINED DEVICE (SCD) TYPE SWITCH.

Where a new switch is required, an SCD can be installed. SCDs are a time-saving method used to install receptacles and switches (Fig. 7.4.4). They were first developed for the manufactured home and recreational vehicle industry, but are also appropriate for site-built houses. In this system, the switch makes a direct attachment to the electrical cable and then, with use of a special tool, is inserted into the wall. The cable is not cut, and an electrical box is not installed. SDCs are not appropriate for direct replacement of conventional light switches.

ADVANTAGES: This device can be installed in a quarter of the time it takes to install conventional switches.

DISADVANTAGES: SCDs can only be used with grounded nonmetallic sheathed cable; this limits their use to houses wired within the last 20 years.

5. INSTALL WIRELESS SWITCHES.

Where a switch needs to be added or relocated, wireless switching can be installed. Wireless control systems utilize infrared light or radio frequency signals to communicate with controls for lighting and electrical devices. A wireless three-way switching kit contains a battery-powered wireless wall switch, and a receiver switch that replaces an existing switch. The wireless switch sends a signal to the receiver switch, allowing a fixture to be controlled from either location. Another system can control lamps or appliances plugged into a receptacle through the use of a receptacle switch module.

ADVANTAGES: Substantial cost savings are realized because rewiring is not required to add a new light switch. This system does not use existing house wiring as a means of communicating, and therefore is not susceptible to other household product crosstalk, which can create erratic behavior.

DISADVANTAGES: This system can cause interference.

6. INSTALL A FIBER-OPTIC SWITCH.

Installing a fiber-optic switch can eliminate the danger of operating a wall switch from a wet location such as a bath or shower. In this system a light pulse is emitted from the switching module, travels along a fiber-optic cable to an optical membrane wall switch, and is reflected back. Pressing the wall switch breaks the light pulse, activating the switching module (Fig. 7.4.5).

ADVANTAGES: Fiber-optic switches will not corrode and fail due to moisture. The risk of electrical shock when operating a switch from a damp or wet location is eliminated.

DISADVANTAGES: The distance between the switch and electrical device is limited to runs of about 100', based on the ability of the fiber-optic cable to conduct light.

7. INSTALL A MOTION-SENSOR SWITCH.

Motion-sensor switches utilize an infrared sensor to detect moving heat sources to activate the switch (Fig. 7.4.6). They can be adjusted to stay on from a few seconds to 20 minutes and will remain on as long as there is movement in the room. Some motion-sensor switches also contain a photocell sensor

FIGURE 7.4.4 SELF-CONTAINED DEVICE TYPE SWITCH

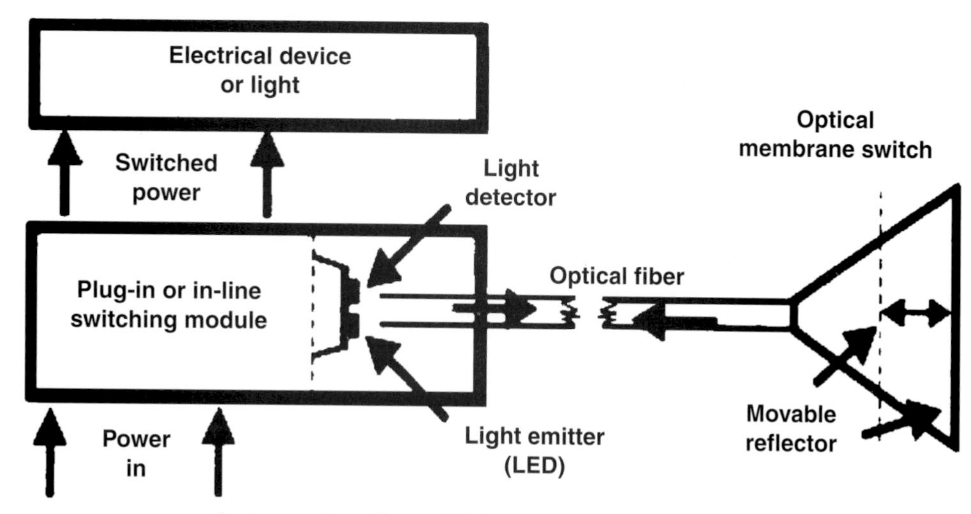

FIGURE 7.4.5 FIBER-OPTIC CONTROL

FIGURE 7.4.6 MOTION-SENSOR SWITCH

that will prevent the switch from turning the lights on when there is enough light already in a space. Some models can be switched from automatic to manual operation at the flip of a switch.

ADVANTAGES: A motion-sensor switch saves energy by turning lights on only when they are needed.

DISADVANTAGES: Motion-sensor switches are more expensive than conventional light switches. Less sensitive sensors can shut off the lights while a room is still occupied and little movement is taking place.

7.5 ELECTRIC BASEBOARD HEATING

ESSENTIAL KNOWLEDGE

Electric baseboard heating has often been used in houses where the installed cost was the primary concern. It may be economical to use where little space heating is needed, such as in a cooling-dominated climate or in a superinsulated house with a small heating load, or in regions with low electric rates. Its advantages include low-cost installation, no indoor combustion, and fast response time to thermostat settings. Since each room has an independent system with its own wiring and thermostat, zoning is flexible. Good-quality units are quiet (except for the minor noise of expanding and contracting metal components), and usually the only maintenance is occasional dusting. Some occupants may choose electric heat to eliminate the possibility of indoor gas leaks.

If electric baseboard heating is used for such reasons as scarce fuel supplies or allergies, there are other electrical resistance systems:

1. Electric thermal storage systems take advantage of lower off-peak electric rates. The storage unit recharges from 11 PM to 7 AM and provides heat when needed, with potentially significant energy savings. The units are more costly than electric baseboards.
2. Electric radiant floors or ceilings include electric wiring embedded in integrated panels, flexible mats, or finish materials. Panelectric and Suncomfort make gypsum boards with embedded wiring for radiant ceilings. Electric radiant wiring boasts immediate response time, but some complain of hot heads from radiant ceilings, and radiant floor installation can be expensive.
3. Modular electric radiant panels are usually installed on walls or ceilings, but sometimes in a floor or kickspace. Attractive and convenient, they are more costly than baseboards without saving energy.
4. Electric hydronic baseboard heaters (Fig. 7.5.1) contain immersion heating elements in an antifreeze solution. Used chiefly in commercial applications, their thermal mass creates an energy-saving lag effect at shutdown, but units are more costly.
5. Electric space heaters are usually supplemented by a fan. They move air quickly but are thicker and a bit more expensive.

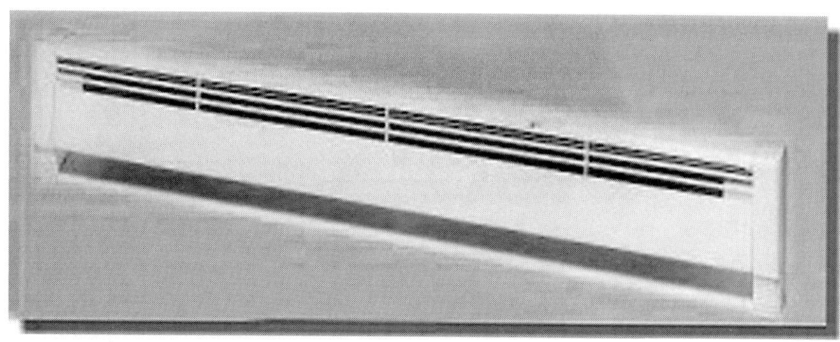

FIGURE 7.5.1 ELECTRIC HYDRONIC BASEBOARD HEATER

6. Electric furnaces contain elements to heat air forced through ducts. The distribution can accommodate air conditioning but the system cannot be zoned as conveniently as electric baseboards. The heating elements are activated individually and sequentially, potentially saving energy during mild weather when full power is not needed.

7. Electric boilers use heating elements to effect a hydronic distribution system. They may be more comfortable than electric furnaces due to the radiating nature of some hydronic systems compared to the drying-out effect of forced air. These systems don't accommodate air conditioning, and energy costs will be similar to electric forced air.

TECHNIQUES, MATERIALS, TOOLS

1. MAINTAIN EXISTING SYSTEM.

Electric resistance systems should be inspected periodically to ensure that components are operating properly and no connections are loose or burned. The fuses or circuit breakers in some electrical systems can easily be checked. Units should be cleaned periodically with a brush or vacuum. There should be adequate clearance from combustibles. Mechanically damaged baseboard heaters should be repaired or replaced.

ADVANTAGES: This system is quiet and has the lowest first cost and lowest maintenance. There are no combustion gases or depressurization.

DISADVANTAGES: The energy costs of electric resistance heating are high in many regions. There has been some concern regarding the health effects of electromagnetic fields (EMFs), which electric resistance heating systems generate during operation. Systems with large open loops create a stronger field than systems that route current back along the same path.

2. REPLACE OLD BASEBOARD UNITS WITH NEW UNITS.

Baseboard heating units are usually installed under windows to heat infiltrating cool air. The unit should sit at least ¾" above the finished floor or carpet to allow cooler air on the floor to reach the radiator fins (Fig. 7.5.2) so it can be heated. The heater should also be installed tightly to the wall to prevent warm air from convecting behind it and streaking the wall with dust particles. The quality of baseboard heaters varies. Cheaper models can be noisy when expanding and contracting and have poor temperature control. New models are not significantly improved from older models, but may be more compact. Units should carry UL and the National Electrical Manufacturer's Association (NEMA) labels. Units 2' to 12' long rate from about 100 to 400 W/ft. Electric hydronic baseboards are rated from 300 to 2,000 W.

FIGURE 7.5.2 ELECTRO-HEAT FIN-CORE UNIT

ADVANTAGES: This is an opportunity to improve the quality or change the size of baseboards.

DISADVANTAGES: Energy costs will not decrease significantly.

3. REPLACE OLD THERMOSTATS WITH NEW THERMOSTATS.

Electric baseboards use on-off thermostats that are classified as line-voltage (120 V) or low-voltage (30 V or less). Low-voltage thermostats, which require a transformer and relay, control temperature more precisely in larger rooms. Electric baseboards are usually controlled with remote wall-mounted thermostats, but many are available with built-in line-voltage thermostats. Built-ins are subject to temperature extremes and therefore do not sense room temperature accurately. Either type should contain a small, internal, anticipation heater, which increases the cycle rate to result in more accurate control of space temperature. An automatic setback thermostat with a clock to control the heater can reduce energy costs.

ADVANTAGES: Appropriate and well-functioning thermostats can maintain comfort zones more accurately.

DISADVANTAGES: Replacement does not significantly affect energy savings.

4. SUPPLEMENT EXISTING ELECTRIC BASEBOARD HEAT WITH ADDITIONAL HEATING SYSTEM.

There are instances in which an electric baseboard system can no longer economically provide for the heating needs of a home (i.e., a seasonal home now used as a full-time residence). One solution is to add an auxiliary heat source to electric baseboard heating, such as a gas furnace and ducts (which can also distribute cooling); a pellet stove; a coal stove (a relatively low-cost fuel, although dirty); an oil stove; a kerosene stove; an unvented gas heater or gas log in the fireplace; a through-wall, Cadet-type heater; or a fan heater. Adding a wood stove can be economical. Chapter 8 discusses such options in detail.

ADVANTAGES: Energy savings can be significantly increased while retaining some advantages of electric baseboard heating.

DISADVANTAGES: There is a dependence upon availability of fuel. Supplemental systems are higher maintenance and are usually less energy efficient than a new electric baseboard system.

5. REPLACE EXISTING BASEBOARD HEAT WITH MORE EFFICIENT SYSTEM.

Conversion to a gas- or oil-fired system may be cost-effective from the standpoint of energy costs. The economics of fuel-switching will vary per house, so an analysis and estimate should be conducted (see Chapter 8). For small rooms, installing radiant electric panels is more energy efficient, provides more even heat distribution, and heats the room faster than electric baseboards.

ADVANTAGES: Conversion can offer substantial energy savings, especially in cooler climates.

DISADVANTAGES: The first costs will be higher than for electric baseboards. A new system may require ductwork. A zoned system will have to be installed to approach the flexibility of electric heating.

7.6 PHONE, COMPUTER, AND TV CABLING

ESSENTIAL KNOWLEDGE

A phone circuit consists of a low-voltage electrical loop. A pair of wires running parallel are necessary to accommodate single-line voice communication. Much of today's analog station wiring (residential telecommunication wiring) of the "plain-old telephone system" (so-called POTS) consists of either two, three, or four (so-called quad wire) insulated conductors running parallel, bunched together, and jacketed (Fig. 7.6.1). These wires were designed to meet minimum analog voice requirements, but interference from motors, power circuits, or crosstalk noise (hearing another phone conversation) is likely, caused by the partial transfer within the cable of a signal from one circuit to another. With the proliferation of phones and modem devices the telecommunications industry developed the economical, unshielded twisted pair (UTP) concept to minimize interference. The paired conductors, usually #24 AWG or #26 AWG, are twisted together at regular intervals (Fig. 7.6.2). In a cable with multiple pairs, this helps isolate and preserve the signals being sent. Some newer telephones and more sophisticated telephone systems will not work properly unless connected to a UTP-type wire.

Each cable has a UL fire rating for the outer insulation jacket (2 or 3). Level 2 is accepted for residential use in many areas and level 3 complies with almost all local building codes. Consult with a building official for local requirements.

In response to the ever-growing wire market, the Telecommunications Industry Association (TIA) and the Electronic Industries Alliance (EIA) established a standard for a generic cabling system that can accommodate many applications. The different telecommunication wires (sometimes called "JK") are rated by category, specifying the number of twists each pair has per unit of length (CAT 1, 2, 3, 4, 5, 5e, 6). The tighter the pairs are twisted together, the higher the performance and signal quality. Today's residential phone systems usually share wires with home computers connected to the Internet or local area network (LAN) and, therefore, not only have to accommodate new digital phone systems but also digital computer transfer signals (ISDN, which will eventually be succeeded by ADSL).

The more bandwidth [indicated in megahertz (MHz)] a wire has, the more data it can transfer. Phone wire performance ranges from the common CAT 1 (1 MHz), and its popular 100-MHz successor CAT 5 UTP, all the way to the 250-MHz CAT 6, at the time this handbook was prepared. Today CAT 5 is widely accepted as the standard for most current and foreseeable residential communication needs. Compared to the 4 megabytes per second (Mbps) data transfer of CAT 1, the 40% higher-priced CAT 5 has a data transfer rate starting at 100 Mbps up to over 150 Mbps, depending on the manufacturer. Some companies market signal enhancers and accelerators as an alternative to cabling upgrades. Such devices have been proven ineffective, however. Shielded wires are not necessary as long as a minimum CAT 3 is used.

The search for the source of problems such as interference and static should start where the phone cable enters the home, namely the network interface device (NID). Cable from this device out to the pole is the phone company's responsibility and is covered by the connection charges for having new or addi-

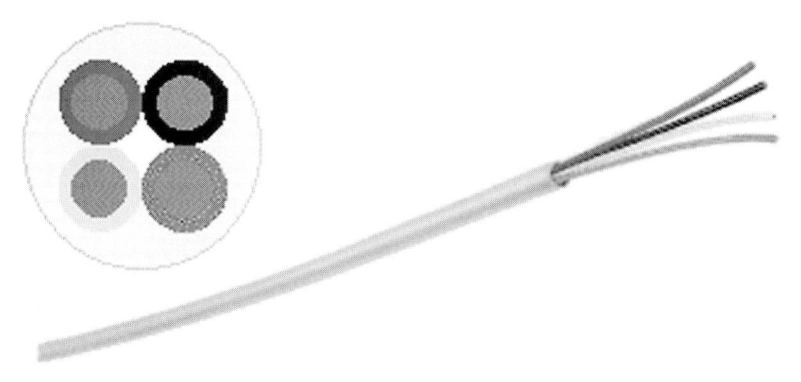

FIGURE 7.6.1 PLAIN-OLD TELEPHONE SYSTEM WIRING

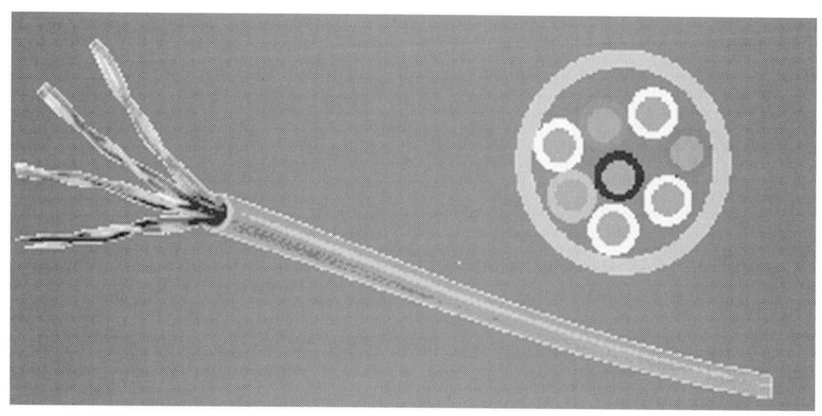

FIGURE 7.6.2 UNSHIELDED TWISTED PAIR WIRING

tional service installed. If there is a dial tone at the NID, the phone company has met its obligation (for test procedures and equipment, see Product Information). Cable from the NID to the telephone sets is the responsibility of the customer, and it is at the customer's discretion how repair work is accomplished. If wiring and cabling problems persist, most phone and TV cable companies provide a maintenance contract where, for a monthly fee, the company takes repair responsibility beyond NID. Phone companies also answer individual service calls but usually bill hourly and may therefore not be as cost-effective as a local installer.

In TV cabling, the common 20-gauge RG-59 coaxial cable (composed of two conductors and several layers of insulation) served the industry when there were only channels 2 though 13 in the TV spectrum. RG-59 is being replaced by the better-performing 18-gauge RG-6 coaxial cable (at the same price). With today's high-frequency channels and the growing demand for digital satellite systems (DSS), high-definition television (HDTV), and interactive services such as video-on-demand and WebTV, homes need cable with greater bandwidth and dependability. RG-6 doubles the capacity of RG-59 and is usually protected by four separate electronic shields (quad shielded) to ensure minimal signal loss and interference. The performance and reliability of RG-6 depends on the manufacturer. Typical aging problems, such as brittle insulation or bare wire, are similar to phone, computer, and electrical wiring.

TECHNIQUES, MATERIALS, TOOLS

1. REPAIR CRACKED INSULATION OR FRAYED WIRE

Crumbling insulation can temporarily be repaired by wrapping the bared wires with electrical tape. However, kinks, crimps, splices, or too much twisting out can ruin the performance of CAT 5 wire and RG-6 cable.

ADVANTAGES: Repairing insulation with electrical tape is significantly less costly than rewiring.

DISADVANTAGES: Compared with rewiring, this fix is short-term. Splicing or using electrical tape significantly decreases signal quality and performance.

2. REWIRE WITHIN EXISTING WALLS AND CEILING.

The most cost-effective way to solve telecommunication wiring problems such as inadequate bandwidth, interference, and deteriorating wiring insulation is to rewire. Unless only single-line analog voice service is necessary, TIA does not recommend using CAT 1 wiring. Currently the trend is to install CAT 5 for both current and future data and voice transmissions. In TV cabling RG-6 replaces the existing coaxial cable RG-59 for the same price. Selecting which wires and terminators in the home are to be updated is critical, since a phone and TV system's performance is only as good as its weakest component. For optimal performance and fire code compliance consult your local telecommunication and TV cabling expert.

It may be possible to use the existing wire as a guide to bring a new cable through the wall. If not, the wire must be snaked through the wall and/or ceiling cavities. Because of possible transmission interference, avoid running telecommunication wiring closer than 2" parallel to electrical wiring, do not cross wires at 90° angles, and do not share bore holes or studs for outlets. If baseboards can be removed, wires can run within the baseboard-wall cavity. Under certain circumstances wires can also be run underneath wall-to-wall carpet.

ADVANTAGES: This is the most effective short- and long-term fix. Most wire problems are eliminated and performance is upgraded.

DISADVANTAGES: Rewiring within existing walls and ceilings is cumbersome and can be costly.

3. REWIRE USING CABLING RACEWAYS

Cabling raceways provide a way to rewire without having to snake through walls or ceilings. The raceways are mounted on interior walls and can encase electrical and communication wires (Fig. 7.6.3). However, telecommunication and TV cabling should not be run in the same raceway as electrical wiring unless the raceway has been designed for dual use. PVC raceways are available in white or wood laminate versions.

ADVANTAGES: Cabling raceways can simplify the task of rewiring and reduce wall penetrations that can compromise a building's thermal performance. Wiring in raceways is accessible for future changes; locations of jacks can also be changed.

DISADVANTAGES: Separate raceways for electrical wiring and telecommunications and TV cabling have to be installed unless the raceways are designed to accommodate both. This fix may not be aesthetically acceptable.

FIGURE 7.6.3 ELECTRICAL AND COMMUNICATION WIRING RACEWAY

7.7 SECURITY SYSTEMS

INTRUSION AND ALARM SYSTEMS

ESSENTIAL KNOWLEDGE

Intrusion detection and alarm systems can share components with smoke and gas detection systems, pipeburst and flood detection systems (see Chapter 8), closed circuit television, child-tracking devices, or keyless entries/access control systems. Sensors can be located on the interior or around the perimeter of the building. When a sensor is activated, it sends a signal to a control panel, which triggers a siren and/or notifies a central monitoring station, automatically dispatching police, fire, or medical help as necessary.

Components of an intrusion detection system include:

1. The control panel is a hidden box that receives information from the sensors and transmits the alarm. Signals can also be sent to a central station, indicating the area of the building in which the alarm originated.
2. The arming station (also called the master station, remote station, or touch pad) turns the system on and off. It is usually a remote keypad with an LED or LCD display, but can be linked to touch screens, panic buttons, TV screens, computers, telephones, or handheld remotes.
3. Sensors send signals to the control panel. Several types of sensors are used for intrusion detection: Magnetic contacts signal when the door or window is opened; glass-break sensors protect windows, sliding glass doors, and skylights, and are activated either by the shock of glass breakage or the sound of breakage; passive infrared sensors detect the presence of an intruder by comparing body heat to ambient temperature in occupied space; motion sensors detect movement in the area of coverage; combination, or dual technology, sensors (Fig. 7.7.1) detect two activities, for example, glass-break and motion detection, which helps prevent false alarms.

TECHNIQUES, MATERIALS, TOOLS

1. INSPECT EXISTING SYSTEM.

Establish a regular inspection schedule to ensure the system performs according to the homeowner's requirements. First, contact the manufacturer to verify it is in business and that the control panel can be repaired or updated. Manufacturers can renew the monitoring contract, which includes periodic checking and repair of the system and may include connecting the system with personnel via the phone line. A contract can limit maintenance costs, especially when a system's reliability is questionable, and can automatically dispatch police should the burglar manage to stop the siren. Locate security loop faults (Fig. 7.7.2) with appropriate testing devices (Fig. 7.7.3). Regularly check backup batteries.

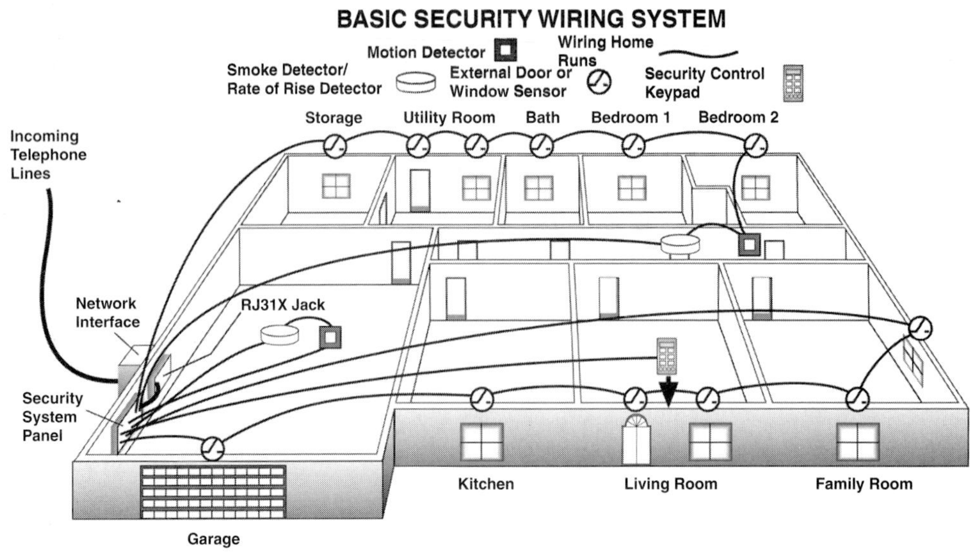

BASIC SECURITY WIRING SYSTEM

Motion Detector
Wiring Home Runs
Smoke Detector/ Rate of Rise Detector
External Door or Window Sensor
Security Control Keypad
Storage
Utility Room
Bath
Bedroom 1
Bedroom 2
Incoming Telephone Lines
Network Interface
RJ31X Jack
Security System Panel
Kitchen
Living Room
Family Room
Garage

FIGURE 7.7.1 COMBINATON SENSORS

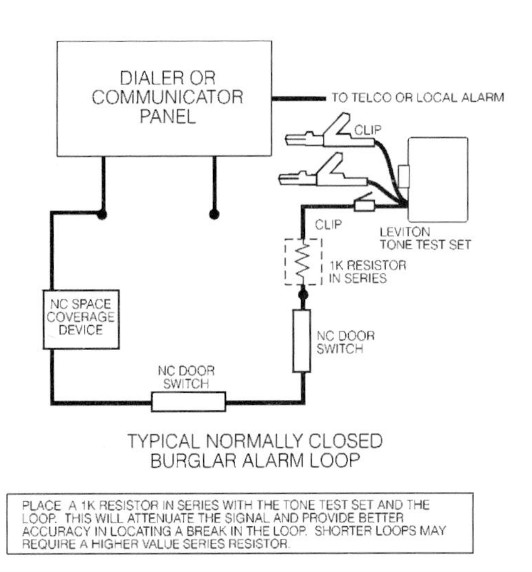

FIGURE 7.7.2 SECURITY LOOP

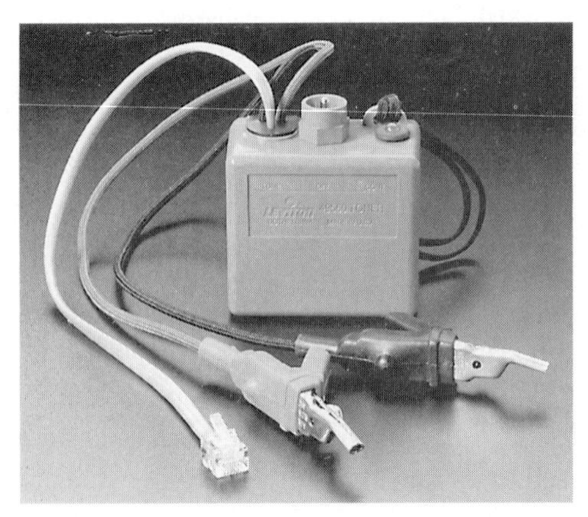

FIGURE 7.7.3 TONE TEST SET

ADVANTAGES: An existing system in good repair may not require many service visits to keep the system in good working order.

DISADVANTAGES: Labor can be costly.

2. REPLACE CONTROL PANEL.

Control panels often need replacement. Most modern security equipment is standardized and compatible with new components. Many installers provide a free control panel with the purchase of a multiyear monitoring contract. Newer control panels have greater reliability and features and reasonable costs.

ADVANTAGES: This is a relatively inexpensive component.

DISADVANTAGES: Labor can be costly.

3. REPLACE SENSORS.

Window and door contacts are the most durable components of the system, which is fortunate, as installation labor can be costly. Old, lead-based foil window sensors are often damaged from thermal expansion and should be upgraded. Conductor cables are often severed during the course of even minor renovation and should be checked for continuity. Various types of sensors, including wireless ones that use radio transmitters, are also available and can replace existing sensors or can be added to them in combination.

ADVANTAGES: Upgrading can reduce the incidence of false alarms or activation failures.

DISADVANTAGES: Rewiring for new sensor locations may be required.

4. REPLACE WIRING.

This is often feasible when the scope of home rehab involves cutting electrical lines. Incorrectly spliced wire, without sufficient solder and electrical tape, is a common cause of false alarms. Attempting to salvage old lines may cost more in labor than installing new wiring. Replace wiring if the insulation is brittle, such as near the contacts where it is exposed. Concealed wiring may require some removal of interior finishes; coordinate with drywall installation.

ADVANTAGES: The incidence of false alarms or activation failures can be reduced.

DISADVANTAGES: Wiring is usually the most costly part of the system.

5. REPLACE ENTIRE SYSTEM.

This may be necessary when, for example, wireless components are installed, which often don't interface with older systems. Systems older than the mid-1980s will need replacement; they used 3-V to 6-V drycell batteries, whereas 12-V rechargeable batteries are now used. The NEC has recently been updated with respect to low-voltage and limited-energy systems. Installing a new security system is an opportunity to integrate other home automation features. Although many installers routinely will run four- or six-conductor cable, it is worth specifying as it provides added flexibility at a small upcharge. Installation contracts for $100 that come with a 3-year, $30/month monitoring contract may be desirable. Without such a contract, installation may be $400+, with service calls at an hourly rate.

ADVANTAGES: Replacement of the entire system affords the opportunity for utilizing the latest security technology and integrating with other advanced wiring systems. Deep discounts are available if purchased with contract.

DISADVANTAGES: Contract payments may be costly.

SMOKE DETECTORS

ESSENTIAL KNOWLEDGE

Smoke and fire detectors have saved thousands of lives since being introduced to the residential market in the 1970s. In spite of the increase in the number of homes since then, the number of fire-related deaths has decreased steadily.

There are three types of detectors: photoelectric, ionization, and heat-sensing. Most photoelectric devices work on the principle of detecting the presence of light due to its scattering by smoke particles. In the absence of smoke, light from an LED passes right through the detection chamber (Fig. 7.7.4) without triggering a response from the light sensor. When smoke enters the chamber, the light scatters onto the photo detector, triggering an alarm. This kind of detector is sensitive to slow smoldering fires that produce large-sized smoke particles. In another type of photoelectric device, the smoke entering the chamber also blocks a light beam, but in this case the reduction in light reaching a photocell sets off the alarm.

Ionization smoke detectors sense fires by relying on the atomic neutralizing property of smoke. The ionization chamber (Fig. 7.7.5) has two plates with a small amount of voltage applied to them. One of the plates has a hole with radioactive material on the other side (approximately 1/5000 gram of Americium-241). The alpha particles generated by the Americium ionize the oxygen and nitrogen atoms. These ionized atoms are attracted to the plates and generate electric current. Smoke particles fill the chamber and attach themselves to the ionized atoms, thus neutralizing them. This disrupts the flow of current in the circuit, which is detected as smoke. This kind of detector is sensitive to fast flaming fires that produce small-sized smoke particles.

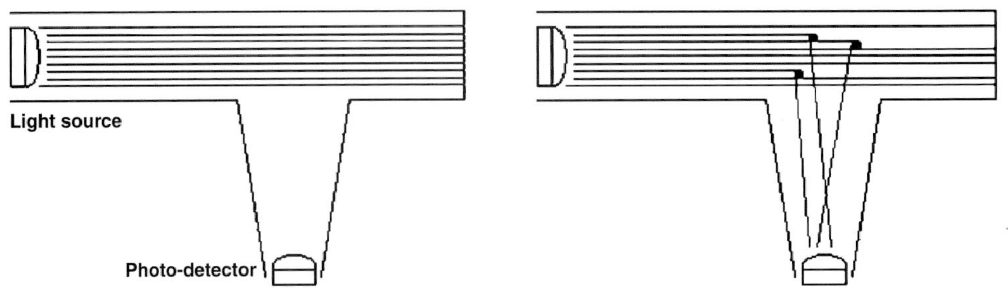

FIGURE 7.7.4 PHOTOELECTRIC SMOKE DETECTION CHAMBER, CLEAR AND PARTICLE FILLED

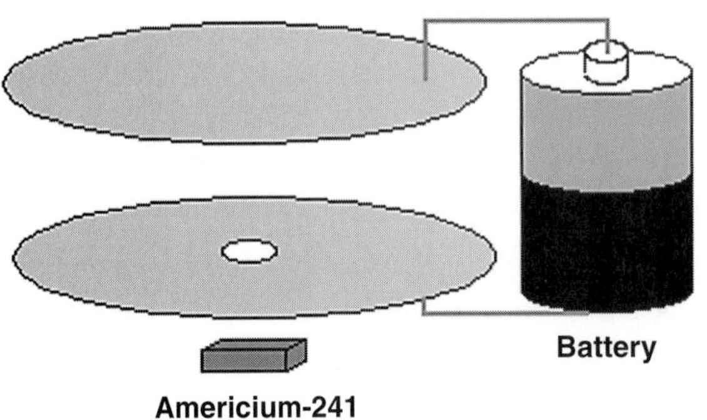

FIGURE 7.7.5 IONIZATION SMOKE DETECTOR COMPONENTS

Heat detectors are available combined with a smoke detector or as a separate product. They are useful where smoke detectors have false alarms, such as in kitchens, as well as in areas that are too hot or too cold for smoke detectors. Heat detectors depend on a fusible link made of lead that distorts with rising temperatures. There are two types of heat sensors. The *fixed* temperature sensor triggers an alarm when it detects a certain air temperature. This is the most common type for residential applications. *Rate-of-rise* sensors depend on a thermostat which detects a percentage temperature rise over a few seconds. Heat detectors either sound their own alarm or a central alarm if part of a system. Heat detectors must be very close to a fire to be set off and are therefore not effective early-warning devices. Heat detectors are long-lived (unless triggered), have few false alarms, and are relatively unaffected by cooking grease. However, it is necessary to install a new unit after the device has sensed a single fire.

Detectors are battery powered or hardwired, or a combination of both. Hardwired detectors are either connected to a 120-V home circuit in series or stepped down to low voltage in a home run to a control box when interconnecting to a low-voltage security system. The NFPA safety standard recommends at least one detector on each level of the house. Ideally, a detector should be located in each bedroom, outside each bedroom area, in the living area, and in the kitchen. It is extremely important to locate the detectors properly to provide good coverage.

Although alarms for the hearing-impaired are not required for residences, the Americans with Disabilities Act and the NFPA suggest strobe light or vibration systems, should they be installed. The vibration system works with a transmitter relayed at the fire panel to send a signal to a bedside receiver that vibrates the bed. Available for new systems or retrofit to existing, vibration systems are more expensive but are considered more likely to wake residents than strobe light systems.

"Smart" fire detectors that minimize false alarms are now in development. Smart detectors use a combination of detection methods and a microprocessor to decide when to alarm. The National Institute of Standards and Technology (NIST) Building and Fire Research Laboratory has developed a flow tunnel to test smart smoke detectors.

TECHNIQUES, MATERIALS, TOOLS

1. MAINTAIN EXISTING SMOKE DETECTOR.

Battery-powered smoke detectors will beep at regular intervals if the battery needs changing. It is imperative that batteries be changed once or twice a year to avoid failed detection. Follow the manufacturer's recommendations for battery size and voltage. Smoke detectors have an aspirated opening to facilitate free flow of indoor air. These openings can get clogged from the accumulation of dust and cooking grease, preventing smoke entry and detection. Detectors should be cleaned regularly to avoid such buildup.Care should be taken when cleaning ionization smoke detectors; the sealed chamber containing radioactive material should not be opened or serviced. Even though there is a small amount of radioactive material in the chamber and a sheet of paper can block the alpha particles generated by it, it is dangerous if inhaled or swallowed. Follow manufacturer's cleaning recommendations.

ADVANTAGES: This is the least-expensive response.

DISADVANTAGES: There is a risk of an old detector not performing properly, even after maintenance.

2. REPLACE EXISTING SMOKE DETECTOR WITH NEW UNIT.

The *1999 National Fire Alarm Code* requires smoke detectors to be replaced every 10 years. During this time they have gone through 3.5 million monitoring cycles and are more likely to fail in the case of a real fire. Although most home fires respond to photoelectric sensors, a combination of sensor types is the best strategy for covering all kinds of fires and contingencies. Using both battery-operated and 120-V units will detect fires during a power outage and if there is a failed battery. It is also recommended that all smoke detectors in a home be interconnected; in the event of a fire in any part of the house, all alarms will activate to alert the occupants. In a household with a hearing-impaired person, visual alarms with strobe lights should be additionally installed. According to the NYC Health Department Radiation Division, there are no federal disposal precautions for ionization smoke detectors.

ADVANTAGES: This is an opportunity to install a reliable, properly specified system.

DISADVANTAGES: This solution is more costly.

CARBON MONOXIDE DETECTORS

ESSENTIAL KNOWLEDGE

Carbon monoxide, known by its chemical name, CO, is a gas generated during most combustion processes. The amount of CO increases with the scarcity of oxygen, leading to partial combustion of carbon, producing CO instead of CO_2 (carbon dioxide). CO is deadly in long-term exposure even in low concentrations. CO forms a chemical bond with hemoglobin in the blood, which is stronger than the bond between oxygen and hemoglobin. Exposure to a high-concentration of CO or a long-term exposure to CO removes oxygen from the bloodstream, leading to possible death.

CO is generated in toxic quantities in homes that have one or more combustion devices such as gas ovens, gas cooking ranges, gas or oil water heaters, gas or oil furnaces, and gas or wood fireplaces with inadequate air for the combustion process. Poor maintenance of combustion devices, cracked heat exchangers, blocked chimneys, leaky furnaces, running automobiles in enclosed spaces, and indoor barbecues can aggravate the CO problem.

Homes without combustion devices have a CO concentration of 0.5 to 5 ppm, which is about the same as outdoors. However, CO levels above 15 to 20 ppm over long periods can be deadly. CO detectors should be able to detect the concentration as well as the duration of exposure to protect the occupants. All CO detectors should be UL 2034 approved, which requires the detector to activate an alarm if the CO level reaches 100 ppm for 90 minutes.

There are presently three CO detection methods commonly available for residential use: (1) Biomimetic sensors mimic blood reaction to CO with color changes sensed by a photocell that triggers an alarm. Sensors must be changed at intervals. (2) Electrochemical sensors react with CO to produce a small charge that corresponds to the CO level. This type of sensor loses sensitivity after approximately 2 years and needs replacement. (3) Solid-state, also called semiconductor or conductometric, sensors measure changes in electrical resistance of a material that absorbs CO.

Infrared (IR) sensors are calibrated to detect specific gases. When used for CO detection, they constantly bombard air samples with IR radiation and then look for the radiation signature of CO molecules. These sensors are accurate and more stable over longer periods than other detection methods. IR gas detection has industrial applications but is too expensive for most residential use. Some of the smart detectors being tested by NIST incorporate CO detection in addition to advanced smoke detection.

TECHNIQUES, MATERIALS, TOOLS

INSTALL A CO DETECTOR.

CO detectors are either battery powered or hardwired into the home's 120-V or low-voltage circuit. Batteries should be replaced every year to avoid failed detection. The best approach is a hardwired detector with a battery backup to ensure proper functioning under all conditions. Ideally, CO detectors should be on each floor of a home, in each sleeping area, and outside any room where combustion devices are located.

ADVANTAGES: CO detectors provide a means for monitoring a dangerous gas.

DISADVANTAGES: The reliability of products varies widely.

LIGHTNING PROTECTION

ESSENTIAL KNOWLEDGE

Lightning is a random, capricious phenomenon that results in more deaths than hurricanes and floods combined and damages approximately 18,000 homes annually in the United States. Lightning strikes the Earth up to 100 times a second. Lightning can carry up to 400,000 A, can produce up to 30 million volts, and can reach temperatures of 50,000°F. Lightning occurs when a buildup of electric potential (usually negative charge) in storm clouds sends downward *leaders* that connect with *streamers* sent up by ground objects of positive charge. Direct effects of lightning are caused by resistance and include ohmic heating, arcing, and fires. Indirect effects are capacitive, inductive, and magnetic behaviors occurring in certain locations, causing electrical surges leading to fire or breakdown of household electronic devices.

Properly designed protection systems (Fig. 7.7.6) have been documented by UL to be over 99% effective in preventing direct damage by lightning. A system consists of five components:

1. Collection devices to direct lightning away from the structure and into the ground (these devices include lightning rods or "air terminals," Faraday's cages, and shielding wires, each made for a specific purpose).
2. Cables to interconnect collection devices and down-lead conductor cables to route lightning between collection devices and down to the ground. Any bends in the cable should not be too sharp (Fig. 7.7.7) to prevent flashover of lightning to nearby objects.
3. Interconnection to metal parts on the roof or within the building.
4. Grounding rods to dissipate lightning into the earth once it has traveled down the building. Single-point (Fig. 7.7.8) or multiple-point grounding systems can be used, depending on ground resistance,

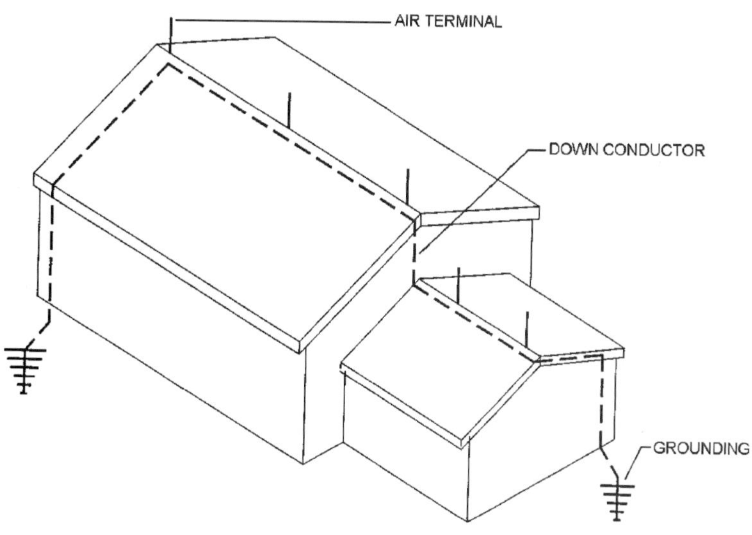

FIGURE 7.7.6 LIGHTNING PROTECTION SYSTEM

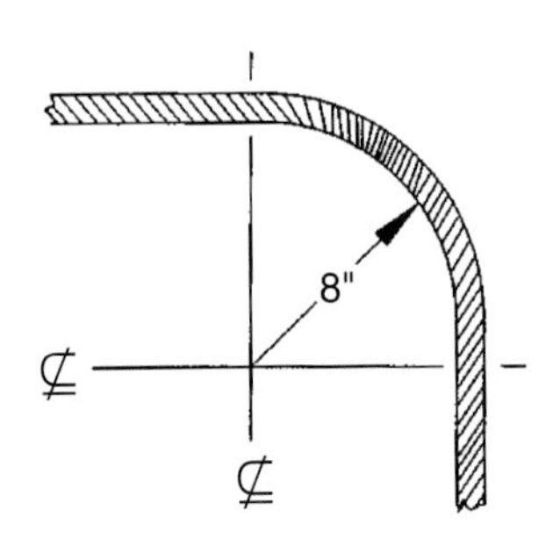

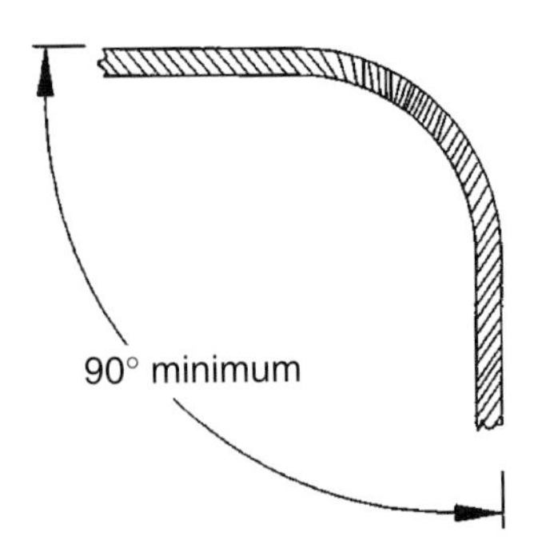

FIGURE 7.7.7 MAXIMUM BENDS IN CABLE

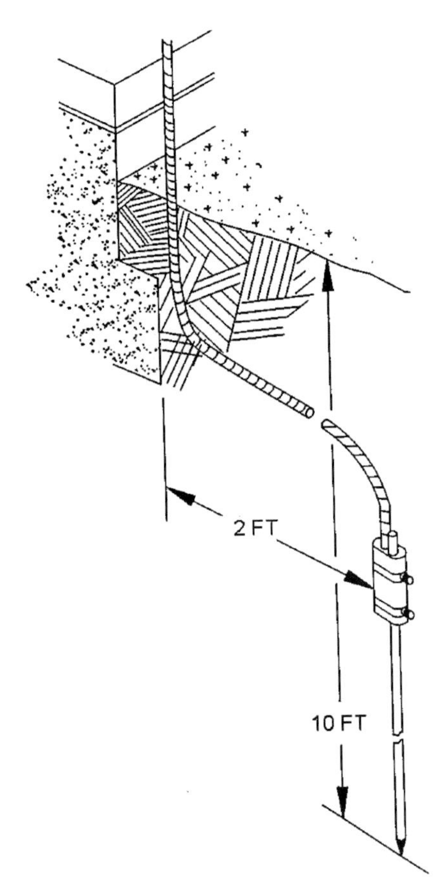

FIGURE 7.7.8 SINGLE-POINT GROUNDING

water table, and soil characteristics. Avoid placing multiple-point grounding rods so close that they allow saturation of the ground; UL requires a minimum 10' separation between multiple grounds. Preferably, each rod should have counterpoise (also called fork-type or crow's foot) radial branches to reduce impedance. The system should share common grounding with electric and telephone ground, and metallic water and gas pipe according to code.

5. Secondary lightning surge arresters on the main electric panel, telephone line entrance, and radio and TV antenna lead-in.

Model and state building codes do not require the installation of lightning protection systems. All lightning codes (UL 96A, NFPA 780, and Lightning Protection Institute Standard of Practice LPI-175) function as guidelines. However, some local codes, such as in Florida, are incorporating requirements for lightning protection to reduce fires resulting from lightning strikes.

TECHNIQUES, MATERIALS, TOOLS

1. MAINTAIN EXISTING LIGHTNING PROTECTION SYSTEM.

The lightning protection system should be inspected at least once every 2 years; utility service companies may have altered the grounding or severed down conductors. Besides a visual check, the inspection should include: (1) A bonding-resistance test, which rates the resistance of each bond along the path of the lightning protection system. Each bond is allowed a maximum of 1 ohm. (2) A resistance-to-earth test, which measures net resistance of the total path from the air terminal to the ground. Resistance-to-earth records should be kept for at least six cycles to note any trends. Do not paint the down conductor, as the increased resistance can cause a fire.

ADVANTAGES: This is the least-expensive option if the existing system is adequate to offset local risk.

DISADVANTAGES: The existing system may not be UL- or code-compliant and may not account for current conditions, such as adjacent tree growth, building additions, or construction or demolition of nearby buildings.

2. INSTALL NEW MATERIALS ON EXISTING LIGHTNING PROTECTION SYSTEM.

Systems are custom-designed, and any existing system can be repaired or added to in a modular way with new components. Replacement components can be semiconcealed if the attic is accessible; down conductors can be routed behind rain leaders or trim boards. Most modifications to existing systems are intended to bring a system up to UL standards or to comply with code changes. The NEC is now in agreement with the *Lightning Protection Code*, which requires that underground service line grounds be connected in series. UL calls for tying electrical, phone, and cable grounds together with gas and plumbing grounds since telephone grounding, often poor, is a common source of damage. Secondary lightning surge arresters, located at electrical and phone service panels, have been code-mandated since the early 1970s for homes with lightning protection. Insurance companies used to offer discounts or credits for UL-approved lightning protection, but since deductibles have become common, homeowners are motivated to install reliable, UL-approved lightning protection to avoid paying deductibles.

Adjacent trees may have grown considerably since the system was installed. Lightning protection, which does not harm the tree, is routed along the trunk and main boughs and is recommended by UL, NFPA, and LPI for valuable trees and those taller than, and within 10' of, any part of the building.

ADVANTAGES: An old system can be updated to perform like new at modest cost. Trees or parts of a structure added after the existing system was installed can be protected.

DISADVANTAGES: If materials or the configuration of the existing system are inferior or incorrect, a new system may offer better protection.

3. INSTALL NEW LIGHTNING PROTECTION SYSTEM.

NFPA 78 includes a Risk Assessment Guide that helps determine whether a lightning protection system is cost-effective for a particular building and location. In general, taller structures relatively isolated from buildings nearby are more likely to be hit, particularly in areas that receive frequent storms. Systems can often be semiconcealed for a modest upcharge, hidden by the roof and building finishes, with only the 10"-high air terminals visible. Reinforcement bars in concrete should not be used for grounding the lightning protection system; lightning strikes produce enough heat to evaporate the moisture in the concrete, causing it to crack. TV and cable TV antennas, though grounded, do not offer lightning protection.

ADVANTAGES: Installing a new lightning protection system can limit injury and property damage, and add value, lasting as long as the home. It is the best opportunity to conceal down conductors.

DISADVANTAGES: This is the most expensive option in the short-term.

SURGE PROTECTION

ESSENTIAL KNOWLEDGE

Surges, or voltage transients, are very high pulses of voltage lasting less than a hundredth of a second. Surges are caused by induction (internal or external) and by power-line transients. Whenever an inductive device, such as a motor, fan, air conditioner, or incandescent bulb is shut off, the electromotive force (EMF) causes a voltage surge. Lightning strikes induce voltage surges by causing nearby metallic objects, wires, and electronic equipment to generate voltage. A typical lightning strike 300' from a house will induce high transient voltage, and a strike near high-tension wires many miles away can surge into many homes. Also, normal utility operation carries surges from load fluctuations, transformer blowouts, and other events.

Though surges and fluctuations presented few problems until the 1980s, modern devices require today's homes to be wired differently. In addition to personal computers (PCs), information technology is being installed in many household devices, such as microwaves, vacuum cleaners, washers, dryers, telephones, and even alarm clocks. Though these microelectronics endow products with additional functions and convenience, they are highly susceptible to common variations in voltage, which can affect their performance and shorten their life.

Homes built before the early 1970s may not have three-prong grounded receptacles, or may have three-prong outlets installed in a two-wire system (see Receptacles, in Section 7.3). Many computers, peripherals, and other electronics are designed to send surges through the third prong to ground; if that path is broken, components can be damaged. The home's wiring may not be supplying power of correct polarity. The ground for the home's wiring may have high impedance, improper neutral-ground bonds, or ground loops that allow unpredictable voltages to remain on the grounding circuit. If any aspect of the wiring is questionable, an electrician should inspect it. Local codes and regulations may influence the extent of surge protection in a rehab since the utilities have some responsibility for such suppression. For example, the phone company should supply a properly grounded network interface box.

Since appliances can also generate transients in a home's electrical system, some electricians recommend keeping sensitive devices plugged into a different circuit than inductive equipment with motors and high-amperage switching loads like portable heaters, electric frying pans, and toasters. A separate ground circuit for electronic equipment may need to be installed if the wiring does not allow separating them from inductive and switching loads.

Beyond the utilities' efforts, transient control should be designed and installed in two stages to suppress all surges. First, whole-house surge protection, or breaker suppression, should be installed at the circuit breaker box or distribution panel to eliminate most incoming surges. It should be able to shunt surges away from power, telephone and modem, and cable lines. Alternatively, telephone surge suppression can be installed as a separate unit, at the telephone connecting block outside the house or apartment building. Second-stage devices at plug outlets protect electronic circuitry from EMF induced inside the house wiring or from other surges. Surge suppressors incorporate a small disc called a metal oxide varister (MOV) that absorbs excess energy. MOVs are rated by diameter, which corresponds to their energy-absorbing ability. The rated life of a cheaper surge suppressor, which may rely on only one or two MOVs, is limited to a few years when exposed to most residential circuits, and less if a series of large transients or lightning occurs. Better-quality units incorporate a more substantial combination of MOVs, zener diodes, avalanche diodes, and/or gas tubes. Surge suppressors should meet UL 1449 Standard, second edition.

Frequently, lightning damage comes through the phone line, and modems are especially vulnerable. The telephone company should have installed a properly grounded network interface box (NIB), which contains a lightning protection device, outside the home. Plug-in communications surge protection devices are available.

Most operational difficulty, however, is caused by power interruptions, also called sags or brownouts, and not power surges or spikes. These sags cause data corruption in PCs. Surge suppressors do not protect against this largest problem in Internet connectivity, but battery-powered uninterruptible power supply units (UPS) do. UPSs come in three general types: (1) Double-conversion UPSs continually run the device through an output circuit converter, powered by a battery connected to the line voltage. (2) Standby supplies connect the device directly to the line voltage, which also keeps the battery charged. When power is lost, the battery-powered inverter is switched on. (3) A ferroresonant supply acts as a power conditioner with a microprocessor-controlled transformer between the device and line voltage. When power failure occurs, the inverter supplies power to the transformer.

TECHNIQUES, MATERIALS, TOOLS

1. INSTALL NEW SURGE SUPPRESSION AT BREAKER PANEL.

Breaker, or hardwired, suppressors (Fig. 7.7.9) protect an entire building, often with better energy-absorbing capacity than plug-in suppressors. They are installed at the main service or at a subpanel, keeping the lead from the panel to the suppressor short and straight. Suppressors are available as series or parallel units. Series protectors are more expensive and require breaking the incoming line. Parallel protectors are usually installed on the load side of the main breaker and are more typical for residences. Breaker suppressor devices should provide protection for line to neutral, line to ground, and neutral to ground; a fuse or circuit breaker for overload protection; thermal overheating protection; and additional protection for telephone line and coaxial cable if needed. The rated clamping voltage should be no more than 280 V, the volt-amp rating should be no less than 750,000 VA, and a response time of 3 to 5 nanoseconds (ns). Equipment liability insurance should be available; some manufacturers offer up to $500,000 coverage with no deductible. The warranty should be either for a lifetime or for at least 5 years; better-quality units last indefinitely.

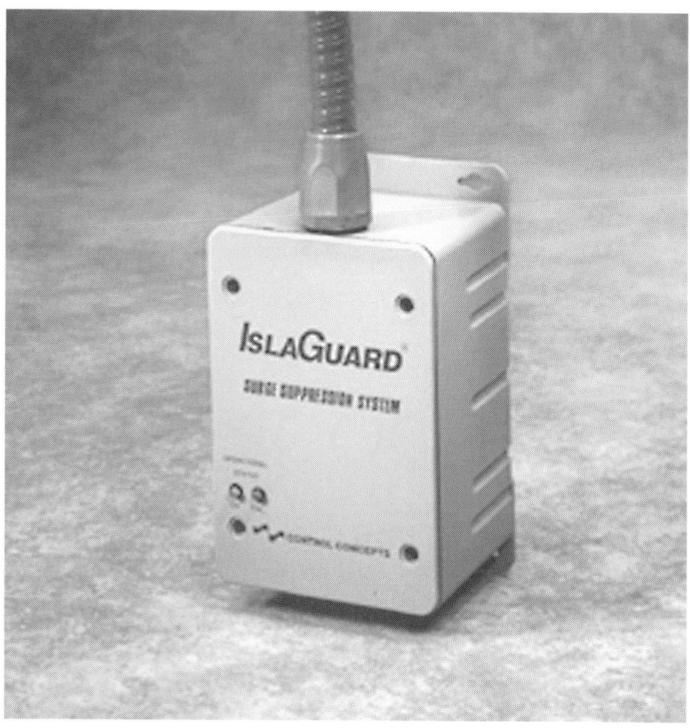

FIGURE 7.7.9 BREAKER SURGE SUPPRESSOR

ADVANTAGES: Installing a new surge suppressor at the breaker panel significantly reduces the load on point-of-use surge suppression.

DISADVANTAGES: Installation requires an electrician. A surge suppressor at the breaker panel does not provide an uninterrupted power supply or protection from internally induced transients and may not include dataline protection.

2. PROVIDE NEW UNINTERRUPTED POWER SUPPLY (UPS).

Many units include outlets dedicated for surge suppression, in addition to combination surge and power protection outlets. However, it may be more economical to buy separate units, as it is less expensive to replace a plug-in surge suppressor with a blown MOV than a UPS. Home office users may consider units with automatic data protection, which connects to the PC with a serial port and software that automatically saves and closes files during a blackout if the user is not present. When selecting a UPS, one should compare sensing time, transfer time, length of backup service, and output quality. It should accommodate a standard size gel-cell battery in an accessible compartment. Batteries must be replaced after 3 to 6 years and should be tested periodically. Standby power sources, as well as generators, are occasionally used but have a longer delay than UPS units, leaving devices less protected. Ferroresonant power conditioners are occasionally used where nearby industry or utility problems might create damaging interference. However, units without batteries do not protect against power loss such as sags and brownouts.

ADVANTAGES: UPSs can provide various levels of point-of-use protection.

DISADVANTAGES: Models with additional features are costly and are generally inadequate without a breaker suppressor.

3. PROVIDE NEW PLUG-IN SURGE SUPPRESSION.

Plug-in suppressors can react quickly and can protect against induced transients. Available as dual-outlet direct plug-in units or strips with 4, 6, 8, or more outlets, many units include modem and fax and/or coaxial surge protection. Units that suppress only dataline surges (or only coaxial surges) are available. Note that most surge suppressors will deliver power after the protection circuitry has been destroyed by a single lightning strike or repeated "hits" by high-energy surges, reducing it to a mere outlet strip. High-quality surge suppressors feature higher joule ratings, lower clamping response times, lower voltage clamping levels, both common-mode and normal-mode protection, and electromagnetic interference (EMI) and RFI filtering of incoming power. Units with these features will be the most costly.

ADVANTAGES: A plug-in surge suppressor inexpensively protects against induction surges.

DISADVANTAGES: A plug-in surge suppressor is relatively short lived and generally inadequate without a breaker suppressor.

GARAGE DOOR OPENERS

ESSENTIAL KNOWLEDGE

An electric garage door must serve two opposite functions: providing automatic entry for residents while maintaining a barrier to would-be burglars. Components of a typical garage door operator (Fig. 7.7.10) include:

1. An electrically reversible motor.

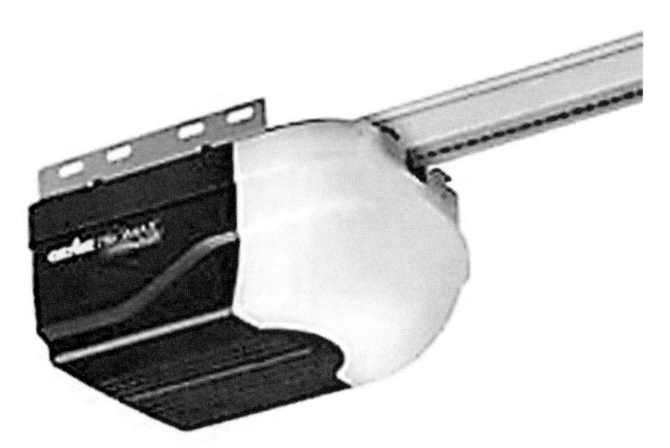

FIGURE 7.7.10 GARAGE DOOR OPERATOR

2. A belt, chain, or screw drive. The belt drive is the quietest, but the neoprene belt material can break. Chains are common but must be lubricated. The screw drive is considered the most durable, though somewhat more difficult to repair if damaged.
3. Limit switches, activated at each end of travel, stop the motor and toggle the state (up or down) of the controller that sets top and bottom positions of the door.
4. Safety stops, or door blockage sensors, detect obstructions and stop or reverse travel. Some types use a twisting motor mount that closes a set of contacts to stop the door.
5. Logic controller relays or a microcontroller allows for essentially three types of door operation: momentary (in which the door operates at one touch of the button), constant or latched hold (which operates the motor only while the button is depressed), and time-latched (which opens or closes the door after a programmed time interval).
6. A remote receiver traditionally tuned to the frequency of the hand unit. A logic program here or in the controller checks the transmission to determine if the codes match. Recent units employ a rolling code or security code or some such pseudorandom code-changing scheme to reduce the chance of interference or code theft. The unit is usually in a box on the wall wired to the motor.
7. Remote controls, transmitters, or hand units have changed with receivers, and the FCC regulates frequency bands dedicated to various types of cordless devices that increasingly fill the airwaves.
8. Light and timer, which is usually a bimetal strip heated to operate a set of contacts. The on time is determined by how long it takes for the strip to cool. The light is activated with the motor and is timed for 3 to 5 minutes. These last approximately 10 years but can be replaced.

Some systems include additional features, such as allowing multiple doors (including non-garage doors) to operate via a single remote, or multiple remotes for several users to operate one or more doors. Some remotes operate lights or appliances in addition to doors. Other remotes range up to 500' with a programmable delay so the driver can enter the garage without stopping. Retrofit transmitter and receiver kits are available that operate at the flick of a headlight. Remote indicators allow residents to see whether the door is open from elsewhere in the house. Some systems come with a "vacation lock switch" to lock out all remote signals until the system is reactivated. Still others automatically push the door back down if someone tries to pry it open. A pick-resistant key switch or electronic lock has an extra measure of security beyond the standard door locking mechanism. Quantum doors are equipped with a locking device that automatically slides into place once the door is closed.

Metal in the vicinity of the receiver can affect remote-control performance. Short-range remotes are becoming more prevalent, particularly as garage doors are now often steel, and insulation may have foil facing. In addition, the FCC has reduced the range of remotes. Receivers should be located away from metal. Another solution is to splice wire onto the existing antenna and have it protrude outside the garage. External antenna kits are available for some models. Windows on the door can also be helpful.

TECHNIQUES, MATERIALS, TOOLS

1. MAINTAIN EXISTING SYSTEM.

A series of accidents has resulted in investigations of automatic door safety. By 1993, laws changed such that UL garage door opener ratings varied from state to state, so systems predating 1990 often do not comply. The laws generally mandate that there be two means of reversing the door: one integral to the circuit board and one means of responding to a physical obstruction in the threshold, such as reverse edge, loop detector, or photocell. If the existing opener complies, maintenance is generally limited to lubricating the chain or screw drive with a light oil recommended by the manufacturer. To maximize the life of the opener, the door itself should function smoothly. Previous, obstructed door operation may have left components bent or broken. Disconnect the door from the opener and try to operate it manually; difficult operation means the opener is also straining.

ADVANTAGES: This is a low-cost alternative.

DISADVANTAGES: Professional inspection is required to offset risk of a noncompliant, damaged, or poorly adjusted system.

2. REPLACE CIRCUIT BOARD AND ADD PHOTOCELL.

Since motors are essentially unchanged, most noncompliant openers can be brought up to code by installing a UL-rated circuit board in the controller that employs safety features, and by adding a photocell, or eye, that will detect when an object is in the threshold. These are available in upgrade kits.

ADVANTAGES: It might be enough to bring an existing system up to code.

DISADVANTAGES: Professional service is required.

3. REPLACE REMOTE CONTROL.

If the remote control is broken, most dealers carry remotes for discontinued models; bring the old one in for a match. Home centers sell universal remote and receiver kits to be installed if the original is lost or is no longer produced, or if additional security or reliability is desired. Standard remote and receiver combinations are vulnerable to burglars recording the entry-code signal sent from the remote. Newer openers (Fig. 7.7.11) randomly alternate between 100 billion different frequencies at each door operation to prevent sound or radiowave interference, and burglars cannot break the code. These openers are recommended over the standard type, in which the user personally selects eight or more fixed DIP switch settings on the receiver and the remote. To eschew the remote, install a keyless entry pad at the garage or at another location.

ADVANTAGES: Both security and reliability will be upgraded.

DISADVANTAGES: The voltage must match that of the transformer, unless a new one is included. State compliance is not ensured.

4. REPLACE EXISTING SYSTEM WITH NEW SYSTEM.

If the existing opener is noncompliant and sufficiently damaged, or parts start failing, complete replacement may be more cost-effective than an overhaul. Modern, low-end openers have plastic gears that wear and eventually no longer engage. These are designed to last 10,000 to 15,000 cycles, for which homeowners can expect 4 to 7 years of trouble-free operation. Higher-end openers, however,

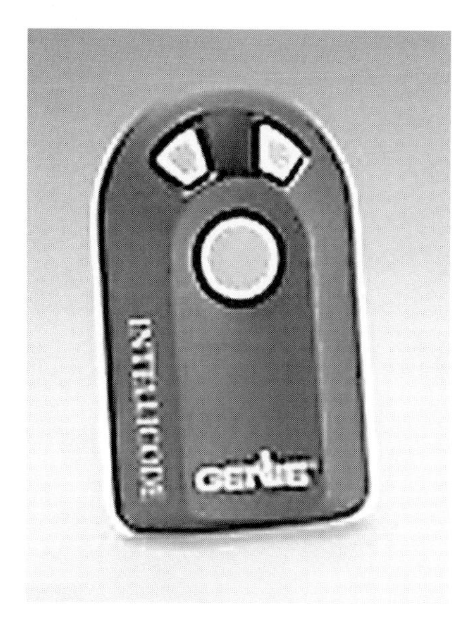

FIGURE 7.7.11 REMOTE-CONTROL OPENER

can last 25 years or more with little maintenance and are cost-effective considering increased life and reliability. Screw drive units have the least moving parts and are considered to be the most reliable.

ADVANTAGES: Replacing the existing system is an opportunity to tailor the system to the residents' lifestyle. It may be cost-effective compared to retrofitting a system that does not meet codes. The whole system will be under warranty.

DISADVANTAGES: This is often the most expensive option.

CH. 8

HOME REHAB HANDBOOK

HVAC AND PLUMBING

Chapter 8
HVAC AND PLUMBING

8.1. HVAC DESIGN AND ENGINEERING
8.2. DISTRIBUTION SYSTEMS
8.3. HEATING
8.4. COOLING
8.5. HEAT PUMPS
8.6. INDOOR AIR QUALITY
8.7. CONTROLS
8.8. FIREPLACES AND CHIMNEYS
8.9. DOMESTIC HOT-WATER HEATING
8.10. PLUMBING DESIGN AND ENGINEERING
8.11. WATER SUPPLY AND DISTRIBUTION SYSTEMS
8.12. DRAIN, WASTE, AND VENT SYSTEMS
8.13. FUEL SUPPLY SYSTEMS
8.14. APPLIANCE VENTS AND EXHAUSTS
8.15. FIRE PROTECTION SYSTEMS

8.1 HVAC DESIGN AND ENGINEERING

HVAC SYSTEMS OVERVIEW

Heating, ventilation, and air-conditioning (HVAC) systems that are properly operating and appropriate for the home are critical for the comfort and safety of the home occupants. Richard Trethewey of *This Old House* states that the home's HVAC systems and plumbing set the foundation on which the family's physical comfort and health depend. Their economic well-being is not only affected by the initial purchase price, but, perhaps more importantly, by the cost of operating and maintaining the systems.

What constitutes an HVAC system is a function of the home, the climate, and the occupants. It is safe to say that all homes have a heating system of some kind, but it is only over the last 20 years that summer air conditioning beyond opening windows became prominent. In 1970, about one-third of new single-family homes had central air conditioning; the figure is now over three-quarters. And, the practice of installing mechanical ventilation systems has only matured in the last 10 years with the emphasis on tighter homes and high-efficiency equipment. Today's HVAC systems can include smart controls, air filtering, humidification, and/or dehumidification (Fig. 8.1.1).

When evaluating the rehabilitation needs of a home's HVAC system, the appropriateness of the current *type* of system for meeting the expectations of today's home occupants needs to be considered.

FIGURE 8.1.1 HVAC SYSTEM COMPONENTS

Home occupants expect to be comfortable in the winter and summer without paying exorbitant energy bills. While a home with electric baseboard heating and passive cooling (i.e., natural ventilation and shading, with no equipment) was fairly common and acceptable 20 years ago, the high winter electric bills due to the increased cost of electricity and poor level of comfort during the summer have diminished its popularity significantly. Even if the baseboard system is in good operating condition, it may be appropriate to consider its replacement.

If it is determined that the type of system is appropriate, the system performance must then be considered. Are there opportunities to improve the performance of the existing system or is it a wiser decision to simply replace the old system with a newer one? Generally speaking, if the heating system equipment is more than 15 years old or the heat pump or air conditioner is more than 10 years old, it should probably be replaced. The energy savings with the higher-efficiency equipment available today will pay for the new equipment within a few years. The typical new gas furnace sold in 1975 had an efficiency of 63%; by 1988, the typical efficiency had increased to 75%; and, in 1997, 86% of the furnaces shipped had efficiencies greater than 80%. Nevertheless, replacement is not always the answer, and there are a number of alternatives for rehabilitating the existing system.

This guideline will review the attributes of many alternatives for rehabilitating HVAC systems. It discusses the advantages and disadvantages of various modifications to existing systems as well as equipment technologies which may be considered for supplementing or replacing the existing system.

As a final note before moving into the discussion of alternatives, it is critically important to understand how HVAC systems interact with other systems throughout the home. A decision concerning the kitchen range exhaust vent can cause the gas furnace to backdraft. Under certain circumstances, an attic ventilator can increase cooling loads rather than decrease them by drawing conditioned air up through the ceiling. When the old natural-draft furnace is replaced with a closed-combustion system, the home's pressure and infiltration rate will be altered, and, as a result, its indoor air quality. Much of the interaction of these systems revolves around the pressurization and depressurization of spaces. Lack of consideration for the effects of various devices on home pressure can result in costly excessive infiltration, damaging condensation in walls, or dangerous levels of carbon monoxide in the home. These issues have been enhanced by today's construction and insulation practices which make homes tighter and therefore easier to pressurize or depressurize. It is important to understand and consider the whole-house implications of each system modification. Contractors who are narrowly focused on a particular trade often do not take all of the interactions within the house into consideration.

REPLACEMENT SYSTEM SIZING

In many instances, the decision will be made to replace the existing heating and cooling system rather than rehabilitate it. The old system may be well beyond its expected life. Many newer systems are more efficient and can quickly pay for themselves in reduced energy bills. The availability of fuels may have changed (e.g., natural gas may now be available) since the system was originally designed and installed.

If the old heating and/or cooling system in the house being rehabilitated is beyond retrofitting and needs to be replaced, there are two primary reasons why it should not simply be replaced with another system of the same size. The old philosophy of "bigger is better" no longer applies. Systems were traditionally oversized, causing them to cycle on and off frequently. Cycling that results from oversizing is inefficient and hard on the equipment. Also, rehab work may also include the addition of more or better insulation and better-performing windows and doors. This will reduce the heating and cooling loads and allow for a smaller-capacity system to be installed.

A design load analysis should be conducted to determine the current heating and cooling capacity needs. There are various methods and levels of sophistication for performing these analyses.

Most equipment vendors are equipped with worksheets or computer software to estimate the appropriate size of the system for the home. They will typically perform a sizing calculation as part of the sales process. While such a service from the dealer is available at no cost, it should be remembered that the dealer is selling equipment, not efficiency. Methods are often oversimplified with factors of safety built in, resulting in oversized equipment. An alternative is to size the system yourself. There is a multitude of books available that provide instructions, data tables, and examples for performing system sizing calculations. It is recommended that calculations be performed more than once with different methods and sources to provide confidence in the results. While sizing the system may cost a modest amount of time, lack of experience by the novice estimator may result in mistakes. Basic estimating techniques may also not properly account for unique aspects of the home. Another alternative is to hire a consultant to size the system. Professional energy specialists and auditors can evaluate the home and provide recommendations on the size and type of equipment. The advantage here is the benefit of an experienced professional who is focused on energy efficiency, but consulting fees may be hefty.

ENERGY SOURCES

ESSENTIAL KNOWLEDGE

The most common energy sources for residential use are natural gas for space and water heating and electricity for cooling. Other heating energy sources include electricity, oil, propane, and the sun. In a few instances, wood or coal may even be the primary heating fuel. Natural gas is a potential energy source for cooling as well.

The most appropriate fuel choice for a particular home depends on its availability, price, and the climate. Some of these factors may have changed since the home was built, and the fuel choice that was made then may not be the most appropriate choice now. Natural gas, a popular choice now, was not readily available prior to World War II and embargoes were placed on new gas services for a period in the 1970s. Consult the local utilities to determine the availability and rate schedules for the energy alternatives. In most instances, the primary energy sources of electricity, natural gas, and oil should be considered. Old coal-fired systems should be converted to a more-efficient and cleaner fuel alternative such as oil or gas. Various energy sources will be discussed.

TECHNIQUES, MATERIALS, TOOLS

1. ELECTRICITY AS AN ENERGY SOURCE.

In regions with a mild heating season, electricity may be the most appropriate fuel for heating and cooling needs. It is the most logical choice for cooling and, if heating loads are small, any higher cost of electricity will not be a severe penalty. And, in some areas such as the northwest where hydroelectric plants provide much of the electricity, it is quite economical. In the case of electric air conditioners and heat pumps, electricity is not used directly as an energy source but indirectly to pump heat from one location to another. For heat pumps in the heating mode, the energy source is the sun warming the air for air-source heat pumps and ground for ground-source heat pumps.

ADVANTAGES: Electric resistance heating systems are inexpensive to install, reliable, quiet, and clean. Electric air conditioners are the most common type of mechanical cooling. Electric heat pumps can serve both heating and cooling needs.

DISADVANTAGES: Electricity is generally the most-expensive fuel choice. Service to the home may need to be upgraded. Electric resistance heating (not heat pumps) is prohibited by code in some areas.

2. NATURAL GAS AS AN ENERGY SOURCE.

In most heating-season–dominated climates, natural gas is the fuel of choice for space and water heating needs. The residential space heating market shares for natural gas are 59%, 72%, and 48%, in the west, midwest, and northeast, respectively. Most of the country's gas processing plants are located in six states: Texas, Louisiana, Oklahoma, Wyoming, Kansas, and New Mexico. Canada is another significant source. The use of natural gas as a heating fuel surged with the post-World War II construction of thousands of miles of pipeline for transportation (Fig. 8.1.2). Transportation costs still make up a large portion of the consumer's price for gas. Major investments in the pipeline system during the 1980s and early 1990s improved the supply to areas in the northeast, west coast, and Florida. With the deregulation of the gas industry, gas prices fell approximately 50% from 1985 to 1991. Market competition has led to innovation and advances in technology for the exploration, extraction, and transportation of natural gas.

ADVANTAGES: Natural gas is generally the least-expensive fuel aside from solar energy. No storage tank is required.

DISADVANTAGES: The initial installation cost may be high if gas is not already supplied to the home. Natural gas is not available in all areas.

3. OIL AS AN ENERGY SOURCE.

Oil is commonly used for heating in cold climates where natural gas is not readily available. It is typically more expensive than gas unless located near a port or refinery or at the end of the natural gas pipeline. The northeast United States is located at the end of the gas pipeline, and gas is not available throughout the region. Heating loads are significant and electricity is expensive. Thus, oil has a 36% market share according to the 1993 census.

ADVANTAGES: Oil is generally less costly than electricity.

DISADVANTAGES: There is the potential for supply shortages and dramatic price fluctuations. A storage tank is required and there are associated environmental concerns and regulations (underground tanks, soil contamination, etc.).

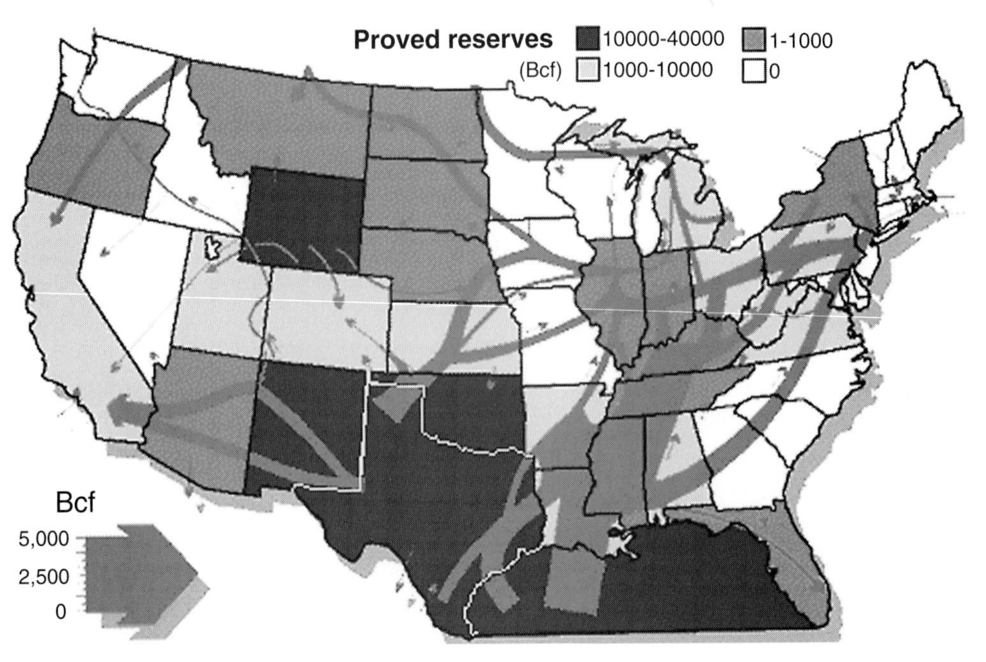

FIGURE 8.1.2 MAJOR NATURAL GAS PIPELINES

4. PROPANE AS AN ENERGY SOURCE.

Propane, or liquid petroleum gas (LPG), is typically an alternative when electricity is unattractive (usually because of price) and natural gas is not available. Propane comes from natural gas processing and crude oil refining and approximately 85% of the country's supply is produced domestically. It is transported in a liquid state by pipeline, rail car, or tank truck to retail markets.

ADVANTAGES: Propane can be used for most gas-fired equipment with only minor modifications.

DISADVANTAGES: An on-site storage tank is required.

5. SOLAR AS AN ENERGY SOURCE.

The use of solar energy to provide a portion of the home's space or water heating is a good option in many geographic areas. Passive systems use building orientation and construction materials to enhance natural processes to collect, store, and distribute heat. Active systems employ pumps and/or fans. Hybrid systems use small pumps or fans to enhance the performance of a passive system. Photovoltaic (PV) systems convert solar energy directly to DC power which is inverted to AC power for home use. Unfortunately, the market advancement of this technology is suffering from the costly and limited production of the ultra-pure silicon wafers which are the main component of the PV system (Fig. 8.1.3).

ADVANTAGES: There is a minimal operating cost, and solar energy is environmentally benign.

DISADVANTAGES: There is an initial cost for design and installation. There are optimum location and aesthetics issues.

FIGURE 8.1.3 PHOTOVOLTAIC PANEL WITH SILICON WAFERS

8.2 DISTRIBUTION SYSTEMS

ESSENTIAL KNOWLEDGE

By the end of World War I, the majority of urban homes and many rural homes were centrally heated by a hot-air, hot-water, or steam distribution system. Today there are four basic methods or media for distributing heat in the home: steam, air, water, and electric resistance. All have unique characteristics which may or may not be desirable for the house being rehabilitated.

Steam is one of the oldest types of central heat distribution systems. A boiler produces steam piped through the house to radiators, which provide concentrated heating surfaces in each room. Such systems are either one-pipe or two-pipe. One-pipe systems require a larger pipe because steam rises to the radiators while condensate returns in the same pipe from the radiators to the boiler. If the single pipe is too small, steam will force condensate back up the pipe, causing noise as steam slugs through pockets of water. Two-pipe systems use smaller pipe but twice as much of it because separate steam supply and condensate return lines are installed. Gravity or a condensate pump may be used to return the condensate to the boiler, depending on the height of the condensate piping relative to the boiler. Two-pipe systems installed after World War I include "steam traps" to prevent steam from getting into the return piping.

Air is by far the most common distribution medium for heating and cooling systems. Forced-air heating systems were used in 63% of the homes in the United States, according to 1993 census data. Original "gravity" air distribution systems relied on natural convection. Heated warm air would rise and distribute itself through the home as cold air fell to be reheated. These systems often resulted in uneven temperatures in the home. As electricity became available to homes, forced-air systems replaced most of the gravity systems. These systems employ forced convection, using an electric fan, to push the hot air through ductwork and supply registers throughout the home and pull cold air through the return (Fig. 8.2.1). Converting a gravity system to a forced-air system involves more than adding a fan to the system. A new ductwork system should be installed with proper duct sizes and register locations to assure the appropriate distribution of warm air throughout the home.

Water, or hydronic, systems are efficient because the higher heat capacity of water requires less pumping energy than fan energy in an air system. They are also inherently safer than steam systems because it is nearly impossible to run a hot-water boiler to dangerous pressures. Old hot-water systems relied on gravity flow. Hot water expands as it is heated, rises to the radiators, and forces the condensed water through the return piping to the boiler. An expansion tank with an overflow pipe discharges any extra water in the system. Today's systems employ a pump and forced flow (Fig. 8.2.2). Hot water moves at a greater speed with better heat transfer efficiency through smaller pipes. In a typical hydronic system, hot water passes through a finned pipe in a baseboard radiator located at the base of the outside wall. Air is warmed and rises by convection to circulate into the room. (They are called radiators, but more heat is delivered via convection than radiation.) True radiant systems with pipes embedded in the floor are also available at an installation cost premium.

Electric resistance distribution systems typically use baseboards (Fig. 8.2.3) but may include wall units or radiant systems using embedded cable or panels. These systems are sometimes referred to as zonal or direct because the primary source of heating or cooling is within the space. There is

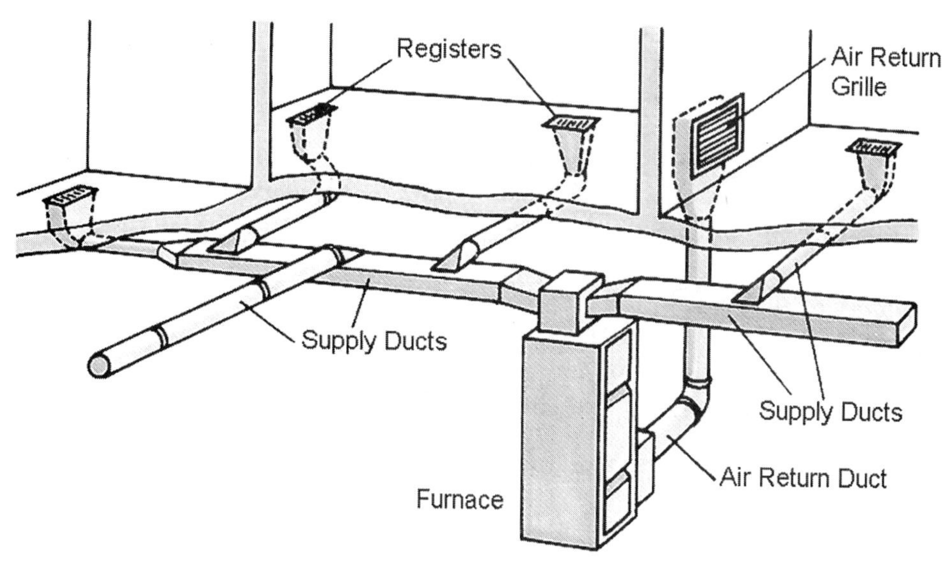

FIGURE 8.2.1 TYPICAL FORCED-AIR DISTRIBUTION SYSTEM

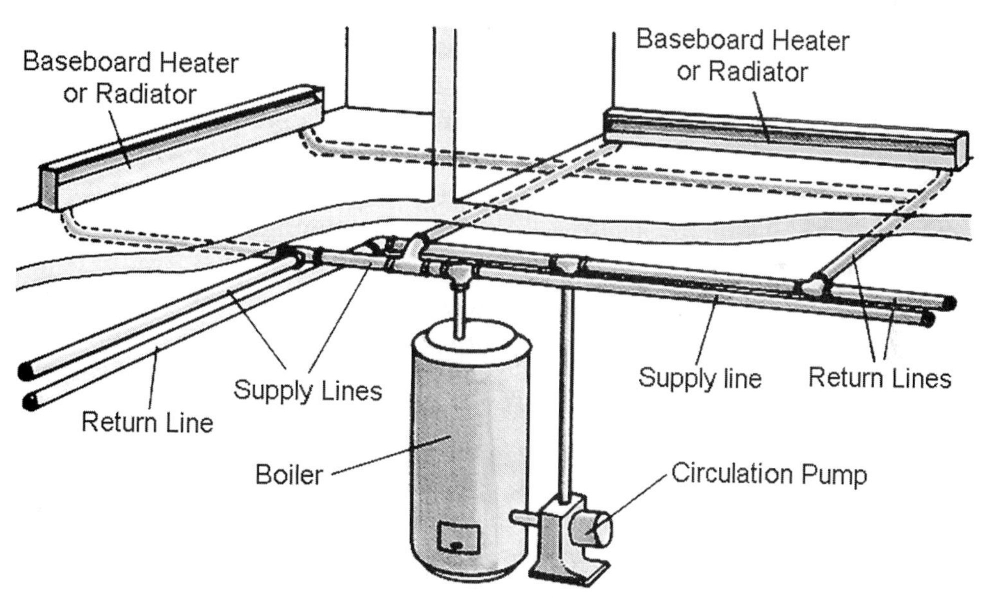

FIGURE 8.2.2 TYPICAL HYDRONIC DISTRIBUTION SYSTEM

no distribution system from a centrally located system. In common electric baseboard systems, the air is heated and relies on convection to distribute heat to the room, similar to hydronic baseboard systems. Electricity can also be used in radiant panel systems which can be surface mounted to existing walls and ceilings. Electric resistance systems are typically inexpensive to install, but generally more expensive to operate because of the high cost of electricity. However, effective use of controls and zoning can reduce operating costs.

Independent of the type of distribution, the existing system in an older home is likely to need major repair or replacement. Corrosion and leaks in ductwork and piping can contribute to inefficiency, poor comfort, and poor indoor air quality. Asbestos insulation may also be present on the old ductwork or pipes. If so, it should be either removed or encased by a professional contractor. If the asbestos insulation is in good condition, encasing it may be more cost-effective than removal. Recommendations regarding the removal and disposal of asbestos are available from the Environmental Protection Agency (EPA) Asbestos Information Hotline: 800-438-2474, or www.epa.gov/iaq/asbestos.html.

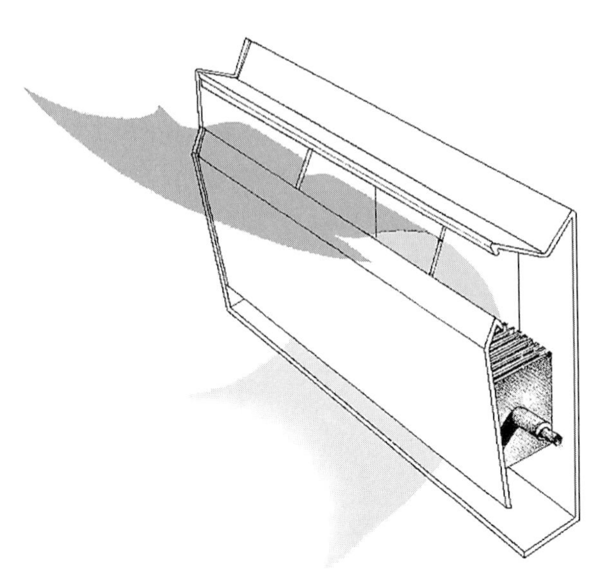

FIGURE 8.2.3 ELECTRIC BASEBOARD

TECHNIQUES, MATERIALS, TOOLS

1. REHAB THE EXISTING FORCED-AIR DISTRIBUTION SYSTEM.

Restoring the existing distribution system will likely involve sealing the supply ductwork to minimize the loss of conditioned air before it reaches the supply registers. According to a Housing and Urban Development (HUD) study, leaky ductwork can increase heating and cooling costs by as much as 30% and cause air pressure problems that result in drafts and uneven room temperatures. Return ductwork should be sealed so that unconditioned air from an attic or crawl space is not pulled into the system. Instead of sealing tape, mastic should be used because it seals better, lasts longer, and is easier to apply. A relatively new system for sealing ducts, particularly useful in retrofit situations where access can be limited, is aerosolized mastic. Ducts are sealed from the inside by pumping aerosolized mastic into the ducts under pressure. The mastic migrates through the system and seals small leaks without coating the interior of the ducts. Note that this works for small openings only. Once sealed, all supply and return ducts that pass through unconditioned space such as attics, crawl spaces, and basements should be insulated to minimize conduction losses in the winter (and gains in the summer, if used for air conditioning). Old registers and grilles that are no longer (or never were) adjustable should be replaced with new adjustable equipment for better air distribution and comfort.

ADVANTAGES: Rehabilitation of the existing forced-air distribution system improved efficiency, comfort, and safety.

DISADVANTAGES: Rehabilitation of a poorly designed system provides minimal benefit.

2. REHAB THE EXISTING HYDRONIC DISTRIBUTION SYSTEM.

For hot-water systems, restoring the existing distribution system will likely involve replacing rusted pipe sections and insulating pipes running through unconditioned spaces. The best type of pipe insulation consists of 3' sections of extruded foam that is slit so it can be applied over the pipe. Valves that are no longer operable should be replaced, as should old baseboard radiators with bent fins or guards. Replacing radiators is probably more an issue of lifestyle and aesthetics. Finned copper radiators work faster and have a higher heat output than old-style cast-iron radiators, but the cast-iron radiators hold more water and give off heat for a longer time. There are three-way thermostatic bypass valves, such as those distributed by Enerjee, which can be retrofitted to existing hydronic baseboard units. When the room temperature sensed at the valve is satisfactory, circulating hot water is bypassed around the finned tube to minimize the heat distributed to the room. This type of valve is intended for use with continuously circulating systems, but could be useful for redistributing the loads on a hydronic system without changing the baseboard units.

For steam systems (Fig. 8.2.4), restoring the existing distribution system will likely involve repairing or replacing radiator valves, which serve as an on/off control, and should either be fully open or closed. Worn valve seats can result in gurgling as steam leaks in, but condensate can't flow back. The stem packing for most radiator inlet valves deteriorates with time; the valves can be repacked with special graphite-impregnated cord. Broken valve handles should be replaced with new handles of insulating plastic. Vents control the rate at which steam enters the radiator (and thus the radiator's output) by regulating the amount of air in and out of the radiator. If the radiator never heats up (never hisses) or steam comes out of the vent (continuously hisses), the vent needs to be replaced. Replacement of vents with the proper speed can also help with unbalanced heat distribution in the home. Vents come in four speeds (very slow, slow, fast, and very fast) or variable. For rooms that tend to overheat, use a slower vent and vice versa. Vacuum vents allow air to escape, but not reenter as the radiator cools. A vacuum is created within the radiator as the steam condenses and hot steam is drawn into the radiator rather than cold air. This can result in a more steady output of heat. Special "packless" airtight control valves need to be installed in conjunction with the vacuum vents. Radiators should be sloped slightly toward the steam valve. If they are not, condensate will likely pool at the far end of the radiator and steam bubbling through will "gurgle." A shim should be placed under the radiator feet farthest from the valve to slope the radiator slightly toward the valve.

ADVANTAGES: This is a low-cost and the least disruptive method.

DISADVANTAGES: Hydronic systems are only suitable for heating.

3. REHAB THE EXISTING ELECTRIC RESISTANCE SYSTEM.

While electric resistance systems are relatively simple with few parts, there may still be the need for rehabilitation. In addition to cleaning, damaged fins or fin guards should be repaired or replaced because they affect the performance of a baseboard.

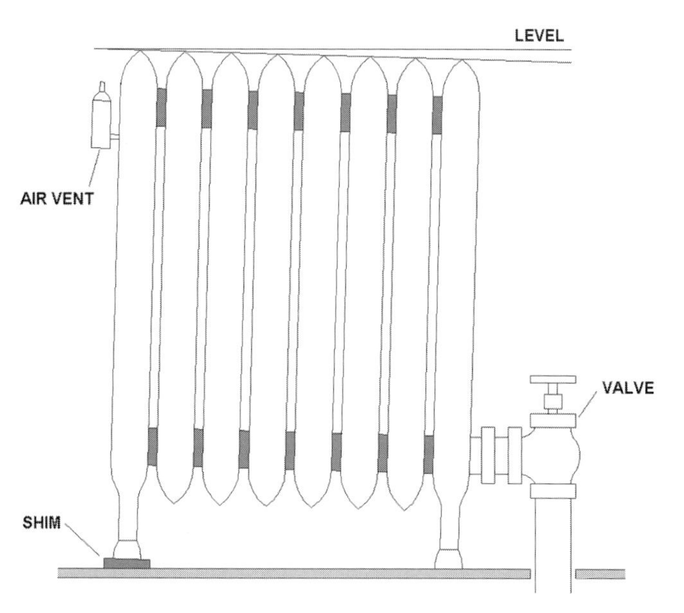

FIGURE 8.2.4 STEAM RADIATOR

ADVANTAGES: Performance of the electric resistance system will be improved.

DISADVANTAGES: Replacement of the entire unit may be simpler than repairing or replacing components.

4. INSTALL A NEW CONVENTIONAL FORCED-AIR DISTRIBUTION SYSTEM.

Conventional forced-air distribution systems are by far the most common type of heating system. However, they are notorious for being the cause of inefficiency and discomfort. Poorly designed and installed ductwork can have high levels of air leakage and poor temperature control. Systems are often designed with poorly insulated ductwork running through unconditioned space. A design that keeps ductwork within the conditioned space avoids many of these problems. Duct losses are reduced because conditioned air is leaking into the conditioned space, and conduction losses are lower because temperature differences between the supply and surrounding air are less. The location of supply outlets depends upon whether the emphasis is on heating or cooling. For heating-dominated systems, standard practice has been to install supply registers along the outside walls under windows to counter cold drafts coming from windows. However, in a tightly constructed home with insulating windows, interior wall registers can be used to save on duct material. For cooling-dominated systems, ceiling supply diffusers are most appropriate. Ducts are usually made of square and/or round sheet metal, but insulated duct board and flex duct are also widely used. Fiberglass duct board is quieter than sheet metal ducts because such ducts attenuate the blower noise that can propagate through the duct system. Care must be taken to avoid problems with condensation, which contributes to mold growth, if the system is used for air conditioning. Owens Corning has introduced EnDuraCoat—a duct board system with an antimicrobial acrylic interior coating to resist the growth of fungi and bacteria.

ADVANTAGES: Conventional forced-air distribution systems are widely used and accepted. They can readily include air conditioning, humidification, or air purification and are fast responding.

DISADVANTAGES: Significant space requirements of ductwork can make installation in an existing structure difficult and costly if the previous distribution system was not air. Duct system leakage can cause inefficiency. Forced-air systems can be noisy.

5. INSTALL A MINI-DUCT HVAC SYSTEM.

There are at least two systems on the market, Unico and SpacePak, that feature small-diameter, flexible ductwork that can pass through studs and joists and snake through narrow openings and around corners (Figs. 8.2.5 and 8.2.6). The typical system delivers less air at higher velocities than conventional forced-air systems; to achieve the same heating or cooling capability, the air is delivered at higher temperatures when heating and lower temperatures when cooling. Air is discharged through plastic collars with 2" diameter holes in the ceiling, floor, or wall. Since air is supplied at more extreme temperatures and higher velocities, these outlets must be strategically placed to avoid blowing directly on occupants. Special sound attenuating tubing is used at the end of each supply run to minimize the noise caused by high air velocities.

ADVANTAGES: Ducts can be installed in tight areas. Lower supply air temperatures may provide better humidity control in the summer.

DISADVANTAGES: Building professionals are unfamiliar with the system. The lower installation cost might be offset by higher equipment cost.

6. INSTALL A NEW HYDRONIC DISTRIBUTION SYSTEM.

There are alternative types of hydronic systems to consider. Systems involving traditional baseboards can be single-pipe or two-pipe, and radiant systems are also an option. All hydronic systems require an expansion tank to compensate for the increase in water volume when it is heated (i.e., the volume

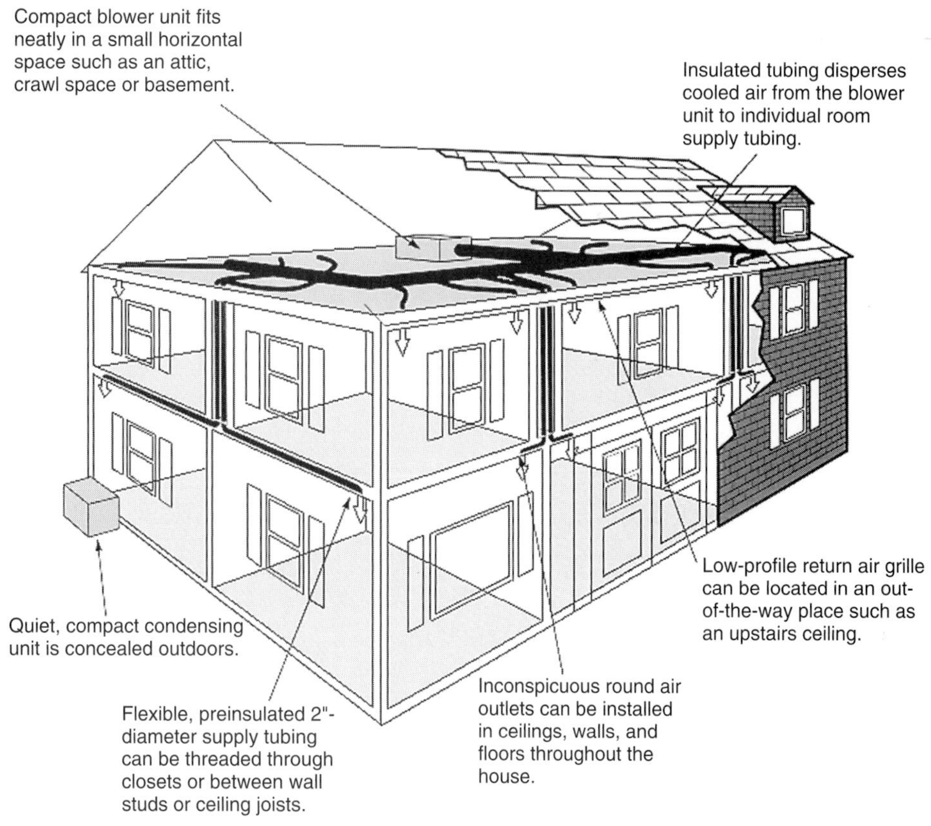

FIGURE 8.2.5 MINI-DUCT DISTRIBUTION SYSTEM

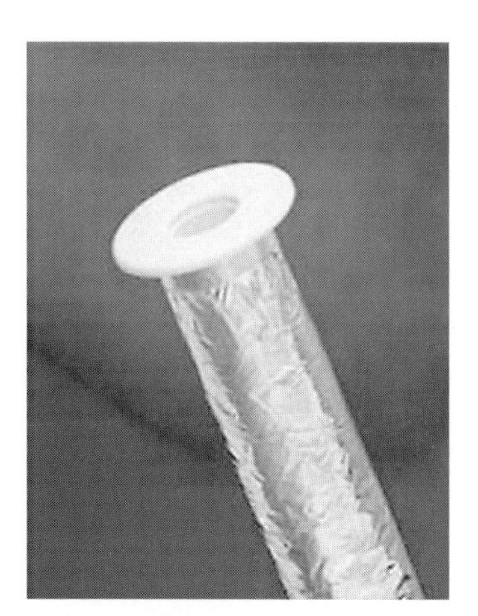

FIGURE 8.2.6 MINI DUCT

of 50°F water increases almost 4% when it is heated to 200°F). Single-pipe hydronic systems are most commonly used in residences. They employ a single pipe with hot water flowing in a series loop from radiator to radiator. The drawback to this arrangement is that the temperature of the water decreases as it moves through each radiator. Thus, larger radiators are needed for those locations downstream in the loop. A common solution to this is multiple loops or zones. Each zone has its own temperature control with circulation provided by a small pump or zone valve in each loop (Fig. 8.2.7). Two-pipe hydronic systems use a pipe for supplying hot water to the radiators and a second pipe for returning the water from the radiators to the boiler. There are also direct- and reverse-return arrangements (Fig. 8.2.8). The direct-return system can be difficult to balance because the pressure drop through the

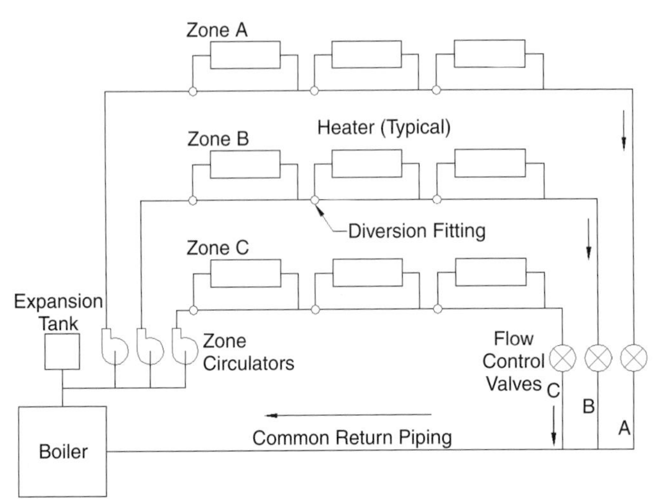

FIGURE 8.2.7 MULTIZONE, SINGLE-PIPE HYDRONIC DISTRIBUTION SYSTEM

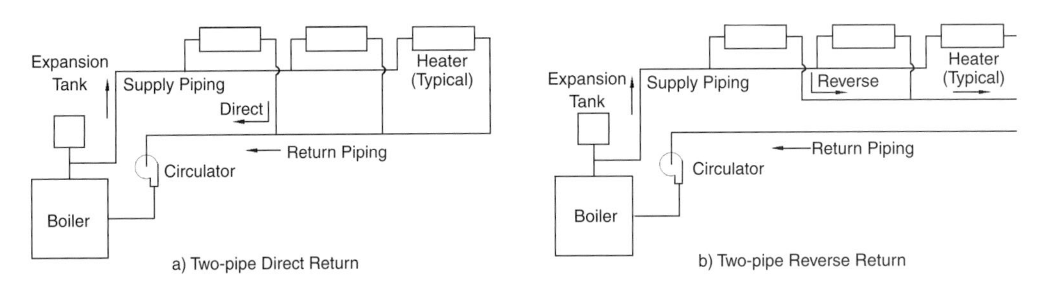

FIGURE 8.2.8 TWO-PIPE HYDRONIC DISTRIBUTION SYSTEMS

nearest-radiator piping can be significantly less than for the farthest radiator. Reverse-return systems take care of the balancing problem, but require the expense of additional piping. Orifice plates at radiator inlets or balancing valves at radiator outlets can also be used to balance the pressure drops in a direct-return system.

Radiant systems run hot water through plastic pipe or tubing typically embedded in floor slabs or under framed floors (Fig. 8.2.9). Systems involving panels that look like baseboard molding or panels that can mount in the wall or ceiling are also available. These systems warm the surrounding objects rather than the air and can generally provide better comfort than baseboard systems. The introduction of cross-linked polyethylene (PEX) tubing to the U.S. market in the 1980s revolutionized the installation of hydronic floor heating with fast installation and longer service life.

There are also radiant cooling systems that involve running cool water through the same pipe or tubing or panels as used for heating. These systems must be designed carefully to ensure that the temperature of the radiant surface (floor, ceiling, or wall panel) remains above the dewpoint of the room air. Otherwise, harmful and potentially dangerous, in the case of a slippery floor, condensation can occur. For this reason, radiant cooling systems typically supplement another type of cooling system.

ADVANTAGES: Small piping or tubing is more adaptable to an existing building structure than ductwork. Hydronic systems are clean, quiet, have fewer heat losses, and can be easily zoned.

DISADVANTAGES: Hydronic systems are only used for heating. Radiant cooling systems are not commonly used and are therefore more difficult to design and usually more costly than other alternatives.

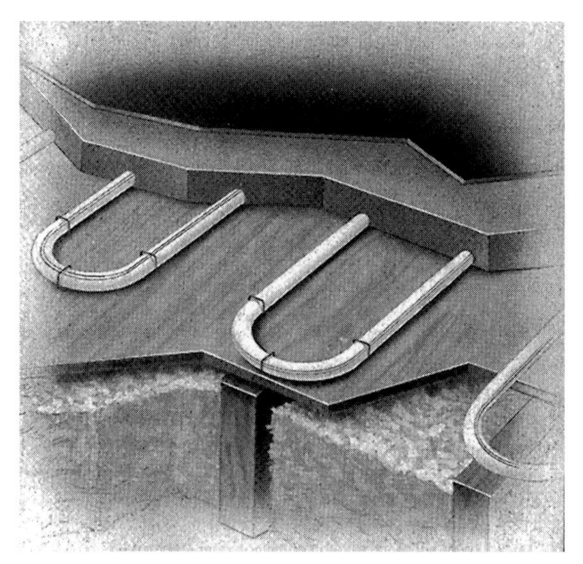

FIGURE 8.2.9 HYDRONIC RADIANT FLOOR HEATING SYSTEM

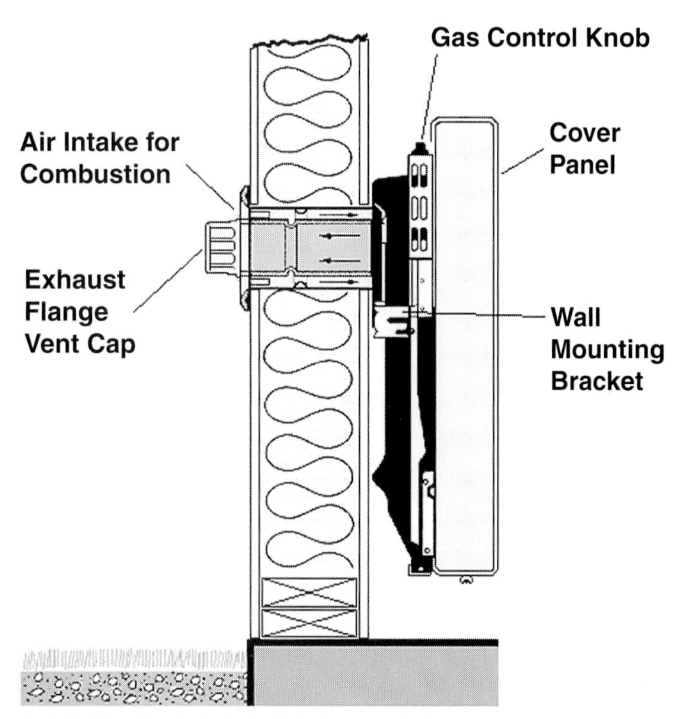

FIGURE 8.2.10 DIRECT-VENT GAS SPACE HEATER

7. INSTALL A ZONAL SYSTEM.

A zonal system utilizes individual in-space heaters in each of the rooms to be heated. Essentially, there is no distribution system with its inherent losses and inefficiencies. Significant savings can be achieved through the use of setback controls and zoning; providing heat when and where it is needed. These heaters can be electric or gas and may employ natural convection, fan-forced convection, and/or radiation to transmit the heat to the space. Some units also include a thermal storage medium to improve comfort.

Gas space heaters are available from several manufacturers in a range of capacities. The units employ sealed combustion, so there is no potential of mixing combustion gases with the internal air via spillage or backdrafting. Direct-vent units operate with efficiencies of 60% to 75%. They mount on an external wall and vent the combustion gases through the wall (Fig. 8.2.10). Power-vented units with efficiencies above 80% can duct the exhaust gases so they have more location flexibility.

Electric heaters are less expensive than gas-fired heaters but electricity is typically the more expensive fuel. Electric baseboard heaters are the most common zonal heater in residential applications. Other electric units include wall, kickspace, floor, and ceiling heaters. Baseboard units typically rely on natural convection while the other types include a fan to augment the convection process. Some units are available with two stages to better match the output to the load. Others include an electric immersion element and a liquid solution hermetically sealed in copper tubing. The thermal capacitance of the liquid serves to modulate the output of the unit.

Radiant systems that use ceiling, wall, or floor panels with electric wiring embedded are also a type of zonal system. These units can be faster responding than hydronic radiant systems.

ADVANTAGES: Zonal systems are inexpensive to install, offer flexibility, and provide zone-control capability.

DISADVANTAGES: The types of zonal systems discussed only provide heating.

8.3 HEATING

ESSENTIAL KNOWLEDGE

There are two basic types of central heating plants: furnaces (used with air distribution systems) and boilers (used with either hydronic or steam systems). Another less-common alternative is to use a domestic hot-water heater to supply hot water to a hydronic coil in an air-handling unit or for a radiant system. These systems are often referred to as combination systems and are gaining in popularity.

Warm-air furnaces (Fig. 8.3.1) are the most common residential heating system in the United States. According to 1993 census data, furnaces are used in 37% of electrically heated homes and 69% of gas-heated homes. If a gas- or oil-fired furnace is presently installed in the house to be rehabilitated, it is probably much less efficient than today's furnaces. The efficiency of most gas- or oil-fired furnaces installed 20 years ago is between 50% and 65%. Standards that went into effect in 1994 require that furnaces sold today be at least 78% efficient. This standard essentially eliminated the use of standing pilots on gas furnaces, and electronic ignition devices and vent dampers became standard. Gas-fired models with efficiencies over 90% and oil-fired models with efficiencies as high as 87% are available today.

Furnaces are available in upflow, downflow, and horizontal configurations. Upflow systems are the most common, with blowers drawing air into the bottom and supplying heated air out the top to the duct system. Downflow systems draw air in at the top and supply heated air out the bottom for floor duct systems. Horizontal systems draw air in one side and supply heated air out the other side. They are used in attics, crawl spaces, below floors, or suspended from ceilings.

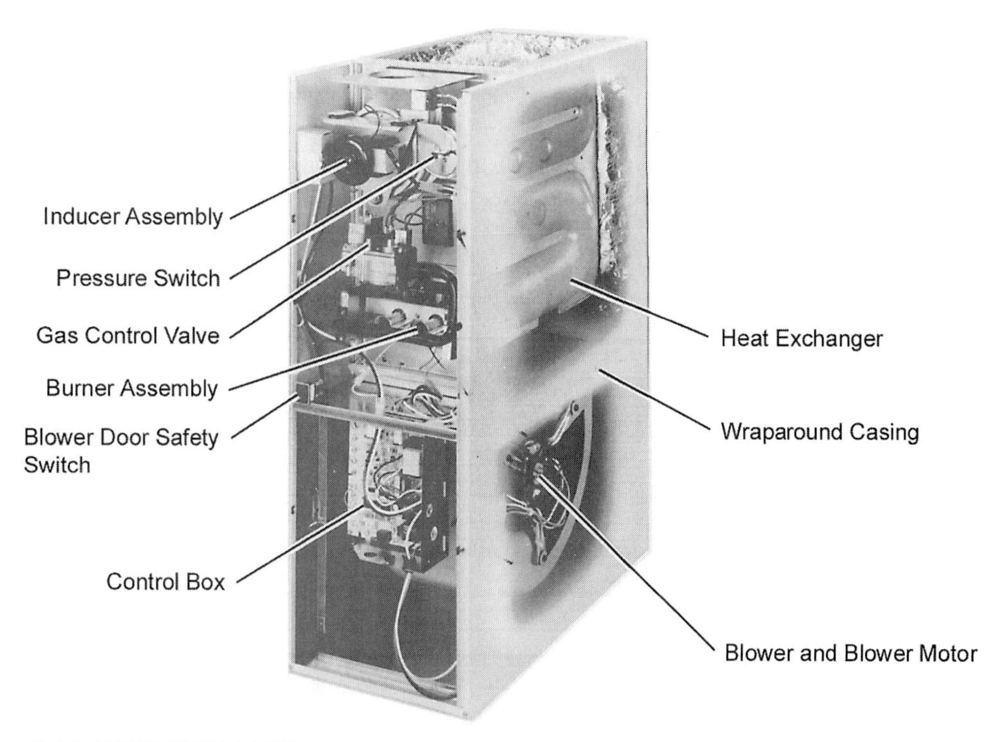

FIGURE 8.3.1 GAS-FIRED FURNACE

According to 1993 census data, boilers are used for heating in 15% of homes across the United States and approximately 45% of those in the northeast (Fig. 8.3.2). Government standards require that today's residential combustion boilers have efficiencies of at least 80%. The Quantum Leap boiler by Dunkirk Radiator has a 95% efficiency. While boiler efficiencies are typically lower than furnace efficiencies, total system efficiency, including distribution system losses, can be higher because duct losses are typically higher than pipe losses.

Electric furnaces and boilers operate with a 100% fuel conversion efficiency. However, distribution system losses often make them less efficient as a system than electric baseboard or other zonal systems. If an electric furnace is presently installed in the home, the integrity of the ductwork system should be examined and improved as a minimum (see Section 8.2). If the electric furnace or boiler is situated in a cold climate, serious consideration should be given to converting it to a gas-fired system to reduce operating costs.

This section will focus on fuel-burning, or combustion, systems. In combustion systems, combustion occurs within a chamber and air circulates over the outside surfaces of a heat exchanger. The circulation air does not come in contact with the fuel or the products of combustion, which are vented to the outside. Traditional combustion systems are natural draft (or atmospheric vent) with a draft hood to carry the combustion gases through the vent to the outdoors. The draft hood has a relief air opening to assure that the exit of the combustion heat exchanger is at atmospheric pressure. Fan-assisted or powered-combustion systems use a small blower to force (forced-draft, power burner) or induce (induced-draft, mechanical draft, power vent) the flue products through the system. Fan-assisted systems do not require a draft hood, resulting in reduced off-cycle losses and improved efficiency.

Backdrafting of combustion appliances occurs when the surrounding area is depressurized and the appliance relies on natural draft to carry the combustion gases through the vent to the outdoors. Backdrafting can be dangerous when deadly combustion gases are drawn into the living space. Depressurization can occur when other devices such as exhaust fans, water heaters, clothes dryers, and fireplaces are drawing air from the space and exhausting it to the outside. The best solution to this problem is the installation of a direct-vent combustion system that uses outside air for combustion and does not have a draft hood. This is referred to as a sealed combustion system because the combustion air is isolated from the indoor air.

The following are techniques for improving the safety, improving the efficiency, and/or reducing the operating costs, of the existing combustion furnace or boiler system. For safety reasons, all modifications should be performed by a qualified mechanic.

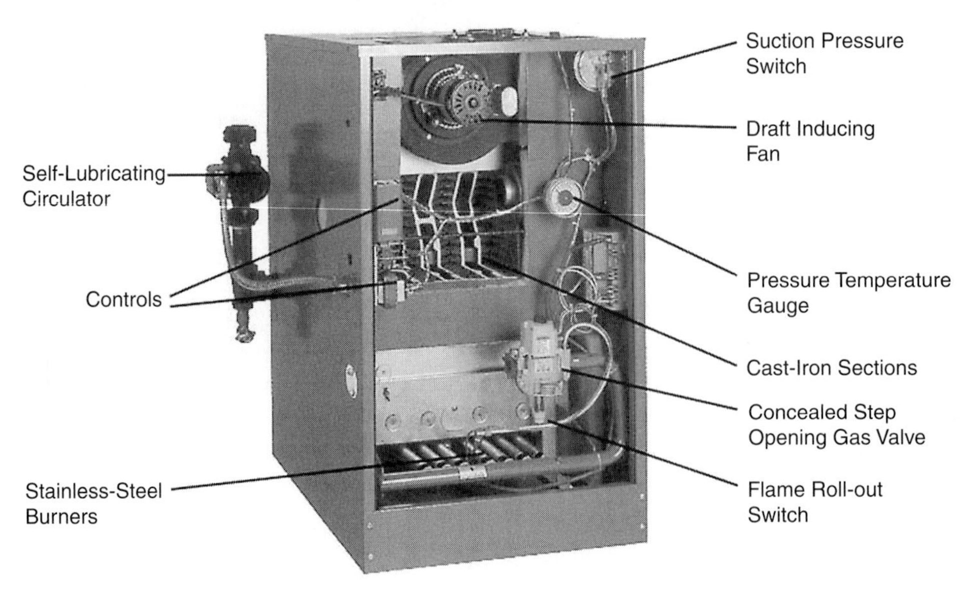

FIGURE 8.3.2 GAS-FIRED BOILER

TECHNIQUES, MATERIALS, TOOLS

1. FOR OIL-FIRED SYSTEMS, INSTALL A FLAME-RETENTION BURNER.

Flame-retention burners have smaller air intakes and a more concentrated flame that enables them to burn hotter and more efficiently.

ADVANTAGES: Combustion efficiency is improved.

DISADVANTAGES: None.

2. INSTALL A VENT DAMPER.

A vent damper is a flapper installed in the flue of a draft hood–equipped natural draft furnace or boiler. (Fan-assisted natural draft systems do not have draft hoods.) When the burners are ignited, the damper is open to allow combustion gases to exhaust up the flue to the outdoors. The damper closes when the burners are shut off to reduce exfiltration of heated air from the house and to prevent residual heat from the heat exchanger from escaping up the flue. The National Appliance Energy Conservation Act (NARECA) standards passed in 1987 made vent dampers a standard feature of today's natural-draft furnaces.

ADVANTAGES: Efficiency is improved if the furnace is located in the conditioned space and is using indoor air for combustion.

DISADVANTAGES: The vent damper is of less value if the combustion air is drawn from outside or unconditioned space.

3. REPLACE THE STANDING PILOT WITH AN ELECTRONIC SPARK IGNITOR.

Old furnaces and boilers employ a standing pilot that consumes a small amount of gas continuously. The NAECA standards passed in 1987 essentially eliminated this energy-wasting practice. Electronic ignitions use an electric spark or a hot surface as an ignition source for the gas mixture when heating is needed.

ADVANTAGES: Efficiency is improved.

DISADVANTAGES: A new ignition system can be costly.

4. FOR OIL-FIRED FURNACES, DOWNSIZE THE BURNER NOZZLE.

The firing rate of an oil-fired furnace is controlled by the pump pressure and the orifice size of the burner injection nozzle. The nozzles are rated in gallons per hour. If the home's heating loads have been reduced or the system was oversized initially, the nozzle size may be reduced to decrease the capacity of the furnace.

ADVANTAGES: Comfort through less system cycling is improved.

DISADVANTAGES: Smaller nozzles become clogged more easily by sediment. Fuel line filters may need to be changed more frequently.

5. INSTALL AN ADVANCED, HIGH-EFFICIENCY GAS FURNACE.

High-efficiency gas furnaces can offer a significant improvement in efficiency. Efficiency ratings between 84% and 89% are not common because acidic corrosive condensate forms at these efficiencies and the modest improvement in efficiency is not worth the added material cost needed to withstand the condensate. Thus, high efficiency means jumping from efficiencies in the low 80s to the low 90s. There are several alternatives available for those who want to install an advanced and efficient furnace system, including pulse and condensing combustion technologies.

Nearly every major furnace manufacturer includes a condensing furnace in their product line. Condensing furnaces are designed to condense the water vapor from the exhaust gases and capture the heat of condensation. This is done by lowering the temperature of the combustion gases with a second heat exchanger in the furnace (Fig. 8.3.3). Condensing furnaces are expensive because they require corrosion-resistant materials, but they can have efficiencies as high as 97%.

Many manufacturers provide furnaces with two-speed or even variable-speed motors. These systems can improve comfort and run more quietly when used with zoning systems, with an air-conditioning system that requires a higher airflow, or when it is desirable to run the fan continuously for ventilation air supply. The Carrier Weathermaker 8000 and Trane XV-90 are two such products.

ADVANTAGES: Operating costs are lower. Higher-efficiency systems have lower-temperature exhaust gases, so PVC piping can be used for venting, a significant advantage if a chimney is not currently in place. The multispeed systems provide better comfort because output follows load more closely. The low-speed setting can provide dehumidification capability with an air conditioning unit.

DISADVANTAGES: The initial cost is higher, and the existing flue may need to be downsized. All fan-assisted furnaces consume more electricity than old furnaces that did not use a fan in the combustion air stream. Local code may require a neutralizer cartridge for condensate disposal. If condensate cannot be drained by gravity, a condensate pump is necessary.

6. INSTALL A COMBINATION SYSTEM.

Combination systems use one combustion device to provide space and water heating. These systems eliminate the gas furnace and its flue by running a hot-water coil from the domestic hot-water heater to the air-handling unit or a hydronic radiant system (Fig. 8.3.4). The Gas Research Institute estimates that nearly a million of these systems have been installed in the United States.

Typically, potable water is used in the air-handler heating coil. For hydronic radiant systems, a separate heat exchanger in the water heater is used to circulate nonpotable water through the radiant system piping. This is a code requirement in most areas.

Water heaters specifically designed for these systems are available. They have an extra set of taps for supply and return of the space heating water and higher recovery rates and direct venting as options.

ADVANTAGES: The furnace and its venting requirements are eliminated.

DISADVANTAGES: There is dealer confusion over sizing and designing of systems.

7. INSTALL AN ADVANCED HIGH-EFFICIENCY BOILER.

Condensing boilers are less common than furnaces because they typically are not compatible with the high return-water temperatures, such as the 160°F of a hydronic baseboard system. They are appropriate for the lower temperatures of radiant systems. However, Dunkirk Radiator produces the Quantum Leap boiler with an efficiency of 95%. This unit, which uses an aluminum heat exchanger, is a condensing boiler even at high return-water temperatures. Condensation at high return-water temperatures is accomplished by heating and saturating the incoming combustion air to raise its dewpoint. This is done with the condensate in an evaporative tower (Fig. 8.3.5).

ADVANTAGES: Operating costs are lower. Higher-efficiency systems have lower-temperature exhaust gases so PVC piping can be used for venting. There are improved safety devices and controls over an old boiler system.

DISADVANTAGES: The initial cost is higher.

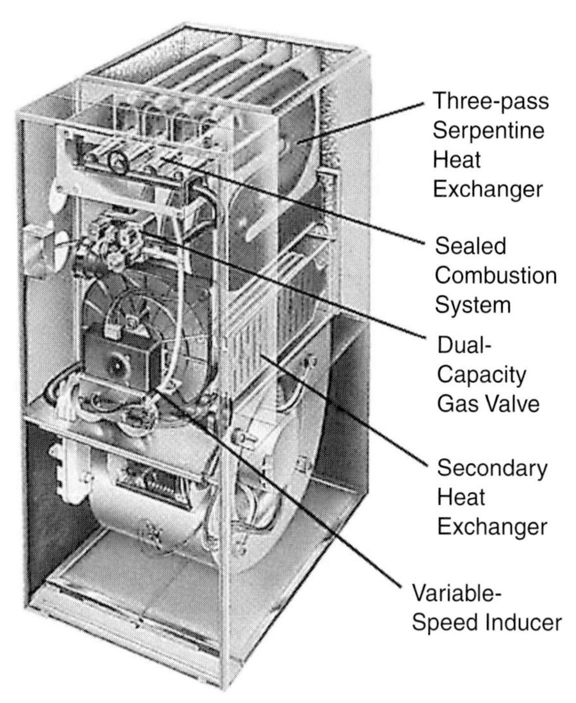

FIGURE 8.3.3 CONDENSING FURNACE

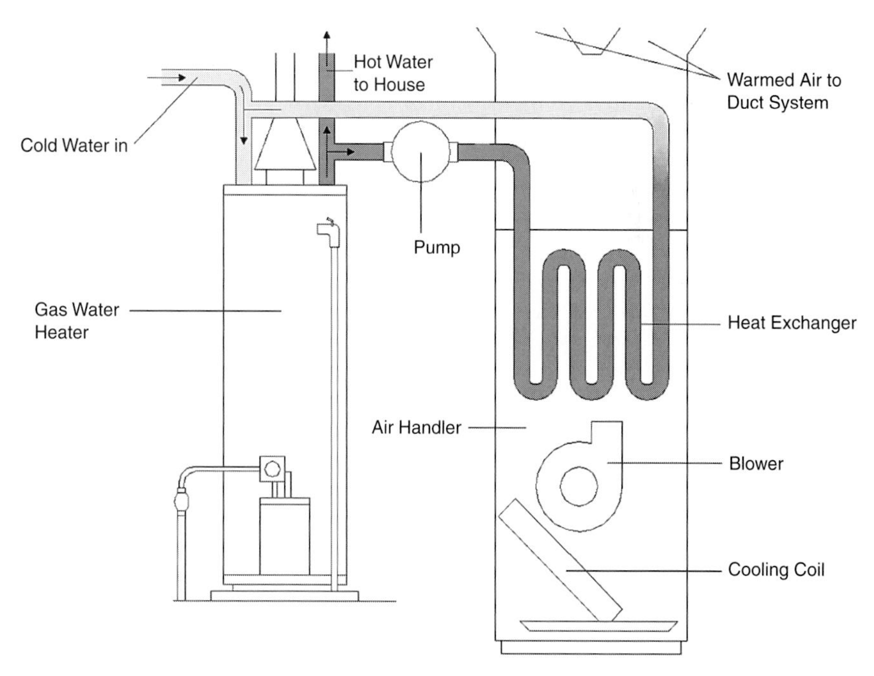

FIGURE 8.3.4 COMBINATION HEATING SYSTEM

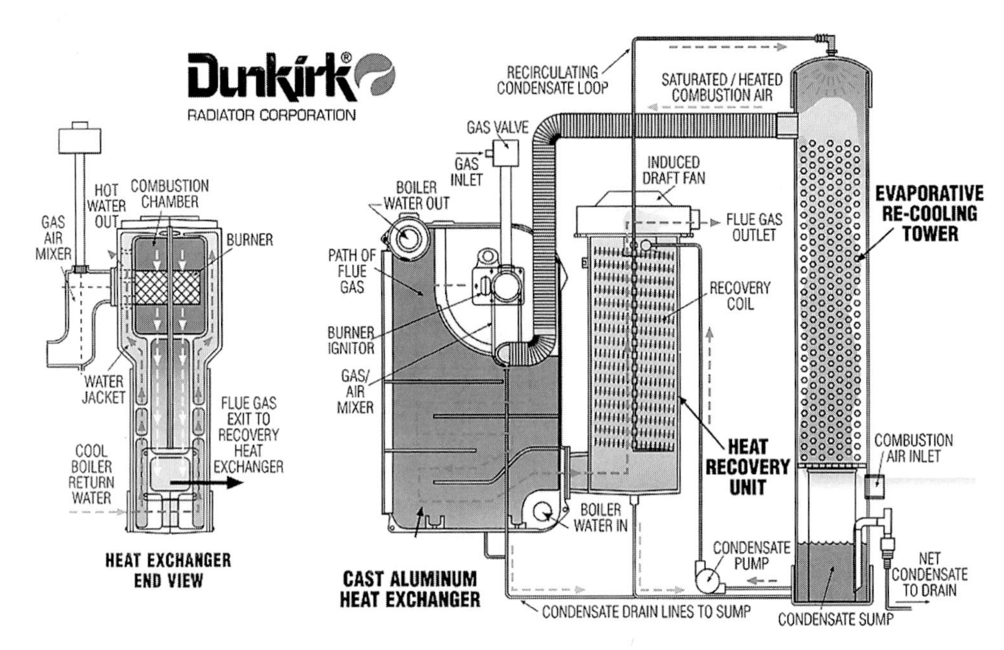

FIGURE 8.3.5 QUANTUM LEAP CONDENSING BOILER

8.4 COOLING

ESSENTIAL KNOWLEDGE

Many old homes relied on passive cooling—the opening of windows and doors, and the provision of shading devices—during the summer months. Homes were designed with windows on opposite walls to encourage cross-ventilation, and large shade trees reduced solar heat gains. This approach is still viable in many areas today, and improved thermal performance (insulating value) windows are available that allow for larger window areas to let in more air in the summer without the heat loss penalty in the winter. However, increased outdoor noise levels, pollution, and security issues make relying on open windows a less-attractive option in some areas today.

An air-conditioning system of some kind may already be installed in the home. It may be a window air conditioner or a through-the-wall unit for cooling one or two rooms, or a central split-system air conditioner or heat pump. In any event, the performance of these systems in terms of providing adequate comfort without excessive energy use should be investigated. The age of the equipment alone will provide some indication. If the existing system is more than 10 years old, replacement should be considered because it is much less efficient than today's systems and nearing the end of its useful life.

The refrigerant commonly used in today's residential air conditioners is R-22. Due to the suspicion that R-22 depletes the ozone layer, manufacturers will be prohibited from producing units with R-22 in 2010. The leading replacements for R-22 are R-134A and R-410A, and new air conditioners are now available with these non-ozone-depleting refrigerants. Carrier's brand name for the R-410A refrigerant is Puron.

The performance measure for electric air conditioners with capacities less than 65,000 British thermal units (Btu) per watt-hour (Wh) is the seasonal energy efficiency ratio (SEER). This is a rating of cooling performance based upon representative residential loads. It is reported in units of Btu of cooling per watt-hour of electric energy consumption including the unit's compressor, fans, and controls. The higher the SEER, the more efficient the system. However, the highest SEER unit may not provide the most comfort. In humid climates, some of the highest SEER units exhibit poor dehumidification capability because they operate at higher evaporator temperatures to attain the higher efficiency. A SEER of at least 10 is required by the NAECA for conventional central split-system air-cooled systems.

Cooling system options vary widely, depending upon the level of control and comfort desired by the homeowner. Fans can increase circulation and reduce cooling loads, but their cooling capability is directly limited by the outdoor conditions, so they may be unsatisfactory in hot climates. Radiant barriers can possibly reduce cooling loads in very hot climates. Evaporative coolers can be a relatively inexpensive and effective method of cooling in dry climates such as the southwest. Electric air conditioning is the answer for those who want to maintain a comfortable indoor temperature and humidity even under the most severe outdoor conditions. Over 75% of new homes in the United States are equipped with some form of central air conditioning; more specifically, 50% of the homes in the northeast, 75% in the midwest, 95% in the south, and approximately 60% in the west. Electric air conditioning filters and removes moisture from the air as well as reducing its temperature. It can be a good investment because, in most parts of the country, the payback is significant when the house is sold.

TECHNIQUES, MATERIALS, TOOLS

1. INSTALL A CIRCULATION FAN.

Air movement can make you feel comfortable even when dry-bulb temperatures are elevated. A circulation fan (ceiling or portable) (Fig. 8.4.1) that creates an airspeed of 150 to 200 feet per minute (ft/min) can compensate for a 4° F increase in temperature.

ADVANTAGES: The same ceiling circulation fan can also be beneficial in the heating season by redistributing warm air that collects along the ceiling.

DISADVANTAGES: A circulation fan can be noisy.

2. INSTALL A POWER ATTIC OR ROOFTOP VENTILATOR.

These units are used to assist the natural flow of air through the attic space (Fig. 8.4.2). Without good ventilation, attic temperatures can exceed 130° F on warm, sunny days. Increasing the attic ventilation results in a cooler attic space, reducing the cooling load on the space below. To estimate the required cubic foot per minute (cfm) rating of an attic ventilator, multiply the attic floor area by a factor of 0.75 and increase the value by 15% if the roof is dark and will absorb heat. The exhausted air is replaced by outside air entering through eave vents.

ADVANTAGES: A ventilator is inexpensive and easy to install.

DISADVANTAGES: A ventilator does not directly cool the living space. If the attic vent area is insufficient, the fan can draw air through the ceiling of the conditioned space and potentially depressurize the space.

FIGURE 8.4.1 CEILING CIRCULATION FAN

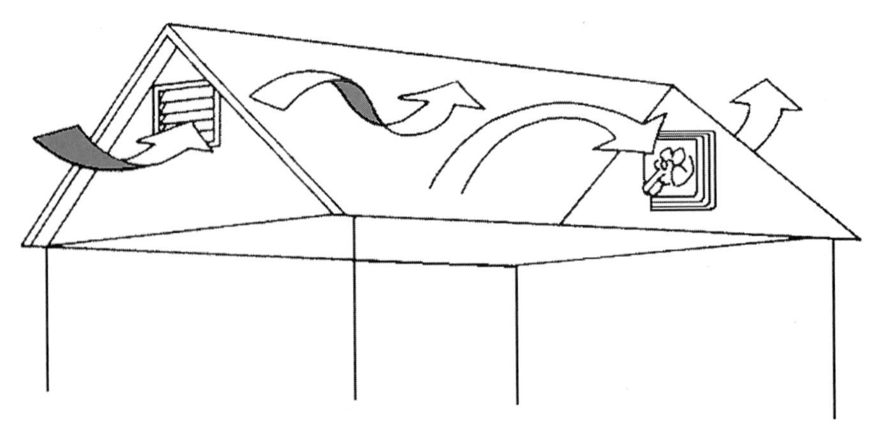

FIGURE 8.4.2 POWERED ATTIC VENTILATOR

3. INSTALL A WHOLE-HOUSE VENTILATOR OR FAN.

Whole-house ventilators are typically installed in the attic floor over a hallway or stairway. They pull hot air from the living space and exhaust it into the attic where it exits through the attic vents. Fresh cooler air is drawn in through open windows. Specific areas of the home can be ventilated by selectively opening and closing doors and windows. The HV1000 unit by Tamarack Technologies (Fig. 8.4.3) uses efficient fan motors to minimize energy use and noise and has an automatic door assembly that provides an airtight seal with an insulating value of R-22. It uses two side-by-side fans rather than one large fan so that the unit can easily fit between 16" or 24" on-center joists with no joist cutting.

ADVANTAGES: The need for mechanical air conditioning in moderate climates can be eliminated.

DISADVANTAGES: A whole-house ventilator or fan is noisy and has limited cooling capability in warm or humid climates. It can depressurize the home if windows are not open. There will be increased heat loss during the winter if the system is not sealed and insulated.

4. INSTALL A RADIANT BARRIER.

In hot climates, where attics can become very hot and air conditioning ducts are often in the attic, a radiant barrier may be beneficial. It can reduce the cooling load on the home and reduce the increase in supply air temperature as conditioned air travels through the supply ducts in the attic before reaching the rooms to be conditioned. Radiant barriers are made of materials that are good at reflecting heat. They reflect the radiant heat emanating from a hot roof and come in a variety of forms, including foil, paint coatings, and chips.

ADVANTAGES: A radiant barrier is fairly simple to install and can reduce the size of the air conditioner needed.

DISADVANTAGES: A radiant barrier can be relatively expensive. Dust can seriously degrade the performance by dulling the reflective surface.

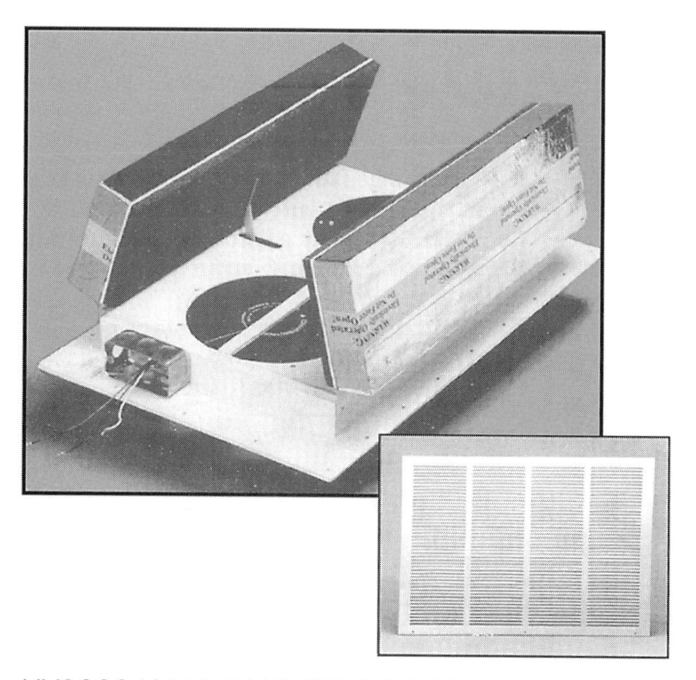

FIGURE 8.4.3 HV1000 WHOLE-HOUSE COOLER

5. INSTALL AN EVAPORATIVE COOLER.

In dry climates such as that in the southwest, an evaporative cooler or "swamp" cooler may provide sufficient cooling. This system cools an airstream by evaporating water into it; the airstream's relative humidity (RH) increases while the dry-bulb temperature decreases. A 95°F, 15% RH airstream can be conditioned to 75°F, 50% RH. The simplest direct systems are centrally located and use a pump to supply water to a saturated pad over which the supply air is blown. Indirect systems use a heat exchanger between the airstream that is cooled by evaporating water and the supply airstream. The moisture level of the supply airstream is not affected as it is cooled. Some companies manufacture a two-stage unit that employs an indirect first stage coupled to a direct second stage (Fig. 8.4.4). Moisture is added to the supply airstream, but not as much as with a direct evaporative cooler.

ADVANTAGES: Evaporative coolers have lower installation and operating costs than electric air conditioning. No ozone-depleting refrigerant is involved. High levels of ventilation are provided because the system typically conditions and supplies 100% outside air.

DISADVANTAGES: Bacterial contamination can result if the system is not properly maintained. The system is only appropriate for dry, hot climates.

6. INSTALL AN ELECTRIC AIR CONDITIONER.

(See Section 8.5 on Heat Pumps as well.)

Electric air conditioners that employ the vapor compression refrigeration cycle are available in a variety of sizes and configurations, ranging from small window units to large central systems. The most common form of central air conditioning is a split system with a warm air furnace (Fig. 8.4.5). The same ductwork is used for distributing conditioned air during the heating and cooling seasons. Supply air is cooled and dehumidified as it passes over an A-shaped evaporator coil. The liquid refrigerant evaporates inside the coil as it absorbs heat from the air. The refrigerant gas then travels through refrigerant piping to the outdoor unit where it is pressurized in an electrically driven compressor, raising its temperature and pres-

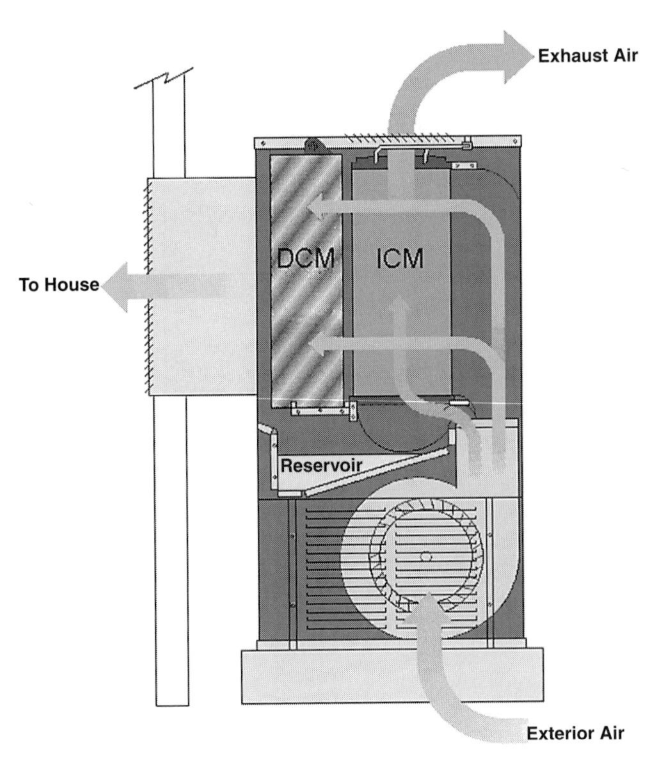

FIGURE 8.4.4 TWO-STAGE EVAPORATIVE COOLER

sure, and returned to a liquid state in the condenser as it releases, or dumps, the heat to the outdoors as a fan draws outdoor air in over the condenser coil. The use of two-speed indoor fans can be advantageous in this type of system because the cooling load can often require higher airflows than the heating load. The lower speed can be used for the heating season and for improved dehumidification performance during the cooling season.

Another electric air-conditioning system configuration is the packaged terminal air conditioner (PTAC) (Fig. 8.4.6). These units are similar to window air conditioners in that they are a single package, but they also provide heat and are designed to be installed through an outside wall. A common application is motel rooms. Their cooling capacities are typically larger than window units and smaller than central systems. Most of the units are equipped with electric heating and can be installed without the need for an HVAC contractor because there is no ductwork or refrigerant piping involved in the installation. Noise can be a problem and the wall penetrations can be a source of uncomfortable air leakage in the winter.

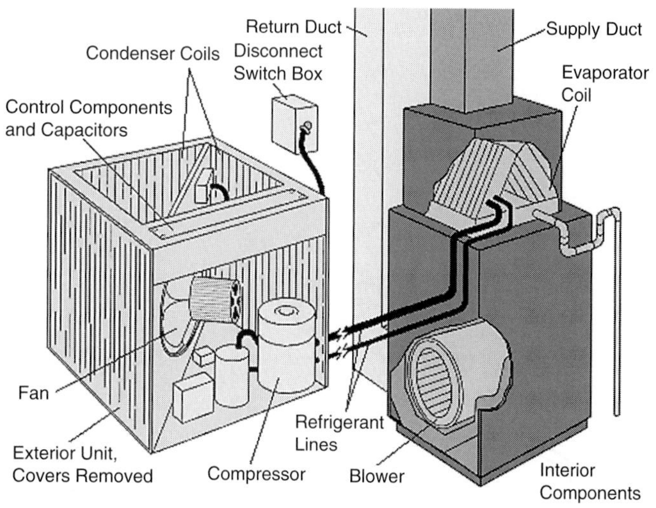

FIGURE 8.4.5 SPLIT-SYSTEM AIR CONDITIONER

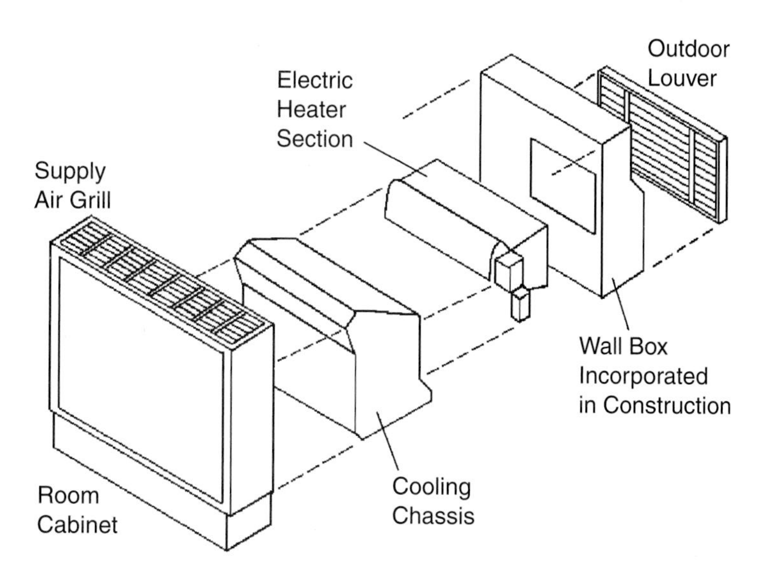

FIGURE 8.4.6 PACKAGED TERMINAL AIR CONDITIONER

ADVANTAGES: This system removes latent heat (moisture) in addition to sensible heat in more humid climates.

DISADVANTAGES: The compressor and condenser fan can be noisy.

7. INSTALL A GAS-FIRED CHILLER.

For the situation where central air conditioning is desired and electricity is very expensive or the cost of additional electrical service is excessive, there are gas-fired alternatives. For some 30 years, Robur Corporation has manufactured small-tonnage absorption products. The 3- and 5-ton Servel systems are air-cooled absorption chillers that utilize ammonia. The system is self-contained in an outdoor unit, and chilled water piping is run to an indoor air handler unit (Fig. 8.4.7).

ADVANTAGES: The operating costs are lower in areas with high electricity prices and low gas prices. The additional electrical service required for a central electric air-conditioning system is avoided. The system does not use an ozone-depleting refrigerant.

DISADVANTAGES: The system has a high initial cost. There is limited market infrastructure and service support. This is a more complex system with greater maintenance requirements.

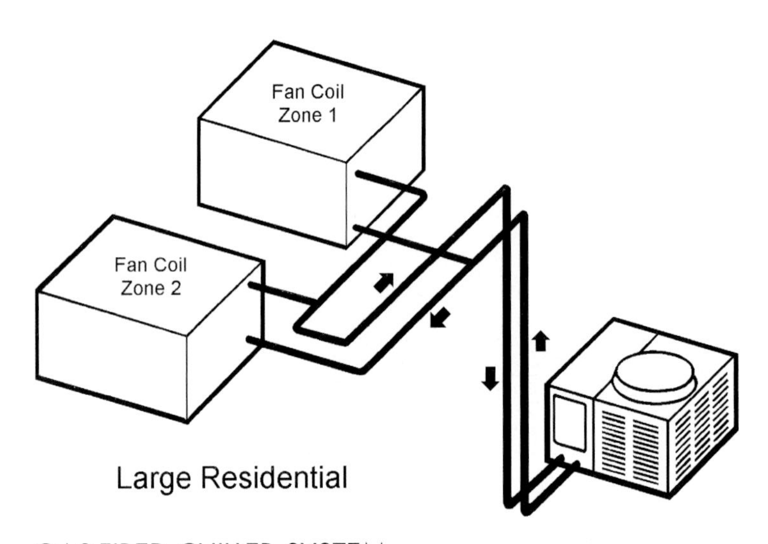

FIGURE 8.4.7 GAS-FIRED CHILLER SYSTEM

8.5 HEAT PUMPS

ESSENTIAL KNOWLEDGE

Heat pumps represent a single system that can provide both heating and cooling. A single source of energy—electricity—is typically used, and vents and chimneys for combustion products can be eliminated. Heat pumps were introduced to the residential market in the 1950s, and early systems had problems with reliability and comfort. Today's heat pump systems are much more reliable and efficient.

The installation of a new heat pump system as part of a home's rehabilitation should be considered if an existing heat pump is nearing the end of its expected life (15 to 20 years) or the existing heating system needs replacement and the addition of air conditioning is desirable. Heat pumps are a logical HVAC system choice in climates with significant cooling loads and *modest* heating loads. If heating loads are small, a less-expensive electric resistance heating and electric air-conditioning system may be more logical. If heating loads are large, a gas furnace and air-conditioner combination may be the optimum choice.

When cooling, heat pumps use the vapor compression refrigeration cycle just like electric air conditioners (see Section 8.4). To provide heating, heat pumps are equipped with reversing and check valves to run the cycle backwards, removing heat from the outdoors and dumping it indoors (Fig. 8.5.1). Unfortunately, as outdoor temperatures drop and heating loads increase, the capacity of a heat pump declines (Fig. 8.5.2). In most climates, a heat pump needs to be equipped with a supplemental heat system, which is typically electric resistance heat.

There is a variety of heat pump types, each with different standards for rating efficiency, but the most common rating terms are seasonal energy efficiency ratio (SEER) for cooling performance and heating season performance factor (HSPF) for heating performance. The SEER is the same rating method used for air-conditioning systems (see Section 8.4). The HSPF is the ratio of total heating output in Btu of a heat pump during its normal annual usage period to the total electric power input in

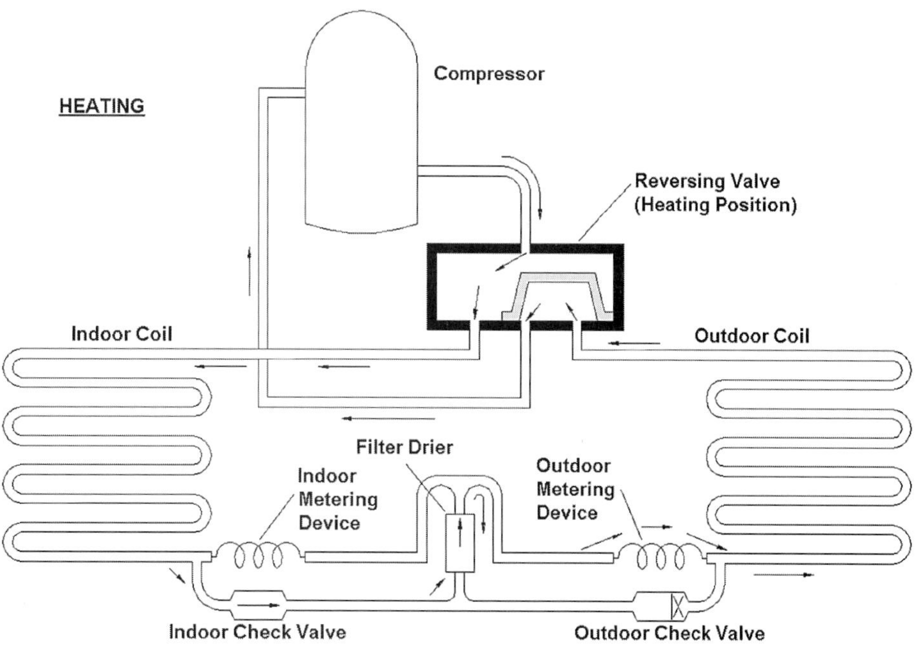

FIGURE 8.5.1 HEAT PUMP CYCLE-HEATING MODE

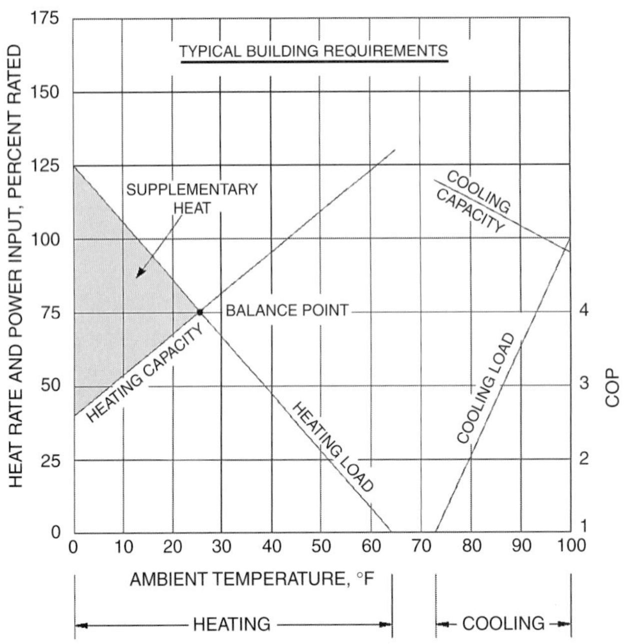

FIGURE 8.5.2 HEAT PUMP CAPACITY AND AMBIENT TEMPERATURE RELATIONSHIP

watt-hours during the same period. It is reported in units of Btu of heating per watt-hour of electric energy consumption, including the unit's compressor, fans, and controls. The higher the HSPF, the more efficient the system. An HSPF of at least 6.8 is required by the NAECA for conventional central split-system air-cooled heat pump systems with capacities of 5 tons or less. However, it should be noted that air-source heat pump heating efficiency varies dramatically with outdoor temperature. If considering a heat pump for a cold climate (i.e., colder than Indianapolis, which is representative of the "typical" used for the rating), a comparison of performance at more severe conditions may be more beneficial than relying on the HSPF comparison. In other words, two heat pumps may have the same HSPF rating, but perform quite differently at cold conditions. Ground-source heat pumps use the coefficient of performance (COP) as an efficiency rating rather than HSPF. However, ground-source heat pumps do not have the varying conditions that air-source units have, so multiplying the COP by 3.4 provides a value that is comparable to the HSPF. Make sure the ground-loop pumping energy is included in the efficiency calculation.

In areas where the cooling load is large relative to the heating load, a heat pump may provide a more comfortable year-round solution than a gas furnace and air conditioner combination. When a furnace is used in combination with an air conditioner in a cooling-load–dominated climate, a significantly oversized furnace may be required to obtain blower performance that is compatible with the size of the cooling coil. Unless the furnace has a variable- or two-speed blower, this will produce short-cycling during the heating season, which compromises comfort.

TECHNIQUES, MATERIALS, TOOLS

1. INSTALL A SPLIT-SYSTEM AIR-SOURCE HEAT PUMP (ASHP).

This is by far the most common residential heat pump system. It is very similar to the split-system air conditioner (see Section 8.4), but operates in the winter as well to provide heating. The indoor unit has a blower, the heat pump coil, and a supplemental heating section that usually contains electric resistance heating elements. Alternatives to using electric resistance heating during cold periods, when the capacity of the heat pump is insufficient, include hydronic coils from the domestic hot-water heater (see Install a Combination System in Section 8.3) or "add-on" heat pumps.

Add-on heat pumps are heat pumps added to a gas furnace. The heat pump operates with declining efficiency as outdoor temperatures drop to the point where it becomes more economical to operate the gas furnace. Two-speed heat pumps with multispeed fans and a two-speed compressor are capable of varying the capacity of the system to better match the load and avoid the inefficiencies of cycling on and off frequently.

ADVANTAGES: This is the most common heat pump system with a well-established service infrastructure.

DISADVANTAGES: Heating performance drops significantly with colder outdoor temperatures.

2. INSTALL A SINGLE-PACKAGE AIR-SOURCE HEAT PUMP.

If the installation of an outdoor compressor-condenser unit is a problem because of noise, aesthetics, or vandalism, the Insider heat pump manufactured by Consolidated Technology Corporation (Fig. 8.5.3) may be a viable solution. This unit, originally designed for manufactured homes, contains all of the components in a single package that is approximately the size of a furnace. A fan is used to draw outdoor air in over the "outside" coil and exhausts it. Another fan draws return air from the space over the "inside" coil and returns the conditioned air to the space.

ADVANTAGES: Installation is easier with no outdoor unit and refrigerant lines.

DISADVANTAGES: There is compressor noise within the living space, because the system generates at 61 decibels.

3. INSTALL A PACKAGED TERMINAL HEAT PUMP (PTHP).

These systems are very similar to PTACs (see Install an Electric Air Conditioner in Section 8.4) except they use an electric heat pump to provide heating rather than relying solely on electric resistance for heating.

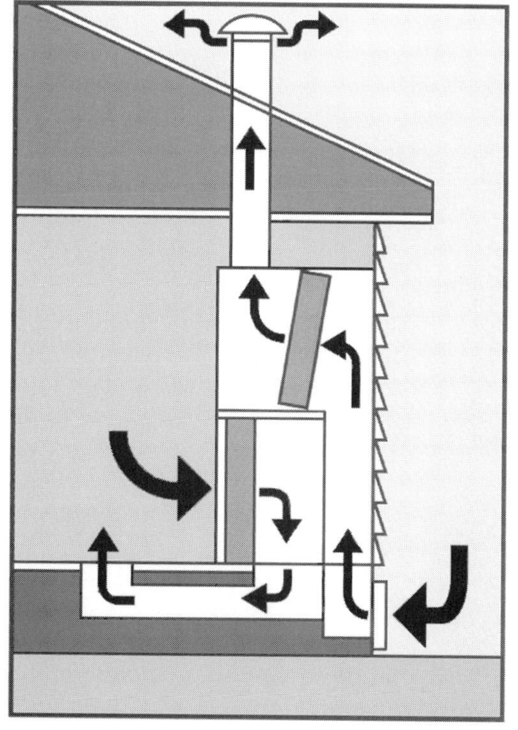

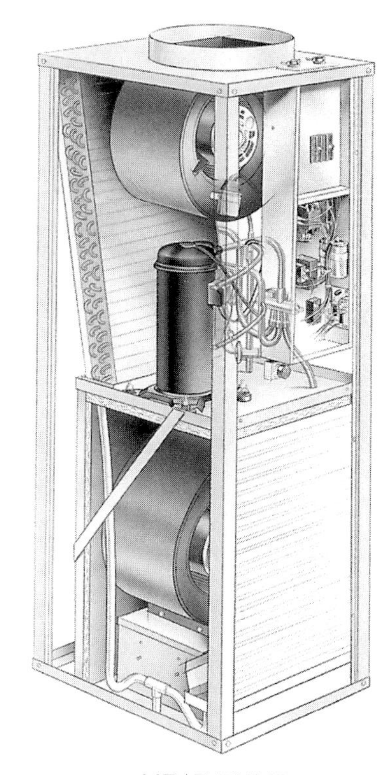

FIGURE 8.5.3 SINGLE-PACKAGE AIR-SOURCE HEAT PUMP HEAT PUMP

ADVANTAGES: No ductwork or refrigerant lines need to be installed on-site. There is zone control capability.

DISADVANTAGES: Units can be noisy, and the wall penetrations can be a source of unwanted infiltration.

4. INSTALL ONE OR MORE DUCTLESS SPLIT SYSTEMS OR MINI SPLITS.

These systems are similar to conventional split systems in that they have an outdoor condenser unit and an indoor evaporator–air handler unit connected by refrigerant piping. However, the indoor unit is located directly in or adjacent to the conditioned space. There is no ductwork required for distributing the air. Several indoor units can be used in conjunction with a single outdoor unit (Fig. 8.5.4).

ADVANTAGES: These systems are easier to install than ducted systems if no ducting is already in place and have zone control capability. They can be linked with motion detectors to minimize operating time.

DISADVANTAGES: These systems operate on recirculated air only and have no means for providing fresh air. Although very popular in Japan and Europe, they are not widely used or manufactured in the United States.

5. INSTALL A GROUND-SOURCE HEAT PUMP (GSHP).

As the name implies, GSHPs use the ground as the source for heat during the winter rather than outdoor air (Fig. 8.5.5). They can be an attractive option to ASHPs in colder climates where the ground temperature is warmer and less variable than the air temperature. In the more common closed-loop design, a ground loop of polybutylene or high-density polyethylene pipe is buried and water or antifreeze solution is pumped through it to absorb heat. The pipe can be buried vertically or horizontally, straight or coiled like a Slinky. Installation will depend on soil conditions, drilling versus trenching costs, and space availability. Open-loop systems pump groundwater from a well through a heat exchanger and then discharge it. They avoid the cost of a buried ground loop, but water quality and

FIGURE 8.5.4 DUCTLESS SPLIT SYSTEM

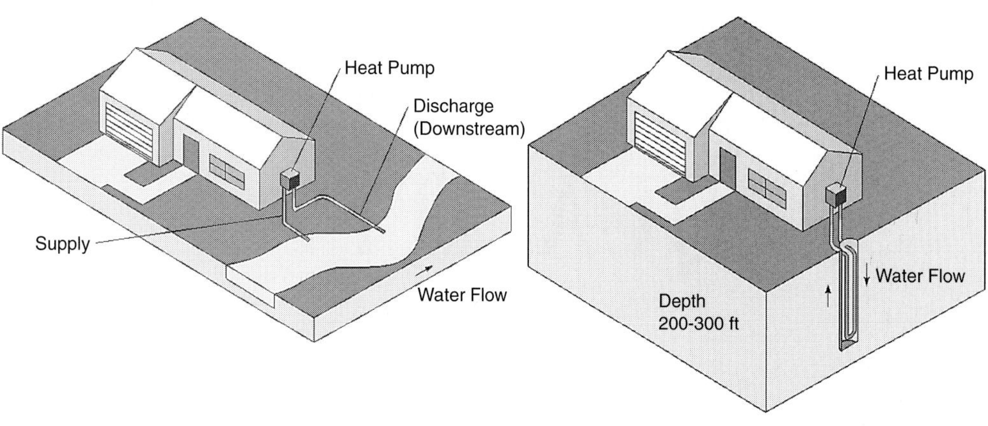

FIGURE 8.5.5 OPEN-LOOP GSHP CLOSED-LOOP GSHP

code issues have to be addressed. Some GSHPs include desuperheaters (see Section 8.9) for water heating as an integral part of the system.

ADVANTAGES: GSHPs are usually more efficient on a seasonal basis than ASHPs because the ground temperature is higher than the air temperature. GSHPs do not need to defrost like ASHPs.

DISADVANTAGES: There is the installation cost of the ground loop.

8.6 INDOOR AIR QUALITY

ESSENTIAL KNOWLEDGE

As homes constructed since the 1980s have become "tighter" in order to reduce infiltration to improve comfort and save energy, concerns about indoor air quality have grown. The list of indoor pollutants is long, but the most common include: formaldehyde fumes from building materials, furniture, curtains, and carpet; organic gases from aerosols, paints, solvents, and air fresheners; carbon monoxide and nitrogen dioxide from unvented appliances such as gas ranges; dust mites, mold spores, and mildew.

Tight homes without a mechanical ventilation system can depressurize when devices such as exhaust fans or clothes dryers operate and remove air from conditioned space. Depressurization can cause back-drafting of combustion appliances and potentially increased radon levels (Fig. 8.6.1).

The simplest way to improve indoor air quality is through a controlled ventilation system. Mechanical ventilation systems are now required in new homes by some building codes, particularly in the northwest and Canada. The recommended amount of ventilation air is a subject of debate by experts in the field. The American Society of Heating, Refrigerating and Air-Conditioning Engineers (ASHRAE) standard 62 recommends 15 cfm of outside air per person or 0.35 air changes per hour (ACH) for residences. Another proposed standard, ASHRAE standard 62.2P, recommends 1 cfm per 100 ft^2 of house area plus 7.5 cfm per occupant.

Ventilation systems assure that adequate fresh air is brought into the home to replace the indoor air that is removed. There are several different mechanical ventilation approaches. In addition to

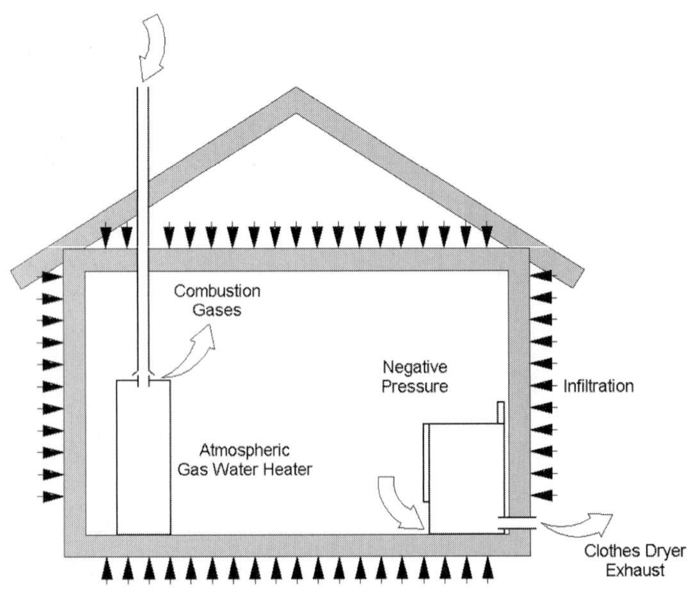

FIGURE 8.6.1 DEPRESSURIZATION-INDUCED BACK-DRAFTING

mechanical ventilation for indoor air quality, humidifiers and dehumidifiers may be installed to control humidity and air cleaners can be used for those sensitive to air contaminants such as dust and pollen.

TECHNIQUES, MATERIALS, TOOLS

1. INSTALL A MECHANICAL VENTILATION SYSTEM.

There are three basic types of mechanical ventilation systems: exhaust, supply, and balanced (Fig. 8.6.2). Exhaust and supply systems typically use a single fan that either pulls air from the home or pushes air into the home in conjunction with intentional air inlets or pressure-relief outlets. These systems either depressurize or pressurize the home. Depressurization can cause problems with combustion appliance back-drafting and potentially increased radon infiltration. Pressurization can promote detrimental moisture infiltration into building materials. To avoid these problems, balanced mechanical ventilation systems typically use at least two fans, supply and exhaust, to maintain a neutral pressure in the home. The most energy-efficient type of balanced mechanical ventilation system is a heat-recovery ventilator or air-to-air heat exchanger (Fig. 8.6.3). These systems include a heat exchanger to exchange heat between the exhaust and supply airstreams. Fresh supply air is preheated in the winter and precooled in the summer. Among the manufacturers of these types of systems is Nutech Energy Systems, which has introduced the Lifebreath Clean Air Furnace (Figs. 8.6.4 and 8.6.5), designed to draw in outside air at a rate that is approximately 10% of the total supply airflow. In regions with significant moisture differences between supply and exhaust airstreams, enthalpy wheels may be used as the heat exchanger. The wheels include a desiccant material on a heat wheel to absorb and release moisture and heat.

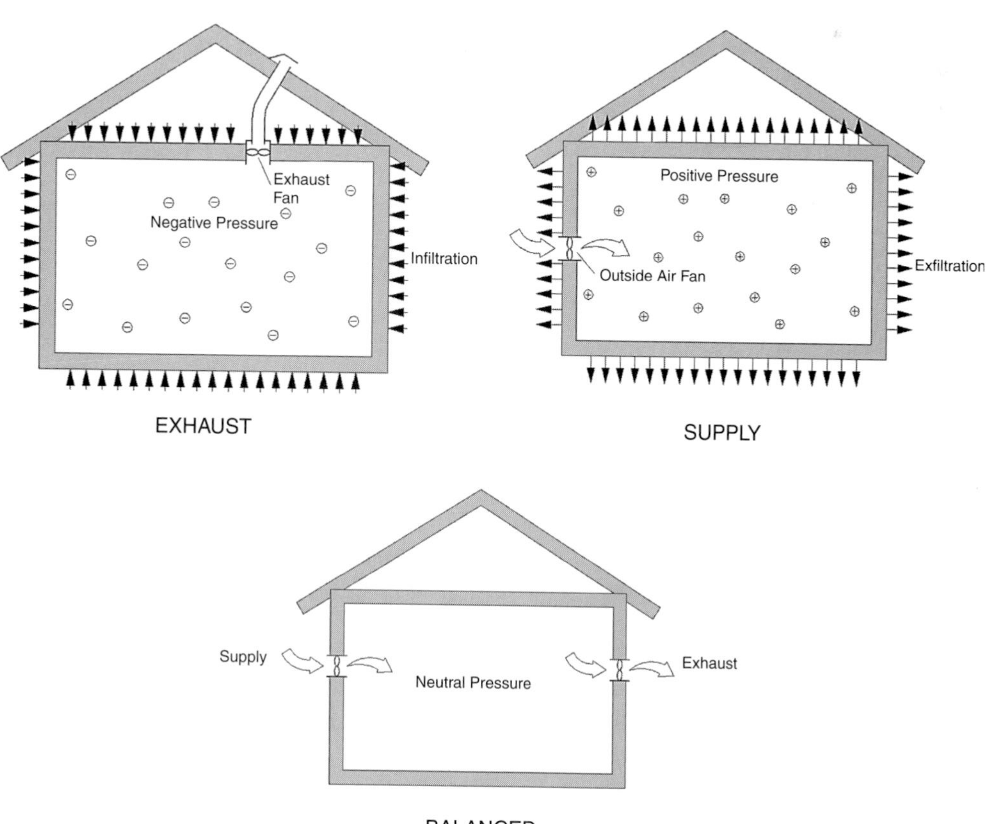

FIGURE 8.6.2 MECHANICAL VENTILATION TYPES

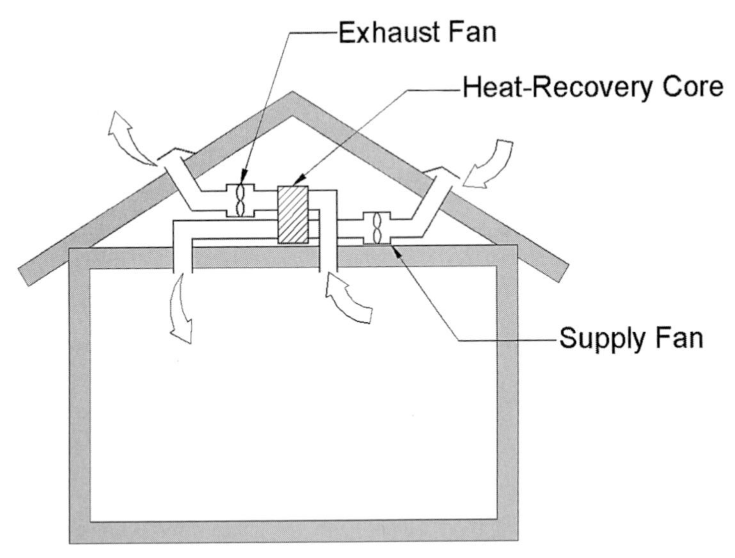

FIGURE 8.6.3 BALANCED SYSTEM WITH A HEAT-RECOVERY VENTILATOR

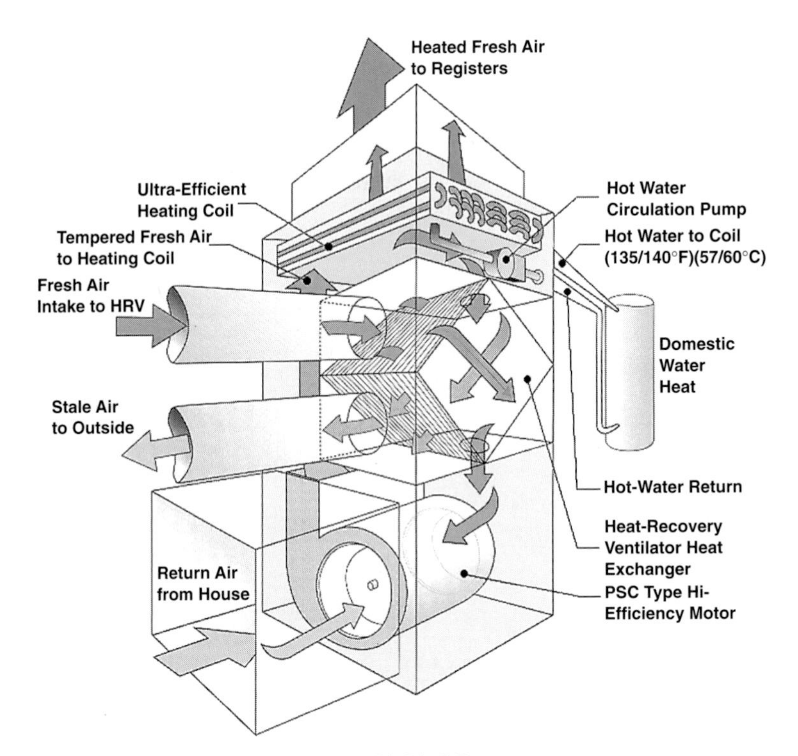

FIGURE 8.6.4 LIFEBREATH CLEAN AIR FURNACE

ADVANTAGES: A mechanical ventilation system provides better health and safety because the supply of an adequate amount of fresh air to the home is assured.

DISADVANTAGES: The initial installation and fan operating cost can be high. Filter maintenance is required. Additional heating and cooling loads are placed on the existing heating and cooling system, but the additional load is minimal with heat- or enthalpy-recovery ventilators.

2. INSTALL A HUMIDIFIER.

Low indoor humidity can be problematic in cold climates during the heating season. Cold winter air holds less moisture than warm summer air (80° F air at 50% RH contains twice as much water

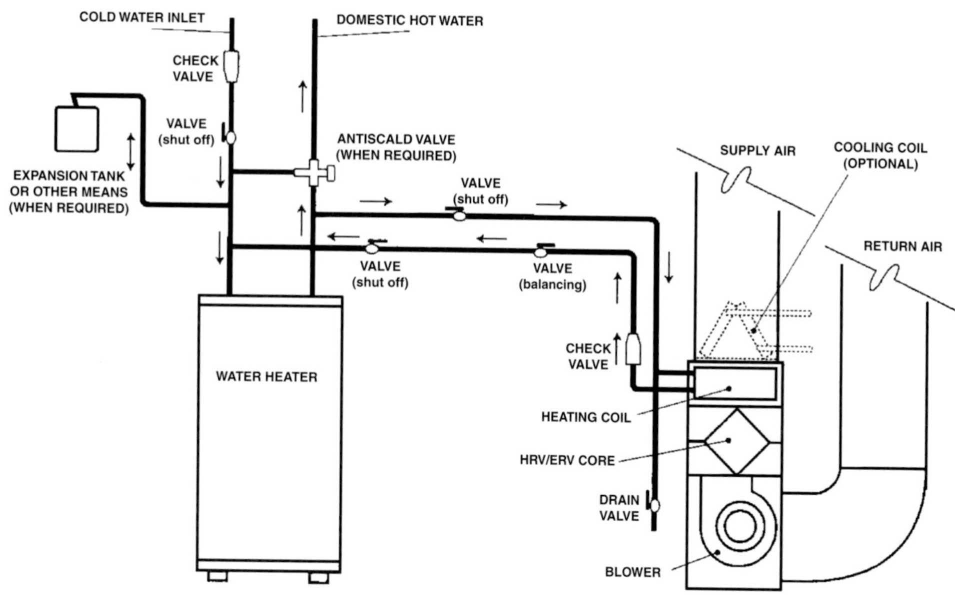

FIGURE 8.6.5 CLEAN AIR FURNACE DIAGRAM

as 42° F air at 100% RH). Low indoor relative humidity can lead to dry skin, nosebleeds, and respiratory problems. An indoor relative humidity of 35% to 50% is recommended. There are three basic humidifier types for residential applications: pan, wetted-element, and steam. Ultrasonic and impeller-type humidifiers, which emit a mist into the air, are discouraged because they require the use of demineralized or distilled water to avoid circulating harmful minerals into the air. Whole-house humidifiers are used in conjunction with a warm-air furnace. Pan-type humidifiers use a pan of water in the furnace plenum. As warm air flows over the pan, water evaporates into the air. A heating element can be used to increase the water temperature and the rate of evaporation, and/or wicking plates can be used to increase the pan's effective surface area (Fig. 8.6.6). Wetted-element humidifiers use a plastic pad that rotates through a reservoir of water (Fig. 8.6.7). As air passes through the pad, water evaporates into the airstream. Steam humidifiers use heating elements to boil the water in a pan. They provide higher humidification output than other humidifier types. Honeywell produces a steam humidifier for residential applications that independently controls the furnace fan to provide humidified air to the space even when the furnace is not running to supply heat. An issue with all types of humidifiers is disposal of mineral deposits that accumulate in the water reservoir. Automatic flushing systems are available, but require a drain and use more water. Humidifiers without a flushing system should be cleaned routinely, every 1 or 2 months, to remove the deposits. Humidistats are typically used to control humidifiers. Research Products has introduced an Aprilaire whole-house humidifier that monitors the outdoor humidity and automatically provides the optimum humidity to the home.

ADVANTAGES: The higher, more comfortable heating season humidity allows lower thermostat settings without sacrificing comfort (for example, air at 69° F and 35% RH provides the same level of comfort as 72° F and 19% RH).

DISADVANTAGES: If improperly controlled, excessive indoor humidity can cause damaging condensation problems. Maintenance is required to remove mineral deposits on nonflushing systems.

3. INSTALL A DEHUMIDIFIER.

High levels of indoor humidity cannot only cause comfort problems, but also health concerns with the potential for increased growth of mold and bacteria. In severe situations, building materials and furnishings can be damaged by mildew.

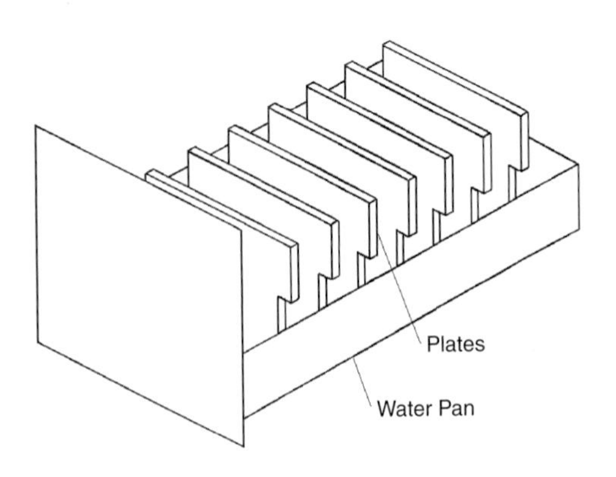

FIGURE 8.6.6 PAN HUMIDIFIER

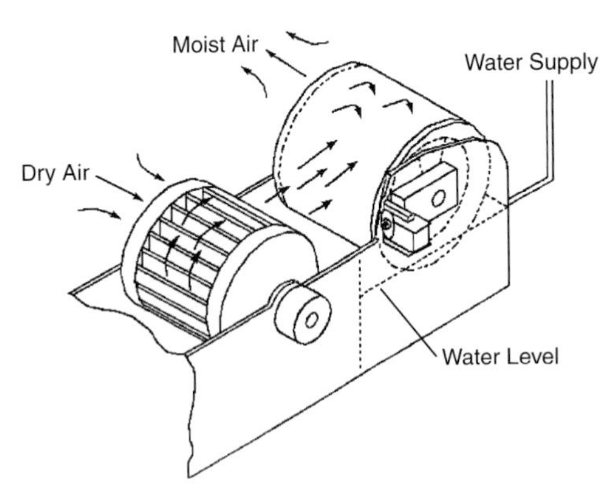

FIGURE 8.6.7 WETTED-DRUM HUMIDIFIER

High levels of humidity can occur in a tight home when internally generated moisture by people, cooking, or showering is not exhausted. Condensation on the inside of cold windows and within wall insulation in the winter can be damaging. Bath fans such as the SmartSense by Nutone automatically activate when high humidity levels are detected.

High humidity levels can also occur in the summer if the ventilation system introduces humid outside air to the conditioned space. Dehumidifiers appropriate for residential applications employ refrigeration to cool the air below its dew point to remove moisture. This process also might cool the air lower than desired. The cool, dry air can be reheated by passing it over the unit's condenser coil. There are several manufacturers of these types of systems, such as Therma-Stor Products' Ultra-Aire Air Purifying Dehumidifier, which combines a dehumidifier with an outdoor air inlet for ventilation air and an optional 95% efficient pleated media filter (Fig. 8.6.8).

ADVANTAGES: Humidity is properly controlled independent of temperature. At lower indoor humidities, thermostat settings can be raised to achieve the same level of comfort.

DISADVANTAGES: A dehumidifier can be costly to install and operate.

4. INSTALL AN AIR FILTER OR CLEANER.

People who are particularly sensitive to air contaminants such as dust and pollen may benefit from a high-efficiency air filter or air cleaner. Air pollutants are either a particulate or a gas, and there are specific types of filters for each category. Particulate pollutants include mold spores, pollen, house dust, animal dander, clothing and furnishing fibers, and dirt. Most filters are designed for particulate removal. Gases include combustion gases, by-products of human and animal metabolism, and volatile

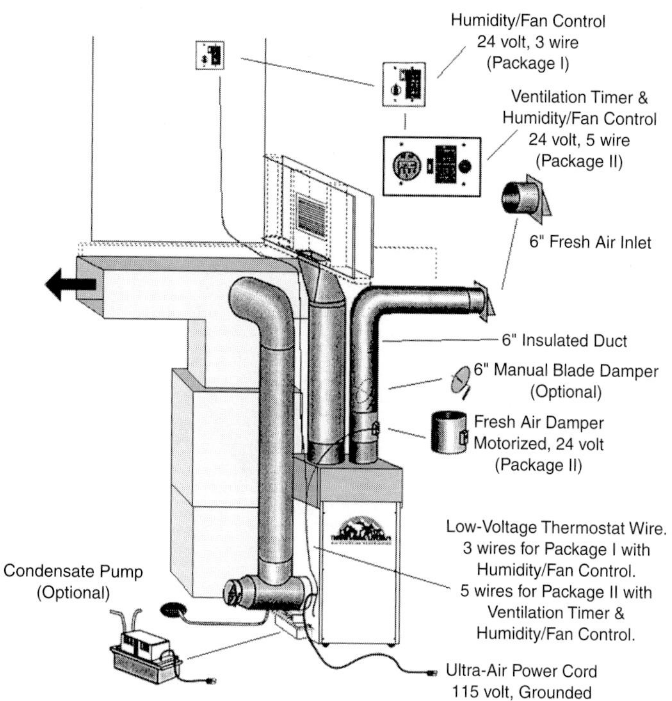

FIGURE 8.6.8 ULTRA-AIRE PURIFYING DEHUMIDIFIER

organic compounds. Adsorption-type filters such as charcoal are used for gas pollutants. Tobacco and wood smoke are both particulates and gases and require a combination of filter types. The standard filter in a furnace or air-conditioning system is typically only good for fairly large particulates. A test for determining the efficiency of a filter is the atmospheric-spot-dust test, which measures particles between 0.3 and 6 microns in size. A standard 1"-thick furnace filter may only be 3% to 5% efficient according to an atmospheric-spot-dust test. The standard furnace/air-conditioner filter is a media filter because it is made of material that is designed to trap particulates. The higher the media filter efficiency, the greater the initial cost and operating cost due to a higher pressure drop and thus fan energy requirements.

Another type of air filter, an electrostatic precipitator, gives particulates a static-electric charge to capture them. The charged particulates are then collected by oppositely charged metal plates. Electrostatic precipitators are often rated at 90% efficiency (atmospheric-spot-dust test), but the metal plates must be cleaned often to maintain efficiency. These types of air filters also produce small amounts of ozone that can then be captured in a subsequent adsorption filter. They do not have the pressure drop problem of media filters. Electrostatic air filters are media filters that use a special fibrous plastic material that becomes charged as air passes through (Fig. 8.6.9). Oppositely charged particles in the airstream then cling to the filter media. They are about 10% to 15% efficient (atmospheric-spot-dust-test), but have a lower pressure drop penalty than a 30% to 60% efficient medium-efficiency media filter.

Complete air cleaning systems will include a particulate filter and an adsorption filter. There are also systems that address the need for ventilation air. Therma-Stor Products manufactures the Filter-Vent System that brings in outside air, combines it with indoor air, and draws the mixed air through a 90% to 95% media filter and an activated carbon filter.

ADVANTAGES: This is the only method of providing pollutant-free indoor air in an area with high levels of outdoor pollutants and/or for those occupants who are especially sensitive.

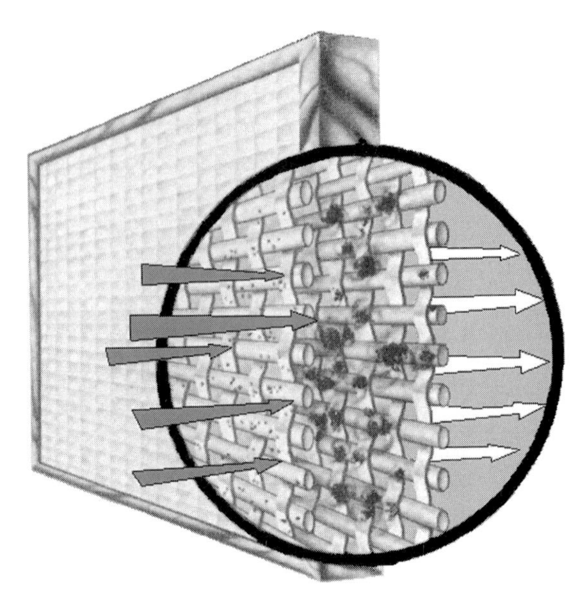

FIGURE 8.6.9 ELECTROSTATIC AIR FILTER

DISADVANTAGES: These systems require high maintenance. Uncleaned media filters produce an excessively high pressure drop. An uncleaned electrostatic precipitator or filter will be ineffective. These systems have high initial and operating costs. Air cleaners alone do not bring in outside air, so they are not a substitute for a ventilation system.

8.7 CONTROLS

ESSENTIAL KNOWLEDGE

A critically important aspect of an HVAC system's ability to efficiently maintain comfort is its controls. Today, nearly all heating and/or cooling systems are controlled by a thermostat. Even factory-built wood fireplaces are available with thermostat control. The sophistication of thermostats and other HVAC system controls has evolved dramatically over the last decade with the advancement of electronic controls. As controls for residential systems are relatively inexpensive, and their technology has advanced significantly in the past few years, it is cost-effective to replace older controls that may not be operating properly.

A thermostat senses the room air temperature and controls the heating system via an electrical switch according to an occupant-selected set-point temperature. Line-voltage thermostats are typically used for zonal electric resistance heating systems and low-voltage thermostats are used for central systems. Thermostats should be located about 5' above the floor where they will sense an air temperature representative of the room or area being controlled. Location near a heat source such as the supply register or heat sink such as a window will result in poor comfort control.

Old-style electromechanical thermostats found in many rehab projects employ either a mercury or snap-action switch, a bimetallic coil, and an anticipator (Fig. 8.7.1). The bimetallic coil expands and contracts as it heats and cools, activating and deactivating the switch that controls furnace or boiler operation. The anticipator is a tiny heater that heats the bimetallic coil to compensate for its slow response to changing air temperature. Adjustment of the thermostat dial establishes the angle or the tension of the bimetallic coil to the contacts. There are versions of these thermostats that include an electric clock to switch the thermostat between two temperature settings to achieve an energy-saving setback condition. An 8-hour overnight setback during the heating season saves 1% to 2% per degree of setback. An 8-hour midday setup during the cooling season saves 1% to 3% per degree of setup.

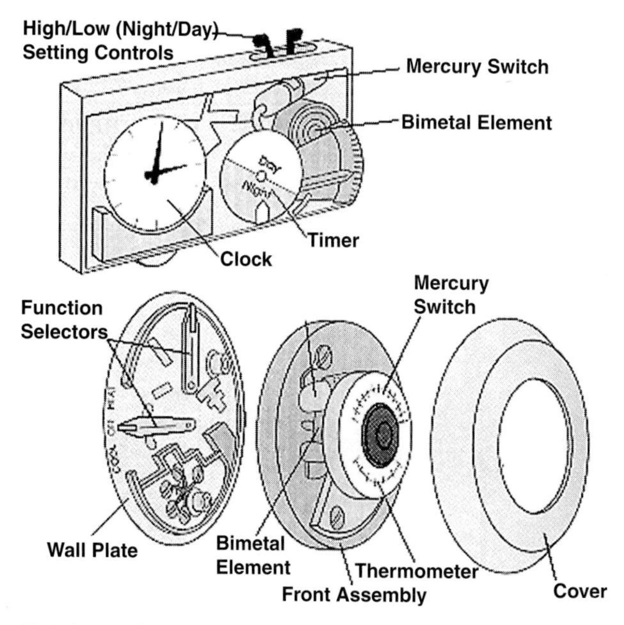

FIGURE 8.7.1 ELECTROMECHANICAL THERMOSTAT

TECHNIQUES, MATERIALS, TOOLS

1. REPLACE THE BIMETAL THERMOSTAT WITH AN ELECTRONIC PROGRAMMABLE THERMOSTAT.

While electromechanical thermostats (Fig. 8.7.2) are lower in cost than electronic programmable thermostats, their performance in terms of deadband and droop is poorer as well. *Deadband* is an indicator of the swing in room temperature between heating/cooling system cycles. *Droop* is an indicator of how well the anticipator is functioning. Thermostats also degrade over time as calibration slips and mechanical parts stick. An old electromechanical thermostat can be cleaned and recalibrated, but a better solution is to simply replace it with a new electronic thermostat. Electronic thermostats employ a thermistor rather than a bimetallic coil for sensing the room air temperature. It is more sensitive and fast responding. Microprocessor programming allows for several different temperature settings throughout the day and for different days of the week. Some units provide for different settings for weekdays and weekends or weekdays, Saturdays, and Sundays. The Lightstat thermostat has the ability to adjust the thermostat setting using the light level in the room as an indicator of whether people are in the room. Honeywell has a thermostat that senses lighting levels and occupancy. Thermostats are also now available as wireless remote-control devices. They may be placed on the coffee table in the center of the room next to the television remote. Home Automation manufactures a line of thermostats that can communicate with home automation systems, personal computers, and utility demand-side management programs. In a region with real-time pricing, the RC-91 model has the ability to display the real-time energy price and the user can then set the thermostat accordingly.

ADVANTAGES: An electronic programmable thermostat provides energy savings by having more capability to set back the temperature during unoccupied or night time hours. It is more accurate and responsive for better comfort control.

DISADVANTAGES: The cost is somewhat high. As with VCRs, there may be some homeowners that have difficulty following the sequential steps to program them properly. Some types of heat distribution systems such as radiant and hydronic baseboard systems are not as responsive as forced-air systems, and the time required to recover from a thermostat setback may be unacceptable.

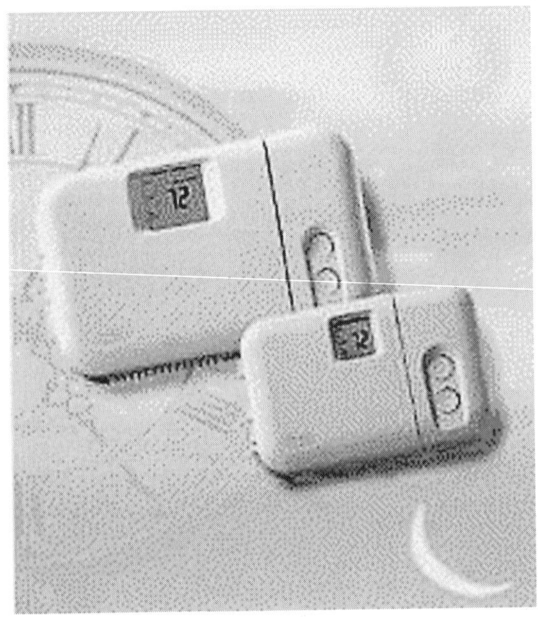

FIGURE 8.7.2 ELECTRONIC PROGRAMMABLE THERMOSTAT

2. INSTALL A HUMIDISTAT.

Typically, heating and cooling systems are designed to maintain comfort in terms of temperature only. Humidity is allowed to fluctuate. In warm humid climates, it may be desirable to control the air-conditioning system to maintain comfort humidity as well as temperature conditions. In colder climates, a humidifier may be installed to maintain higher and more comfortable indoor humidity levels. In either situation, the installation of a humidistat (or dehumidistat, depending on the situation) is appropriate. The location of the humidistat can vary. It can be in the room next to, or integral with, the thermostat or in the return-air duct. The humidistat works in a similar manner to the thermostat. It senses humidity and turns the air conditioner or humidifier on or off based upon a humidity set point. For two-speed air conditioners, the dehumidistat may trigger the system to go to low speed for enhanced dehumidification capability.

ADVANTAGES: The humidistat provides better control of humidity for improved comfort and air quality.

DISADVANTAGES: Calibration needs to be checked periodically. When controlling an air-conditioning system in humid climates, the humidistat can cause overcooling.

3. INSTALL A CONTROL SYSTEM THAT SENSES OUTDOOR CONDITIONS AND ADJUSTS ACCORDINGLY.

More-sophisticated control systems incorporate outdoor temperature sensors as well as indoor sensors for better control and operating efficiency. For heat pump systems equipped with auxiliary electric resistance heat, an outside temperature sensor is used to lock out the resistance heat when the outdoor temperature is above a specific setting. This prevents the resistance heat from unnecessarily coming on to quickly recover from a nighttime setback. Outdoor temperature sensors are also used to adjust the water temperature in hydronic systems to the lowest temperature necessary to meet the home's heating load. For the control of humidifiers in cold climates, Honeywell Perfect Climate Comfort Center Control System's winter humidity control senses outdoor temperature and resets the indoor humidity setting accordingly to prevent condensation on windows. The Honeywell system can be customized for the insulating value of the home's windows. Aprilaire also has a humidistat that controls according to outdoor temperature but does not have the window customization capability.

ADVANTAGES: The control system reduces operating costs for the heat pump auxiliary heat lock-out control.

DISADVANTAGES: The initial cost is high.

4. INSTALL VENTILATION SYSTEM CONTROLS.

There is a variety of methods for controlling mechanical ventilation systems. On the simple, low-cost end of the spectrum are the manual controls. These include a simple on/off switch. Additional features might be a timer that shuts the fan off automatically after a certain period of time or a speed control that allows the user to select the ventilation airflow. The manual controls rely upon the user recognizing when ventilation is needed and taking the appropriate control action. The average human nose is more sensitive to air pollutants than most sophisticated electronic sensors. Another set of control options that are more sophisticated and more expensive are automatic. These include time clocks programmed by the user. Honeywell's Trol-A-Temp Timed Make-Up Air Control (TMAC) or Timed Ventilation Control (TVC) are simple timers that control a motorized damper that allows outside air into the central HVAC system (Fig. 8.7.3). The unit will turn on the HVAC blower if it is not already running. Automatic controls also include a variety of demand-controlled ventilation (DCV) strategies. The DCV methods rely upon a sensor in the space or return duct to detect when additional ventilation air is needed. The sensing parameter can be motion, humidity, carbon dioxide, or a mixture of gaseous pollutants such as tobacco smoke, cooking odors, and VOCs. For the most part, the manual controls tend to be most appropriate for local exhaust fans and the more-sophisticated automatic controls are appropriate for whole-house ventilation systems.

ADVANTAGES: Manual controls are relatively inexpensive and reliable. Automatic controls do not rely upon the user.

DISADVANTAGES: Manual controls rely upon the user for proper control. Automatic controls are more expensive and can fall out of calibration resulting in either insufficient or excessive ventilation air quantities.

5. INSTALL A FORCED-AIR ZONING SYSTEM.

Forced-air zoning systems utilize thermostats in conjunction with motorized dampers in individual supply ducts to different control zones (Fig. 8.7.4). By varying the airflow in response to the zone

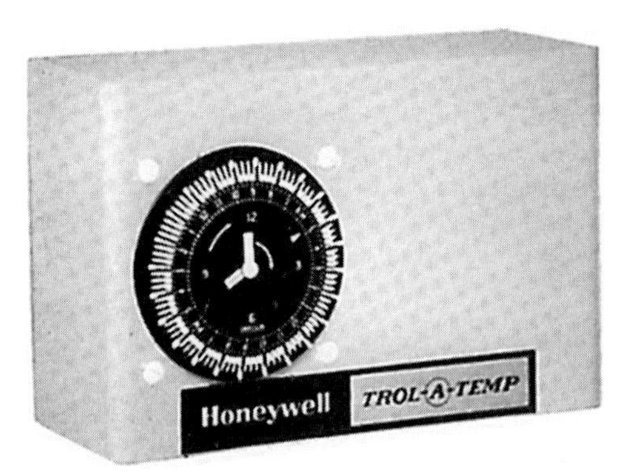

FIGURE 8.7.3 VENTILATION CONTROLLER MOTORIZED OUTSIDE AIR DAMPER

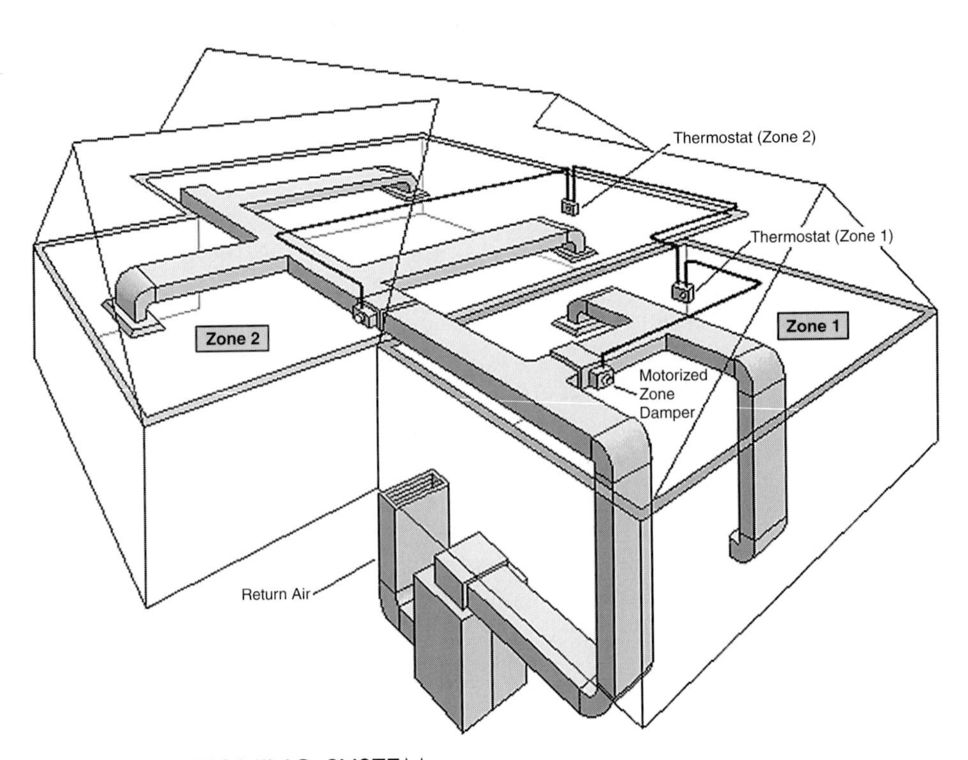

FIGURE 8.7.4 FORCED-AIR ZONING SYSTEM

thermostat, a single air-conditioning and/or heating system can be used to meet the varying needs of multiple zones. These systems are most easily installed in a new installation, but can be retrofitted into an existing system. Care must be taken to either bypass extra air or oversize the ductwork to assure that central system airflow does not drop below the design minimum when only one or two zones are calling for air and all other zones are closed. There are several manufacturers of this type of system including Research Products Corporation, Carrier, Honeywell/Trol-A-Temp, and Durozone.

ADVANTAGES: A forced-air zoning system provides energy savings and improved comfort.

DISADVANTAGES: The initial cost is high. Systems need to be carefully designed to assure that minimum airflow requirements are met under all operating conditions.

8.8 FIREPLACES AND CHIMNEYS

ESSENTIAL KNOWLEDGE

In recent years fireplaces have captured a renewed interest as a supplemental heating source for the home. In a rehabilitation project, rehabilitating an existing fireplace or installing a new fireplace may be under consideration. The rehabilitation of an existing fireplace cannot only add to the appearance of the room, but also provide for safe and efficient space heating. The installation of a new, efficient fireplace to supplement an existing system, which may be inefficient and/or undersized, may be an attractive alternative to replacing the existing system.

Traditional wood-burning fireplace designs are not efficient, can produce emissions that are harmful to the environment, and, if poorly designed or maintained, can be dangerous. In fact, some western municipalities in the United States have passed ordinances prohibiting the installation of a conventional fireplace for environmental reasons. EPA-approved, factory-built wood and gas fireplaces, wood stoves, or gas-fired logs are acceptable alternatives which should be considered during a rehabilitation project.

If constructing or significantly reconstructing a fireplace or installing a stove, building codes should be carefully reviewed for minimum requirements for materials, heights, and clearances. Local code requirements apply even if the existing chimney is used. If an old, unused fireplace is found encased in plaster during the course of a rehabilitation project, it can be reconstructed by a competent mason who knows and follows recommended design practices. Many early fireplaces and chimneys smoked too much due to flaws in their design. This poor performance may be the reason that the fireplace had been abandoned. If an old, closed-off chimney is to be rehabilitated, make sure that it still runs through the roof to an appropriate height and that openings for other stove pipes have not been cut into it. Old chimneys were often used to serve multiple stoves with a single flue.

Conventional fireplaces (Fig. 8.8.1) provide heat primarily by radiation. Thus, the amount of brick masonry, surface area exposed to the fire, its distance from the fire, and the size of the fire determine the amount of heat provided. Rumford-style fireplaces feature angled side walls, a shallow depth, and a high opening. These fireplaces tend to be more efficient than conventional designs because they radiate more heat and draw less room air up the chimney. Air-circulating fireplaces (Fig. 8.8.2) capture heat from the back of the firebox by circulating room air through brick baffles or steel plates. This warmed air can then be circulated by a fan to spaces to provide heat by convection.

The primary function of a chimney flue is to exhaust combustion wastes—carbon dioxide, nitrogen, sulfur dioxide, water vapor, and carbon monoxide. A second function is to create a draft that pulls the air over the fire. For a wood-burning fire, the draft must be strong. To accomplish this, the fireplace has a sloped back and a stepped throat controlled by a damper. Each combustion heat source (e.g., furnace and fireplace) needs its own flue, but a single chimney can contain several flues from multiple fireplaces and/or a furnace (Fig. 8.8.3). Two combustion appliances such as a water heater and a furnace can share a flue. In this instance, problems can arise if one of those appliances is replaced with a direct-vent unit because the flue is too large for the remaining appliance. Condensation of combustion gases can occur in the chimney, resulting in corrosion. This can also occur with the installation of a higher-efficiency chimney-vented combustion appliance, particularly in a cold climate.

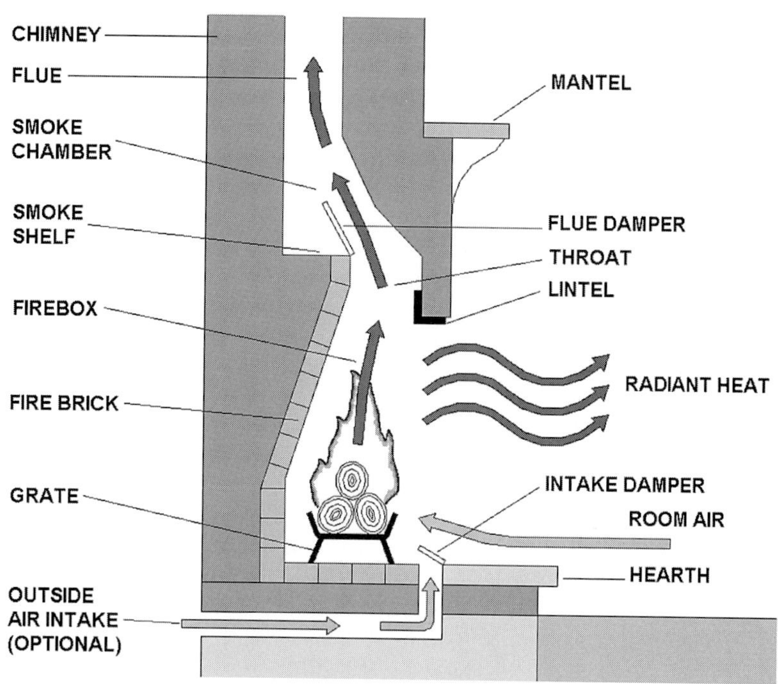

FIGURE 8.8.1 CONVENTIONAL WOOD-BURNING FIREPLACE

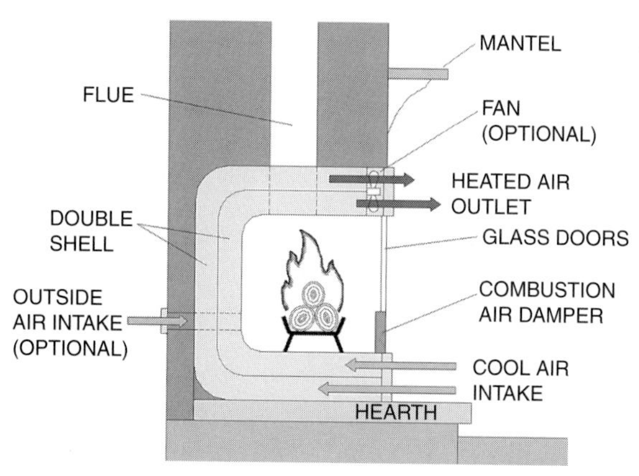

FIGURE 8.8.2 AIR-CIRCULATING FIREPLACE

Beyond replacing the brick and mortar of an old disintegrating chimney, clearing several inches of debris from above the damper, and clearing other blockages in the chimney flue, several additional steps may be necessary to assure that an old fireplace and chimney are in safe working condition. Alternatives or enhancements to the fireplace such as stoves and inserts can also improve the efficiency of the system.

TECHNIQUES, MATERIALS, TOOLS

1. REBUILD OR REPOINT THE CHIMNEY.

The mortar between the bricks of an old chimney is likely to be in need of renewal. It was not intended to last as long as the bricks. The replacement process is known as repointing or tucking. Whatever kind of mortar was used initially should be used for repointing so as to have the same expansion and compression characteristics under varying weather conditions. In some instances, the condition of the chimney may have deteriorated to a point beyond repair by repointing and relining. When the mortar

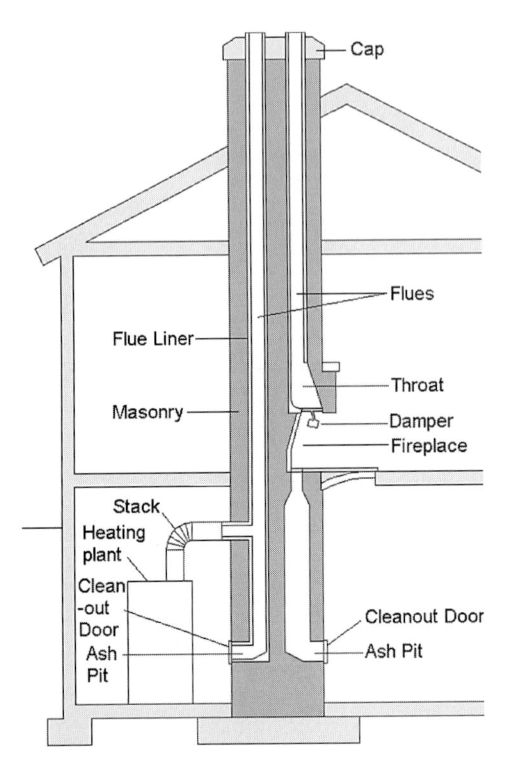

FIGURE 8.8.3 MULTIFLUE CHIMNEY

crumbles easily and/or bricks are loose, removal and replacement should be considered. It may be possible to reuse the original bricks. Creosote is a carcinogen, so a respirator should be worn when rehabilitating an old chimney, and the work area should be enclosed in plastic sheeting to protect the rest of the house.

ADVANTAGES: Safety and the integrity of the chimney will be improved.

DISADVANTAGES: The process can be expensive because it is time-consuming and requires the expertise of a mason.

2. INSTALL A CHIMNEY CAP.

Rain mixing with soot and fireplace gases produces a destructive acid. Thus, all chimneys should have a rain cap to prevent this deterioration (Fig. 8.8.4). Rain caps can also minimize downdrafts, and unwanted animals. They can also function as spark arrestors to keep sparks from igniting nearby combustible materials.

ADVANTAGES: A chimney cap is fairly inexpensive and extends the life of the chimney.

DISADVANTAGES: A chimney cap is difficult to install on high chimneys.

3. RELINE THE CHIMNEY.

Liners, which are now required by code, prevent the heat from a chimney fire from causing a house fire. Older chimneys may be unlined (pre-1910), lined with mortar that has deteriorated, or have old tile liners that are cracked. They also often served more than one fireplace and had more than one flue in them with a wall of brick as a divider. This brick divider is likely to be in very poor condition, and tumbling bricks from this divider can block the flue.

The solution to all of these problems is to reline the chimney. There are different methods for relining a chimney.

FIGURE 8.8.4 CHIMNEY CAP

■ *Install a metal pipe.* Metal pipe is available in 3′ sections from local heating suppliers. Stainless steel should be used if the flue will be used for furnace gases, which are very corrosive. Seal around the pipe at the bottom and top of the chimney with lightweight mortar. Flexible pipe is also available for nonstraight chimneys. Caution is advised if the chimney has a weak draft. Reducing the cross-section area of the flue with a metal pipe will only worsen the situation.

■ *Install a tile lining.* This is required by code in some areas if the chimney is to be used with wood fires. Although more expensive, a tile liner is more resistant to corrosion and the high temperatures of a chimney fire due to creosote buildup.

■ *Use a patented process to line the chimney with a cementitious mix.* The patented process is only available through franchised dealers. A vinyl or rubber hose can be inserted and inflated (Fig. 8.8.5). Lightweight concrete is then poured around the hose, which is deflated and removed after the concrete sets.

ADVANTAGES: Relining the chimney improves the safety and integrity of the chimney.

DISADVANTAGES: The size of the flue can be reduced, thus reducing the draft to an insufficient level.

4. INSTALL AN EXTERIOR AIR SUPPLY FOR COMBUSTION AND DRAFT AIR. Conventional fireplaces draw room air that has been heated by the home's primary heating system and exhausts it out the chimney. This air is replaced by infiltration of cold outside air through cracks and openings in the building envelope. In tightly constructed homes or in homes that are already exhausting air with fans and dryers, the fireplace may be starved for air and dump smoke into the room because of insufficient draft. A solution to these problems is to provide a means for using outside air for combustion.

There are three basic components to an exterior supply air system for a fireplace: intake, passageway, and inlet (Fig. 8.8.1). The intake is typically located on an outside wall or the back of the fireplace, but can be in a crawl space, attic, or other unheated space. Many codes will not allow location of an inlet within a garage because of the potential presence of fuel fumes. A passageway or duct connects the intake to the inlet. It is usually insulated to reduce heat loss. The inlet introduces the outside air to the firebox. A damper is necessary to control the volume and direction of airflow.

Glass doors are typically installed to prevent indoor air from entering the firebox and going up the chimney. Unfortunately, the tempered glass that is so often used is not a good transmitter of infrared radiation so the radiant heat from the fire itself is significantly reduced.

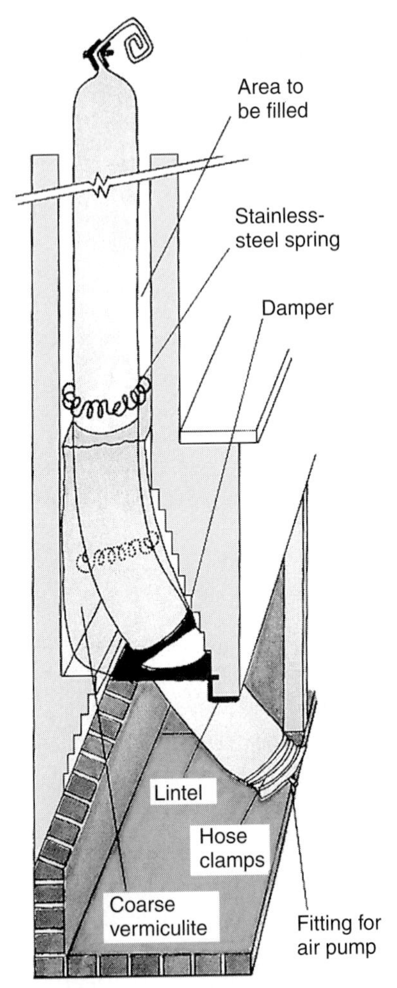

FIGURE 8.8.5 CHIMNEY RELINING

ADVANTAGES: The exfiltration of heated conditioned-space air is minimized.

DISADVANTAGES: An exterior air supply may be difficult to install in an existing fireplace.

5. INSTALL AN ENVIRONMENTALLY FRIENDLY GRATE.

The specially designed EcoFire Super-Grate, marketed by Andiron Technologies, can be installed to reduce the harmful emissions of a conventional fireplace. This stainless-steel grate replaces the wrought-iron grate that holds wood logs. It is attached to a fan which blows air through holes in the tubes of the grate. This air results in an extremely hot fire which has lower emission levels. The heat-reflecting shield which is attached improves heating efficiency by increasing the radiant heat output.

ADVANTAGES: An environmentally friendly grate can be easily retrofited to existing fireplaces.

DISADVANTAGES: An environmentally friendly grate is relatively expensive and still subject to wood-burning bans.

6. INSTALL A GAS LOG SET OR FIREPLACE.

A gas log set is primarily a decorative appliance. It includes a grate holding ceramic logs, simulated embers, a gas burner, and a variable flame controller. These sets can be installed in most existing fireplaces. There are two principal types: vented and unvented. Vented types require a chimney flue for exhausting the gases. They are only 20% to 30% efficient, and most codes require that the flue be welded open, which results in an easy exfiltration path for heated room air. Unvented types operate like the burner on a gas stove, and the combustion products are emitted into the room. They are more efficient because no heat is lost up the flue and most are equipped with oxygen depletion sensors, but they are banned in some states, including Massachusetts and California.

Gas fireplaces incorporate a gas log set into a complete firebox unit with a glass door (Fig. 8.8.6). Some have built-in dampers, smoke shelves, and heat-circulating features that give them the capability to provide both radiant and convective heat. Units can have push-button ignition, remote control, variable heat controls, and thermostats. Gas fireplaces are more efficient than gas logs with efficiencies of 60% to 80%. Many draw combustion air in from the outside and are direct vented, eliminating the need for a chimney (Fig. 8.8.7). Some of these units are wall-furnace rated.

There are also electric fireplaces which provide the ambience of a fire and, if desired, a small amount of resistance heat. These units have no venting requirements.

ADVANTAGES: There are no ashes or flying sparks that occur with wood-burning fireplaces. Their use is not affected by wood burning bans imposed in some areas when air-quality standards are not met. Direct-vented gas or electric models eliminate the need for a chimney.

DISADVANTAGES: The cost for equipment and running the gas line can be high. An existing masonry chimney cannot be used with a gas fireplace.

7. INSTALL A WOOD STOVE, FIREPLACE INSERT, OR ADVANCED FIREPLACE.

There are wood-burning equipment alternatives that offer an improvement over a conventional wood-burning fireplace that may only be 10% efficient. These include wood stoves, fireplace inserts, and advanced fireplace cores. The performance of wood-burning systems varies dramatically with the type of equipment, the type of wood being burned, the wood's moisture content, and the way it is maintained and operated.

Wood stoves without air controls, such as Franklin stoves, have efficiencies of 20% to 30%. Stoves with controlled air inlets into primary and secondary combustion areas can have efficiencies as high as 55%. Advanced designs can have efficiencies as high as 75%. The more efficient systems require much less excess air for combustion and produce lower levels of incomplete combustion products which produce creosote.

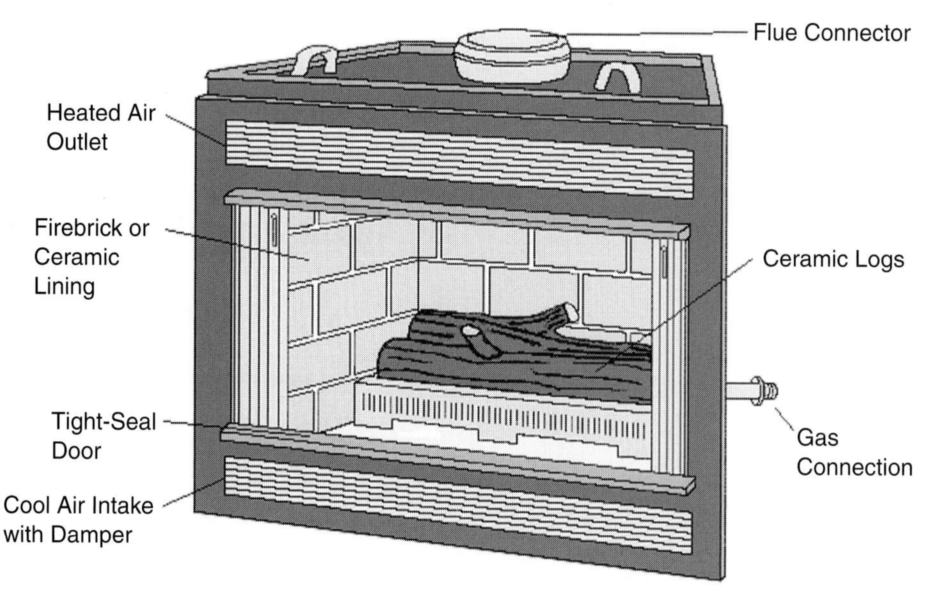

FIGURE 8.8.6 GAS FIREPLACE

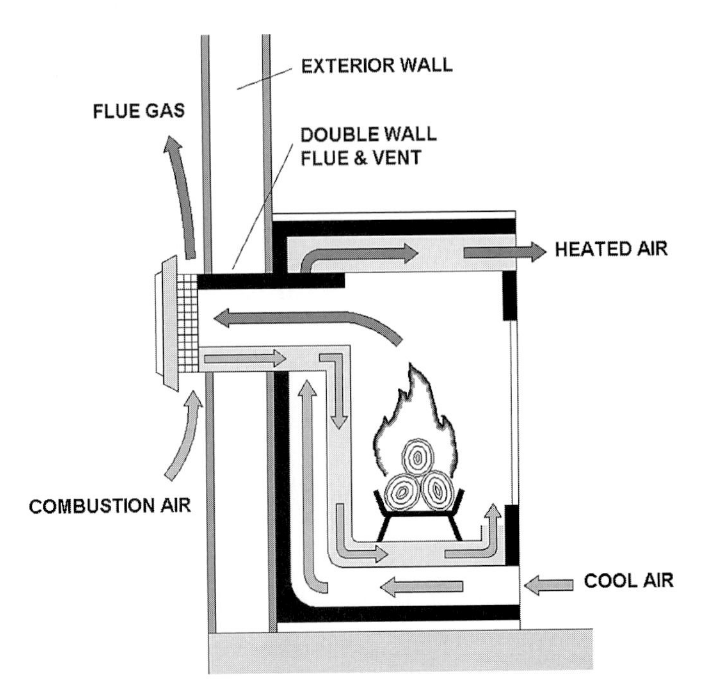

FIGURE 8.8.7 DIRECT-VENT GAS FIREPLACE

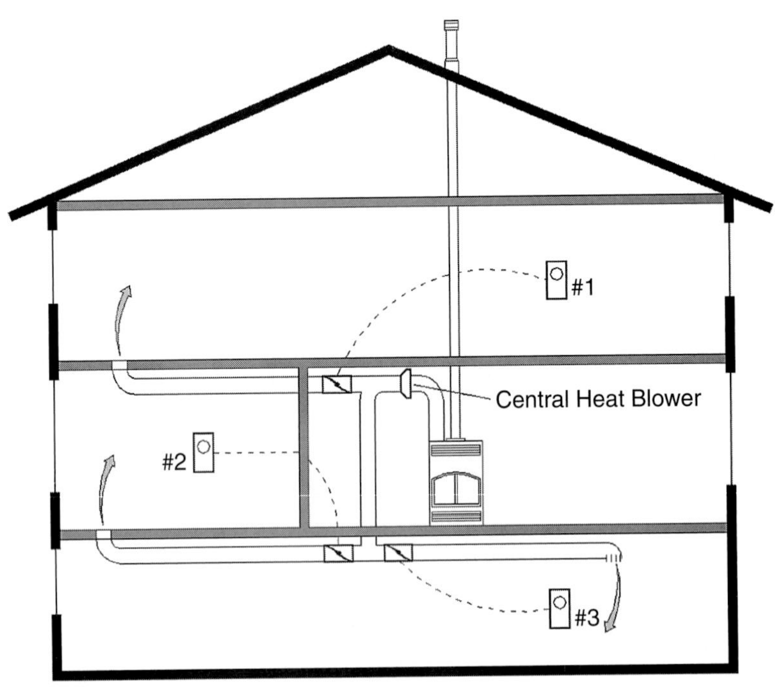

FIGURE 8.8.8 DUCTED FIREPLACE

Wood-burning fireplace inserts are designed to fit into existing fireplaces to improve their heating performance. Some stoves sit on an existing fireplace hearth and vent up the chimney.

In response to some western-state ordinances banning the installation of a traditional fireplace, manufacturers responded with fireplaces that meet the EPA's standards for wood stoves. These units are as efficient as the advanced wood stoves. They are airtight with gasketed doors and a pyro-ceramic glass window that allows the infrared heat from the flame into the room. The advanced fireplace has an insulated casing to reduce heat loss through the outside wall of the house. A squirrel-cage fan is used to draw room air in and around the casing to pick up additional convective heat and supply it to the room. Heat output is controlled by the amount of outside air intake for combustion. Some even allow for short duct runs to distribute heat via natural convection to isolated rooms. RSF Energy produces a fireplace system that can supply a whole-house duct system with an in-line blower and thermostats (Fig. 8.8.8).

The Rumford-style fireplace designed by Jim Buckley is one of the only masonry fireplaces that meets most air-quality standards. A fireplace kit is available for the construction of a masonry fireplace based upon the traditional Rumford design. The kit includes a one-piece curved clay throat, a clay flue tile liner, a smoke chamber, a stainless-steel damper, and optional glass doors.

ADVANTAGES: These systems have higher efficiencies than a conventional wood-burning fireplace.

DISADVANTAGES: These systems can be subject to wood-burning bans when local air-quality standards are not met. Maintenance to prevent the dangerous buildup of creosote is required.

8.9 DOMESTIC HOT WATER HEATING

ESSENTIAL KNOWLEDGE

Depending on whether the home is air conditioned or not, water heating is the second or third largest energy expense in the home. Traditionally, water heating accounts for approximately 14% of the utility bill. As space heating and cooling loads decrease and system efficiencies increase, water heating represents a greater portion of the home's energy bill.

The predominant design for water heaters is the storage type (Fig. 8.9.1). These units hold heated water in a thermostatically controlled storage tank. Tanks in residential applications usually have storage capacities ranging from 30 to 80 gallons. The tank may have a combustion burner in the bottom with a flue running up through the center of the tank, electric resistance heating elements immersed in the water, or a heat exchanger circulating fluid heated by another source such as a boiler.

When deciding whether or not the domestic water heating system needs rehabilitation, the following questions should be answered:

- Does the present system have several years left on its expected life?
- Is the time it takes to deliver hot water to faucets acceptable?
- Is there sufficient hot water?

In a rehabilitation situation, the answer to any of these questions could be no. Poor maintenance can shorten the life of water heaters. Bathrooms added over the years may be remote from the tank location, or the addition of clothes and dish washers may have increased hot-water demands beyond the original design.

The life of storage-type, water heaters is typically 10 to 15 years due to the corrosion of the tank. Maintenance, such as replacing the anode and cleaning sediment from the bottom of the tank, can extend its life, but this is rarely done. Thus, if the storage tank is more than 10 years old, it should probably be replaced. The month and year that the tank was built is usually encoded in its serial number.

Demand, or instantaneous-type, water heaters can be located closer to the fixture to minimize the waiting time for hot water. These units eliminate the tank and its associated losses altogether.

If purchasing a new water heater, selecting a system of the proper size and recovery rate is important to ensure that all hot-water demands are met. A water heater of insufficient capacity will result in cold showers, but a water heater that is too large wastes energy. The water heater size is determined by the first hour rating (FHR), which is the amount of hot water (in gallons) that can be produced in 1 hour. The FHR is not only a function of tank size, but also recovery rate, which is a measure of how quickly the incoming cold water can be heated. Gas water heaters have higher recovery rates than electric units. Thus, for the same FHR, the gas water heater requires a smaller tank than an electric water heater.

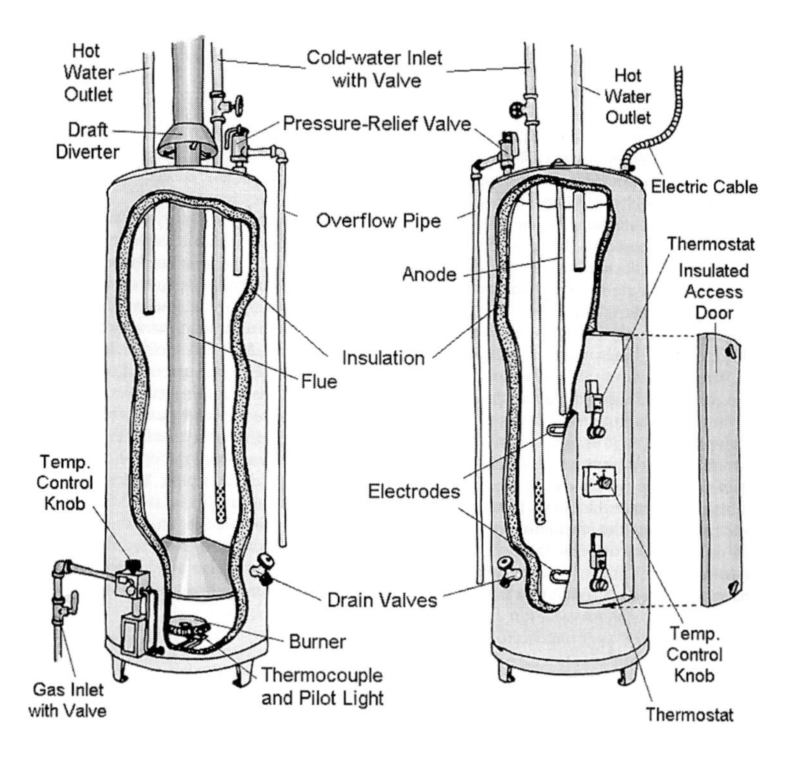

FIGURE 8.9.1 STORAGE-TYPE WATER HEATERS

In addition to the FHR, storage-type water heaters are given an energy factor (EF) rating. This is a seasonal efficiency rating that takes into account the water heater's recovery efficiency, standby losses, and energy input. Recovery efficiency is the ratio of the amount of heat that is absorbed by the water to the amount of heat input.

TECHNIQUES, MATERIALS, TOOLS

1. IMPROVE THE EFFICIENCY OF THE EXISTING STORAGE-TYPE WATER HEATER.

In lieu of purchasing a new, more-efficient water heater, there are a few relatively easy and inexpensive ways to improve the efficiency of the existing system. These are essentially the same methods that manufacturers have used to improve the efficiency of today's tank water heaters.

Insulate the tank and pipes. Older storage-type water heaters can benefit from the simple installation of an insulating jacket or blanket. This reduces the heat loss from the tank to the surrounding area. Be careful not to cover thermostats, drains, flues, or combustion air inlets. Insulating the pipes reduces the losses from the hot water as it flows through the pipes to the faucet. The split foam rubber type of insulation is effective and easy to install.

Install anticonvection valves or loops. These devices are installed on the hot-water inlet and outlet pipes to prevent the convection of hot water up the pipes from the tank when in the standby mode. There are numerous types. Some are based on a simple ball-type check valve. A loop in the piping serves the same purpose (Fig. 8.9.2).

ADVANTAGES: These methods are inexpensive and easy to do.

DISADVANTAGES: Tank blankets are not as effective as internal insulation because certain areas must be left exposed for access and venting purposes. The useful life of these measures is limited to the

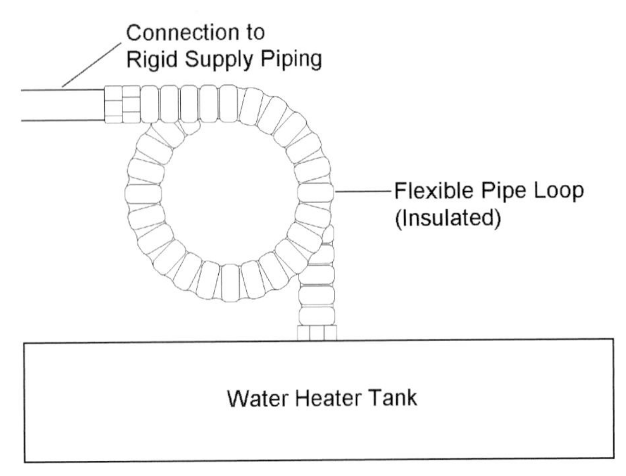

FIGURE 8.9.2 ANTICONVECTION LOOP

remaining life of the water heater. They will not be necessary for most new water heaters because high levels of internal tank insulation and anticonvection devices are standard features.

2. INSTALL AN INDIRECT STORAGE WATER HEATER.

In an older home with a hydronic heating system supplied by a boiler, a tankless coil may be the method for heating domestic hot water. This is a coil located within the boiler. There is no storage of hot water. This approach is suitable in the winter when the boiler is operating for space heating, but is inefficient in the summer because the boiler must start and stop frequently just to provide domestic hot water. An alternative is to install an indirect storage water heater (Fig. 8.9.3). Heating fluid from the boiler is circulated through a coil inside the storage tank. The boiler is still used throughout the year for domestic hot water, but it comes on less frequently in the summer because it responds to a drop in storage tank temperature and not every demand for hot water.

ADVANTAGES: An indirect storage water heater improves system efficiency.

DISADVANTAGES: Additional floor space for the storage tank is required.

3. INSTALL A NEW ELECTRIC RESISTANCE STORAGE WATER HEATER.

Although this is usually the most expensive method for heating water, it is the second most common type of water heater after gas storage units. If hot-water requirements are relatively low, this may be the most practical choice. New electric storage water heaters have higher levels of insulation than the old versions. Some are all-plastic, which do not need anodes and come with lifetime, never-leak warranties. Electric storage water heaters can be located almost anywhere because there are no combustion air and venting issues. Time clocks can be used to prevent the resistance elements from operating during peak electric charge periods where time-of-use rates are in effect.

ADVANTAGES: Electric storage water heaters have a lower initial cost than gas storage water heaters. The installation location is flexible.

DISADVANTAGES: Operating costs are high.

4. INSTALL A NEW GAS STORAGE WATER HEATER.

New gas storage water heaters have better tank insulation, improved baffle designs, lower pilot burner inputs, and new combustion chamber configurations than older versions. Baffles regulate the flow of combustion air up through the flue, and new designs increase the transfer of heat from the flue gases to the water, increasing system efficiency. They also reduce convective air movement and heat loss up the flue during standby periods.

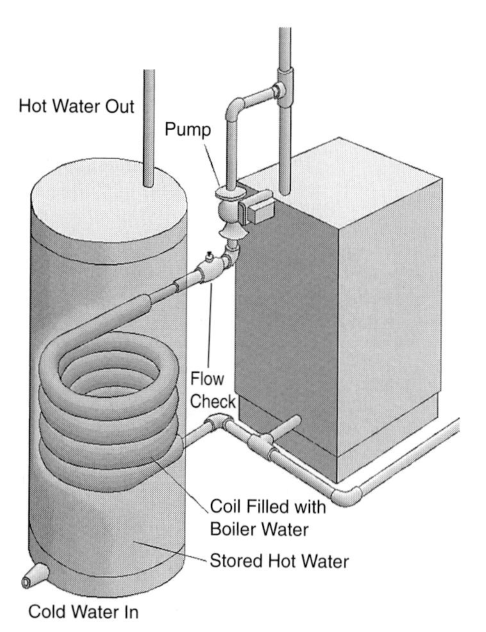

FIGURE 8.9.3 INDIRECT STORAGE WATER HEATER

As with gas furnaces, there are several venting options. Standard units are atmospheric vented with the vertical flue that is often tied into the same chimney flue as the gas furnace. Power-vented units use a fan to assist the venting of the combustion gases. These provide more location flexibility because longer vertical or horizontal vent pipes can be used. They also cannot back-draft while the burner is operating. However, they still use indoor air for combustion and now require electricity to operate. Direct-vented units are also available that draw outside air in for combustion and vent directly out the wall.

A gas water heater that avoids the problems of venting entirely is the Seahorse by Gas-Fired Products Incorporated. This unit is a gas-fired heat exchanger with a pump that is placed in an insulated box on the outside wall of the home and plumbed into a storage tank inside the home which may be the old electric water heater (Fig. 8.9.4). Going a step further is the Weather-Pro by American Water Heater. This is a gas water heater that can be installed outside, including the 50-gal tank. Its distribution is currently limited to southern states where freeze protection is not an issue.

ADVANTAGES: The operating cost is low.

DISADVANTAGES: The initial cost is higher than that for electric resistance storage-type water heaters. Venting requirements restrict location flexibility. Power-vented units require electricity to operate and have had occasional problems with nuisance shutdowns when their pressure safety switches have mistaken windy conditions for blocked vents.

5. INSTALL A DEMAND WATER HEATER.

In situations where space for a 20- to 50-gal storage tank is limited or the wait for hot water to a tap remote from the storage tank is excessive, the installation of a demand water heater may be appropriate. Also called tankless, instantaneous, and point-of-use, demand water heaters heat the water as it is called for. There is no storage tank. Some point-of-use units may not be truly instantaneous or tankless because they employ a small 2- to 4-gal storage tank (Fig. 8.9.5).

Electric demand water heaters heat the water as it passes over a resistance element. The power requirements limit these units to water flow rates for a single sink or low-flow shower. One of the largest electric demand water heaters is the Seisco RA-28. Rated at 28 kW, it is capable of supplying 2.5 gallons per minute (gpm) at a 78°F temperature rise. Electric demand water heaters typically modulate their output by using multiple heating elements. Less expensive, fixed output units do not allow for much variation in water flow.

FIGURE 8.9.4 SEAHORSE OUTSIDE GAS-FIRED WATER HEAT EXCHANGER

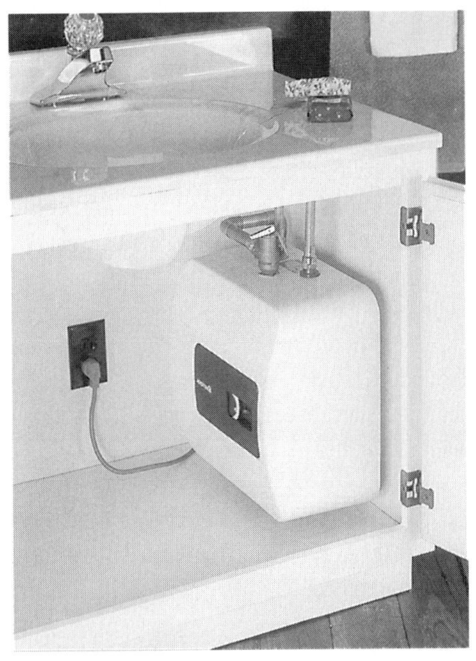

FIGURE 8.9.5 POINT-OF-USE WATER HEATER

Gas demand water heaters employ a modulating burner to supply hot water at a specific preset temperature. Their output is sufficient to satisfy the hot-water demands of an entire home. They must be mounted on an external wall for venting purposes.

ADVANTAGES: Demand water heaters provide location flexibility due to their small size. They have reduced standby losses and longer life because the tank corrosion issue is eliminated. When located near the point of use, they are water conserving because there is less cold water to go down the drain prior to the arrival of hot water.

DISADVANTAGES: While demand water heaters can deliver hot water for an indefinite period of time, the in-flow rate (gallons per minute) may not be sufficient. Most units cannot serve multiple tap requirements simultaneously. Initial costs are higher than for storage-type units. Larger gas lines or more power are required than for storage-type units. Service support may be less than for the more common storage-type units.

6. SUPPLEMENT THE WATER HEATER WITH A PREHEATING OR HEAT RECOVERY SYSTEM.

The heating requirements of the water heater can sometimes be supplemented by recovering waste heat from other processes or capturing solar energy. One of the simplest approaches is a tempering tank. This is a second uninsulated tank located in a warm or sunny area and connected in series with the primary water heater tank. Cold water first enters the tempering tank where it warms up to the surrounding air temperature. Solar water heating, both passive and active systems, can be used to preheat water and during some times of the year can meet all of the water heating needs.

Hot-water desuperheaters, such as the DS06 unit by Trevor-Martin Corp., are hot refrigerant-to-water heat exchangers on the refrigerant line of an air conditioner or heat pump. Such a unit is installed after the compressor, but before the condenser, to remove the superheat from the refrigerant vapor and transfer it to the domestic hot water (Fig. 8.9.6). The amount of water heating provided by these systems is a function of the air conditioning usage since they only provide heat when the air conditioner or heat pump is operating.

Another method of heat recovery is gravity film exchange (GFX). This is a heat exchanger between the water waste pipe and the domestic hot water tank cold water inlet (Fig. 8.9.7). It consists of a section of 3" or 4" copper drainpipe with a coil of 1/2" or 3/4" tubing wrapped around it. At times of high hot water use such as showering, the water going down the drain is still hot. This device recovers some of that heat and preheats the cold water flowing into the tank. There are no controls or moving parts.

ADVANTAGES: These systems are relatively inexpensive, require little or no maintenance, and capture heat energy that would otherwise be wasted. Desuperheaters can improve the HVAC system cooling efficiency slightly.

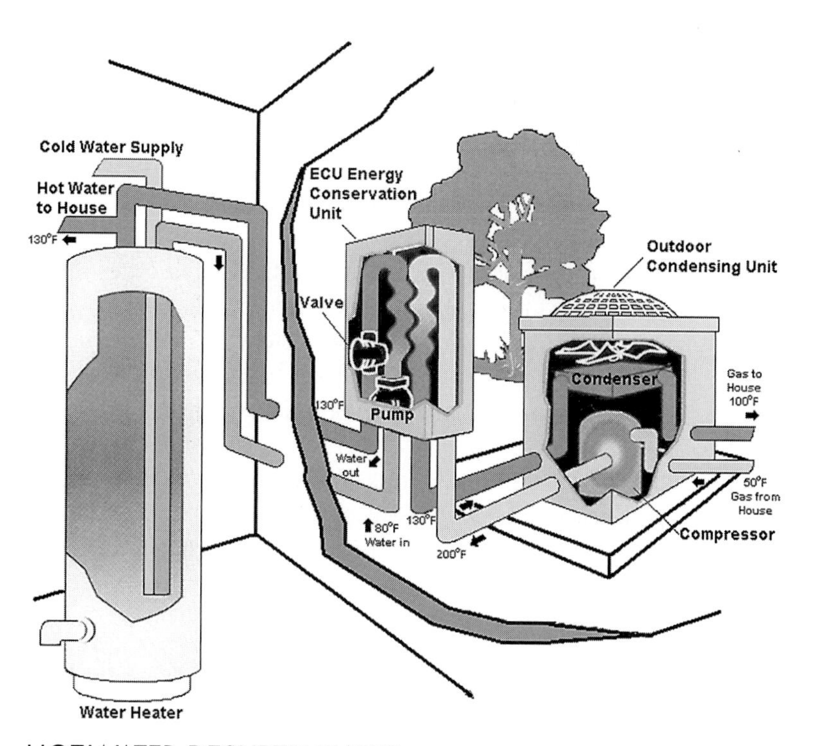

FIGURE 8.9.6 HOT-WATER DESUPERHEATER

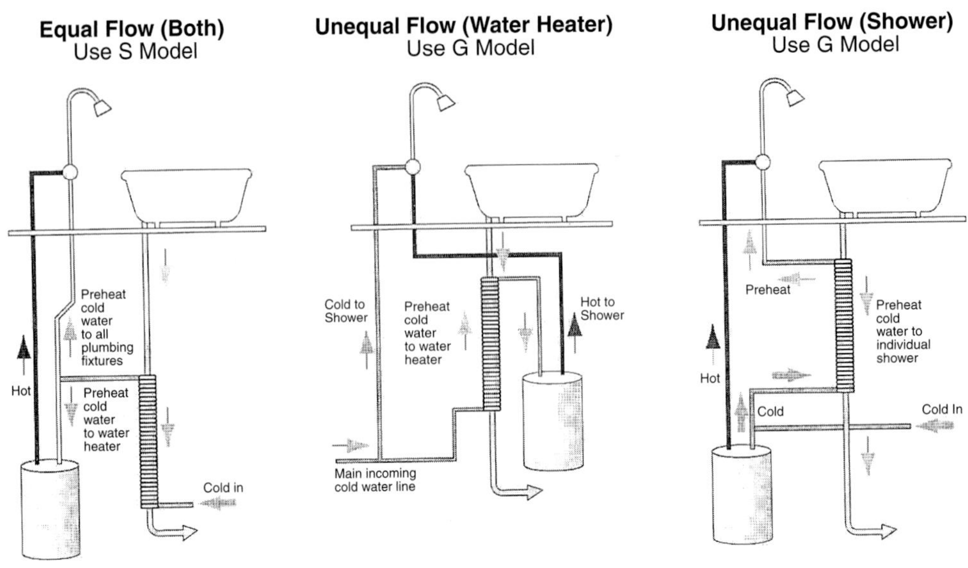

FIGURE 8.9.7 GFX HEAT RECOVERY SYSTEM

DISADVANTAGES: Desuperheaters only provide a benefit when the heat pump or air conditioner is operating, and they can decrease the heating capacity of a heat pump system. GFX devices only recover heat when water is draining out at the same time that makeup water is flowing into the tank (e.g., water is flowing directly from the spigot or showerhead down the drain). Building codes or building officials in some areas do not allow GFX devices.

7. INSTALL A HEAT PUMP WATER HEATER (HPWH).

If electricity is the only water heating fuel alternative and the domestic hot-water requirements are relatively high (e.g., four or more occupants), a heat pump water heater can be considered to reduce operating costs. It can typically provide hot water at one-half to one-third the energy use of an electric resistance water heater. Rather than heating the water directly by electric resistance, these heat pumps use electricity to drive a vapor compression cycle that moves heat from the surrounding air to the tank water. HPWHs can be integral systems with the compressor-evaporator unit sitting on top of the storage tank or as separate units requiring a pump and flow loop (Fig. 8.9.8). Integral units avoid the need for a pump and control loop, but the separate unit design can be retrofitted to the home's existing tank.

HPWHs are most appropriate in warm climates because they are typically located in basements and garages, where freeze damage is not a concern. The efficiency of the heat pump also drops at low air temperatures. HPWHs cool and dehumidify the air surrounding the evaporator section like an air conditioner. This can be beneficial in a conditioned space in the summer, but detrimental in the winter. The evaporator section cannot be in a confined space where the surrounding air is not mixed with warmer air.

HPWHs do not have the quick recovery of standard water heaters. A larger tank could be used to meet peak demands, but the more common solution is to install an electric resistance heating element in the tank to handle the peak demand periods. This defeats some of the efficiency benefits of the HPWH.

In cooler climates or where a ventilation system is employed, an exhaust air heat pump water heater (EAH PWH) may be appropriate. This system, manufactured by DEC/Therma-Stor, captures heat from the air before it is exhausted.

ADVANTAGES: Lower operating costs than an electric resistance system.

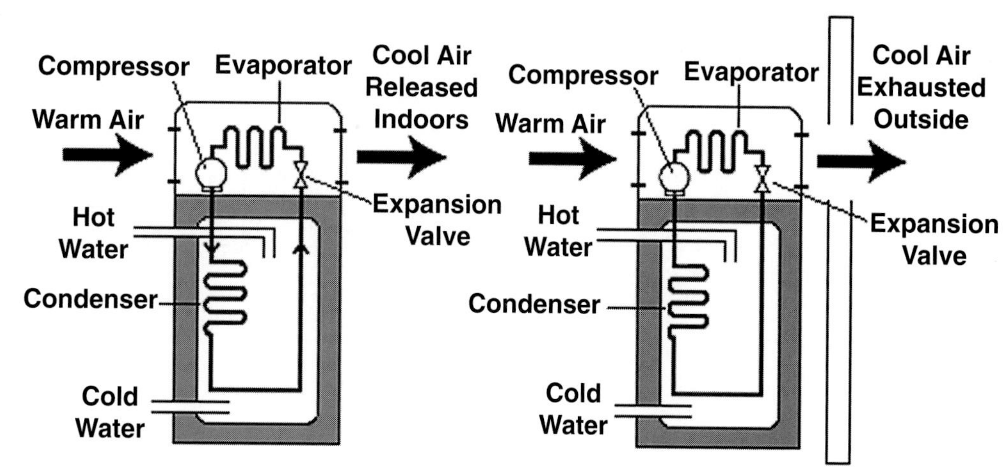

FIGURE 8.9.8 HEAT PUMP WATER HEATER

DISADVANTAGES: The initial cost is higher. There are currently only a few U.S. manufacturers of these heaters, and there is little market infrastructure in place. Finding qualified, experienced installation and service technicians may be difficult. Installation of these heaters requires expertise in water heater service as well as refrigerant handling regulations. HPWHs need to be located where the compressor noise will not be a problem. They require regular maintenance similar to that for an air conditioner.

8.10 PLUMBING DESIGN AND ENGINEERING

ESSENTIAL KNOWLEDGE

Properly sized and performing plumbing systems are vital for the comfort, convenience, health, and safety of home occupants. Rehabilitation reasons can range from obvious failures of a system, such as leaks, corrosion, and stoppage, to not-so-obvious failures, such as leaks of dangerous gas from sewers or combustion appliances. Plumbing systems can include water supply and distribution; drain, waste, and vent systems; fuel supply and storage; appliance venting and exhausting; fire protection systems; and gray water systems. Plumbing fixtures themselves are covered in another volume, as are septic systems.

There are a few major factors that drive the decision to rehab plumbing systems: how well the system meets the needs of occupants, how the system performs and complies with code requirements, how possible rehabilitation might save resources, and the cost and feasibility to rehab. The scope of rehab work can vary from simple repair, to removal and replacement of systems, to adding new ones. Fire protection systems and fuel systems are examples of added systems.

Materials, designs, and codes have changed and evolved over the years. Code-complying original installations may now be below standard or dangerous. For example, an old plumbing fixture such as a bathtub could be filled with contaminated water that could drain back into potable water supply because the fill spout is below an overflow drain. Older toilets used much more water, and older shower heads used more water and energy than those that now comply with the Energy Policy Act and the National Plumbing Standards. While fixture concerns are not addressed here, they drive decisions to change systems and components. Existing vents might not work with a new, more efficient appliance that produces cooler and high-moisture vent products. Materials previously approved may be hazardous to occupant health or prone to failure, such as lead in water systems, which is a well-publicized hazard. Certain connectors and pipe in polybutylene water systems are known to fail. Asbestos insulation in many old plumbing installations is a hazard. Recommendations regarding the removal and disposal of asbestos are available from the Environmental Protection Agency (EPA) Asbestos Information Hotline: 800-438-2474.

Designs and installations previously deemed to be state-of-the-art may not work well, and in some instances could be dangerous. Old drain, waste, and vent systems are examples. Without proper venting, sewer gases can create unhealthy conditions. New materials and designs can alleviate flaws. There are positive system developments as well. Research and, subsequently, codes have permitted water pipe supply sizes to be downsized for branches. The house's present system, if in good condition, may be able to accommodate more fixtures. There are many new materials and upgrades to old materials and methods. There is a plethora of connectors, valves, reducers, and adaptors to allow change from one material to another.

As design, research, and new materials have evolved, so have codes. One national code, the *International Residential Code* for one- and two-family dwellings, was first promulgated in draft form and issued in 1998. It was updated for an April 2000 release. This code covers most residential plumbing systems. The International Association of Plumbing and Mechanical Officials (IAPMO) nevertheless still publishes its *Dwelling Requirements of the Uniform Plumbing Code*.

This guide will review attributes of many systems and materials, along with advantages and disadvantages. These attributes are not necessarily comprehensive and readers are advised to undertake research of individual systems, products, installation recommendations, warranties, and code acceptance for their own locale. This guide cannot foresee the scope of a project or predict costs, but systems components are evaluated as more or less costly than others.

8.11 WATER SUPPLY AND DISTRIBUTION SYSTEMS

ESSENTIAL KNOWLEDGE

There are many causes for rehabilitation maladies that affect water supply and distribution systems. Leaks, poor water quality, poor supply of hot or cold water, insufficient pressure, noise, vibration, and presence of hazardous materials can all be driving forces.

Water distribution in today's homes originates from a one-pipe delivery system with pipes and fittings of lead. Ancient Romans had lead water pipes, and in fact the term *plumbing* is derived from the Latin *plumbum*, meaning lead. Ancient Egyptian artifacts show use of copper water piping. A variety of pipes and fittings could be present in any rehab project, including brass, copper, iron, and steel. Water piping materials and fittings have changed in popularity and code acceptance. Many types are approved for use and can be used in combination if electrogalvanic and grounding continuity are properly addressed. Galvanized iron and galvanized steel piping (with some yellow and red brass and copper) was quite popular prior to World War II, but were replaced by copper tubing postwar and by plastics from the 1970s. Iron and steel pipes suffered corrosion and scale buildup in certain water types, especially in hot-water lines. Copper tubing was approved by ASTM B88 in 1932 and is the same material used today. Copper has had some problems when used with aggressive, corrosive water.

There is a variety of plastic materials approved for use. Some plastic flexible water piping materials have had problems with splitting and fitting failure. One material used in the 1970s and 1980s, polybutylene fittings, failed in many homes with disastrous results. A new hybrid piping material, KITEC by IPEX, is a composite pipe made of aluminum laminated between interior and exterior plastic layers. It claims national code approval, easy workability, sound dampening, and is corrosion-proof with good flow rates.

Water conservation has become mandated in recent years. Toilets, faucets, and showerheads must now often meet lower water use standards. Consumers are more cognizant of excess water use and energy cost for heating water that is wasted. Energy-recovery devices are available and growing in popularity. Insulation, better system design, and solar heating options can also be explored.

Water quality and other health concerns are national and local issues. Filtering water systems are becoming more popular, installed below the kitchen sink or counter. Galvanized iron piping has corrosion problems that are often exacerbated by adverse water conditions and heat. Lead water services for houses were quite common, and lead in water continues to be a problem. Lead solder for copper joining has been banned. Jacksonville, Florida, has "aggressive, corrosive water" and has banned the use of copper in domestic water systems. Asbestos was a popular insulation material for piping and must be carefully removed or encapsulated.

TECHNIQUES, MATERIALS, TOOLS

The water system should be thoroughly inspected and tested. Depending on the reason for rehabilitation, the water distribution system can be repaired, modified, or replaced. Some reasons include presence of lead piping, failed polybutylene joints or pipes, low water pressure, corroded or broken pipes in under slab or concealed locations, or condensation on cold-water piping due to deteriorated or missing insulation. Codes may require renovations to any plumbing system to conform to current code without requiring the entire system to comply. Failed systems can be ripped out and replaced entirely or in part with the same or differing materials. Many adapters and couplings are available for these purposes. Note that water distribution systems have historically been used for grounding electrical systems. Please keep this in mind during rehab work, and restore or provide for this important safety requirement.

Design requirements for water distribution systems are water volume, water pressure, number of fixture units, the total water demand, the height of fixtures above the water supply (static head), frictional loss due to distance and piping material, and the developed length of the pipe and fittings. Code requirements may vary from one municipality to another, but water supply fixture limits and minimum branch sizes are mostly the same. These are some code requirements to satisfy proper system performance: minimum (15 psi) at the highest plumbing fixture, and maximum (80 psi) system pressures are required for the water source. Tanks and pumps increase pressure, and pressure regulators and relief or vacuum valves decrease pressure. Piping design for single-family houses is usually done by a plumber with approval by the plumbing inspector. Professional engineers typically are never involved with single-family plumbing systems. A good guide to design is *Plumbing a House* by plumber Peter Hemp.

1. IMPROVE WATER DISTRIBUTION.

Water distribution systems and house sizes have changed over the years from small homes with limited systems to large ones with elaborate systems. Long delays for hot water at remote fixtures created a shift from single-pipe systems and introduced recirculating hot-water systems. These loop systems can be expensive and use large amounts of energy in water heating and pumping if not properly designed. The draining of cold water awaiting hot water in the single-pipe system is resource-depleting and time-consuming. NIBCO has a Just Right modified loop product that uses natural convection to circulate hot water. A check valve and passive recirculating line forming the "forgotten" system can be easily installed for a similar passive recirculation system. The electric tracer wire system has a single-pipe hot-water self-regulating supply system that solves many loop problems. MetLund D'Mand System is suitable for rehabilitation and adaptation to existing plumbing. A remote sink has a pump and control system installed on the fixture angle stops. A push-button calls for hot water, and the pump recirculates the water, which saves energy and water (Fig. 8.11.1).

Manifold distribution systems are now recognized by major codes. The introduction of flexible plastic tubing like cross-linked polyethylene allows easy distribution of small-diameter, joint-free branches to individual fixtures. A larger supply line is connected to the manifold. The systems claim faster hot-water delivery, balanced flow, easily accessible control valves, and no water hammer. This is an appropriate solution that can be installed around an existing failed system, with phased replacement of fixtures and branches. A classic distribution concern is water leaks and resultant damage. Aqua-Stop offers a water leak detection system (Fig. 8.11.2). Sensors are placed on the floor near potential leak sources, such as toilets, tubs, dishwashers, and water heaters. The sensors will shut the main supply valve if a leak occurs and sound an alarm. System sensors can be hard-wired or remote.

ADVANTAGES: Responsiveness of water distribution throughout the home is improved.

DISADVANTAGES: Some systems can be costly. Local codes need to be consulted to determine whether such systems are permitted.

2. INSTALL NEW COPPER PIPING.

Copper is a very popular material and has been used since the 1930s. Joint solders previously contained some percentage of lead. Federal law changed to prohibit lead solder in potable water systems. Joints are made by soldering, brazing, and two-component adhesives. Soldering can be done

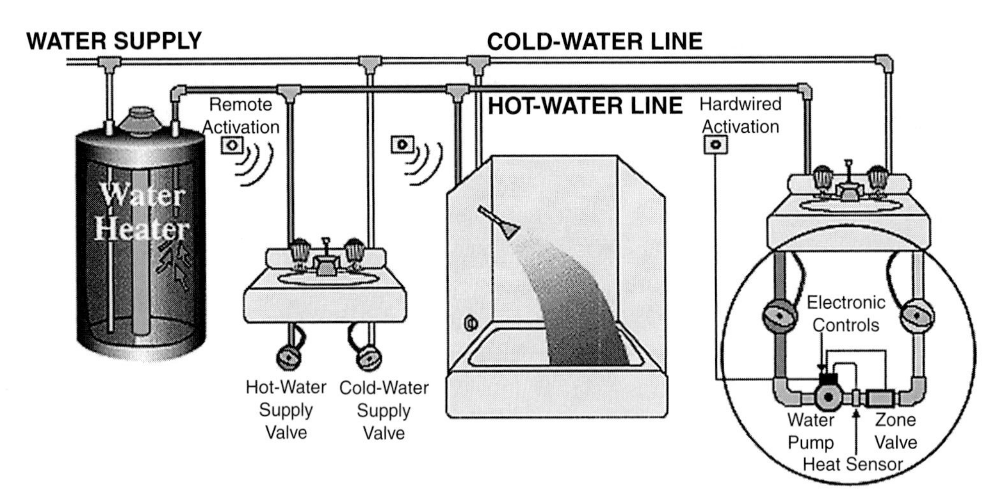

FIGURE 8.11.1 METLUND D'MAND RECIRCULATION SYSTEM

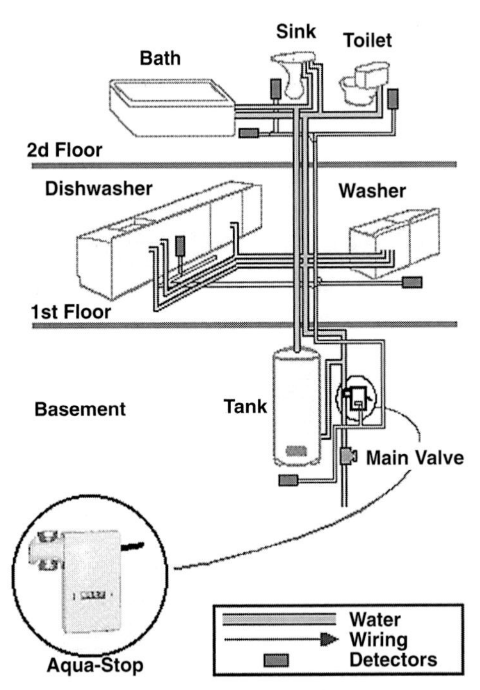

FIGURE 8.11.2 AQUA-STOP INSTALLATION

by electric resistance tools and by torches. Tees can now be mechanically pulled from continuous tubing with new tools. A new copper fitting has recently been introduced from Europe: patented in 1934, the *integral soldering* fitting has been a standard in England (Fig. 8.11.3). IMI Yorkshire now imports American-sized integral soldering fittings, which ensure the proper amount of solder is provided. The solder is in the middle of the fitting, which provides a better joint and saves labor. Copper tubing is inserted into each end of a fitting, applied heat melts the solder ring, and the two tubes are fused together.

ADVANTAGES: Nearly every code approves. It has a long track record, relatively easy installation, and a limited 50-year warranty. It is corrosion resistant with some water types, needs less support than plastic pipe, is fire resistant, and comes in coils and tubes.

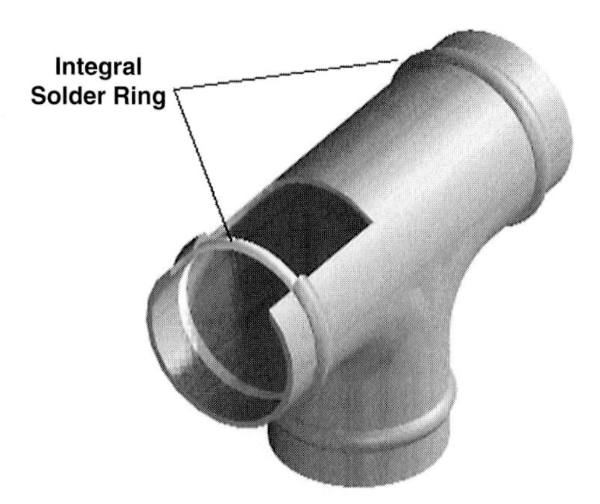

FIGURE 8.11.3 INTEGRAL SOLDERING FITTING

DISADVANTAGES: The cost is high. Some aggressive water attacks copper. Normally a flame is present when joining piping. The system is noisy, conducts heat, and needs insulation. Condensation may occur, and tubing can kink.

3. INSTALL CHLORINATED POLYVINYL CHLORIDE (CPVC) PIPING.

Used since the 1950s, CPVC is approved by most codes for both hot- and cold-water distribution. BF Goodrich, inventor of CPVC, reformulated it in 1992 to be more workable in cold weather—an earlier drawback. CPVC is popular in new construction applications above and below slabs.

ADVANTAGES: CPVC has a lower cost and easy installation and joining. There are no corrosion concerns and no water hammer. Conductance losses are reduced, which saves energy and reduces condensation. The product comes in coils and tubes.

DISADVANTAGES: More support is needed than for copper or steel piping, A 24-hour joint cure time is recommended. Expansion capacity for piping is specified by all codes. This is a relatively recent product (first used in 1959). Installation during extremely hot temperatures can result in later joint failure.

4. INSTALL CROSS-LINKED POLYETHYLENE (PEX) PIPING.

Cross-linked polyethylene (PEX) flexible thin wall tubing had its beginnings in Europe in the 1970s and has been in use in the United States since the 1980s for radiant heating systems and hot- and cold-water distribution systems. Most codes approve its use, and some allow ⅜" branches. Long PEX coils are used in the new manifold water distribution systems and eliminate expensive joints.

ADVANTAGES: PEX piping is low cost and relatively easy to install. It is highly flexible, with greater water flow. Joints and repairs can be made wet. The tubing has memory and will return to its old shape. Kinks can be removed by heat and water hammer is eliminated. PEX piping has improved freeze-resistance with better insulation than metallic piping and is easily repaired. It seems less vulnerable to nail punctures.

DISADVANTAGES: Not all codes approve PEX piping. It should not be exposed to sunlight, needs room for expansion, and needs more support than copper or steel piping.

8.12 DRAIN, WASTE, AND VENT SYSTEMS

ESSENTIAL KNOWLEDGE

There are a number of indications of the need to rehabilitate the drain, waste, and vent (DWV) systems: leaks; clogged, slow, or nonworking drains and odors. The nature of these problems can be complex, so thorough investigation is advised. Leaks and slow drainage may reveal corrosion in waste pipes. Previous rehabs might have introduced dissimilar materials that corrode through electrogalvanic action, e.g., brass and iron fittings joined. Drains need to be sloped to remove wastewater and solids. Building or pipe settlement can change the pipe slope, and the system may not work properly or may fail entirely. Odors can be a warning of serious problems. A toxic, explosive blend of gases could be present, caused by something as simple as a dry trap or as serious as a nonexistent vent system. Vents protect against back pressure and siphoning and provide system air circulation. Cross-connections are also a concern. Waste system design has evolved over time, and older approved methods did not recognize that siphoning could occur between potable and contaminated water (Fig. 8.12.1). Local codes should be checked before proceeding with rehabilitation.

Drainage systems are basically gravity designed with venting introduced to assure the system performs properly. System capacities are now better understood, and codes allow smaller pipes.

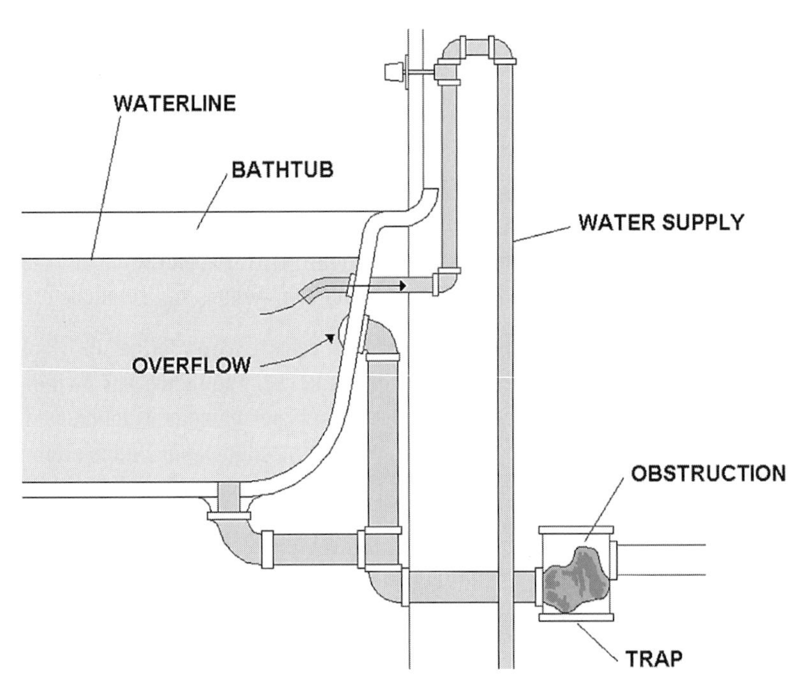

FIGURE 8.12.1 POTABLE WATER CONTAMINATION

The older, existing system to be rehabbed may potentially allow for greater expansion. The National Plumbing Standards was adopted in 1992 to mandate water-saving toilets, shower heads, and faucets. New materials, methods, and devices are now used along with old standards. Cast-iron piping in a bell-and-spigot configuration was a common material and has been replaced since the 1970s by plastics. Now hubless cast iron is regaining some upscale market share due to its sound-dampening properties.

Depending on the reason for rehabilitation, systems can be repaired, modified, or replaced. Proper analysis of the existing system and its condition is crucial. Various nondestructive diagnostic tools and services are available if opening up walls or ceilings or access under slabs for inspection is not an option. Small-pipe TV inspection and electronic leak detection systems are examples.

Drainage system materials used in residential construction include cast-iron, galvanized, and nongalvanized steel; galvanized wrought iron; lead; copper; brass; vitrified clay pipe; and plastics—acrylonitrile butadiene styrene (ABS-DWV) and polyvinyl chloride (PVC-DWV). Rehab projects on houses built after 1970 might involve plastic DWV pipe and fittings. Adapter fittings allow use of a variety of materials if desired. Cast iron is a better sound insulator but more costly, so blended systems are popular. Cast iron is used in sound-sensitive areas and plastic in nonsensitive areas. Existing materials can be combined with new or other materials, provided electrogalvanic or code concerns are addressed.

There are many new code-approved drainage systems and designs that can be considered. If the system is to be modified or replaced, a new design could be in order. New designs might be adapted to solve problems with the existing system. Drainage systems are calculated to handle a certain load, and fixtures are assigned values known as fixture units. These units are added to determine pipe sizes. Code tables state maximum unit capacity and lengths for drains and vents and minimum pipe sizes. Local codes should be consulted.

Vent strategies have evolved from conventional to common vent, wet venting, waste stack venting, and circuit venting. Various strategies evolved to minimize pipe runs and roof penetrations. A common vent connects two fixtures. The wet vent pipe doubles as a waste pipe and is increased in size. Wet vents may be horizontal (Fig. 8.12.2), vertical (Fig. 8.12.3), or combined (Fig. 8.12.4). Waste stack venting requires all fixture drains to connect separately to it (Fig. 8.12.5). Size requirements apply to these alternate methods. Circuit venting can connect a maximum of eight fixtures to a horizontal drain (Fig. 8.12.6). Slope and size requirements apply here as well. In cold regions, vents penetrating roofs may experience frost or snow closure; a pipe at least 3" in diameter is required for the last portion of the vent.

TECHNIQUES, MATERIALS, TOOLS

1. INSTALL CAST-IRON DWV PIPING.

Cast iron was originally used in bell-and-spigot configurations with lead and oakum joints and upgraded to Tyler or rubber compression joints. Hubless pipe and fittings joined with elastomeric and stainless-steel couplings are most popular currently. Cast iron can rust and is coated with an asphaltic compound. It almost disappeared as a residential waste pipe material in the 1970s, but is still used.

ADVANTAGES: Cast-iron DWV piping is accepted by all codes. It is fireproof; is a good sound insulator; is stronger, requiring less support; is long lasting; can be buried in a trench; and needs no protection from nail punctures. Three-inch pipe fits in a standard 2x4 wall without furring. No expansion allowance is needed. Hubless pipe is easy to disassemble.

DISADVANTAGES: Cast-iron DWV pipe has a high cost and weight, requires more labor to install, and is not easy to cut.

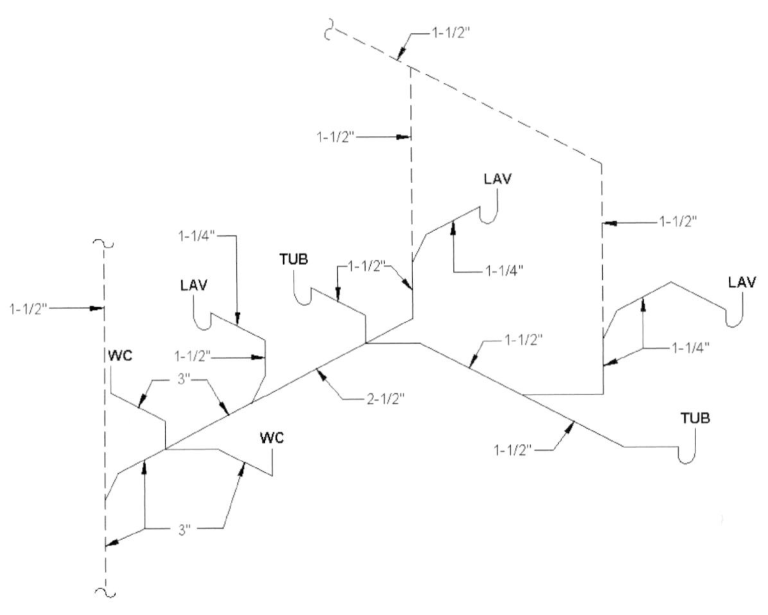

FIGURE 8.12.2 TYPICAL HORIZONTAL WET VENTING

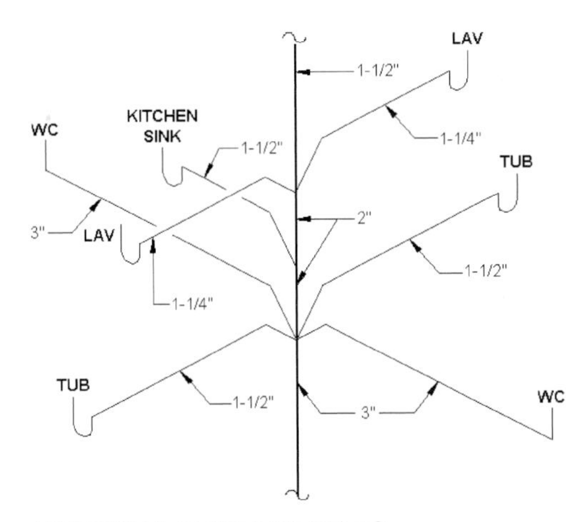

FIGURE 8.12.3 VERTICAL WET VENTING

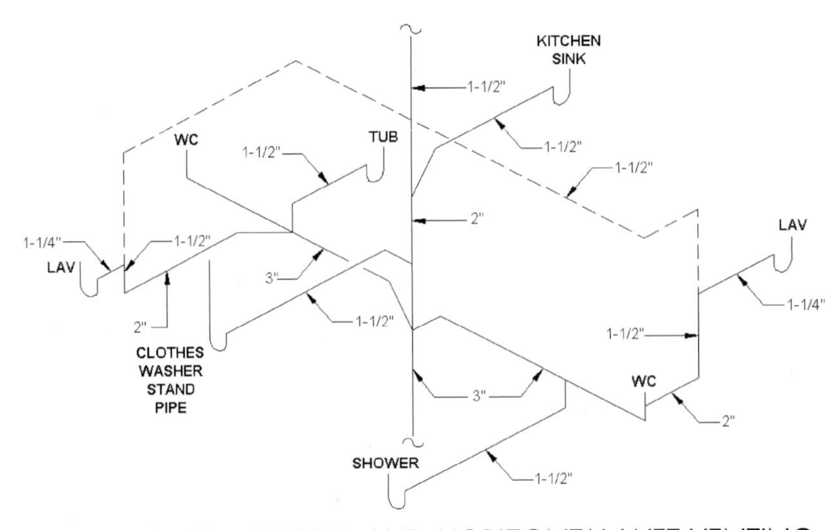

FIGURE 8.12.4 COMBINATION VERTICAL AND HORIZONTAL WET VENTING

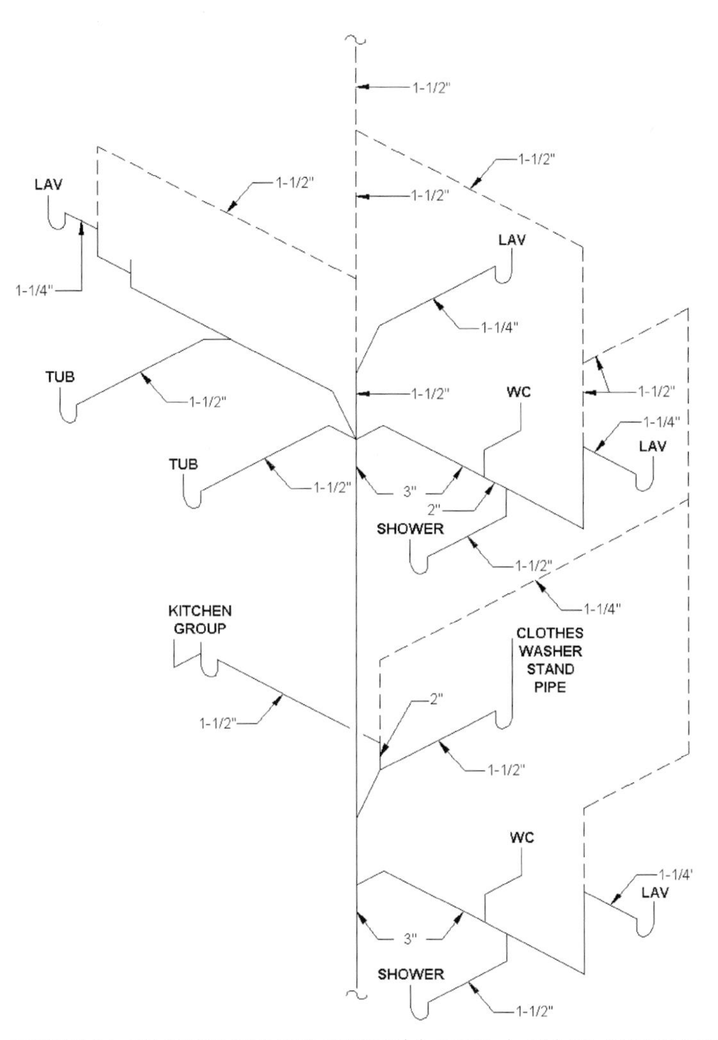

FIGURE 8.12.5 TYPICAL SINGLE-STACK SYSTEM FOR A TWO-STORY DWELLING

2. INSTALL COPPER DWV PIPING.

Copper DWV is found in many older systems. It is also more costly than cast-iron piping. The torch needed for joining copper creates a potential fire hazard, but new joining techniques are available. For example, an electric resistance joint tool can be used. Copperbond is a UL-approved two-component adhesive that replaces soldering or brazing.

ADVANTAGES: Copper DWV piping is accepted by all codes. It has a thin wall, is lightweight, and is easy to cut and assemble. Three-inch pipe and fittings fit in a 2x4 wall. Copper DWV piping is fireproof, and less labor is required than for cast-iron piping.

DISADVANTAGES: Copper DWV piping costs more than cast-iron piping, transmits noise but not as much as plastic does, and requires nail protection plates. An open flame is typically required for securing joints.

3. INSTALL ACRYLONITRILE BUTADIENE STYRENE (ABS-DWV) AND POLYVINYL CHLORIDE (PVC-DWV) PIPING.

Plastics—ABS-DWV and PVC-DWV—are the predominant materials now used in rehab and new residential construction. In various regions of the country, plumbing professionals seem to have preferences for one material over the other—typically the materials do not share a given market equally. Reduced cost and labor are reasons for popularity. Most codes accept these materials. New York State code will not allow plastic in underground residential use.

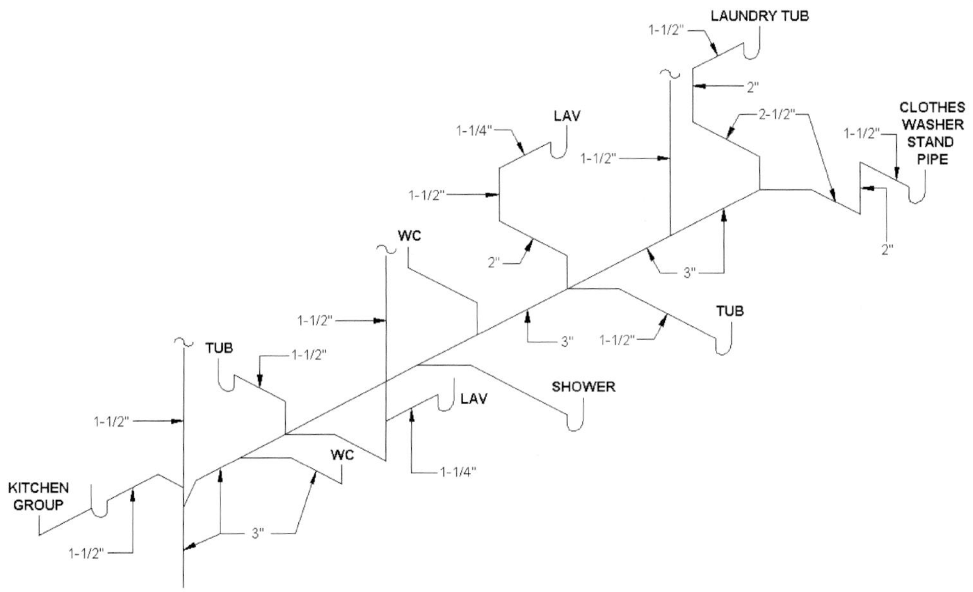

FIGURE 8.12.6 CIRCUIT VENT WITH ADDITIONAL NONCIRCUIT VENTED FIXTURES

ADVANTAGES: This type of piping is low cost, less labor intensive, easy to cut and assemble, and lightweight.

DISADVANTAGES: There are fumes present with solvents and primers. The material transmits noise, expands slightly, and needs more support. Care is needed with underground installation, and protection is needed for nail punctures. ABS burns on its own, while PVC burns only with a flame present. ABS deteriorates in sunlight.

4. INSTALL PUMPS AND EJECTORS TO DISCHARGE WASTE.

Slow or stopped waste drains may indicate that the drainage connections do not meet gravity requirements in rehab situations. Adding a pump might solve marginal or nonperforming lines. Also, new plumbing fixtures might be located so that gravity will not expel waste (i.e., basement fixtures). Sewage ejectors, pumps, or grinder pumps may be required to lift discharge. Specific code requirements exist on pipe sizing, venting, electrical connections, and audiovisual alarms. These devices should only be used when absolutely necessary because of cost, access requirements, pump failure, and power failures. The units must be well sealed and designed for removal and replacement. Check with local authorities on placement; exterior locations are typically most desirable.

ADVANTAGES: Plumbing fixtures can be below gravity connection points.

DISADVANTAGES: Pumps and ejectors are costly. Power failures will stop the system. A monitoring system and alarm are needed. Easy access is required for unit replacement.

5. INSTALL GRAY WATER AND HEAT-RECOVERY DEVICES.

Rehabilitation requires consideration for resource savings alternatives. Low-flow fixtures and shower heads to save water and energy are now mandated by the Energy Policy Act. Gray-water collection systems are a sustainable choice for any rehab project because they recycle water from bathtubs, showers, lavatories, and clothes washing machines. The *Uniform Plumbing Code* allows gray water to be collected and used for irrigation. As a conservation method or in areas with restricted water use, graywater systems may be appropriate. However, cross-contamination is a concern with gray water. Check your local code for design requirements and approval. Heat-recovery devices for waste hot water are new, energy-saving options. One wastewater heat-recovery device is the GFX. Shower, laundry, bath, dishwasher, and sink waste hot water flows through a copper DWV section wrapped with a copper water pipe feeding the water heater (Fig. 8.12.7). The device is installed vertically in the main drain and works best in basement applications, but nonbasement applications are available. Most codes approve its use.

ADVANTAGES: These devices save resources. Gardens in arid locations can be landscaped with gray water. GFX devices save energy.

DISADVANTAGES: These devices are costly. There are cross-contamination concerns, and storage containers are needed. GFX devices need to be located in the basement for best results. It may be difficult to obtain these devices.

6. INSTALL AN AIR ADMITTANCE VALVE.

An air admittance value is a relatively new device approved by some codes that can aid plumbing rehab. Some plumbing fixtures are difficult to vent due to location—kitchen island sinks are the classic example. An air admittance valve placed above the trap eliminates all other vent piping (in some circum-

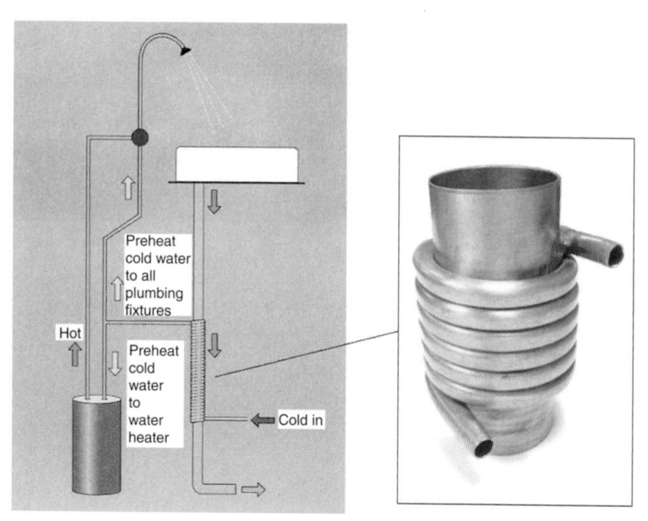

FIGURE 8.12.7 GFX WASTEWATER HEAT RECOVERY SYSTEM

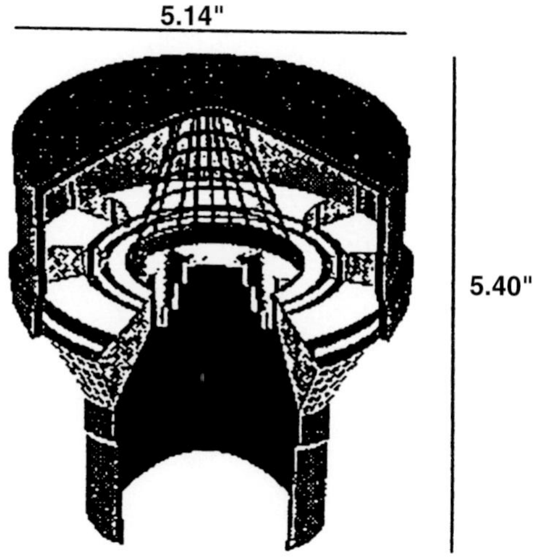

FIGURE 8.12.8 MAXI-VENT®

stances; generally, at least one vent in the system is required to terminate above the roof line). Studor Inc. invented the technology, and such vents are popular in Europe. Air admittance valves allow air to enter but prevent gases from escaping (Fig. 8.12.8). Valves must be accessible and installed above the fixture served (Fig. 8.12.9).

ADVANTAGES: Air admittance valves allows venting in difficult locations and saves venting pipe and materials.

DISADVANTAGES: Less costly, spring-activated air admittance valves were used in manufactured homes, and spring corrosion caused a high failure rate. Diaphragm designs of Studor and others do not have the corrosion problem. Not all codes allow air admittance valves.

FIGURE 8.12.9 AIR ADMITTANCE VALVE BEHIND SINK

8.13 FUEL SUPPLY SYSTEMS

ESSENTIAL KNOWLEDGE

There are three predominant residential fuel systems—natural gas, propane, and fuel oil. Rehabilitation might be necessary in the event of leaks, odors, corrosion, system damage, or appliance change. Gas—manufactured from coal or derived naturally from underground sources—has been used in houses since the mid-1880s. Lighting and cooking were the first uses of gas, with space and water heating added later. Propane (liquefied petroleum gas) became a residential fuel source in the 1920s and has approximately 5 million residential customers currently. Oil became a source in the same period, but the 22,000 houses built in Levittown, New York starting in 1947 gave oil a big boost. There were 16 million residential customers pre–oil embargo (1973), and 12 million thereafter.

As with any plumbing system, the existing condition and code compliance should be determined. Natural gas system designs depend upon utility supply pressure, heating value of the gas, appliance needs, and distances from supply to appliances. Negotiating a higher service pressure from the utility allows for a smaller internal distribution system. The gas load center concept is a relatively new fuel system innovation. The concept is similar to electrical distribution with a central panel and branch circuits. Gas lines come to a manifold at a panel, and valved branch runs connect to gas appliances. Flexible gas tubing with fewer joints allows the concept to work easily. In some homes in Japan, portable cooking appliances and ventless portable space heaters are plugged into quick-connect gas outlets. The trend is likely to grow here as well. MB Sturgis makes quick-connect gas outlets for barbeques and other appliances. Another trend concerns seismic safety; Los Angeles now requires seismic activity-triggered shutoff valves for gas systems in new homes. Flexible gas tubing, now available, may be a safe alternative because it is less likely to shear or rupture between joints.

Propane and oil systems usually have a fuel delivery source that provides service as well. System designs are per local and national codes. These two fuels require a storage tank in addition to the delivery source. The propane tank is owned by the service provider and must be outside the residence. Tanks can be below grade, but typically are above. Ten feet from the house exterior is a typical distance for tank location, but tanks can be located closer to the house. Underground connections are typical, with polybutylene frequently used as a new underground material for piping.

Oil storage tanks can be indoors or outdoors. Existing underground tanks should be surveyed for corrosion—a major problem, with potential environmental impacts due to leaks. Indoor tanks require venting, and a blocked vent line causes filling problems. Oil systems may have a single supply pipe or a loop.

Carbon monoxide (CO) from fuel combustion is a safety concern, and certain manufacturers and associations recommend installing sensors near potential CO sources. Some first-generation CO detectors were too sensitive and alarmed at nondangerous levels. New detectors are available now with appropriate alarm thresholds.

TECHNIQUES, MATERIALS, TOOLS

The following are the most common fuel supply system piping and fittings. Consult your local code agency, utility, or supplier for other materials and methods.

1. INSTALL STEEL PIPE AND FITTINGS.

This is the traditional material used for residential fuel systems, along with cast-iron pipe. The majority of systems use threaded joints, but welded joints are also common in older homes (which may make systems rehab more complicated). It is the most popular natural gas piping material in new construction. The rigid nature of steel pipe might make rehab projects difficult. Flexible tubing, if allowed by your local code, usually is easier for installation in rehab work.

ADVANTAGES: Steel pipe and fittings are approved by all codes. They have a low material cost, require fewer supports, and are nailproof.

DISADVANTAGES: Installation of steel pipe and fittings is labor-intensive, and sophisticated tools are required. Steel pipe is rigid, is difficult to modify, is heavy, and can corrode. The pipe should not be used for support when exposed (as clothes racks, etc.) because of pipe damage/breakage.

2. INSTALL COPPER PIPE AND FITTINGS.

This is the traditional material used today in residential propane and fuel oil installations. It is allowed by some utilities and codes for natural gas fuel supply and is painted yellow to distinguish it from water piping. Sulphur found in some gas corrodes copper and internal tinning is required in the tubing. See Section 8.11 for connection techniques.

ADVANTAGES: Copper pipe and fitting are approved by most codes. Copper pipe is flexible, faster, and easier to install; allows for long runs and fewer fittings; is lightweight; corrosion-resistant; easy to modify; easy for manifold systems; easy appliance hookup.

DISADVANTAGES: Copper pipe fittings are not allowed by some utilities for natural gas. Copper piping can be easily confused with water piping. Some gas is corrosive to copper. Nail plates are needed to prevent puncture. More pipe supports are needed than for steel.

3. INSTALL CORRUGATED STAINLESS-STEEL TUBING (CSST).

Developed initially in Japan, CSST was first manufactured in the United States in 1988. It is now approved by most major codes. CSST is used in residential and commercial work with natural gas and propane. It is gaining market acceptance, with five U.S. manufacturers now producing CSST, and features mechanical couplings that vary per manufacturer. Plumbers require training before installations. CSST is widely used in manifold distribution systems.

ADVANTAGES: CSST has wide code approval, is flexible and lightweight, is faster and easier to install, allows for long runs with fewer fittings, and is easy to modify. Installation can be performed by one person. CSST is corrosion-proof, is easy to use with manifold systems, and allows for easier appliance hookup.

DISADVANTAGES: CSST is a new material and is still not approved by some codes. There is a higher material cost, specialized tools are required for assembly, nail plates are needed to stop punctures, and more support is needed than for steel.

8.14 APPLIANCE VENTS AND EXHAUSTS

ESSENTIAL KNOWLEDGE

Combustion appliances require venting, and certain appliances require exhausting. Rehab work on vents and exhausts may be needed because of failure of equipment, venting, or structure. Venting failures may be deadly or cause health problems and are difficult to ascertain. Carbon monoxide deaths are often in the news, and detectors are becoming more common. Venting failure indicators might include corrosion, smoke (with oil equipment), soot, back-drafting, appliance operation failure, excess moisture, or visual damage.

Combustion appliances can be fueled by natural gas, propane (liquified petroleum gas), or oil. Traditionally, appliances were designed with a draft hood or diverter to vent combustion gases. Vents were sized based on tables indicating maximum capacities for certain vent area, height, lateral (horizontal run), and material. Traditional venting materials are masonry (lined and unlined), cement asbestos pipe, and single-walled and multiwalled metal pipe.

The 1987 U.S. National Appliance Energy Conservation Act (NAECA) increased minimum efficiency requirements for residential gas and oil-fired appliances, because vents will not perform properly if they are too large or too small. The annual fuel utilization efficiency (AFUE) became at least 78% for gas appliances and 80% for oil. New appliances meeting these and higher AFUE ratings produce fewer and cooler combustion products. Condensation and corrosion can be a problem because moisture can deteriorate chimneys through freeze-thaw action and corrode metal vents. New high-efficiency, self-condensing appliances produce such cool, moist exhaust that they can be vented using drain system materials and need to be sloped to drain the moisture.

In response to higher-efficiency appliances, the *National Fuel Gas Code* (NFPA 54) in 1992 changed venting design requirements with concern for minimum and maximum flue capacities. Certain fan-assisted combustion system appliances were included to overcome venting problems. Oil appliance venting design standards have never been published but are expected in NFPA standard 31.

Some noncombustion appliances require exhausts. Codes require clothes dryer exhausts, and gas ranges should be exhausted (but are not required to be). Dryer failure indicators may be excess moisture, clothes not drying, high energy bills, smoke, or odors. Lint and grease buildup can create fire hazards, while blocked systems or combustion by-products can create health hazards.

The physical condition of the existing venting and exhaust system should be checked, especially to ensure the passageway is properly lined, clear, and free of obstructions. If the vent or chimney was previously used for solid or liquid fuel, it should be cleaned. The failure indicators mentioned above need to be reviewed and system operation verified. Ages of combustion appliances should be matched to the vent system to determine if a new efficient appliance has been installed without an appropriate flue modification. As mentioned above, improper venting and flue damage may result. Existing oil appliances might have had their older burners replaced with the more efficient retention-head oil burner. This could require a flue modification because new levels of water and acid concentration could be present. Typically corrosion-resistant materials and/or heat loss reduction are introduced to ensure proper draft and reduce condensation. Some codes require the existing chimney or vent be brought into code conformance if a new appliance is connected.

A properly designed vent controls draft and removes flue gases. When selecting a replacement vent or checking compliance, factors such as appliance draft, configuration, size, heat, and condition; construction of surroundings; building height; material selection; and code requirements should be required. Charts in codes show selection, clearance from combustible material, and vent termination requirements. Gas appliances are classified in four categories that allow different listed vent products. Category I has nonpositive vent static pressure with vent gas temperature that avoids excessive condensation production in the vent. A Type B vent is listed and labeled for this category. Other categories require special vents. Additional requirements exist for multiple appliance vents. Direct-vented appliances have through-wall designs. To ensure intake air and combustion gas outlets are in proper locations, one guide is shown (Fig. 8.14.1).

TECHNIQUES, MATERIALS, TOOLS

1. INSTALL A NEW EXHAUST FOR A CLOTHES DRYER.

Installing a new clothes dryer or modifying an existing dryer installation may be part of the rehab project. These appliances, whether gas or electric, require exhausting to the home's exterior. Lint buildup may lead to fire or health problems, making exhausting imperative. The International Residential Code for One and Two Family Dwellings (IRC) has specific requirements for exhausts regarding duct size, length, and construction standards. A maximum length of 25' has been established based upon a compilation of appliance manufacturers' recommendations. Bends reduce the length allowed, but there are exterior-mounted powered exhausts available. The local code should be checked for approvals. Most dryers exhaust approximately 150 to 200 cfm of air. The IRC requires makeup air provisions if dryers exhaust more than 200 cfm. With bath exhaust fans commonly exhausting 50 cfm of air, and range hoods 150 cfm, adding a clothes dryer exhaust may cause an assortment of problems, such as backdrafting of combustion appliances or fireplaces. Bath and kitchen exhausts might not exhaust during dryer operation.

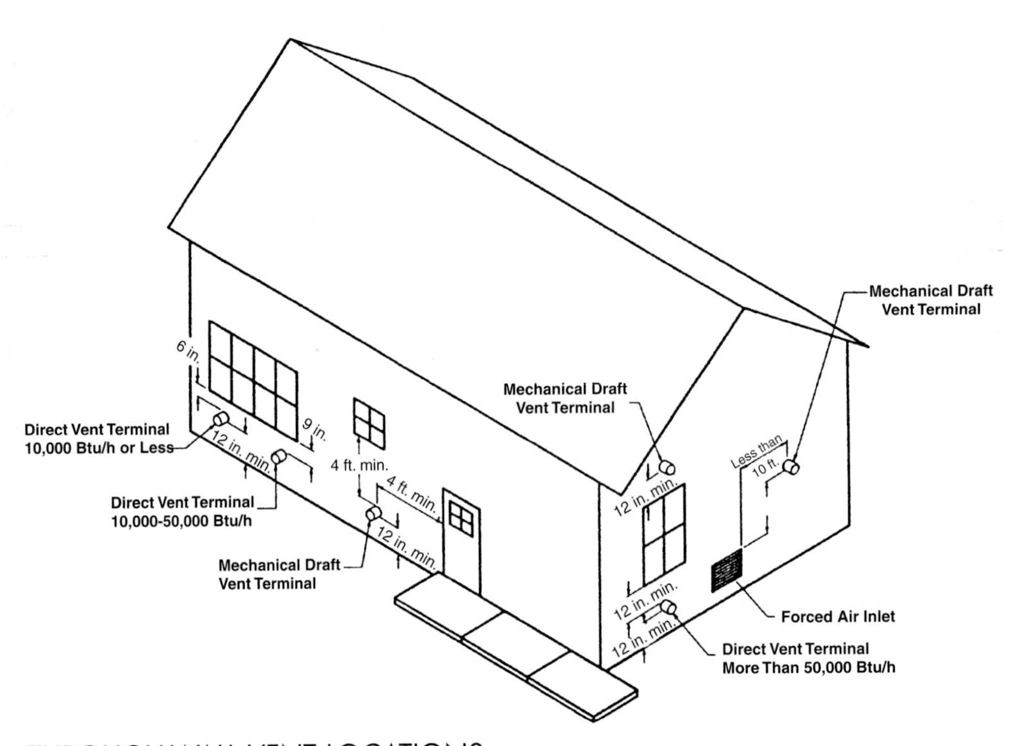

FIGURE 8.14.1 THROUGH-WALL VENT LOCATIONS

ADVANTAGES: A clothes dryer exhaust can ensure fire safety with dryer use, improve indoor air quality, and reduce lint and moisture inside the home.

DISADVANTAGES: A clothes dryer exhaust may induce back-drafting of combustion appliances and fireplaces and may inhibit function of other exhausts in the home.

2. INSTALL A NEW EXHAUST FOR A RANGE.

Range exhausts typically are not required by code, relying on windows for exhaust or nonventing filters for cleaning air. Range exhausts are recommended for indoor air quality and safety issues, particularly for gas ranges, as emissions of carbon monoxide, nitrous oxide, and others are present. Fouled burners can produce dangerous levels of benzene, indicated by a yellow flame. Moisture is a product of gas combustion as well, with a gallon of moisture released per 100,000 Btu/h input (⅛ gallon per hour per average burner). Kitchen exhausts can be recirculating or circulating. Recirculating "exhaust" hoods do not exhaust and move approximately 40 cfm of air through a filter and then back to the kitchen. Filters are either charcoal or activated carbon coated fiber, and require maintenance. Circulating exhausts vary from 150-cfm average two-speed hoods to downdraft exhausts of 500 cfm or more. Manufacturers provide similar duct openings so equipment can be replaced (typically 3 ¼" x 10" or 7" round). Duct length is critical, and manufacturers' recommendations should be followed. Smooth metal duct material is preferred over flexible metal. An equivalent length adjustment must be made for flex.

ADVANTAGES: A range exhaust allows for better indoor air quality and mitigates fire hazard.

DISADVANTAGES: Recirculating hoods do not address combustion products and are not recommended.

8.15 FIRE PROTECTION SYSTEMS

ESSENTIAL KNOWLEDGE

According to the National Fire Protection Association (NFPA), there were 406,500 residential fires and 3,360 deaths in the United States in 1997. Domestic fires are low-probability but high-consequence events. Experts indicate that most fires occur in kitchens, family or living rooms, and bedrooms, with the highest loss of life in bedrooms and family and living rooms. Fires reportedly burn hotter and faster than 25 years ago because of newer materials in home furnishings.

This Old House magazine recommends adding sprinkler heads during rehab to water systems—a "some protection is better than none" philosophy. However, fire protection engineers say an added system may not perform adequately because there are no hydraulic calculations to verify that the system will work. No valves or alarms are available to tell when the sprinklers might accidentally fail. A loop is needed with sprinkler heads so that standing water in the pipes will not be a concern—potable water that does not circulate can become contaminated. The Director of Operation Life Safety of the International Association of Fire Chiefs suggests adding fire sprinklers to various areas rehabbed, essentially adding a system on an "installment plan" basis. Other professionals take an all-or-nothing approach to sprinkler systems. Adding an automatic fire sprinkler system can add to life safety and property protection. Automatic fire sprinkler systems are required in some rehab construction in certain jurisdictions. The local code should be consulted.

NFPA Section 13D governs residential sprinkler systems for one- and two-family construction. Residential systems are relatively new with NFPA 13D, adopted in 1980. Complex automatic fire protection systems are commonly installed in commercial, institutional, and high-rise construction. Residential systems are designed to be less complex. Life safety is the top priority in residential applications, compared to saving property in commercial systems. Still, requirements are deemed too extreme and costly by many. Opponents feel more effective solutions could be enacted. NFPA 13D requires sprinklers in most rooms with flows to allow occupants 10 minutes escape time. The multipurpose system combining plumbing and fire sprinklers is newly approved by the NFPA and could be considered. Insurance companies may provide a 5% to 20% yearly insurance rebate to homeowners with sprinkler systems. Canadian insurers may provide a 35% to 40% rebate.

Automatic fire sprinkler systems are designed to provide minimum water flow from heads for a specified escape time. NFPA 13D requires at least one automatic water supply. For homes not on a public water system, a captured water supply large enough for a 10-minute flow from two heads is required. Pools, tanks (elevated or with electric pumps), or tanks pressurized with nitrogen propellant may be used.

Residential sprinklers are separate wet systems, compared to wet or dry standpipe systems found in commercial construction. System supply configurations and connections vary in type and complexity and are usually determined by local code. NFPA 13D requires sprinklers in most rooms, except baths, closets, attics, and garages. Heads in nonhabitable spaces provide a system with better coverage, but this is more difficult and costly, particularly in rehab work. Wet systems also need protection for freezing temperatures. A fire sprinkler professional, familiar with the local code requirements, should design the system.

Piping materials include welded and seamless steel, wrought steel, copper, and chlorinated polyvinyl chloride (CPVC) and polybutylene (PB). All are common materials used in residential construction. Copper can be used in any area, but CPVC must be used in concealed areas or adjacent to smooth ceilings. CPVC fire sprinkler pipe is a different product than water supply pipe with different code approvals, and the two cannot be combined in most cases. Both can be used with the same fittings but have varying wall thicknesses. For example, BF Goodrich BlazeMaster CPVC has a pipe wall thickness of SDR 13.5 IPS versus Flow Guard Gold with SDR 11 CTS.

Cross-linked polyethylene, multipurpose tubing system for sprinklers has been submitted for UL approval, (approval appears likely), which would lower the 175-psi pressure requirement to 130. Smaller pipe runs, ½" versus ¾" or 1" would be featured with multiple service connections to each sprinkler head. The material would be a very cost-effective and easy product to install in rehabs. Check with your local code official.

Sprinkler heads vary in design and type. They can be ceiling or sidewall mounted, concealed or exposed, and in metallic or prepainted finishes. Quick-response residential sprinkler heads contain a fusible link, pellet, or frangible glass bulb. All must be nationally listed and approved. There are many heads available with widely ranging performance ratings (k factor). Coverage from 144 to 400 ft^2 per head is possible depending on available flow rates and system pressure. There is a Consumer Products Safety Commission national action filed along with various suits against Central's Omega sprinkler heads. The heads when tested failed to work at the designated pressure, but most worked at higher pressures. If the rehab project includes these heads, the International Association of Fire Chiefs recommends you have Omega heads removed and replaced.

Other system elements include a riser and components that control and monitor for the flow of water to the system. The riser components may vary depending on design and code requirements. Typically, they consist of a check valve that lets water flow in only one direction, a pressure gauge, a pressure relief circuit and drain, a flow switch, and a test valve. The riser is connected to the supply. If the water supply is not public, a tank, pond, or source other than a well will be needed. Self-contained systems are available with electric pumps or pneumatic pressure systems.

If running distribution piping within finished walls is not possible, DecoShield makes a UL-listed cover and support system. The product was designed for surface mounting retrofit systems in existing buildings and would be appropriate for most rehabs (Fig. 8.15.1).

TECHNIQUES, MATERIALS, TOOLS

1. INSTALL A STAND-ALONE RESIDENTIAL SPRINKLER SYSTEM.

The stand-alone system is separate from a house's water distribution system and is the most common. A main control valve from the water source to the domestic and fire systems, pressure gauges, check valves, water flow detectors, and drain and test connections are typical.

FIGURE 8.15.1 DECOSHIELD SYSTEM

ADVANTAGES: System water flow activates an alarm. These systems are approved by all codes and can be connected to the fire station, providing more control of system (flushing and maintenance).

DISADVANTAGES: These systems can be deactivated by accident. More materials and labor are required and installation is difficult.

2. INSTALL A MULTIPURPOSE RESIDENTIAL SPRINKLER SYSTEM.

The NFPA issued a Technical Interim Amendment on October 28, 1998, approving multipurpose or combined systems for Section 13D. This means the fire sprinkler and domestic water distribution system can be combined. Most plumbing codes prohibit such systems, but Highland Springs, California, and Dupont, Washington, are two of many locales that allow them.

ADVANTAGES: These systems have a lower cost than stand-alone systems and cannot be deactivated accidentally. Materials and labor are reduced, and these systems are easier to install.

DISADVANTAGES: These systems do not include an alarm and are not allowed by some plumbing codes. Contaminated water is possible if there are no loops in the system. Future additions to the house's water system (i.e., water softeners and backflow prevention) could hurt the calculated effectiveness.

3. INSTALL A COPPER-PIPED SYSTEM.

Copper has been in use for sprinklers since the 1930s and remains the most popular piping material for stand-alone and multipurpose systems. Approval of alternative materials that cost less and are easier to install might change the use of copper piping. The integral soldering fitting described in Section 8.11 can speed construction.

ADVANTAGES: These systems are code-approved in any installation, are commonly used in other plumbing systems, and are relatively easy to install. They are fabricated by soldering or brazing using conventional equipment or electric resistance tools. They are compact within thin walls, and tees can be pulled mechanically.

DISADVANTAGES: These systems are costly and more rigid than plastic. Brazing or soldering are potentially dangerous. The pipe may be subject to condensation and are susceptible to theft.

4. INSTALL A CPVC-PIPED SYSTEM.

CPVC water piping has been used in single-family housing since 1959. A reformulation in the 1990s solved cold weather workability problems. CPVC for fire sprinklers is a different product and was introduced in 1986. Code requires CPVC to be installed behind a thermal barrier in most cases, except with quick response sprinklers in special cases.

ADVANTAGES: These systems are low cost, somewhat flexible, and more easily installed with cut and glue/weld fittings. They have better insulation properties than copper.

DISADVANTAGES: Use is limited to specific areas, more support is required and transitions are needed to other non-PVC sections. This product will char during a fire.

CH. 9

HOME REHAB HANDBOOK

SITE WORK

Chapter 9
SITE WORK

9.1. DECKS, PORCHES, AND FENCES
9.2. PAVED DRIVEWAYS, WALKS, PATIOS, AND MASONRY WALLS
9.3. UNDERGROUND CONSTRUCTION
9.4. LANDSCAPING

9.1 DECKS, PORCHES, AND FENCES

DECK AND PORCH STRUCTURE

ESSENTIAL KNOWLEDGE

The structure of a wood deck or porch is subject to moisture and rot conditions because of its exposure to the elements. Water absorbed by the wood structure increases the material's moisture content, causing rot and decay, compromising the material's strength. Among the indications of wood rot are cracked or sagging floorboards, peeling paint (which provides a moisture barrier), and discolored or soft areas of the wood that should be cut and removed to halt rot spread. Such moisture and rot problems can easily spread to the structure of the house itself where it comes into contact with decks and porches, if remedial action is not taken. The best prevention for rot in wood is to use treated wood and to seal it periodically. Rehabilitating the structure of a wood deck or porch may require removing the rotted portions of the structure and replacing them. Another option is to remove the rotted material and fill with an epoxy. If the deckboards are damaged from weathering, you can sand the surface to make them smooth. Countersink the deck nails or screws below the surface with a nail set. Using a belt sander, sand in the direction of the grain and keep the sander moving to avoid gouges in the deck boards.

TECHNIQUES, MATERIALS, TOOLS

1. REPAIR MOISTURE-DAMAGED POSTS OR COLUMNS.

The most efficient and effective correction to rotting wood posts and columns is to remove the damaged material and replace it with new material. If you are removing deck posts or columns, support the deck with a jack post, house jack, or crossed timbers under the rim joist. Place a board between the ground and the jack as well as between the jack and the rim joist to prevent damaging the joist. To support the roof for removing porch posts or columns, use ground-to-eave bracing of doubled 2x4s or 2x6s. Nail a cleated top plate to the roof framing, and nail the bracing to the top plate in front of the cleat. Nail a cleat to the bottom plate, and nail the bracing behind the cleat. Stakes behind the bottom plate prevent the bracing from sliding, keeping bracing as vertical as possible.

Cut a lap joint at least 18" long in the existing post, above the damaged area. To make the crosscut, set a circular saw to exactly half the depth of the post. Then rip from both sides down the center. Use a chisel to clean up the portion of the joint where the circular saw blade does not reach. Install three staggered bolts through the posts to hold them together, with the top bolt 3" from the old post and the bolts spaced 6" apart. Make sure that the new member has a preservative treatment retention greater than the other deck members not coming into contact with the ground.

ADVANTAGES: This repair saves material and time because only part of a post is replaced.

DISADVANTAGES: Conditions that cause rot should be corrected before repair work is done.

2. REPAIR MOISTURE-DAMAGED JOISTS.

If only the top of a joist needs repair, leave the joist in place. Apply a thick coat of sealer-preservative to the damaged joist, let dry, and then apply a second coat. Treat all sides of the new partner reinforcing joist (2x4 or 2x6 of pressure-treated lumber) with clear sealer-preservative, and let dry. Align the top edges of the partner and old joist. Partner the new piece by nailing it to the damaged joist tightly with 16d galvanized nails or deck screws (Fig. 9.1.1). Treat the ends of the new joist with sealer-preservative where it was cut.

Wood grades are important for structural members. Quality grades run in descending order. Choose grade no. 1 for railings and benches, and grade no. 2 or BTR for decking. Also check the grade stamp for the letters KDAT (kiln-dried after treatment) which is more dimensionally stable than air-dried wood. Use hot-dipped galvanized, aluminum, or stainless-steel (ring-shanked) nails, screws, and other fasteners.

ADVANTAGES: This repair saves materials and time because only a portion of the joist is replaced.

DISADVANTAGES: This repair may be difficult if access under the deck is limited. In such cases, the deck boards can be removed to access the joists, and then reinstalled after the repair. If all of the rot in a joist has not been cut out and removed, the rot can easily spread to the rest of the joist.

3. REPAIR ROTTED LEDGER BOARDS.

The ledger board is the joist attached to the house. Very often, the ledger will start to rot from water trapped between it and the house, unless it is properly flashed (Fig. 9.1.2). This rot then can spread to the ends of the joists that rest on the rotted ledger board. You will need to replace the ledger board and trim the rotten ends off the joists (Fig. 9.1.3).

Remove deck boards over the area of rotted joists and ledger. Run a 2x6 on edge under the joists and support it with 4x4s. Remove the rotted ledger, and trim all of the joists the same length with a reciprocating or circular saw. Coat the ends with wood preservative. Cut two or more new ledger boards the same size as the old material from pressure-treated lumber. Sandwich the ledger boards together, and bolt them to the house, using the old ledger board for the bolt hole spacing. Seal around the ledger with caulking compound. If the house has siding, install flashing under the siding and over the ledger (Fig. 9.1.2). Attach joists to the sandwiched ledger with joist hangers and replace deck boards.

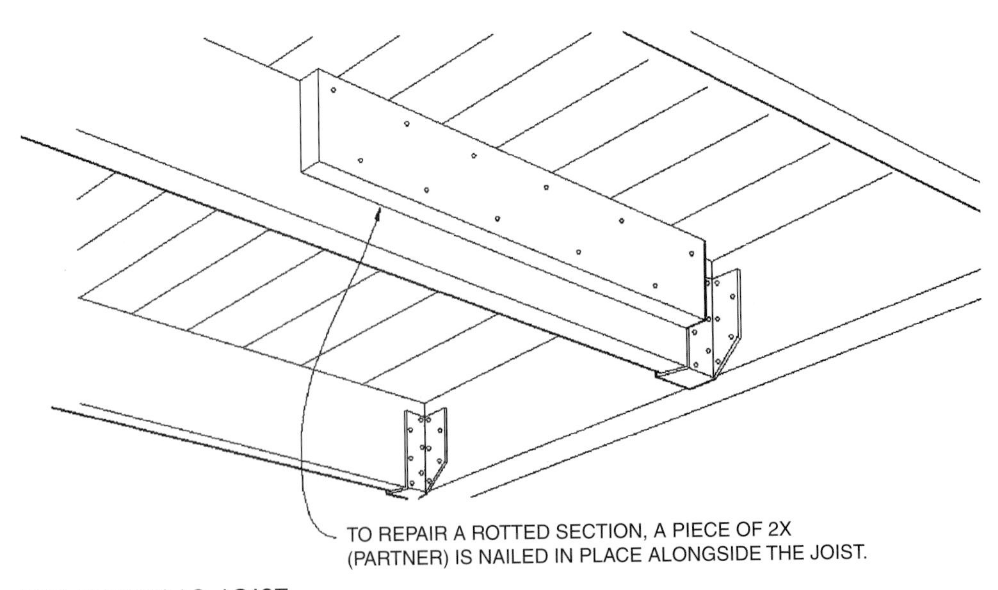

FIGURE 9.1.1 REINFORCING JOIST

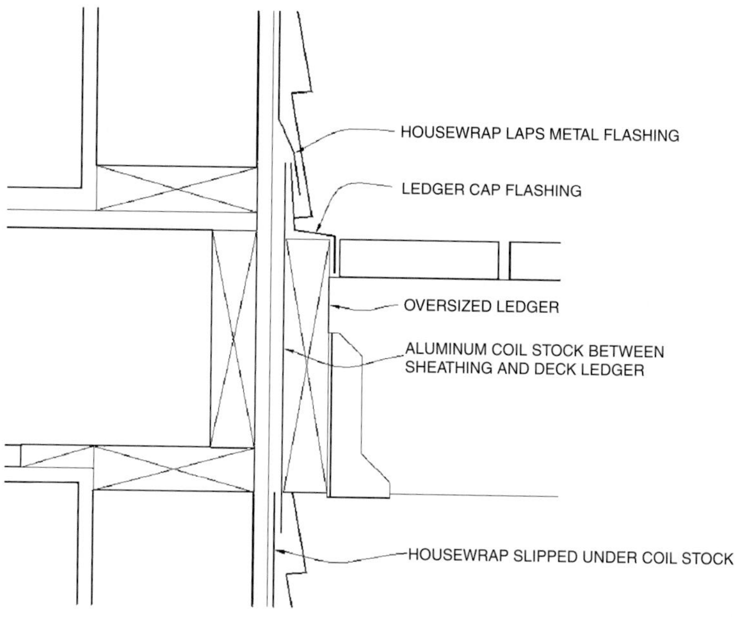

FIGURE 9.1.2 LEDGER FLASHING

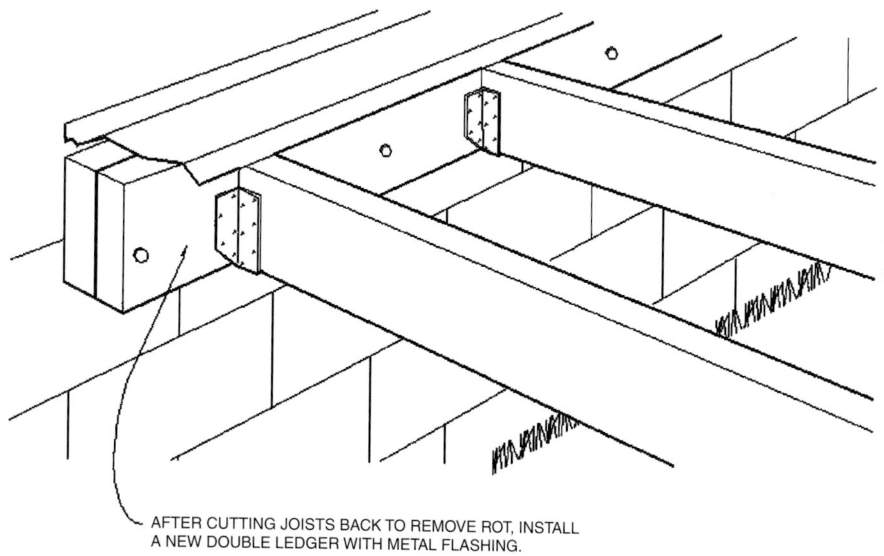

FIGURE 9.1.3 REPLACEMENT LEDGER BOARDS

If you cannot remove the ledger board, use an epoxy wood hardener or filler. To repair a partially rotted ledger board, drill several ⅛" holes throughout the area and inject the mixed liquid epoxy hardener. At any damaged corners, chisel away to the good wood and slightly overfill with liquid epoxy wood filler. After it is dry, sand and shape the patch.

ADVANTAGES: This repair saves materials and time because only a portion of the deck is replaced.

DISADVANTAGES: Ledger board may be difficult to repair if access under the deck is limited.

4. STABILIZE ROTTING MATERIAL WITH EPOXY.

An alternative to replacing entire structural members or portions of members is to use epoxy restoration products. If the rot is not extensive, products such as LiquidWood or WoodEpox can be used to restore the member's structural integrity. Such repairs can be achieved by two different products: consolidants for wood that is intact, and a puttylike filler material for areas that are missing or require removal because they are beyond repair. Epoxy consolidants penetrate and bind with the wood fibers while preventing further decay. They are either poured or brushed on the surface in liquid form, and bond with the wood fibers to create a filler with greater strength than wood that is water resistant. The material cures in a few minutes to a few hours (depending on the extent of the repair). Consolidants may be used as a primer for the application of epoxy putty filler.

ADVANTAGES: Epoxy fillers are preferable to solvent-based wood fillers, which may shrink as they cure or become loose as materials expand and contract at different rates. Epoxy fillers can be worked as wood, and are a low-cost solution.

DISADVANTAGES: The repair of damaged structural members will not address the cause or progressive deterioration of adjacent materials.

REPAIR OR REPLACE WOOD DECKING MATERIALS

ESSENTIAL KNOWLEDGE

Wood decking that has discolored because of mildew can be cleaned and restored. Untreated or unsealed decking exposed to the elements can split, cup, splinter, or check. The best solution is to replace these decking materials with new materials, either matching the original materials or with new decking materials that have greater resistance to degradation. Alternative materials that are rot-resistant require very little maintenance and offer a longer life cycle. For example, new types of plastic lumber can eliminate the problems with wood rot and may be more cost-effective over the life of a structure, even though they typically cost more than treated wood. An array of new decking materials lasts longer and reduces maintenance. Other choices include exotic hardwoods that are naturally insect resistant. In addition, there is a variety of nonwood products, such as engineered vinyl systems and plastic-wood composites, all available in different shapes, widths, grades, and treatments.

TECHNIQUES, MATERIALS, TOOLS

1. CLEAN MILDEW-STAINED WOOD DECKING.

You can remove stains and weathering with a deck brightener (a wood bleach), trisodium phosphate (TSP), or restain the deck. Afterwards, apply a wood sealer to protect the finish. It may be easier to bleach out the individual stains than to stain the entire deck. Make sure that the wood deck brightener that you use is compatible with your type of wood. Test a small, inconspicuous portion of the deck before applying to a large area.

ADVANTAGES: This is a relatively easy, low-cost process.

DISADVANTAGES: Bleached "hot spots" can occur if the product is not carefully applied.

2. REPAIR OR REPLACE EXISTING DECKING.

Splintered deck boards can be sanded or turned over so that the underside is used as the new surface. If sanding, countersink or remove the screws prior to sanding. Should a deck board need to be removed, get underneath the deck and hammer up on the boards to loosen them, or use a pry bar. If access from below is not possible, bore holes into the good wood adjacent to the damaged area. Connect these holes with saw cuts, using a keyhole saw or electric jigsaw, avoiding the joists below, and remove the decking. Cut out any sections of bad decking up to the joist. Treat any remaining rot on the existing joist with a fungicide. Then install a short piece of 2x4 (support cleat) even with the top edge of the joist to support the new decking (Fig. 9.1.4). Cut the new decking to size and nail it into place, using two nails at each joint for 2x4s and three nails for 2x6s. To keep the 2x4 cleat flush against the underside of the new decking, if access from below is possible nail the cleat loosely to the joist and, after decking is installed, push the cleat tight against the deck underside and then nail into place. If all the decking must be replaced due to moisture damage, ensure that the new decking slopes away from the house so that water will drain.

ADVANTAGES: This repair saves material and time because only deck boards that are rotted are replaced.

DISADVANTAGES: Rot spreads throughout wood from the inside out; a cursory inspection may not reveal areas of rot that need removal. Conditions contributing to rot should be corrected before replacing material.

3. REPLACE WITH CERTIFIED SUSTAINABLY HARVESTED WOOD.

Typical lumber harvesting puts stress on the ecosystem and animal habitats of the forest and nearby streams used to transport the lumber. The Forest Stewardship Council (FSC) has established environmental standards for forestry operations, which require a logging company to avoid the overharvesting of timber, minimize waste of felled trees, protect threatened and endangered species, and control erosion. In addition, there are two other lumber certification groups: Scientific Certification Systems and the Rainforest Alliance's SmartWood Program. The FSC has accredited both groups to certify that wood has met FSC standards. To date, SmartWood and SCS have certified more than 40 U.S. forestry operations, covering nearly 5 million acres. Consult with forest certification organizations for certified lumber suppliers and Co-Op America, which also maintains a database.

ADVANTAGES: Use of sustainably harvested wood benefits the local ecosystem.

DISADVANTAGES: Sustainably harvested lumber may not be readily available. Contact national suppliers of certified wood for their nearest retail outlets (some may even supply directly).

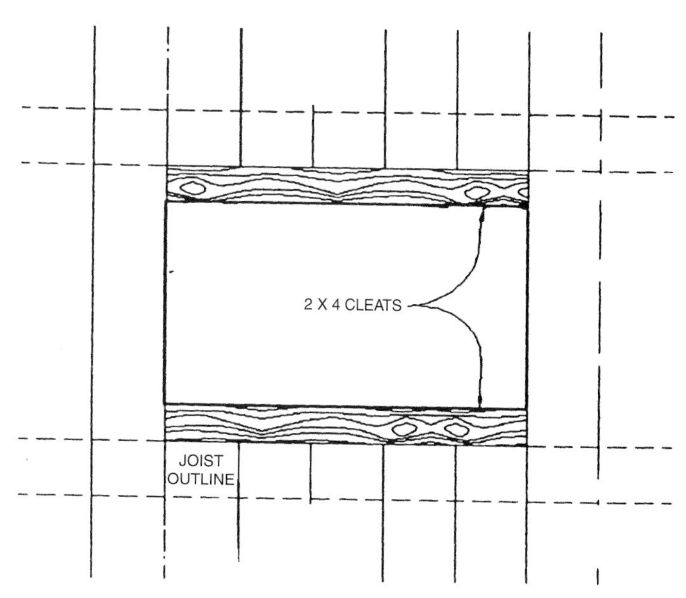

FIGURE 9.1.4 SUPPORT CLEATS FOR NEW DECKING

4. REPLACE WITH REDWOOD AND CEDAR.

Cedar deck boards are easily available, while redwood may require several days for special orders. For redwood, use a kiln-dried deck heart or A deck common grades. Clear heart, clear, B heart, and B grade redwood are economical choices. For cedar, clear all-heart is the premium choice. Good, lower-cost alternatives include appearance grade and, for dry areas, No. 1 select tight knot (STK). Use hot-dipped galvanized aluminum or ring-shanked stainless-steel nails, screws, and other fasteners.

ADVANTAGES: Redwood and cedar heartwood has a glowing color, straight grains, and natural resistance to rot and insects. Boards are easy to cut and can be left untreated to weather naturally. Service life is easily more than 20 years.

DISADVANTAGES: Both woods are costly. Redwood sapwood is the newer wood closer to the bark; it may rot when exposed to moisture for sustained periods of time. Cedar sapwood breaks down quickly in moist conditions.

5. REPLACE WITH TROPICAL HARDWOODS.

Among tropical hardwoods are Ipe, Ironwood, Cambara, and Pau Lope. Ipe is a durable, teaklike tropical hardwood widely available sold as Ironwood. Ipe is twice as strong as oak and is more durable than redwood and cedar. It has a service life of 40 years or more and is resistant to insects and decay. Ironwood is backed by a 25-year transferable warranty. Another tropical hardwood option is Cambara, which is not as durable, hard, or stable as Ipe, and requires more maintenance. However, Cambara is also knot-free and less costly than Ipe. Cambara is similar to cedar or Meranti in durability and requires sealing. Pau Lope is a long-lasting material that requires virtually no maintenance. It is five times harder than pine, cedar, and redwood; resists splitting and splintering; and is free of toxic preservatives and arsenic. Pau Lope has the highest rating from the U.S. Department of Agriculture for insect- and decay-resistance; is priced comparable to cedar; and requires almost no maintenance. The material comes with a 20-year limited warranty.

ADVANTAGES: All grades are virtually knot-free, and tight grain patterns make tropical hardwoods impervious to water.

DISADVANTAGES: These materials are often costly. Because of their hardness, they require predrilling for fasteners. Ipe is available in most standard dimensions for decking, but delivery can take up to 3 weeks.

6. REPLACE WITH PLASTIC-WOOD COMPOSITES.

Virtually indestructible, plastic-wood composites blend 30% to 50% recycled plastic with wood fibers for skid resistance and stainability. Composite lumber is low-maintenance, and resists rot, insects, and ultraviolet (UV) rays. It is also splinter-free and therefore easy to work with. Composite lumber weathers to a light gray and can be painted or stained, although protective sealers are not required. Galvanized fasteners are recommended. Both ChoiceDek and DuraWood deck-and-rail systems are formed to allow running wires within posts or deck boards. Nexwood is a structural recycled plastic/rice hull composite decking material. Trex, TimberTech, and DuraWood EX come with 10-year warranties, while ChoiceDek and DuraWood PE are backed for 20 years.

ADVANTAGES: Materials are nearly maintenance-free and contain nontoxic elements.

DISADVANTAGES: Contractors have to become familiar with different load characteristics and fastening techniques. Because composites lack the strength of solid wood, they require more material for framing. Some building inspectors may be unfamiliar with the materials. Some composite lumber has a plastic appearance, and some colors fade over time. During construction, sawdust and shavings should be collected in a drop cloth because they are not biodegradable. Not all composite lumber can span 16" or 24" joist spacing. Therefore, a closer joist layout may be needed, which increases costs. Some building codes may not allow composite lumber; check local code requirements.

7. REPLACE WITH VINYL DECK SYSTEMS.

Vinyl deck systems typically include deck boards, rails, spindles, and fascia. Three major manufacturers of vinyl deck systems are Kroy, DreamDeck, and EZ Deck, which offer a variety of plank widths and limited, lifetime warranties. These materials cut and work much like wood, but are lighter. Vinyl decking is framed conventionally, but special clips are screwed into stringers and boards are hammered into the clips. Skid-resistant planks provide an added measure of safety. Vinyl decks weather a bit during the first few months after installation. To reduce fading, most high-end vinyl contains more UV inhibitors than even high-end siding. Color-fast, no-fade materials available from manufacturers such as EZ Deck hold up better under exposure from UV light. Planks can be cut to length with a circular saw; plank ends are covered with vinyl caps. Proprietary strip systems are screwed to joists with galvanized or stainless-steel screws, and then planks snap into place.

ADVANTAGES: These are low-maintenance materials that require no sealers or finishes and are free of splinters and cracks. Planks have good spanning ability and resist UV rays if treated at the factory. Fasteners can be completely hidden once planks are installed.

DISADVANTAGES: These materials are relatively costly and must often be ordered through distributors. Vinyl can fade and become brittle with age unless specially treated at the factory; all vinyl eventually loses its gloss. Sawdust from vinyl decking is not biodegradable, so it must be collected.

PRESERVATIVE-TREATED WOOD

ESSENTIAL KNOWLEDGE

Typically, any wood within 6" of finish grade should be factory-treated or have natural rot and insect resistance, such as heartwood of cedar, redwood, black locust, and Ipe. The three primary methods of manufacturer-treated wood are (1) creosote pressure-treated wood, (2) pentachlorophenol pressure-treated wood, and (3) inorganic arsenic-based pressure-treated wood. Copper napthenate, zinc napthenate, and tributyltin oxide wood treatments can be site-applied. All of these treatment processes involve dangerous chemicals, and you should use material treated according to American Wood Preservers Association standards. While pressure-treated "green" Southern yellow pine is used in 80% of all decks, there are several alternatives even within the pressure-treated (PT) lumber category. Pressure-treated lumber should be used only in well-ventilated areas. Rubber or vinyl gloves should be worn when handling the material, and a dust mask and goggles when cutting it (use sizes and lengths of treated wood that require no cutting, if possible). Newly cut surfaces should be protected with wood preservative solution applied according to the manufacturer's directions. Penta-treated wood should not be used where people, pets, or plants are likely to come into contact with it.

Typically, decks should not be painted since paints and solid-color stains cannot hold up to the severe weather exposure from sun and rain. As small cracks develop in the paint layer on the deck's surface, water can pass through the crack and under the paint layer, and the paint will soon begin to peel. Also, the typical wood species used for pressure-treated lumber is southern pine. This type of wood does not hold paint well since it expands and contracts. The U.S. Forest Products Lab has shown that semi-transparent stains and clear finishes actually last longer on chromated copper arsenate (CCA) pressure-treated wood than wood with other types of pressure treatment, since the chromium in the CCA treatment protects the wood surface from UV light degradation. Paints and solid-color stains will perform well on pressure-treated wood that is used in an upright position (on fences, for instance), but only when the wood has been cleaned and is thoroughly dry before painting.

Since penetrating finishes do not form a film, they can withstand sun and rain and are much more suitable for decks. Decks should be finished either every year or two with a penetrating semi-transparent stain or a penetrating clear finish designed for use on decks.

TECHNIQUES, MATERIALS, TOOLS

1. USE WOOD WITH FACTORY-APPLIED PRESERVATIVES.

Chromated copper arsenate (CCA) is one of the most popular insecticides and wood-preservative treatments. The chemicals are inert within the material and offer protection from moisture and decay fungi. However, the chemicals do not penetrate all the way into the heartwood. It is recommended that a sealer be applied to the cut ends of CCA-treated wood and be reapplied to the whole member periodically. Southern yellow pine is usually treated with CCA. Use wood with a CCA density of 0.40 for decks and joists and 0.60 for posts. When using CCA-treated lumber, the designation LP-2 is for above ground use and LP-22 is for locations with soil contact.

Ammoniacal copper quaternary (ACQ) is a new wood preservative that is less toxic than CCA and performs similarly to CCA, but does not contain arsenic, chromium, or other EPA-classified hazardous preservatives, which are components in CCA-treated woods. The treated wood has a longer life than redwood or cedar and a more consistent quality than plastic-wood products. ACQ weathers to a natural brown tone and can be used in environmentally sensitive settings. The wood scraps can be disposed of by ordinary trash collection.

Treated wood may be disposed of by bringing it to a licensed sanitary landfill or wrapped in paper and included with other household garbage. Treated wood should not be burned. Wood treated with any of the commonly used wood preservatives when burned produces highly toxic compounds in the smoke or ashes. Treated wood from construction sites may be burned only in commercial or industrial incinerators and boilers in accordance with state and federal regulations. If replacing decking, consider factory pressure-treated wood with a water repellent, such as Hickson's Thompsonized PT (pressure treated) or Osmose's Armor All PT.

ADVANTAGES: Economical and plentiful, pressure-treated wood lasts approximately 15 years if properly treated with a water repellent every 2 years and is widely available.

DISADVANTAGES: The dominant pressure-treated species is southern yellow pine, which will check and splinter as it dries. Pressure-treated wood should be carefully chosen for structures that will be used by children, as chemical residues may cause health problems.

2. APPLY NONTOXIC FINISHES, STAINS, AND CLEANERS.

Pressure treating makes a wood surface more porous and susceptible to moisture damage. Penetrating sealers protect decking from moisture, and some also block damaging UV rays. Sealers usually need to be reapplied every year or two. Over time the wood surface may become rough, and small cracks are likely to develop. Eventually, nails and fasteners will loosen and deck boards may develop checks. Urethane, shellac, latex, epoxy, enamel, and varnish are acceptable sealers for pentachlorophenol-treated wood. Low-VOC, low-odor sealers are less harmful to the environment and work just as well. The low-VOC formulations are suitable for decks in areas near water and wetlands. For long-term protection, wood surfaces can be coated with a water-resistant paint or stain. Water repellents and water-repellent preservatives slow the uptake of water and help keep wood dry. The only difference between these finishes is that water-repellent preservatives include a fungicide or mildewcide. Both contain a 10% to 20% binder such as varnish resin or drying oil (linseed or tung oil), a solvent, and a substance that repels water (wax or waxlike chemical). Prior to finishing, old paint must be removed and the surface sanded thoroughly in order for the paint to adhere. A primer may be necessary depending on the type of paint (follow application instructions).

ADVANTAGES: Wood preservatives help maintain the structural integrity of the material.

DISADVANTAGES: Wood preservatives should not be applied within 150' of a water source or wetlands area, or near a drinking water well.

3. APPLY BORATE-BASED PRESERVATIVES.

Borate products are available for site application. For preservation of wood not exposed to water, boron is a water-soluble preservative that automatically moves into areas where high moisture content has led to decay and insect infestations. It provides relatively nontoxic protection against wood-destroying beetles, fungus, and termites (upwards of 20 to 50 years) and helps increase the fire resistance of the wood. Borates can be applied by spraying, brushing, injection, and with the use of foggers. Boron penetrates the wood through osmosis. The treatment works best if the wood is slightly damp (this can be accomplished by slightly wetting the wood prior to application). The ideal treatment is to submerge fresh-sawn lumber in a vat of boron solution for a while, though it is regularly used on existing structures to stop or prevent insects and rot. In very wet climates, it may be necessary to retreat the wood every few years. Wood just treated with borates should be protected from rain or snow. Nisus sells a water-repellent concentrate, Co-Pel, that can be mixed in with the Bora-Care for partially exposed timbers. U.S. Forest Products Laboratory research suggests that a water repellent should be reapplied to the exposed surface of the wood every 2 to 3 years. If this is done, there should be enough borate remaining in the wood to provide adequate protection.

ADVANTAGES: Borate-based preservatives are relatively nontoxic and relatively easy to apply in a variety of ways.

DISADVANTAGES: Borates will leave a slight whitish discoloration.

STAIRS AND HANDRAILS

ESSENTIAL KNOWLEDGE

Handrails are required on at least one side of all stairs with three or more risers and on ramps with a slope of 1:12 or more. The recommended height of handrails is a minimum of 30" and a maximum of 38". The recommended perimeter of a handrail is 4" to 6 ¼". Wood handrails and their supporting posts and wood stairs are subject to rot from standing water and infiltration of water into fasteners. Some rotted portions of the material can be replaced without having to replace the entire handrail, supporting post, or stair treads. Rotting problems can be serious and must be corrected before replacing the material.

TECHNIQUES, MATERIALS, TOOLS

1. ADD SUPPORT TO EXISTING RAILING POSTS.

If railing posts are nailed in place, they can be renailed or stabilized with lag bolts for additional support. Bore pilot holes for the lag bolts, and counterbore holes in the posts so that the lag bolt head and washer will be recessed below the railing post surface. Renailing handrails to posts will secure the handrails too.

ADVANTAGES: Renailing or bolting provides more support at a relatively low cost.

DISADVANTAGES: Bolting posts requires more work than renailing.

2. REPLACE EXISTING WOOD RAILINGS.

Wood rot usually occurs where two or more pieces of wood intersect or where wood is fastened. When replacing a portion of a handrail that has rotted, cut out the rotted section using a diagonal scarf joint. After measuring a new piece from treated lumber, use waterproof glue and galvanized screws to attach the new piece. If the railing is attached with a bracket, tighten all of the screws in the bracket. If the screws cannot be tightened and are loose, use longer screws in the holes or install a new bracket with holes in new positions (Fig. 9.1.5). If the handrail is separated from the post, fasten a 2x4 block under the handrail onto the post for support, or use a metal angle bracket that is fastened to both the handrail and the post. For a bottom handrail that has come loose from its post, use a metal T-connector plate fastened to the end of the handrail and at the side of the post.

ADVANTAGES: Replacing only the rotted section of wood conserves time and materials.

DISADVANTAGES: Handrails may be rotted in several areas, thus requiring replacement of several sections. Multiple sections may be unsightly compared to one continuous handrail.

3. REPLACE DAMAGED WOOD TREADS OR RISERS.

To remove a damaged tread on exterior stairs (Fig. 9.1.6) saw through the middle and remove the pieces from the stringers with a prybar. Cut a new tread from pressure-treated lumber, matching existing tread depth and thickness. Treat the cut ends with preservative. Attach the tread to the stringers with 16d galvanized nails or galvanized deck screws (deck screws will secure the treads more tightly to the stringers). If risers are damaged, remove the tread first to access the risers and replace.

ADVANTAGES: Replacing only damaged treads is time- and resource-efficient and is a low-cost fix.

DISADVANTAGES: If treads are subject to rot, the problem should be corrected before replacing material.

4. REPLACE DAMAGED STRINGERS.

If wood stair stringers are partially rotted, they should be replaced. Replace with pressure-treated wood of the same size (usually 2x10 or 2x12). Use the old stringer as a pattern for the new stringer. Stringers should rest on a pressure-treated sleeper if the wooden stair ends on a concrete pad. Stringers and cleats should be bolted to the carriage (Fig. 9.1.7).

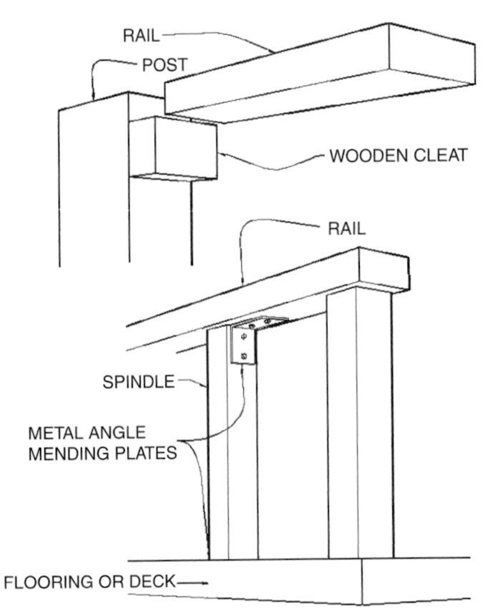

FIGURE 9.1.5 WOOD RAIL SUPPORT

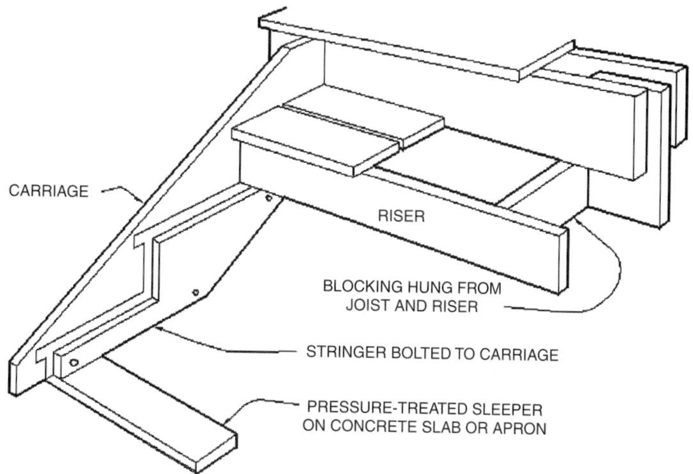

FIGURE 9.1.6 EXTERIOR STAIR ELEMENTS

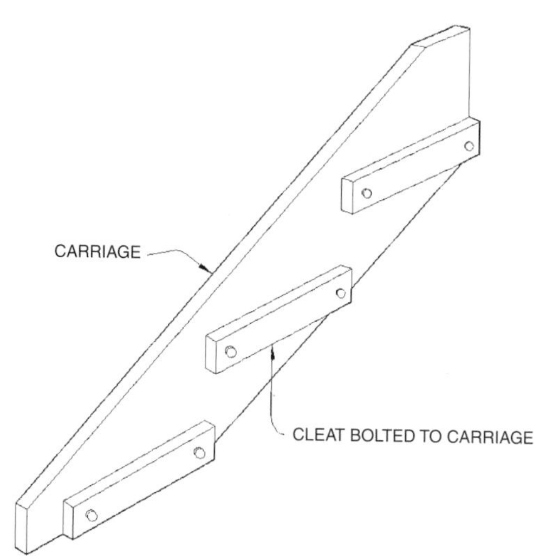

FIGURE 9.1.7 BOLTED CLEATS ON CARRIAGE

ADVANTAGES: Replacing only the damaged stringer is time- and resource-efficient and is a low-cost fix.

DISADVANTAGES: If stringers are subject to rot, the problem should be corrected before replacing material.

WOOD FENCES AND RETAINING WALLS

ESSENTIAL KNOWLEDGE

While some fences are meant to decorate lawns and gardens, others are meant to create barriers and privacy. A well-maintained fence will contribute to both privacy and yard maintenance. Posts are the vertical elements of a fence that are anchored into the ground, and rails are the horizontal members. Fences will collapse if the post supports are inadequate. This can result from rotted post bases, or post bases can overturn if not set deep enough or set with insufficient concrete. Wood fence posts are usually guaranteed for 20 years if professionally installed. Fence posts should be inspected regularly for rot and damage. At the point where the post meets the ground, if the wood is not painted, a coat of wood preservative should be applied yearly. Paint or preservatives should be applied to the entire fence every few years.

TECHNIQUES, MATERIALS, TOOLS

1. STRENGTHEN AN EXISTING WOOD FENCE POST.

If fence posts are wobbly, dig down along one side of the post to check for rot. If the lower section does not flake or feel spongy, the post may only need to be reset in its hole. If physical inspection reveals that many posts are rotted, entire replacement may be warranted. If a wooden post is loose, support can be added by driving pressure-treated 1x4 stakes into the ground on either side of the post (Fig. 9.1.8). Apply a wood preservative to the stakes prior to putting them into the ground. Then nail the stakes to the post. If the post has been pushed upwards by frost heaves, drive the post back into the ground with a mallet or sledge hammer or anchor the post in concrete. Tighten fasteners that are popping out between the post and rails, or add fasteners for additional strength. Wobbly posts set in concrete can be strengthened by pounding the concrete into the ground with a sledge hammer, and then pouring at least 6" of concrete over the area. To plumb a leaning post, rails, and screening, drive a ground stake on the opposite side of the direction of lean. Attach an eye screw to the middle of the leaning post and, using a winch and cable attached to the eye screw, tighten the cable until the fence is plumb.

ADVANTAGES: This is a quick and easy way to fix a fence post and is a low-cost remedy. Salvageable pieces can be restored to make this approach more resource-efficient.

DISADVANTAGES: This is not a long-term solution. Posts with rotted ends should be replaced.

2. REPLACE A WOOD FENCE POST.

Fence posts that need to be replaced can be removed from the ground by first removing the screen and rail boards from the post. To replace only the bottom portion of a rotted post, cut the post at ground level while it is attached to the fence rails, and then remove soil from around the post bottom. Remove

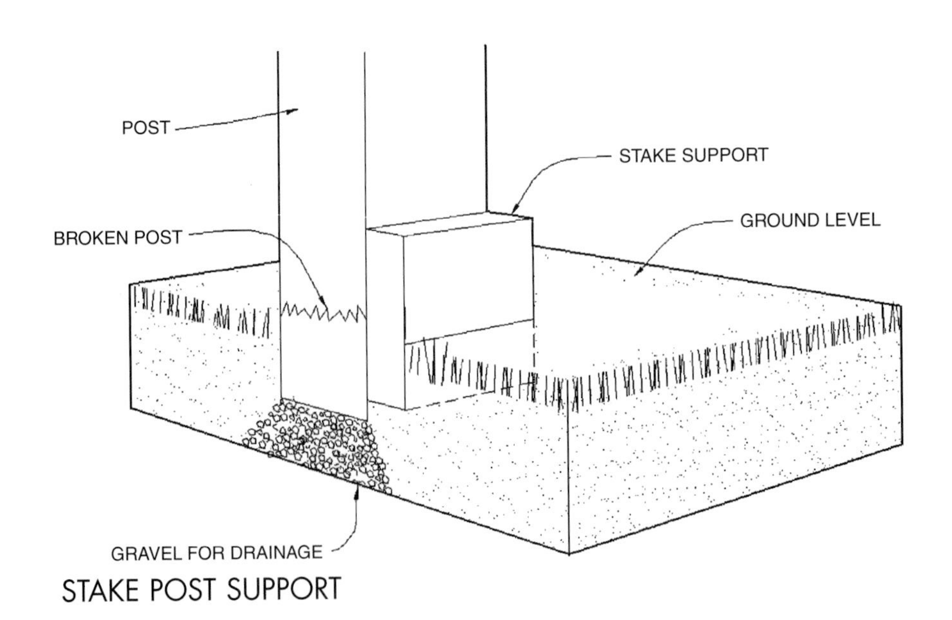

FIGURE 9.1.8 STAKE POST SUPPORT

the rotted portion of the post. Cut a brace or "sister" post approximately 5' long from pressure-treated lumber suitable for in-ground contact. Redwood and cedar will rot in moist ground, so use a galvanized angle iron set in the concrete, bolted to the wood post above ground. Treat with a wood preservative before placing the new post into the hole. Angle cut the top of the new brace post for water runoff. The new brace post should rest on a bed of gravel and extend 2' to 3' out of the ground (Fig. 9.1.9). Fasten it to the old post with two or three ½" carriage bolts. Make sure the post is plumb, and fill the hole with rocks or broken pieces of brick to make it secure.

To anchor a new post in concrete, add the ready-mix concrete and water after the post has been inserted into an 8"- to 12"-diameter hole. Add gravel to the hole up to the base of the post. Holding the post vertically in the hole, pour about 3" of concrete mix into the bottom of the hole. Add water and tamp. Add more concrete mix, water, and tamp it again, until the post hole is full.

ADVANTAGES: This is a relatively easy and low-cost remedy.

DISADVANTAGES: Even if treated with a preservative, moisture in the ground will wick up into the post and can lead to rot. Setting the new wood post in concrete will help prevent the flow of moisture from the ground into the new post.

3. STRENGTHEN AN EXISTING WOOD FENCE RAIL.

Usually a rail begins to rot where it is attached to a post, due to water infiltration. Add a pressure-treated corner block under the rail at the post and attach it to both with galvanized fasteners (similar to the technique for supporting a porch rail, Fig. 9.1.5). If the rail is rotted between two posts, cut the affected section out and scab a wood connector piece to the outside of the rail using galvanized nails or two ½" carriage bolts. If the area of rot is extensive throughout the rail, attach a full-length rail to the damaged rail with galvanized screws every 20". Secure the new rail to the posts with cornerblocks or corner angle braces.

ADVANTAGES: Replacing only the portion of the rail that is damaged is time- and resource-efficient.

DISADVANTAGES: Repair work may show differences in materials.

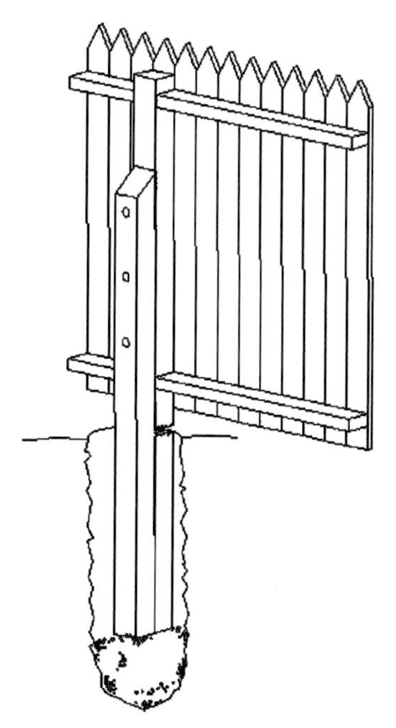

FIGURE 9.1.9 BRACE OR "SISTER" POST

4. REPLACE AN EXISTING WOOD FENCE RAIL.

To replace rotted rails, first remove the screen boards. This is best done by splitting the rail in two before removing the screen boards. Make sure to mark the order of the screen boards prior to removal. Remove any extra nails in the screen or rail boards. Cut a piece of pressure-treated lumber and test the new rail piece for fit between the two posts. Paint the new rail before installation. Level the rail between the posts and fasten it with nails or a metal hanger. Replace screen boards in their original order.

ADVANTAGES: Replacing only part of the fence saves time and material.

DISADVANTAGES: Repair work may show differences in materials.

5. REPLACE DAMAGED WOOD FENCE SCREENING (PICKETS).

Fence screening that has become loose from the rails should be renailed or fastened with new nails or screws. If the screening needs replacement, remove the old boards and apply a wood preservative or finish to the new boards before installing them. Apply a wood preservative to all cuts in treated lumber. If the boards are attached to the rails with cleats, remove one cleat and then slip the board out without removing the other cleat. When reattaching the screen boards, check to be sure that the boards are vertical before nailing them to the rails. Wood screen boards will sometimes rot when water collects under them at grade. A remedy is to install a drainage path by digging a shallow trench under the fence and then filling it with gravel or crushed stone.

ADVANTAGES: This is a relatively easy, low-cost repair.

DISADVANTAGES: Conditions contributing to rot should be corrected before repair.

6. STRENGTHEN A SAGGING WOOD GATE.

A sagging wood gate may be difficult or impossible to close. Replace the hinges on the gate if they are rusty or loose. If the post that the gate is hinged to is not vertical, it can be plumbed (see option 1, Strengthen an Existing Wood Fence Post) or replaced (see option 2, Replace a Wood Fence Post). If the gate itself is out of plumb and sags, it can be squared up with a turnbuckle (Fig. 9.1.10). Install an eye screw on the gate near the top hinge. Position another eye screw on the gate diagonally across from the first, near the bottom of the gate on the latch side. Attach wire to each end of the turnbuckle (the turnbuckle's screw and threads should be just barely engaged). Cut the wires so that the total length of the turnbuckle and both wires is equal to the distance between the two eye screws plus 8".

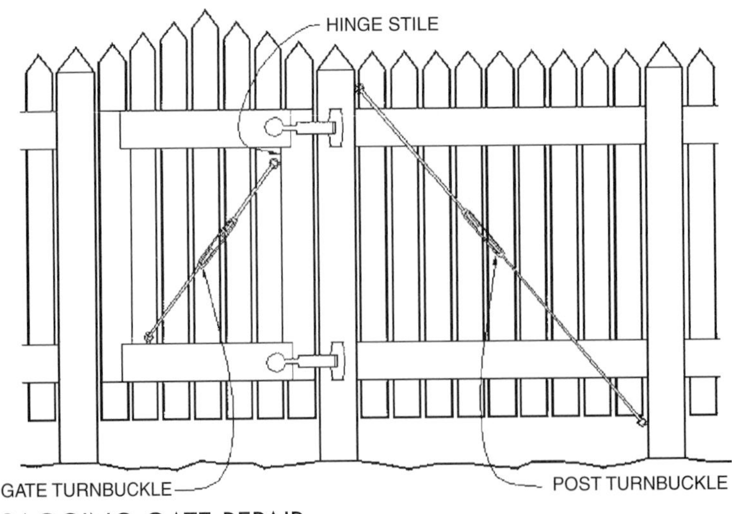

FIGURE 9.1.10 SAGGING GATE REPAIR

Insert each wire end into an eye screw, pull the wire taut, and wrap the remaining 4" of wire around the eye screw. Tighten the wire by turning the turnbuckle with pliers until the gate is plumb. An alternative method is to remove the gate and lay it on a flat area, square it, and then attach a wooden diagonal cross-brace on the back of the gate with galvanized screws or nails.

ADVANTAGES: This is a relatively easy, low-cost repair.

DISADVANTAGES: If the gate sags because of rotted material, it may be necessary to completely replace it.

7. REALIGN A TIMBER RETAINING WALL.

Timbers and railroad ties in a retaining wall can become misaligned due to lateral forces pushing the wall out. One technique is to carefully remove the timbers one by one and restack them (it is best to do this during a dry spell so that the retained earth has little water). After the timbers are realigned and restacked, they can be pinned in place by driving long gutter spikes through the edges of the ties. For more stability, drive two 1x4 or 2x4 stakes into the ground behind the retaining wall and then nail the stakes to the ties. If you cannot access the back of the wall, bore two holes into the top of each timber as it is restacked, and then align the holes using a reinforcing rod. Drive the rod into the ground so that the top of the rod will be flush with the top of the wall. If all the timbers are removed, restack them on a 6" trench filled with gravel for drainage.

ADVANTAGES: Materials are conserved if the timbers can be reused.

DISADVANTAGES: Over time, timbers will likely become misaligned again unless they are pinned together.

9.2 PAVED DRIVEWAYS, WALKS, PATIOS, AND MASONRY WALLS

PAVED DRIVEWAYS, WALKS, AND PATIOS

ESSENTIAL KNOWLEDGE

For walks and driveways, the two main paving materials are asphalt and concrete. With proper maintenance these materials should withstand years of use. These paving materials are both susceptible to freeze-thaw cycles in temperate regions, which can cause asphalt and concrete to crack and crumble. The use of road salts to melt snow in cold climates contributes to the deterioration of concrete. It is important to maintain paving in areas where the temperatures vary from one season to another. Heat can cause a concrete slab to expand and buckle, and extreme cold can heave and crack a slab. Ice can cause the surface of concrete to chip and spall.

It is recommended that concrete slabs be inspected every spring. Check the edges of the slab for erosion. If a sidewalk or driveway slopes, water erosion problems along the edges of the slab are common. Water will begin to wash away the borders until it undercuts the gravel base beneath the edges of the concrete slab, thus leaving the slab unsupported. Check the edges along the slab, and refill any eroded or washed out areas with gravel, dirt, or sod.

Concrete paving is usually 4" of concrete over a 4" gravel base. Asphalt paving is 2" of asphalt over 4" of gravel base. Precast concrete pavers are also available and are less costly than either concrete or asphalt. Asphalt is a good choice for driveways and sidewalks. It works best in climate areas that do not have freeze-thaw cycles. Asphalt costs less than concrete and can last almost as long.

There are basically three delivery methods for concrete: ready mix from a plant, premixed concrete by the bag, or premix with a power mixer. Ready mix from a plant will have portland cement, gravel, sand, a water-reducing agent, an air-entrainment agent, and water. A standard mix will reach 3,000-psi compressive strength in 28 days. For a stronger mix more portland cement is added. Since each component is weighed precisely by computer, the concrete from a plant will be a predictable mix. The water-reducing agent allows a lower volume of water to be added, thus reducing the amount of concrete shrinkage. The aggregate size can be anywhere from ¼" to 1", unless specified otherwise. The amount concrete shrinks decreases as the aggregate size increases. The air-entrainment agent helps the finished concrete to resist freeze/thaw cycles. Additionally, polypropylene fibers can be added to reduce cracking of the concrete.

Premixed concrete by the bag sold at retail building suppliers should conform to the ASTM C-387 standard, yielding a minimum of 3,500 psi after 28 days. Some manufacturers offer fiber reinforcing, for crack reduction, available in premixed bags. The aggregate size in premixed bags ranges from $\frac{3}{8}$" to $\frac{9}{16}$". The premixed bags will probably not include either the water-reducing agent or the air-entrainment agent. Each bag will yield about one-half cubic foot.

TECHNIQUES, MATERIALS, TOOLS

1. SEAL ASPHALT PAVEMENT.

The life of the asphalt pavement can be greatly extended with the yearly application of a sealer. The driveway should be sealed after filling potholes and cracks. This will help the patch adhere to the existing asphalt and not pop out when expansion or contraction occurs. The sealer will also cover over all crack and hole repairs to give a uniform, even appearance.

ADVANTAGES: Sealer will help preserve the surface and prevent small, hairline cracks from growing larger and letting water seep in under the slab, causing potholes.

DISADVANTAGES: A good sealer should be used, as some sealers are merely cosmetic.

2. REPAIR CRACKS IN ASPHALT.

Cracks greater than $\frac{1}{8}$" cannot be properly repaired with sealer and should be repaired with a crack filler designed specifically for the purpose. Asphalt crack filler is available in a caulking tube.

ADVANTAGES: Cracks are easily filled; this is a low-cost remedy.

DISADVANTAGES: After cracks are filled, the surface should be sealed.

3. REPAIR HOLES IN ASPHALT.

As water seeps through cracks in asphalt pavement, it undermines the gravel and earth bed beneath the surface. This will cause the asphalt to cave in and form a pothole. Cold patch or black top patch should be used for pothole repair. The hole should be cleaned out of all loose asphalt pieces and undercut (Fig. 9.2.1) so that the patch will bond and not pop out. Follow manufacturer's instructions, filling the hole and tamping it to compress the material.

ADVANTAGES: Depending on the extent of the patch, this is a relatively low-cost repair.

DISADVANTAGES: The pothole may return if gravel is washed out below the asphalt surface.

4. REMOVE STAINS IN CONCRETE PAVING.

Stains on concrete surfaces may be caused by both organic and inorganic chemicals or by growth of fungus. Effective stain removal begins with matching the cleanser to the stain. There are two general methods of removing stains: physical and chemical. A combination of the two may be needed. Physical methods include sand blasting, grinding, steam cleaning, brushing, scouring, or blow torch. Wire brushes should not be used, as the wire can become embedded in the concrete and cause rust stains.

Chemical solvents dissolve stains or react with them to form a compound that will not show the stain. Solvents can be applied directly on the surface. They can also be applied by soaking a cotton

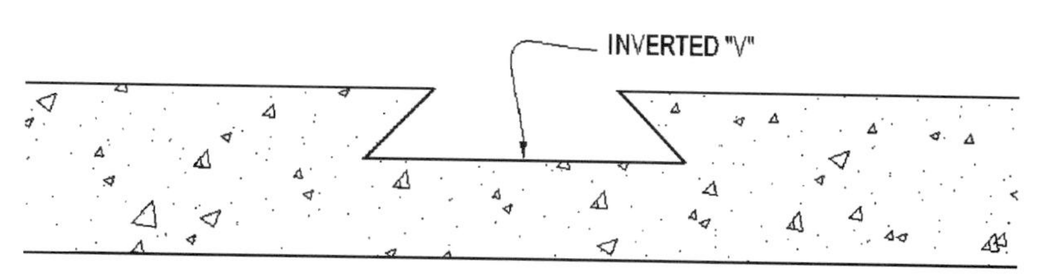

FIGURE 9.2.1 UNDERCUT HOLE FOR PATCH

cloth and placing it on the concrete. This is done to keep just enough chemical at the surface of the concrete to complete the chemical reaction. Mix the solvent and a finely ground powder such as whiting (calcium carbonate), hydrated lime (calcium hydroxide), talc, or diatomaceous earth and spread it on the stained area in a ¼" to ½" layer.

If the slab is not oil stained, clean it with a strong solution of trisodium phosphate (TSP). Scrub food stains with hot water and soap (for tough food stains, cover the spot with a wad of cheesecloth soaked in a solution of one part glycerin, two parts rubbing alcohol, and four parts water). Keep the cheesecloth in place until the stain is absorbed.

For motor-oil stains, saturate them with paint thinner, and then cover with cat litter. Leave the litter in place overnight. Repeat the process if the stain persists, and then scrub with a strong cleaner. Finally, apply full-strength chlorine bleach, wait 15 minutes, and rinse it off with water. Another option is to use a premixed concrete-degreasing solution or a cleaner-degreaser. Another technique is to sprinkle dry portland cement over motor oil stains until all the oil is absorbed, and then remove the powdered cement.

Rust stains can be removed with an acid solution of 1 lb of oxalic acid in 1 gal of water. Carefully apply the solution to the stain, let sit for 2 hours, and then rinse with water. Old stains may require several treatments.

ADVANTAGES: There are many concrete cleaners and degreasers which are less toxic than acids.

DISADVANTAGES: The cause of the stain may be difficult to determine; several stain treatments may be needed.

5. SEAL CONCRETE PAVING.

Water under a concrete slab may freeze and expand causing the surface to crack. Sealers protect the concrete from stains, producing fine dust particles that enter the house, and damage due to water penetrating the surface of the slab. Concrete sealers can provide a surface that is resistant to scaling from the use of de-icers. Rust and oil usually do not penetrate sealed concrete and are more difficult to remove from unsealed concrete. After you have patched or resurfaced the concrete slab, apply at least one coat of a water-repellent sealer. Clear concrete sealers can be applied with a roller, brush, or sprayer.

ADVANTAGES: Sealers prevent oil and antifreeze from being absorbed by the concrete.

DISADVANTAGES: Concrete sealers are not impervious to grease that saturates the surface. They have to be reapplied every few years depending on the product and the conditions. No matter how good a sealer is, it will eventually wear off and be ineffective.

6. REPAIR CRACKS IN CONCRETE.

Liquid cement crack filler, available in a caulk tube, should be used to seal the crack. Clean old caulk or patching products from cracks before refilling with a new patching material. Cracks up to ⅜" wide can be repaired using either a mix of latex concrete bonding liquid and portland cement, available as a repair kit, or a ready-mixed concrete patch product available in a caulk tube. For narrow cracks, use concrete caulk, a ready-mixed liquid repair product for better penetration. Small holes, or "pops," can be filled with portland cement. Concrete bonding liquid provides a better bond between the old concrete and the new patch than does plain water.

Before repairs, oil and grease stains should be cleaned. Using a mason's or cold chisel, break away loose or cracked concrete to create a V-cut. Clean loose material from the crack, and then remove any loose dust or concrete particles. The damaged area should not have any loose dust that could prevent proper bonding. Apply a thin layer of bonding adhesive to the entire repair to help keep the repair material adhered to the crack. Trowel the latex patching compound into the crack, making it even with the surrounding surface. For large patches, after cleaning, pour sand into the crack to within ½" of the surface. Prepare sand-mix concrete by adding a concrete fortifier. Trowel the mixture into the crack and smooth.

ADVANTAGES: Repairing cracked concrete prevents moisture from damaging reinforcing; small crack repair is relatively easy to accomplish.

DISADVANTAGES: Large crack repair may be a significant undertaking; new concrete may be needed.

7. REPAIR A SUNKEN CONCRETE WALK OR DRIVE.

To level a small portion of a sunken concrete slab such as a sidewalk, first bore two 1"-diameter holes, 1' apart, through the lower side of the slab. Then pry the slab up with a lever until it is level. Mix cement, sand, and water to pour into the holes. For good drainage along the slab edges, redirect water away from these areas by filling any accessible voids beneath the concrete with gravel.

ADVANTAGES: This is a relatively easy way to fill the void beneath a slab to prevent further sinking.

DISADVANTAGES: If erosion of gravel and soil beneath the slab continues, the surface will sink again.

8. PATCH CONCRETE.

For large areas of damage or spalling that cannot be repaired as cracks, remove any loose concrete particles with a hammer and cold chisel or stiff bristle brush, and vacuum. Mix the concrete patching material and use a knife or trowel to spread the material over the hole, or use a pourable concrete patch material (Fig. 9.2.2). Overfill the patch area, and then use the patching tool to compress and level the patching material.

ADVANTAGES: This is a relatively simple, low-cost repair, provided the damaged area is not extensive.

DISADVANTAGES: The repair should be done in weather that is warm but not too hot.

9. REPAIR CRUMBLING CONCRETE.

Repair small broken areas of a slab, curb, or step with standard concrete or a concrete patching product (Fig. 9.2.3). If you use standard concrete, apply a concrete bonding adhesive (commonly available) to bind the old concrete to the new. Apply the patching compound according to the manufacturer's instructions. If the damaged area is deep or large, bore holes through the slab and install some reinforcing bars into the existing concrete to tie it into the patched concrete.

FIGURE 9.2.2 POURABLE CONCRETE PATCH

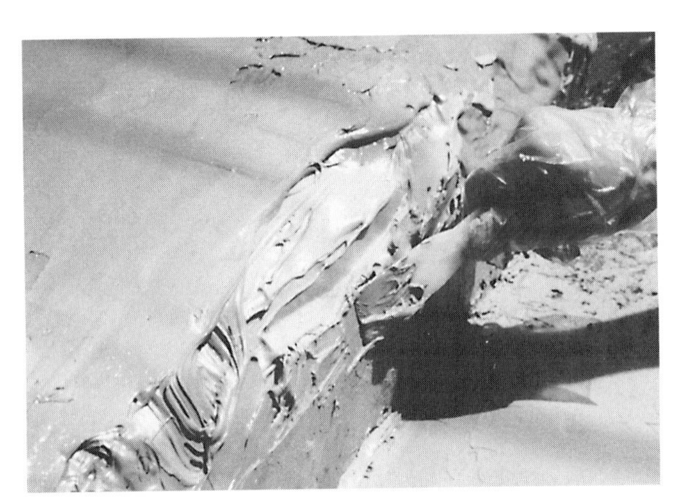

FIGURE 9.2.3 CONCRETE PATCH PASTE

ADVANTAGES: This is a relatively low-cost, easy repair if the damaged area is not extensive.

DISADVANTAGES: The repair should be made when there are warm air and ground temperatures.

10. RESURFACE CONCRETE.

If the concrete slab or steps are badly peeled or extensive repairs are needed, the surface can be renewed with a resurface product. Evidence of concrete surface damage is scaling and flaking of the concrete. This usually occurs from de-icing chemicals or water that has seeped into the concrete and frozen. To repair a scaled surface, apply a thin cement overlay system, which includes both a patching compound for deep holes and a thin coat as a resurfacer (Fig. 9.2.4). A concrete contractor can apply a special-formula thin overlay, such as A-300 Pourable Outdoor Concrete Topping from Ardex Engineered Cements. Pattern-stamped overlays may be used to match an existing finish texture. First, the overlay concrete is poured on, and then it can be colored to imitate brick or stone. As the concrete hardens, a pattern is pressed onto the surface, leaving finishes that replicate brick, slate, or stone.

ADVANTAGES: Resurfacing is less labor intensive than pouring a new slab.

DISADVANTAGES: Resurfacing material should be applied when there are warm air and ground temperatures.

11. LAY NEW CONCRETE NEXT TO EXISTING.

To add a small area of concrete next to existing material, it is important that the new footings for the slab addition match the footings under the old slab. Matching the new footings to the old ensures that winter frost will not heave the slab unequally. Usually a driveway slab is not attached to the house and is laid on a prepared grade at ground level. If a carport or garage is attached to the house, the footings must extend to the local frost line.

To prepare the area for a slab extension, remove all dirt and grass from the area. Compact the soil, since slabs poured on uncompacted soil will settle. A 3" to 6" gravel base spread over the compacted soil should extend 4" past the edge of the new slab. The concrete will be better supported by a deeper gravel base. Install a layer of 6-mil plastic sheeting with overlapping, and wire reinforcing mesh over the gravel base. On top of this are placed 2x4 forms, which should be level with the surface of the existing slab. Install an expansion strip at the joint between the old and new slabs, to allow for movement. Use an edger to form a neat groove at control joints. After pouring the new slab, raise the reinforcing wire so it is in the center of the concrete slab. Wet the concrete

FIGURE 9.2.4 CONCRETE RESURFACING

three times a day for 5 days; remove the forms after 10 days. The new slab should not be driven on for about 30 days.

ADVANTAGES: This is the most practical and cost-effective way to repair large areas of damaged concrete slab.

DISADVANTAGES: Concrete mix used for the extension slab should be as similar as possible to the concrete mix used for the original slab (for fewer expansion and contraction problems).

12. REPAIR OR REPLACE DAMAGED CONCRETE STEPS.

To repair damaged concrete steps, first patch any holes and fill any cracks. Enlarge small holes to 1" depth, minimum, and then clean out hole with a stiff brush or compressed air. Rinse and let dry. Apply a liquid bonding agent before filling the hole with new concrete or apply a concrete patching compound. For cracks, enlarge the crack to get a good bond and fill with concrete patching compound or expansive mortar. If there are signs of water seepage in the steps, use hydraulic cement for patching. If the nose of a step is chipped, place a board in front of the nosing along the front edge of the riser (Fig. 9.2.5). Using heavy stones to hold this form in place, fill with concrete and let cure before removing the form. If a concrete stair corner has broken off, use a latex-based ready-mix or a sand-cement-epoxy mix. Precast concrete steps are inexpensive and can be installed quickly.

ADVANTAGES: Repairing concrete steps by patching is easier than replacing them or repouring them.

DISADVANTAGES: Concrete patches may not hold if the repair area has not been cleaned properly before repair.

13. REPAIR STEEL HANDRAILS.

Steel handrails are most often used on concrete or masonry steps. If the post and handrail are in good condition but the post is loose, unscrew the post at the bottom and remove it from the bracket. Pull out the metal bracket if it needs to be replaced. To do this, enlarge the area in the concrete where the bracket was with a masonry drill. Clean out the hole with water and let it dry. Using masonry anchoring cement or an epoxy, install a new bracket into the hole. Wait 24 hours before screwing the metal post into the new bracket.

ADVANTAGES: This is a relatively simple, low-cost repair.

DISADVANTAGES: Railings with significant rust should be replaced.

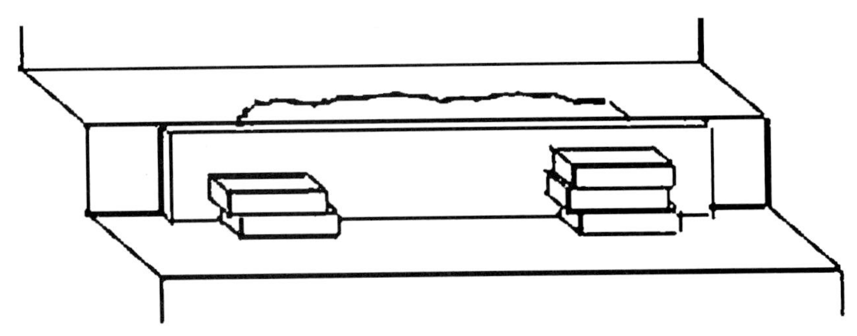

FIGURE 9.2.5 CONCRETE STEP REPAIR

14. REPAIR BRICK OR STONE STEPS.
Check the mortar of the brick or stone steps. If the mortar is loose, but the bricks are tight, remortar the bricks. Remove the old mortar with a chisel. Clean out the joints. Place concrete into the joints with a tuck point trowel. After the joints are packed with concrete, smooth the joint with the tuckpoint trowel. Should the bricks come loose, remove any concrete on them with a chisel. Wash the bricks and soak in a bucket of water. Lay a mortar bed equal in height to the rest of the steps or at least ½". Align the soaked bricks, and then lightly tap them into the fresh mortar with a hammer. Next, place mortar in the areas between the bricks or stones. Trowel smooth the joints between the bricks.

ADVANTAGES: This is a relatively easy way to renew masonry steps.

DISADVANTAGES: Bricks or stones on steps may come loose again through frost heave.

MASONRY WALLS

ESSENTIAL KNOWLEDGE

Brick and masonry unit walls may have steel reinforcing bars that extend from the footing up into the body of the wall. Rough or uncut stone walls are usually laid up with mortar only. Drystack walls use no mortar at all. These walls occur over much of the Eastern United States, and often appear to be little more than a running pile of rocks separating pasture from woodlands.

Impact collisions with a masonry wall can cause anything from minor surface dents to the collapse of the entire wall section. Even if the wall has not been damaged outside, it can crumble from within. If the mortar or concrete is poorly mixed, it can break down and material can begin to crumble. Occasionally brick or block will deteriorate from acid rain or poor material construction.

Masonry retaining walls can be made of stone, poured-in-place and reinforced concrete, or reinforced brick. A newer type of retaining wall made with interlocking concrete blocks can be laid up without mortar or specialized skills. These walls can go to great heights because they are engineered with plastic wire basket structures, called geogrids, that attach to the block at calculated intervals to give the wall an effective width of 4' to 6' or more. Retaining walls are commonly found on properties with sloping yards or on hilly sites that require level ground. Because they can provide support for a level ground surface, they help to slow down surface runoff and help control erosion. The wider the wall, the greater resistance it has to sliding and overturning.

Retaining walls often fail when the soil behind the wall holds back too much water, making the lateral soil load much greater. More often, they fail from the expansive forces of freezing water trapped behind the wall. This lateral load causes cracking and, eventually, wall failure. Holes or drain pipes placed in the wall to relieve water buildup often silt up from fine particles leached from the soil above and become ineffective. Repairing retaining walls may entail altering the soil as well as making structural improvements to the wall.

A hybrid fence-wall system made of precast concrete is available as either a fence or wall. It is offered as a fence (an open split-rail system made of concrete) or a weight-bearing 1 ¾"-thick wall. The walls are built by setting notched concrete posts into concrete footings and sliding 8" interlocking panel sections on top of one another until the desired wall height is achieved. Cast into molds with finishes that resemble rough sawn wood planks or brick panels, this reinforced precast concrete system needs virtually no maintenance.

TECHNIQUES, MATERIALS, TOOLS

1. REPOINT EXISTING MASONRY WALL.

As the mortar weakens in old masonry walls, it begins to crumble and fall out of the joints. This problem is typical in construction before 1930, as the mortar then used was a lime putty. Water aids the lime in leaching out of the mortar, and the mortar crumbles as it loses strength. To test existing mortar, poke an ice pick into the joint. If the joint starts to crumble or the mortar appears to be sandy, repointing may be necessary. Clean out the old mortar joints with a chisel, being careful not to damage the brick or stonework. Make sure all of the crumbling mortar is removed until solid mortar is found. Remove all loose pieces, and use a hose with a pressure nozzle to clean out the joints. Repoint the joints with the same mortar type, unless the old mortar was of the lime-based type. In that case, use lime putty but use 20% portland cement in the mortar mix. The color can be adjusted by adding colored sand. After mixing the mortar, place it on a flat board. Using a joint filler, push the mortar from the board into the clean, wet joint. After the mortar has stiffened, use a jointer as a shaping tool to match the old joint profiles.

If a brick has come loose from the mortar, chisel out all of the mortar around the brick. To remove the brick it may be split with a chisel or a brick set. If necessary, trim the new brick (possibly ¾") so it will fit into the hole. Apply mortar into the clean void and then push the brick in. Scrape away excess mortar and repoint gaps in the joint. Smooth the joint profiles with a jointer. (If the wall is to be pressure washed, do so before repointing the joints.)

ADVANTAGES: Repointing old joints provides a more uniform appearance and prevents moisture from entering the wall.

DISADVANTAGES: Repointing an entire brick wall face can be labor intensive. It may be easier to repoint only the joints that are crumbling, and then apply a sealer or stucco over the face to prevent moisture from entering the wall.

2. RESURFACE AN EXISTING MASONRY WALL.

To repair a crack in a reinforced concrete wall, clean out the break with a cold or brick chisel. Cut away crumbling mortar until hard concrete is reached. Mix the concrete mortar to a thick consistency. Wet the crack, fill it with mortar mix, and then smooth the surface with a trowel.

To repair a hole, apply the first scratch coat 8 to 9 days before the final coat. Attach a new wire lath over the damaged area to any existing lath. After wetting the area, apply a scratch coat from ⅜" to ½" thick, with most of the stucco behind the lath. When the area begins to dry, score the scratch coat horizontally with a rakelike scarifier tool. After scratching the scratch coat, keep the area damp for 48 hours. Apply a second layer that comes within ⅛" of the final surface and keep the area damp for another 48 hours. Let the area cure for 4 to 5 more days. Dampen the area and apply the finish coat, with a pigment if desired. It is important to keep the finish coat damp for 2 days after application.

For scaling of depths to ¾", you should use a special repair mortar and a sealer. For scaling depths up to 1 ¾" or 6" in diameter, use a bonding bridge, special repair mortar, and a sealer. Clean the crack with a wire brush. Chip out a V-shaped groove at the crack line and dampen the area. Patch with three parts fine sand to one part cement. Keep the area damp for 48 hours. There are stuccos made for resurfacing masonry that are readily available at masonry stores. If sealing a brick face wall, use a silicone sealer that allows water vapor trapped in the brick to escape. Apply when the temperature is over 50°F and the wall has been dry for at least 7 days. After spraying the sealer onto the wall, smooth out any runs with a brush.

ADVANTAGES: Repairing only the damaged parts of a wall is less labor intensive and less costly.

DISADVANTAGES: If the wall has some scaling problems, it is best to resurface the entire wall.

3. REPLACE PARTS OF EXISTING MASONRY WALL.

If stones, bricks, blocks, or mortar are falling out of the wall, it may be shifting due to the additional weight, water in the soil, poor drainage, or ground movement. Clean out the old mortar joints with a chisel, being careful not to damage the solid areas. Remove all crumbling mortar and loose pieces, and then flush with a water hose. Fill large voids inside the wall with chunks of rock or stone. If bricks, blocks, or stones are loose from the old mortar, they should be reset into the new mortar. Pieces of asphalt expansion strips or fiberglass insulation can also be used. The edges of the filler material should be recessed from the surface of the wall for the new mortar. Mix the cement to a thick consistency. Mist the area to be repaired with water. Pack the mortar tight into the joint and smooth the surface of the joint for water drainage.

ADVANTAGES: Repairing only the damaged parts of a wall is less labor intensive and less costly.

DISADVANTAGES: The problem of the wall damage should be fully addressed before repairs.

4. GOOD RETAINING WALL DRAINAGE.

Retaining walls should have weep holes to prevent damage from moisture building up behind the wall. When rebuilding a brick, block, or stone wall, leave some of the vertical joints in the first course unmortared as weep holes. For an existing wall, drill 2" weep holes into the bottom of the wall just above grade every 2' to 3', and line them with plastic tubing, such as a piece of hose. If possible, backfill behind the wall with gravel or crushed stone for better drainage. If the wall has no weep holes and it is not possible to bore through the walls, dig out the earth behind the wall and install a french drain. Geotextile fabric around a perforated drain pipe behind the wall will promote water flow from around the wall. In addition, storm drains may be needed at the upper level of the retaining wall to help surface water to drain away from the wall instead of flowing down behind it. Place the covered perforated drain in a layer of crushed rock, gravel, or free-draining river sand. The perforated pipe needs to be pitched ¼" per foot in the gravel trench.

ADVANTAGES: Good drainage can extend the life of the wall.

DISADVANTAGES: It may be very difficult to excavate behind the wall to install a drainage pipe and trench.

5. REPAIR A MASONRY RETAINING WALL.

Since retaining walls most often fail due to unseen forces, repair techniques will vary. In most cases of structural failure the wall needs to be excavated for some, if not all, of its distance. If physical inspection reveals that backfilled material is similar to surrounding poor-draining soil, then this soil should be removed. The wall's specific engineering requirements should be determined by a qualified designer or engineer. If the wall is salvageable, the repair will be completed prior to the replacement of new granular backfill, drain pipe, and filter fabric.

ADVANTAGES: This repair may be cheaper than rebuilding if the existing wall can be reused. It may have less impact on trees and plants than constructing a new wall.

DISADVANTAGES: Existing saved portions may have unseen defects. This repair could be more costly if some work has to be redone after placement.

6. REPLACE A MASONRY RETAINING WALL.

If a masonry retaining wall is failing beyond repair, consider rebuilding the wall completely. If the new wall is similar to the existing wall, site excavation work should be less costly than if an entirely new wall were built. A good retaining wall permits surface water to drain away slowly. Once the wall is built, it is backfilled with draining gravel material 2' to 3' out from the wall up to within 6" of final grade. A layer of filter fabric should be placed between the gravel and the native soil, with a drain pipe at the bottom of the gravel backfill to permit easy drainage of water away from the wall. This drain pipe may also be filter wrapped to inhibit silt blockage.

In colder climates the earth may be protected from freezing by placing a 2" layer of rigid polystyrene against the inside face of the foundation wall to keep the earth's natural warmth from being lost to the cold face of the exposed wall (and should reduce the likelihood of damage from the freeze-thaw cycle).

ADVANTAGES: A new wall will last indefinitely if properly constructed and maintained.

DISADVANTAGES: A new wall can be very costly; shrubs and trees may be lost.

9.3 UNDERGROUND CONSTRUCTION

WELLS

ESSENTIAL KNOWLEDGE

A drilled well consists of a hole bored into the ground, with the upper part lined with casing that prevents the collapse of the borehole walls and prevents surface or subsurface contaminants from entering the water supply. It also provides a housing for a pumping mechanism and for the pipe that moves water from the pump to the surface. The casing should have a drive shoe attached to the bottom to prevent damage during well driving. Below the casing is an intake through which water enters the well. This may be an open hole in solid bedrock or screened and gravel-packed, depending upon the geologic conditions. After drilling, it is necessary to remove fine material remaining from the drilling process so that water can more readily enter the well by using compressed air (blowing), bailing, jetting, surging, or pumping. After proper disinfection, the well is capped and has an air vent, to provide sanitary protection until it is hooked into the house's plumbing system. Inspection of the well should reveal causes of poor water quality and restricted water flow.

TECHNIQUES, MATERIALS, TOOLS

1. REMEDY POOR-QUALITY WATER.

If water quality is poor, revealed either by taste or testing, the water can be sealed off from the surrounding soil. One method is to install an additional casing inside the original casing and grout into place. If the water quality remains unsatisfactory, or if original well construction defects cannot be remedied, the well may need to be abandoned. It must be completely sealed prior to drilling a new well, in order to prevent cross-contamination between sites.

ADVANTAGES: Installing additional casing may be more cost-effective than drilling another well.

DISADVANTAGES: Adding another casing inside the well may not solve the problem if the water supply itself is contaminated.

2. REMEDY IMPROPERLY SIZED WELL.

A well that is too small may lead to water usage problems, such as poor or limited simultaneous use of a sink and washer. Well size should be based on peak demand. A day's use may be concentrated into a period of 1 to 2 hours, often in different areas of the house at the same time (laundry, bathroom, and lawn). A conservative estimate is that a home will need 3 to 10 gpm to meet all the needs of two to four people. If the water supply is insufficient, another well may need to be drilled in a different location.

ADVANTAGES: Drilling an additional well lets each well alternate so as not to deplete the water sources.

DISADVANTAGES: This can be costly.

3. PLUG ABANDONED WELLS AND BOREHOLES NEAR DRINKING WELL.

Potential groundwater contamination sources are abandoned wells and boreholes that penetrate aquifers or breach a zone that is a barrier to contaminant migration. Cross-contamination can occur in abandoned wells or boreholes when two or more aquifers are penetrated and a seal has not been placed or is no longer intact between the zones to prevent the water from mixing. The potential for direct groundwater contamination exists in abandoned wells and boreholes along the borehole if an adequate surface seal was not installed or the surface seal has deteriorated. Using a geotechnical fabric around the outside of the abandoned well will keep fine particles out. Another method is to fill unused or "dry" wells with cement grout, bentonite chips, or native soil. Most local codes allow any person to abandon a well, but it is recommended that a registered well driller do the work.

ADVANTAGES: Installing a filter around the abandoned well will only work for a while; filling and plugging the well will keep it from cross-contaminating nearby drinking wells.

DISADVANTAGES: The majority of states have regulations specifying decommissioning (plugging) methods for abandoned water wells or boreholes; check local codes.

ON-SITE WASTEWATER TREATMENT

ESSENTIAL KNOWLEDGE

On-site wastewater treatment is usually provided by cesspools, septic systems, or aerobic treatment units. Cesspools, drywells, or seepage pits (Fig. 9.3.1) are the oldest forms of on-site sewerage systems, often still in use in older homes. They are simply a single hole in the ground loosely blocked up with locally available materials (stone, brick, block, or railroad ties) and capped either with ties covered with a layer of old steel roofing or a cast-in-place concrete lid with a cleanout hole near the center. Household waste water enters the cesspool, and the liquid portion is absorbed into the ground. When the soil plugged, a

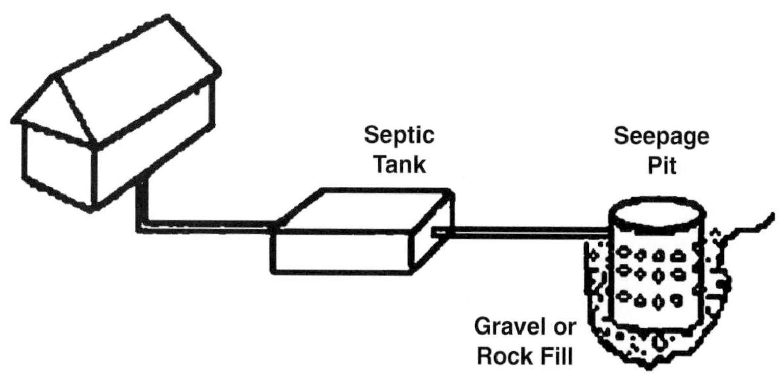

FIGURE 9.3.1 SEEPAGE PIT SYSTEM

new cesspool was added. Sometimes installers placed an elbow or a tee in the outlet pipe from the first cesspool, creating a baffle to hold back the floating greases and scums. The first tank is pumped out to maintain system operation.

A septic system's main components are a watertight chamber that all household wastewater enters for settling and anaerobic digestion of greases and solids, a distribution box, drainage tile, and a drainage or leaching field to dissipate the waste. Older septic tanks were made of asphalt-coated steel, while today tanks are made of concrete, fiberglass, or plastic, with a set of baffles. Most tanks have an inspection hatch at both the inlet and the outlet. Some have a third hatch in between for pumping access. Steel tanks often have one round lid that covers the entire tank. Access locations should be recorded and/or marked.

Septic tanks should be pumped approximately every 3 years of normal operation. They should not be treated with any additives and should be protected from harmful chemicals used in many homes, which can inhibit the anaerobic process. Septic systems can usually last 15 to 40 years or longer, depending on their design and maintenance. For example, even a relatively new system can fail if it is located in poor soil, is undersized, or is not properly installed or maintained. If soil or site conditions are not suitable for a conventional drainfield, an alternative system might be used.

In aerobic treatment systems, wastewater is mixed with air, promoting bacterial digestion of organic wastes and pathogens. The biological breakdown of wastes in a septic tank's anaerobic (oxygen-deprived) environment is relatively slow. Aerobic units are more expensive and require significantly more maintenance than conventional septic systems. However, they provide good wastewater treatment on homesites that are otherwise unsuitable for development because the soil type, depth, or area is inadequate for on-site treatment. The effluent from an aerobic unit can be discharged into a soil absorption system.

The exact locations of system components are not obvious, because they are below ground. If the location of the system is not in the home's records, check with a pumper's records or the county's Environmental Health Division. If no record of the location exists, look in the basement for the sewer pipe leaving the house and note the direction in which it goes through the wall. In houses without basements, go to the outside of the house, along the wall closest to the lowest bathroom. Outside, probe with a steel rod by gently tapping into the ground, starting 5' from where the sanitary sewer leaves the house, or dig 10' to 20' away from the house in the direction of the house sewer line. The septic tank is usually within 2' of the ground surface. The distribution box and drainfield are usually located downslope from the septic tank. In some cases, wastewater is pumped to a drainfield uphill from the septic tank. Check with the state's septic design standards on slopes (greater than 15% are common) that are unacceptable.

To prevent contamination of water supplies, the drainfield should be set back (Fig. 9.3.2) from any wetland, shoreline, stream bed, or drinking water well (check with local building officials). A well is better protected if the system is downhill from it.

TECHNIQUES, MATERIALS, TOOLS

1. REPAIR THE SEPTIC SYSTEM.

If the septic system appears to be blocked, check the washing machine outlet, floor drain, bathtub, or remove the cleanout plug very carefully to avoid a flood to see if there is water backed up, through any of the drains which are below the level of a toilet. If no water is backed up, then the problem is likely with a toilet or plumbing. Tree roots are a common cause of blockage. If roots are blocking one or more pipes, the roots should be cut out and the joints sealed where roots entered the pipe. Settling, breaking, crushing, and pulling apart are problems due to poor installation. Freezing, plugging at joints, corrosion, or decomposition are other sources of blockage. Insulating, replacing, releveling the pipes, sealing joints, and properly backfilling will resolve most problems. Failed dry wells can sometimes be excavated and repacked with crushed stone to create a new soil surface for absorption.

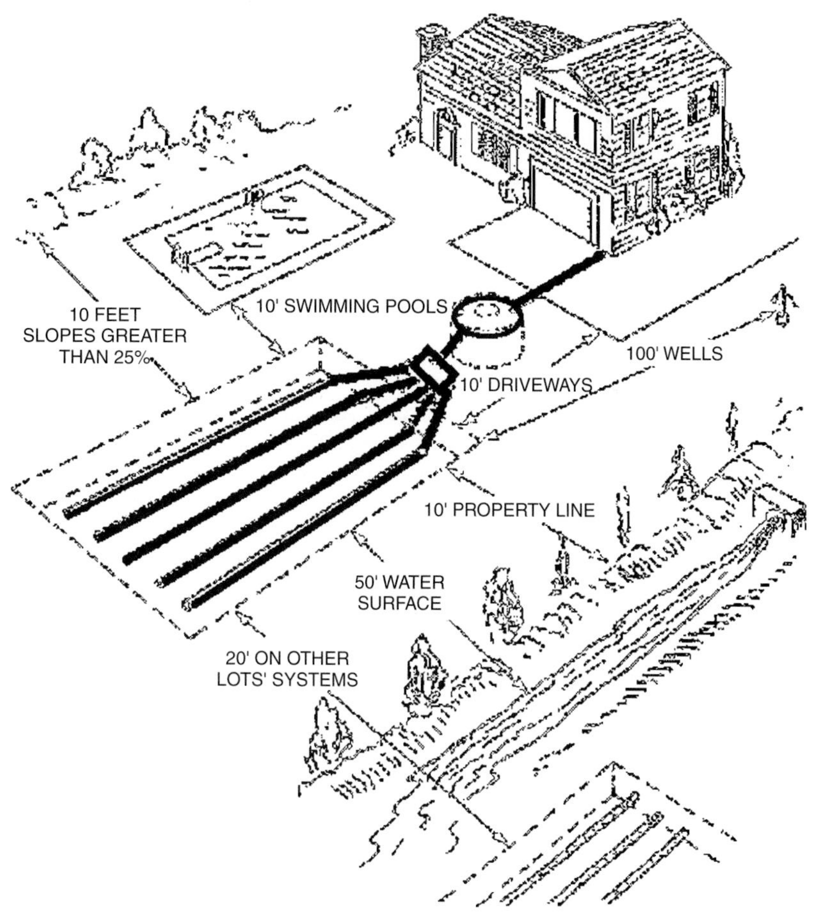

FIGURE 9.3.2 SETBACKS FOR SEPTIC SYSTEM

ADVANTAGES: Repairing a system is far less costly than installing a new one.

DISADVANTAGES: Depending on the problem, repairs can be costly.

2. PUMP OUT THE SEPTIC SYSTEM.

Regular pumping of the septic tank (Fig. 9.3.3) is the most important factor in system maintenance. As more solids accumulate in the tank, particles are more likely to flow out of the tank and into the drainfield. The septic tank needs to be pumped if the sum of the solid layers (sludge plus scum) takes up more than half of the tank capacity, the top of the sludge layer is less than 1' below the outlet baffle or tee, or the bottom of the scum layer is within 3" of the bottom of the outlet baffle (or top of the outlet tee). Most septic tanks need to be pumped every 3 to 5 years, depending on the tank size, and the amount and type of solids entering the tank. This should be done by a septic system maintenance company.

ADVANTAGES: The cost of pumping a septic tank ($250 to $300) is far less than the expense of replacing a drainfield clogged by escaping solids ($3,000 to $12,000, depending on site conditions, the size of the home, and the type of system).

DISADVANTAGES: Pumping out the septic system may not fix the problem.

3. REJUVENATE THE LEACHING FIELD.

A leaching field that is compacted or plugged will fill up with effluent and cause the system to fail. Instead of installing a new system, the leaching field soil can be "fractured" through a process called

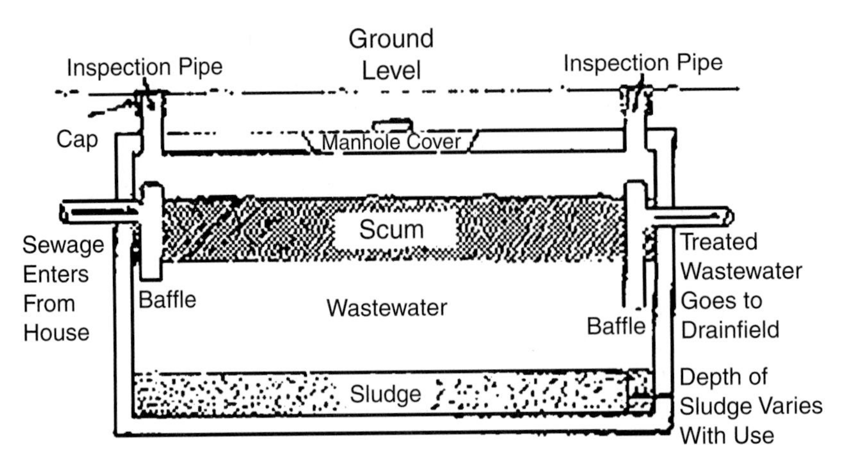

FIGURE 9.3.3 SEPTIC TANK COMPONENTS

terra-lifting. A hollow tube is pushed into the soil, which is then pumped full of air, creating fissures in the soil. The fissures will permit the leaching field to drain and aerobic bacteria to flourish. A leaching field can be treated through terra-lifting in a few hours, depending on size.

ADVANTAGES: Terra-lifting is a quick, relatively low-cost solution and does not require digging.

DISADVANTAGES: It may not remedy problem if blockage is extensive or leaching field is exhausted.

4. INSTALL A NEW SEPTIC SYSTEM.

Both the septic tank and drain field must be located according to setbacks from surrounding elements (Fig. 9.3.2), designed for the maximum occupancy of the home, and have adequate capacity to treat all the wastewater generated in your house, even at times of peak use. Installing low-flow toilets and water-saving faucets (see Chapter 6) will reduce the size of the system needed. Each state sets procedures for calculating wastewater flow and sizing on-site treatment systems. There are various rules of thumb for calculating the volume of wastewater from a single-family home: 100 to 200 gallons per bedroom per day multiplied by the number of bedrooms in the home; or 75 gallons per person per day. The septic tank should be large enough to hold two days' worth of wastewater, as this is long enough to allow solids to settle out by gravity. The addition of a bathroom, bedroom, or water-using appliance (such as a Jacuzzi, dishwasher, or water softener) to the home may require expanding your system. Typically, a new three-bedroom home is equipped with a 1,000-gal tank. A two-compartment tank or a second tank in series can improve sludge and scum removal and help prevent drainfield clogging (Fig. 9.3.4).

ADVANTAGES: A new system is more efficient and will alleviate problems with old systems if properly designed and installed.

DISADVANTAGES: Lot size may not support the installation of a new system. A new system can be costly depending on design, size, and other constraints.

WATER AND SEWER LINES

ESSENTIAL KNOWLEDGE

Water lines must be in good repair to deliver water without risk of contamination. Sewer lines must carry waste out of the house without risk of contamination to the soil or water outside the home. Water supply lines usually deliver water at pressures that range from as low as 15 psi up to 120 psi. To help

FIGURE 9.3.4 DRAINFIELD INSTALLATION

prevent them from freezing, municipal water and well-water supply lines are buried in trenches at or below the frost line. In older homes supply lines may be ¾" to 1" galvanized or copper pipe or (in very old homes) lead. In houses constructed since the 1970s, supply lines are in the range of 1" to 2" copper or polyethylene coiled pipe, suitable for burying and approved for drinking water.

In-ground supply pipes seldom need purging. If problems occur with water quantity or pressure, the pipe may be leaking or it may be clogged with sedimentation or scale. If water pressure drops at sinks and showers and no leaks are evident in the house, it could be caused by a leaking supply pipe. If the water meter is outside the house, there could be a leak from the meter to the house. If the first few feet of water supply line can be unearthed and inspected, the remedy may be apparent. Old, small, heavily scaled pipe may suggest that replacement is in order. If the pipe does not look decrepit, spot repair may be sufficient. Accessing and replacing water lines is often complicated by surface paving, walls, trees, and other obstructions, and these obstacles can greatly affect construction costs. Unfortunately, fixing pipes with poor water pressure or tainted water may require digging up the entire line.

Sewer lines run from the house to public sewers or private septic systems. Because the pipe carries wastes that come from within the house, sewer lines do not need to run below the frost line. Slow draining waste could mean pipe restriction due to internal clogging, roots, or a damaged line. Problems with sewer lines can be assessed with snakes to check for blocks; some professionals use snakes with cameras that offer video images within the pipe. Most sewer pipe failure results from ground settlement caused by improper backfilling or, in rare cases, ground settlement.

TECHNIQUES, MATERIALS, TOOLS

1. REPAIR THE WATER LINE.

If a leak is detected, once it is located and fixed water pressure should improve in the house. Follow the pipe supplier's recommendations and local plumbing regulations for repairing the type of water pipe found. Avoid low-cost repair strategies that are not cost-effective when considering the cost to dig up pipe. After pipe repair, care should be taken to properly backfill below and above the pipe with non-settling material to assure proper pipe support underground.

ADVANTAGES: Repairing only damaged sections of pipe can save time and money; if leaks can be detected, extensive excavation can be avoided and there is less disruption to surrounding grounds.

DISADVANTAGES: All problem sections of the pipe may not be uncovered. Fixing a section of the pipe will not solve systemwide problems, such as defective joints or pipe deterioration.

2. REPLACE THE WATER LINE.

If the water line is of questionable condition and extensive landscaping is planned, it may make sense to remove and replace the entire line. When replacing supply lines, care should be taken to provide adequate bearing support by placing gravel into the trench before laying new pipe and replacing soil. If supply pipe runs are more than a few lengths, polyethylene tubing may be most cost-effective, as it is available in 200' coiled lengths. Lines can also be replaced using a "trenchless" method that does not require excavation. A hardened steel splitting head is pulled through the old pipe. A new pipe is attached to the back of the splitting head (Fig. 9.3.5) and follows the path of the old line.

3. REPAIR THE SEWER LINE.

Unlike water lines, sewer lines can be snaked to locate problems. Because sewer lines regularly carry warm water from the house, they are usually laid only a few feet from the surface of the earth and are thus easier to find. Poor flowing sewer pipes can be clogged with roots or nondissolving household waste. Sewer snakes, power snakes, or roto-rooting can unclog lines, though the condition may return if the source of the problem is not corrected. Once a clogged or broken line is located, the line should be excavated back to sound, undamaged pipe where a new section can be spliced in cleanly. Older homes will have 4" cast-iron sewer lines. Newer homes will have PVC schedule SDR 35 plastic pipe. In either case, the replacement pipe will be the PVC 35 and the joint can be made using a Fernco soft coupling, which can join the same or similar pipe of roughly the same diameter. These joints should last as long as the pipe and permit splicing old to new, even of

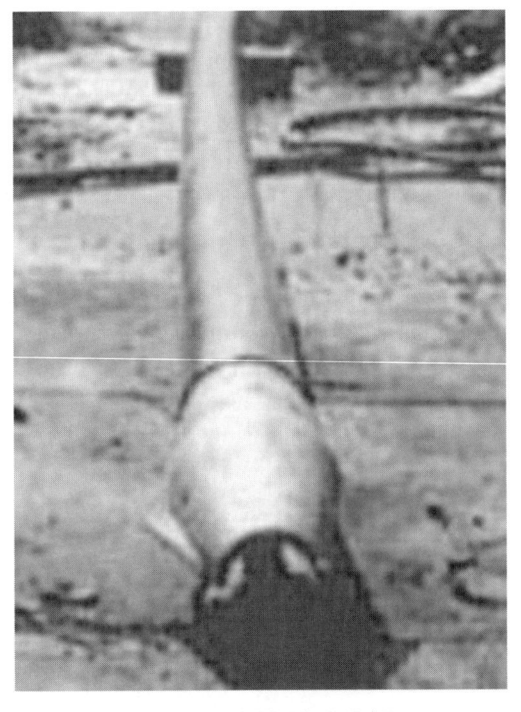

FIGURE 9.3.5 TRENCHLESS SPLITTING HEAD

different materials. Once repairs are complete, the trench should be backfilled with gravel or similar material to assure stability.

ADVANTAGES: Since sewer lines will usually last many years without degradation, fixing a part of the line should return the line to dependable, like-new condition (if the cause of the problem is also fixed). Repairing the line minimizes disruption to property and conserves materials and cost.

DISADVANTAGES: Care must be taken with repair work; must be backfilled properly.

4. REPLACE THE SEWER LINE.

Replacing sewer lines involves essentially the same processes as repair. In addition to digging a new trench and backfilling with good bearing material, care needs to be given to maintain a slope of ⅛" per foot from the house to the outlet elevation in the street. New lines should be made with 4" round PVC schedule SDR 35 pipe, which comes in 13′ lengths. The pipe is joined together by a hub and gasket system that snaps together to form a permanent joint without special tools.

ADVANTAGES: This is the best approach if a large addition or other site excavation requires disturbing an existing sewer line; long-term solution.

DISADVANTAGES: Full replacement can be disruptive to existing property and could threaten the health of trees and shrubs in the vicinity.

UNDERGROUND STORAGE TANKS

ESSENTIAL KNOWLEDGE

Underground storage tanks are commonly used for home heating oil. It is recommended that buried tanks be tested, which typically costs about $500. Specialists and some oil companies have equipment to test buried tanks for leaks. Both simple pressure-testing and sophisticated electronic testing are used on residential tanks. Periodic tank testing for a small, buried residential tank is typically less cost-effective than tank relocation.

Testing for water in the tank is simple and can be done by any service person using a simple chemical paste on a probe. A lot of water in the tank may indicate leaks. Underground fuel storage tanks usually fail from rust perforation due to several effects of water inside the tank including, in the case of heating oil, the combination of water with sulphur in the fuel. External rust, unless very heavy, is not highly correlated with internal rust. There are also mandatory requirements for abandoning buried old and unused tanks.

TECHNIQUES, MATERIALS, TOOLS

REMOVE OR ABANDON AN UNDERGROUND TANK.

If a tank is not to be used, or testing indicates a leaking tank, it should be removed or abandoned. Proper abandonment procedures involve pumping out remaining fuel, confirming that there has been no leakage, and filling the tank with an approved filler or removing it entirely. One should check local building codes for regulations and methods of proper removal (Fig. 9.3.6) of an underground storage tank. It should be replaced with a double-walled or aboveground tank.

FIGURE 9.3.6 TANK REMOVAL

ADVANTAGES: Removal eliminates problems with leaky, unsafe tanks. Removing a tank allows certification that the tank was not leaking, mitigating future claims against the property owner and negative impacts on resale value.

DISADVANTAGES: Tanks can be costly to remove and replace, depending on size, location, and soil contamination (if any).

9.4 LANDSCAPING

LANDSCAPE CARE

ESSENTIAL KNOWLEDGE

Trees and plants add beauty to a home and also help regulate and replenish the natural environment. Healthy growing plants and trees (Fig. 9.4.1) improve air quality by scrubbing out dust and increasing oxygen levels. Over time, soil conditions can be improved by mulching and composting plant material that is generated by trees, lawns, and gardens. Trees and plants are also a soil binder that prevents soil erosion. Water quality is protected when runoff is moderated, and water that percolates through soil into aquifers is cleaned. Water that is filtered by leaves and foliage has some of its impurities scrubbed out.

Planted too close to buildings (Fig. 9.4.2), trees may cause foundation damage because roots can cause cracks or wall failure. Overhanging limbs may fall and damage roofs or cause gutters to clog from twigs and leaves that can block water drainage. Foundation plantings may promote mold and fungus growth on siding that is shrouded from the sun by dense evergreen branches.

Buildings can reduce the amount of nutrients available to plants, so trees especially need to be properly sized and located to assure health. Trees and shrubs may have been planted too close to other plants or objects, and if they are still healthy, they may be transplanted to a more suitable location. Feeding plants and trees can improve plant vitality, and pruning may also help by reducing the nutrient load required by the tree. If possible, a licensed arborist should be consulted for specific recommendations for each tree or plant species being disturbed. Large trees that are damaged or require trimming of higher branches require an experienced tree service (Fig. 9.4.3). Properly planted and nourished, most shrubs and trees will not require constant maintenance.

Trees have root systems that usually extend far beyond the drip edge of branches. This makes roots very susceptible to damage from trenching or excavation. Where possible, utility lines should be tunneled under trees rather than trenched through.

When the natural slope of yards and gardens is changed intentionally, it can result in softer or steeper grades achieved by regrading to desired effect. A berm may serve as a visual screen to block a noisy eyesore such as a highway, or help to slow or redirect surface water runoff.

Berms and other open areas of a yard are often covered with turf grasses that are not native to the environment. Most ornamental lawn grasses contain mixes of Kentucky bluegrass that like alkaline soils and lots of moisture and sunlight. Since much of the continental United States does not naturally have these conditions, keeping sod healthy and vigorous is an ongoing and costly endeavor requiring regular fertilizer, watering during dryspells, regular weeding, and cutting. Specialty lawn treatments are controversial because they use pesticides and chemicals, traces of which can be found in water and air. At certain levels these chemicals are known to be detrimental to the environment. The reader interested in more information about local lawn care should inquire directly to local providers and applicators who may have formulations specially designed for a specific area and who will know of applications that may be hazardous.

FIGURE 9.4.1 HEALTHY PLANTINGS

FIGURE 9.4.2 TREE TOO CLOSE TO HOUSE

TECHNIQUES, MATERIALS, TOOLS

1. REPAIR TREES AND SHRUBS.

Tree and shrub growth can be stunted by reduced access to water or nutrients, which can cause the plants to lose leaves or entire limbs and branches. Restoration should be attempted in phases to see what works before trying alternative measures. After consulting local professionals or nurseries for

FIGURE 9.4.3 TRIMMING HIGH LIMBS

advice, an approach might be to apply recommended fertilizer to the root base of the tree or shrub. If plant malaise persists the following year, consider limb thinning using proper tools (Fig. 9.4.4) or some other approach that is recommended by a local professional. This will reduce the nutrient load on the root system. Mature but sickly landscaping may require special applications that are unique to the species and site. In such cases, consultation with a landscape professional may be necessary to restore the plant to health.

ADVANTAGES: Appropriate action keeps plantings healthy.

DISADVANTAGES: Palliatives for plant distress or disease may delay inevitable new plant stock or improved landscape design. Money spent for repair or restoration may not be cost-effective because it may mean less money is available for a later replacement phase.

2. PROTECT TREES AND SHRUBS DURING CONSTRUCTION.

When building or remodeling around mature trees and shrubs it is necessary to take preventive steps to avoid harming healthy species to be preserved. The first step is to inventory what is worth saving and to prioritize which plants are more important. Flag important trees, branches, and shrubs before starting, and then hire a small-scale firm to do the digging (the excavator-operator of the small firm will likely be the same person who does the initial estimate and job appraisal and should remember the special care that was agreed upon for the job). Install a fence around building areas and areas for parking and make the rest off limits.

Care should then be taken with digging in the vicinity of saved trees. In general, keep trenches at least 15′ from the center of tree trunks or, if not possible, consider boring a tunnel directly under the trunk a few feet below the surface. Avoid raising or lowering the grade, or even scraping the surface above tree and shrub roots or excessively treading over the earth. All these activities can weaken or kill the root system. Most root systems are within 18″ of grade with large feeder roots permeating the top 6″ of soil, and these must be protected.

If construction will cut across roots, consider using piers instead of a continuous trench. When stockpiling soil, avoid piling onto root areas of saved trees. If unavoidable, consider using a tile aeration system so that oxygen is not cut off from root tops. When digging is required across tree root centers, dig from a radial position with a vertical chopping motion so that roots will be cut cleanly. Avoid ripping up lateral root structures, which could be affected by conventional digging. Broken roots should be neatly trimmed with a pruner or saw and kept from drying or freezing while excavation is open.

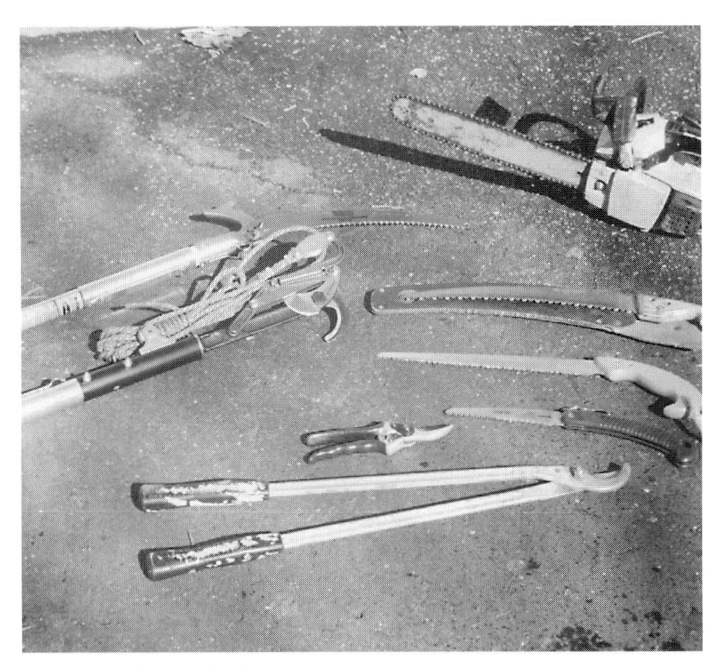

FIGURE 9.4.4 PRUNING TOOLS

ADVANTAGES: Protecting trees allows mature plantings to be retained.

DISADVANTAGES: Unless carefully protected, trees and shrubs could be lost.

3. REMOVE TREES OR SHRUBS.

Sometimes trees or shrubs must be removed. When cutting down trees, whenever possible avoid digging out the trunk. Dig down around the base of the tree and remove the portion of the stump that is above ground. Digging or grubbing out trunks can kill roots from other trees and shrubs in the area.

ADVANTAGES: Removing trees or shrubs can extend the useful life of plant species and is less disruptive and less costly than replacement.

DISADVANTAGES: Work can be extensive and costly.

4. REPLACE TREES OR SHRUBS.

When choosing to replace existing trees and shrubs that cannot be maintained or have become diseased or rotted, consider replacing successful species with the same species. If considering upgrading or altering the landscape with new vegetation, evaluate environmental factors such as soils, climate, slope, and hydrology. Consider the effect the new trees will have on views, screening, shade, wildlife habitat, color, and scent (if any).

When replacing trees or planting new specimens, always chose native species if possible. As first cost, they are usually cheaper to purchase than exotic species and will thrive in the environment without pesticides, fertilizers, or special watering, which makes them a best buy in life-cycle costs too. Choose an exotic, nonnative species only after concluding that you can provide the proper soil preparation and necessary additional maintenance that they will require.

ADVANTAGES: Affords the chance to solve long-term problems and make improvements that will be beneficial to the yard or landscape for years to come. Opportunity to replace nonnative species with native species.

DISADVANTAGES: Replacing mature trees with well-established nursery stock can be very expensive. Replacing with smaller, more affordable trees can leave property looking empty or disfigured and requires patience for plants to develop into mature specimens.

5. INSTALL LIGHTNING PROTECTION FOR TREES.
According to the National Recreation Association, the 10 species most likely to be struck by lightning are, in order: oak, elm, pine, tulip, poplar, ash, maple, sycamore, hemlock, and spruce. Since the expense of removing a large, lightning-struck tree (Fig. 9.4.5) may approach that of installing a lightning protection system, it is economically prudent to protect large or valuable trees from lightning strikes. In addition, protecting nearby trees from lightning helps protect the house itself, as electric current may otherwise travel through the tree to the house, seeking a ground. If the house features lightning protection, it should share a common ground with the tree.

Similar to that for a building, lightning protection for a tree (Fig. 9.4.6) consists of these copper components: air terminal points (rods); down conductor cables; copper fasteners; and adequate grounding, usually 10' rods driven into the earth. Grounding is buried while terminals are positioned at high points and cables are installed along the trunk and main boughs with fasteners driven through the bark. Since specialized equipment is required, installation and testing should be performed by a professional lightning protection company to comply with Underwriters' Laboratories, National Fire Protection Association, and Lightning Protection Institute standards. The National Arborists Association is updating its own lightning protection standards for trees, which promises to be most comprehensive. For further information, refer to the section on "Lightning Protection" in Chapter 7.

ADVANTAGES: Systems can protect large or prize trees from lightning damage. Does not harm tree.

DISADVANTAGES: Improperly installed or poorly grounded systems may not provide protection.

ENERGY-EFFICIENT AND SUSTAINABLE LANDSCAPING

ESSENTIAL KNOWLEDGE

In a variety of climates, trees and other plants can have great effect on house comfort in both summer and winter. Properly selected, trees can shade harsh summer sun and permit winter rays to shine into

FIGURE 9.4.5 TREE LIGHTNING STRIKE

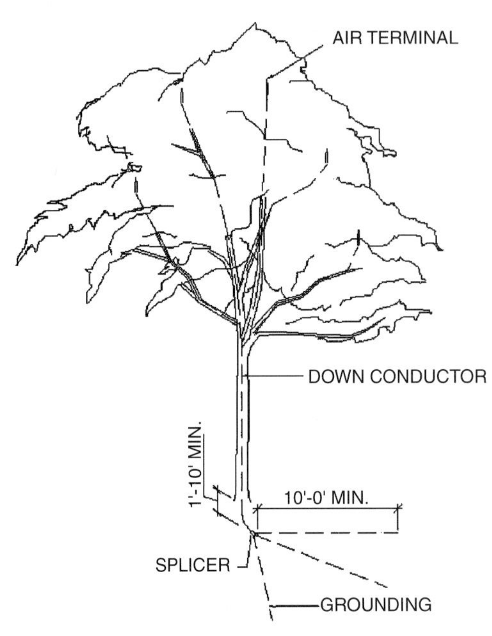

FIGURE 9.4.6 LIGHTNING PROTECTION FOR TREES

southfacing glass. Trees naturally aid in summer cooling by photosynthesis: They absorb solar energy and store it as molecular bonds in wood fiber; they transpire water vapor that causes cooling as it evaporates; they provide shade and temperature differentials which promote breezes; and trees absorb carbon dioxide that would otherwise add to global warming. Trees and shrubs can also aid as wind shields where exposure to wind is a problem.

TECHNIQUES, MATERIALS, TOOLS

1. LOCATE TREES AND SHRUBS FOR SHADE.

In addition to awnings and covered porches, carefully placed trees and shrubs can keep a home much cooler in summer. Trees, shrubs, and vines all provide shade to moderate summer heat gain. Such plants also create a cool microclimate that can dramatically reduce the temperature by as much as 9°F in the surrounding area. During photosynthesis, large amounts of water vapor escape through leaves, cooling the passing air. The dark, coarse leaves absorb solar radiation. Deciduous trees (trees that lose their leaves each year) offer one of the best ways to cut solar gain, by shading roofs, walls, windows, and air-conditioning units. Trees can reduce cooling loads by 20% to 40%. Deciduous trees also allow winter sun to penetrate and warm the house. The northeast–southeast and northwest–southwest sides of the home are the best locations for shade tree placement. In regions where it is hot year-round, trees can be planted to shade the southern exposures. Shrubs planted a few feet away from the house can provide extra shade without obstructing air currents. Vine trellises can shade windows or a wall of the house, and should be located away from the wall to allow air to circulate. Vegetation too close to the house can trap heat and make the air around the house even warmer.

ADVANTAGES: Carefully placed trees are an effective way to promote energy conservation for homes through natural means.

DISADVANTAGES: It may take many years for trees and shrubs to reach an effective size.

2. LOCATE TREES AND SHRUBS AS WIND BUFFERS.

Trees and shrubs can be used to mitigate the effects of cold in the winter. Trees and bushes can buffer winds, mitigating cold air infiltration, which helps to lower heating costs. Evergreens, such as pine and spruce, should be densely planted for windbreaks, which can reduce wind speeds by 80%, over a distance equal to twice the height of the windbreak. Trees and shrubs can also be used to create loosely planted windbreaks that are more effective in cutting winds over greater distances. Loosely planted windbreaks, made up of conifers and decidious trees, can slow wind speeds by 40% over a distance five times the height of the windbreak.

ADVANTAGES: Trees and shrubs are an effective way to promote energy conservation for homes through natural means.

DISADVANTAGES: It may take many years for trees and shrubs to reach an effective size.

3. WATER PLANTS WITH CONSERVATION IN MIND.

Watering of trees, shrubs, and ground cover should take place in the early morning or the evening hours when temperatures are lower and winds are at a minimum to help reduce water loss due to evaporation. Water slowly for long periods of time (drip water is best) to encourage deep rooting. If possible, a drip irrigation system should be used. As a result, plants will have great reservoirs of soil area for drawing moisture and anchoring themselves. Avoid overwatering, as this will kill vegetation. Use moisture-conserving mulch, such as ground bark or other composted material, to cool soil and reduce evaporation.

ADVANTAGES: Conserves water; promotes healthy plants.

DISADVANTAGES: Drip irrigation systems can be costly.

4. PROMOTE SUSTAINABLE LANDSCAPING.

Landscape choices should be made that promote sustainability. For example, use low-maintenance, drought-resistant plants, shrubs, and ground cover to reduce water consumption. Avoid the use of chemical fertilizers and pesticides on lawns and landscaping elements. Native plant species often require little or none of these treatments. Avoid landscaping materials that leach pollutants into soils, groundwater, or nearby streams. For example, creosote or CCA-treated landscaping ties can be replaced with nontreated woods or with newer, less toxic pressure-treated timbers. Use reclaimed water (from retention ponds or cisterns, for example) for landscape irrigation.

ADVANTAGES: Conserves water and natural resources.

DISADVANTAGES: Alternative plantings may not be readily available.

5. REPLACE LAWN WITH LOW-IMPACT VEGETATION.

There are alternatives to a manicured lawn that are attractive and do not require fertilizer, special watering, and seldom or never need cutting. Buffalo grass is one such alternative to sod that may be suitable in many central and western U.S. locations. Another alternative to the turf lawn is moss. Many varieties of moss will provide green, maintenance-free ground cover that looks good in all seasons and, once established, needs no tending, feeding, watering, or cutting and can actually benefit from light foot traffic.

ADVANTAGES: Conserves water and maintenance costs; eliminates run-off of chemicals into streams and wells.

DISADVANTAGES: May not be an esthetically acceptable alternative.

FURTHER READING

CHAPTER 1: FOUNDATIONS

EXISTING FOUNDATION OVERVIEW

CONCRETE

Building Movement and Joints, EB086, PCA, 1982.

Casorso, Greg, "Foundation Replacement," *Journal of Light Construction*, Sept. 1, 1994.

Concrete Construction Handbook, McGraw-Hill, 1993.

Design and Control of Concrete Mixtures, EB001, PCA, 1994 (rev.).

Guide to Residential Cast-in-Place Concrete Construction, ACI 332R-84, American Concrete Institute, 1984.

Residential Concrete, National Association of Home Builders Research Center, 1994.

CONCRETE BLOCK

Concrete Masonry Handbook for Architects, Engineers, and Builders, Portland Concrete Association, 1991.

Design of Concrete Masonry Foundation Walls, TEK 15-1A, National Concrete Masonry Association, 1994.

Homeowners Guide to Building with Concrete, Brick and Stone, SP038, PCA, 1988.

NCMA Guide for Home Owners and Home Builders on Residential Concrete Masonry Basement Walls, National Concrete Masonry Association, 1994.

Recommended Practice for Laying Concrete Block, Portland Cement Association, 1993.

Strength Design of Reinforced Concrete Masonry Foundation Walls, TEK 15-2A, National Concrete Masonry Association, 1994.

BASEMENT FLOORS

Concrete Slab Surface Defects: Causes, Prevention, Repair, IS177, Portland Cement Association, 1987.

Guide for Concrete Floor and Slab Construction, 302.IR, American Concrete Institute, 1996.

Residential Concrete, National Association of Home Builders Research Center, 1994.

Resurfacing Concrete Floors, IS144, Portland Cement Association, 1996.

Slabs On Grade, CCS-1, American Concrete Institute, 1994.

PERMANENT WOOD FOUNDATION SYSTEMS

Labs, Ken, et al., *Building Foundation Design Handbook*, prepared for the U.S. Dept. of Energy by the Underground Space Center, University of Minnesota, 1988.

Permanent Wood Foundations, Red Deer, Alberta, Canada: Sure-West Publishing 1998.

Permanent Wood Foundation Design & Construction Guide, Southern Pine Council, 1995.

PREFABRICATED FOUNDATIONS

NAHB Research Center, "Precast Concrete Stud Foundation Wall," *New Building Products*, vol. 2, no. 5, August/September 1996, p. 1.

SURFACE AND SUBSURFACE DRAINAGE

Investigating, Diagnosing & Treating Your Damp Basement, Canada Mortgage and Housing Corp., 1992.
Stormwater Control to Prevent Basement Flooding, Canada Mortgage and Housing Corp., March 1992.

FOUNDATION DRAINAGE

Carmody, John, et al., *Builder's Foundation Handbook*, prepared for the U.S. Dept. of Energy by the Underground Space Center, University of Minnesota, 1991.
Investigating, Diagnosing & Treating Your Damp Basement, Canada Mortgage and Housing Corp., 1992.
Methods of Constructing Dry, Fully Insulated Basements, Canada Mortgage and Housing Corp., 1991.

DAMPROOFING AND WATERPROOFING

Carmody, John, et al., *Builder's Foundation Handbook*, prepared for the U.S. Dept. of Energy by the Underground Space Center, University of Minnesota, 1991.
Development of an Interior Dampproofing Strategy to Prevent Basement Wall Condensation During Curing, Canada Mortgage and Housing Corp., 1994.
Flannigan, B., and P. R. Morey, *Control of Moisture Problems Affecting Biological Indoor Air Quality*, International Society of Indoor Air Quality and Climate, 1996.
Investigating, Diagnosing & Treating Your Damp Basement, Canada Mortgage and Housing Corp., 1992.
Joints in Walls Below the Ground, CR059 PCA, 1982.
Kerbal, Michael T., *Waterproofing the Building Envelope*, McGraw-Hill, 1993.
Methods of Constructing Dry, Fully Insulated Basements, Canada Mortgage and Housing Corp., 1991.
Moisture Problems, Canada Mortgage and Housing Corp., 1995.
Preventing Water Penetration in Below-Grade Concrete Masonry Walls, TEK 19-3A, NCMA, 1994.
Preventing Wet Basements, CTT, vol. 17, no. 1, PL 961, PCA, 1996.
Trtechsel, Heinz R., ed., *Moisture Control in Buildings*, American Society for Testing and Materials, 1994.

INSULATING BELOW-GRADE WALLS

Carmody, John, et al., *Builder's Foundation Handbook*, prepared for the U.S. Dept. of Energy by the Underground Space Center, University of Minnesota, 1991.
Concrete Masonry Handbook, Portland Cement Association, 1991.
Fine Homebuilding on Foundations and Masonry, Taunton Press, 1990.

Labs, Ken, et al., *Building Foundation Design Handbook*, prepared for the U.S. Dept. of Energy by the Underground Space Center, University of Minnesota, 1988.
"Preventing Termite Damage," *Lite-Form International Technical Bulletin,* August 1996.
"Should You Insulate the Basement?" *Journal of Light Construction*, June 1992.

INSULATING CRAWL SPACES

Carmody, John, et al., *Builder's Foundation Handbook*, prepared for the U.S. Dept. of Energy by the Underground Space Center, University of Minnesota, 1991.
Labs, Ken, et al., *Building Foundation Design Handbook*, prepared for the U.S. Dept. of Energy by the Underground Space Center, University of Minnesota, 1988.

INSULATING SLABS

Concrete Floors on Ground, EB075, Portland Cement Association, 1997.
Design Guide for Frost-Protected Shallow Foundations, 2d ed., NAHB Research Center, Inc., 1996.
Frost Protected Shallow Foundations, NAHB Builder's Press, 1994.
Slabs on Grade, American Concrete Institute, 1994.

VENTILATING BASEMENT SPACES

Clean Air Guide, Canada Mortgage and Housing Corp., 1993.
Flannigan B., and P. R. Morey, *Control of Moisture Problems Affecting Biological Indoor Air Quality*, International Society of Indoor Air Quality and Climate, 1996.
Indoor Air Quality and Climate: Investigation, Evaluation and Remediation, Indoor Air Technologies, Ottawa, ON, Canada KIVOW2.
Lstiburek, Joseph, with John Carmody, *Moisture Control Handbook*, U.S. Dept. of Energy, 1991.
Trechsel, Heinz R., ed., *Moisture Control in Buildings*, ASTM, 1994.

VENTILATING CRAWL SPACES

"Investigation of Crawl Space Ventilation and Moisture Control Strategies for British Columbia Homes," Canada Mortgage and Housing Corp., 1991.
Lstiburek, Joseph, with John Carmody, *Moisture Control Handbook*, U.S. Dept. Of Energy, 1991.
"Recommended Practices for Controlling Moisture in Crawl Spaces," Technical Data Bulletin, vol. 10, no. 3, January 1994, ASHRAE Symposium.
Trechsel, Heinz R., ed., *Moisture Control in Buildings*, ASTM, 1994.

VENTILATION FOR SOIL GASES

Crosbie, Michael J., "Selected Detail: Radon Abatement," *Progressive Architecture*, November 1994, p. 125.
Guide to Radon Control, Canada Mortgage and Housing Corp., 1990.
"Model Standards and Techniques for Control of Radon in New Residential Buildings," March 1994.
"Radon-Resistant Construction Techniques for New Residential Construction: Technical Guidance," U.S. Environmental Protection Agency, 1991.

SHORING, UNDERPINNING, AND REPAIR

Handy, Richard C., Ph.D., *The Day the House Fell*, American Society of Civil Engineering Press, 1995.
Litchfield, Michael W., *Renovation: A Complete Guide*, Prentice-Hall, 1991.
Wray, Warren K., ed., *So Your House is Built on Expansive Soil*, American Society of Civil Engineering Press, 1995.

CRACKS IN WALLS AND SLABS

Causes, Evaluation and Repair of Cracks in Concrete Structures, publication 244.1 R-93, American Concrete Institute, 1993.
Concrete Repair and Maintenance Illustrated, Peter H. Emmons, R.S. Means, 1993.
Concrete Repair Guide, Report of ACI Committee 546-R97, American Concrete Institute.
Guide for Evaluation of Concrete Structures Prior to Rehabilitation, publication 364.1 R-94, American Concrete Institute, 1994.
"Preventing Building Joint Sealant Failures," *Concrete Technology Today*, vol. 15, no. 2, PCA, 1994.
Who's Who in Concrete Repair: Membership Directory, International Concrete Repair Institute, October 1995.

COATINGS AND FINISHES

Gozdan, Walter J., "Specifying Quality Architectural Paints," *Construction Specifier*, July 1996, pp. 44–47.
Painting Concrete, 15134, Portland Cement Association, 1992.
Smead, Tom, "Preventing Masonry Failures with Proper Coating Choices," *Construction Specifier*, July 1996, pp. 48–53.
Water Repellents for Concrete Masonry Walls, TEK 19-1, National Concrete Masonry Association, 1995.

CHAPTER 2: EXTERIOR WALLS

EXTERIOR WALL OVERVIEW

Bock, Gordan, "The Structure of Wood Frame Homes," *Old-House Journal*, March/April 1992.

WOOD-FRAME SEISMIC RESISTANCE

Crosbie, Michael J., *Buildings at Risk: Wind Design Basics for Practicing Architects*, Washington, D.C.: American Institute of Architects, 1998.
A Guide to Strengthening and Repairing Your Home Before the Next Earthquake, developed by the Governor's Office of Emergency Services, State of California, and the Federal Emergency Management Agency, revised May 1995. Association of Bay Area Governments (ABAG) Publications, P.O. Box 2050, Oakland, CA 94604-2050.
How the City of San Leandro Can Help Strengthen Your Home for the Next Big Earthquake in the Bay Area (publication includes prescriptive details). The City of San Leandro Development Services, Building Regulation Division, 835 East 14th St., San Leandro, CA 94577; 510-577-3405.

Hyman, Harris, P.E., "Bracing Walls Against Racking," *Journal of Light Construction*, April 1987.
An Ounce of Prevention: Strengthening Your Wood Frame House for Earthquake Safety: A Do-It-Yourself Program. Video and book. Governor's Office of Emergency Services, State of California, 1993. Association of Bay Area Governments (ABAG) Publications, P.O. Box 2050, Oakland, CA 94604-2050.
Prescriptive Seismic Strengthening Plan for Wood Frame Residential Structures, Simpson Strong-Tie Co., 1996.
The San Francisco Bay Area—On Shaky Ground. Association of Bay Area Governments, 1995 (multi-media CD-ROM, 1996). Association of Bay Area Governments (ABAG) Publications, P.O. Box 2050, Oakland, CA 94604-2050.
Scoggins, John, "Tying Down the House," *Journal of Light Construction*, September 1997.
Structural Strengthening for Seismic Conditions (Video 1997), Simpson Strong-Tie® Co.

WOOD-FRAME WIND RESISTANCE

Crosbie, Michael J., *Buildings at Risk —Wind Design Basics for Practicing Architects*, Washington, D.C.: American Institute of Architects, 1998.

MOISTURE DETERIORATION

Lichfield, Michael W., *Renovation: A Complete Guide,* 2d ed., New York: Prentice Hall, 1991.
Nash, George, *Renovating Old Houses*, Newtown, CT: Taunton Press, 1996.
Poore, Patricia, ed., *The Old House Journal Guide to Restoration*, New York: Dutton, 1992.
Rehabilitation of Wood-Frame Homes, USDA, Forest Service, Handbook No. 704, February 1998.
"Saving House Sills," "Structural Repair Under Old Floors," *Old House Journal*, March/April 1992.
"The Structure of Wood-Frame Houses," *Old House Journal*, March/April 1992.

MITIGATING INSECT DAMAGE

The Approved Reference Procedures for Subterranean Termite Control, National Pest Control Associations (NPCA), 1991.
Mallis, Arnold, *Handbook of Pest Control*, 8th ed., Mallis Handbook & Technical Training Company, 1997.

FIRE DAMAGE TO WOOD FRAMING

Evaluation, Maintenance, and Upgrading of Wood Structures, American Society of Civil Engineers.
"Research Sheds New, Unfavorable Light on Ozone Generators," IEQ Strategies, P.O. Box 129, Center Strafford, NH 03815-0129; 603-664-6942; www.cutter.com/energy/.
NIDR—Guidelines for Fire and Smoke Damage Repair, National Institute for Disaster Restoration (a division of the Association of Specialists in Cleaning and Restoration); 1997; 410-729-9900; www.ascr.org.
Zlotnik, Clifford B., *Odor Removal Manual,* Vol. I, Unsmoke Systems, Inc.
Zlotnik, Clifford B., *Restoration Technology,* Vol. I, Unsmoke Systems, Inc.

CLEAN EXISTING MASONRY WALLS

"Cleaning Brick Masonry," BIA Technical Note #20, Rev. 11, November 1990.
Mark, Robert C., "The Cleaning and Waterproof Coating of Masonry Buildings," *Preservation Briefs*, No. 1, National Park Service, Washington, D.C., 1975.
"Moisture Resistance of Brick Masonry," BIA Technical Note #7F, January 1987.
Sivinski, Valerie, "Gentle Blasting—New Methods of Abrasive Cleaning," *Old House Journal*, July/Aug. 1996.

APPLY COLORLESS PROTECTIVE COATINGS

"Colorless Coatings for Brick Masonry," BIA Technical Note #6A, April 1995.

REPOINT EXISTING MASONRY WALLS

"Mortars for Brick Masonry," BIA Technical Note #8, revised August 1995.
"Repointing (Tuckpointing) Brick Masonry," BIA Engineering and Research Document #622.

REPAIR EXISTING MASONRY WALLS

"Anchored Brick Veneer Wood Frame Construction," BIA Technical Note #28, revised August 1991.
"Brick Masonry Cavity Walls—Detailing," BIA Technical Note #21B, January 1987.
"Brick Veneer Existing Construction," BIA Technical Note #28A, September 1988.
"Guideline on the Rehabilitation of Walls, Windows, and Roofs," U.S. Department of Housing and Urban Development, 1986.
"Guidelines for Residential Building Systems Inspection," U.S. Department of Housing and Urban Development, 1986.
Mark, Robert C., *Preservation Briefs 1: The Cleaning and Waterproof Coating of Masonry Buildings*, U.S. National Park Service, Nov. 1975.
Mark, Robert C., Patterson Tiller, and James S. Askins, *Preservation Briefs 2: Repoint Mortar Joints in Historic Buildings*, U.S. National Park Service, September 1980.
Swanson, Rob, "Brick Veneer Basics," *Journal of Light Construction*, June 1994.
Thomas, Steve, "Getting Started with Brick Veneer," *Journal of Light Construction*, November 1997.

SHEATHING

House Building Basics, APA-Engineered Wood Association, APA Form X461, 1997.
Oriented Strand Board, APA-Engineered Wood Association, APA Form W410, April 1996.
OSB Performance by Design, Structural Board Association, 1997/98.
Residential & Commercial Design/Construction Guide, APA–Engineered Wood Association, APA Form E30, April 1996.

VAPOR RETARDERS

ASHRAE Handbook of Fundamentals, Chapters 22–24. Atlanta: American Society of Heating, Refrigeration and Air Conditioning Engineers, 1997.

AIR INFILTRATION BARRIERS

Cushman, Ted, ed., "Can Moisture Beat Housewrap?", *Journal of Light Construction*, June 1997, p. 9.
Fishett, Paul, "Housewrap vs. Felt," *Journal of Light Construction*, November 1998.
Greenlaw, Bruce, "Sizing Up Housewraps," *Fine Homebuilding*, October/November, no. 91, 1994, p. 42.
"Housewraps or Building Paper? No Perfect Answer," *Energy Design Update*, July 1998.
"Housewrap Manufacturers Prescribe New Details for Windows and Doors," *Energy Design Update*, August 1998.

INSULATION

ASHRAE, 1997 Handbook of Fundamentals, Inch-Pound Edition, Chapter 22: Thermal and Moisture Control in Insulated Assemblies—Fundamentals; Chapter 23: Thermal and Moisture Control in Insulated Assemblies —Applications; Chapter 24: Thermal and Water Vapor Transmission Data, American Society of Heating, Refrigerating and Air-Conditioning Engineers, Inc., Atlanta, GA; 404-636-8400; www.ashrae.org.
Cushman, Ted, "Fiberglass vs. Cellulose: Making the Choice," *Journal of Light Construction*, September 1995, pp. 27–31.
Energy Source® Catalog; Iris Communications, Inc.; 800-346-0104.
Energy Star Fact Sheets; Builder Guides; U.S. Environmental Protection Agency (EPA), Atmospheric Pollution Prevention Division; 888-STAR-YES.
Home Energy's Consumer Guide to Insulation; www.homeenergy.org/eehem/96/9609insulation.html.
"Insulation Fact Sheet," August 1997, Document DOE/CE-0180.
"Loose-Fill Insulations," May 1995, Document DOE/GO-10095-060.
Residential Energy Efficiency Database (REED); http://139.142.36.88/reed/index.htm.
Sawnee Energy Library; /www.energydepot.com/sawnee/library/library.cfm.
U.S. Department of Energy (DOE) Energy Efficiency and Renewable Energy Clearing House (EREC); 800-363-3732; e-mail: doe.erec@nclinc.com; http://erecbbs.nclinc.com.
Wilson, Alex, "Insulation Materials: Environmental Comparisons," *Environmental Building News*, vol. 4, no. 1, January/February 1995.
Wilson, Alex, "Insulation Comes of Age," *Fine Homebuilding*, no. 100, February/March 1996.

VINYL SIDING

"Application Instructions," Vinyl Siding Institute, the Society of Plastics Industry Inc., 1994.
"Cleaning of Vinyl Siding," Vinyl Siding Institute, the Society of Plastics Industry Inc., 1994.
"Fire Properties," Vinyl Siding Institute, the Society of Plastics Industry Inc., 1994.
"Installation Guide for Vinyl Siding and Accessories," Wolverine/Certainteed, 1998; 888-838-8100; www.vinylsiding.com.
Schamback, George, "Vinyl Siding," *The Journal of Light Construction*, June 1997, pp. 27–31.

WOOD SHINGLES AND SHAKES

Cedar Shake and Shingle Bureau Design and Application Manual for Exterior and Interior Walls, Cedar Shake and Shingle Bureau, 1991.
Cedar Shake and Shingle Bureau Membership Directory and Buyers Guide, Cedar Shake and Shingle Bureau, January 1998.
A Guide to Southern Pine Shakes, Kenner, LA: Southern Forest Products Association, 1994.
The Sovebec Guide to Installing Eastern White Cedar Shingles, Charney, Canada: Sovebec, Inc.

SOLID WOOD SIDING

Certified Kiln Dried Siding Applications, Novato, CA: California Redwood Association, 1995; 888-225-7339; www.calredwood.org.

Designer's Handbook, Vancouver, BC: Western Red Cedar Lumber Association, November. 1997; 604-684-0266; www.wrcla.org.

Guidelines for Installing and Finishing Wood and Hardboard Siding Over Foam Sheathing, American Forest and Paper Association; 202-463-2700; www.afandpa.org.

Installing Cedar Siding, Western Red Cedar Lumber Association; 604-684-0266; www.wrcla.org.

Natural Wood Siding—Technical Guide, Portland, OR: Western Wood Products Association, May 1998; 503-224-3930; www.wwpa.org.

Redwood Architectural Guide, Novato, CA: California Redwood Association, 1998; 888-225-7339; www.calredwood.org.

Redwood Lumber Grades and Uses, Novato, CA: California Redwood Association, December. 1995; 888-225-7339; www.calredwood.org.

Specifying Cedar Siding, Vancouver, BC: Western Red Cedar Lumber Association, May 1998; 604-684-0266; www.wrcla.org.

Using Redwood Siding Over Rigid Foam Insulation, Novato, CA: California Redwood Association, September. 1994; 888-225-7339; www.calredwood.org.

HARDBOARD SIDING

ANSI/AHA 135.6-1990 Hardboard Siding, Palatine, IL: American Hardboard Association, 1998.

Maintenance Tips for Hardboard Siding, Palatine, IL: American Hardboard Association, 1997.

Recommended Basic Application Instructions for Hardboard Siding, Palatine, IL: American Hardboard Association, 1994.

Today's Hardboard, Palatine, IL: American Hardboard Association, February 1998.

ENGINEERED WOOD SIDING

Application Instructions for Smart Panel™ & Smart Lap™ Siding, Louisiana-Pacific, March 1998.

PLYWOOD PANEL SIDING

303 Siding Manufacturing Specifications, APA-Engineered Wood Association, March 1997.

Performance Rated Siding, APA-Engineered Wood Association, April 1996.

Residential Design and Construction Guide, APA-Engineered Wood Association, April 1996.

FIBER-CEMENT SIDING

Frane, David, "On Site with Fiber-Cement Siding," *Journal of Light Construction*, January 1998.

EXTERIOR INSULATION AND FINISH SYSTEMS

Culpepper, Steven, "Synthetic Stucco," *Fine Homebuilding*, October/November, 1996.

EIFS Design Handbook, CMD Associates, 1800 Westlake Avenue North, Suite 203, Seattle, WA 98109; 206-285-6811; www.eifs.com/aboutcmd.htm.

EIFS New Construction Inspection Manual, CMD Associates, 1800 Westlake Avenue North, Suite 203, Seattle, WA 98109; 206-285-6811; www.eifs.com/aboutcmd.htm.

EIFS Restoration Guide, Dow Corning #62-510B096; 517-496-6000.

"EIFS Wall Weathers 75-mph Wet Spray Test with No Sealers," *Energy Design Update*, October 1998, pp. 10–12.

Fisette, Paul, "Housewraps vs. Felt," *Journal of Light Construction*, November 1998.

Installation Checklist, Exterior Insulation Manufacturer's Association (EIMA), 3000 Corporate Center Drive, Suite 270, Morrow, GA 30260; 800-294-3462; www.eifsfacts.com.

Minkovich, Russ, "Installing Water-Managed Synthetic Stucco," *Journal of Light Construction*, September. 1998. PN

"Sill Sentry Designed to Treat Leaking Windows," *Energy Design Update*, June 1998, pp. 12–13.

STUCCO

Oriental Stucco, U.S. Gypsum Co., P682, April 1993.

Portland Cement Plaster Stucco Manual, Portland Cement Association, No. EB049, 1996.

Webber, Ron, "Patching Stucco," Ron Webber, *Journal of Light Construction*, September 1997.

EXTERIOR TRIM

Paul Fisette, "Alternatives to Solid Wood Exterior Trim," Building Materials and Wood Technology Program, University of Massachusetts at Amherst, www.umass.edu/bmatwt/index.html.

The Secretary of the Interior's Standards for Rehabilitation and Guidelines for Rehabilitating Historic Buildings, Washington: U.S. Dept. of the Interior, National Park Service Preservation Assistance Division, 1990.

SEALANTS AND CAULKS

Brendenberg, Al, "Caulking," *Today's Homeowner*, www.todayshomeowner.com/todayarticles/paint/04.97.62.html.

Garskof, Josh, "Caulking About the Weather," *Old-House Journal*, November/December 1996.

Greenlaw, Bruce, "Caulks and Sealants," *Fine Homebuilding*, June/July 1990.

Nicastro, David H., and Joseph P. Solinski, "Premature Sealant Failure, *The Construction Specifier*, April 1997.

Zavitz, Brian, "Silicone Caulking Basics," *Fine Homebuilding*, August/September 1997.

PAINT AND OTHER FINISHES

Berney, James, Dan Greenough, and Doug Kelly, "Latex Enamel Problems and Solutions," *Journal of Light Construction*, January 1997.

"Finishes for Exterior Wood—Selection, Application, and Maintenance," U.S. Department of Agriculture, Forest Service, Forest Products Laboratory, Forest Products Society, Madison, WI; 608-231-2152.

Lemley, Brad, "The Art of Painting," *This Old House*, June 1998.

CHAPTER 3: ROOFS

TYPICAL FRAMING ERRORS

"Advanced Framing—Techniques, Troubleshooting and Structural Design," *Journal of Light Construction*, 1992.

Bock, Gordon, "The Structure of Wood Frame Homes," *Old-House Journal*, March/April 1992.

Litchfield, Michael W., *Renovation: A Complete Guide, 2d ed.*, New York: Prentice Hall, 1991.

Nash, George, *Renovating Old Houses*, Newtown, CT: Taunton Press, 1996.

Poore, Patricia, ed., *The Old House Guide to Restoration*, New York: Dutton, 1992.

Troubleshooting Guide to Residential Construction, The Journal of Light Construction, 1997.

Wood-Frame Construction Problems, Canadian Mortgage and Housing Corporation, 1989, revised 1995.

WOOD-FRAME WIND RESISTANCE

Building in High Wind and Seismic Zones—Design Concepts, APA—The Engineered Wood Association, 1997.

Buildings at Risk: Wind Design Basics for Practicing Architects, American Institute of Architects, 1735 New York Avenue, NW, Washington, D.C. 20006-5292.

Coastal Construction Manual, Federal Emergency Management Agency (FEMA), 500 C Street, SW, Washington, D.C. 20006; 800-480-2520; www.fema.gov; February 1986 (being revised).

Is Your Home Protected from Hurricane Disaster—A Homeowner's Guide to Hurricane Retrofit, Institute for Business and Home Safety, 73 Tremont Street, Suite 510, Boston, MA 02108-3910; 617-292-2003; www.ibhs.org; 1997.

Standard for Hurricane Resistant Residential Construction—SSTD 10-96, Southern Building Code Congress International, Inc., 900 Montclair Road, Birmingham, AL 35213-1206; 205-591-1853.

Wood Frame Construction Manual for One and Two-Family Dwellings—High Wind Edition, 1995, SBC, American Forest & Paper Association—American Wood Council, 1111 19th Street, NW, Washington, D.C. 20006; 202-463-2700; www.afandpa.org.

WOOD-FRAME SEISMIC RESISTANCE

Buildings at Risk: Seismic Design Basics for Practicing Architects, AIA/ACSA Council on Architectural Research, National Hazards Research Program, 1735 New York Avenue, NW, Washington, D.C. 20006, 1994.

A Guide to Strengthening and Repairing Your Home Before the Next Earthquake, Developed by the Governor's Office of Emergency Services, State of California, and the Federal Emergency Management Agency, revised May 1995. Association of Bay Area Governments (ABAG) Publications, P.O. Box 2050, Oakland, CA 94604-2050; www.abag.ca.gov.

How the City of San Leandro Can Help Strengthen Your Home for the Next Big Earthquake in the Bay Area (publication includes prescriptive details). The City of San Leandro Development Services, Building Regulation Division, 835 East 14th Street, San Leandro, CA 94577; 510-577-3405.

An Ounce of Prevention: Strengthening Your Wood Frame House for Earthquake Safety: A Do-It-Yourself Program, video and book. Governor's Office of Emergency Services, State of California, 1993. Association of Bay Area Governments (ABAG) Publications, P.O. Box 2050, Oakland, CA 94604-2050; www.abag.ca.gov.

Prescriptive Seismic Strengthening Plan for Wood-Frame Residential Structures, Simpson Strong-Tie Company, Inc; www.strongtie.com.

The San Francisco Bay Area—On Shaky Ground. Association of Bay Area Governments, 1995 (multimedia CD-ROM, 1996). Association of Bay Area Governments (ABAG) Publications, P.O. Box 2050, Oakland, CA 94604-2050; www.abag.ca.gov.

Scoggins, John, "Tying Down the House," *Journal of Light Construction*, September 1997.

FIRE DAMAGE

Evaluation Maintenance and Upgrading of Wood Structure, American Society of Civil Engineers.

NIDR—Guidelines for Fire and Smoke Damage Repair, National Institute of Disaster Restoration (a division of the Association of Specialists in Cleaning and Restoration), 1997; 301-684-4411; www.ascr.org.

SHEATHING

House Building Basics, APA Form X461, APA—The Engineered Wood Association, 1997.

Oriented Strand Board, APA Form W410, APA—The Engineered Wood Association, April 1996.

Residential & Commercial Design/Construction Guide, APA Form E30, APA—The Engineered Wood Association, 1996.

Roof Alterations and Renovations, APA Form M325, APA—The Engineered Wood Association, 1991.

FLASHING

Architectural Sheet Metal Manual, Sheet Metal and Air Conditioning Contractor's National Association (SMACNA).

Copper in Architecture, Copper Development Association, Inc.

Copper and Common Sense, Revere Copper Products, Inc., P.O. Box 300, Rome, NY 13442-0300.

Flashings and Weatherings, Lead Industries Association, www.leadinfo.com.

Gutters, Flashing and Roof Details, video, Copper Development Association, Inc.

Old House Journal—Guide to Restoration, Old House Journal, 1992.

Rheinzink Roofing and Wall Cladding, Rheinzink Canada Ltd., www.rheinzink@rheinzink.com.

Roofing and Cladding, Lead Industries Association, www.leadinfo.com.

Roofing and Flashing Problems, Canada Mortgage and Housing Corporation, rev. 1995.

SSINA Designer Handbook and Standard Practices for: Stainless Steel Roofing, Fascias and Copings, Specialty Steel Industry of NA, February 1995.

Steep Roofing Manual, National Roofing Contractors Association, 1996.

Troubleshooting Guide to Residential Construction, Journal of Light Construction, 1997.

Working Safely with Sheet Lead, Lead Industries Association, www.leadinfo.com.

UNDERLAYMENTS AND MOISTURE BARRIERS

The NRCA Residential Steep-Roofing Manual, National Roofing Contractors Association, 1997.

Residential Steep-Slope Roofing Materials Guide, National Roofing Contractors Association, 1997.

INSULATION

ASHRAE 1997 Handbook of Fundamentals, Inch-Pound Edition, Chapter 22: Thermal and Moisture Control in Insulated Assemblies—Fundamentals; Chapter 23: Thermal and Moisture Control in

Insulated Assemblies—Applications; Chapter 24: Thermal and Water Vapor Transmission Data, American Society of Heating, Refrigerating and Air-Conditioning Engineers, Inc., Atlanta, GA; 404-636-8400; http//www.ashrae.org.
Cushman, Ted, "Fiberglass vs. Cellulose: Making the Choice," *Journal of Light Construction*, September 1995, pp 27–31; Builderburg Partners, Ltd., Washington, D.C.; 800-552-1951.
Energy Source® Catalog, Iris Communications, Inc.; 800-346-0104.
Energy Star Fact Sheets; Builder Guides; U.S. Environmental Protection Agency (EPA), Atmospheric Pollution Prevention Division; 888-STAR-YES.
Home Energy's Consumer Guide to Insulation; http://www.homeenergy.org/eehem/96/9609insulation.html.
Insulation Fact Sheet, August 1997, Document DOE/CE-0180.
Loose-Fill Insulations, May 1995, Document DOE/GO-10095-060.
Residential Energy Efficiency Database (REED), Information Technology Specialists, Inc.; 403-892-3594; http://139.142.36.88/reed/index.htm.
Sawnee Energy Library; http://www.energydepot.com/sawnee/library/library.cfm.
Southface Energy Institute, P.O. Box 5506, Atlanta, GA 30307; 404-525-7657.
U.S. Department of Energy (DOE) Energy Efficiency and Renewable Energy Clearing House (EREC); 800-363-3732. email: doe.erec@nclinc.com; http://erecbbs.nclinc.com.
Wilson, Alex, "Insulation Comes of Age," *Fine Homebuilding*, no. 100, February/March 1996, Newtown, CT: The Taunton Press, Inc.; 800-283-7252.
Wilson, Alex, "Insulation Materials: Environmental Comparisons;" *Environmental Building News*, vol. 4, No. 1, January/February 1995, West River Communications, Inc., Brattleboro, VT; 802-257-7300.

WOOD SHINGLES AND SHAKES

Cedar Shake and Shingle Bureau Design and Application Manual for New Roof Construction, Cedar Shake and Shingle Bureau, 515 116th Avenue, NE, Suite 275, Bellevue, WA 98004; 425-453-1323; www.cedarbureau.org.
Cedar Shake and Shingle Bureau Membership Directory and Buyer's Guide, Cedar Shake and Shingle Bureau, 515 116th Avenue, NE, Suite 275, Bellvue, WA 98004; 425-453-1323; www.cedarbureau.org.
"A Guide to Southern Pine Shakes," Southern Forest Products Association, P.O. Box 641700, Kenner, LA; 504-443-4464; www.southernpine.com.
Hamilton, Patricia, "Long Lasting Wood Roofs," *Journal of Light Construction*, May 1997.
"Wood Shingle Roofs—Best Job Specifications and Care," *Old House Journal*, May/June 1990, pp. 34–41.

ASPHALT SHINGLES

"Algae Discoloration of Roofs," ARMA Technical Bulletin #217- RR-89.
"Blue Sky Construction Guidelines," Town of Southern Shores, NC.
"Builder Tips—How to Minimize the Buckling of Asphalt Composition Shingles," APA, the Engineered Wood Association, Form No. K310F/revised April 1994; www.apawood.org/buildertips/buckling.html.
Certainteed Shingle Application Manual, Certainteed Corp., P.O. Box 860, Valley Forge, PA 19482, 3d ed., 1997–98.
"Cold Weather Recommendations for Application of Asphalt Roofing Shingles," ARMA Technical Bulletin #225-RR-97.
"Direct Application of Asphalt Shingles Over Insulation or Insulated Decks," ARMA Technical Bulletin #211-RR-86.

"Recommendations for Application of Asphalt Shingles on Steep Slopes and Mansard Construction," ARMA Technical Bulletin #213-RR-87.

"Reroofing: Tear-Off vs. Re-Cover," ARMA Technical Bulletin #223-RR-96.

Residential Asphalt Roofing Manual, Asphalt Roofing Manufacturers Association, 1997, www.asphalt roofing.org.

Residential Steep-Slope Roofing Materials Guide, National Roofing Contractors Association; www.roofonline.org.

"Shingles and Siding: How to Know When You Need Them—Costs, Product Tests," *Consumer Reports*, August 1997, P.O. Box 2015, Yonkers, NY 10703-9015.

The Steep Roofing Manual, National Roofing Contractors Association, www.roofonline.org.

"Ventilation and Moisture Control for Residential Roofing," ARMA Technical Bulletin #209-RR-86.

LOW-SLOPE ROOFING

"Everything You Always Wanted to Know About Flexible Membrane Roofing, but Didn't Know Who to Ask," *The Roofing Specifier*, July 1998.

Flexible Membrane Roofing: A Professional's Guide to Low-Slope Membrane Roofing Construction Details, 4th ed., NRCA Specifications, Single-Ply Roofing Institute, 1998.

Fricklas, Richard C., and C. W. Griffen, *Manual of Low-Slope Roofing Systems*, 3d ed., McGraw-Hill, 1995.

Hardy, Steve, *Time-Saver Details for Roof Designs*, McGraw-Hill, 1998.

Manual for Inspection and Maintenance of Built-up and Modified Bitumen Roof Systems: A Guide for Building Owners, NRCA/ARMA, 1996.

The Manual of Low-Slope Roof Systems, NRCA, 1996.

Modified Bitumen Roofing Systems Design Guide, ARMA, 1997.

The NRCA Roofing and Waterproofing Manual, 4th ed., NRCA, 1996.

Quality Control Guidelines for the Association of Thermoset Single-Ply Roof Membrane, NRCA, March 1996.

Repair Manual for Low-Slope Membrane Roof Systems, ARMA/NRCA/SPRI, 1997.

Residential Asphalt Roofing Manual, ARMA, revised 1997.

SPRI/NRCA Manual of Roof Inspection, Maintenance, and Emergency Repair for Existing Single-Ply Roofing Systems, NRCA/SPRI, April 1992.

METAL ROOFING

"Alternative Roofing Materials," *Journal of Light Construction*, August 1997.

Architecture Sheet Metal Manual, Sheet Metal and Air Conditioning Contractors National Association (SMACNA); www.smacna.org.

Copper in Architecture Handbook, Copper Development Association, Inc.; www.copper.org.

The Metal Roofing Manual, National Roofing Contractors Association, 1996.

Prepainted Metal Roofs, Video, National Coil Coaters Association, 1996.

Roof It With Steel, American Iron & Steel Institute; 800-79- STEEL.

SLATE

Crosbie, Michael J., "A Chip Off the Old Slate," *Construction Specifier,* November 1997, pp. 49–54.

Jenkins, Joseph, *The Slate Roof Bible*, Jenkins Publishing, P.O. Box 607, Grove City, PA 16127; 800-689-3233, 1997.

Laying Slate, National Slate Association (defunct). Available from Buckingham-Virginia Slate Corp. and other slate manufacturers.

The Repair, Replacement and Maintenance of Historic Slate Roofs, Preservation Brief #29, Technical Preservation Service, National Park Service; www2.cr.nps.gov.

Stearns, Brian, Alan Stearns, and John Meyer, *The Slate Book – How to Design, Specify, Install, and Repair a Slate Roof*, Vermont Slate and Copper Services, Inc., 888-766-4273, 1998.

The Steep Roof Manual, National Roofing Contractors Association, 1996.

CLAY, CONCRETE, FIBER-CEMENT, AND COMPOSITE TILE

"Alternative Roofing Materials," *Journal of Light Construction*, August 1997, pp. 33–37.

"Choosing Roofing," *Fine Homebuilding*, January 1995, Vol. 92, pp. 46–51.

The NRCA Steep Roofing Manual, National Roofing Contractors Association, Rosemont, IL.

Polson, Mary Ellen, "Clay in Context," *Old Home Journal*, April 1998, pp. 39–43.

"Roofing Materials: A Look at the Options for Pitched Roofs," *Environmental Building News*, vol. 4, no. 4, July/August 1995.

"Stu's Tile Talk," Northern Roof Tile Sales, Blasdell, NY.

"The Top Choice," National Tile Roofing Manufacturer's Association, Eugene, OR.

GUTTER AND LEADER SYSTEMS

"Architectural Sheet Metal Manual," Sheet Metal and Air Conditioning Contractors National Association (SMACNA), www.smacna.org.

"Copper in Architecture Handbook," Copper Development Association, Inc., www.copper.org.

"Gutters, Flashing and Roof Details," video, Copper Development Association, Inc., www.copper.org.

CHAPTER 4: WINDOWS AND DOORS

EXISTING WINDOW AND DOOR OVERVIEW

ASHRAE Handbook of Fundamentals, Atlanta, GA: American Society of Heating, Refrigerating and Air Conditioning Engineers, 1997.

O'Bannon, James, and Andre Grieco, "Energy-Efficient Window Retrofits: Install with Care," *Home Energy*, January/February 1997.

Carmody, John, Stephen Selkowitz, & Lisa Heschong, *Residential Windows*, New York: W.W. Norton, 1996.

Wardell, Charles, "Shopping for Wood Windows," *Journal of Light Construction*, June 1994.

"Choosing Replacement Windows," *Journal of Light Construction* (New England Edition), February 1993.

DeKorne, Clayton, "Shopping for Entry Doors" *Journal of Light Construction* (New England Edition), December 1991.

Designing Low-Energy Buildings: Passive Solar Strategies and Energy-10 Software, Passive Solar Industries Council, Washington, D.C., Passive Solar Industries Council, 1996.

Engstrom, Paul, and Jeanne Huber, "Worrisome Windows," *This Old House*, January/February 1997.
Fisette, Paul, "Understanding Energy-Efficient Windows," *Fine Homebuilding*, February/March 1998.
Guidelines for the Evaluation and Control of Lead-Based Paint Hazards in Housing, Washington, D.C.: U.S. Department of Housing and Urban Development, 1995.
Jester, Thomas C., *Twentieth Century Building Materials: History and Conservation*, New York: McGraw-Hill, 1995.
Kolle, Jefferson, "Taking a Look at Windows," *Fine Homebuilding*, No. 97.
MacDonald, Marylee, "Shopping for Replacement Windows," *Journal of Light Construction*, June 1989.
McCluney, R., *Choosing the Best Window for Hot Climates*, Florida Solar Energy Center, 1993.
New York Landmarks Conservancy, *Repairing Old and Historic Windows: A Manual for Architects and Homeowners*, New York: John Wiley & Sons, 1992.
Residential Window and Door Installation Guide, Association of Window and Door Installers, 11300 U.S. Highway 1, Suite 400, North Palm Beach, FL 33408-3296.
Shapiro, Andrew M., and Brad James, "Creating Windows of Energy-Saving Opportunity," *Home Energy*, September/October 1997.
Steven Winter Associates, *The Passive Solar Design and Construction Handbook,* New York: John Wiley & Sons, 1998.
Steven Winter Associates, *Accessible Housing by Design: Universal Design Principles in Practice*, New York: McGraw-Hill, 1997.
Wilson, Alex, "Windows: Looking through the Options," *Environmental Building News*, March/April 1996.
Wilson, Alex, "Window Technology Update," *Journal of Light Construction* (New England Edition), December. 1991.
Ziegner, Rich, "Choosing a Front Door," *Fine Homebuilding*, no. 90.
Zingman-Leith, Elan, "Historic Metal Windows: Their Architectural History," *Old-House Journal*, November, 1986.

GLAZING

Carmody, John, Stephen Selkowitz, & Lisa Heschong, *Residential Windows*, New York: W.W. Norton, 1996.
Guide to Avoiding Glass Failure, Association of Industrial Metallizers Coaters and Laminators, November 1996.
Jester, Thomas C., *Twentieth Century Building Materials: History and Conservation*, New York: McGraw-Hill, 1995.
"Low-E Glass—Why the Coating Is Where It Is," *Energy Design Update*, March 1990, pp. 5–7.
"No Pane, No Gain (Window Technology: Part One)," *Popular Science*, June 1993, pp. 92–98.
"Through the Glass Darkly," *Popular Science*, July 1993, pp. 80–87.

WINDOW FRAMES AND REPLACEMENT UNITS

Carmody, John, Stephen Selkowitz, & Lisa Heschong, *Residential Windows*, New York: W.W. Norton, 1996.
New York Landmarks Conservancy, *Repairing Old and Historic Windows: A Manual for Architects and Homeowners*, New York: John Wiley and Sons, 1992.
Richardson, Barry, *Wood Preservation*, 2d ed., Chapman to Hall, 1993.
Wood Protection Guidelines: Protecting Wood From Decay Fungi & Termites, Wood Protection Council, National Institute of Building Science, 1993.

STORM WINDOWS AND SCREENS

Greenlaw, Bruce, "What's The Difference—Door and Window Screening: Aluminum or Fiberglass?" *Fine Homebuilding*, no. 97, September 1995, p.120.

SKYLIGHTS

Carmody, John, Stephen Selkowitz, and Lisa Heschong, *Residential Windows*, New York: W.W. Norton, 1996.

Cotton, J. Randall, "Skylights: The Design and Upkeep of Old Fashioned Rooftop Windows," *Old House Journal*, July/August 1992, pp.42–46.

Milner, Donna, "Skylights: Design & Installation Basics," *Journal of Light Construction* (New England Edition), May 1989, pp. 47–51.

Rieger, Ted, "Tubular Daylighting for Sun Lovers," *Home Energy*, January/February 1997, pp. 9–10.

Wagner, John, "Skylight Options and Accessories," *Journal of Light Construction*, April 1996, pp. 47–50.

Wilk, Sanford, "Skylights: Daylights & Dollars," *Roofer Magazine*, January 1997.

DOORS & FRAMES

Borel, William, "Silent Woods," *Doors and Hardware*, August 1997, pp. 43–44.

Fishburn, Christine, "Door Prize," *Remodeling Magazine*, December 1997, p. 78.

"Fiberglass Bulkhead Door," *Fine Homebuilding*

Kearns, Steve, "Ordering and Installing Prehung Doors," *Fine Homebuilding*, April/May 1992, pp. 62–65.

Koscos, Jim, "Safer, More Secure Garage Doors," *Today's Homeowner Magazine*, http://index.today homeowner.com/todayarticles/door/11.95.16.html.

Leeke, John, "Open Door Policy—Troubleshooting Interior Doors," *Old House Journal*, January/February 1997, pp. 40–43.

Phillips, Bill, "New Break-In Resistant Exterior Doors," *Today's Homeowner Magazine*, www.todays homeowner.com.

Poore, Jonathan, "How to Fix Old Doors," *The Old-House Journal*, June 1986, pp. 222–227.

Screen Manufacturers Assn., 2850 South Ocean Blvd., #114, Palm Beach, Fl, 33480; 561-533-0991.

Ziegner, Rich, "Choosing a Front Door," *Fine Homebuilding*, August/September 1994, pp.40–45.

CASING AND TRIM

Fisette, Paul, "Alternatives to Solid Wood Exterior Trim," Building Materials and Wood Technology Program, University of Massachusetts at Amherst; www.umass.edu/bmatwt/index.html.

The Secretary of the Interior's Standards for Rehabilitation and Guidelines for Rehabilitating Historic Buildings, Washington: U.S. Dept. of the Interior, National Park Service Preservation Assistance Division, 1990.

HARDWARE

Bowers, Dan, "Upgrading Door Locks and Deadbolts," http://index.todayshomeowner.com/todayarticles/door/06.95.30.html.

Katz, Gary M., "Installing Locksets," *Fine Homebuilding*, February/March 1993, pp. 40–45.

NFPA 80-1990, Standard for fire doors and windows, Quincy, MA: National Fire Protection Association.

Swearingen, David, "Locks & Alarms: Old House Security," *The Old-House Journal*, December 1986, pp. 472–475.

Taylor, Timothy, and Michael J. Crosbie, "Natural Selection," *Doors and Hardware*, March 1997, pp. 45–53.

FLASHING

Door and Window Installation–Builder's Series, Canada Mortgage and Housing Corporation, 1995.
Fisette, Paul, "Details That Keep Walls Tight," Building Materials and Wood Technology, University of Massachusetts at Amherst, www.umass.edu/bmatwt/walls.html.
"Flashing Details," *Remodeling*, mid-February 1997, p. 46.
Larson, James R., "How to Avoid Common Flashing Errors," *Fine Homebuilding*, April/May 1998.
Thallon, Rob, *Graphic Guide to Frame Construction*, Newtown, CT: Taunton Press, 1991.
"Window Nail Flange Air Sealing Gasket," *Energy Design Update*, September 1996, p. 11.

CAULKING AND WEATHER STRIPPING

"ANSI A156.22 Door Gasketing Systems," American National Standards Institute, available from the Builders Hardware Manufacturers Association.
Brendenberg, Al, "Caulking," *Today's Homeowner*, http://index.todayshomeowner.com/todayarticles/paint/04.97.62.html.
"Building Sealing and Ventilation, Green Seal's Choose Green Report," March 1997.
Garskof, Josh, "Caulking about the Weather," *Old-House Journal*, November/December 1996.
Greenlaw, Bruce, "Caulks and Sealants," *Fine Homebuilding*, June/July 1990.
Katz, Gary, "Retrofit Weatherstripping with Silicone Bead," *Journal of Light Construction*, November 1993.
Nicastro, David H., and Joseph P. Solinski, "Premature Sealant Failure," *The Construction Specifier*, April 1997.
Rose, Helen, "Choosing Gasketing Products," *The Construction Specifier*, April 1997.
Sanger, John, "The Theory of Caulk," http://www.paintstore.com/book/pwc/PWC_Archives/caulk/103.html.
Brian Zavitz,"Silicone Caulking Basics," *Fine Homebuilding*, August/September 1997.
Old House Journal, "Weatherstripping Entry Doors," December 1986, pp. 470–471.

SHUTTERS AND AWNINGS

Carmody, John, Stephen Selkowitz, and Lisa Heschong, *Residential Windows*, New York: W.W. Norton, 1996.
Lemley, Brad, "Serious Shutters," *This Old House*, July/August 1996, pp. 88–91.

CHAPTER 5: CEILINGS, PARTITIONS,FLOORS, AND STAIRS

FLOOR AND CEILING STRUCTURE

Bauer, Eric, "Steelwork in Wood Frames," *Journal of Light Construction*, April 1992.
Bock, Gordon, "The Structure of Wood Frame Homes," *Old-House Journal*, March/April 1992.
Borgemeister, Peter, "Structural Repair Under an Old Floor," *Old-House Journal*, March/April 1992.

Evaluation, Maintenance, and Upgrading of Wood Structure, American Society of Civil Engineers, 1982.
Fire Resistance Design Manual —Sound Control, Gypsum Association, 15th ed., 1997.
Glued Laminated Timber Appearance Classifications for Construction Applications, APA/EWS Y110, April 1998.
Glulams in Residential Construction, APA-Engineered Wood Systems, December 1995.
Gold Bond^R Gypsum Construction Guide, 5th ed., National Gypsum Company, 1998.
Gypsum Construction Handbook, 3d ed., United States Gypsum Company, 1994.
Hamilton, Patricia, "On Site with Parallam," *Journal of Light Construction*, December 1995.
Handling, Installing and Bracing MPC'ed Wood Trusses, HIB-91, Truss Plate Institute.
Harris, David A., *Noise Control Manual for Residential Buildings*, New York: McGraw-Hill, 1997.
Hyman, Harris, "Taking the Bounce Out of Floors and Beams," in *Advanced Framing Techniques, Troubleshooting and Structural Design*, Journal of Light Construction, 1992.
Litchfield, Michael W., *Renovation: A Complete Guide* (2d Ed.), New York: Dutton, 1992.
Nash, George, *Renovating Old Homes*, Newtown, CT: Taunton Press, 1996.
Nelson, Sherman, P.E., "Structural Composite Lumber," in *Engineered Wood Products—A Guide for Specifiers, Designers and Users*, Madison, WI: PFS Research Foundation, pp. 147–172.
NIDR—Guidelines for Fire and Smoke Damage Repair, National Institute Disaster Restoration (a division of the Association of Specialists in Cleaning and Restoration); 1997; (301) 684-4411; www.ascr.org.
Noise-Rated Systems, APA—Engineered Wood Association, 1994.
Poore, Patricia, ed., *The Old House Journal Guide to Restoration*, New York: Dutton, 1992.
Product and Application Guide, APA-Engineered Wood Systems, June 1995.
Rehabilitation of Wood-Framed Houses, U.S. Department of Agriculture, Forest Service, Agriculture Handbook No. 704, February 1998.
"Research Sheds New, Unfavorable Light on Ozone Generators," *IEQ Strategies*, P.O. Box 129, Center Strafford, NH 03815-0129; 603-664-6942; www.cutter.com/energy/.
Utterbach, David, "Wall and Floor Framing," in *Troubleshooting Guide to Residential Construction*, Journal of Light Construction, 1997.
What We Learned by Framing the American Dream, Wood Truss Council of America, 1996.
"Wood Framing," in *Trouble Shooting Guide to Residential Construction*, Journal of Light Construction, 1997, pp. 34–38.
Zlotnik, Clifford B., *Odor Removal Manual,* vol. I, Unsmoke Systems, Inc., 1997.
Zlotnik, Clifford B., *Restoration Technology,* vol. I, Unsmoke Systems, Inc, 1996.

SHEATHING SUBFLOORING AND UNDERLAYMENTS

"Application Instructions to Hardboard Underlayment," *Basic Hardboard Products*, American Hardboard Association, July 1994.
Builder Tips: Steps to Construct a Solid, Squeak-free Floor System, APA Form Q300, APA—The Engineered Wood Association, P.O. Box 11700, Tacoma, WA 98411-0700; (206) 565-6600; www.apa.org.
Data File: Installation and Preparation of Plywood Underlayment for Resilient Floor Covering, APA Form L335.
Design/Construction Guide: Residential and Commercial, APA Form E30.
Handbook for Ceramic Tile Installation, Tile Council of America, Inc., 1999.
Proper Selection and Installation of APA Plywood Underlayment, APA Form R3400/Rev. October 1995/0010.
Source List: Plywood Underlayment for Use Under Resilient Finish Flooring, APA Form L330.
"Subfloors and Underlayments," *Armstrong Guaranteed Installation Systems*, Armstrong World Industries, Inc., 1995.

FINISH FLOORING

Acids and Acid Cleaning—Friend or Foe, Field Report, Tile Council of America, 1985.
American Institute of Architects, *Environmental Resource Guide*, New York: John Wiley & Sons, Inc., 1997.
Armstrong Guaranteed Installation Systems, Lancaster, PA: Armstrong World Industries, Inc., 1998.
Bollinger, Don, *Hardwood Floors—Laying, Sanding and Finishing*, Newtown, CT: Taunton Press, 1990.
Caring for Your Ceramic Tile, Tile Council of America, Inc., 1996.
Carpet: Installation and Maintenance for Maximum Performance, Carpet and Rug Institute, 1998.
Cleaning and Repointing Old Tile Installations, Tile Council of America, 1985.
CRI-104 Installation Guidelines, Carpet and Rug Institute, 1996.
Finishing Hardwood Flooring, National Oak Flooring Manufacturers Association, March 1998.
Flooring Grade Brochure, National Oak Flooring Manufacturers Association, March 1998.
General Restoration Guidelines for Water-Damaged Carpet, Carpet and Rug Institute, 1998.
Handbook for Ceramic Tile Installation, Tile Council of America, Inc., 1997.
How to Hire a Carpet-Cleaning Professional, Carpet and Rug Institute, 1998.
Installing Hardwood Flooring, National Oak Flooring Manufacturers Association, March 1998.
Installing Vinyl Flooring, Chaska, MN: Hometime, 1994.
Maintaining a Tile Installation, Field Report, Tile Council of America, 1986.
Problems Causes Cures, National Wood Flooring Association, May 1991.
Standard Reference Guide for Professional On-Location Cleaning of Installed Textile Floor Covering Material, Institute of Inspection, Cleaning and Restoration Certification, 1991.
Water and Wood—How Moisture Affects Wood Flooring, National Wood Flooring Association, 1996.
Wood Floor Care Guide, National Oak Flooring Manufacturers Association, November 1997.
Wood Flooring—A Lifetime of Beauty, National Wood Flooring Association.
Wood Species Used in Wood Flooring, National Wood Flooring Association, Technical Publication A200, 1994.

FINISH WALLS AND CEILINGS

Caring for Your Historic House, Washington, D.C.: Heritage Preservation and National Park Service, 1998.
Dixon, Mark, *House Painting Inside and Out*, Newtown, CT: Taunton Press, 1997.
Drywall Application, Canadian Mortgage and Housing Corporation; 1995.
Ferguson, Myron R., *Drywall-Professional Techniques for Walls and Ceilings*, Newtown, CT: Taunton Press, 1996.
Fire Resistance Design Manual, 15th ed., Gypsum Association, 1997.
Gypsum Construction Guide, National Gypsum Company; www.national-gypsum.com; 1999.
Gypsum Construction Handbook, U.S. Gypsum Company; www.usg.com, 1995.
Hamburg, H. R., and W. M. Morgans, *Hess's Paint Film Defects: Their Causes and Cure*, Chapman and Hall, 1979.
How to Install, Finish and Repair SHEETROCK Brand Gypsum Panels, U.S. Gypsum Company, 1994.
"Rx for Latex Paint Problems," *Journal of Light Construction*, January, 1997.
Shivers, Natalie, *Respectful Rehabilitation: Walls and Molding*, Washington, D.C.: The Preservation Press, 1990.
"Stripping Wallpaper," *Old-House Journal*, December, 1997.
Troubleshooting Guide to Residential Construction, Richmond, VT: Editors of the *Journal of Light Construction*, 1997.
Wallpaper in America, New York: Barra Foundation/Cooper Hewitt Museum, 1980.
Weismantel, Guy E., *Paint Handbook*, New York: McGraw-Hill, 1981.

STAIRS

Dixon, Jed, "Trimming Out Stairs," *Journal of Light Construction*, March 1999.
Floors and Stairways, Alexandria, VA: Editors of Time-Life Books, 1995.
Frane, David, "Sturdy Site-Built Stairs," *Journal of Light Construction*, Volume 14, Number 2, November 1995.
Gilbert, Rob Dale, "Trouble-Free Stairs," *Journal of Light Construction*, April 1997.
Lipinski, Edward R., "Finding and Fixing Stair Problems," *New York Times*, November 1998.
McLearn, Bill, "Building Stairs with Stock Parts," *Journal of Light Construction*, vol.9, no. 5, February 1991.
Nunn, Richard V., *Home Improvement Home Repair*, Creative Homeowner Press, 1980.
Schuttner, Scott, *Basic Stairbuilding*, Newtown, CT: Taunton Press, Inc., 1990.
Tully, Gordon, "Stiffening Stairway Railings," *Journal of Light Construction*, vol. 8, no. 10, July 1990.
Wardell, Charles, "New Curve to Stairway Design," *Journal of Light Construction*, vol. 12, no. 5, February 1994.
Zepp, Don, "Foolproof Handrail Layout," *Journal of Light Construction*, Volume 15, Number 1, October 1996.

CHAPTER 6: KITCHENS AND BATHS

KITCHENS AND BATHS OVERVIEW

Cheever, Ellen, CKD, CBD, ASID, *Bathroom Industry Technical Manual;* Vol. 3, *Bathroom Equipment and Materials*, National Kitchen and Bath Association, 1997.
Cheever, Ellen, CKD, CBD, ASID, *Bathroom Industry Technical Manual;* Vol. 4, *Bathroom Planning Standards and Safety Criteria*, National Kitchen and Bath Association, 1997.
Cheever, Ellen, CKD, CBD, ASID, *Bathroom Industry Technical Manual;* Vol. 5, *Design Principles for Bathroom Planning*, National Kitchen and Bath Association, 1997.
Cheever, Ellen, CKD, CBD, ASID, *Kitchen Industry Technical Manual;* Vol. 3, *Kitchen Equipment and Materials*, National Kitchen and Bath Association, 1996.
Cheever, Ellen, CKD, CBD, ASID, *Kitchen Industry Technical Manual;* Vol. 4, *Kitchen Planning Standards and Safety Criteria*, National Kitchen and Bath Association, 1996.
Cheever, Ellen, CKD, CBD, ASID, *Kitchen Industry Technical Manual;* Vol. 5, *Design Principles for Kitchen Planning*, National Kitchen and Bath Association, 1996.
Germer, Jerry, *Kitchen and Bathroom Carpentry: Installation Techniques*, National Kitchen and Bath Association, 1997.
McDonald, MaryLee; Ellen Cheever, CKD, CBD, ASID; and Nicolas Geragi, CKD, CBD, *Bathroom Industry Technical Manual;* Vol. 1, *Building Materials, Construction, and Estimating for the Kitchen and Bathroom*, National Kitchen and Bath Association, 1997.
McDonald, MaryLee; Ellen Cheever, CKD, CBD, ASID; and Nicolas Geragi, CKD, CBD, *Bathroom Industry Technical Manual;* Vol. 2, *Bathroom Mechanical Systems*, National Kitchen and Bath Association, 1997.
McDonald, MaryLee; Ellen Cheever, CKD, CBD, ASID; and Nicolas Geragi, CKD, CBD, *Kitchen Industry Technical Manual;* Vol. 1, *Building Materials, Construction, and Estimating for the Kitchen and Bathroom*, National Kitchen and Bath Association, 1996.
Peterson, Mary Jo, CKD, CBD, CHE, *Universal Kitchen and Bath Planning*, New York: McGraw-Hill, 1998.
"The Principles of Universal Design," The Center for Universal Design, North Carolina State University, April 1, 1997; www.design.ncsu.edu/cud/pubs/udprinciples.html.

Stoeppelwerth, Walter W., *Kitchen and Bathroom Installation Manual;* Vol. 1, National Kitchen and Bath Association, 1994.

Stoeppelwerth, Walter W., *Kitchen and Bathroom Installation Manual;* Vol. 2, *Installation Project Management*, National Kitchen and Bath Association, 1995.

SURFACE MAINTENANCE AND REPAIRS

Baker, Paula, AIA; Erica Elliot, MD; and John Banta, *Prescriptions for a Healthy House*, InWord Press, 1998.

Bower, Lynn Marie, *The Healthy Household*, Healthy House Institute, 1995.

Kimball, Herrick, *Refacing Cabinets; Making an Old Kitchen New,* Taunton Press, 1997.

Lipinski, Edward, R., "Steps to Refacing Kitchen Cabinets," *The New York Times*, July 19, 1998.

HARDWARE TO MAXIMIZE ACCESS AND FUNCTION

"1999 Buyer's Guide," *Kitchen and Bath Business*, February 1999.

"Annual Directory and Buyer's Guide," *Kitchen and Bath Design News*, February 1999.

Peterson, Mary Jo, CKD, CBD, CHE, *Universal Kitchen and Bath Planning*, McGraw-Hill, 1998.

ACCESSORIES TO MAXIMIZE ACCESS AND IMPROVE STORAGE

"1999 Buyer's Guide," *Kitchen and Bath Business*, February 1999.

"Annual Directory and Buyer's Guide," *Kitchen and Bath Design News*, February 1999.

Cheever, Ellen, CKD, CBD, ASID, *Kitchen Industry Technical Manual;* Vol. 4, *Kitchen Planning Standards and Safety Criteria*, National Kitchen and Bath Association, 1996.

Universal Kitchen and Bath Planning, Peterson, Mary Jo, CKD, CBD, CHE, McGraw-Hill, 1998.

REPLACE OR ADD CABINETRY

"1999 Buyer's Guide," *Kitchen and Bath Business*, February 1999.

"Annual Directory and Buyer's Guide," *Kitchen and Bath Design News*, February 1999.

Baker, Paula, AIA; Erica Elliot, MD; and John Banta, *Prescriptions for a Healthy House*, InWord Press, 1998.

Bower, Lynn Marie, *The Healthy Household*, Healthy House Institute, 1995.

Cheever, Ellen, CKD, CBD, ASID, *Kitchen Industry Technical Manual;* Vol. 3, *Kitchen Equipment and Materials,* National Kitchen and Bath Association, 1996.

Cheever, Ellen, CKD, CBD, ASID, *Kitchen Industry Technical Manual;* Vol. 4, *Kitchen Planning Standards and Safety Criteria,* National Kitchen and Bath Association, 1996.

SURFACE MAINTENANCE AND REPAIRS

"1999 Buyer's Guide," *Kitchen and Bath Business*, February 1999.

Alseth, James, "Repairing Stone Utilizing the Overfill Grind Technique," *Dimensional Stone*, August 18, 1998.

"Annual Directory and Buyer's Guide," *Kitchen and Bath Design News,* February 1999.

Bower, Lynn Marie, *The Healthy Household,* Healthy House Institute, 1995.

Cheever, Ellen, CKD, CBD, ASID, *Bathroom Industry Technical Manual;* Vol. 3, *Bathroom Equipment and Materials*, National Kitchen and Bath Association, 1997.

Cheever, Ellen, CKD, CBD, ASID, *Kitchen Industry Technical Manual;* Vol. 3, *Kitchen Equipment and Materials*, National Kitchen and Bath Association, 1996.
Hamilton, Gene, and Katie Hamilton, *Home Improvement for Dummies*, IDG Books Worldwide, 1998.
Hufnagel, James A., *Bathrooms: How To,* Hometime Video Publishing, Inc., 1998.

IMPROVE INDOOR AIR QUALITY

Baker, Paula, AIA; Erica Elliot, MD; and John Banta, *Prescriptions for a Healthy House*, InWord Press, 1998.
Bower, Lynn Marie, *The Healthy Household*, Healthy House Institute, 1995.
Goldbeck, David, *The Smart Kitchen,* Ceras Press, 1994.

IMPROVE BACKSPLASH AND COUNTERTOP SEAMS

Baker, Paula, AIA; Erica Elliot, MD; and John Banta, *Prescriptions for a Healthy House*, InWord Press, 1998.
Bower, Lynn Marie, *The Healthy Household*, Healthy House Institute, 1995.

MAXIMIZE ACCESS AND FUNCTION

Peterson, Mary Jo, CKD, CBD, CHE, *Universal Kitchen and Bath Planning*, McGraw-Hill, 1998.

IMPROVING FUNCTIONAL COUNTERTOP SPACE

McIlvain, Jess, AIA, CCS, CSI, and Joe Ranzoni, *Installation of Ceramic and Stone Tiles, and Dimension Stone Countertops*, National Kitchen and Bath Association, 1997.

REPLACE OUTDATED OR NONFUNCTIONING APPLIANCES

"1999 Buyer's Guide," *Kitchen and Bath Business*, February 1999.
"Annual Directory and Buyer's Guide," *Kitchen and Bath Design News*, February 1999.
Builder Appliances, www.builderappliances.com
Cheever, Ellen, CKD, CBD, ASID, *Kitchen Industry Technical Manual;* Vol. 3, *Kitchen Equipment and Materials*, National Kitchen and Bath Association, 1996.
Goldbeck, David, *The Smart Kitchen*, Ceres Press, 1994.
Moris, Allison Murray, "Repair or Replace; Weighing the Options," *Women's Day, Kitchens and Bath Special Interest Publication*, Winter 1998, p. 22.
Sullivan, Bruce, "Kitchens; A Smorgasbord of Energy Savings," *Home Energy Magazine*, Reprint.
No Regrets Remodeling, Energy Auditor & Retrofitter, Inc., 1997.

IMPROVE ACCESS AND FUNCTION WITHIN WORK CENTERS

Cheever, Ellen, CKD, CBD, ASID, *Kitchen Industry Technical Manual;* Vol. 3, *Kitchen Equipment and Materials*, National Kitchen and Bath Association, 1996.

Cheever, Ellen, CKD CBD ASID, *Kitchen Industry Technical Manual;* Vol. 4, *Kitchen Planning Standards and Safety Criteria*, National Kitchen and Bath Association, 1996.

Wylde, Margaret A., Ph.D.; Deany Hillhouse; Joe Austin; Kristen Suslar Guinn, *ProMatura Enabling Products Sourcebook 2*, National Kitchen and Bath Association, 1995.

Peterson, Mary Jo, CKD, CBD, CHE, *Universal Kitchen and Bath Planning*, McGraw-Hill, 1998.

UPGRADE APPLIANCE APPEARANCE

Krengel, James, CKD, CBD, IIDA, "Thinking Creatively About Appliances," in *Kitchens by Professional Designers*, Book IX, pp. 24–29.

IMPROVE RESOURCE AND ENERGY EFFICIENCY OF EXISTING APPLIANCES

Baker, Paula, AIA; Erica Elliot, MD; and John Banta, *Prescriptions for a Healthy House*, InWord Press, 1998.

Goldbeck, David, *The Smart Kitchen,* Ceres Press, 1994.

Krengel, James, CKD, CBD, IIDA, "Thinking Creatively About Appliances," in *Kitchens by Professional Designers*, Book IX, pp. 24–29.

No Regrets Remodeling, Energy Auditor & Retrofitter, Inc., 1997.

Sullivan, Bruce, "Kitchens; A Smorgasbord of Energy Savings," *Home Energy Magazine*, Reprint.

INSTALL RESOURCE- AND ENERGY-EFFICIENT APPLIANCES

No Regrets Remodeling, Energy Auditor & Retrofitter, Inc., 1997.

Sullivan, Bruce, "Kitchens; A Smorgasbord of Energy Savings," *Home Energy Magazine*, Reprint.

SURFACE MAINTENANCE AND REPAIRS

"1999 Buyer's Guide," *Kitchen and Bath Business,* February 1999.

"Annual Directory and Buyer's Guide," *Kitchen and Bath Design News,* February 1999.

Bower, Lynn Marie, *The Healthy Household*, Healthy House Institute, 1995.

Cheever, Ellen, CKD, CBD, ASID, *Bathroom Industry Technical Manual;* Vol. 3, *Bathroom Equipment and Materials*, National Kitchen and Bath Association, 1997.

Cheever, Ellen CKD, CBD, ASID, *Kitchen Industry Technical Manual;* Vol. 3, *Kitchen Equipment and Materials*, National Kitchen and Bath Association, 1996.

This Old House Journal Guide to Restoration, Penguin Books, 1992.

WATER PURIFICATION

Baker, Paula AIA; Erica Elliot, MD; John Banta, *Prescriptions for a Healthy House,* InWord Press, 1998.

Bandon, Alexandra, "Shopping Smart; On the Waterfront," *American Homestyle and Gardening*, March 1999, p. 40.

Bower, Lynn Marie, *The Healthy Household*, Healthy House Institute, 1995.

McDonald, Marylee; Nicholas Geragi, CKD, CBD, NCIDQ; Ellen Cheever, CKD, CBD, ASID, *Kitchen Industry Technical Manual;* Vol. 2, *Kitchen Mechanical Systems,* National Kitchen and Bath Association, 1996.

WATER CONSERVATION

Cheever, Ellen M., CKD, CBD, ASID, *Bathroom Industry Technical Manual;* Vol. 3, *Bathroom Equipment and Materials*, National Kitchen and Bath Association, 1997.
Designing and Remodeling Bathrooms, Ortho Books, 1990.
Goldbeck, David, *The Smart Kitchen*, Ceres Press, 1994.
No Regrets Remodeling. Energy Auditor & Retrofitter, Inc., 1997.
Johnson, David, "Remodeling Bathrooms; Let the Energy Savings Flow," *Home Energy Magazine*, Reprint.

MAXIMIZE ACCESS AND FUNCTION

"1999 Buyer's Guide." *Kitchen and Bath Business*, February 1999.
"Annual Directory and Buyer's Guide." *Kitchen and Bath Design News,* February 1999.
Cheever, Ellen, CKD, CBD, ASID, *Bathroom Industry Technical Manual;* Vol. 3, *Bathroom Equipment and Materials,* National Kitchen and Bath Association, 1997.
Peterson, Mary Jo, CKD, CBD, CHE, *Universal Kitchen and Bath Planning,* McGraw-Hill, 1998.
Wylde, Margaret A., Ph.D.; Deany Hillhouse; Joe Austin; Kristen Suslar Guinn, *ProMatura Enabling Products Sourcebook 2,* National Kitchen and Bath Association, 1995.

SURFACE MAINTENANCE AND REPAIRS

"1999 Buyer's Guide." *Kitchen and Bath Business*, February 1999.
Alseth, James, "Repairing Stone Utilizing the Overfill Grind Technique," *Dimensional Stone*, August 18, 1998, p. 29.
"Annual Directory and Buyer's Guide." *Kitchen and Bath Design News,* February 1999.
Cheever, Ellen, CKD, CBD, ASID, *Bathroom Industry Technical Manual;* Vol. 3, *Bathroom Equipment and Materials*, National Kitchen and Bath Association, 1997.
Hufnagel, James A., *Bathrooms: How To*, Hometime Video Publishing, Inc., 1998.
Hamilton, Gene, and Katie, *Home Improvement for Dummies,* IDG Books Worldwide, 1998.

MOISTURE CONTROL

ANSI Specification for the Installation of Ceramic Tile, Tile Council of America.
Byrne, Michael, "Durable Substrates for Thinset Tile," *Journal of Light Construction*, August 1996, p. 40.
Cheple, Marilou, and Pat Huelman, "Moisture Control in Bathrooms," *Home Energy Magazine*, March/April 1998.
Handbook for Ceramic Tile Installation, Tile Council of America.
McDonald, MaryLee; Nick Geragi, CKD, CBD; and Ellen Cheever, CKD, CBD, ASID, *Bathroom Industry Technical Manual;* Vol. 2, *Bathroom Mechanical Systems*, National Kitchen and Bath Association, 1997.
No Regrets Remodeling, Energy Auditor & Retrofitter, Inc., 1997.
Pennebaker, E. David, "Build It Right the First Time," *EEBA Excellence*, Winter 1998, pp. 10–12.
Pennebaker, E. David, "Tile Wall Failures," *EEBA Excellence,* Winter 1998, pp. 13, 17.

MAXIMIZE ACCESS AND FUNCTION

Cheever, Ellen M., CKD, CBD, ASID, *Bathroom Industry Technical Manual;* Vol. 3, *Bathroom Equipment and Materials,* National Kitchen and Bath Association, 1997.

Peterson, Mary Jo, CKD, CBD, CHE, *Universal Kitchen and Bath Planning*, McGraw-Hill, 1998.
Wylde, Margaret A., Ph.D.; Deany Hillhouse; Joe Austin; Kristen Suslar Guinn, *ProMatura Enabling Products Sourcebook 2*, National Kitchen and Bath Association, 1995.

WATER CONSERVATION

Cheever, Ellen M., CKD, CBD, ASID, *Bathroom Industry Technical Manual;* Vol. 3, *Bathroom Equipment and Materials*, National Kitchen and Bath Association, 1997.
Horner, Thomas, "Water Resource Efficiency; The Performance Contracting Method," *ASPE Plumbing Engineer*, reprinted by Water Management, www.watermgt.com.
No Regrets Remodeling, Energy Auditor & Retrofitter, Inc., 1997.

MAINTENANCE AND REPAIRS

Cheever, Ellen, CKD, CBD, ASID, *Bathroom Industry Technical Manual;* Vol. 3, *Bathroom Equipment and Materials*, National Kitchen and Bath Association, 1997.
Hamilton,Gene, and Katie, *Home Improvement for Dummies*, IDG Books Worldwide, 1998.

WATER CONSERVATION

Baz, Jose, "Flushing 1.6 gpf; How 1.6 gpf Technologies Impact Today's Plumbing Conservation Goals," Sloan Valve Company.
"Choosing a Toilet," *Fine Homebuilding*, www.taunton.com.
Johnson, David, "Remodeling Bathrooms; Let the Energy Savings Flow," *Home Energy Magazine*, Reprint.
"Less Wasteful Flushing," *Environmental Building News*, vol. 8 no. 3, March 1999, www.ebuild.com.

MAXIMIZE ACCESS AND FUNCTION

Cheever, Ellen, CKD CBD ASID, *Bathroom Industry Technical Manual;* Vol. 3, *Bathroom Equipment and Materials*, National Kitchen and Bath Association, 1997.
Peterson, Mary Jo, CKD, CBD, CHE, *Universal Kitchen and Bath Planning*, McGraw-Hill, 1998.

CHAPTER 7: ELECTRICAL AND ELECTRONIC SYSTEMS

ELECTRICAL AND ELECTRONIC SYSTEMS OVERVIEW

Cauldwell, Rex, *Wiring a House*, Newtown, CT : Taunton Press, 1996.
Editors of the *Journal of Light Construction*, *Troubleshooting Guide to Residential Construction*, Builderburg Group, Inc., 1997.
Editors of Time-Life Books, *Home Repair and Improvement—Advanced Wiring*, Alexandria, VA: Time-Life, 1998.
Editors of Time-Life Books, *Home Repair and Improvement—Basic Wiring*, Alexandria, VA: Time-Life, 1994.
Enterprise Foundation, *A Consumer's Guide to Home Improvement, Renovation, and Repair*, New York: John Wiley & Sons, Inc., 1990.

The Guide to Low-Voltage and Limited Energy Systems, National Electrical Contractors Association, 1999.
Landfield Group archive of publicly available material: www.faqs.org/faqs/electrical-wiring.
Nash, George, *Renovating Old Houses*, Newtown, CT: Taunton Press, 1992.
National Electrical Code 1999, National Fire Protection Association, 1998.
Poore, Patricia, ed., *The Old-House Journal Guide to Restoration*, New York: Dutton, 1992.
Richter, Herbert P., and W. Creighton Schwan, *Practical Electrical Wiring,* 17th ed., New York: McGraw-Hill Book Company, 1996.
Shapiro, David E., *Old Electrical Wiring Maintenance and Retrofit*, New York: McGraw-Hill, 1998.
Trimmer, H. William, B.S., Butterworth Heinemann, *Understanding and Servicing Alarm Systems*, 2d ed., 1989.
Whelchel, Harriet, ed., *Caring for Your Historic House,* Heritage Preservation, 1998.

SERVICE PANELS

Cauldwell, Rex, "The Electric Panel," *Journal of Light Construction*, October 1993.
Roberts, Sam, "Panel Discussion: Assessing the Safety of Old-House Electric Panels," *Old-House Journal*, January/February 1996.

WIRING OVERVIEW

Cauldwell, Rex, "Plug-In Electrical Testers," *Journal of Light Construction*, December 1995.
Kenny, Sean, "Rewiring Old Houses," *Journal of Light Construction*, April 1999.
Kenny, Sean, "Upgrading Kitchen Wiring," *Journal of Light Construction*, May 1997.
Shapiro, David E., "Working with Old Wiring," *Fine Homebuilding*, December/January 1999, pp. 112–115.

ALUMINUM WIRING

The Aluminum Wire Information Website, www.mhv.net/~dfriedman/aluminum.htm.
"Aluminum Wiring in Residential Properties: Hazards & Remedies," www.inspect-ny.com/aluminum.htm.

RECEPTACLES

Cauldwell, Rex, "Installing Electrical Boxes and Receptacles," *Fine Homebuilding*, June/July, 1996.
Kardon, Redwood, "Converting 2-Prong to Grounded 3-Prong Receptacles," www.codecheck.com/NEC_hnd_2prong/250_50_commentary.html.
Rhodes, J. Michael, "Don't Guess at GFCI Regs!," *Electrical Contractor Magazine*, May 1996.
Shapiro, David E., "More 'Groundless' Worries," *Electrical Contractor Magazine*, April 1999.

INTERIOR LIGHTING

#1 Lighting, Home Energy Brief, Rocky Mountain Institute, 1994.
"Lighting Your Life, a Home Lighting Guide," American Lighting Association, 1992; www.americanlightingassoc.com/resources/index.cfm.
Maker, Tim, "Dedicated Fixtures for Compact Fluorescents," *Journal of Light Construction*, January 1994.

Maril, Nadja, *American Lighting 1840-1940*, Schiffer Publisher, 1995.
Ridenour, Jeff, "Low Voltage Lighting," *Journal of Light Construction*, May 1996.
"White LED Lighting," Partnership for Advancing Technology in Housing, Technology Inventory, www.nahbrc.org/ToolBase/pandt/tech/abstracts/elecab11.html.

EXTERIOR LIGHTING

"Full Cut-Off Outdoor Light Fixtures," Partnership for Advancing Technology in Housing, Technology Inventory; www.nahbrc.org/homebase/index.html.
"Outdoor Photovoltaic Lighting," Energy Efficiency and Renewable Energy Network (EREN), U.S. Department of Energy; www.eren.doe.gov/consumerinfo/refbriefs/db8.html.
Stockton, Martha Tuzson, "A Guide to Exterior Lighting," *Fine Homebuilding*, February/March 1998.

CONTROLS

Home Automation Systems, The Energy Efficiency and Renewable Energy Network (EREN)/ U.S. Department of Energy; www.eren.doe.gov/consumerinfo/refbriefs/ad7.html.
Rosenberg, Paul, "Sell Energy Technology to Reduce Long Term Costs," *Electrical Contractor Magazine*, May 1995.

ELECTRIC BASEBOARD HEATING

1996 ASHRAE Handbook, HVAC Systems and Equipment, Atlanta, GA: ASHRAE, 1996.
Bobenhausen, William, *Simplified Design of HVAC Systems*, New York: John Wiley & Sons, Inc., 1994.
"Saving Energy with Electric Resistance Heating," Energy Efficiency and Renewable Energy Network, U.S. Department of Energy; www.eren.doe.gov/erec/factsheets/elecheat.html.
Trewethey, Richard, with Don Best, *This Old House: Heating, Ventilation, and Air Conditioning*, Little, Brown, 1994.

PHONE, COMPUTER, AND TV CABELING

Better Homes and Gardens Online Encyclopedia, www.bhglive.com/homeimp/wiring.
"Wired For the Future," *Builder*, September 1998, www.BuilderOnline.com/frmArtFront/1,1071,` 1~322~323~854~1`,00html.

INTRUSION AND ALARM SYSTEMS

Bernard, Robert, *Intrusion Detection Systems*, 2d ed., Butterworth Heinemann, 1988.
Capel, Vivian, *Security Systems & Intruder Alarms*, 2d ed., Butterworth Heinemann, 1999.
Cumming, Neil, *Security: A Guide to Security System Design and Equipment Selection and Installation*, 2d ed., Butterworth Heinemann, 1994.
The Guide to Low-Voltage and Limited Energy Systems, National Electrical Contractors Association, 1999.
Honey, Gerard, *Electronic Security Systems Pocket Book*, Butterworth Heinemann, 1999.
Residential and Light Commercial Installation Practices (TIA-570 Compliance), Leviton Manufacturing Co., Inc., 1998; www.levitontelcom.com/pdf/strategies.pdf.

Safe & Sound, Your Guide to Home Security, National Burglar and Fire Alarm Association.
"To Foil a Thief," *Consumer Reports*, September 1998.
Trimmer, H. William, B.S., *Understanding and Servicing Alarm Systems*, 3d ed., Butterworth Heinemann, 1999; www.bhusa.com.

SMOKE DETECTORS

"Cleaning and Testing Smoke Detectors," *Security Professional*, August 1999.
Guide to Low-Voltage and Limited Energy Systems, National Electrical Contractors Association, 1999.
"How Smoke Detectors Work," www.howstuffworks.com/smoke1.htm.
NFPA 72: National Fire Alarm Code, 1999 Edition, Chapters 1, 7, 8, National Fire Protection Association.
NFPA 101: Life Safety Code, National Fire Protection Association.
Safe & Sound, Your Guide to Home Security, National Burglar and Fire Alarm Association.
Trimmer, H. William, B.S., *Understanding and Servicing Alarm Systems*, 3d ed., Butterworth Heinemann, 1999, www.bhusa.com.
What You Should Know About Smoke Detectors, U.S. Consumer Product Safety Commission, January 1985, www.ag.ohio-state.edu/~ohioline/aex-fact/0690_1.html.
"Working Knowledge: Smoke Detectors," *Scientific American*, April 1997, www.sciam.com/0497issue/0497working.html.

CARBON MONOXIDE DETECTORS

NFPA 720: Recommended Practice for the Installation of Household Carbon Monoxide (CO) Warning Equipment, 1998 Edition, National Fire Protection Association.
"GRI Technical Report: Performance Testing of Residential CO Alarms," Gas Research Institute, 1999.
"Sleeping Safely," *Consumer Reports*, November 1996, www.consumerreports.org.
"UL 2034-98: Carbon Monoxide Alarm (CO) Standards," Underwriters' Laboratories, 1998.

LIGHTNING PROTECTION

NFPA78: Lightning Protection Code, National Fire Protection Association.
NFPA780: Standard for Installation of Lightning Protection Systems, 1997 Edition, National Fire Protection Association.
LPI-175 Installation Standard, Lightning Protection Institute; www.lightning.org.
LPI-176 Components Standard, Lightning Protection Institute.
Towne, H.M., "Lightning: Its Behavior and What to Do About It," United Lightning Protection Association, Inc., 1956.
UL96A: Lightning Protection Standards, Underwriters' Laboratories.

SURGE PROTECTION

NFPA70, National Electrical Code Article 280: Surge Arresters, National Fire Protection Association.
UL 1449, Surge Suppression Standards, 2d ed., Underwriters' Laboratories.

GARAGE DOOR OPENERS

Goldwasser, Samuel M., "Notes on the Troubleshooting and Repair of ..." *Sci. Electronics. Repair Frequently Asked Questions*, http://repairfaq.org/sam, 1999.

"UL 325 Standard: Federal Laws for Reversing Mechanisms and Sensing Edges," Underwriters' Laboratories.

CHAPTER 8: HVAC AND PLUMBING

HVAC SYSTEMS OVERVIEW

Bower, John, *Understanding Ventilation: How to Design Select, and Install Residential Ventilation Systems*, The Healthy House Institute, Bloomington, IN 47408; 1995; www.hhinst.com/UVnews.html.

Building Energy Efficiency, OTA-E-518 p. 20. U.S. Congress, Office of Technology Assessment, U.S. Government Printing Office, Pittsburgh, PA 15250; May 1992, 202-512-1800; www.ota.nap.edu/gpo.html.

Trethewey, Richard, with Don Best, *This Old House Heating, Ventilation, and Air Conditioning*, Little, Brown and Company, 1994; www.pbs.org/wgbh/thisoldhouse/show/rich.html.

REPLACEMENT SYSTEM SIZING

Bobenhausen, William, *Simplified Design of HVAC Systems*, New York: John Wiley & Sons, Inc., March 1994; www.swinter.com/company/publications.html.

Cooling and Heating Load Calculation Manual, American Society of Heating, Refrigerating and Air-Conditioning Engineers, Inc., 1791 Tullie Circle, N.E., Atlanta, GA 30329; 800-527-4723; www.ashrae.org.

Consumers' Directory of Certified Efficiency Ratings for Residential Heating and Water Heating Equipment, Gas Appliance Manufacturers Association, Arlington, VA, October 2000; www.gamanef.org/consumer/certification/certdir.htm.

Directory of Certified Unitary Products, Air Conditioning and Refrigeration Institute, 4301 North Fairfax Drive, Suite 425, Arlington, VA 22203, January 1, 2001; 703-524-8800; www.ariprimenet.org.

Residential Load Calculation Manual J, Air Conditioning Contractors of America, Arlington, VA 22206; 703-575-4477; www.acca.org/catalog/product.asp.

ENERGY SOURCES

Bobenhausen, William, *Simplified Design of HVAC Systems*, New York: John Wiley & Sons, Inc.; March 1994; www.swinter.com/company/publications.html.

Passive Solar Design Strategies: Guidelines for Home Building, Sustainable Buildings Industry Council, 1331 H Street NW, Suite 1000, Washington, D.C. 20005; 202-628-7400 x209; www.sbi-council.org/soft.

DISTRIBUTION SYSTEMS

Bobenhausen, William, *Simplified Design of HVAC Systems*, New York: John Wiley & Sons, Inc., 1994; www.swinter.com/company/publications.html.

Nash, George, *Renovating Old Houses*, Newton, CT: Taunton Press, 1992.

Poore, Patricia, ed., *The Old-House Journal Guide to Restoration*, New York: Penguin Books, 1992.

Siegenthaler, John, *Modern Hydronic Heating for Residential and Light Commercial Buildings*, Delmar Publishers, 1995.

HEATING

1996 ASHRAE Systems and Equipment Handbook, ASHRAE, 1791 Tullie Circle, NE, Atlanta, GA 30329; 800-527-4723; www.ashrae.org.

Consumers' Directory of Certified Efficiency Ratings for Residential Heating and Water Heating Equipment, Gas Appliance Manufacturers' Association, 1995, available from ETL Testing Laboratories, Inc.; 607-753-6711; www.gamanet.org/consumer/certification/certdir.htm.

Santucci, Robert M., Brooke C. Stoddard, and Peter Werwath, *A Consumer's Guide to Home Improvement, Renovation & Repair*, The Enterprise Foundation, John Wiley & Sons, Inc., 1995.

COOLING

1996 HVAC Systems and Equipment Handbook, American Society of Heating, Refrigerating and Air-Conditioning Engineers, Inc. (ASHRAE), Atlanta, GA; 800-527-4723; www.ashrae.org.

Bobenhausen, William, *Simplified Design of HVAC Systems*, New York: John Wiley & Sons, Inc., 1994; www.swinter.com/company/publications.html.

Trethewey, Richard, *This Old House Heating, Ventilation, and Air Conditioning*, Little, Brown and Company, 1994.

HEAT PUMPS

1996 HVAC Systems and Equipment Handbook, Atlanta, GA: American Society of Heating, Refrigerating and Air-Conditioning Engineers, Inc.

ARI Directory of Certified Applied Air-Conditioning Products, Arlington, VA: Air-Conditioning & Refrigeration Institute.

Closed-Loop/Ground-Source Heat Pump Systems: Installation Guide, Stillwater, OK: International Ground-Source Heat Pump Association, 1988; 800-626-4747; www.igshpa.okstate.edu/Publications/Brochures/Brochures.html.

INDOOR AIR QUALITY

Bower, John, *Understanding Ventilation*, Bloomington, IN: The Healthy House Institute, 1995; www.hhinst.com.

ASHRAE Standard 62-1989, Ventilation for Acceptable Indoor Air Quality, Atlanta, GA: American Society of Heating, Refrigerating and Air-Conditioning Engineers, 1989.

Certified Home Ventilating Products Directory, Home Ventilating Institute, May 2001, Division of AMCA, 30 West University Drive, Arlington Heights, IL 60004; 847-394-0150; www.hvi.org/directory.

EPA's Consumer Guide to Radon Reduction, Environmental Protection Agency, August 1992 (402-K92-003), 800-490-9198; www.epa.gov/iaq/radon/pubs/consguid.html.

CONTROLS

Automatic and Programmable Thermostats (DOE/GO-10097-375), March 1997, DOE Energy Efficiency and Renewable Energy, Washington, D.C.; www.eren.doe.gov/erec/factsheets/thermo.pdf.

Bower, John, *Understanding Ventilation*, Bloomington, IN: The Healthy House Institute, 1995; www.hhinst.com.

Comfort, Air Quality, and Efficiency by Design, Manual RS, Air Conditioning Contractors Association, 1999, 80 pp.; www.acca.org/catalog/category.asp?cid=3.

FIREPLACES AND CHIMNEYS

"Back to the Future Fireplace," *This Old House, no.10,* January/February 1997, pp. 74–79; www.pbs.org/wgbh/thisoldhouse/magazine/no10.html.

Brick Institute of America (BIA) Technical Notes on Brick Construction. Available on-line at www.bia.org.:

Technical Note 19—Residential Fireplace Design, January 1993; www.bio.org/BIA/technotes/t19.htm.

Technical Note 19A—Residential Fireplaces—Details and Construction, May 1980; Reissue Jan. 1988; www.bio.org/BIA/technotes/t19a.htm.

Technical Note 19B—Residential Chimneys—Design and Construction, June 1980. Reissue April 1988; www.bio.org/BIA/technotes/t19b.htm.

Labine, Clem, and Carolyn Flaherty, *The Old-House Journal Compendium*, Woodstock, NY: The Overlook Press, 1980; www.oldhousejournal.com.

The Fireplace Book: Idea Book of Fireplace Design, The Aberdeen Group, vol. 1, June 1991; vol. 2, December 1995.

Fireplaces & Wood Stoves, Time Life Books, August 1997.

Hufnagel, James A., *The Stanley Complete Step-by-Step Book of Home Repair and Improvement*, New York: Simon & Schuster, 1993; www.simonsays.com.

Nash, George, *Renovating Old Houses*, Newton, CT: Taunton Press, 1992, 206 pp.; www.taunton.com/books/th/renoldhs.

"Water Damage and Your Masonry Chimney," Indianapolis, IN: Chimney Safety Institute of America; 800-536-0118; www.csia.org/home/water.html.

DOMESTIC HOT-WATER HEATING

"Energy Efficient Water Heating," Western Area Power Administration; www.es.wapa.gov/pubs/files/water heating.pdf.

Residential Heat Pump Water Heaters, Federal Technology Alert, September, 1995; 800-DOE-EREC; www.pnl.gov.

"Water Heaters and Energy Conservation—Choices, Choices!," *Home Energy Magazine Online*, May/June 1996; www.homeenergy.org/9398.html or http://homeenergy.org/archive/hem.dis.anl.gov/eehem/96/9605contents.html.

PLUMBING DESIGN AND ENGINEERING

1996 ASHRAE Handbook—HVAC Systems and Equipment, Atlanta, GA: ASHRAE, 1996.

The Enterprise Foundation, *The Consumer's Guide to Home Improvement, Renovation & Repairs*, New York: John Wiley & Sons, 1995.

Hemp, Peter, *Plumbing a House*, Newtown, CT: Taunton Press, 1994.
International Residential Code (Draft), International Code Council, Falls Church, VA 22041, 1998.
Nash, George, *Renovating Old Houses*, Newtown, CT: Taunton Press, 1996.
Poore, Patricia, *The Old House Journal Guide to Restoration*, New York: E.P. Dutton Books, 1992.

WATER SUPPLY AND DISTRIBUTION SYSTEMS

1996 ASHRAE Handbook—HVAC Systems and Equipment, Atlanta, GA: ASHRAE, 1996.
"The Forgotten Hot Water Recirculating System," *PM Engineer*, May 1998.
Hemp, Peter, *Plumbing a House*, Newtown, CT: Taunton Press, 1994.
Journal of Light Construction, March 1997.
Nash, George, *Renovating Old Houses*, Newtown, CT: Taunton Press, 1996.
Poore, Patricia, *The Old House Journal Guide to Restoration*, New York: E.P. Dutton Books, 1992.

DRAIN, WASTE, AND VENT SYSTEMS

1997 Dwelling Requirements of the Uniform Plumbing Code, Walnut, CA: IAPMO; 909-595-8449.
Directory of Water Conserving Plumbing Products, Walnut, CA: IAPMO, updated monthly; 909-595-8449.
The Enterprise Foundation, *The Consumer's Guide to Home Improvement, Renovation & Repairs*, New York: John Wiley & Sons, 1995.
International Residential Code (Draft), Falls Church, VA: International Code Council, 1998.
Hemp, Peter, *Plumbing a House*, Newtown, CT: Taunton Press, 1994.
Poore, Patricia, *The Old House Journal Guide to Restoration*, New York: E.P. Dutton Books, 1992.

FUEL SUPPLY SYSTEMS

1996 ASHRAE Handbook—HVAC Systems and Equipment, Atlanta, GA: ASHRAE, 1996.
Burkhardt, Charles A., *Domestic Oil Burners*, New York: McGraw-Hill, 1961.
National Fuel Gas Code (NFPA 54), International Approval Services, Cleveland, OH 44131, 1996; www.gasweb.org/gasweb.

APPLIANCE VENTS AND EXHAUSTS

1996 ASHRAE Handbook—HVAC Systems and Equipment, Atlanta, GA: ASHRAE, 1996.
Burkhardt, Charles A., *Domestic Oil Burners*, New York: McGraw- Hill, 1961.
E Source Technology Atlas Series: Residential Appliances, E Source, Inc., Boulder, CO, 80302,1996; 303-440-8500; www.esource.com.
National Fuel Gas Code (NFPA 54), International Approval Services, Cleveland, OH 44131, 1996.

FIRE PROTECTION SYSTEMS

Automatic Sprinkler Systems Handbook, Quincy, MA: National Fire Protection Association, 1997.
"The Economics of Fast Response Residential Sprinkler Systems," *Fire Journal*, May 1985.
"Installation and Sprinkler Systems in One- and Two-Family Dwellings and Manufactured Homes" Quincy, MA: National Fire Protection Association, Section 13D, 1996.

"Sprinkle, Sprinkle Little Star," *This Old House*, January 1998.
"Sprinkler Codes A'Changing," *Journal of Light Construction*, December 1989.

CHAPTER 9: SITE WORK

DECK AND PORCH STRUCTURE

Beckstrom, Robert J., and John Reed, *Ortho's Home Improvement Encyclopedia* (Revised), Columbus, OH: Ortho Books, 1994, 512 pp.; www.ortho.com/content/books/5620.cfm.
Leeke, John, "Epoxy Repairs for Exterior Wood," *Practical Restoration Reports*, Historic Home Works, Portland, ME 04103; 207-773-2306; www.historichomeworks.com/HHW/reports/reports.htm,10 pp.
Leeke, John, "Exterior Wood Columns," *Practical Restoration Reports,* Historic Home Works, Portland, ME 04103; 207-773-2306; www.historichomeworks.com/HHW/reports/reports.htm, 17 pp.
Leeke, John, "Exterior Woodwork Details," *Practical Restoration Reports*, Historic Home Works, Portland, ME 04103; 207-773-2306; www.historichomeworks.com/HHW/reports/reports.htm, 16 pp.
Schuttner, Scott, *Building & Designing Decks*, Newtown, CT: Taunton Press, 1993, 154 pp.

REPAIR OR REPLACE WOOD DECKING MATERIALS

"Plastic Lumber and Other Recycled Plastic Products," King County Environmental Purchasing Program, October 2000, Seattle, WA; 800-325-6165; www.metrokc.gov/procedure/green/plastic.htm.
"Recycled Plastic Lumber," *Environmental Building News*, vol. 2, no. 4, July/August, 1993, pp. 1–17; *Environmental Building News*, Brattleboro, VT 05301; 802-257-7300; www.building green.com.
"State of Recycled Plastic Lumber," Plastic Lumber Trade Assoc., November 2000, PTLA Akron, OH; 330-762-1963; www.plasticlumber.org
McDonald, Kent, and Robert Falk, *Wood Decks: Materials, Construction, and Finishing*, Madison, WI: Forest Products Society, 1994, 94 pp; http://shop.forestprod.org.
WoodWise Consumer Guide, 2001 Edition, Co-op America, Washington, D.C. 20006; 800-58-GREEN; www.WoodWise.org.

PRESERVATIVE-TREATED WOOD

"Disposal: The Achilles' Heel of CCA-Treated Wood," *Environmental Building News*, vol. 6, no. 3, March 1997; www.buildinggreen.com/features/tw/treated_wood.html.
"Pressure-Treated Wood: How Bad Is It and What Are the Alternatives?" *Environmental Building News*, vol. 2, no. 1, 1993, pp. 1–13; www.buildinggreen.com.
Williams, R. Sam, and William Feist, *Selection and Application of Exterior Stains for Wood*, Madison, WI: USDA Forest Service, Forest Products Laboratory, Report FPL-GTR-106, January 1999; http://www.fpl.fs.fed.us/documnts/FPLGTR fplgtr106.pdf.
Williams, R. Sam and William Feist, *Water Repellents and Water-Repellent Preservatives for Wood*, USDA Forest Service, ForestProducts Laboratory, Report FPL-GTR-109, Madison, WI, January 1999; http://www.fpl.fs.fed.us/documnts/FPLGTR fplgtr109.pdf.

STAIRS AND HANDRAILS

Litchfield, Michael, *Renovation: A Complete Guide*, 2d ed., Englewood Cliffs, NJ: Prentice Hall, 1991.

WOOD FENCES AND RETAINING WALLS

Hufnagel, James A., *The Stanley Complete Step-by-Step Book of Home Repair and Improvement*, New York: Simon & Schuster, 1993.

Swift, Penny, and Janek Szymanowski, *Build Your Own Walls & Fences*, London: New Holland Publishers Ltd., 1997.

PAVED DRIVEWAYS, WALKS, AND PATIOS

Allen, Benjamin W., and Ben Allen, eds., *Step-By-Step Masonry & Concrete*, Better Homes & Gardens Books, September 1997.

Building Concrete Walks, Driveways, Patios, and Steps. Portland Cement Association (PCA), 5420 Old Orchard Road, Skokie, IL 60077; 800-868-6733 or 847-966-6200; http://www.portcement.org/pdf_files/IS209.pdf.

Building Quality Concrete Driveways, Video, published jointly by NRMCA and Portland Cement Association (PCA), 1990, 10 min; 800-868-6733.

Concrete for Small Jobs, Portland Cement Association (PCA), Skokie, IL 60077; 800-868-6733 or 847-966-6200; www.portcement.org/concrete.htm.

Dobrowolski, Joseph A., and Joseph J. Waddell, eds., *Concrete Construction Handbook,* 3d ed., McGraw-Hill Handbooks, 1993; www.ConcreteNetwork.com/bookstore/concrete.htm.

"The Concrete Pavement Restoration Guide: Procedures for Preserving Concrete Pavements" (TBO20P), American Concrete Pavement Association, Skokie, IL 60077, 1998, 24 pp.; 847-966-2272; www.pavement.com/CartConfig/keywords/CPR.htm.

Guide for Selecting and Specifying Concrete Repair Material (No. 03733*)* and *Guide for Selecting and Specifying Surface Preparation for Sealers, Coatings, and Membranes* (No. 03732), Sterling, VA: International Concrete Repair Institute (ICRI).

Kosmatka, Steven H., *Finishing Concrete Slabs with Color and Texture*, Portland Cement Association (PCA), Skokie, IL 60077; 800-868-6733 or 847-966-6200; www.portcement.org and www.Concrete Network.com/bookstore/decoconcrete.htm, 1991.

Perkins, Philip H., *Concrete Floors, Finishes and External Paving*, http://www.ConcreteNetwork.com/book store/concrete.htm.

Portland Cement Association, *The Homeowner's Guide to Building with Concrete, Brick and Stone*, Emmaus, PA: Rodale Press, 1998.

Residential Concrete, 3d ed., American Concrete Institute (ACI) International, 1999.

MASONRY WALLS

Concrete Masonry Handbook for Architects, Engineers, Builders, Portland Cement Association (PCA), Skokie, IL 60077; 800-868-6733 or 847-966-6200; http://www.portcement.org/info_resources.

Foundations and Masonry, National Concrete Masonry Association, Herndon, VA 20171; 703-713-1900; http://www.ncma.org/.

Guide to Concrete Masonry Residential Construction in High-Wind Areas, Portland Cement Association (PCA), Skokie, IL 60077; 800-868-6733 or 847-966-6200; http://www.portcement.org/concrete.htm, 1997.

Home Masonry Repairs & Projects, Black and Decker Home Improvement Library, Creative Publishing International, Minnetonka MN 55343, January 1996; 800-328-0590 or 612-936-4700.

"The Keys to Historic Masonry Restoration," *Building Design & Construction*, February 1, 1997; www.bdc.mag.com.

Nunn, Richard V., *Home Improvement-Home Repair*, Upper Saddle River, NJ: Creative Homeowner Press, May 1981.

Ortho's Home Improvement Encyclopedia, San Ramon, CA: The Solaris Group, 1994; www.ortho.com/content/books/5620.cfm.

Recommended Practices for Laying Concrete Block, Portland Cement Association (PCA), Skokie, IL 60077, 1993; 800-868-6733 or 847-966-6200; http://www.portcement.org/info_resources.

Segmental Retaining Wall Installation Guide, National Concrete Masonry Association, Herndon, VA 20171, 703-713-1900; http://www.ncma.org/.

Swift, Penny, and Janek Szymanowski, *Build Your Own Walls and Fences,* New Holland Publishers, distributed by Sterling Publishing Co., New York, NY, September 1997.

WELLS

Alth, Max, and Charlotte Alth, and S. Blackwell Duncan (contributor), *Wells and Septic Systems*, 2d ed., Tab Books, 1991.

Brassington, Rick, *Finding Water: A Guide to the Construction and Maintenance of Private Water Supplies*, 2d ed., New York: John Wiley & Sons, 1995.

Woodson, R. Dodge, *Builder's Guide to Wells and Septic Systems*, New York: McGraw-Hill, 1997.

ON-SITE WASTEWATER TREATMENT

ANSI/NSF Standard 40: Residential Wastewater Treatment Systems, Ann Arbor, MI: NSF International, 800-NSF-MARK; www.nsf.org/wastewater.

Kahn, Lloyd, Blair Allen, Jullie Jones, and Peter Aschwanden, *The Septic System Owner's Manual*, Bolinas, California: Shelter Publications, 1999; www.shelterpub.com.

Septic System Owner's Guide (#PC-6583-OF2), University of Minnesota Extension Service, St. Paul, MN 55108; 800-876-8636; www.bae.umn.edu/~septic.

WATER AND SEWER LINES

"Onsite Domestic Sewage Disposal Handbook," MU Extension—Missouri University—Columbia, Columbia, MO 65211; 800-292-0969; http://muextension.missouri.edu/xplor/envqual.

Septic Tank/Absorption Field Systems: A Homeowner's Guide to Installation and Maintenance, 800-292-0969; http://muextension.missouri.edu/xplor/envqual.

Trenchless Technology Magazine, 1770 Main Street, P.O. Box 190, Peninsula, OH 44264; 330-467-7588; fax: 330-468-2289; www.ttmag.com; www.trenchlessonline.com/magazine.html.

UNDERGROUND STORAGE TANKS

"Catalog of EPA Materials on Underground Storage Tanks," (EPA-510-B-00-001), EPA, January 2000; 800-424-9346; www.epa.gov/swerust1/pubs/index.htm.

"Homeowners Guide to Fuel Storage," Verbank, NY: Agway Energy Products, November 1990; 888-AGWAY24; www.agwayenergy.com.

"Which UST Systems Are Federally Regulated," EPA, September 30, 1998; 800-424-9346; www.epa.gov/swerust1/faqs/index.htm.

LANDSCAPE CARE

"How to Prune Trees," U.S. Department of Agriculture Forest Service, 1992 Folwell Ave., St. Paul Field Office, St. Paul, MN 55108; 651-649-5243.
"How to Recognize Hazardous Defects in Trees," www.na.fs.fed.us/spfo/pubs/howto.htm.
"Installation Requirements for Lightning Protection Systems, UL96A," July 8, 1998, 46 pp., Northbrook, IL 60062-2096; 888-853-3503; www.ul.com/info/standard.htm.
"Rx For Wounded Trees," www.na.fs.fed.us/spfo/pubs/misc/woundedtrees/rxwounded.htm.
"Tree Care Tips & Most Frequently Asked Questions," Manchester, NH: National Arborists Association, 800-733-2622; http://www.natlarb.com/tips.htm.

ENERGY-EFFICIENT AND SUSTAINABLE LANDSCAPING

Daniels, Stevie, *The Wild Lawn Handbook: Alternatives to the Traditional Front Lawn,* IDG Books Worldwide, 1997.
Ellefson, C., *Xeriscape Gardening: Water Conservation for the American Landscape,* New York: Macmillan Publishing Company, 1992.
"Environmental Landscaping," Naval Facilities Engineering Service Center-Environmental Services, April 1996; http://enviro.nfesc.navy.mil/p2Library/cd/docs/afdoc/april96a.htm.
Horn, Brad, "Landscaping Gets Back to Nature," *Environmental Design & Construction,* May/June 1998; www.edcmag.com.
"Landscaping for Energy Efficiency," DOE/GO-10095-046, The Energy Efficiency and Renewable Energy Clearinghouse (EREC), April 1995; http://www.eren.doe.gov/erec/factsheets/landscape.html.
Moffat, A. S., M. Schiler, and the staff of Green Living, *Energy-Efficient and Environmental Landscaping,* South Newfane, VT: Appropriate Solutions Press, 1995; 240pp.; 802-348-7441.
Moffat, A. S., and M. Schiler, *Landscaping Design that Saves Energy*, New York: William Morrow and Company, Inc., 1991.
Nelson, W. R., *Landscaping for Energy Conservation*, available from the Building Research Council, College of Fine and Applied Arts, University of Illinois at Urbana-Champaign, One East Saint Mary's Road, Champaign, IL 61820, 1991.
Sternberg, Guy, and Jim Wilson, *Landscaping with Native Trees*, New York: Houghton Mifflin Co., 1995.
"Trees, Saving Energy Naturally," Southern California Edison; 800-952-5062.
"Xeriscape Plant Guide," Denver Water, 1996; www.water.denver.co.gov/bookstore/bookstore.html.

HOME REHAB TECHNIQUES CHECKLIST

CHAPTER 1: FOUNDATIONS

BASEMENT FLOORS

❑ Provide a new floor slab or replace a portion of the existing slab.
❑ Pour a new fully bonded floor slab over the existing slab.
❑ Provide a new unbonded floor slab over the existing slab.

CRAWLSPACE FLOORS

❑ Pour a crawlspace floor slab of lightweight concrete.

PERMANENT WOOD FOUNDATION SYSTEMS

❑ Replace portion of existing foundation with permanent wood foundation system.

PREFABRICATED FOUNDATIONS

❑ Replace damaged foundation sections with structural insulated panel foundation system.
❑ Replace damaged foundation sections with precast concrete panel system.

SURFACE AND SUBSURFACE DRAINAGE

❑ Grade away from the house.
❑ Provide a "ground roof" around the perimeter of the house.
❑ Create swales to channel water away from foundation.
❑ Terrace slope to reduce water flow.
❑ Provide and maintain roof gutters and leaders.
❑ Provide trench or soil trap drains.
❑ Provide good drainage under basement window areaways.
❑ Replace backfill adjacent to house with free-draining material.

FOUNDATION DRAINAGE

❑ Install an interior perimeter baseboard "gutter" drainage system.
❑ Install a sump pump.
❑ Replace the basement slab and install interior foundation drains.
❑ Excavate the exterior foundation wall and install and repair or replace the existing drainage system.

DAMPPROOFING AND WATERPROOFING

❑ Apply crystallization products or cementitious coatings to foundation wall interior.
❑ Apply cementitious coatings to foundation wall exterior.
❑ Apply asphaltic coatings without modifiers to foundation wall exterior.
❑ Apply acrylic or other approved polymer sealers to foundation wall exterior or slab surface.
❑ Apply polyethylene sheet below floor slab.
❑ Apply an asphaltic-base product to the foundation wall exterior.
❑ Apply a rubberized asphalt coating to the foundation wall exterior or slab surface.
❑ Apply an asphalt-modified urethane coating to the foundation wall exterior.
❑ Apply a urethane coating system to the foundation wall exterior.
❑ Apply a rubber-based coating to the foundation wall exterior.
❑ Apply a clay-based waterproofing system to the foundation wall exterior.

INSULATING BELOW-GRADE WALLS

❑ Apply exterior insulation.
❑ Install interior insulation covering the entire wall from floor to ceiling.

INSULATING CRAWLSPACES

❑ Insulate vented crawlspaces.
❑ Insulate unvented crawlspaces.

INSULATING SLABS

❑ Provide exterior insulation for slabs with stem walls.
❑ Exterior insulation for slabs with grade beams.
❑ Insulate basement slabs from inside.

VENTILATING BASEMENT SPACES

❑ Provide direct-ventilation (sealed combustion) boilers or furnaces.
❑ Provide a fan with a motorized damper to induce outside air into the boiler or furnace room only when the equipment fires.
❑ Provide a screened, open vent to allow outside air into the basement near the furnace or boiler.
❑ Separate the basement air from the air in the rest of the house.
❑ Treat the basement as contiguous, conditioned space open to the rest of the house.
❑ Provide a ventilated "room within a room."

VENTILATING CRAWLSPACES

❑ Naturally ventilate through required sized openings in the foundation wall.
❑ Mechanically ventilate the unconditioned crawlspace.
❑ Mechanically ventilate the semiconditioned crawlspace.

VENTILATION FOR SOIL GASES

❑ Provide sub-slab ventilation of soil gasses in basement spaces.
❑ Provide foundation wall depressurization.
❑ Provide basement heating and ventilation, separate basement air from that of the rest of the house, and pressure the space to keep out oil gasses.
❑ Provide ventilation of soil gases from crawlspaces.
❑ Provide a ventilated "room within a room."

SHORING, UNDERPINNING, AND REPAIR

- ❑ Stabilize and underpin settled foundation with reinforced concrete piers.
- ❑ Stabilize or raise settled foundation with steel minipiers.
- ❑ Stabilize or raise settled foundation with helical piers.
- ❑ Stabilize or raise settled foundation by "pressure grouting" or "mud grouting."
- ❑ Stabilize or raise foundation with compaction grouting.
- ❑ Stabilize, waterproof, or raise foundation or slab with various highly specialized grouting techniques.
- ❑ Underpin masonry or concrete stem wall with enlarged footing.
- ❑ Buttress stone foundation walls.
- ❑ Replace the foundation.

CRACKS IN WALLS AND SLABS

- ❑ Repair cracked and outwardly displaced foundation wall by exterior jacking.
- ❑ Stabilize cracked and inwardly bowed foundation wall by use of "earth anchor" and wall plate.
- ❑ Stabilize cracked and inwardly bowed foundation wall by use of helical screw anchor.
- ❑ Repair wall cracks with conventional grouting techniques.
- ❑ Repair cracked wall with epoxy injections.
- ❑ Repair cracked wall with urethane injections.

COATINGS AND FINISHES

- ❑ Apply cementious coatings.
- ❑ Apply elastomeric coatings.
- ❑ Apply paint coatings.

CHAPTER 2: EXTERIOR WALLS

WOOD FRAME SEISMIC RESISTANCE

- ❑ Attach sill to foundation with anchor bolts.
- ❑ Attach sill, joist, or stud to foundation with side brackets or straps.
- ❑ Reinforce cripple walls.
- ❑ Provide secure load path from roof to foundation.

WOOD FRAME WIND RESISTANCE

- ❑ Reinforce connections of wood-frame walls to foundations.
- ❑ Reinforce wood-frame walls for shear resistance.
- ❑ Reinforce connections of wood-frame walls to first floor.
- ❑ Reinforce connections of wood-frame walls to roof trusses.
- ❑ Reinforce connections of wood-frame walls to roof overhangs.

REPAIR MOISTURE DETERIORATION

- ❑ Repair sill with built-up lumber.
- ❑ Replace large sections or the entire sill.
- ❑ Repair portions of the structure with epoxy.

MITIGATE INSECT DAMAGE

- ❑ Remove cellulous material near structure or crawlspace.
- ❑ Eliminate contact of wall materials with soil.
- ❑ Provide access for termite inspection.
- ❑ Eliminate dirt-filled porches and steps.
- ❑ Provide termite shields.
- ❑ Use pressure-treated lumber.

REPAIR WOOD-FRAME FIRE DAMAGE

- ❑ Assess damage to structure.
- ❑ Remove char with scraping or abrasion.
- ❑ Clean and deodorize damaged materials.

CLEAN EXISTING MASONRY WALLS

- ❑ Clean with brush by hand.
- ❑ Clean with pressurized water.
- ❑ Clean by abrasive blasting.

APPLY COLORLESS PROTECTIVE COATINGS

- ❑ Apply film coating to masonry wall.
- ❑ Apply penetrating coating to masonry wall.

REPAIR EXISTING MASONRY WALLS

- ❑ Replace missing masonry units.
- ❑ Repair damaged masonry units.
- ❑ Stabilize wall with replacement ties.

WALL SHEATHING

- ❑ Repair existing wall sheathing.
- ❑ Replace existing wall sheathing with oriented strand board.
- ❑ Replace existing wall sheathing with plywood.
- ❑ Replace existing wall sheathing with fiberboard sheathing.
- ❑ Replace existing wall sheathing with gypsum sheathing.
- ❑ Replace existing wall sheathing with paperboard sheathing.
- ❑ Replace existing wall sheathing with fiber-cement sheathing.
- ❑ Replace existing wall sheathing with foam insulating sheathing.

VAPOR RETARDERS

- ❑ Apply vapor retarder paint coating.
- ❑ Install treated paper or foil vapor retarder.
- ❑ Install clear polyethylene vapor retarder.
- ❑ Install black polyethylene vapor retarder.
- ❑ Install cross-laminated or fiber-reinforced polyethylene vapor barrier.

AIR INFILTRATION BARRIERS

- ❑ Install housewrap over existing sheathing.
- ❑ Install housewrap over new sheathing.

INSULATION

- ❑ Install batt insulation.
- ❑ Install encapsulated fiberglass insulation.
- ❑ Install blown-in loose-fill insulation in closed stud spaces.
- ❑ Install blown-in or sprayed on insulation in open stud spaces.
- ❑ Install rigid wall insulation.
- ❑ Install radiant barrier.
- ❑ Install structural insulated panel wall.

VINYL SIDING

- ❑ Remove stains from existing siding.
- ❑ Repair existing siding.
- ❑ Replace/cover existing siding with new vinyl siding.

METAL SIDING

- ❑ Maintain existing metal siding.
- ❑ Repair metal siding.
- ❑ Replace existing siding with steel siding.
- ❑ Replace existing siding with aluminum siding.

WOOD SHINGLES AND SHAKES

- ❑ Replace individual cedar shingles.
- ❑ Reside with new cedar shingles and shakes.

SOLID WOOD SIDING

- ❑ Repair/replace damaged bevel wood siding.
- ❑ Replace deteriorated or damaged siding with new Western red cedar bevel siding.
- ❑ Replace deteriorated or damaged siding with new Western redwood bevel siding.
- ❑ Replace deteriorated or damaged siding with new quartersawn spruce or pine bevel siding.
- ❑ Replace deteriorated or damaged siding with new nonbevel wood siding.

HARDBOARD SIDING

- ❑ Repair existing hardboard lap siding.
- ❑ Replace existing siding with new hardboard lap siding.
- ❑ Replace existing siding with new hardboard panel siding.

ENGINEERED WOOD SIDING

- ❑ Replace damaged or deteriorated siding with new L-P Smart Lap siding.
- ❑ Replace damaged or deteriorated siding with new L-P Smart Panel and EZ Panel engineered wood panels.

PLYWOOD PANEL SIDING

- ❑ Repair existing plywood siding.
- ❑ Replace existing siding with new plywood panel siding.

FIBER-CEMENT SIDING

❑ Repair existing fiber-cement siding.
❑ Replace existing siding with fiber-cement lap siding.
❑ Replace existing siding with fiber-cement panel siding.

EXTERIOR INSULATION AND FINISH SYSTEMS (EIFS)

❑ Repair existing EIFS .
❑ Install an EIFS barrier system.
❑ Install an EIFS moisture drainage system.

STUCCO

❑ Patch existing stucco.
❑ Install a stucco exterior wall finish.

EXTERIOR TRIM

❑ Patch existing wood trim with epoxy filler.
❑ Install new solid wood trim.
❑ Install new laminated veneer trim.
❑ Install new engineered wood trim.
❑ Install new wood/thermoplastic trim.
❑ Install new fiber-cement trim.
❑ Install new polymer trim.

SEALANTS AND CAULKS

❑ Prepare surface, remove existing sealants.
❑ Install sealant.

PAINT AND OTHER FINISHES

❑ Maintain existing coated surfaces.
❑ Prepare previously coated surfaces.
❑ Apply paint to new or existing wood or wood-based composite materials.
❑ Apply oil-based penetrating stains to new or existing wood or wood-based material.
❑ Apply solid color acrylic stains to new or existing wood or wood-based material.
❑ Apply specialty coatings.

CHAPTER 3: ROOFS

WOOD-FRAME WIND RESISTANCE

❑ Reinforce existing structure with metal connectors, straps, and additional fasteners.
❑ Reinforce existing roof sheathing to rafter/truss connection with adhesives.
❑ Reinforce existing roof-to-wall connection with Kevlar straps.

STRUCTURAL DECAY

❑ Control moisture intrusion.

FIRE DAMAGE

❑ Restore fire-damaged roof elements.

SHEATHING

❑ Repair existing roof sheathing.
❑ Replace damaged existing roof sheathing with plywood.
❑ Replace damaged existing roof sheathing with oriented strand board.
❑ Replace damaged existing roof sheathing with tongue-and-groove wood decking.
❑ Replace damaged existing roof sheathing with fiberboard sheathing/decking.

ROOF FLASHING

❑ Repair existing flashing.
❑ Install new copper or lead-coated copper flashing.
❑ Install new aluminum flashing.
❑ Install new galvanized steel flashing.
❑ Install new Galvalume sheet metal flashing.
❑ Install new stainless steel flashing.
❑ Install Rheinzink flashing.
❑ Install new lead flashing.
❑ Install new roll roof flashing.
❑ Install new ice and water barrier membrane.

UNDERLAYMENTS AND MOISTURE BARRIERS

❑ Replace existing roofing felts with new asphalt-saturated felts.
❑ Replace existing roofing felts with new reinforced underlayment.
❑ Replace existing roofing felts with new ice and water barrier.

INSULATION

❑ Install batt insulation at ceiling level.
❑ Install loose-fill insulation at ceiling level.
❑ Install batt insulation at roof level in cathedral ceilings.
❑ Install encapsulated fiberglass insulation.
❑ Install blown-in loose-fill insulation into closed rafter spaces at roof level.
❑ Install blown-in or sprayed on insulation into open rafter spaces at roof level.
❑ Install rigid insulation below the roof structure.
❑ Install rigid insulation above the roof structure.
❑ Install a radiant barrier.
❑ Install a structural insulated panel roof.

WOOD SHINGLES AND SHAKES

❑ Repair existing cedar shingles.
❑ Install new cedar shingles.
❑ Install new cedar shakes.
❑ Install new Southern yellow pine shakes.

ASPHALT SHINGLES

❑ Repair existing asphalt shingles.
❑ Install new asphalt shingles.

LOW-SLOPE ROOFING

❑ Repair existing built-up roof membrane.
❑ Replace existing roof with built-up roofing membrane.
❑ Repair existing modified bitumen membrane.
❑ Replace existing roof with modified bitumen membrane.
❑ Repair existing thermoset and thermoplastic single-ply roofing membrane.
❑ Replace existing roofs with new thermoset and thermoplastic single-ply roofing membranes.

METAL ROOFING

❑ Repair existing metal roofs.
❑ Repair existing roof with new standing seam metal roof.
❑ Replace existing roofs with metal shakes and tiles.
❑ Replace existing roof with flat metal shingles.

SLATE ROOFING

❑ Repair existing slate roofs.
❑ Replace existing slate roof with new slate.

CLAY, CONCRETE, FIBER-CEMENT, AND COMPOSITE TILE ROOFING

❑ Replace existing clay or concrete tiles.
❑ Install new clay roof tile.
❑ Install new concrete roof tile.
❑ Install new fiber-cement roof tiles.
❑ Install new composite roof shingles.
❑ Install new plastic composite roof tiles.

GUTTER AND LEADER SYSTEMS

❑ Repair existing gutter systems.
❑ Install new wood gutters and leaders.
❑ Install new steel gutters and leaders.
❑ Install new K-style aluminum gutters.
❑ Install new K-style copper gutters.
❑ Install new half-round copper and aluminum gutters.
❑ Install new vinyl (PVC) gutters.
❑ Replace/provide gutter screens/guards.

CHAPTER 4: WINDOWS AND DOORS

GLAZING

❑ Clean and polish existing damaged glazing.
❑ Apply tinted or reflective film to existing window glazing.
❑ Apply safety film to existing window glazing.
❑ Install insulated glass inserts.
❑ Replace existing glazing with better performing glazing.

WINDOW FRAMES AND REPLACEMENT UNITS

- ❑ Replace existing window units.
- ❑ Replace existing window sash and track.
- ❑ Install new (secondary) window unit within existing window frame.
- ❑ Install replacement sills.
- ❑ Replace existing damaged wood with epoxy consolidants and fillers.
- ❑ Adjust window or door frame with shim screws.

STORM WINDOWS AND SCREENS

- ❑ Install interior (fixed/removable) storm windows.
- ❑ Install exterior (operable) storm windows.
- ❑ Repair or replace screen material.
- ❑ Install exterior sun screening devices.

SKYLIGHTS

- ❑ Repair or replace existing metal skylights.
- ❑ Replace existing skylights with new units.
- ❑ Install a tubular skylight.

PRIMARY ENTRY DOORS

- ❑ Repair existing door with traditional materials.
- ❑ Replace existing door with new door slab.
- ❑ Replace existing door with secondary door frame.
- ❑ Replace existing door with a new prehung door.

GARAGE AND BULKHEAD DOOR

- ❑ Repair existing bulkhead or garage door.
- ❑ Replace bulkhead or garage door.

STORM AND SCREEN DOORS

- ❑ Repair existing storm door unit.
- ❑ Replace existing storm door unit.

INTERIOR DOORS

- ❑ Repair existing interior doors.
- ❑ Replace existing interior doors.

CASING AND TRIM

- ❑ Repair existing wood trim.
- ❑ Install solid/engineered wood trim.
- ❑ Install fiber cement/plastic/polymer trim.
- ❑ Install modular/preassembled trim.

HARDWARE

- ❑ Repair/replace existing hardware with original components.
- ❑ Replace existing hardware with new units.

FLASHING

❑ Install sheet metal flashing.
❑ Install vinyl (PVC) flashing.
❑ Install self-adhering membrane or tape flashing.

CAULKING, SEALANTS, AND WEATHERSTRIPPING

❑ Install sealant materials.
❑ Install new or replacement interlocking weatherstripping.
❑ Install new or replacement resilient (compression or sliding) weatherstripping.
❑ Install new or replacement door thresholds.

SHUTTERS AND AWNINGS

❑ Repair existing shutters and awnings.
❑ Replace with new shutters and awnings.

CHAPTER 5: PARTITIONS, CEILINGS, FLOORS, AND STAIRS

FLOOR/CEILING STRUCTURAL ASSEMBLIES

❑ Check for excessive notching or cutting of beams or joists.
❑ Use guidelines for cutting, notching, or boring joists.
❑ Reinforce existing main beam by adding support.
❑ Reinforce existing beams or joists.
❑ Transfer load to existing joists.

ALTERNATIVES TO SOLID LUMBER FOR FLOOR FRAMING

❑ Repair existing trusses.
❑ Replace existing joists with new metal plate connected trusses.
❑ Replace existing structure with glulams.
❑ Replace existing structure with wood I-joists.
❑ Replace existing structure with laminated veneer lumber.
❑ Replace existing structure with parallel strand lumber.
❑ Replace existing structure with laminated strand lumber.
❑ Replace existing structure with steel floor joists.

MOISTURE DETERIORATION

❑ Repair existing joists with epoxy consolidants.
❑ Repair/replace existing joists with supporting structure.

FIRE DAMAGE TO FLOOR FRAMING

❑ Restore fire-damaged structural members.
❑ Salvage fire-damaged glulams.

SOUND CONTROL

❑ Reduce airborne and impact sounds in floor/ceiling systems.

DAMAGED SHEATING/SUBFLOORING

❑ Repair deteriorated or damaged floor sheathing.

FLOOR SQUEAKS

❑ Apply additional surface fasteners.
❑ Refasten floor sheathing to joist from below with lumber strip and screws.
❑ Refasten floor sheathing to joist with specialty fasteners.
❑ Renail existing bridging.
❑ Add new bridging.
❑ Adjust joist hangers.
❑ Adjust ductwork within floor cavity.

UNDERLAYMENTS

❑ Repair existing underlayment.
❑ Replace existing underlayment.

WOOD FLOORING

❑ Maintain wood flooring.
❑ Repair damaged wood floors.
❑ Stabilize and repair moisture-damaged wood floors.
❑ Sand wood floors.
❑ Refinish wood flooring with water-based urethanes.
❑ Refinish wood flooring with oil-modified urethanes.
❑ Refinish wood flooring with moisture-cured urethanes.
❑ Refinish wood flooring with "Swedish" finishes.
❑ Refinish wood flooring with oil finishes.
❑ Refinish wood flooring with waxes.
❑ Replace or recover wood floors.

VINYL SHEET FLOORING AND TILE

❑ Remove stains from sheet vinyl and vinyl tile flooring.
❑ Repair vinyl floor covering.
❑ Replace vinyl floor tile.
❑ Remove resilient tile and sheet goods.
❑ Install resilient vinyl tile.
❑ Install resilient sheet flooring.

CERAMIC TILE

❑ Maintain/restore ceramic tile.
❑ Replace loose, cracked, or damaged ceramic tile.

CARPET AND RUGS

❑ Maintain carpet.
❑ Restore water-damaged carpet.
❑ Repair stained carpet.
❑ Restore carpet from smoke damage.
❑ Install carpet over damaged floors.

PLASTER AND DRYWALL

❑ Refasten bowing or deflecting plaster.
❑ Refasten bowing or deflecting drywall.
❑ Repair cracks and holes in plaster.
❑ Repair small holes, cracks, dents, and popped nails in drywall.
❑ Repair medium holes.
❑ Repair large cracks (1/8" to 1/4").
❑ Repair torn drywall panel face paper.
❑ Repair large holes or water-damaged areas with U.S. Gypsum repair kit.
❑ Repair large sections of drywall.
❑ Laminate walls and ceilings with new drywall.
❑ Repair cracked or "alligatored" walls and ceilings with fiber-glass mats.
❑ Apply replacement drywall to damaged wall sections.
❑ Apply corner bead and joint compound to new drywall.
❑ Use alternative taping tools.

PAINTS AND WALL COVERINGS

❑ Treat stains on painted walls.
❑ Clean paper and vinyl wall coverings.
❑ Repair paper and vinyl wall coverings.
❑ Remove paper and vinyl wall coverings.

MOLDINGS AND TRIM

❑ Repair or replace damaged trim.
❑ Patch damaged trim.

REPAIRING STAIR TREADS AND RISERS

❑ Use graphite powder to fix squeaky treads and risers.
❑ Refasten tread from above.
❑ Refasten tread from below.
❑ Use wood wedges to fix squeaky stairs.

REPLACING STAIR TREADS AND RISERS

❑ Replace treads and risers.

SAGGING CARRIAGES

❑ Reinforce a sagging carriage.

DAMAGED OR BROKEN BALUSTERS

❑ Strengthen existing balusters.
❑ Replace filleted balusters.
❑ Replace doweled balusters.
❑ Replace dovetailed balusters.

PREFABRICATED STAIRS

❑ Install a prefabricated stair.
❑ Use a prefabricated stringer for stair construction.

ATTIC LADDERS

❑ Install an attic ladder.

CHAPTER 6: KITCHENS AND BATHS

CABINET SURFACE MAINTENANCE AND REPAIRS

❑ Touch-up or patch surfaces.
❑ Refinish surfaces.
❑ Reface fronts and sides.
❑ Replace cabinet doors and drawer fronts.

HARDWARE TO MAXIMIZE ACCESS AND FUNCTION

❑ Replace pulls or knobs.
❑ Replace hinges.
❑ Upgrade to specialty bifold hinges to improve access.
❑ Upgrade to specialty retractable door hinges.
❑ Install upswinging hinges.
❑ Upgrade to drop-down door hinges.
❑ Replace or upgrade to bottom-mount drawer glides.
❑ Replace or upgrade to side-mount drawer glides.
❑ Replace or upgrade to full-extension drawer glides.

ACCESSORIES TO MAXIMIZE ACCESS AND IMPROVE STORAGE

❑ Install a step stool.
❑ Improve corner storage.
❑ Replace fixed shelves with adjustable/roll-out shelves.
❑ Install specific purpose accessories.
❑ Install backsplash storage accessories.
❑ Install height-adjustable storage.

REPLACE OR ADD CABINETRY

❑ Determine cabinet condition.
❑ Supplement existing cabinetry.
❑ Select cabinetry with inset doors.
❑ Select cabinetry with offset doors.
❑ Select cabinetry with full overlay doors.
❑ Improve indoor quality with cabinetry.

COUNTERTOP SURFACE MAINTENANCE AND REPAIRS

❑ Replace laminate surfaces.
❑ Repair cultured stone.
❑ Repair aggregate stone.
❑ Repair solid surfaces.

❑ Repair or replace ceramic tile and grout.
❑ Repair or replace stone.
❑ Repair or replace wood or butcher block.
❑ Replace with stainless steel.
❑ Repair or replace concrete surfaces.

IMPROVE INDOOR AIR QUALITY

❑ Reduce particleboard outgassing.
❑ Cover all edges of substrate.
❑ Use low or nonformaldehyde substrate.

IMPROVE BACKSPLASH AND COUNTERTOP SEAMS

❑ Fill seams in laminate countertops.
❑ Fill seam between countertop and backsplash.
❑ Integrate backsplash with countertop.

MAXIMIZE ACCESS AND FUNCTION

❑ Install counters with eased or beveled corners.
❑ Add different counter heights.
❑ Install adjustable height counter.

IMPROVING FUNCTIONAL COUNTERTOP SPACE

❑ Install small appliances under wall cabinets.
❑ Install a pull-out work surface.
❑ Install a heat-resistant counter material.
❑ Install a rolling table or cart.

ENHANCE BACKSPLASH WITH SURFACE MATERIAL

❑ Install metallic laminate backsplash.
❑ Install decorative ceramic tile backsplash.
❑ Install stone backsplash.
❑ Install other materials.

REPLACE OUTDATED OR NONFUNCTIONING APPLIANCES

❑ Replace refrigerator.
❑ Replace range.
❑ Replace cooktop.
❑ Replace or add a range ventilation system.
❑ Replace a wall oven.
❑ Replace microwave.
❑ Replace dishwasher.
❑ Replace food waste disposer.

IMPROVE ACCESS AND FUNCTION WITHIN WORK CENTERS

❑ Consider refrigerator location and relocate if necessary.
❑ Consider cooktop and separate oven rather than range.
❑ Consider cooktop location and relocate if necessary.
❑ Consider optional oven placement.

- ❑ Consider microwave location and relocate if necessary.
- ❑ Consider dishwasher location and relocate if necessary.

UPGRADE APPLIANCE APPEARANCE

- ❑ Paint appliance exterior.
- ❑ Install panels.
- ❑ Create a built-in look for the refrigerator.
- ❑ Recess refrigerator.

IMPROVE RESOURCE AND ENERGY EFFICIENCY OF EXISTING APPLIANCES

- ❑ Consider appliance adjacencies and air temperature.
- ❑ Provide ventilation and circulation.
- ❑ Provide adequate ventilation to exterior.

INSTALL RESOURCE- AND ENERGY-EFFICIENT APPLIANCES

- ❑ Consider replacing the refrigerator.
- ❑ Consider replacing the dishwasher.
- ❑ Consider replacing the oven.
- ❑ Consider replacing the washer and dryer.

SINK AND LAVATORY SURFACE MAINTENANCE AND REPAIRS

- ❑ Refinish cultured stone.
- ❑ Repair vitreous china.
- ❑ Repair enameled steel.
- ❑ Repair or refinish enameled cast iron.
- ❑ Maintain stainless steel.
- ❑ Repair solid surface.

WATER PURIFICATION

- ❑ Install a carbon filter adapter or faucet.
- ❑ Install an under-counter carbon filtration system.
- ❑ Install an under-counter reverse osmosis and carbon combination filtration system.

SINK AND LAVATORY WATER CONSERVATION

- ❑ Install low-flow faucet.
- ❑ Retrofit a faucet aerator.
- ❑ Replace fittings.

MAXIMIZE SINK ACCESS AND FUNCTION

- ❑ Install diverter for spray.
- ❑ Install scald protection device at faucet.
- ❑ Install gooseneck or high arc spout.
- ❑ Install instant hot and instant cold dispensers.
- ❑ Install pot-filler faucet.
- ❑ Install pedal valve controls.
- ❑ Consider shape, size, and number of sinks.
- ❑ Install touchless faucet control.

TUB AND SHOWER SURFACE MAINTENANCE AND REPAIRS

- ❑ Repair fiberglass tub and surround.
- ❑ Repair acrylic tub, base, and surround.
- ❑ Repair cultured stone base and surround.
- ❑ Install a liner over existing tub or shower base.
- ❑ Repair enameled steel tub.
- ❑ Repair or refinish enameled cast iron tub.
- ❑ Repair solid surface base and surround.
- ❑ Repair or replace ceramic tile and grout.
- ❑ Repair stone base and surround.

MOISTURE CONTROL

- ❑ Improve ventilation.
- ❑ Reduce moisture at recessed lights.
- ❑ Manage water penetration at tile.
- ❑ Change installation of tile substrate at tub/shower seam.

MAXIMIZE TUB AND SHOWER ACCESS AND FUNCTION

- ❑ Install scald protection devices.
- ❑ Relocate controls.
- ❑ Install built-in seat to improve transfer.
- ❑ Install blocking for grab bars.
- ❑ Use a "no threshold" shower base.

TUB AND SHOWER WATER CONSERVATION

- ❑ Replace fittings.
- ❑ Install high-efficiency low-flow showerhead.

TOILET AND BIDET MAINTENANCE AND REPAIRS

- ❑ Repair vitreous china surface.
- ❑ Repair water seepage.
- ❑ Repair run-on toilet.

TOILET AND BIDET WATER CONSERVATION

- ❑ Install water-limiting devices.
- ❑ Install a gravity-assisted toilet.
- ❑ Install a pressure-assisted toilet.
- ❑ Install a wall-hung urinal.

MAXIMIZE TOILET AND BIDET ACCESS AND FUNCTION

- ❑ Install special seat to increase seat height.
- ❑ Increase height of toilet seat.
- ❑ Install blocking for grab bar.
- ❑ Install wall-hung toilets at desired height.
- ❑ Install toilet/bidet combination.

CHAPTER 7: ELECTRICAL AND ELECTRONIC SYSTEMS

SERVICE PANELS

❑ Install new panel with greater capacity.
❑ Install a home automation-ready service panel.
❑ Install a new panel with same capacity but more circuits.
❑ Install a subpanel.
❑ Install a dual or half-sized breaker.
❑ Install "Type S" adapters in existing fuse box.
❑ Install ground fault circuit breaker.
❑ Install an arc fault circuit interrupter (AFCI).

OLD WIRING

❑ Rewire the electrical system.
❑ Install wiring raceways.
❑ Repair insulation.

ALUMINUM WIRING

❑ Replace existing wiring.
❑ Splice wire using AMP COPALUM connectors.
❑ Replace wire using "Scotchlok" twist-on connectors.

RECEPTACLES

❑ Replace existing wiring.
❑ Install a new ground conductor.
❑ Connect receptacle to the ground receptacle box.
❑ Install a receptacle-type ground fault circuit interrupter.

INTERIOR LIGHTING

❑ Rewire existing fixture.
❑ Install a new, period reproduction fixture.
❑ Install additional light fixtures.
❑ Install a dimmer switch.

EXTERIOR LIGHTING

❑ Rewire existing fixture.
❑ Install a new, period reproduction fixture.
❑ Install photovoltaic (PV) lighting.
❑ Install a full cut-off fixture.
❑ Install white LED lighting.

CONTROLS

❑ Install a new, period reproduction switch.
❑ Install a dimmer with built-in noise filter.

❑ Install a lamp debuzzing coil (LDC).
❑ Install a self-contained device (SCD) type switch.
❑ Install wireless switches.
❑ Install a fiber-optic switch.
❑ Install a motion-sensor switch.

ELECTRIC BASEBOARD HEATING

❑ Maintain existing system.
❑ Replace old baseboard units with new units.
❑ Replace old thermostats with new thermostats.
❑ Supplement existing electric baseboard heat with additional heating system.
❑ Replace existing baseboard heat with more efficient system.

PHONE/COMPUTER/TV CABLING

❑ Repair cracked insulation or frayed wiring.
❑ Rewire within existing walls and ceilings.
❑ Rewire using cabling raceways.

INTRUSION AND ALARM SYSTEMS

❑ Inspect existing system.
❑ Replace control panel.
❑ Replace sensors.
❑ Replace wiring.
❑ Replace entire system.

SMOKE DETECTORS

❑ Maintain existing smoke detector.
❑ Replace existing smoke detector with new unit.

CARBON MONOXIDE DETECTORS

❑ Install a CO detector.

LIGHTNING PROTECTION

❑ Maintain existing lightning protection system.
❑ Install new materials on existing lightning protection system.
❑ Install new lightning protection system.

SURGE PROTECTION

❑ Install new surge suppression at breaker panel.
❑ Provide new uninterrupted power supply (UPS).
❑ Provide new plug-in surge suppression.

GARAGE DOOR OPENERS

❑ Maintain existing system.
❑ Replace circuit board and add photocell.
❑ Replace remote control.
❑ Replace existing system with new system.

CHAPTER 8: HVAC/PLUMBING

ENERGY SOURCES

❑ Consider electricity as an energy source.
❑ Consider natural gas as an energy source.
❑ Consider oil as an energy source.
❑ Consider propane as an energy source.
❑ Consider solar as an energy source.

DISTRIBUTION SYSTEMS

❑ Rehab the existing forced-air distribution system.
❑ Rehab the existing hydronic distribution system.
❑ Rehab the existing electric distribution system.
❑ Install a new conventional forced-air distribution system.
❑ Install a mini-duct HVAC system.
❑ Install a new hydronic distribution system.
❑ Install a zonal system.

HEATING

❑ For oil-fired systems, install a flame retention burner.
❑ Install a vent damper.
❑ Replace the standing pilot with an electronic spark ignitor.
❑ For oil-fired furnaces, downsize the burner nozzle.
❑ Install an advanced, high-efficiency gas furnace.
❑ Install a combination system.
❑ Install an advanced high-efficiency boiler.

COOLING

❑ Install a circulation fan.
❑ Install a power attic or rooftop ventilator.
❑ Install a whole-house ventilator or fan.
❑ Install a radiant barrier.
❑ Install an evaporative cooler.
❑ Install an electric air conditioner.
❑ Install a gas-fired chiller.

HEAP PUMPS

❑ Install a split-system air-source heat pump (ASHP).
❑ Install a single-package air-source heat pump.
❑ Install a packaged terminal heat pump (PTHP).
❑ Install one or more ductless split systems or mini-splits.
❑ Install a ground-source heat pump (GSHP).

INDOOR AIR QUALITY

❑ Install a mechanical ventilation system.
❑ Install a humidifier.
❑ Install a dehumidifier.
❑ Install an air filter or cleaner.

CONTROLS

- ❑ Replace the bimetal thermostat with an electronic programmable thermostat.
- ❑ Install a humidistat.
- ❑ Install a control system that senses outdoor conditions and adjusts accordingly.
- ❑ Install ventilation system controls.
- ❑ Install a forced-air zoning system.

FIREPLACES AND CHIMNEYS

- ❑ Rebuild or repoint the chimney.
- ❑ Install a chimney cap.
- ❑ Reline the chimney with metal pipe.
- ❑ Reline the chimney with tile.
- ❑ Reline the chimney with a cementitious mix.
- ❑ Install an exterior air supply for combustion and draft air.
- ❑ Install an environmentally friendly grate.
- ❑ Install a gas log set or fireplace.
- ❑ Install wood stove, fireplace insert, or advanced fireplace.

DOMESTIC HOT WATER HEATING

- ❑ Improve the efficiency of the existing storage-type water heater.
- ❑ Install an indirect storage water heater.
- ❑ Install a new electric resistance storage water heater.
- ❑ Install a new gas storage water heater.
- ❑ Install a demand water heater.
- ❑ Supplement the water heater with a preheating or heat recovery system.
- ❑ Install a heat pump water heater (HPWH).

WATER SUPPLY AND DISTRIBUTION SYSTEMS

- ❑ Improve water distribution.
- ❑ Install new copper piping.
- ❑ Install chlorinated polyvinyl chloride (CPVC) piping.
- ❑ Install cross-linked polyethylene (PEX) piping.

DRAIN, WASTE, AND VENT SYSTEMS

- ❑ Install cast iron drain, waste, and vent (DWV) piping.
- ❑ Install copper DWV piping.
- ❑ Install acrylonitrile butadiene styrene (ABS-DWV) and polyvinyl chloride (PVC-DWV piping.
- ❑ Install pumps and ejectors to discharge waste.
- ❑ Install gray water and heat recovery devices.
- ❑ Install an air admittance valve.

FUEL SUPPLY SYSTEMS

- ❑ Install steel pipe and fittings.
- ❑ Install copper pipe and fittings.
- ❑ Install corrugated stainless-steel tubing (CSST).

APPLIANCE VENTS AND EXHAUSTS

- ❑ Install a new exhaust for a clothes dryer.
- ❑ Install a new exhaust for a range.

FIRE PROTECTION SYSTEMS

- ❑ Install a stand-alone residential sprinkler system.
- ❑ Install a multipurpose residential sprinkler system.
- ❑ Install a copper piped system.
- ❑ Install a CPVC-piped system.

CHAPTER 9: SITE WORK

DECK/PORCH STRUCTURE

- ❑ Repair moisture-damaged posts or columns.
- ❑ Repair moisture-damaged joists.
- ❑ Repair rotted ledger boards.
- ❑ Stabilize rotting material with epoxy.

REPAIR/REPLACE WOOD DECKING MATERIALS

- ❑ Clean mildew-stained wood decking.
- ❑ Repair or replace existing decking.
- ❑ Replace with certified sustainably harvested wood.
- ❑ Replace with redwood and cedar.
- ❑ Replace with tropical hardwoods.
- ❑ Replace with plastic-wood composites.
- ❑ Replace with vinyl deck systems.

PRESERVATIVE-TREATED WOOD

- ❑ Use wood with factory-applied preservatives.
- ❑ Apply nontoxic finishes, stains, and cleaners.
- ❑ Apply borate-based preservatives.

STAIRS AND HANDRAILS

- ❑ Add support to existing railing posts.
- ❑ Replace existing wood railings.
- ❑ Replace damaged wood treads or risers.
- ❑ Replace damaged stringers.

WOOD FENCES AND RETAINING WALLS

- ❑ Strengthen an existing wood fence post.
- ❑ Replace a wood fence post.
- ❑ Strengthen an existing wood fence rail.
- ❑ Replace an existing wood fence rail.
- ❑ Replace damaged wood fence screening (pickets).
- ❑ Strengthen a sagging wood gate.
- ❑ Realign a timber retaining wall.

PAVED DRIVEWAYS, WALKS, AND PATIOS

- ❑ Seal asphalt pavement.
- ❑ Repair cracks in asphalt.
- ❑ Repairs holes in asphalt.
- ❑ Remove stains in concrete paving.

- ❑ Seal concrete paving.
- ❑ Repair cracks in concrete.
- ❑ Repair a sunken concrete walk or drive.
- ❑ Patch concrete.
- ❑ Repair crumbling concrete.
- ❑ Resurface concrete.
- ❑ Lay new concrete next to existing.
- ❑ Repair or replace damaged concrete steps.
- ❑ Repair steel handrails.
- ❑ Repair brick or stone steps.

MASONRY WALLS

- ❑ Repoint existing masonry wall.
- ❑ Resurface an existing masonry wall.
- ❑ Replace parts of existing masonry wall.
- ❑ Ensure good retaining wall drainage.
- ❑ Repair a masonry retaining wall.
- ❑ Replace a masonry retaining wall.

WELLS

- ❑ Remedy poor water quality.
- ❑ Remedy improperly sized well.
- ❑ Plug abandoned wells and boreholes near drinking well.

ON-SITE WASTEWATER TREATMENT

- ❑ Repair the septic system.
- ❑ Pump out the septic system.
- ❑ Rejuvenate the leaching field.
- ❑ Install a new septic system.

WATER AND SEWER LINES

- ❑ Repair the water line.
- ❑ Replace the water line.
- ❑ Repair the sewer line.
- ❑ Replace the sewer line.

UNDERGROUND STORAGE TANKS

- ❑ Remove/abandon an underground tank.

LANDSCAPE CARE

- ❑ Repair trees and shrubs.
- ❑ Protect trees and shrubs during construction.
- ❑ Remove trees or shrubs.
- ❑ Replace trees or shrubs.
- ❑ Install lightning protection for trees.

ENERGY-EFFICIENT AND SUSTAINABLE LANDSCAPING

- ❑ Locate trees and shrubs for shade.
- ❑ Locate trees and shrubs as wind buffers.
- ❑ Water plants with conservation in mind.
- ❑ Promote sustainable landscaping.
- ❑ Replace lawn with low-impact vegetation.

PRODUCT INFORMATION

CHAPTER 1: FOUNDATIONS

BASEMENT FLOORS

There are numerous companies that manufacture self-leveling toppings and other concrete repair products. Some of those with national distribution include:

Abatron, Inc., 5501 95th Ave., Kenosha, WI 53144; 800-445-1754; www.albatron.com.

Ardex, Inc., 400 Ardex Park Drive, Aligneippa, WI 15001; 412-264-4240; www.ardex.com.

Bonsal Co., P.O. Box 241148, Charlotte, NC 28224; 800-334-0784; www.bonsal.com.

Chemrex, Inc. (Thoro Products), 889 Valley Park Drive, Shakopee, MN 55379; 800-433-9517; www.chemrex.com.

Dayton Superior Concrete Accessories, 721 Richard St., Miamisburg, OH 45342; 800-745-3700; www.datonsuperior.com.

Larsen Products Corp., 8264 Preston Ct., Jessup, MD 20794; 800-633-6668; www.larsenproducts.com.

Laticrete International, Inc., 1 Laticrete Park North, Bethany, CT 06524-3423; 800-243-4788; www.laticrete.com.

Master Builders, Inc. (Div. of Construction Chemicals), 23700 Chagrin Blvd., Cleveland, OH 44122; 800-628-9990; www.masterbuilders.com.

Maxxon Corp. (formerly Gyp-Crete Corp.) 920 Hamel Rd., Hamel, MN 55340; 800-356-7887; www.maxon.com.

Quick Crete Products Corp., 3097 Presidential Drive, Suite 30340, Atlanta, GA 30329; 800-282-5828; www.quickcrete.com.

Sika Corp., 201 Polito Ave., Lyndhurst, NJ 07071; 800-933-7452; www.sika.com.

Sonneborn Building Products (now a part of Chemrex, Inc., see above).

Sto Concrete Restoration Division, 3800 Camp Creek Parkway Bldg. 1400, Suite 120, Atlanta, GA 30331; 800-221-2397; www.stocorp.com.

Tamms Industries Inc., 3835 State Route 72, Kirkland, IL 60146; 800-862-2667.

WR Meadows, Inc., 300 Industrial Drive, Hampshire, IL 60140; 800-342-5976; www.wrmeadows.com.

CRAWL SPACE FLOORS

Neutocrete, 564 Danbury Rd., New Milford, CT, 06776; 860-354-8500; www.neutocrete.com.

PREFABRICATED FOUNDATIONS

Insulspan, P.O. Box 38, Blissfield, MI 49228; 800-726-3510.

Kistner Concrete Products, Inc. 8713 Read Road P.O. Box 218 East Pembroke, New York 14056-0218; 716-894-2267, www.kistner.com.

Superior Walls of America, Ltd., 937 East Earl Road, New Holland, PA 17557; 800-452-9255; www.superiorwalls.com.

FOUNDATION DRAINAGE

American Wick Drain Corp., 1209 Airport Road, Monroe, NC 28105; 800-242-WICK; www.american-wick.com.

Basement De-Watering Systems™, 162 East Chestnut St., P.O. Box 160, Canton, IL 61520; 800-331-2943; www.bdws.com.

Drainboard, Roxul, Inc., 551 Harrop Dr., Milton, Ontario L9T 3H3 Canada; 800-265-6878.

Enkadrain, Colbond-USA, div. of Akzo Nobel Geosynthetics Co., P.O. Box 1057 Sand Hill Road, Enka, NC 28728; 800-365-7391; www.colbond-usa.com.

Hancor, Inc., 401 Olive St., P.O. Box 1047, Findlay, OH 45839; 800-537-9520; www.hancor.com.

Home Guard, HiLo Industries (a division of Zoeller Co.), 3649 Cane Run Road, Louisville, KY 40211; 502-778-0234; 888-968-3309; www.hiloind.com.

INSUL-DRAIN™, Owens-Corning, One Owens Corning Parkway, Toledo, OH 43659; 800-GET-PINK; www.owens-corning.com.

Sanford Irrigation, 444 East Highway 79, Elbow Lake, MN 56531; 218-685-4344; www.sanfordirrigation.com.

Styrofoam PERIMATE, Dow Chemical Co., 2040 Willard H. Dow Center, Midland, MI, 48674; 800-441-4369; www.dow.com/styrafoam/index.html.

System Platon, Big "O" Inc., 33 Centennial Road, Orangeville, Ontario L9W 1R1 Canada; 800-265-7622; www.systemplaton.com.

WARM-N-DRI, Koch Materials Co., 800 Irving Wick Drive, Heath, OH 43056; 800-379-2768; www.tuff-n-dri.com.

WaterGuard™ by Basement Systems, Inc., 60 Silvermine Road, Seymour, CT 06483; 800-768-0935; www.basementsystems.com.

DAMPROOFING AND WATERPROOFING

Aquafin, Inc., P.O. Box 1440, Columbia, MD 21044; 888-482-6339; www.vandexus.com.

Bonsal Co., P.O. Box 241148, Charlotte, NC 28224-1148; 800-334-0784; www.bonsal.com.

Concrete Restoration, Materials and Application Field Guide, STO Concrete Restoration Division, P.O. Box 44609 Atlanta, GA 30331; 800-321-2397; www.stocorp.com.

Gaco Western Inc., P.O. Box 88698, Seattle, WA 98138-2698; 800-456-GACO; www.gaco.com.

Grace Construction Products, W.R. Grace & Co., 62 Whittemore Ave., Cambridge, MA 02140-9901; 800-354-5414; www.grace.com.

Griffolyn, Div. of Reef Industries, Inc., P.O. Box 750250, Houston, TX 77275; 800-231-6074; www.reefindustries.com.

Karnak Corp., 330 Central Ave., Clark, NJ, 07066; 800-526-4236; http://karnakcorp.com.

Koch Materials Co., 800 Irving Wick Drive, Heath, OH 43056; 800-379-2768; www.kochmaterials.com.

Permaquik Corp., 6178 Netherhart Road, Mississauga, Ontario L5T 1K4 Canada; 905-564-6100; www.permaquik.com.

Poly-Wall International, 408 Red Cedar Street, Suite 6, Menomonie, WI 54751; 800-846-3020; www.polywall.com.

"Rub-R-Wall," Rubber Polymer Corp., 1135 West Portage Trail Ext., Akron, OH 44313; 800-860-7721; www.rpcinfo.com.

Sonneborn, ChemRex, Inc., 889 Valley Park Dr., Shakopee, MN 55379; 800-433-9517; www.chemrex.com.

ThoroSeal, ChemRex, 889 Valley Park Drive, Shakopee, MN 55379; 800-243-6739; www.chemrex.com/thoro.

Tremco, Inc., 3735 Green Road, Beachwood, OH 44122; 800-321-6412; www.tremcosealants.com.

Volclay Waterproofing, Colloid Environmental Technologies Co., 1500 W. Shure Dr., Arlington Heights, IL 60004; 800-527-9948; www.cetco.com.

Xypex Chemical Corp., 13731 Mayfield Pl., Richmond, British Columbia, V6V 2G9 Canada; 888-443-7922; www.hi-dri.com.

INSULATING BELOW-GRADE WALLS

DFI Pultruded Composites, Inc., 1600 Dolwick Drive, Erlanger, KY 41018; 606-282-7300; www.dfi-inc.com.

INSUL-DRAIN™, Owens Corning, One Owens Corning Parkway, Toledo, OH 43659; 800-GET-PINK; www.owenscorning.com.

Perform Guard, Advance Foam Plastics, 24000 Hwy 7, P.O. Box 246, Excelsior, MN 55331; 800-255-0176; www.advancefoam.com.

Styrofoam PERTMATE and Styrofoam Wallmate, Dow Chemical Co., 2040 Willard H. Dow Center, Midland, MI, 48674; 800-441-4369; www.dow.com/styrafoam/index.htm.

TUFF-N-DRI, Koch Materials Co., 4111 East 37th Street North, Witchita, KS 67220; 800-DRYBSMT; www.tuff-n-dri.com.

INSULATING CRAWL SPACES

Refer to "Insulating Below-Grade Walls."

INSULATING SLABS

Refer to "Insulating Below-Grade Walls."

VENTILATING BASEMENT SPACES

ECHO System, Indoor Air Technologies, Inc., P.O. Box 22038, Sub 32, Ottawa, Ontario, Canada K1V 0W2; 800-558-5892; www.indoorair.ca.

VENTILATING CRAWL SPACES

CellarSaver, Tamarack Technologies, Inc., 11A Patterson's Brook Road, PO Box 490, West Wareham, MA 02576; 800-222-5932; www.tamtech.com.

SHORING, UNDERPINNING, AND REPAIR

A.B. Chance Company, 210 North Allen St., Centralia, MO 65240; 573-682-8414; www.abchance.com.

Atlas Systems, Inc., 1026-B South Powell Road, Independence, MO 64056; 800-325-9375; www.atlassys.com.

Hayward Baker Geotechnical Engineers and Drilling Contractors, 1875 Mayfield Rd., Odenton, MD 21113; 410-551-8200, 800-888-4292; www.haywardbaker.com.

Perma-Jack of Kansas, 9066 Watson Road, St. Louis, MO 63126, 800-843-1888; www.permajack.com.

CRACKS IN WALLS AND SLABS

3M Construction Markets Division, 3M Center Building, 304-1-01, St. Paul, MN 55144-1000; 800-3M-HELPS; www.3m.com/market/construction.

A.B. Chance Company, 210 North Allen St., Centralia, MO 65240; 573-682-8414.

Atlas Systems, Inc., 3114 Weatherford Road, Independence, MO 64055; 800-325-9375; www.atlassys.com.

Abatron, Inc., 5501 95th Avenue, Kenosha, WI, 53144; 800-445-1754; www.albatron.com

DeNeef Construction Chemicals, P.O. Box 1219, Waller, TX 77484; 409-372-9185, 800-732-0166; www.deneef.com.

Green Mountain International, 235 Pigeon Street, Waynesville, NC 28786; 800-942-5151; www.mountaingrout.com.

GRIP-TITE Anchored Walls, Inc., 1962 Highway 92, Winterset, IA 50273; 800-221-4699; www.anchoredwalls.com.

Prime Resins, Inc., 2381 Rockaway Industrial Blvd., Conyers, GA 30012; 800-321-7212; www.primeresins.com.

W.R. Bonsal 8201 Arrowridge Blvd., Charlotte, NC 28224-1148; 800-334-0784.

COATINGS AND FINISHES

Sherwin-Williams Company, 101 Prospect Avenue, N.W., Cleveland, OH 44115; 800-321-8194; www.sherwinwilliams.com.

W.R. Bonsal Company, P.O. Box 241148, Charlotte, NC 28224-1148; 800-334-0784; www.bonsal.com.

CHAPTER 2: EXTERIOR WALLS

WOOD-FRAME SEISMIC RESISTANCE

Earthquake Resistant Construction Connectors, Simpson Strong-Tie® Co., 4637 Chabot Drive, Suite 200, Pleasanton, CA 94588; 800-999-5099; www.strongtie.com.

United Steel Products Co. (USP), 703 Rogers Drive, Montgomery, MN 56069; 800-328-5934; www.unitedsteelproducts.com.

WOOD-FRAME WIND RESISTANCE

High-Wind-Resistant Construction Connectors, Simpson Strong-Tie Co., Inc., 4637 Cabot Drive, Suite 200, Pleasanton, CA 94588; 800-999-5099; www.strongtie.com.

High-Wind Retrofit of Wood Trusses or Rafters to Masonry or Concrete Walls, Simpson Strong-Tie Co., Inc., 4637 Cabot Drive, Suite 200, Pleasanton, CA 94588; 800-999-5099; www.strongtie.com.

REINFORCING EXISTING MASONRY WALL CONSTRUCTION

High-Wind Retrofit of Wood Trusses or Rafters to Masonry or Concrete Walls, Simpson Strong-Tie Co., Inc., 4637 Cabot Drive, Suite 200, Pleasanton, CA 94588; 800-999-5099; www.strongtie.com.

MOISTURE DETERIORATION

Abatron, Inc, Wood Restoration Systems, 5501-95th Avenue, Kenosha, WI 53144; 800-445-1754; www.abatron.com.

Preservation Resource Group, P.O. Box 1768, Rockville, MD 20849-1768; 301-309-2222.

FIRE DAMAGE TO WOOD FRAMING

Unsmoke Systems, Inc., 1135 Braddock Avenue, Braddock, PA 15104; 800-332-6037; www.unsmoke.com.

REPAIR EXISTING MASONRY WALLS

DRAINAGE MESH

Mortar Net™ USA Ltd., 541 S. Lake Street, Gary, IN 46403; 800-664-6638; www.mortarnet.com.

BRICK TIES, ACCESSORIES, AND STABILIZATION SYSTEMS

Dur-o-Wal®, Inc., 3115 N. Wilke Road, Suite A, Arlington Heights, IL 60004; 800-323-0090; www.dur-o-wal.com.

Heckman Building Products, Inc., 4260 Westbrook Drive, Suite 120, Aurora, IL 60504; 800-621-4140; 877-851-8400; www.heckmannbuildingprods.com.

Helifix®, 110 Maple Creek Road, Concord, Ontario L4K-184 Canada; 888-992-9989; www.helifix.com; www.block-lock.com.

Hohman and Bainard, Inc. 30 Rasons Court, P.O. Box 5270, Hauppauge, NY 11788-0270; 800-645-0616; www.H-B.com.

SHEATHING

PLYWOOD AND OSB

APA-The Engineered Wood Association, P.O. Box 11700, Tacoma, WA 98411-0700; 253-565-6600; www.apawood.org.

OSB

Structural Board Association, 45 Sheppard Avenue East, Suite 412, Willowdale, Toronto, Ontario M2N 5W9 Canada; 416-730-9090; www.sba-osb.com.

RIGID FOAM INSULATION

Celotex Building Products, P.O. Box 31602, Tampa, FL 33631-3602; 800-CELOTEX; www.celotex.com.

Dow Chemical Company, Styrofoam Brand Products, 2020 Willard H. Dow Center, Midland, MI 48674; 800-258-2436; www.dow.com/styrafoam.

Owens Corning, One Owens Corning Parkway, Toledo, OH 43659; 800-354-PINK or 800-GET-PINK; www.owenscorning.com.

Tenneco Building Products, 2907 Log Cabin Drive, Smyrna, GA 30080; 800-241-4402; www.tennecobuilding.com.

RADIANT BARRIER AND PAPER BOARD PRODUCTS

Energy-Brace™ reflective sheathing; Fiber-Lam, Inc., P.O. Box 2002, Doswell, VA 23047; 804-876-3135.

Ludlow Coated Products; www.ludlow.com.

Thermo-ply™ reflective sheathing; Simplex Products Division, P.O. Box 10, Adrian, MI 49221; 517-263-8881.

VAPOR RETARDERS

Owens Corning, Fiberglass Tower, Toledo, OH 43659; 800-GET-PINK (kraft and foil-faced batt insulation); www.owenscorning.com.

Raven Industries, P.O. Box 5107, Sioux Falls, SD 57117-5107; 800-635-3456 (Rufco Moisture/Vapor Barriers); www.ravenind.com.

Reef Industries, Inc., P.O. Box 750250, Houston, TX 77275-0250; 800-231-6074 (Griffolyn Reinforced Vapor Barriers); www.reefindustries.com.

Sto-Cote Products, Inc., 218 South Road, Genoa City, WI 53128; 800-435-2621 (Tu-Tuf products).

AIR INFILTRATION BARRIERS

The Celotex Corp., P.O. Box 31602, Tampa, FL 33631; 800-CELOTEX (Tuff Wrap); www.celotex.com.

DuPont, Chesnut Run Plaza, Building 705 Room 2ES, Center Road, Wilmington, DE 19805; 800-441-7515; (Tyvek Homewrap); www.tyvekconstruction.com; www.dupont.com.

Owens Corning, Fiberglass Tower, Toledo, OH 43659; 800-GET-PINK (Pinkwrap); www.owenscorning.com.

Raven Industries, P.O. Box 5107, Sioux Falls, SD 57117-5107; 800-635-3456 (Rufco Wrap); www.ravenind.com.

Simplex Products Division, P.O. Box 10, Adrian, MI 49221-0010; 800-345-8881 (R-Wrap, Barricade Building Paper); www.ludlowcp.com.

INSULATION

BATT, SPRAY-ON, AND LOOSE-FILL INSULATION

American Rockwool, Inc., P.O. Box 880, Spring Hope, NC 27882; 919-478-5111; www.amerrock.com.

Ark-Seal International, 2190 S. Kalamath Street, Denver, CO 80223; 800-525-8992; www.arkseal.com.

Building Products Division, The Celotex Corporation, P.O. Box 31602, Tampa, FL 33631; 800-CELOTEX; www.celotex.com.

Cellulose Insulation Manufacturers Association, 136 S. Keowee Street, Dayton, OH 45402; 937-222-2462; www.cellulose.org.

CertainTeed Corporation, 750 E. Swedesford Rd., Valley Forge, PA 19482; 800-523-7844; www.certainteed.com.

GreenStone Industries Inc., 809 Westhill Street, Suite 4, Charlotte, NC 28208-9124; 800-666-4824; www.greenstone.com.

Icynene Inc., 5805 Whittle Road, Suite 110, Mississauga, Ontario L42 2J1 Canada; 800-946-7325; www.icynene.com.

International Cellulose Corp., 12315 Robin Blvd., Houston, TX 77245-0006; 800-444-1252; www.internationalcellulose.com.

Johns Manville Corporation (formerly Schuller International Inc.), 717 17th St., Denver, CO 80202; or P.O. Box 5108, Denver, CO 80217-5108; 800-654-3103; www.jm.com.

Knauf Fiber Glass, 240 Elizabeth St., Shelbyville, IN 46176; 800-825-4434; www.knauffiberglas.

Owens Corning, One Owens Corning Parkway, Toledo, OH 43659; 800-354-PINK or 800-GET-PINK; www.owenscorning.com.

Par-Pac™, 27 Main Street West, Swanley, NH 03446; 877-937-3257; www.parpac.com.

Rock Wool Manufacturing Co., 203 N. 7th St., Leeds, AL 35094; 800-874-7625.

Sloss Industries Corporation, 3500 35th Avenue North, Birmingham, AL 35207; 205-808-7803; www.sloss.com.

U.S. Fiber, Inc., 809 Westhill Street, Charlotte, NC 28208; 800-666-4824; www.usgf.com.

Western Fiberglass Group, 6955 Union Park Center, Suite 580, Midvale, UT 84047; 801-562-9558.

STRUCTURAL INSULATED PANELS

Structural Insulated Panel Association, 3413 56th Street, Suite A, Gig Harbor, WA 98335; 253-858-7472; www.sips.org.

RIGID FOAM INSULATION

Celotex Building Products, P.O. Box 31602, Tampa, FL 33631-3602; 800-CELOTEX; www.celotex.com.

Dow Chemical Company, Styrofoam Brand Products, 2020 Willard H. Dow Center, Midland, MI 48674; 800-258-2436; www.dow.com/styrofoam.

Expanded Polystyrene Molders Association (EPSMA), 2128 Espey Court, Suite 4, Crofton, MD 21114; 800-607-3772, www.epsmolders.org.

Johns Manville Corporation (formerly Schuller International Inc.), 717 17th St., Denver, CO 80202; or P.O. Box 5108, Denver, CO 80217-5108; 800-654-3103; www.jm.com.

Owens Corning, One Owens Corning Parkway, Toledo, OH 43659; 800-354-PINK or 800-GET-PINK; www.owenscorning.com.

Polyisocyanurate Insulation Manufacturer's Association (PIMA), 1331 F Street NW, Suite 975, Washington, DC 20004; 202-624-2709; www.pima.org.

Tenneco Building Products, 2907 Log Cabin Dr., Smyrna, GA 30080; 800-241-4402; www.tennecobuildingprod.com.

RADIANT BARRIER PRODUCTS

Radiance™ Low-e interior paint, ChemRex, 889 Valley Park Drive, Shakopee, MN 55379; 800-433-9517; www.chemrex.com.

TYCO-Plastic and Adhesives; 800-345-8881; www.ludlowcp.com.

VINYL SIDING

Foam-Core™—International Paper, P.O. Box 1839, Statesville, NC 28687-1839; 800-438-1701; www.ipaper.com; www.gatorfoam.com.

Progressive Foam Products, Beach City, Ohio; 800-860-3636; www.progressivefoam.com.

TechWall™—VIPCO, 6753 Chestnut Ridge Road, Beach City, OH 44608; www.crane-plastics.com.

Vinyl Siding Institute, 1801 K Street, Suite 600K, Washington, D.C. 20006; 888-FORVSI-1; www.vinyl-sid
ing.org. (A complete list of member companies, their Web sites and products, can be obtained from the Vinyl Siding Institute.)

METAL SIDING

ABC Seamless, 3001 Feichner Drive, Fargo, ND 58103; 800-732-6577; www.abcseamless.com.

Alcoa Building Products, P.O. Box 57, 1501 Michigan St., Sidney, OH 45365-0057; 800-962-6973; www.alcoahomes.com.

Alside, 373 State Road, P.O. Box 2010, Akron, OH 44309; 800-257-4335; www.edcoproducts.com.

EDCO, 30350 Edison Road, New Carlisle, IN 46552-0960; 800-298-2628; www.edcoproducts.com.

Norandex/Reynolds Distribution Co., 8450 S. Bedford Road, Macedonia, OH 44056; 330-468-2200; www.norandex.com; www.reynoldsbp.com.

Reynolds Building Products, One Norandex Place, Macedonia, OH 44056; 330-468-2200; www.reynoldsbp.com.

United States Seamless, Inc., 2001 1st Ave. N., P.O. Box 2426, Fargo, ND 58108-2426; 800-SIDE ME 2; www.usseamless.com.

WOOD SHINGLES AND SHAKES

Cedar Breather™, Benjamin Opdyke, Inc., 199 Precision Drive, Horsham, PA 19044; 800-346-7655; www.opdyke.com/cbindex.htm.

Cedar Shake & Shingle Bureau, P.O. Box 1178, Sumas, WA 98295; 604-462-8961; www.cedarbureau.org.

Sovebec Eastern White Cedar, Sovebec, Inc., 9201 Centre Hospitalier Blvd., Charny, Quebec, Canada G6x 1L5; 418-832-1456; www.sovebec.com.

SOLID WOOD SIDING

Granville Manufacturing Co., Granville, VT 05747; 802-767-4747; quartersawn spruce and pine clapboard siding; www.woodsiding.com.

Siding 2000, prefinished cedar siding: Coastal Forest Products, 451 South River Road, P.O. Box 10898, Bedford, NH 03110; 800-932-WOOD; www.coastalforestproducts.com.

Skookum Lumber Co., Box 1398, Olympia, WA 98507-1398; 360-352-7633 (cedar siding); www.skookum.com.

Step Saver Siding: Factory Primed Western Red Cedar; PPG Industries, Inc., 1 PPG Place, Pittsburgh, PA 15272; 800-441-9695; www.ppgaf.com.

HARDBOARD SIDING

American Hardboard Association, 1210 W. Northwest Highway, Palatine, IL 60067; 847-934-8800; www.ahardbd.org; www.hardboard.org.

Hardboard Manufacturers:

AL Specialty Products, 10115 Kincey Ave., Suite 15U, Hunterville, NC 28075; 800-566-2282.

Collins Products, LLC, P.O. Box 16, Klamath Falls, OR 97601; 800-417-3674; www.collinswood.com.

Forestex Co., P.O. Box 68, Forest Grove, OR 97116; 503-357-2131; www.stimsonlumber.com.

Georgia Pacific Corp., 133 Peachtree St., NE, Atlanta, GA, 30303; 800-284-5347; www.gp.com.

Masonite Corp., 1 South Wacker Drive, Suite 3600, Chicago, IL 60606; 800-323-4591; www.masonite.com.

Temple-Inland Forest Products Corp., P.O. Drawer N, Diboll, TX 75941; 800-231-6060; www.temple.com.

ENGINEERED WOOD SIDING

Louisiana-Pacific, 111 SW Fifth Avenue, Portland, OR 97204; 800-648-6893; www.LPcorp.com (Smart Lap™, Smart Panel™, and EZ Panel™ siding).

PLYWOOD PANEL SIDING

Breckenridge Siding™: Roseburg Forest Products, P.O. Box 1088, Roseburg, OR 97420; 800-859-6998; www.rfpo.com.

Duratemp Plywood Siding™: Stimson Lumber Company, 520 Southwest Yamhill, Suite 325, Portland, OR 97204; 800-445-9758; www.stimsonlumber.com.

Georgia-Pacific Corporation, P.O. Box 105605, Atlanta, GA 30348-5605; 800-284-5347; www.gp.com.

Guardian Siding™: Simpson Timber Company, 204 East Railroad Ave., Shelton, WA 98584; 800-782-9378; www.simpson.com.

International Paper 6600 L.B.J. Freeway, Dallas, TX 75240; 800-527-5907.

Louisiana-Pacific, 111 S.W. Fifth Avenue, Portland, OR 97204; 800-231-1292; www.lpcorp.com.

U.S. Forest Industries, Inc., P.O. Box 820, Medford, OR 97501; 800-541-6906.

FIBER-CEMENT SIDING

FIBER-CEMENT MANUFACTURERS

AL Specialty Products, 10115 Kincey Ave., Suite 150, Hunterville, NC 28075; 800-566-2282; www.lpcorp.com (lap and panel siding).

Cemplank Inc. (successor sales arm of Eternit, Inc.), Excelsior Industrial Park, P.O. Box 99, Blandon, PA 19510-0099; 888-327-0723 (Cemplank™ smooth, rough-sawn, and wood grain lap siding; Cempanel™ smooth, stucco, and wood grain vertical siding panel); www.cemplank.com.

James Hardie Building Products, Inc., 26300 La Alameda, Suite 250, Mission Viejo, CA 92691; 800-9-HARDIE; www.jameshardie.com (Hardiplank™ smooth, rough sawn, and wood grain lap siding; Hardipanel™ smooth, stucco, wood grain vertical siding panels; Shingleside™ fiber-cement shingle; Hardisoffit™, Harditrim™).

Maxitile, Inc., 17141 S. Kingview Ave., Larson, CA 90746; 800-338- 8453 (MaxiPlank™ smooth and wood grain lap siding; MaxiPanel™ smooth, wood grain, V-groove, stucco panels; MaxiTrim™); www.maxibuildingproducts.com.

Temple-Inland Forest Products Corp., Inc., P.O. Drawer N, 303 Temple Dr., Diboll, TX 75941; 800-231-6060; www.temple.com.

CUTTING TOOLS

Pacific International Tool & Shear, P.O. Box 1604, Kingston, WA 98346; 800-297-7487; www.snappershear.com.

PATCHING PRODUCTS

Macklanberg Duncan™, P.O. Box 25188, Olkahoma City, OK 73125; 800-654-8454; www.mdteam.com.

VINYL TRIM AND VENT ACCESSORIES

Tamlyn and Sons, 13623 Pike Road, Stafford, TX 77477-5103; 800-334-1676; www.tamlyn.com.

EXTERIOR INSULATION AND FINISH SYSTEMS

Dow Corning Corporation, P.O. Box 994, Midland, Michigan 48686-0994; 517-496-6000; www.dowcorning.com.

Dryvit Energy Systems, Inc., One Energy Way, P.O. Box 1014, West Warwick, RI 02893; 800-556-7752; www.dryvit.com.

EIFS Industry Members Association (EIMA), 3000 Corporate Center Drive, Suite 270, Morrow, GA 30260; 800-294-3462; www.eifsfacts.com.

Finestone (Simplex Products Div.), 1149 Treat Street, Adrian, MI 49221-0010; 517-263-8881.

Omega Products Corp., P.O. Box 1869, Orange, CA 92668; 800-600-6634; www.omega-products.com.

Parex, Inc., P.O. Box 189, Redan, GA 30074; 800-537-2739; www.parex.com.

Pleko Systems International, Inc., P.O. Box 98360, Tacoma, WA 98498; 206-472-9637; www.pleko.com.

Retro Tek, 3876 Hoepker Road, Madison, WI 53718; 800-225-9001.

Senergy Inc., 3550 St. Johns Bluff Road, South Jacksonville, FL 32224-2614; 800-221-9255; www.senergyelfs.com.

Stuc-o-Flex International, Inc., 17639 N.E. 67th Court, Redmond, WA 98052; 800-305-1045; www.stucoflex.com.

TEC Incorporated, 315 South Hicks Road, Palatine, IL 60067; 847-358-9500; www.tecspecialty.com.

Universal Polymers, Inc., 319 N. Main Street, Springfield, MO 65804; 800-752-5403.

US Gypsum Company, 125 South Franklin Street, Chicago, IL 60606; 800-USG-4YOU; www.usg.com.

W.R. Bonsal Co., P.O. Box 241148, Charlotte, NC 28224-1148; 800-334-0784; www.bonsal.com.

EXTERIOR TRIM

RESTORATION PRODUCTS

Abatron, Inc., 5501 95th Avenue, Kenosha, WI 53144; 800-445-1754; www.albatron.com.

Conservation Services, 8 Lakeside Trail, Kinnelon, NJ 07045; 973-838-6412.

Gougeon Bros., Inc., P.O. Box 908, Bay City, MI 48707; 517-684-7286; www.gougeon.com.

Preservation Resource Group, P.O. Box 1768, Rockville, MD 20849-1768; 301-309-2222; www.prginc.com.

LAMINATED VENEER LUMBER

South Coast Lumber Co., 815 Railroad Ave., P.O. Box 670, Brookings, OR 97415; 541-469-4177; www.socomi.com.

ENGINEERED WOOD

American Hardboard Association, 520 N. Hicks Rd., Palatine, IL 60067; 312-934-8800, www.ahardb.org.

Forestrim™, Stimpson Lumber, P.O. Box 68, Forest Grove, OR 97116; 503-357-2131; www.stimpsonlumber.com.

Prime Trim™, Georgia Pacific Corp., 133 Peachtree St., NE, Atlanta, GA 30303; 404-652-4000; www.gp.com.

Protrim™, AL Specialty Products, 10115 Kincey Ave., Suite 150; www.lpcorp.com.

SmartTrim™, Louisiana-Pacific Corp., 1 East First St., Duluth, MN 55802; 800-648-6893; www.lpcorp.com.

Trim Craft™, Temple Inland Forest Products, P.O. Box N, Biboll, TX 75941; 800-231-6060; www.temple.com/tpgl.html.

WOOD AND THERMOPLASTIC COMPOSITE

Crane Plastics Co., P.O. Box 1047, Columbus, OH 43216; 800-366-8472, www.vinyl-siding.com.

Durawood PE, The Eaglebrook Companies, 2600 W. Roosevelt Rd., Chicago, IL 60601; 312-491-2500.

FrameSaver™, BMS, P.O. Box 631247, 1124 Bennet Clark Rd., Nacogdoches, TX 75963; 409-569-8211.

FIBER CEMENT

Cem-Trim™, Cemplank, Excelsior Industrial Park, P.O. Box 99, Blandon, PA 19510-0099; 888-327-0723; www.cemplank.com.

Harditrim™, James Hardie Building Products, 26300 Los Alameda, Suite 250, Mission Viejo, CA 92691; 888-J-HARDIE; www.jameshardie.com.

Maxitile, Inc., 849 East Sandhill Ave., Carson, CA 90746; 800-338-8453; www.maxibuildingproducts.com.

Temple Inland Forest Products, P.O. Box N, Biboll, TX 75941; 800-231-6060; www.temple.com.

POLYMER

Fypon, 22 West Pennsylvania Ave., Stewartstown, PA 17363; 800-537-5349; www.fypon.com.

Outwater Plastic Industries, P.O. Box 347, Woodridge, NJ 07075; 800-835-4400; www.outwater.com.

Style-Mark, Inc., 960 West Barre Road, Archibold, OH 43502; 800-446-3040; www.style-mark.com.

SEALANTS AND CAULKS

AC Products, 172 East La Jolla St., Placentia, CA 92870; 800-238-4204.

AEG, 3 Shaw's Cove, P.O. Box 6003, New London, CT 06320-1777.

ChemRex/PL Adhesives & Sealants, 889 Valley Park Drive, Shakopee, MN, 55379; 800-433-9517; www.chemrex.com.

DAP, 2400 Boston St., Suite 200, Baltimore, MD 21224; 800-543-3840; www.dap.com.

Dow Corning Corporation, Midland, MI 48686-0994; 517-496-6000; www.dowcorning.com.

Franklin International, Construction Adhesives and Sealants, 2020 Bruck St., Colombus, OH 43207; 800-877-4583; www.franklini.com.

GE Silicones, 260 Hudson River Road, Waterford, NY 12188; 800-255-8886; www.gesilicones.com.

Insta-Foam Products, Inc., Dow Chemical Company, 2020 Willard H. Dow Center, Midland, MI 48674; 800-258-2436; www.dow.com.

Macco Adhesives, 925 Euclid Ave., Cleveland, OH 44115; 800-634-0015; www.liguidnails.com.

Macklanburg-Duncan, P.O. Box 25188, Oklahoma City, OK 73125; 800-654-8454; www.mdteam.com.

Miracle Adhesives, TAC Intl, Air Station Industrial Park, Rockland, MA 02370; 800-503-6991; www.tacint.com.

NPC Sealants, 1208 South 8th Ave., P.O. 645, Maywood, IL 60153; 800-654-1042.

OSI Sealants, 7405 Production Dr., Mentor, OH, 44060; 800-321-3578; www.osisealants.com.

Polytite, 324 Ridge Avenue, Cambridge, MA 02140; 800-776-0930; www.polytite.com.

Red Devil, 2400 Vaux Hall Rd., Union, NJ 07083; 800-4-A-DEVIL; www.reddevil.com.

Resource Conservation Technology, 2633 N. Calvert Street, Baltimore, MD 21218; 410-366-1146.

CHAPTER 3: ROOFS

WOOD-FRAME WIND RESISTANCE

High Wind Resistant Construction Connectors, Simpson Strong-Tie® Company, Inc., 4637 Chabot Drive, Suite 200, Pleasanton, CA 94588; 800-999-5099; 1995; www.strongtie.com.

ITW Foamseal, 2425 North Lapeer Road, Oxford, MI 48371; 248-628-2587; www.itwfoamseal.com.

New Necessities, Inc., 5710 Pebblebrook Trail, Gainesville, GA 30506; 770-844-9438.

WOOD-FRAME SEISMIC RESISTANCE

Earthquake Resistant Construction Connectors, 1995, Simpson Strong-Tie® Company, 4637 Chabot Drive, Suite 200, Pleasanton, CA 94588; 800-999-5099; www.strongtie.com.

SHEATHING

APA–The Engineered Wood Association, P.O. Box 11700, Tacoma, WA 98411-0700; 253-565-6600; www.apawood.org.

Homasote Company, Box 7240, W. Trenton, NJ 08628; 800-257-9491; www.homasote.com.

FLASHING

Alcoa Building Products, P.O. Box 57, 1501 Michigan Street, Sidney, OH 45365-0057; 800-962-6973; www.alcoahomes.com (aluminum building products).

Bethlehem Steel Corporation; 1170 Eighth Ave., Bethlehem, PA 18016-7699; 800-352 5700; www.bethsteel.com/spgalval.html (information on galvanized and Galvalume™ products).

H. Bixon & Co., P.O. Box 1198, New Haven, CT 06505; 203-777-7445 (supplier of sheet lead).

Heckman Industries, P.O. Box 250, Mill Valley, CA 94942; 800-841-0066 (SBS rubberized asphalt in roll form).

Revere Copper Products, Inc., One Revere Park, Rome, NY 13442-0300; 800-950-1776; www.reverecopper.com (prepatinated copper and other copper products).

Rheinzink Canada Ltd., 4595 Tillicum Street, Burnaby, B.C. V5J 3J9 Canada; 604-291-8171; www.rheinzink.com (zinc flashing, roofing, siding and gutter systems).

Specialty Steel Industry of North America (SSINA), 3050 K St., N.W., Washington, DC 20007; 800-982-0355; www.ssina.com (information on stainless steel).

Tamlyn and Sons; 13623 Pike Road, Stafford, TX 77477; 800-334-1676; www.tamlyn.com (roofing accessories).

W.R. Grace & Co., 62 Whittemore Ave., Cambridge, MA 02140; 800-444-6459; www.graceconstruction.com (ice and water barriers; see Underlayments and Moisture Barriers for a complete list of ice and water barrier manufacturers).

UNDERLAYMENTS AND MOISTURE BARRIERS

ALCO Shield Ice & Water Protector, ALCO-NVC, Inc., P.O. Box 14001, Detroit, MI 48214; 800-323-0029; www.alconvc.com.

Arctic Seal #170, Herbert Malarkey Roofing Co., P.O. Box 17217, Portland, OR 97217; 800-546-1191; www.malarkey.com.

Celoguard shingle underlayment, Celotex Corp., P.O. Box 31602, Tampa, FL 33631; 813-873-1700; www.celotex.com.

Eave & Valley Shield, Globe Building Materials, Inc., 2230 Indianapolis Blvd., Whiting, IN 46394; 800-456-5649; www.globebuilding.com.

Flashband, Andek Corp., P.O. Box 392, Moorestown, NJ 08057; 888-88ANDEK; www.andek.com.

Miradri WIP 100 and 200, TC Miradri, 2170 Satellite Boulevard, Suite 350, Duluth, GA 30097-4074; 888-464-7234; www.miradi.com.

Moisture Guard Plus, TAMKO Roofing Products, Inc., P.O. Box 1404, 220W 4th St., Joplin, MO 64802; 417-624-6644; 800-841-1923.

Stormaster DG Ice & Water Protection, Atlas Roofing Corp., 1775 The Exchange #160, Atlanta, GA 30339; 770-952-3170; 800-254-2852; www.atlasroofing.com.

Stormshield/Black Diamond Base Sheet, GS Roofing Products Co., P.O. Box 868, Valley Forth, PA 11482; 972-580-5600; 800-274-8530; www.gsroof.com.

Typar® 30, Twinpak, Inc., 369 Elgin Street, Branford, Ontario N3T 5V6, Canada.

VycorTM Ice and Water Shield® and VycorTM Ultra, W.R. Grace & Company, 62 Whittemore Avenue, Cambridge, MA 02140; 800-444-6459; www.gcp-grace.com.

Weather Rock polyethylene surface and granulated surface, Owens Corning, One Owens Corning Parkway, Toledo, OH 43659; 800-GET-PINK; www.owens-corning.com.

Weather Watch and Stormguard waterproof underlayment, GAF Materials Corp., 1361 Alps Rd., Wayne, NJ 07470; 800-ROOF-411; www.gaf.com.

WinterguardTM, Certainteed Corp., Roofing Products Group, P.O. Box 1100, 1400 Union Meeting Road, Blue Bell, PA 19422; 800-322-3060; www.certainteed.com.

INSULATION

BATT, SPRAY-ON, AND LOOSE FILL INSULATION

American Rockwool, Inc., P.O. Box 880, Spring Hope, NC 27882; 919-478-5111; www.amerrock.com.

Ark-Seal International, 2190 S. Kalamath, Denver, CO 80223; 800-525-8992; www.arkseal.com.

Building Products Division, The Celotex Corporation, P.O. Box 31602, Tampa, FL 33631; 800-CELO-TEX; www.celotex.com.

CertainTeed Corporation, 750 E. Swedesford Rd, Valley Forge, PA 19482; 800-523-7844; www.certainteed.com.

GreenStone Industries Inc., 809 Westhill Street, Suite 4, Charlotte, NC 28208; 800-666-4824; www.greenstone.com.

Icynene Inc., 5805 Whittle Road, Suite 110, Mississauga, Ontario L42 2J1 Canada; 800-946-7325; www.icynene.com.

International Cellulose Corp., 12315 Robin Blvd., Houston, TX 77245-0006; 800-444-1252; www.internationalcellulose.com.

Johns Manville Corporation (formerly Schuller International Inc.), 717 17th St., Denver, CO 80202; or P.O. Box 5108, Denver, CO 80217-5108; 800-654-3103; www.jm.com.

Knauf Fiber Glass, 240 Elizabeth St., Shelbyville, IN 46176; 800-200-0802; www.knauffiberglass.com.

Owens Corning, One Owens Corning Parkway, Toledo, OH 43659; 800-354-PINK or 800-GET-PINK; www.owens-corning.com.

Rock Wool Manufacturing Co., 203 N. 7th St., Leeds, AL 35094; 800-874-7625.

Sloss Industries Corporation, 3500 35th Avenue North, Birmingham, AL 35207; 205-808-7803; www.sloss.com.

U.S. Fiber, Inc., 809 Westhill Street, Charlotte, NC 28208; 800-666-4824; www.us-gf.com.

Western Fiberglass Group, 6955 Union Park Center, Suite 580, Midvale, UT 84047; 801-562-9558.

STRUCTURAL INSULATED PANELS

Structural Insulated Panel Association, 3413 56th Street, Suite A, Gig Harbor, WA 98335; 253-858-7472; www.sips.org.

RIGID FOAM INSULATION

Celotex Building Products, P.O. Box 31602, Tampa, FL 33631; 800-CELOTEX; www.celotex.com.

Dow Chemical Company, Styrofoam Brand Products, 2020 Willard H. Dow Center, Midland, MI 48674; 800-258-2436; www.dow.com/styrafoam.

Johns Manville Corporation (formerly Schuller International Inc.), 717 17th St., Denver, CO 80202; or P.O. Box 5108, Denver, CO 80217-5108; 800-654-3103; www.jm.com.

Owens Corning, One Owens Corning Parkway, Toledo, OH 43659; 800-354-PINK or 800-GET-PINK; www.owens-corning.com.

Tenneco Building Products, 2907 Log Cabin Dr., Smyrna, GA 30080; 800-241-4402; www.tennecobuildingprod.com.

RIGID FOAM COMPOSITE NAILBASE PANELS

Celotex Building Products, P.O. Box 31602, Tampa, FL 33631-3602; 800-CELOTEX; www.celotex.com.

Cornell Corporation, P.O. Box 338, Cornell, WI 54732; 715-239-6411.

Homasote Company, Box 7240, W. Trenton, NJ 08628; 800-257-9491; www.homasote.com.

Johns Manville Corporation (formerly Schuller International Inc.), 717 17th St., Denver, CO 80202; or P.O. Box 5108, Denver, CO 80217-5108; 800-654-3103; www.jm.com.

RADIANT BARRIER PRODUCTS

Radiance™ Low-E interior paint, ChemRex, 889 Valley Park Drive, Shakopee, MN 55379; 800-433-9517; www.chemrex.com.

Super-R™ radiant barriers, Innovative Insulation Inc., 6200 W. Pioneer Parkway, Arlington, TX 76013; 800-825-0123; www.radiantbarrier.com.

TechShield™, Louisiana-Pacific Corp., 111 SW 5th Ave., Portland, OR 97204; 800-648-6893; www.lpcorp.com.

WOOD SHINGLES AND SHAKES

Cedar Breather™ and Roll Vent for Cedar, Benjamin Opdyke, Inc., 199 Precision Drive, Horsham, PA 19044; 800-346-7655; www.opdyke.com.

Ice & Water Shield™, W.R. Grace & Co., Construction Products, 62 Whitemore Avenue, Cambridge, MA 02140; 617-876-1400; 800-472-2391; www.graceconstruction.com.

Winterguard™, Certainteed Corp., Roofing Products Group, P.O. Box 860, Valley Forge, PA 19482; 800-523-7844; www.certainteed.com.

PARTIAL LIST OF CSSB MEMBER AND NONMEMBER SUPPLIERS OF WOOD SHINGLES

A complete list of the approximately 608 supplier and installer members of the Cedar Shake and Shingle Bureau is found in the *Cedar Shake and Shingle Bureau's Membership Directory and Buyer's Guide*.

Amaraut Wood Products, Inc., P.O. Box 1008, 4935 Boyd Road, Arcarta, CA 95521; 707-822-4849 (redwood shingles and shakes).

Ambrook Industries Ltd., 17360 Frazer Dyke Road, Pitt Meadows, B.C., Canada V3Y 1Z1; 604-465-5657 (western red cedar, wholesale and retail).

Clarke Group Marketing, Inc., P.O. Box 515, Sumas, WA 98295; 800-963-3388; www.cedarplus.com (largest manufacturer of cedar shake and shingle products).

Sovbec, Inc., 9201 Center Hospilalier Blvd., Chary, Quebec, Canada G6X 1L5; 418-832-6181 (largest supplier of Eastern white cedar, representing 26 mills).

Teal Cedar Products, 17897 Trigg Road, Surrey, B.C., Canada V3T 5J4; 604-581-6161 (western red cedar and yellow cedar; wholesale only; good source of information about shingle and shake products).

Watkin Sawmills, P.O. Box 314, Sumas, WA 98295; 800-663-8301 (western red cedar; wholesale and retail).

ASPHALT SHINGLES

Atlas Roofing Corp., 1775 The Exchange, #160, Atlanta, GA 30339; 800-955-1476; www.atlasroofing.com.

Bird Inc., 1077 Pleasant St., Norwood, MA 02062; 800-BIRD-INC; www.certainteed.com.

BPCO, P.O. Box 3177, Wayne, NJ 07474-3177.

Celotex Corp., P.O. Box 31602, Tampa, FL 33631; 800-CELOTEX (phone); 813-873-4080 (fax); www.celotex.com.

ELK, 14643 Dallas Parkway, Suite 1000, Dallas, TX 75240; 972-851-0400; www.elcor.com.

GAF Materials Corp., 1361 Alps Rd., Wayne, NJ 07470; 800-766-3411; www.gaf.com.

Georgia-Pacific Corp., 4300 Wildwood Parkway, Suite 300, Atlanta, GA 303TK; 800-839-2588; www.gp.com.

Globe Building Materials, Inc., 2230 Indianapolis Blvd., Whiting, IN 46394; 219-473-4500.

GS Roofing Products Co., 5525 MacArthur Blvd., Suite 900, Irving, TX 75038; 972-580-5600; www.gsroof.com; www.certainteed.com.

Herbert Malarkey Roofing Co., P.O. Box 17217, Portland, OR 97217; 800-545-1191 (phone); 503-283-5405 (fax); www.malarkey.com.

IKO Manufacturing, Inc., 120 Hay Rd., Wilmington, DE 19809; 800-IKO-ROOF; www.iko.com.

Owens Corning, One Owens Corning Parkway, Toledo, OH 43659; 800-438-7465 (phone); 419-248-7354 (fax); www.owenscorning.com.

PABCO Roofing Manufacturers, 1718 Thorne Rd., Takoma, WA 98421; 800-286-0498; www.pabcoroofing.com.

ICE AND WATER BARRIERS

For a more comprehensive list of ice and water barriers, see the previous section, "Underlayments and Moisture Barriers."

Ice & Water Shield™, W.R. Grace & Co. Construction Products, 62 Whitemore Avenue, Cambridge, MA 02140; 800-472-2391; www.graceconstruction.com.

Winterguard™, Certainteed Corp., Roofing Products Group, P.O. Box 860, Valley Forge, PA 19482; 800-286-0498; www.pabcoroofing.com.

ALGAE MITIGATION

3M™, Algae Block™, System, 3M Industrial Mineral Products Division, 3M Center 225-2N-07, P.O. Box 33225, St. Paul, MN 55133-3225; 800-447-2914; www.3M.com.

LOW-SLOPE ROOFING

MODIFIED BITUMEN MEMBRANE SUPPLIERS

GAF Materials Corporation, 1361 Alps Road, Wayne, NJ 07470; 973-628-3000; www.ispcorp.com; www.gaf.com.

Henry Co. (formerly Monsey-Bakor), 336 Cold Stream Road, Kimberton, PA 19442; 800-523-0268; www.henry.com.

Siplast, 222 W. Las Colinas Boulevard #1600, Irving, TX 75039; 800-922-8800; www.siplast.com.

Tamko Roofing Products, P.O. Box 1404, Joplin, MO 64802; 800-641-8935; www.tamko.com.

THERMOPLASTIC MEMBRANE SUPPLIERS

Bondcote Roofing Systems, 4090 Pepperell Way, Dublin, VA 24084; 800-368-2160; 706-882-3410; www.bondcote.com.

Canadian General Tower Limited, 52 Middleton Street, Cambridge, Ontario N1R ST6; 519-623-1630.

Cooley Engineered Membranes, Inc., P.O. Box 939, 50 Esten Avenue, Pawtucket, RI 02862-0939; 800-444-4023; 401-724-0490; www.cooleygroup.com.

Duro-Last Inc., 525 Morley Drive, Saginaw, MI 48601; 800-248-0280, www.duro-last.com.

Sarnafil Inc., 100 Dan Road, Canton, MA 02021; 800-451-2504; www.sarnafilus.com.

Seaman Corporation, 1000 Venture Boulevard, Wooster, OH 44691-9360; 330-262-1111; 800-927-8578; www.seamancorp.com.

THERMOSET MEMBRANE SUPPLIERS

Carlisle Syntec Inc., 1285 Ritner Highway, P.O. Box 7000, Carlisle, PA 17013-0925; 800-4-SYNTEC; www.carlisle-syntec.com.

THERMOPLASTIC AND THERMOSET MEMBRANE

ERSystems, 50 Medina Street, P.O. Box 56, Loretto, MN 55357-0056; 800-403-7747; www.ersystems.com.

Genflex Roofing Systems, 1722 Indian Wood Circle, Suite A, Maumee, OH 43537; 800-443-4272; www.genflex.com.

JPS Elastomerics Corp., Stevens Roofing Systems, Nine Sullivan Road, Holyoke, MA 01040-2800; 800-621-ROOF; www.stevensroofing.com.

Kelly Energy Systems, Inc., P.O. Box 2583, Waterbury, CT 06723; 800-537-7663; 203-575-9220.

Verisco, Inc., 3485 Fortuna Drive, Akron, OH 44312; 800-992-7662; www.versico.com.

MODIFIED BITUMEN, THERMOSET, AND THERMOPLASTIC SUPPLIERS

Firestone Building Products, 525 Congressional Boulevard, Carmel, IN 46032-5607; 800-428-4511; www.firestonebpco.com.

Johns Manville Corporation, Roofing Systems Group, 717 17th Street, P.O. Box 5108, Denver, CO 80217; 800-654-3103; www.jm.com.

METAL ROOFING

Atas International, Inc., 6612 Snowdrift Road, Allentown, PA 18106; 800-468-1441; www.atas.com (s-tile and shake panels, metal shingles, standing seam).

Berridge Manufacturing Co., 1720 Maury Street, Houston, TX 77026; 800-231-8127; www.berridge.com (full line of metal roofing products).

Custom-Bilt Metals, 9845 Joe Vargus Way, So. El Monte, CA 91733; 800-826-7813, www.custombiltmetals.com (full line of roofing products, 15 branches nationwide).

Dura-Loc Roofing Systems Limited, Box 220, RR 2, Courtland, Ontario, Canada N0J 1E0; 800-265-9357; www.duraloc.com (s-tile and shake panels).

Met-Tile, Inc., 1745 Monticello Court, P.O. Box 4268, Ontario, CA 91761; 800-899-0311; www.met-tile.com/roof; www.met-tile.com (s-tile panels).

Perfection Co., 8512 Industry Park Drive, P.O. Box 1524, Piqua, OH 45356; 888-788-2427 ("Country Manor" aluminum shake).

Revere Copper Products, Inc., One Revere Park, Rome, NY 13440-5561; 800-490-1776; www.reverecopper.com.

Rheinzink Canada, Ltd., 4560 Dawson Street, Burnaby, British Columbia, Canada V5J 3J9; 604-291-8171 (zinc standing seam roofing).

Vail Metal Systems, LLC, P.O. Box 2030, Edwards, CO 81632; 888-245-6385, www.vailmetal.com/roofing.

COMPREHENSIVE LISTS OF MANUFACTURERS

There are many manufacturers of metal-standing seam roofs and metal tiles. A list of manufacturers and their products can be found in the referenced publications.

Product Capability Directory, National Coil Coaters Association, 1997.

Residential Steep-Slope Roofing Material Guide, National Roofing Contractors Association, 1997.

SLATE

MANUFACTURERS AND SUPPLIERS

Buckingham-Virginia Slate Corp., 1 Main St., P.O. Box 8, Arvonia, VA 23004-0008; 800-235-8921; www.bvslate.com.

Greenstone Slate, P.O. Box 134, Pultney, VT, 05764; 888-592-7684; www.greenstone.com.

Hilltop Slate Co., Rt. 22A, P.O. Box 201, Middle Granville, NY 12849; 518-642-2270; www.hilltopslate.com.

Newfoundland Slate Co. Inc., 8800 Shepard Avenue East, Scarborough, ON, Canada, M1B5R4; 800-975-2835.

Structural Slate Co., 222 E. Maine St., Pen Argyl, PA 18072; 610-862-4145; 800-67-SLATE; www.structuralslate.com.

DISTRIBUTORS AND INSTALLERS

Allied Building Products, 15 East Union Avenue, East Rutherford, NJ 07073; 800-541-2198; www.alliedbuilding.com (distributor).

Durable Slate Co., 1050 N. 4th St., Columbus, OH 43201; 800-666-7445; www.durableslate.com (new and salvaged slate and tile distributor and installer).

Evergreen Slate Co., 68 East Potter Avenue, Granville, NY 12832; 518-642-2530; www.evergreenslate.com (distributor).

New England Slate Co., 1385 U.S. Route 7, Pittsford, VT 05763; 888-NE-SLATE; www.neslate.com (distributor of United States., Quebec, and Newfoundland tile).

Roof Tile & Slate Co., 1209 Carroll, Carrolltown, TX 75006; 800-446-0220; www.claytile.com (new and salvaged slate distributor).

Slate International, Inc., 15106 Marlboro Pike, Upper Marlboro, MD 20772; 800-343-9785 (distributor).

SNOW GUARDS AND SLATE ACCESSORIES

Berger Building Products Corp., 805 Pennsylvania Ave., Feasterville, PA 19053; 800-523-8852; www.bergerbros.com.

Sieger Snow Guard Co., #63 Ziegler Road, Leesport, PA 19533; 610-926-2074 (heavy-duty snow guards).

Vermont Slate & Copper Services, Inc., P.O. Box 430, Stowe, VT 05672-0430; 888-766-4273; www.vermontslateandcopper.com (heavy-duty snow guards).

SLATE RIDGE VENTS

Petersen Aluminum Corporation, 9060 Junction Drive, Annapolis Junction, MD 20701; 800-PAC-CLAD; www.pac-clad.com.

VENTED ATTACHMENT SYSTEM

Slate International, 15106 Marlboro Pike, Upper Marlboro, MD 20877; 800-343-9875 (Fastrack for Slate™).

CLAY, CONCRETE, FIBER-CEMENT, AND COMPOSITE TILE

National Tile Roofing Manufacturers Association, P.O. Box 40337, Eugene, OR 97404; 541-689-0366.

CLAY TILE

Altusa Tile, 6645 N.W. 77th Ave., Miami, FL 33166; 800-54-TILE-1; www.altusa.com.

Ameri-Clay Roof Tile, 2905 47th Ave., North, St. Petersburg, FL 33714-3131; 813-522-3900; www.americlay.com.

Celadon, P.O. Box 309, New Lexington, OH 43764-0309; 800-235-7528; www.celadonslate.com.

Gladding, McBean, P.O. Box 97, 601 7th Street, Lincoln, CA 95648; 800-776-1133; www.gladdingmcbean.com.

Ludowici Roof Tile, P.O. Box 69, 4757 Tile Plant Road, New Lexington, OH 43764; 800-945-8453; www.ludowici.com; www.certainteed.com.

Maruhachi Ceramics of America, Inc. (MCA), 1985 Sampson Avenue, Corona, CA 91719; 800-736-6221; www.mca-tile.com.

U.S. Tile Company, 909 West Railroad St., Corona, CA 92882-1906; 909-737-0200; www.ustile.com.

SALVAGED AND NEW CLAY TILE

Renaissance Roofing, Box 5024, Rockford, IL 61125; 800-699-5695; www.claytileroof.com.

The Roof Tile and Slate Co., 1209 Carroll, Carrollton, TX 75006; 800-446-0220; www.claytile.com.

Tile Roofs Inc., 12056 S. Union Ave., Chicago, IL 60623; 708-479-4366.

Tilesearch, P.O. Box 1694; Roanoke, TX 76262; 817-491-2444; www.tilesearch.com.

The Tile Man, Inc., 520 Vaiden Rd., Louisberg, NC 27549; 919-853-6923; www.thetileman.com.

CONCRETE TILE

Boral Life Tile, Inc., 4685 MacArthur Court, Suite 300, Newport Beach, CA 92660; 800-LIFETILE; www.lifetile.com.

Columbia Concrete Products, Ltd., 8650 130 Street, Surrey, B.C. V3W 1G1 Canada; 604-596-3388; www.crooftile.com.

Eagle Roofing Products Co., 3546 North Riverside Ave., Rialto, CA 92376; 800-300-EAGLE; www.eagleroofing.com.

Entegra Roof Tile Corp., 1201 N.W. 18th St., Pompano Beach, FL 33069; 800-586-7663.

Metro Roof Tile, Inc., 11501 N.W. 117th Way, Medley, FL 33178; 305-863-0021.

Monier Lifetile LLC, 1 Park Place, Suite 900, Irvine, CA 92714; 714-756-1605; www.monierlifetile.com.

Pioneer Roofing Tile, Inc., 10650 Poplar Ave., Fontana, CA 92337; 909-350-4238; 800-411-TILE; www.pioneertile.com.

Vande Hey-Raleigh, 1665 Bohm Drive, P.O. Box 263, Little Chute, WI 54140; 800-236-8453; www.vhr-roof-tile.com.

Westile, 8331 W. Carder Court, Littleton, CO 80125; 800-433-8453; www.westile.com.

FIBER-CEMENT TILE

Eternit, Inc., 610 Corporate Drive, Reading, PA 19605; www.eternit.usa.com.

FireFree, 580 Irwin Street 100, San Rafeal, CA 94901; 888-990-3388; www.firefree.com.

James Hardie Building Products, 26300 La Alameda, Suite 250, Mission Viejo, CA 92691; 800-942-7343; www.JamesHardie.com (Hardishakes®).

Louisiana-Pacific Corp., 111 S.W. Fifth Ave., Portland, OR 97204; 800-579-8401; www.lpcorp.com (Nature Guard®).

Northern Roof Tile Sales Co., P.O. Box 275 Millgrove, Ontario L0R 1V0 Canada; 905-627-4035; www.northernrooftiles.com (Cembrit B7).

COMPOSITE TILE

American Sheet Extrusion Corporation, 1618 Lynch Road, Evansville, IN 47711; 800-347-3390 (Perfect Choice® plastic shake).

Crowe Industries, Ltd, 116 Burris Street, Hamilton, Ontario L8M 2J5; 905-529-6818; www.authentic-roof.com (Authentic Roof® plastic slates).

Owens Corning, One Owens Corning Parkway, Toledo, OH 43659; 800-766-3464; www.owens-corning.com, (MiraVista™ shakes).

Re-New Wood, P.O. Box 1093, Waggoner, OK 74467; 800-420-7576; www.ecoshake.com (ECO-Shakes® wood and plastic shake).

FOREIGN-MADE TILE

HG Roofing, P.O. Box 406, Lakeville, IN 46536; 219-784-2006.

Northern Roof Tile Sales Co., P.O. Box 275, Millgrove, Ontario L0R 1VD Canada; 905-627-4035; www.northernrooftiles.com.

GUTTER AND LEADER SYSTEMS

WOOD GUTTERS

Blue Ox Millworks, 1X Street, Eureka, CA 95501-0847; 800-248-4259; www.blueoxmill.com (redwood gutters).

ALUMINUM GUTTERS

Alcoa Building Products, Inc., 2600 Campbell Road, Sidney, OH 46365; 800-962-6973; www.alcoahomes.com (stick systems).

Berger Building Products, 805 Pennsylvania Blvd., Feasterville, PA 19053; 800-523-8852, www.bergerbros.com.

Classic Gutter Systems, 5621 East "D.E." Avenue, P.O. Box 2319, Kalamazoo, MI 49004; 616-382-2700; www.classicgutters.com (half-round aluminum gutters and accessories).

Custom-Bilt Metals (corporate office), 9845 Joe Vargas Way, S. El Monte, CA 91733; 800-826-7813; www.custombiltmetals.com.

STEEL GUTTERS AND ACCESSORIES

Berger Building Products, 805 Pennsylvania Blvd., Feasterville, PA 19053; 800-523-8852; www.bergerbros.com.

Custom-Bilt Metals (corporate office), 9845 Joe Vargas Way, S. El Monte, CA 91733; 800-826-7813; www.custombiltmetals.com.

Klauer Manufacturing, P.O. Box 59, Dubuque, IA 52004-0059; 319-582-7201; www.klaur.com (stick systems).

COPPER GUTTERS AND ACCESSORIES

Berger Building Products, 805 Pennsylvania Blvd., Feasterville, PA 19053; 800-523-8852, www.bergerbros.com.

Classic Gutter Systems, 5621 East "D.E." Avenue, Kalamazoo, MI 49004; 616-382-2700; www.classicgutters.com(half-round gutters and accessories).

VINYL (PVC) GUTTER SYSTEMS

Bemis Manufacturing Co., 300 Mill Street, Sheboygan Falls, WI 53085-0901; 800-558-7651, www.bemismfg.com, (K-style and "U" shaped).

Genova Products, Inc., 7034 East Court Street, P.O. Box 309, Davison, MI 48423-0309; 800-521-7488, www.genovaproducts.com, (K-style and "U"-shaped gutters and vinyl gutter guards).

GSW Thermoplastic Co., 1994 A Woodward Avenue #361, Bloomfield Hills, MI 48302; 800-662-4479; www.gsothermo.com (K-style and "U"-shaped gutters and accessories).

Plastmo Vinyl Raingutters, 8246 Sandy Court #B, Jessup, MD 20794; 800-899-0992; www.plastmo.com (K-style and half-round systems).

Tilt 'N Clean, Rain Gutter Systems™, P.O. Box 1231, Pottstown, PA 19464; 800-454-TILT; www.tiltnclean.com (half-round gutter systems that can rotate for cleaning from the ground).

GUTTER GUARD PRODUCTS

FLO-FREE™, FLO-FREE Gutter Co., 225 West Avenue, Springfield, PA 19064; 888-543-4484; www.flofree.com.

Gutter Helmet™, American Metal Products, 8601 Hacks Cross Road, Olive Branch, MS 38654; 888-4-HELMET; www.gutterhelmet.com.

Gutter ProTech™, Absolute Gutter Protection, P.O. Box 568, Woodbury, NJ 08096-7568; 800-283-7791; www.gutterprotech.

Gutter Topper™, Gutter Topper Ltd., P.O. Box 349, Amelia, OH 45102; 800-915-5888; www.guttertopper.com.

Waterfall®, Crane Plastics Holding Company, P.O. Box 1047, Columbus, OH 43216-1047; 800-307-7780, www.crane-plastics.com.

GUTTERLESS DEFLECTING SYSTEMS

Rainhandler™, Savetime Corporation, 2710 North Avenue, Bridgeport, CT 06604; 800-942-3004; www.rainhandler.com.

CONCEALED GUTTER HANGERS

White Oak Manufacturing, LLC, 125 Kingcourt, New Holland, PA 17557; 800-245-4086.

SHINGLE OIL

Chevron Products Company, Global Lubricants, 575 Market Street, San Francisco, CA 94105; 415-894-7700; www.chevron.com.

GUTTER LINERS

Heckman Industries, 405 Spruce Street, Mill Valley, CA 94941; 800-841-0066.

SNOW GUARDS AND ROOFING ACCESSORIES

Berger Building Products, 805 Pennsylvania Blvd., Feasterville, PA 19053; 800-523-8852; www.bergerbros.com.

Vermont Slate and Copper Services, Inc., P.O. Box 430, Stowe, VT 05672; 888-SNOGARD; www.alpinesnowguards.com.

CHAPTER 4: WINDOWS AND DOORS

EXISTING WINDOW AND DOOR OVERVIEW

Andersen Windows, 100 4th Avenue North, Bayport, MN 55003-1096; 800-426-4261; www.andersenwindows.com (wood, vinyl clad, and composite wood windows and doors).

Caradco, P.O. Box 920, Rantoul, IL 61866; 800-238-1866; www.caradco.com (wood, aluminum clad, and vinyl windows and doors).

CertainTeed Corporation, P.O. Box 860, Valley Forge, PA 19482; 800-233-8990; www.certainteed.com (vinyl windows and doors).

Comfort Line Inc., 5500 Enterprise Boulevard, Toledo, OH 43612; 800-522-4999; www.comfortline.com (wood, composite wood, vinyl, and fiberglass windows and doors).

Hope's Windows, Inc., 84 Hopkins Avenue, P.O. Box 580, Jamestown, NY 14702-0580; 716-665-5124; www.hopeswindows.com [metal (steel) framed windows].

Hurd Millwork Co., 575 South Whelen Avenue, Medford, WI 54451; 800-223-4873; www.hurd.com (wood, aluminum clad, and vinyl windows and doors).

Jeld-Wen Window Products (Pozzi, Wenco, Norco, Summit, Caradco), P.O. Box 1329, Klamath Falls, OR 97601-0268; 800-877-9482; www.jeld-wen.com (wood, aluminum. clad, composite wood, vinyl, and fiberglass windows and doors).

Kolbe and Kolbe Millwork Co., Inc., 11th Avenue, Wausau, WI 54401-5998, 715-842-5666, www.kolbe-kolbe.com (wood and aluminum. clad windows and doors).

Marvin Windows, Attn.: Architecture Dept., P.O. Box 100, Warroad, MN 56763; 800-346-5128; www.marvin.com (wood, aluminum. clad, composite wood, and fiberglass windows and doors).

Milgard Windows, 1010 54th Ave. E., Tacoma, WA 98424; 800-645-4273; www.milgard.com (wood, vinyl, and fiberglass windows and doors).

Omniglass, 1205 Sherwin Road, Winnipeg R3H 0V1 Canada; 204-987-8522; www.omniglass.com (composite wood and fiberglass windows).

Pella Corporation, 102 Main Street, Pella, IA 50219; 800-847-3552; www.pella.com (wood and aluminum. clad windows and doors).

Seekircher Steel Window Repair, 2 Weaver Street, Scarsdale, NY 10583; 914-725-1904; www.design-site.net/seekirch.htm (metal framed window repair).

Torrance Steel Window Co., Inc., 1819 Abalone Ave., Torrence, CA 90501; 310-328-9181; www.torrancesteel.com (metal framed windows).

Weather Shield Manufacturing, Inc., One Weather Shield Plaza, P.O. Box 309, Medford, WI 54451; 800-222-2995; www.weathershield.com (wood, aluminum. clad, vinyl clad, and vinyl windows and doors).

GLAZING

3M, 3M Center Div. Specified Construction Products 225-4S-08, St. Paul, MN 55144-1000; 800-480-1704; www.3M.com/market/construction (applied performance film products).

Bi-Glass Systems, Inc., 108 Wadesworth Rd., Ducksbury, MA 02332; 800-729-0742; www.bi-glass.com.

Cardinal IG, 12301 Whitewater Drive, Minnetonka, MN 54343; 800-843-1484; www.mhtc.net/~cardinal (glass manufacturer).

Co-Ex Corp., 5 Alexander Drive, Wallingford, CT 06492; 800-888-5364; www.co-excorp.com (single- and multiple walled polycarbonate).

CP Films, Solutia Inc., P.O. Box 5068, Martinsville, VA 24115; 800-223-4385; www.cpfilms.com (applied performance film products).

Hy-Lite Block Windows, 101 California Ave., Beaumont, CA 92223; 800-827-3691; www.hy-lite.com.

IBP Glass Block Grid System, Acme Building Brands, P.O. Box 425, Fort Worth, TX 76107; 800- 932-2263; www.ibpglassblock.com (aluminum frame glass block).

MSC Specialty Films Inc., 13770 Automobile Blvd., Clearwater, FL 33777-1430; 800-282-9031; www.solarguard.com (applied performance film products).

Pilkington Building Products North America, P.O. Box 799, 811 Madison Ave., Toledo, OH 43697; 419-247-3731; www.pilkington.com/building (glass manufacturer).

PPG Industries, Inc., One PPG Place, Pittsburgh, PA 15272; 412-434-3131; www.ppg.com (glass manufacturer).

Southwall Technologies, 1029 Corporation Way, Palo Alto, CA; 800-365-8794; www.southwall.com (applied performance film products).

Viracon, 800 Park Drive, P.O. Box 248, Owatonna, MN 55060; 800-533-2080; www.viracon.com (laminated glass).

WINDOW FRAME AND REPLACEMENT UNITS

Abatron, Inc., 5501 95th Ave, Kenosha, WI 53144; 800-445-1754; www.albatron.com (wood repair products).

American Windowsill Corp., P.O. Box 1454, Orem, UT 84059-1454; 801-426-4303; http://vpp.com/easysills (retrofit vinyl window sills).

Amerimax Building Products; 3950 Medford Dr., Loveland, CO 80538; 972-701-4900 (vinyl replacement units).

Caradco, 201 Evans Drive, P.O. Box 920, Rantoul, IL 61866; 217-893-4444; www.cardaco.com (wood sash only kits).

CertainTeed Corporation, P.O. Box 860, Valley Forge, PA 19482; 800-233-8990; www.certainteed.com/pro/windows (vinyl replacement units).

Chelsea Building Products, Customer Service, 565 Cedar Way, Oakmount, PA 15139; 800-424-3573; www.chelseabuildingproducts.com (vinyl replacement units).

Gougeon Bros. Inc., P.O. Box 908, Bay City, MI 48707; 517-684-7286; www.gougeon.com (wood repair products).

Harvey Industries, Inc., 1400 Main Street, Waltham, MA 02451; 800-9-HARVEY; www.harveyind.com (vinyl replacement units).

Kolbe and Kolbe, 1323 S. Eleventh Avenue, Wausau, WI 54401-5998; 800-477-8656; www.kolbe-kolbe.com/s-rpdh.html (wood sash only kits).

Marvin Windows, P.O. Box 100, Warroad, MN 56763; 800-346-5128; www.marvin.com (wood sash only kits).

Milgard Windows, 1010 54th Ave. E., Tacoma, WA 98424; 800-645-4273; www.milguard.com/consumer_retrofit.html (vinyl replacement units).

Pella Corporation, 102 Main Street, Pella, IA 50219; 800-847-3552; www.pella.com (wood replacement units).

Preservation Resource Group, Inc., P.O. Box 1768, Rockville, MD 20849-1768; 800-774-7891; www.prginc.com (wood repair products).

U.S. Borax, 26877 Tourney Rd., Valencia, CA 91355; 661-287-5400; http://borax.com.

Weather Shield Manufacturing, Inc., P.O. Box 309, Medford, WI 54451; 800-222-2995; www.weathershield.com (wood sash only kits).

STORM WINDOWS AND SCREENS

Allied Window, Inc., 2724 W. Mc Micken Avenue, Cincinnati, Ohio 45214; 800-445-5411; www.invisiblestorms.com (interior and exterior storm windows).

Innerglass Window Systems, 15 Herman Drive, Simsbury, CT 06070; 800-743-6207; www.stormwindows.com (interior storm windows).

Harvey Industries Inc., 1400 Main Street, Waltham, MA 02451; 800-9-HARVEY; www.harveyind.com.

McNichols Company, 45 Powers Road, Westford, MA 01866; 800-237-3820; www.mcnichols.com (screen material).

MD Building Products, P.O. Box 25188, Oklahoma City, OK 73125; 800-654-8454; www.mdteam.com (screen and storm window components and materials).

Perma-Glas Mesh; P.O. Box 220, Dover, OH, 44622; 800-762-6694; www.permaglas-mesh.com (screen material).

Petit Industries Inc., (MagnaSeal Magnetic Interior, window system, screens)374 South Street, Biddeford, ME 04005; 800-947-3848; www.petitindustries.com.

Phifer Wire Products, Inc., P.O. Box 1700, Tuscaloosa, AL 35403-1700; 800-633-5955; www.phifer.com (screen material).

Winstrom Manufacturing, 70 North St., Park Forest, IL 60466; 708-748-8200; www.winstromwindows.com.

SKYLIGHTS

Andersen Windows, 100 Fourth Ave. North, Bayport, MN 55003-1096; 800-426 4261 (skylights).

Fisher Skylights, Inc., 5005 Veterans Memorial Hwy., Holbrook, NY 11741; 631-563-4001; www.fischerskylights.com.

Pella Corporation, 102 Main Street, Pella, IA 50219; 800-847-3552; www.pella.com (skylights).

Roto Frank of America, P.O. Box 599, Research Park Chester, CT 06412; 800-243-0893; www.roto-roofwindows.com (skylights).

Solatube International, Inc., 2210 Oak Ridge Way, Vista, CA 92083; 800-966-7652; www.solatube.com.

The Sun Pipe Company, Inc., P. O. Box 5760, Elgin, IL 60121; 800-844-4786; www.sunpipe.com.

Sun Tunnel Skylights, 786 McGlincey Lane, Campbell, CA 95008; 800-369-3664; www.suntunnel.com.

Velux-America, P.O. Box 5001, Greenwood, SC 29648; 800-888-3589; www.velux-america.com (skylights).

Wagner Roofing Company, 4909 46th Ave., Hyattsville, MD 20781; 301-927-9030; www.wagnerroofing.com.

Wasco Products, Inc., P.O. Box 351, Sanford, Maine 04073; 800-388-0293; www.wascoproducts.com (skylights).

DOORS AND FRAMES

Acadia Windows & Doors, 9611 Pulaski Park Drive, Baltimore, MD 21220; 800-638-6084; www.acadiawindows.com (vinyl doors).

Benchmark, General Products Company, Inc., P.O. Box 7387, Fredericksburg, VA 22404; 800-755-3667; www.benchmarkdoors.com (prehung, knocked-down, split jamb, and replacement system steel doors and frames).

FrameSaver™, BMS, P.O. Box 631247, Nacogdoches, TX 75963; 800-599-9349; www.bmslp.com (composite wood-frame material).

GRK Canada Ltd., R.R. #1-1499 Rosslyn Rd., Thunder Bay, Ontario P7C 4T9; 800-263-0463; www.grk-canada.com (fasteners).

Jeld-Wen (Challenge, IWP, Nord), 3303 Lakeport Blvd., Klamath Falls, OR 97601; 800-877-9482; www.jeld-wen.com (wood, steel, and fiberglass doors).

Morgan Manufacturing, 228 West 6th Ave., Oshkosh, WI 54903; www.jeld-wen.com (wood doors).

Peachtree Doors, Inc., P.O. Box 5700, Norcross, GA 30091-5700; 888-888-3814; www.peach99.com (steel, fiberglass, and carbon doors).

Pease Industries, 7100 Dixie Highway, Fairfield, OH 45014-8001; 800-888-6677; www.peasedoors.com (wood, steel, and fiberglass doors).

Simpson Door Co., 400 Simpson Ave., McCleary, WA 98557; 800-356-4423 (outside WA), 800-451-1087 (in WA) (wood doors).

Stanley Door Systems, 1000 Stanley Drive, New Britain, CT 06050; 800-647-8145; www.stanley-works.com (steel and fiberglass doors).

Therma-Tru Corp., 1687 Woodland Drive, Maumee, OH 43537; 800-537-8827; www.thermatru.com (steel and fiberglass doors).

Weather Shield Manufacturing, Inc., 1 Weather Shield Plaza, P.O. Box 309, Medford, WI 54451; 800-477-6808; www.weathershield.com (steel and wood doors).

GARAGE DOORS

Clopay Corp., 312 Walnut St., Suite 1600, Cincinnati, OH 45202; 800-2CLOPAY; www.clopaydoor.com.

Designer Doors, Inc., 283 Troy Street, River Falls, WI 54022; 800-241-0525; www.designerdoors.com (traditional appearance doors).

General American Door Co., 5050 Baseline Rd., Montgomery, IL 60538; 800-323-0813; www.gadco.com.

Martin Door Manufacturing, Inc., P.O. Box 27437, Salt Lake City, UT 84127-0437; 800-973-9310; www.martindoor.com.

Overhead Door Corp., 6750 LBJ Fwy., Dallas, TX 75240; 800-929-DOOR; www.overheaddoor.com.

Raynor Garage Doors, P.O. Box 448, Dixon, IL 61021-0448; 800-4RA-YNOR; www.raynor.com.

Stanley Door Systems, 1000 Stanley Drive, New Britain, CT 06050; 800-647-8145; www.stanley-works.com.

BULKHEAD DOORS

The Bilco Company, P.O. Box 1203, New Haven, CT 06505; 800-854-9724; www.bilco.com.

Cole Sewell Corporation, 2288 University Avenue, St. Paul, MN 55114; 800-328-6596.

Emco Forever Doors, P.O. Box 853, Des Moines, Iowa 50304-0853; 800-777-3626; www.forever.com.

Gerkin Windows and Doors, P.O. Box 3203, Sioux City, IA 51102; 800-475-5061; www.gerkin.com.

The Gordon Corp., 170 Spring St., Southington, CT 06489; 800-333-4564; www.gordoncellardoor.com.

Hid-N-Screen Inc., 131 Golden Days Drive, Casselberry, FL 32707; 407-339-1527; www.hidnscreen.com.

International Paper (CraftMaster), Masonite, Molded Products Group, One South Wacker Drive, Chicago, IL 60606; 800-446-1649; www.masonite.com.

Jeld-Wen, Inc. (Doorcraft, Elite, Nord, Yakima,), P.O. Box 10266, Portland, OR 97210-9879; 800-877-9482; www.jeld-wen.com.

Kaylien Corp., 8520 Railroad Ave., Santee, CA 92071; 800-748-5627; www.kayliendoors.com (fiberglass interior).

Larson Manufacturing Company, 2333 Eastbrook Drive, Brookings, SD 57006; 800-334-1328; www.larsondoors.com.

Madawaska Doors, P.O. Box 938, Barry's Bay, Ontario, Canada, KOJ 1BO; 800-263-2358; www.madawaska-doors.com.

MD Building Products, P.O. Box 25188, Oklahoma City, OK 73125; 800-654-8454; www.mdteam.com (screen and storm window components).

Morgan Manufacturing, P.O. Box 2446, Oshkosh, WI 54903-2446; 800-678-2975; www.morgandoors.com (wood doors).

New Tech Doors, Phoenix Door Manufacturing, Inc., 2652 Dow Ave., Tustin, CA 92780-7208; 800-622-0688; www.phoenixdoor.com.

Palmer River Products, Inc., 97 Broad Common Road, Bristol, RI 02809; 401-253-1711.

CASING AND TRIM

WOOD RESTORATION

Abatron, Inc., 5501 95th Avenue, Kenosha, WI 53144; 800-445-1754; www.albatron.com.

Gougeon Bros. Inc., P.O. Box 908, Bay City, MI 48707; 517-684-7286; www.gougeon.com.

ENGINEERED WOOD TRIM

Durawood United States Plastic Lumber Ltd., 2600 W. Roosevelt Rd., Chicago, IL 60608; 888-733-2546; www.uspl-ltd.com.

FrameSaver ™, Burns, Morris, Stewart, LP, P.O. Box 631247, Nacogdoches, TX 75963; 800-657-2239; www.bmslp.com.

Forestrim ™, Stimpson Lumber Co., P.O. Box 68, Forest Grove, OR 97116; 800-445-9758; www.stimpsonlumber.com.

Pacific Wood Laminates, P.O. Box 670, Brookings, OR 97415; 541-469-4177; www.socomi.com.

Prime Trim ™, Georgia Pacific Corp., 133 Peachtree St., NE, Atlanta, GA 30303; 404- 652-5127; www.gp.com.

Protrim™, ABT Co., P.O. Box 509, Middlebury, IN 46540; 800-521-4259; www.lpcorp.com.

SmartStart™, Louisiana-Pacific Corp., P.O. Box 16657, Duluth, MN 55816; 800-648-6893; www.lpcorp.com.

FIBER CEMENT TRIM

Cemtrim™, Cemplank Inc., Excelsior Industrial Park, P.O. Box 99, Blandon, PA 19510-0099; 888-327-0723; http://cemplank.com.

Harditrim™, James Hardie Industries, 26300 La Alameda, Suite 250, Mission Viejo, CA 92691; 888-J-HARDIE; www.jameshardie.com/hardtrim.htm.

Maxitrim™, Maxtile, Inc., 849 E. Sandhill Avenue, Carson, CA 90746; 800-338-8453; www.maxibuildingproducts.

Temple Inland Forest Products, P.O. Drawer N, Diboll, TX 75941; 800-231-6060; www.temple.com.

POLYMER TRIM

Flex Trim Industries Inc., 210 Citrus, Redlands, CA 92373; 800-356-9060; www.flextrim.com.

Focal Point Architectural Products, 3006 Anaconda Drive, Tarboro, NC 27886; 800-662-5550; www.focalpointap.com.

Fypon Molded Millwork, 22 West Pennsylvania Ave., Stewartstown, PA 17363; 800-537-5349; www.fypon.com.

Ornamental Mouldings, P.O. Box 4068, Archdale, NC 27263; 800-779-1135; www.ornamentalmouldings.com.

Zago Manufacturing Co., Inc., 190 Murray Street, Newark, NJ 07114; 973-643-6700; www.flexibletrim.com.

PLASTIC TRIM

ABTCO, Div. of Lousiana Pacific, P.O. Box 16657, Duluth, MN 55816; 800-521-4259; www.lousianapacific.com.

Duraflex, Resinart Corp., 1621 Placentia Ave., Costa Mesa, CA 92627; 800-258-8820; www.resinart.com (flexible trim products).

Easy Sills, American Windowsill Corp., P.O. Box 1454, Orem, UT 84059-1454; 801-426-4303; http://vpp.com/easysills (retrofit vinyl window sills).

The James Wood Company, Box 3547, 2916 Reach Rd., Williamsport, PA 17701; 570-326-3662.

Outwater Plastic Industries, 4 Passaic Street, Wood-Ridge, NJ 07075; 888-688-9283; www.outwater.com.

HARDWARE

Ball and Ball Hardware Reproductions, 463 W. Lincoln Highway, Exton, PA 19341; 800-257-3711; www.ballandball-us.com.

Barry Supply Co., 36 West 17th Street, New York, NY 10011; 212-242-5200; www.blainewindow.com (window replacement hardware).

Blaine Window Hardware, Inc., 17319 Blaine Dr., Hagerstown, MD 21740; 800-678-1919.

Cirecast, Inc., 1790 Yosemite Ave., San Francisco, CA 94124; 415-822-3030; www.cirecast.com.

Crown City Hardware Co., 1047 N. Allen Ave., Pasadena, CA 91104-3298; 800-950-1047; ww.crowncityhardware.com.

Dawson's Supply, Inc., P.O. Box 570415, Houston, TX 77257-0415; 800-816-0750; www.dawsonsupply.com.

GRK Canada Ltd., R.R. #1-1499 Rosslyn Rd., Thunder Bay, Ontario, P7C 4T9 Canada; 800-263-0463; www.grk-canada.com.

G-U Hardware, Inc., 11761 Rock Landing Drive, Suite M-6, Newport News, VA 23606-4235; 800-927-1097; www.g-u.com.

Knape & Voght Mfg. Co., 2700 Oak Industrial Drive, Grand Rapids, MI, 49505; 800-253-1561; www.knapeandvoght.com.

Kwikset Corporation, 19701 Da Vinci Street, Lake Forest, CA 92610; 800-854-3151; www.kwikset.com.

Johnson Hardware, L.E. Johnson Products, Inc., 2100 Sterling Avenue, Elkhart, IN 46516; 800-837-5664; www.johnsonhardware.com.

Monarch Hardware, Div. of Ingersoll-Rand, P.O. Box 548, Shepherdsville, KY, 40165; 800-826-5792; www.monarchhardware.com/index.html.

Piersons Co., 8 W. Brookhaven Road, Suite A, Brookhaven, PA 19015; 800-446-9111; www.piersons.com (hardware supplier with search service).

Pullman Mfg. Corp., 77 Commerce Dr., Rochester, NY 14580; 716-334-1350 (spring counterbalances).

Resource Conservation Technology, Inc., 2633 North Calvert Street, Baltimore, MD 21218; 410-366-1146.

Schlage, Lock Co. Customer Service Dept., 1010 Santa Fe, Olathe, KS 66051; 800-847-1864; www.schlagelock.com.

Stanley Hardware, The Stanley Works, 1000 Stanley Drive, New Britain, CT 06053; 800-622-4393; www.stanleyworks.com.

Truth Hardware, 700 West Bridge Street, Owatonna, MN 55060; 800-866-7884.

U-Change Lock Industries, Inc., 1640 West Highway 152, Mustang, OK, 73064; 800-253-5625; www.u-change.com.

Weiser Lock, 6700 Weiser Lock Drive, Tucson, AZ 85746; 800-677-LOCK; www.powerbolt.com.

Weso Industrial Products, Inc., P.O. Box 47, Lansdale, PA 19446; 800-445-5681; www.wescomfg.com.

Yale Security Group, 1902 Airport Rd., Monroe, NC, 28110; 800-438-1951; www.yalesecurity.com.

FLASHING

AFCO Products Inc., 44 Park Street, Somerville, MA 02143; 800-397-2687; www.afcoproducts.com (copper, aluminum, vinyl).

Benjamin Obdyke Inc., 65 Steamboat Drive, Warminster, PA 18974; 800-523-5261; www.obdyke.com.

Galvalume, Bethlehem Steel Corporation, 511 Northpoint Blvd., Sparrows Point, MD 21219; 800-521-4789; www.bethsteel.com.

Heckman Industries, P.O. Box 250, Mill Valley, CA 94942; 800-841-0066; www.deckseal.net (Deck Seal, self-adhesive aluminum flashing).

Owens Corning, One Owens Corning Parkway, Toledo, OH 43659; 800-438-7465; www.owenscorning.com.

Polytite, Dayton Superior, 2564 Kohnle Drive, Miamisburg, OH 45342; 800-776-0930; www.polytite.com.

Protecto Wrap Company, 2255 South Delaware, Denver, CO 80223; 800-759-9727; www.protectowrap.com.

Specialty Steel Industry of North America, 3050 K St., N.W., Washington, D.C. 20007; 800-982-0355; www.ssina.com.

Tamlyn and Sons, 13623 Pike Road, Stafford, TX 77477; 800-334-1676; www.tamlyn.com.

Tyvek HomeWrap, DuPont Non-Woven, 974 Centre Road, Wilmington, DE 19805-0705; 800-448-9835; www.tyvek.com.

W.R. Grace & Co., Waterproofing Division, 6051 W. 65th Street, Bedford Park, IL 60638; 800-558-7066; www.graceconstruction.com (Perm-A-Barrier wall seam tape).

CAULKING AND WEATHER STRIPPING

Accurate Metal Weatherstripping Co. Inc., 725 S. Fulton Ave., Mount Vernon, NY 10550; 800-536-6043.

AEG, 3 Shaw'S Cove, P.O. Box 6003, New London, CT 06320-1777.

Dow Chemical Co., 1881 W. Oak Parkway, Marietta, GA 30062; 800-800-3626; www.flexibleproducts.com.

Macklanburg-Duncan, 4041 North Santa Fe, Oklahoma City, OK 73118; 800-654-8454; www.mdteam.com.

Pemko Co., 4226 Transport Street, Ventura, CA 93006; 800-283-9988; www.pemko.com.

Polytite, Dayton Superior, 2564 Kohnle Drive, Miamisburg, OH 45342; 800-776-0930; www.polytite.com.

Prazi™ Drill Mate, Prazi U.S.A., P.O. Box 1165, Plymouth, MA 02362; 800-262-0211; www.praziusa.com.

Resource Conservation Technology, 2633 N. Calvert Street, Baltimore, MD 21218; 410-366-1146.

Schlegal Systems, P.O. Box 23197, Rochester, NY 14692; 800-828-6237; www.schlegal.com.

Zero International, Inc., 415 Concord Ave., Bronx, NY 10455-4898; 800-635-5335; http://users.rcn.com/zero.

SHUTTERS AND AWNINGS

TRADITIONAL WOOD SHUTTERS

Cobblestone Mill Woodworks, Inc., 325 Wilbanks Drive, Ball Ground, GA 30107; 800-591-4597; www.cobblestone-mill.com.

Kestrel Mfg. Co., #9 East Race Street, Stowe, PA 19464; 800-494-4321; www.DIYShutters.com.

Vixen Hill Manufacturing Company, P.O. Box 389, Elverson, PA 19520; 800-423-2766; www.vixenhill.com.

METAL AND FABRIC AWNINGS

Americana Building Products by Hindman Manufacturing Co., P.O. Box 1290, Salem, IL 62808; 800-851-0865; www.americana.com.

Somfy Systems, Inc., 47 Commerce Drive, Cranbury, NJ 08512; 800-647-6639; www.somfysystems.com (control mechanisms).

Sunbella Glen Raven, Inc., P.O. Box 6107, Elberton, GA 30635; 800-433-1748; www.glenraven.com.

STORM RESISTANT SHUTTERS

ASI Building Products, 2021 North 40th Street, Tampa, FL 33605; 800-282-6624; www.asibp.com.

Roll-A-Way, Inc., 10597 Oak Street N.E., St. Petersburg, FL 33713; 800-683-9505; www.roll-a-way.com.

CHAPTER 5: PARTITIONS, CEILINGS, FLOORS, AND STAIRS

FLOOR AND CEILING STRUCTURE

STRUCTURAL CONNECTORS

Wood Construction Connectors, Simpson Strong-Tie Co., Inc., 4120 Dublin Blvd., Suite 400, Dublin, CA 94568; 800-999-5099; www.strongtie.com.

WOOD RESTORATION SYSTEMS

Abatron, Inc. Wood Restoration Systems, 5501 95th Avenue, Department HP, Kenosha, WI 53144; 800-445-1754; www.abatron.com.

BCI® Joists, Versa-Lam® and Versa-Rim®, Boise Cascade Corporation, P.O. Box 50, Boise, Idaho 83728-0001; 208-384-6161; www.bc.com.

Membership and Product Directory, 1996, APA-Engineered Wood Association, P.O. Box 11700, Tacoma, WA 98411-0700; 253-565-6600; www.apawood.org.

North American Steel Framing Alliance, 1726 M Street, Suite 601, Washington, D.C. 20036-4523; 202-785-2022; www.steelframingalliance.com.

Parallam®, Microllam®, TimberStrand®, TJI® joist; Trus Joist MacMillan, P.O. Box 60, Boise, ID 83707; 800-628-3997; www.tjm.com.

Preservation Resource Group, P.O. Box 1768, Rockville, MD 20849-1768; 301-309-2222; www.prginc.com.

Steel Floors, LLC, 9251 E. 104th Ave., Henderson, CO 80640; 303-804-9700; www.steelfloorsllc.com.

StrucJoist®, StrucLam®, Glulam and E-Z Rim®, Willamette Industries, Inc., 1300 SW 5th Ave., Suite 3800, Portland, OR 97201; 800-887-0748; www.wii.com.

Unsmoke Systems, Inc., 1135 Braddock Avenue, Braddock, PA 15104; 800-332-6037; www.unsmoke.com.

Wood Care Systems, 751 Kirkland Avenue, Kirkland, WA 98033; 800-827-3480; www.woodcaresystems.com.

SHEET CORK

Badger Cork, 26112 110th Street, P.O. Box 25, Trevor, WI 53179; 800-255-2675.

W.E. Cork, P.O. Box 276, Exeter, NH 03833; 800-666-CORK; www.wecork.com.

SOUND CONTROL MATS AND ISOLATION PADS

Acoustic Surfaces, Inc., 123 Columbia Court North, Chaska, MN 55318; 800-854-6047; www.asi-stop.com.

Colband, Inc., P.O. Box 1057, Enka, NC 28728; 800-365-7391; www.colbond-usa.com.

Kinetic Noise Control, Inc., 6300 Trelan Place, P.O. Box 655, Dublin, OH; 614-889-0480; www.kineticnoise.com.

SHEATHING AND (SUBFLOORING) AND UNDERLAYMENTS

E&E Consumer Products, 7200 Miller Drive, Warren, MI 48092; 810-978-3800; www.eeeng.com.

O'Berry Enterprises, Inc., 664 Exmoor Court, Crystal Lake, IL 60014; 800-459-8428; www.oberry-enterprises.com.

FINISH FLOORING

HARDWOOD MANUFACTURERS

Buyer's Guide—Directory of Companies, National Oak Flooring Manufacturers Association, 1998 (list of Association members and their products); www.nofma.org.

PREFINISHED AND LAMINATED WOOD FLOORS

Amtico International, Inc., 6480 Roswell Rd., Atlanta, GA 30328; 800-370-7324; www.amtico.com.

Anderson Hardwood Floors, P.O. Box 1155, Clinton, SC 29325; 864-833-6250; www.andersonfloors.com.

Armstrong World Industries, Inc., P.O. Box 3001, Lancaster, PA 17604-3001; 800-292-6308; www.armstrong.com.

Bruce Hardwood Floors, P.O. Box 660100, Dallas, TX 75248; 800-527-5903; www.brucehardwoodfloors.com.

BurkeMercer Flooring Products, P.O. Box 1240, Eustis, FL 32727-1240; 800-447-8442; www.burkemercer.com.

Congoleum, P.O. Box 3127, Mercerville, NJ, 08619; 609-584-3000; www.congoleum.com.

Fritz Industries, Inc., P.O. Drawer 17040, Dallas, TX, 75217; 972-285-5471; www.fritztile.com.

Harris-Tarkett Hardwood Floors, P.O. Box 300, Johnson City, TN 37605-0300; 800-842-7816; www.harristarkett.com.

Hartco Flooring Co., P.O. Box 4009, Oneida, TN 37841; 423-569-8526; www.hartcoflooring.com.

Marley Floors Inc., P.O. Box 553, Tuscumbia, AL 35674; 800-633-3151; www.marleyfloors.com.

Robbins Hardwood Flooring, 25 Whitney Drive, Suite 106, Milford, OH 45150; 800-733-3309; www.hiflooring.com/products/robbins.htm.

U.S. EPA Indoor Air Quality Information Clearinghouse, P.O. Box 37133, Washington, D.C. 20013-7133; 800-438-4318; www.epa.gov/iaq.

Zickgraf Hardwood Company, P.O. Box 1149, Franklin, NC 28734; 800-243-1277; www.zickgraf.com.

FINISH WALLS AND CEILINGS

REPAIR PRODUCTS

Homax Products, Inc., P.O. Box 5643, Bellingham, WA 98225-7363; 800-729-9029; www.homaxproducts.com.

Nu-Wal® Restoration Systems (a fiberglass-acrylic system for interior and exterior walls), Specification Chemical Inc.®, 824 Keeler Street, Boone, Iowa 50036; 800-247-3932; www.spec-chem.com.

Permaglas-Mesh® and FibraTape® (self-adhesive fiberglass tape, patches and rolls), Permaglas-Mesh, P.O. Box 220, Dover, Ohio 44622; 800-762-6694; www.permaglas-mesh.com.

Sheetrock® All-in-One Drywall Repair Kit, US Gypsum Company, P.O. Box 806278, Chicago, IL 60606-4678; 800-874-4968; www.usg.com.

CORNER BEADS AND ACCESSORIES

Arvid's Woods, 2500 Hewitt Ave., Everett, WA 98201; 800-627-8437.

Benjamin Moore & Co., 51 Chestnut Ridge Rd., Montvale, NJ 07645; 201-573-9600; www.benjaminmoore.com.

Duron Paints & Wallcoverings, 10406 Tucker St., Beltsville, MD 20705; 800-723-8766; www.duron.com.

Fibratape® metal corner tape, Permaglas-Mesh®, P.O. Box 220, Dover, Ohio 44622; 800-762-6694; www.permaglas-mesh.com.

Focal Point, P.O. Box 93327, Atlanta, GA 30377-0327; 800-662-5550; www.focalpointap.com.

Fuller O'Brien, 925 Euclid Ave., Cleveland, OH 44115; 888-265-6753; www.fullerpaint.com.

Fypon Molded Millwork, 22 West Pennsylvania Ave., Stewartstown, PA 17363; 800-537-5349; www.fypon.com.

Martin Senour Paints, Cleveland, OH; 800-MSP-5270; www.martinsenour.com.

Maya Romanoff, 1730 West Greenleaf, Chicago, IL 60626; 312-465-6909; www.mayaromanoff.

MDC Wallcoverings, 1200 Arthur Ave., Chicago, IL 60007; 847-437-4000; www.mdcwallcoverings.com.

No-Coat® prefinished drywall corners, Drywall Systems International, P.O. Box 5937, Bend, OR 97708; 800-662-6281; www.no-coat.com.

Sheetrock®, paper-faced metal drywall bead and trim, US Gypsum Company, P.O. Box 806278, Chicago, IL 60680-4124; 800-874-4968; www.usg.com.

The Sherman-Williams Company, Cleveland, OH 44115; 800-321-8194; www.sherwin-williams.com.

StraitFlex® drywall corner tape, Conform International, Inc., 11632 Fairgrove Ind. Blvd., Maryland Heights, MO 63043; 314-692-8999; www.straitflex.com.

Trim-Tex, Inc., rigid vinyl drywall accessories, 3700 West Pratt Avenue, Lincolnwood, IL 60645; 847-679-3000; www.trim-tex.com.

STAIRS

REPLACEMENT TREADS AND RISERS

Auciello Iron Works, Inc., 560 Main St., Hudson, MA 01749; 978-568-8382; www.aiw-inc.com.

Carlisle Restoration Lumber, 1676 Route 9, Stoddard, NH 03464; 800-595-9663; www.wideplank-flooring.com.

Coffman Stairs, 1000 Industrial Rd., Marion, VA 24354; 540-783-7251; www.coffmanstairs.com.

Designed Stairs, 1480 East Sixth St., Sandwich, IL 60548; 877-478-2477; www.designedstairs.com.

Holbrook Lumber Co. Inc., P.O. Box 5229, Albany, NY 12205; 518-489-4708; www.holbrook-lumber.com.

L.J. Smith Stair Systems, 35280 Scio-Bowerston Rd., Bowerston, OH 44695; 614-269-2221; www.ljsmith.net.

Mylen Stairs, Inc., 650 Washington St., Peekskill, NY 10566; 800-431-2155; www.mylen.com.

Precision Ladders, LLC, 5727 Superior Dr., Morristown, TN 37814; 800-225-7814; www.precisionladders.com.

Spaulding Craft Inc., 1053 Harbor Lake Dr., Safety Harbor, FL 34695; 727-725-2057; www.floridacolumns.com.

Stairways, Inc., 4166 Pinemont, Houston, TX 77018; 800-231-0793; www.stairwaysinc.com.

Steel Floors, LLC, 9251 E. 104th Ave., Henderson, CO 80640; 303-804-9700; www.steelfloorsinc.com

Vintage Lumber, 1 Council D., P.O. Box 104, Woodsboro, MD 21798; 800-499-7859; www.vintagelumber.com.

CHAPTER 6: KTICHENS AND BATHS

SURFACE MAINTENANCE AND REPAIRS

Elmer's Adhesives, Division of Borden, Inc., 180 E. Broad St., Columbus, OH 43215; 800-648-0074; 218-436-2482 (fax).

Kampel Ent., Inc., Wellsville, PA 17365-0157; 717-432-9688; 717-432-5601 (fax); www.kampelent.com.

Kitchen Tune-Up Enterprises, 813 Circle Drive, Aberdeen, SD 57401; 800-333-6385; 605-225-1371 (fax); www.kitchentuneup.com.

HARDWARE TO MAXIMIZE ACCESS AND FUNCTION

Hafele, P.O. Box 4000, 3901 Cheyenne Drive, Archdale, NC 20263; 336-889-2322; 800-423-3531; 336-431-3831 (fax); www.hafeleonline.com.

Julius Blum, Inc., 7733 Old Plank Road, Stanley, NC 28164; 800-438-6788; 704-827-0799 (fax); www.blum.com.

Kiwi Connection, Ezyfold Hinge, 82 Shelburne Center Road, Shelburne, MA 01370; 413-625-2854; 413-625-6014 (fax).

ACCESSORIES TO MAXIMIZE ACCESS AND IMPROVE STORAGE

Accessible Designs Adjustable Systems, AD-AS, 2728 South Cole Road, Boise, ID 83709; 800-208-2020 (phone); 208-362-8009 (fax).

Feeny Mfg., P.O. Box 191, 6625 North State Rd. 3 North, Muncie, IN 47303; 765-288-8730; 765-288-0851 (fax); www.kv.com.

Hafele, P.O. Box 4000, 3901 Cheyenne Drive, Archdale, NC 20263; 336-889-2322; 800-423-3531; 336-431-3831 (fax); www.hafeleonline.com.

Hinge-It Corp., 3999 Millersville Rd., Indianapolis, IN 46205; 800-284-4643; 317-542-9514; 317-542-9524 (fax); www.hingit.com.

Rev-a-Shelf, P.O. Box 99585, 2409 Plantside Dr., Jeffersontown, KY 40299; 800-626-1126; 505-499-5835; 502-491-2215 (fax); www.rev-a-shelf.com.

REPLACE OR ADD CABINETRY

Ampco, Highway 1N, Rosedale, MS 38769; 662-759-3521; 662-759-3721 (fax).

Arctic Metal Products Corp., 507 Wortman Ave., Brooklyn, NY 11208; 718-257-5277; 718-257-5452 (fax).

Cervitor Kitchens Inc., 10775 Lower Azusa Road, Almonte, CA 91731; 800-523-2666; 826-443-0184; 626-443-0400 (fax).

Dwyer Products Corp., 418 N. Calumet Ave., Michigan City, IN 46360; 800-348-58508; 219-874-5236; 219-874-2823 (fax); www.dwyerkitchens.com.

Heritage Custom Kitchens, 215 Diller Ave., New Holland, PA 17557; 717-354-4011; 717-355-0169 (fax).

Neff Kitchen Manufacturers, 6 Melanie Drive, Brampton, Ontario, Canada L6T 4K9; 800-268-4527; 905-791-7770; 905-791-7788 (fax); www.neffweb.com.

Neil Kelly Naturals Collection, Neil Kelly Signature Cabinets, 804 N. Alberta, Portland, OR 97217; 503-288-6345; 503-288-7464 (fax); www.neilkelly.com.

St. Charles Manufacturing Co., 1611 E. Main St., St. Charles, IL 60174; 708-584-3800; 708-584-3992 (fax).

SURFACE MAINTENANCE AND REPAIRS

Avonite, Inc., 1945 Hwy. 304 Belen, NM 87002; 800-428-6648; 505-864-7790 (fax); www.avonite.com.

Block-Tops, Inc., 4770 E. Wesley Dr., Anaheim, CA 92807; 714-779-0475; 714-779-2284 (fax).

Custom Copper and Brass, Inc., 420 Rt. 46 E., Fairfield, NJ 07001; 973-227-9334; 973-575-6499.

Dupont Corian, Barley Mill Plaza, P.O. Box 80012, Building 12, Wilmington, DE 19880; 800-426-7426; 302-992-2855 (fax); www.corian.com.

Formica Corp., 10155 Reading Rd., Cincinnati, OH 45241; 513-786-3400; 513-786-3024 (fax); www.formica.com.

Get Real Surfaces, 121 Washington Street, Poughkeepsie, NY 12601; 845-452-3988; 845-483-9580 (fax).

John Boos & Co. 315 S. 1st St., Effingham, IL 62401; 217-347-7701; 217-347-7705 (fax); www.effingham.net/johnboos/.

Marble Renewal, 6807 W. 12th Street, Little Rock, AR 72204; 501-663-2080; 501-663-2401 (fax); www.marblerenewal.com.

Nevemar Int'l Paper, 8339 Telegraph Rd., Odenton, MD 21134; 800-638-4380; 410-551-0357 (fax); www.nevemar.com.

Pionite Decorative Surfaces, One Pionite Rd, Auburn, ME 04211; 800-746-6483; 207-748-9111; 207-784-0392 (fax); www.pionite.com.

Silestone, Cosentino USA, 10707 Corporate Drive #136, Stafford, TX 77477; 800-291-1311; 281-494-7277, 281-494-7299 (fax); www.cosentinousa.com.

Soupcan, Inc, Counter:culture Division, 1500 South Western Ave., Chicago, IL 60608; 312-243-6928; 312-243-6958 (fax); www.soupcan.com.

Taylor Wood-Craft, Inc., P.O. Box 245 Malta, OH 43758; 614-962-3741.

WilsonArt, International, Inc., 2400 Wilson Pl., Temple, TX 76504; 800-433-3222; 254-207-7000; 254-207-2545 (fax); www.wilsonart.com.

IMPROVE INDOOR AIR QUALITY

SierraPine Ltd, Medite Division, 2151 Professional Drive, Roseville, CA 95661; 888-633-7477; 916-772-3415; www.sierrapine.com.

IMPROVE BACKSPLASH AND COUNTERTOP SEAMS

Kampel's Enterprises, Inc., 8930 Carlisle Road, Wellsville, PA 17365-0157; 800-778-7006; 717-432-9688; 717- 432-5601 (fax).

MAXIMIZE ACCESS AND FUNCTION

Accessible Designs Adjustable Systems, ADAS, 94 North Columbus Road, Athens, OH 45701, 740-593-5240; 740-593-7155 (fax); www.ad-ad.com.

Auton Motorized Systems, P.O. Box 802320, Velencia, CA 91380-2320; 661-257-9282; 661-295-5638 (fax); www.auton.com.

Hafele, P.O. Box 4000, 3901 Cheyenne Drive, Archdale, NC 20263; 336-889-2322; 800-423-3531; 336-431-3831 (fax); www.hafeleonline.com.

Pressalit, 6615 West Boston Street, Chandler, AZ 85226; 480-961-5353; 480-961-8787 (fax).

IMPROVING FUNCTIONAL COUNTERTOP SPACE

Hafele, P.O. Box 4000, 3901 Cheyenne Drive, Archdale, NC 20263; 336-889-2322; 800-423-3531; 336-431-3831 (fax); www.hafeleonline.com.

Household Appliances, Inc., (formerly Black and Decker Household Appliances, 6 Armstrong Rd., Shelton, CT 06484; 800-231-9786; www.blackanddecker.com.

ENHANCE BACKSPLASH WITH SURFACE MATERIALS

Abet Laminati, 60 Sheffield Ave., Englewood, NJ 07631; 800-228-2238; 201-541-0700; 201-541-0701 (fax).

Crossville, 349 Sweeney Drive, Crossville, TN 38557; 931-484-2110; 931-484-8418 (fax); www.crossvilleceramics.com.

Formica Corp., 10155 Reading Rd., Cincinnati, OH 45241; 513-786-3400; 513-786-3024 (fax); www.formica.com.

Lamin-Art, 1330 Mark St., Elk Grove, IL 60007; 800-323-7624; 847-860-0246 (fax); www.laminart.com.

WilsonArt, International, Inc., 2400 Wilson Pl., Temple, TX 76504; 800-433-3222; 254-207-7000; 254-207-2545 (fax); www.wilsonart.com.

REPLACE OUTDATED OR NONFUNCTIONING APPLIANCES

In-Sink-Erator, 4700 21st Street, Racine, WI 53406; 800-558-5700; 414-554-5432; 414-554-3639 (fax); www.insinkerator.com.

U.S. Environmental Protection Agency ENERGY STAR Program, 401 M Street, SW (6202J), Washington, D.C. 20460; 888-STAR-YES (782-7937); www.energystar.gov.

IMPROVE ACCESS AND FUNCTION WITHIN WORK CENTERS

Fisher & Paykel Appliances Inc., 22982 Alcade Dr. #201, Laguna Hills, CA 92653; 888-936-7872; 949-829-8865; 949-829-8699 (fax); www.fisherpakel.com.

Frigidaire Home Products, 6000 Perimeter Dr., Dublin, OH 43017; 614-792-4100; 614-792-4079 (fax); www.frigidaire.com.

GE Appliances, Appliance Park, Louisville, KY 40225; 800-626-2000; 502-452-0352 (fax); www.ge.com/appliances.

Refrigerator Drawers, Sub-Zero Freezer Co., Inc., 4717 Hammersley Rd., Madison, WI 53711; 800-222-7820; 608-271-2233; 608-271-7471 (fax); www.subzero.com.

IMPROVE RESOURCE AND ENERGY EFFICIENCY OF EXISTING APPLIANCES

Fantech, Inc., 1712 Northgate Blvd., Sarasota, FL 34234; 800-747-1762 (phone); 800-487-9915 (fax); www.fantech-us.com.

U.S. Environmental Protection Agency ENERGY STAR® Program, 401 M. Street SW (6202J), Washington, D.C. 20460; 888-STAR-YES (782-7937); www.energystar.gov.

INSTALL RESOURCE- AND ENERGY-EFFICIENT APPLIANCES

Frigidaire Home Products, 6000 Perimeter Dr., Dublin, OH 43017; 614-792-4100; 614-792-4079 (fax); www.Frigidaire.com.

GE Appliances, Appliance Park, Louisville, KY 40225; 800-626-2000; 502-452-7876; 502-452-0352 (fax); www.ge.com/appliances.

Jenn-Air Products, 403 West 4th St. North, Newton, IA 50208; 800-536-6247; 317-545-2271 (fax); www.jennair.com.

Maytag Corp., 403 W. 4th St. Newtown, IA 50208; 888-462-9824; 641-792-7000; 641-787-8264 (fax); www.maytag.com.

Northland Kitchen Appliances. P.O. Box 400, Greenville, MI 48838; 800-223-3900; 616-754-0970 (fax).

U.S. Environmental Protection Agency ENERGY STAR Program, 401 M. Street SW (6202J), Washington, D.C. 20460; 888-STAR-YES (782-7937); www.energystar.gov.

Wolf Range Co., 19600 S. Alameda St., Compton, CA 90221; 310-637-3737; 310-637-7931 (fax).

SURFACE MAINTENANCE AND REPAIRS

Abatron, Inc., 5501 95th Ave., Kenosha, WI 53144; 262-653-2000; 262-653-2019 (fax); www.abatron.com.

Avonite, Inc., 1945 Hwy. 304 Belen, NM 87002; 800-428-6648; 505-864-7790 (fax); www.avonite.com.

Dupont Corian, Barley Mill Plaza, P.O. Box 80012, Building 12, Wilmington, DE 19880; 800-426-7426; 302-992-2855 (fax); www.corian.com.

Gibraltar, WilsonArt, Int'l Inc., 2400 Wilson Pl., Temple, TX 76504; 800-433-3222; 254-207-7000; 254-207-2545 (fax); www.wilsonart.com.

Surell, Formica Corp., 10155 Reading Rd., Cincinnati, OH 45241; 513-786-3400; 513-786-3024 (fax); www.formica.com.

WATER PURIFICATION

Everpure, Inc., 660 Blackhawk Dr., Westmont, IL 60559; 800-323-7873; 630-654-1115 (fax); www.everpure.com.

Franke Triflow, Franke Kitchen Systems Division, 3050 Campus Dr., Suite 500, Hatfield, PA 19440; 800-626-5771, 215-822-5873 (fax); www.franke.com.

GE SmartWater, GE Appliances, Appliance Park, Louisville KY 40225; 800-626-2000; 502-452-0352 (fax); www.ge.com/appliances.

Moen Puretouch, Moen, Inc., 25300 Al Moen Dr., N. Olmstead, OH 44070; 800-289-6636; 440-962-2000; 440-962-2770 (fax); www.moen.com.

Price Pfister Teledyne Water Pik, Price Pfister, Inc., 13500 Paxton St., Pacoima, CA 91333-4518; 818-896-1141; 818-897-0097 (fax); www.PricePfister.com.

Ultraflo Corp., P.O. Box 2294, 310 Industrial Ln., Sandusky, OH 44870; 800-760-5629; 419-626-8182; 419-626-8183 (fax); www.ultraflo.com.

WATER CONSERVATION

Neoperl, Inc., 171 Mattatuck Heights, Waterbury, CT 06705; 203-756-8891; 203-754-5868; www.neoperl.com.

Resource Conservation, Inc., 39 Mapletree Ave., Stamford, CT 06906; 800-243-2862; 203-964-0600; 203-324-9352 (fax).

MAXIMIZE ACCESS AND FUNCTION

Aqua Touch, 800 Ellis St., Glassboro, NJ 08028; 800-220-3036; 856-881-7890; 856-881-7938.

Chicago Faucets; 2100 Clearwater Drive, Des Plaines, IL 60018; 847-803-5000; 847-298-3101 (fax); www.chicagofaucet.com.

Franke, Instant Hot/Instant Cold, Franke Kitchen Systems Division, 3050 Campus Dr., Suite 500, Hatfield, PA 19440; 800-626-5771; 215-822-5873 (fax); www.franke.com.

Geberit Mfg., Inc., P.O. Box 2008, 1100 Boone Dr., Michigan City, IN 46360; 800-225-7217; 219-879-4466; 219-872-8003 (fax); www.geberit.com.

Grohe America Inc., 241 Covington Dr., Bloomingdale, IL 60108; 630-582-7711; 630-582-7722 (fax); www.groheamerica.com.

In-Sink-Erator; 4700 21st St., Racine, WI 53406; 800-558-5700; 414-554-5432; 414-554-3639 (fax); www.insinkerator.com.

KWC Faucets, Inc., 1770 Corporate Drive, #580, Norcross, GA 30093; 877-592-3287; 678-334-2121; 678-334-2128 (fax); 770-248-1608 (fax); www.kwcfaucets.com.

Pedal Valves, Inc., 13625 River Rd., Luling, LA 70070; 800-431-3668; 504-785-9997; 504-785-0082 (fax); www.pedalvalve.com.

ScaldSafe, Resource Conservation, Inc., 39 Mapletree Ave., Stamford, CT 06906; 800-243-2862; 203-964-0600; 203-324-9352 (fax).

Speakman Sensor Flo, P.O. Box 191 Wilmington, DE 19899; 800-537-2107; 800-977-2747 (fax).

SURFACE MAINTENANCE AND REPAIRS

Flitz International Ltd., 821 Mohr Ave., Waterford, WI 53185; 800-558-8611; 414-534-5898; 414-534-2991 (fax).

Marble Renewal, P.O. Box 56349, Little Rock, AR 72215-6349; 888-664-7866; 501-663-2080; 501-663-2401 (fax); www.marblerenewal.com.

Re-Bath Corp., 1055 Country Club Drive, Mesa, AZ 85210-4613; 800-426-4573; 480-844-1575; 480-833-7199 (fax).

Stone Industry.com, 1014 Makani Rd., Pukalani, HI 96768; 808-572-1222; 808-572-6886 (fax); www.stoneindustry.com.

Worldwide Refinishing Systems, 1020 N. University Parks Dr., Waco, TX 76707; 254-745-2444; 254-745-2590; www.wwrefinishing.com.

MOISTURE CONTROL

Dens-Shield, Georgia-Pacific, 133 Peachtree St. NE, Atlanta, GA 30303; 404-652-4000; 404-652-4732 (fax).

Durock, U.S. Gypsum Industries, 125 S. Franklin St., Chicago, IL 60606; 800-621-9622; 312-606-4093 (fax).

Hardibacker, James Hardie Building Products, 10901 Elm Ave., Fontana, CA 92337; 800-942-7343; 909-355-0690 (fax).

Mapei Corp., 530 Industrial Drive West, Chicago, IL 630-293-5800; 630-293-5079 (fax).

Waterproof membranes, The Noble Co., P.O. Box 350, Grande Haven, MI 49417-0350; 800-878-5788 (phone); 800-272-1519 (fax); 231-799-8008 (phone); 231-799-8850 (fax); www.noblecompany.com.

Wonderboard, Custom Building Products, 13001 Seal Beach Blvd., Seal Beach, CA 90740; 800-282-8786; 800-200-7765 (fax).

MAXIMIZE ACCESS AND FUNCTION

Hewi Inc., 2851 Old Tree Drive, Lancaster, PA 17603, 717-293-1313; 877-439-4462; 717-293-3270 (fax).

Otto Bock Rehabilitation, 3000 Xenium Lane Minneapolis, MN 55441; 800-328-4058; 800-962-2549 (fax); www.ottobock.com.

Pinnacle Innovations Inc., 8 Martin Ave., South River, NJ 08882; 732-257-6900; 732-257-6926 (fax); www.wingits.com.

PlumbingProducts; www.plumbingproducts.com.

Resource Conservation, Inc., 39 Mapletree Ave., Stamford, CT 06906; 800-243-2862; 203-964-0600; 203-324-9352 (fax).

WATER CONSERVATION

Grohe America Inc., 241 Covington Dr., Bloomingdale, IL 60108; 630-582-7711; 630-582-7722 (fax); www.groheamerica.com.

Hansgrohe, 465 Ventura Drive, Cumming, GA 30040; 770-844-7414; 770-844-0236 (fax); www.hansgrohe-usa.com.

Resource Conservation, Inc., 39 Mapletree Ave., Stamford, CT 06906; 800-243-2862; 203-964-0600; 203-324-9352 (fax).

MAINTENANCE AND REPAIRS

Abatron, Inc., 5501 95th Ave., Kenosha, WI 53144; 262-653-2000; 262-653-2019 (fax); www.abatron.com.

Resource Conservation, Inc., 39 Mapletree Ave., Stamford, CT 06906; 800-243-2862; 203-964-0600; 203-324-9352 (fax).

WATER CONSERVATION

Briggs Industries, P.O. Box 71077, Charleston, SC 29415; 800-888-4458; 800-627-4450 (fax); www.briggsplumbing.com.

The Fuller Group, 3461 Summerford Court, Marietta, GA 30062; 770-565-8539; 770-565-4197 (fax); www.aquasaver.com.

Grohe Built-in tank, 241 Covington Dr., Bloomingdale, IL 60108; 630-582-7711; 630-582-7722 (fax); www.groheamerica.com.

MisterMiser Urinal, 4901 N. 12th, Quincy, IL 62301; 888-228-6900; 217-228-6906 (fax).

Niagra Conservation Corp., 45 Horsehill Rd., Cedar Knolls, NJ 07927; 800-831-8383; 201-829-1400 (fax); www.niagraconservation.com.

Resource Conservation, Inc., 39 Mapletree Ave., Stamford, CT 06906; 800-243-2862; 203-964-0600; 203-324-9352 (fax).

Sloan Flushmate, Sloan Valve Company, 10500 Seymour Ave., Franklin Park, IL 60131; 800-875-9116; 800-447-8329 (fax).

MAXIMIZE ACCESS AND FUNCTION

Argenta Trading and Consulting Corp., 7930 NW 36 St., Ste. 23-1351, Miami Springs, FL 33166; 888-462-4288; 305-883-1212; 305-883-1911 (fax).

Geberit Mfg., Inc., P.O. Box 2008, 1100 Boone Dr., Michigan City, IN 46360; 800-225-7217; 219-879-4466; 219-872-8003 (fax); www.geberit.com.

Hewi Inc., 2851 Old Tree Drive, Lancaster, PA 17603, 717-293-1313; 877-439-4462; 717-293-3270 (fax).

Lubidet, 1980 S. Quebec Street, Denver, CO 80231; 303-368-4555; 800-582-4338; 303-368-0812 (fax); www.lubidet.com.

Low Flow Plumbing Fixtures, Plumbing Manufacturers Institute (PMI); 847-884-9764; 847-884-9775 (fax); www.pmihome.org.

Otto Bock Rehab, 3000 Xenium Ln., Minneapolis, MN 55441; 800-328-4058; 800-962-2549 (fax); 612-553-9464; 612-519-6150; www.ottobock.com.

Pinnacle Innovations Inc., 181 West Clay Avenue, Roselle Park, NJ 07204; 877-894-6448; 908-259-8922 (fax); www.wingits.com.

Toto, 1155 Southern Road, Morrow, GA 30260; 770-282-8686; 770-968-8697 (fax); www.totousa.com.

CHAPTER 7: ELECTRICAL AND ELECTRONIC SYSTEMS

SERVICE PANELS

Cutler-Hammer, 1000 Cherington Parkway, Moon Township, PA 15108; 800-525-2000; Advanced Power Center™ System; www.ch.cutler-hammer.com/apc.

The Electrical Outlet, Inc., 5905 Lamar Street, Arvada, CO 80003; 800-227-5731 (obsolete breakers and other equipment); www.electricaloutlet.com.

GE Industrial Systems, 41 Woodford Avenue, Plainville, CT 06062; 860-747-7110; www.geindustrialsystems/markets/residential.

Romac, 7400 Bandini Blvd., Commerce, CA 90040; 800-77-ROMAC (obsolete breakers and other equipment); www.romacsupply.com.

Siemens Energy & Automation, 3333 Old Milton Parkway, Alpharetta, GA 30005; 800-241-4453; www.sea.siemens.com/reselec.

Square D Company, 1415 South Roselle Road, Palatine, IL 60067; 847-397-2600; www.squared.com.

WIRING OVERVIEW

Carlon, 25701 Science Park Drive, Cleveland, OH 44122; 800-322-7566 (surface raceways moldings); www.carlon.com.

Wiremold Company, 60 Woodlawn Street, West Hartford, CT 06110; 800-621-0049; www.wiremold.com (surface raceways moldings).

ALUMINUM WIRING

3M Electrical Products, 6801 River Place Blvd, Austin, TX 78726; 800-245-3573; www.mmm.com/elpd.

AMP Inc., P.O. Box 3608, Harrisburg, PA 17105-3608; 800-522-6752; 800-522-6752; www.amp.com.

RECEPTACLES

Cooper Wiring Devices/Eagle Electric Manufacturing Co., Inc., 45-31 Court Square, Long Island City, NY 11101; 800-366-6789; www.eagle-electric.com.

Leviton Manufacturing Co., 59-25 Little Neck Parkway, Little Neck, NY 11362-2591; 800-824-3005; www.leviton.com.

Pass & Seymour/Legrand, P.O. Box 4822 Syracuse, NY 13221-4822; 800-611-7277; www.passandseymour.com (self-contained devices).

INTERIOR LIGHTING

Automated Voice Systems, Inc., 17059 El Cajon Avenue, Yorba Linda, CA 92686; 714-524-4488; www.mastervoice.com.

B&P Lamp Supply, Inc., 843 Old Morrison Highway, McMinnville, TN 37110; 931-473-9248 (replacement parts for old fixtures); www.bplampsupply.com.

Holly Solar Products, P.O. Box 864, Petaluma, CA 94953; 800-622-6716; www.hollysolar.com.

Kim Lighting, P.O. Box 60080, City of Industry, CA 91716-0080; 626-968-5666; www.kimlighting.com.

LEDtronics, Inc., 23105 Kashiwa Court, Torrance, CA 90505; 800-579-4875; www.ledtronics.com.

The Lighting Resource, P.O. Box 48345, Minneapolis, MN 55448-0343; 952-939-1717; www.lightresource.com.

Midwest Lamp Parts Company, 3534 North Spaulding Avenue, Chicago, IL 60618-5576; 773-539-0628 (replacement parts for old fixtures); www.midwestlamp.com.

Old-House Journal's Restoration Directory, Hanley-Wood, Inc., One Thomas Circle, NW, Suite 600, Washington, D.C. 20005; 202-452-0800; www.remodeling.hw.net/frmRestDir.

W.N. de Sherbinin Products, Inc., POB 63 Hawleyville, CT 06440-0063; 800-458-0010 (replacement parts for old fixtures); www.wndesherbinin.com.

EXTERIOR LIGHTING

LIGHTING-INC.COM, 1236 Wood Station Place, St. Louis, Missouri 63021; 314-225-7042; www.lighting-inc.com/searchman.html.

Old-House Journal's Restoration Directory, Hanley-Wood, Inc., One Thomas Circle, NW, Suite 600, Washington, D.C. 20005; 202-402-0800; www.remodeling.hw.net/frmRestDir.

Nightscaping, 1705 E. Colton Ave., Redlands, CA 92374; 800-544-4840; www.nightscaping.com.

Solar Outdoor Lighting, Jade Mountain Inc., P.O. Box 4616, Boulder, CO 80306-4614; 800-442-1972; www.jademountain.com/lightingProducts/outdr.html.

CONTROLS

Classic Accents, Inc., P.O. Box 1191, Southgate, MI 48195; 800-245-7742 (push-button switches); www.classicaccents.net.

Cooper Wiring Devices/Eagle Electric Manufacturing Co., Inc., 45-31 Court Square, Long Island City, NY 11101; 800-366-6789; www.eagle-electric.com.

Fiberswitch Technologies, 2511 N. Plaza Drive, Rapid City, SD, 57702; 800-811-9370; www.rapid-net.com/fiberswitch.

Leviton Manufacturing Co, 59-25 Little Neck Parkway, Little Neck, NY 11362-2591; 800-323-8920; www.leviton.com.

Lutron Electronics Co., Inc., 7200 Suter Rd., Coopersburg, PA 18036; 800-523-9466; www.lutron.com.

Sensorswitch, Inc., 900 Northrop Rd., Wallingford, CT 06492; 800-PASSIVE; www.sensorswitch.com.

ELECTRIC BASEBOARD HEATING

Cadet Manufacturing Co., P.O. Box 1675, Vancouver, WA 98668; 800-442-2338; www.cadetco.com.

Electro-Heat, Inc., P.O. Drawer D, Allegan, MI 49010; 616-673-6688; www.electroheatinc.com.

Enerjoy by SSHC, Inc., P.O. Box 769, Old Saybrook, CT 06475; 800-544-5182; www.sshcinc.com.

Markey Engineered Products, 470 Beauty Spot Rd. East, Bennetsville, SC 29512; 843-479-4006; (Qmark) www.qmarkmeh.com.

Radiant Electric Heat, Inc., 3695 North 126th Street-Unit N, Brookfield, WI 53005; 800-774-4450; www.electricheat.com.

Runtal North America, Inc., 187 Neck Rd., Ward Hill, MA 01835; 800-526-2621; www.runtalnorthamerica.com.

PHONE, COMPUTER, AND TV CABELING

Leviton Manufacturing Company Inc., 59-25 Little Neck Parkway, Little Neck, NY 11362-2591; 800-323-8920; www.leviton.com.

Residential Cabling System, Molex PN 8 Executive Drive, Hudson, NH 03051; 800-866-3827; www.residentialcabling.com.

INTRUSION AND ALARM SYSTEMS

ADEMCO, 165 Eileen Way, Syosset, NY 11791, 800-645-7568; www.ademco.com.

AMP Incorporated, P.O. Box 3608, Harrisburg, PA 17105-3608; 800-522-6752; www.amp.com.

Applied Future Technologies, Inc., 11615 West 75th Avenue, Arvada, CO 80005; 800-790-3353; www.appliedfuture.com.

Axlon Electronics Corp., 7F, No. 356, Sec. 5 Nan-King E. Road, Taipei, ROC Taiwan 866-2-26983336; www.axlon.com.

Caddx Controls, Inc., 1420 N. Main Street, Gladewater, TX 75647; 800-727-2339; www.caddx.com.

C&K Systems, Inc., 170 Michael Dr., Syosset, NY 11791; 800-573-0154; www.cksys.com.

Digital Security Controls Ltd., 3301 Langstaff Rd., Concord, Ontario L4K 4L2 Canada; 888-888-7838; www.dscsec.com.

First Alert Professional Security Systems, 175 Eileen Way, Syosset, NY 11791; 800-793-5949; www.firstalertpro.com.

Home Automation, Inc., 5725 Powell St., Suite A, New Orleans, LA 70123; 800-229-7256; www.homeauto.com.

Interactive Technologies, Inc., 2266 Second St. No., North St. Paul, MN 55109; 800-777-5484; www.ititechnologies.com.

Napco Security Systems, Inc., 333 Bayview Ave., Amityville, NY 11701; 800-645-9445; www.napcosecurity.com.

UStec, 100 Rawson Road, Suite 205, Victor, NY 14564; 716-924-1740; 800-836-2312; www.ustecnet.com.

X-10 (USA) Inc., 91 Ruckman Road, Box 420, Closter, NJ 07624-0420; 800-526-0027; www.execulink.com/~hometech/x10menu.htm.

SMOKE DETECTORS

Chemtronics, Kidde-Fenwal, Inc., 400 Main St., Ashland, MA 01721; 800-496-8383; www.chemtronics.com.

Edwards Signaling, 90 Fieldstone Ct., Cheshire, CT 06410-1212; 203-699-3300; www.edwards-signals.com.

ESL/Sentrol, 12345 SW Leveton Dr., Tualatin, OR 97062-9938; 800-547-2556; www.sentrol.com.

Gentex Corp., 10985 Chicago Dr., Zeeland, MI 49464; 800-436-8391; www.gentex.com.

MTI Industries, Inc., 31632 N. Ellis Dr., No. 301, Volo, IL 60073; 800-383-0269; www.safe-t-alert.com.

Silent Call Communications Corp., P.O. Box 868, Clarkston, MI 48347; 800-572-5227; www.silent-call.com (vibration systems).

System Sensor, 3825 Ohio Ave., St. Charles, IL 60174; 800-SENSOR2; www.systemsensor.com.

CARBON MONOXIDE DETECTORS

Aim Safety USA Inc., 1624 Headway Circle, Austin, TX 78754; 800-275-4246; www.aimsafeair.com.

Chemtronics, Kidde-Fenwal, Inc., 400 Main St., Ashland, MA 01721; 800-496-8383; www.chemtronics.com.

ESL/Sentrol, 12345 SW Leveton Dr., Tualatin, OR 97062; 800-547-2556; www.sentrol.com.

First Alert, 3901 Liberty St. Rd., Aurora, IL 60504; 800-323-9005; www.firstalert.com (BRK Brands included).

Macurco Gas Protectors, 3946 So. Mariposa St., Englewood, Co 80110; 303-781-4062; www.macurco.com.

MTI Industries, Inc., 31632 N. Ellis Dr., No. 301, Volo, IL 60073; 800-383-0269; www.safe-t-alert.com.

LIGHTNING PROTECTION

Advanced Lightning Technology, Inc., 122 Leesley Lane, Argyle, TX 76226; 800-950-7933; www.advancedlightning.com.

Automatic Lightning Protection, 7548 West Bluefield Ave., Glendale, Arizona 85308; 800-532-0990; www.lightningrod.com.

Independent Protection Company, 1607 South Main St., Goshen, IN 46526; 800-860-8388; www.ipclp.com.

Lightning Master Corp., 1351 N. Archuras Ave., Clearwater, FL 33765; 800-749-6800; www.lightningmaster.com.

Stormin Protection Products, 10749 63rd Way N., Pinellas Park, FL 33782; 888-471-1038; http://members.tripod.com/~storminprotection/index-14.html.

SURGE PROTECTION

Advanced Protection Technologies, 14450 58th N., Clearwater, FL 34620; 800-237-4567; www.apttvss.com.

American Power Conversion, 132 Fairgrounds Road, West Kingston, RI 02889; 800-800-4272; www.apcc.com.

Best Power, P.O. Box 280, Necedah, WI 54646; 800-356-5794; www.bestpower.com.

Control Concepts, 328 Water St., P.O. Box 1380, Binghamton, NY 13902-1380; 800-288-6169; www.control-concepts.com.

Leviton Manufacturing Co. Inc., 59-25 Little Neck Parkway, Little Neck, NY 11362-2591; 800-824-3005; www.leviton.com.

Panamax Inc., 150 Mitchell Blvd., San Rafael, CA 94903-2057; 800-472-5555; www.panamax.com.

Protek Devices, 2929 S. FairLane, Tempe, AZ 85282; 602-431-8101; www.protek-tvs.com.

Square D Co., 1415 S. Roselle Rd., Palatine, IL 60067; 847-397-2600; www.squared.com.

Stormin Protection Products, 10749 63 Way N., Pinellas Park, FL 33782; 888-471-1038; http://members.tripod.com/~storminprotection/index-14.html.

Tripp Lite Power Protection, 1111 W. 35th St., Chicago, IL 60609; 773-869-1111; www.tripplite.com.

Volt-Guard Inc., 400 23 St. So., St. Petersburg, FL 33712; 800-237-0769; www.voltguard.com.

GARAGE DOOR OPENERS

Allstar Corp., P.O. Box 240, Downingtown, PA 19335; 877-441-9300; www.allstarcorp.com.

The Chamberlain Group, Inc., 845 Larch Ave., Elmhurst, IL 60126; 800-282-6225; www.chamberlaingroup.com.

The Genie Co., 22790 Lake Park Blvd., Alliance, OH 44601-3498; 800-354-3643; www.geniecompany.com.

Quantum, Wayne-Dalton Corp., One Door Drive, P.O. Box 67, Mount Hope, OH 44660; 800-827-3667; www.wayne-dalton.com.

CHAPTER 8: HVAC AND PLUMBING

DISTRIBUTION SYSTEMS

Burnham Radiant Heating Co., P.O. Box 3079, Lancaster, PA 17603; 717-481-8400; www.burnham.com.

Empire Comfort Systems, Inc., 918 Freeburg Avenue, Belleville, IL 62222-0529; 800-851-3153; www.empirecomfort.com.

En Dura Coat, duct board, Owens Corning World Headquarters, One Owens Corning Parkway, Toledo, OH 43659; 800-GET-PINK; www.owenscorning.com/around/sound/products/ducts.asp.

Enerjee, 24 S. Lafayette Ave., Morrisville, PA 19067; 215-295- 0557; www.enerjee.com.

Radiant Technology, 11A Farber Drive, Bellport, NY 11713; 800-784-0234; www.radiant-tech.com.

SpacePak, Mestek, Inc., 260 N. Elm Street, Westfield, MA 01085; 413-564-5530; www.spacepak.com.

SSHC, Inc., P.O. Box 769, Old Saybrook, CT 06475; 800-544-5182; www.sshcinc.com.

Unico, Inc., 7401 Alabama Ave., St. Louis, MO 63111-9906; 800-527-0896; www.unicosystem.com.

Wirsbo Company, 5925 W. 148th Street, Apple Valley, MN 55124; 800-321-4739; www.wirsbo.com.

HEATING

Apollo HydroHeat & Cooling, A Division of State Industries, Inc., 500 Lindahl Parkway, Ashland City, TN 37015; 800-365-8170 x4210; www.stateind.com.

Burnham Corporation, P.O. Box 3079, Lancaster, PA 17603; 717-481-8400; www.burnham.com.

Carrier North American Operations, P.O. Box 4808, Carrier Parkway, Syracuse, NY 13221-4808; 800-227-7437; www.carrier.com.

Dunkirk Radiator Corporation, 85 Middle Road, Dunkirk, NY 14048; 716-366-5500; www.dunkirk.com.

Lennox Industries, 2100 Lake Park Boulevard, Richardson, TX 75080; 214-497-5000; www.lennox.com.

Rheem Manufacturing, Air Conditioning Division, P.O. Box 17010, Fort Smith, AR 72917-7010; 800-548-RHEEM; www.rheemac.com.

Trane Company, 3600 Pammel Creek Road, LaCrosse, WI 54601; 608-787-3111; www.trane.com.

York International Corporation, 631 South Richland Avenue, York, PA 17403; 717-771-7890; www.york.com.

COOLING

AdobeAir, Master Cool, 500 South 15th Street, Phoenix, AZ 85034; 602-257-0060; www.adobeair.com.

Carrier North American Operations, P.O. Box 4808, Carrier Parkway, Syracuse, NY 13221; 315-432-6000; www.carrier.com.

Rheem Manufacturing, Air Conditioning Division, P.O. Box 17010, Fort Smith, AR 72917-7010; 800-548-RHEEM; www.rheemac.com.

Robur Corporation, 2300 Lynch Road, Evansville, IN 47711; 812-424-1800; www.robur.com.

SunAmp Power Company, 800-MR-SOLAR; sales@sunamp.com (email).

Tamarack Technologies, Inc., P.O. Box 490, West Wareham, MA 02576; 800-222-5932; www.tamtech.com.

Trane Company, 3600 Pammel Creek Road, LaCrosse, WI 54601; 608-787-3111, www.trane.com.

York International Corporation, 631 South Richland Avenue, York, PA 17403; 717-771-7890; www.york.com.

HEAT PUMPS

Carrier North American Operations, P.O. Box 4808, Carrier Parkway, Syracuse, NY 13221; 315-432-6000; www.carrier.com.

Consolidated Technology Corporation, 4601 Eisenhower Avenue, Alexandria, VA 22304; 703-370-8700; www.ctctheinsider.com.

Enviromaster International, LLC, 5780 Success Drive, Rome, NY 13440; 800-228-9364; www.enviromaster.com.

Trane Company, 3600 Pammel Creek Road, LaCrosse, WI 54601; 608-787-3111, www.trane.com.

WaterFurnace International, Inc., 9000 Conservation Way, Fort Wayne, IN 46809; 800-934-5667; www.waterfurnace.com.

York International Corporation, 631 South Richland Avenue, York, PA 17403; 717-771-7890; www.york.com.

INDOOR AIR QUALITY

American Aldes Ventilation Corporation, 4537 Northgate Court, Sarasota, FL 34234-2124; 800-255-7749; www.american aldes.com.

Aprilaire Automatic Humidifier, Research Products Corporation, 1015 East Washington Avenue, Madison, WI 53703; 800-545-2219; www.resprod.com/aa.html.

Honeywell, Inc., MN10-1461, 1885 Douglas Drive N., Golden Valley, MN 55420; 800-328-5111; www.honeywell.com.

Nutech Energy Systems, Inc., 511 McCormick Blvd., London, Ontario N5W 4C8 Canada; 519-457-1904; www.lifebreath.com.

NuTone, 4820 Red Bank Road, Cincinnati, OH 45227; 800-543-8687; www.nutone.com.

Therma-Stor Products, DEC International, Inc., P.O. Box 8050, Madison, WI 53708; 800-533-7533; www.thermastor.com.

Trion, Inc., 101 McNeil Road, Sanford, NC 27330; 800-884-0002; www.trioninc.com.

CONTROLS

Broan-NuTone, Inc., 926 West State Street, Hartford, WI 53027; 800-637-1453; www.broan.com.

Carrier North American Operations, P.O. Box 4808, Carrier Parkway, Syracuse, NY 13221; 315-432-6000; www.carrier.com.

DuroZone, Duro Dyne Corp., 130 Route 110, Farmingdale, NY 11735; 800-899-3876; www.duro-dyne.com.

Home Automation, Inc., HAI, 5725 Powell Street, Suite A, Metairie, LA 70123; 800-229-7256; www.homeauto.com.

Honeywell, Inc., Perfect Climate Comfort Center Control System, MN10-1461, 1885 Douglas Drive N., Golden Valley, MN 55422; 800-828-8367; http://content.honeywell.com:80/Home/ac-automated_control/hc.htm.

Lightstat, Inc., 22 W. West Hill Road, Barkhamsted, CT 06063; 800-292-2444; www.lightstat.com.

PerfectTemp, Research Products Corp., P.O. Box 1467, Madison, WI 53701; 800-334-6011; www.resprod.com.

Robert Shaw, Uni-Line Division Invensis, P.O. Box 2000, Corona, CA 92878-2000; 800-304-6563; www.robertshaw.com.

Tamarack Technologies, Inc., 11A Patterson's Brook Rd., West Wareham, MA 02576; 800-222-5932; www.tamtech.com.

Trol-A-Temp, Honeywell, MN10-1461; 1885 Douglas Drive N., Golden Valley, MN 55422; 800-828-8367; www.trolatemp.com.

FIREPLACES AND CHIMNEYS

Andiron Technologies, Inc., 2995 Woodside Road, Suite 400-PMB 226, Woodside, CA 94062; 888-4-ECOFIRE; www.EcoFire.com.

Bramec Corp., 403 Highway 105, P.O. Box 9, N. Sioux City, SD 57049; 800-843-9974; www.bramec.com.

Buckley Rumford Co., 1035 Monroe Street, Port Townsend, WA 98368; 800-447-7788; www.rumford.com.

ChimCap, 120 Schmitt Blvd., Farmingdale, NY 11735; 800-262-9622; www.chimneys.com/chimcap.

Heatilator, Inc., 1915 West Saunders Street, Mt. Pleasant, IA 52641; 800-843-2848; www.heatilator.com.

Heat-N-Glo, Div. of Hearth Technologies, 20802 Kensington Blvd., Lakeville, MN 55044; 888-743-2887; www.heatnglo.com.

Industrial Chimney Company/RSF Energy, 400 J.F. Kennedy, St. Jerome, Quebec J74 4B7, Canada; 450-565-6336; www.icc-rsf.com.

Majestic Products Company, 410 Admiral Blvd., Mississauga, Ontario L5T 2N6 Canada; 800-525-1898; www.majesticproducts.com.

Superior Clay Corp., P.O. Box 352, Uhrichsville, OH 44683; 888-254-1905; www.superiorclay.com.

Temco Fireplace Products, 301 S. Perimeter Park Drive, Suite 227, Nashville, TN 37211; 800-671-9394.

DOMESTIC HOT-WATER HEATING

Addison Products Company, P.O. Box 607776, Orlando, FL; 407-292-4400; www.americanwater-heater.com.

Bosch AquaStor, Controlled Energy Corp., 340 Mad River Park, Waitsfield, VT 05673; 800-642-3199; www.controlledenergy.com.

DEC/International/Therma-Stor, P.O. Box 8050, Madison, WI 53708; 800-533-7533, www.thermastor.com.

DS06 (HotTop), Trevor-Martin Corp., 4151 112th Terrace, Clearwater, FL 33762; 800-875-1490; www.trevormartincorp.com.

GFX, Doucette Industries, 701 Grantley Road, York, PA 17403; www.doucetteindustries.com.

Sea Horse, Gas-Fired Products Incorporated, P.O. Box 36485, Charlotte, NC 28236; 704-372-3485; www.gasfiredproducts.com.

Seisco RA-28, Seisco (Tankless Water Heating System), 223 W. Airtex, Houston, TX 77090; 888-296-9293; www.seisco.com.

Weather-Pro, American Water Heater, P.O. Box 1597, Johnson City, TN 37605; 800-999-9515.

WATER SUPPLY AND DISTRIBUTION SYSTEMS

BF Goodrich, Flow Guard Gold (CPVC), Cleveland, OH 44141; 800-864-4851; www.flowguardgold.com.

Copper Development Association, New York, NY; 800-CDA-DATA; www.copper.org.

IPEX, Inc. (KITEC), Englewood, CO 80112; 800-473-9808; www.ipexinc.com.

Metlund, Costa Mesa, CA; 800-METLUND; www.metlund.com.

NIBCO "Just Right," Elkhart, IN 46516; 800-234-0227; www.nibco.com.

SPARCO Anti-Scald Valve, Warwick, RI 02886; 401-738-4290; www.sparco-inc.com.

US Brass, Brass PEX, Dallas, TX; 800-872-7277; www.usbrass@zurn.com.

Yorkshire Fittings, USA, Brentwood, TN 37027; 615-309-8669.

DRAIN, WASTE, AND VENT SYSTEMS

Cast Iron Soil Pipe Institute, Chattanooga, TN 37421; 423-892-0137; www.cispi.org.

Charlotte Pipe & Foundry (cast iron), Charlotte, NC 28235; 800-438-6091; www.charlottepipe.com.

Studor, Inc. Air Admittance Valves, Dunedin, FL 34698; 800-447-4721, www.studor.com.

WaterFilm Energy, Inc., P.O. Box 48, Oakdale, NY 11769; 516-758-6271; www.oikos.com/gfx.

FUEL SUPPLY SYSTEMS

Copper tube, Copper Development Association, New York, NY; 800-CDA-DATA; www.copper.org.

Gas Tite, TiteFlex Corporation (CSST), P.O. Box 90054, Springfield, MA 01139; 800-662-0208; www.titeflex.

P.G.P., Parker Hannifin Corp., Paraflex Division, Ravenna, OH 44266; 800-4-PARFLEX, www.parker.com/fcg.

TracPipe, OmegaFlex, Exton, PA 19341; 800-671-8622; www.omegaflex.com.

WARDFLEX Ward Manufacturing, Blossburg, PA 16912; 800-248-1027; www.wardmfg.com.

FIRE PROTECTION SYSTEMS

BF Goodrich BlazeMaster, Cleveland, OH 44141; 800-331-1144; www.bfgoodrich.com.

Creative Systems, Inc., DecoShield, Jamesville, WI 53545; 608-757-0717.

Reliable Automatic Sprinkler Co., Mt. Vernon, NY 10552; 800-431-1588; www.reliablesprinkler.com.

CHAPTER 9: SITE WORK

DECK AND PORCH STRUCTURE

ConServ Epoxy/Housecraft Associates, 7 Goodale Road, Newton, NJ 07860; 973-579-1112; www.conservepoxy.com.

DeckWorks Construction, 1511 Avondale Dr., Norman, OK 73069; 888-297-1455; www.deckworks.com.

Gougeon Bros., Inc., P.O. Box 908, Bay City, MI, 48707-0908; 989-684-7286; www.gougeon.com.

Life Seal, Life Industries Corp., 2081 Bridgeview Drive, P.O. Box 71789, Charleston, SC 29415; 843-566-1225; www.lifeindustries.com (caulks with UV inhibitors).

LiquidWood (wood consolidant) and WoodEpox (adhesive wood-replacement paste), solvent-free liquid epoxy systems by Abatron Inc., 5501 95th Avenue, Kenosha, WI 53144; 800-445-1754; http://www.abatron.com.

REPAIR OR REPLACE WOOD DECKING MATERIALS

All-Coast Forest Products, Inc. (Pau Lope Hardwood Decking, CafeFree Plastic Lumber), 250 Asti Road, Cloverdale, CA 95425; 800-767-2237; www.all-coast.com.

Arch Wood Protection, Inc., 1955 Lake Park Drive, Suite 250, Smyrna, GA 30080; 770-801-6600; www.WolmanizedWood.com.

Brock Deck & Brock Dock (100% virgin PVC decking), Triple Conference Royal Crown Ltd., P.O. Box 360, Milford, IN 46542-0360; 800-488-5245; www.royalcrownltd.com.

Cabot Wood Care Products, Attn.: Marketing Dept. (wood cleaner, water proofing, stains), 100 Hale Street, Newburyport, MA 01950; 800-US-STAIN; www.cabotstain.com.

Carefree Decking (100% recycled plastic decking), SmartDeck® Systems (composite lumber made from recovered wood and recycled plastic), DuraWood EX (crosslinked wood fiber/ polymer composite made from 70% recovered and recycled wood waste and recycled milkjugs with hollow centers can hide cabeling within deck), and DuraWood PE (100% postconsumer polyethylene plastic lumber), USPL Corp., 2300 Glades Road, Suite 440 W., Boca Raton, FL 33431; 561-394-3511; http://www.uspl-ltd.com.

Certified Forest Products Council, 14780 SW Osprey Drive, Suite 285, Beaverton, OR 97007; 503-590-6600; www.certifiedwood.org.

Certified sustainable harvested pressure-treated lumber source list, SmartWood Program of Rainforest Alliance, Goodwin-Baker Building, 61 Millet Street, Richmond, VT 05477; 802-434-5491; www.smartwood.org.

ChoiceDek, Advanced Environmental Recycling Technologies (AERT) 914 N. Jefferson, Springdale, AK 72765; 800-951-5117; www.choicedek.com.

Co-op America's National Green Pages, 1612 K Street NW, Suite 600, Washington, D.C. 20006; 800-58-GREEN; www.coopamerica.org.

DeckWorks Construction, 1511 Avondale Dr., Norman, OK 73069; 888-297-1455; www.deckworks.com.

DreamDeck (extruded from 100% vinyl), Thermal Industries, Inc., 301 Brushton Avenue, Pittsburgh, PA 15221; 800-245-1540; www.dreamdeck.com.

EnviroWood (100% postconsumer and postindustrial scrap plastic), Enviro Products, P.O. Box 2714, Gulf Shores, AL 36547; 334-955-1490.

Forest Stewardship Council-U.S. (FSC), 1155 30th Street NW, Suite 300, Washington, D.C. 20007; 877-372-5646; www.fscus.org.

Greenheart-Durawoods, Inc., P.O. Box 279, Bayville, NJ 08721; 800-783-7220; www.paulope.com.

Greenseal, 1001 Connecticut Avenue, Suite 827, Washington D.C. 20036-5525; 202-872-6400; www.greenseal.org.

Heritage Vinyl Products' Teck Deck (vinyl decking system has planks that snap onto boards with screws which are hidden by the planks), 1576 Magnolia Drive, Macon, MS 39341; 800-763-5143 x2944; www.heritagevinyl.com.

Iron Woods Ipe and Cambara wood, (insect-resistant species hardwoods), Timber Holdings Ltd., 2400 West Cornell, Milwaukee, WI 53209; 414-445-8989; www.ironwoods.com.

Kroy Building Products, P.O. Box 636, York, NE 68467; 800-933-5769; www.kroybp.com.

MAXiTUF (100% recycled plastic lumber), Resco Plastics, Inc., 1170 Newport Avenue, Coos Bay, OR 97420; 541-269-5485; www.rescoplastics.com.

Nexwood (structural recycled plastic, rice hulls composite decking), Composite Technology Resources, Ltd., 7655 Newman Blvd., Suite 308, Lasalle, Quebec H8N 1X7 Canada; 888-763-9966; www.nexwood.com.

Pau Lope by SKC Ltd., General Woodcraft, Inc., 531 Broad Street, New London, CT 06320; 860-444-9663 or 7524; www.paulopedecking.com.

Penofin (wood sealers), Performance Coatings, Inc., P.O. Box 1569, Ukiah, CA 95482; 800-736-6346; www.penofin.com.

Scientific Certification Systems, 139 Harrison Street, Suite 400, Oakland, CA 94612; 510-832-1415; www.scs1.com.

Smartwood Program, Rainforest Alliance, 65 Bleecker Street, New York, NY 10012; 888-MY-EARTH; www.rainforest-alliance.org, www.smartwood.org.

TimberTech Ltd. (wood/plastic resin composite hollow profile boards fit together in a T&G design without exposed fasteners), Crane Plastics P.O. Box 182880, Columbus, OH 43218-2880; 800-307-7780; www.timbertech.com.

Trex Easy Care Decking (approximately 50% wood fiber and 50% plastic), Trex Company, 20 S. Cameron Street, Winchester, VA 22601; 800-289-8739; www.trex.com.

Wood Guard, ISK Biocides, 416 East Brooks Road, Memphis, TN 38109; 800-238-2523; www.woodguard.com.

ZCL Composites, Inc., Pultronex Corp., 111 Fairhope Street, Forest City, NC 28043; 828-286-1515; www.ezdeck.com.

PRESERVATIVE-TREATED WOOD

ACQ preserve and built-in water repellent (marketed as PreservePlus), Chemical Specialties, Inc., 200 East Woodlawn Road, Charlotte, NC 28217; 800-421-8661; www.treatedwood.com.

Auro Borax Wood Impregnation No. 111, Auro Products, Sinan Company, P.O. Box 857, Davis, CA 95616-0857; 530-753-3104; www.dcn.davis.ca.us/go/sinan/.

BORA-CARE (termiticide, insecticide, and fungicide concentrate) and Timber Borate Treatment, Nisus Corporation, 215 Dunavant Drive, Rockford, TN 37853; 800-264-0870; www.nisuscorp.com.

Impel Rods (site-applied borate rods), Ultrawood (water-repellent treated wood), and Preserve (arsenic-free treated wood), Chemical Specialties, Inc., 200 East Woodlawn Road, Charlotte, NC, 28217; 800-421-8661; www.treatedwood.com.

NatureSeal (low-VOC waterproofing wood and concrete and brick primer/sealer), Seal-N-Protect, 431 Barn Swallow Lane, Allentown, PA 18104; 610-366-7931; www.sealnprotect.com.

Osmose Advance Guard (Borate Pressure Wood), P.O. Box 16657, Duluth, MN 55816-9930; 800-580-4296; www.smartguard.lpcorp.com.

Osmose Weathershield (water repellent, wood sealers, wood stains, wood water repellents, wood brighteners), Osmose Inc., 980 Ellicott Street, Buffalo, NY 14209; 800-877-POLE; www.osmose.com.

Pro-Tech 2000 (low-VOC soy-based wood and concrete sealer) Nycon MidWest (NMW) Inc., 8402 East 33rd Street, Indianapolis, IN 46226; 800-253-4237, 317-898-0292; www.nycon.com.

STAIRS AND HANDRAILS

See Preservative-Treated Wood.

PAVED DRIVEWAYS, WALKS, AND PATIOS

Abocrete (solvent-free, water-borne, resurfacing system), AboWELD55-1 (repairing concrete stairs without forms), WoodEpox and LiquidWood (to fill and repair, can be sawed, nailed, and painted), Abatron, Inc., 5501 95th Ave., Kenosha, WI 53144; 800-445-1754 or 262-653-2000, www.abatron.com.

ARDEX CD (resurface old, spalled or worn concrete surfaces), ARDEX A-300 (pourable outdoor concrete topping), ARDEX Poly-Top Polymer Concrete Patching Compound Ardex Inc., 400 Ardex Park Dr., Aliquippa, PA 15001; 412-604-1200; www.ardex.com.

CIA-Gel Epoxy (multipurpose, solvent-free, low-odor, low-toxicity, nonshrink formula for anchoring), Covert Operations, 1940 Freeman Avenue, Long Beach, CA 90804; 800-827-7229 or 562-986-4212; www.covertoperationsinc.com.

ConSeal (masonry patches and sealants), H & C (concrete and masonry waterproofing sealer), H & C Shield Plus (concrete stain), H&C Stains, Sherwin-Williams, 101 Prospect Avenue, Suite 1460, Cleveland, Ohio 44115; 800-867-8246; www.hc-concrete.com and www.sherwin.com.

EliteCrete Systems, Inc. (acrylic polymer concrete overlay products for decorative concrete resurfacing and concrete repair/resurfacing), P.O. Box 96, Valparaiso, IN 46383; 888-323-4445 or 219-465-7671; www.elitecrete.com.

Endur-o-Seal (low-VOC water-borne concrete sealer) and EPO-TOXY (solvent-free concrete repair epoxy compound), Lone Star Epoxies, P.O. Box 121, Rowlett, TX 75030-0121; 972-475-2501; www.lsepoxies.com (CR60 Super Bond Repair Cement).

Gardner-Gibson Co. 4161 E. 7th Ave., Tampa, FL 33605; 888-SHURSTIK; www.gardnerasphalt.com (foundation coatings and sealants including driveway-maintenance products, caulking and repair compounds).

Green Mountain International, Inc., 235 Pigeon Street, Waynesville, NC 28786; 800-942-5151; www.mountaingrout.com/ (polyurethane grout systems-USDA approved).

Homex 300 (concrete expansion joint filler and light duty forming material made from 100% post-consumer recycled newsprint. Contains no formaldehyde or asbestos), Homasote Co., P.O. Box 7240, West Trenton, NJ 08628-0240; 800-257-9491; www.homasote.com.

INCRETE (stamped-in-place decorative concrete system), STAIN-CRETE (deep penetrating concrete stain for existing concrete), SPRAY-DECK (decorative nonskid concrete overlay system for existing concrete surface), STONE-CRETE (poured-in-place decorative concrete wall system for sound barrier walls, retaining walls), Increte, 1611 Gunn Highway, Odessa, FL 33556; 800-752-4626; www.increte.com.

Macco Adhesives, 925 Euclid Avenue, Cleveland, OH 44115; 800-634-0015; www.liquidnails.com.

PC-CRETE 300 (two-component, solvent-free, exterior-use polyurethane for repair of cracks and divots in concrete and masonry substrates), POLYGLAZE AR-SF (solvent-free, single-component, urethane polyurea topcoat for waterproof concrete decks), Polycoat Products, 14722 Spring Avenue, Santa Fe Springs, CA 90670; 562-802-8834; www.polycoatusa.com.

Perma•Crete Perma-Bond Crack Repair and concrete resurfacing system. PermaCrete, Nashville, TN 37211; 800-60-PERMA; www.permacrete.com.

Sika Corp. (adhesives, sealants, masonry, and repair products for concrete), 201 Polito Avenue, Lyndhurst, NJ 07071; 800-933-SIKA; www.sikaUSA.com.

Sonneborn, ChemRex Inc.(high-performance adhesives and sealants, wall and floor coatings, waterproofing membranes, concrete-repair products and stucco systems), 889 Valley Park Drive, Shakopee, MN 55379, 952-496-6000; 800-433-9517; www.chemrex.com.

Soy-Clean (biodegradable soy-based concrete cleaner, asphalt cleaner, graffiti remover, herbicide stain remover, paint stripper, driveway cleaner), Soy Environmental Products Inc., Kansas City, KS; 8855 N. Black Canyon Hwy., Suite 2000, Phoenix, AZ 85021; 602-674-5500; www.ia-usa.org/k0164.htm and www.soyclean.com.

Specco Industries, Inc., 13087 Main Street, Lemont, IL 60439; 800-441-6646 or 630-257-5060; www.specco.com (concrete and masonry products including bonders, cleaners, coatings, epoxies, patching, sealers, surface treatments and water repellents).

Symons Corporation, 200 E. Touhy Avenue, Des Plaines, IL 60018, 847-298-3200; http://www.symons.com (line of concrete repair and surfacing products and concrete forms).

VOCOMP-20 (water emulsion acrylic curing and sealing compound), SEALTIGHT ASPHALT (Expansion Joint filler), W.R. Meadows, Inc., 300 Industrial Park Drive, P.O. Box 338, Hampshire, IL 60140-0338; 800-342-5976.

MASONRY WALLS

AboCrete (fills cracks, anchors posts and railings, structural patching, resurfacing and restoration. It has no solvents or volatiles), AboWeld 55-1 (rebuild broken risers, corners, or treads in stairs), Abatron, Inc., 5501 95th Avenue, Kenosha, WI 53144; 800-445-1754 or 262-653-2000, www.abatron.com.

Armorloc (interlocking concrete blocks and concrete erosion control systems), Armortec, 3260 Pointe Parkway, Suite 200, Norcross, GA 30092; 800-305-0523; http://www.Armortec.com.

Ashford Formula (concrete, stone and masonry product for curing, sealing, dustproofing), Curecrete Chemical Co. Inc., Springville, UT 84663; 800-998-5664; www.ashfordformula.com.

Cherokee Sanford Brick Co., 1600 Colon Road, Sanford, NC 27330; 800-277-2700; www.cherokeesanford.com.

Crumpler Plastic Pipe, Inc. (corrugated plastic culverts, catch basins, and drain pipes), Highway 24 W., P.O. Box 2068, Roseboro, NC 28382; 800-334-5071; http://www.cpp-pipe.com.

Cunningham Brick Co., Inc., 701 N. Main Street, Lexington, NC 27292; 800-672-6181; www.cunninghambrick.com.

Doublewall Corporation (interlocking precast gravity wall systems), 59 East Main Street, Plainville, CT 06062; 860-747-3412.

Duogard II (low-VOC water emulsion concrete form release),VOCOMP-20 (low-VOC water emulsion acrylic curing and sealing compound), W.R. Meadows, Inc., 300 Industrial Drive, P.O. Box 338, Hampshire, Elgin, IL 60140-0338; 800-342-5976.

GEOBLOCK (Porous 50% postconsumer recycled plastic porous pavement system with interlocking grids), GEOWEB (engineered HDPE, expanded honeycomb-like matrix that is filled with earth to provide stable retaining wall systems with variable slope), Presto Products Co., Geosystems Porducts, 670 N. Perkins Street, P.O. Box 2399, Appleton, WI 54913; 800-548-3424 or 414-739-9471; www.prestogo.com.

GeoStone (segmental retaining walls that do not need mortar or concrete footing, uses clip system of earth reinforcement), GeoStone Retaining Wall Systems, Inc., P.O. Box 325, Westover, AL 35185; 800-GEO-990; www.geostone.com.

Grasscrete (porous paving system used for driveways), Custom Rock Concrete Wall System Custom Rock International, 1156 Homer Street, St. Paul, MN 55116; 800-637-2447; (flexible form liner system used to achieve a look of a stone wall), Bomanite Corp., P.O. Box 599, Madera, CA 93639-0599; 800-854-2094 or 559-673-2411; www.bomanite.com and www.custom-rock.com.

Grasspave^2 (manufactured from 100% recycled HDPE for porous paving applications) and Gravelpave^2 (porous geotextile filter fabric backing to hold small aggregate particles in place, for gravel fill instead of sand and turf), Invisible Structures, Inc., 20100 E. 35th Drive, Aurora, CO 80011-8160, 800-223-1510 or 303-373-1234; http://www.invisiblestructures.com/

Keystone Retaining Wall Systems Inc., 4444 W. 78th Street, Minneapolis, MN 55435; 800-891-9791; http://www.keystonewalls.com/.

Stonhard, Dex-O-Tex (low-VOC epoxies that can be applied in patterns like terrazzo), Isoset (water-based adhesives), Ashland Specialty Polymers & Adhesives Division, 5200 Blazer Parkway, Dublin, OH 43017; 614-790-4159; www.ashchem.com.

TIRECRETE (wire-mesh reinforced porous exterior rubber surfacing for roadways and driveways made from 50% recycled tires), Zeller International, Main Street, P.O. Box 2, Downsville, NY 13755; 800-722-USFX; www.zeller-int.com/polsol-01.htm.

Versa-Lok Retaining Wall Systems, R.I. Lampus Co., (motarless retaining wall) P.O. Box 167, Springdale, PA 15144; 724-274-5035; 800-770-4525; www.versa-lok.com.

ON-SITE WASTEWATER TREATMENT

Infiltrator Systems Inc., 6 Business Park Road, P.O. Box 768, Old Saybrook, CT 06475; 800-718-2754; www.infiltratorsystems.com (Chamber Septic Systems).

K-87 Soap Digester Septic System Treatment for Drywells, K-67 Bacterial Drain and Trap Cleaner, K-57 Septic Tank and Cesspool Cleaner from Roebic Laboratories, Inc.; 25 Connair Road, P.O. Box 927, Orange, CT 06477; 203-795-1283; www.roebic.com.

Orenco Systems, Inc., 814 Airway Avenue, Sutherlin, OR 97479; 800-348-9843; www.orenco.com (wastewater treatment systems, shallow gravelless drainfields).

Septic Protector, (effluent filters) 14622 268th Ave., Zimmerman, MN 55398; 888-873-6504; www.septicprotector.com.

Terralift International, Inc. (soil fracturing), 104 E. Main Street, Box 532, Stockbridge, MA 01262; 413-298-4272; www.terraliftinternational.com.

WATER AND SEWER LINES

AAA Trenchless (replace problem pipe without major excavation), 205 22nd Street, Sacramento, CA 95816; 916-325-9992; www.aaatrenchless.com.

American Leak Detection, 888 Research Drive, Suite 100, Palm Springs, CA 92262; 800-755-6697; fax: 760-320-1288; www.leakbusters.com.

UNDERGROUND STORAGE TANKS

Advanced Environmental, 11 Virginia Road, White Plains, NY 10603; 914-761-8020; www.thetankspecialists.com.

Envirotube (PVC-coated copper tubing for fuel lines), LDC EnviroCorp., 94 McClellan Ave., Mineola, NY 11501; 516-248-7233; www.hvacweb.com.

Highland (Residential Heating System Tanks) Tank, 99 W. Elizabethtown Rd., Manheim, PA 17545; 717-664-0600; www.highlandtank.com.

Northeast Environmental, Inc., 225 Valley Pl., Suite B, Mamaroneck, NY 10543; 877-574-TANK; www.northeastenvironmental.com (tank removals and installations).

LANDSCAPE CARE

Advanced Lightning Technology, 122 Leesley Lane, Argyle, TX 76226; 800-950-7933; 940-455-7300; http://www.advancedlightning.com/contact.html (see catalog 700 series).

Automatic Lightning Protection, 7548 West Bluefield Avenue, Glendale, AZ 85308; 800-532-0990; www.lightningrod.com.

A Climber's Guide to Hazard Trees, Manchester, NH: National Arborists Association; 800-733-2622.

Guide for Plant Appraisal, Council of Tree and Landscape Appraisers, Manchester, NH: National Arborists Association.

Independent Protection Company, Inc., 1607 South Main Street, Goshen, IN 800-860-8388; 219-533-4116; www.ipclp.com.

Lightning Master Corp., 1351 N. Arcturas Avenue, Clearwater, FL 33765; 800-749-6800; www.LightningMaster.com.

Lightning Protection Specialists, 635 Clara Drive, Eads, TN 38028; 901-867-8948; www.lightningspecialists.com.

Pruning, Trimming, Repairing, Maintaining, and Removing Trees, and Cutting Brush—Safety *Requirements* (ANSI Z133.1-1994), New York: American National Standards Institute, 1994; 202-293-8020; www.ansi.org.

Stormin Protection Products, Inc., 10749 63rd Way N., Pinellas Park, FL 33782; 888-471-1038; www.storminprotection.com.

Tree, Shrub, and Other Woody Plant Maintenance—Standard Practices (ANSI A300-1995), New York: American National Standards Institute, 1995.

PROFESSIONAL ORGANIZATIONS

ADAPTIVE ENVIRONMENTS CENTER
374 Congress Street
Suite 301
Boston, MA 02210
617-695-1225
www.adaptenv.org

ADHESIVE AND SEALANT COUNCIL
7979 Old Georgetown Road
Suite 500
Bethesda, MD 20814
301-986-9700
www.ascouncil.org

AIR CONDITIONING AND REFRIGERATION INSTITUTE (ARI)
4301 North Fairfax Drive, Suite 425
Arlington, VA 22203
703-524-8800
www.ari.org

AIR CONDITIONING CONTRACTORS OF AMERICA, INC. (ACCA)
2800 Shirlington Road, Suite 300
Arlington, VA 22206
703-575-4477
www.acca.org

AMERICAN ARCHITECTURAL MANUFACTURER'S ASSOCIATION
1827 Walden Office Square, Suite 104
Schaumberg, IL 60173-4268
847-303-5664
www.aamanet.org

AMERICAN ASSOCIATION OF RETIRED PERSONS
601 E Street, NW
Washington, DC 20049
800-424-3410
www.aarp.org

AMERICAN CONCRETE INSTITUTE
P.O. Box 9094
Farmington Hills, MI 48333
248-848-3700
www.aci-int.org

AMERICAN CONCRETE PAVEMENT ASSOCIATION
5420 Old Orchard Road, Suite A-100
Skokie, IL 60077-1057
847-966-2272
www.pavement.com

AMERICAN COUNCIL FOR AN ENERGY-EFFICIENT ECONOMY
1001 Connecticut Avenue, NW, Suite 801
Washington, DC 20036
202-429-0063
www.aceee.org

AMERICAN FIBERBOARD ASSOCIATION
AMERICAN HARDBOARD ASSOCIATION
1210 W. Northwest Highway
Palatine, IL 60067
847-934-8800
www.afiberboard.org
www.ahardb.org

AMERICAN FOREST & PAPER ASSOCIATION
111 19th Street, NW, Suite 800
Washington, DC 20036
202-463-2700
www.afandpa.org

AMERICAN GALVANIZERS ASSOCIATION
6881 South Holly Circle
Englewood, CO 80112
800-468-7732
www.galvanizeit.org

AMERICAN GAS ASSOCIATION
400 N. Capitol Street, NW
Washington, DC 20001
202-824-7000
www.aga.org

AMERICAN INSTITUTE OF ARCHITECTS
1735 New York Ave, NW
Washington, DC 20006
202-626-7300
www.aia.org

AMERICAN INSTITUTE OF TIMBER CONSTRUCTION
7012 S. Revere Parkway, Suite 140
Englewood, CO 80112
303-792-9559
www.aitc-glulam.org

AMERICAN IRON AND STEEL INSTITUTE
1101 17th St., NW, Suite 1300
Washington, DC 20036-4700
800-79-STEEL
www.steel.org

AMERICAN LIGHTING ASSOCIATION
P.O. Box 420288
Dallas, TX 75342-0288
800-274-4484
www.americanlightingassoc.com

AMERICAN NATIONAL STANDARDS INSTITUTE
11 West 42nd Street, 13th Floor
New York, NY 10036
212-642-4900
www.ansi.org

AMERICAN NURSERY AND LANDSCAPE ASSOCIATION (ANLA)
1250 I Street, NW, Suite 500
Washington, DC 20005-3922
202-789-2900
www.anla.org

AMERICAN OCCUPATIONAL THERAPY ASSOCIATION, INC.
4720 Montgomery Lane
Bethesda, MD 20814
301-652-2682
www.aota.org

AMERICAN PORTLAND CEMENT ALLIANCE
1225 Eye Street, NW, Suite 300
Washington, DC 20005
202-408-9494
www.apca.com

AMERICAN SOCIETY FOR CONCRETE CONTRACTORS
38800 Country Club Drive
Farmington Hills, MI 48331
800-877-2753
www.ascconc.org

AMERICAN SOCIETY FOR TESTING AND MATERIALS
100 Barr Harbor Drive
West Conshohocken, PA 19428-2959
610-832-9585
www.astm.org

AMERICAN SOCIETY OF CIVIL ENGINEERS
1801 Alexander Bell Drive
Reston, VA 20191-4400
800-548-2723
www.asce.org

AMERICAN SOCIETY OF HEATING, REFRIGERATION AND AIR CONDITIONING ENGINEERS, INC.
1791 Tullie Circle, NE
Atlanta, GA 30329
404-636-8400
www.ashrae.org

AMERICAN SOCIETY OF INTERIOR DESIGNERS
608 Massachusetts Avenue, NE
Washington, DC 20002
202-546-3480
www.asid.org

AMERICAN SOCIETY OF LANDSCAPE ARCHITECTS (ASLA)
636 I Street, NW
Washington, DC 20001-3736
202-898-2444
202-898-1185 (fax)
www.asla.org

AMERICAN SOCIETY OF PLUMBING ENGINEERS (ASPE)
3617 Thousand Oaks Boulevard, Suite 210
Westlake, CA 91362-3649
805-495-7120
www.aspe.org

AMERICAN SOCIETY OF SANITARY ENGINEERING (ASSE)
901 Canterbury, Suite A
Westlake, OH 44145
440-835-3040
www.asse-plumbing.org

AMERICAN UNDERGROUND CONSTRUCTION ASSOCIATION
3001 Hennepin Avenue S.
Suite D202
Suite 248
Minneapolis, MN 55408
612-825-8933
www.avca.org

AMERICAN WATER WORKS ASSOCIATION (AWWA)
6666 W. Quincy Avenue
Denver, CO 80235
303-794-7711
www.awwa.org

AMERICAN WOOD COUNCIL
P.O. Box 5364
Madison, WI 53705-5364
800-890-7732
www.awc.org

AMERICAN WOOD PRESERVERS ASSOCIATION
P.O. Box 5690
Granbury, TX 76049-0690
817-326-6300
www.awpa.com

AMERICAN WOOD PRESERVERS INSTITUTE
2750 Prosperity Avenue
Suite 550
Fairfax, VA 22031-4312
800-356-AWPI
www.awpi.org

APA-THE ENGINEERED WOOD ASSOCIATION
P.O. Box 11700
Tacoma, WA 98411-0700
253-565-6600
www.apawood.org

ARCHITECTURAL WOODWORK INSTITUTE
1952 Isaac Newton Square
Reston, VA 20190
703-733-0600
www.awinet.org

ASPHALT ROOFING MANUFACTURERS ASSOCIATION
4041 Powder Mill Road, Suite 404
Calverton, MD 20705
301-231-9050
www.asphaltroofing.org

ASSOCIATED SHEET METAL & ROOFING CONTRACTORS
1 Regency Drive
P.O. Box 30
Bloomfield, CT 06002
860-243-3977

ASSOCIATION OF HOME APPLIANCE MANUFACTURERS
1111 19th Street NW, Suite 402
Washington, DC 20036
www.aham.org

ASSOCIATION OF SPECIALISTS IN CLEANING AND RESTORATION
10830 Annapolis Junction Road, Suite 312
Annapolis Junction, MD 20701-1120
301-604-4411

ASSOCIATION OF THE WALL AND CEILING INDUSTRIES INTERNATIONAL
803 West Broad Street, Suite 600
Falls Church, VA 22046
703-534-8300
www.awci.org

ASSOCIATION OF WINDOW AND DOOR INSTALLERS
11300 US Highway 1, Suite 400
North Palm Beach, FL 33408-3296
561-691-6224
www.awdi.com

BRICK INDUSTRIES ASSOCIATION
11490 Commerce Park Drive
Reston, VA 20191-1525
703-620-0010
www.bia.org

BUILDERS HARDWARE MANUFACTURERS ASSOCIATION
355 Lexington Avenue, 17th Floor
New York, NY 10017
212-297-2122
www.buildershardware.com

BUILDING CONSERVATION INTERNATIONAL
1901 Walnut Street
Suite 902
Philadelphia, PA 19103
215-568-0923

BUILDING OFFICIALS & CODE ADMINISTRATORS (BOCA) INTERNATIONAL, INC.
4051 West Flossmoor Road
Country Club Hills, IL 60478
708-799-2300
www.bocai.org

BUILDING SEISMIC SAFETY COUNCIL
NATIONAL INSTITUTE OF BUILDING SCIENCES
1090 Vermont Avenue, NW, Suite 700
Washington, DC 20005
202-289-7800
www.bssconline.org

BUILDING STONE INSTITUTE
P.O. Box 5047
White Plains, NY 10602-5047
914-232-5725

CALIFORNIA REDWOOD ASSOCIATION
405 Enfrente Drive, Suite 200
Novato, CA 94949
888-225-7339
www.calredwood.org

CANADA MORTGAGE AND HOUSING CORPORATION
HOUSING INFORMATION CENTRE
700 Montreal Road
Ottawa, ON, Canada K1A 0P7
800-668-2642
www.cmhc-schl.gc.ca/chic-ccdh/en

CANADIAN CONSTRUCTION MATERIALS CENTRE
INSTITUTE FOR RESEARCH IN CONSTRUCTION
NATIONAL RESEARCH COUNCIL OF CANADA
Building M-24, 1500 Montreal Road
Ottawa, ON, Canada K1A 0R6
613-993-6189
www.cistinrc.ca.irc (for IRC);
www.nrc.ca/ccmc

CARPET AND RUG INSTITUTE
P.O. Box 2048
Dalton, GA 30722
800-882-8846
www.carpet-rug.com

CEDAR SHAKE AND SHINGLE BUREAU
P.O. Box 1178
Sumas, WA 98295
604-820-7700
www.cedarbureau.org

CELLULOSE INSULATION MANUFACTURERS ASSOCIATION
136 S. Keowee Street
Dayton, OH 45402
937-222-2462
www.cellulose.org

CENTER FOR UNIVERSAL DESIGN
NORTH CAROLINA STATE UNIVERSITY
Box 8613
Raleigh, NC 27695
800-647-6777
www.design.ncsu.gov

CERAMIC TILE DISTRIBUTORS ASSOCIATION
800 Roosevelt Road, Building C, Suite 20
Glen Ellyn, IL 60137
800-938-2832
www.ctdahome.org

CERAMIC TILE INSTITUTE OF AMERICA
12061 Jefferson Boulevard
Culver City, CA 90230-6219
310-574-7800
www.ctioa.org

COMPOSITE PANEL ASSOCIATION
18922 Premiere Court
Gaithersburg, MD 20879
301-670-0604
www.pbmdf.com

CONCRETE FOUNDATIONS ASSOCIATION OF NORTH AMERICA
107 First Street W.
PO Box 204
Mount Vernon IA 52314
319-895-6940
www.cfawalls.org

CONCRETE REINFORCING STEEL INSTITUTE
933 N. Plum Grove Road
Schaumburg, IL 60173
847-517-1200
www.crsi.org

CONCRETE SAWING AND DRILLING ASSOCIATION
6089 Frantz Road
Suite 101
Dublin, OH 43017
614-798-2252
www.csda.org

CONSORTIUM FOR ENERGY EFFICIENCY
One State Street, Suite 1400
Boston, MA 02109
617-330-9755
www.ceeformt.org

CONSTRUCTION SPECIFICATIONS INSTITUTE
99 Canal Center Plaza, Suite 300
Alexandria, VA 22314
800-689-2900
www.csinet.org

COPPER DEVELOPMENT ASSOCIATION
260 Madison Avenue
New York, NY 10016
800-232-3282
www.copper.org

COUNCIL OF AMERICAN BUILDING OFFICIALS
5203 Leesburg Pike, Suite 708
Falls Church, VA 22041
703-931-4533
www.intlcode.org

DOOR AND HARDWARE INSTITUTE
14150 Newbrook Drive, Suite 200
Chantilly, VA 20151-2223
703-222-2010
www.dhi.org

DRYWALL, LATH AND PLASTER ASSOCIATION
2286 N. State College Boulevard
Fullerton, CA 92831
760-837-9094
www.tsib.org

EDISON ELECTRIC INSTITUTE
701 Pennsylvania Avenue, NW
Washington, DC 20004-2696
800-EEI-4688
www.eei.org

ELECTRIC POWER RESEARCH INSTITUTE
3412 Hillview Avenue
Palo Alto, CA 94303
800-313-3774
www.epri.com

ELECTRONICS INDUSTRIES ALLIANCE
2500 Wilson Boulevard
Arlington, VA 22201-3834
703-907-7500
www.eia.org.

ENERGY AND ENVIRONMENTAL BUILDING ASSOCIATION
10740 Lyndale Avenue South
Suite 10W
Bloomington, MN 55420-5615
952-881-1098
www.eeba.org

ENERGY EFFICIENCY AND RENEWABLE ENERGY CLEARINGHOUSE (EREC)
P.O. Box 3048
Merrifield, VA 22116
800-DOE-EREC
e-mail: doe.erec@nciinc.com
www.eren.doe.gov/consumerinfo

ENERGY-EFFICIENT BUILDING ASSOCIATION
NORTHCENTRAL TECHNICAL COLLEGE
1000 Campus Drive
Wausau, WI 54401-1899
888-682-7144
www.northcentral.tech.wi.us

ENGINEERED WOOD ASSOCIATION
7011 S. 19th Street
P.O. Box 11700
Tacoma, WA 98411
253-565-6600
www.apawood.org

THE ENTERPRISE FOUNDATION
10227 Wincopin Circle, Suite 500
Columbia, MD 21044
800-624-4298
www.enterprisefoundation.org

EVAPORATIVE COOLING INSTITUTE
P.O. Box 3ECI
Las Cruces, NM 88003
505-646-3948
www.evapcooling.org

EXPANDED POLYSTYRENE
MOLDERS ASSOCIATION (EPMA)
2128 Espy Court, Suite 4
Crofton, MD 21114
800-607-3772
www.espmolders.org

FLOOR COVERING INSTALLATION
CONTRACTORS ASSOCIATION
7439 Millwood Drive
West Bloomfield, MI 48322
248-661-5018
www.fcica.com

FLORIDA SOLAR ENERGY CENTER (FSEC)
1679 Clearlake Road
Cocoa, FL 32922
321-638-1000
www.fsec.ucf.edu

FOREST PRODUCTS LABORATORY
U.S. DEPARTMENT OF AGRICULTURE
One Gifford Pinchot Drive
Madison, WI 53705-2398
608-231-9200
www.fpl.fs.fed.us/

FOREST PRODUCTS SOCIETY
2801 Marshall Court
Madison, WI 53705-2295
608-231-1361
www.forestprod.org

GAS APPLIANCE MANUFACTURERS ASSOCIATION
2107 Wilson Boulevard, Suite 600
Arlington, VA 22201
Phone: 703-525-7060
www.gamanet.org

GAS RESEARCH INSTITUTE
1700 South Mount Prospect Road
Des Plaines, IL 60018
847-768-0500
www.gri.org

GAS TECHNOLOGY INSTITUTE
1700 S. Mount Prospect Road
Des Plaines, IL 60018-1804
847-768-0500
www.gri.org

GEOTHERMAL HEAT PUMP CONSORTIUM
701 Pennsylvania Ave., NW
Washington, D.C. 20004-2696
202-508-5500
www.ghpc.org

GREEN SEAL
1001 Connecticut Ave. NW, Suite 827
Washington, DC 20036-5525
202-872-6400
www.greenseal.org/

GYPSUM ASSOCIATION
810 1st Street, NE, Suite 510
Washington, DC 20002
202-289-5440
www.gypsum.org

HARDWOOD MANUFACTURERS ASSOCIATION
400 Penn Center Boulevard, Suite 530
Pittsburgh, PA 15235
800-373-9663
www.hardwood.org

HARDWOOD PLYWOOD & VENEER ASSOCIATION
P.O. Box 2789
Reston, VA 20195-0789
703-435-2900
www.hpva.org

HEARTH PRODUCTS ASSOCIATION
1601 North Kent Street, Suite 1001
Arlington, VA 22209
703-522-0086
www.hearthassociation.org

HOME AUTOMATION ASSOCIATION
1444 I Street NW, Suite 700
Washington DC 20005
202-712-9050
email: 75250.1274@compuserve.com
www.homeautomation.org

HOME VENTILATING INSTITUTE
DIVISION OF THE AIR MOVEMENT AND CONTROL ASSOCIATION INTERNATIONAL, INC.
30 West University Drive
Arlington Heights, IL 60004
847-394-0150
www.hvi.org

IDEA CENTER
University of Buffalo
Buffalo, NY 14214
716-829-3485, Ext. 329
www.arch.buffalo.gov/idea

ILLUMINATING ENGINEERING SOCIETY (IES)
120 Wall Street, 17th Floor
New York, NY 10005
212-248-5000
www.iesna.org

INSTITUTE OF ELECTRICAL AND ELECTRONICS ENGINEERS
IEEE SERVICE CENTER
445 Hoes Lane, P.O. Box 1331
Piscataway, NJ 08855-1331
800-678-4333
www.ieee.org

INSTITUTE OF INSPECTION, CLEANING AND RESTORATION CERTIFICATION
2715 E. Mill Plain Boulevard
Vancouver, WA 98661
360-693-5675
www.iicrc.org

INSULATING CONCRETE FORMS ASSOCIATION
1807 Glenview Road, Suite 203
Glenview, IL 60025
847-657-9730
www.forms.org

INTERNATIONAL ASSOCIATION OF PLUMBING AND MECHANICAL OFFICIALS (IAPMO)
20001 E. Walnut Drive South
Walnut, CA 91789-2825
909-595-8449
www.iapmo.ort

INTERNATIONAL ASSOCIATION OF STONE RESTORATION AND CONSERVATION
30 Eden Alley, Suite 301
Columbus, OH 43215
614-461-5852

INTERNATIONAL CAST POLYMER ASSOCIATION
8201 Greensboro Drive, Suite 300
McLean, VA 22102
703-610-9005
www.icpa-hq.com

INTERNATIONAL CODE COUNCIL
5203 Leesburg Pike, Suite 600
Falls Church, VA 22041
703-931-4533
www.intlcode.org

INTERNATIONAL CONCRETE REPAIR INSTITUTE (ICRI)
1323 Shepard Drive, Suite D
Sterling, VA 21064-4428
703-450-0116
703-450-0119 (fax)
www.icri.org

INTERNATIONAL CONFERENCE OF BUILDING OFFICIALS
5360 S. Workman Mill Road
Whittier, CA 90601
800-336-1963
www.icbo.org

INTERNATIONAL FIRESTOP COUNCIL
25 N. Broadway
Tarrytown, NY 10591
914-332-0040
www.firestop.org

INTERNATIONAL GROUND SOURCE HEAT PUMP ASSOCIATION
490 Cordell South
Oklahoma State University
Stillwater, OK 74078
800-626-4747
www.igshpa.okstate.edu

INTERNATIONAL INSTITUTE FOR LATH AND PLASTER
820 Transfer Road
St. Paul, MN 55114-1406

INTERNATIONAL INTERIOR DESIGN ASSOCIATION
341 Merchandise Mart
Chicago, IL 60654
312-467-1950
www.iida.org

INTERNATIONAL MASONRY INSTITUTE
The James Brice House
42 East Street
Annapolis, MD 21401
410-280-1305
www.imiweb.org

INTERNATIONAL ORGANIZATION FOR STANDARDIZATION
1, rue de Varembe
Case postale 56
CH-1211 Geneva 20, Switzerland
41-22-749-01-11
http://iso.ch/welcome.html

INTERNATIONAL STAPLE AND TOOL ASSOCIATION (ISANTA)
512 West Burlington Avenue, Suite 203
La Grande, IL 60525
312-644-0828
www.senco.com/isanta/isanta2.html

INTERNATIONAL WINDOW FILM ASSOCIATION
P.O. Box 3871
Martinsville, VA 24115-3871
540-666-4932
http://iwfa.com

ITALIAN TRADE COMMISSION
Tile Department
499 Park Avenue
New York, NY 10022
212-980-1500
www.italtrade.com

KITCHEN CABINET MANUFACTURERS ASSOCIATION
1899 Preston White Drive
Reston, VA 20191
703-264-1690
www.kcma.org

LAMINATING MATERIALS ASSOCIATION
116 Lawrence Street
Hillsdale, NJ 07642
201-664-2700

LAWRENCE BERKELEY NATIONAL LABORATORY
BUILDING TECHNOLOGIES DEPARTMENT
ENVIRONMENTAL ENERGY TECHNOLOGIES DIVISION
LAWRENCE BERKELEY LABORATORY
Berkeley, CA 94720
510-486-6845
http://eandelbl.gov/BT.html

LEAD INDUSTRIES ASSOCIATION
13 Main Street
Sparta, NJ 07871
973-726-5323
www.leadinfo.com

LIGHTING RESEARCH CENTER
21 Union Street
Troy, NY 12180-3352
518-687-7100
www.lrc.rpi.edu/

LIGHTNING PROTECTION INSTITUTE
3335 N. Arlington Heights Road, Suite E
Arlington Heights, IL 60004
800-488-6864
www.lightning.org

MAPLE FLOORING MANUFACTURERS ASSOCIATION
60 Revere Drive, Suite 500
Northbrook, IL 60062
847-480-9138
www.maplefloor.org

MARBLE INSTITUTE OF AMERICA
30 Eden Alley, Suite 301
Columbus, OH 43215
614-228-6194
www.marble-institute.com

MASON CONTRACTORS ASSOCIATION OF AMERICA
1910 S. Highland Ave., Suite 101
Lombard, IL 60148
630-705-4200
800-536-2225
www.masoncontractors.com

THE MASONRY SOCIETY
3970 Broadway, Suite 201-D
Boulder, CO 80304-1135
303-939-9700
www.masonrysociety.org

METAL ROOFING SYSTEMS ASSOCIATION
1300 Sumner Avenue
Cleveland, OH 44115-2851
216-241-7333
www.taol.com/mbma

NAHB REMODELORS COUNCIL
1201 15th Street, NW
Washington, DC 20005-2800
202-822-0212
800-368-5242 ext. 216
www.nahb.net/remodelor_working

NAHB RESEARCH CENTER
400 Prince George's Blvd.
Upper Marlboro, MD 20774
800-638-8556
www.nahbrc.org

NATIONAL ALARM ASSOCIATION OF AMERICA
P.O. Box 3409
Dayton, OH 45401
800-283-6285
www.naaa.org

NATIONAL ARBOR DAY FOUNDATION (NADF)
100 Arbor Avenue
Nebraska City, NE 68410
402-474-5655
www.arborday.org

NATIONAL ARBORISTS ASSOCIATION (NAA)
2 Perimeter Road, Unit 1
Manchester, NH 03103
800-733-2622
www.natlarb.com

NATIONAL ASSOCIATION OF ARCHITECTURAL METAL MANUFACTURERS
8 S. Michigan Avenue, Suite 1000
Chicago, IL 60603-3305
312-332-0405
www.naamm.org

NATIONAL ASSOCIATION OF OIL HEAT SERVICE MANAGERS (NAOHSM)
P.O. Box 67
E. Petersburg, PA 17520
888-552-0900
www.naohsm.org

NATIONAL ASSOCIATION OF THE REMODELING INDUSTRY
780 Lee Street, Suite 200
Des Plaines, IL 60016
847-298-9200
www.nari.org

NATIONAL BURGLAR AND FIRE ALARM ASSOCIATION
8300 Colesville Road, Suite 750
Silver Spring, MD 20910
301-585-1855
www.alarm.org

NATIONAL COIL COATERS ASSOCIATION
401 N. Michigan Avenue
Chicago, IL 60611
312-321-6894
www.coilcoaters.org

NATIONAL CONCRETE MASONRY ASSOCIATION
2302 Horse Pen Road
Herndon, VA 20171-3499
703-713-1900
www.ncma.org

NATIONAL ELECTRICAL CONTRACTORS ASSOCIATION
3 Bethesda Metro Center, Suite 1100
Bethesda, MD 20814
301-657-3110
www.necanet.org

NATIONAL ELECTRICAL MANUFACTURERS ASSOCIATION (NEMA)
1300 North 17th Street, Suite 1847
Rosslyn, VA 22209
703-841-3200
www.nema.org

NATIONAL FENESTRATION RATING COUNCIL
1300 Spring Street, Suite 500
Silver Spring, MD 20910
301-589-6372
www.nfrc.org

NATIONAL FIRE PROTECTION ASSOCIATION
1 Batterymarch Park
Quincy, MA 02269-9101
800-344-3555
www.nfpa.org

NATIONAL FRAME BUILDERS ASSOCIATION
4840 W. 15th Street, Suite 1000
Lawrence, KS 66049-3876
800-844-3781
www.postframe.org

NATIONAL GLASS ASSOCIATION
8200 Greensboro Drive
McClean, VA 22102
703-442-4890
www.glass.org

NATIONAL INSTITUTE OF BUILDING SCIENCES
1090 Vermont Avenue, NW, Suite 700
Washington, DC 20005
202-289-7800
www.nibs.org

NATIONAL KITCHEN AND BATH ASSOCIATION
687 Willow Grove Street
Hackettstown, NJ 07840
800-843-6522
908-852-1695
www.nkba.org

NATIONAL MULTI HOUSING COUNCIL
1850 M Street, NW, Suite 540
Washington, DC 20036-5803
202-974-2300
www.nmhc.org

NATIONAL OAK FLOORING MANUFACTURERS ASSOCIATION
P.O. Box 3009
Memphis, TN 38173-0009
901-526-5016
www.nofma.org

NATIONAL PAINT AND COATINGS ASSOCIATION
1500 Rhode Island Avenue, NW
Washington, DC 20005
202-462-6272
www.paint.org

NATIONAL PARTICLEBOARD ASSOCIATION
18928 Premiere Court
Gaithersburg, MD 20879-1569
301-670-0604
www.pbmdf.com

NATIONAL PROPANE GAS ASSOCIATION
1600 Eisenhower Lane, Suite 100
Lisle, IL 60532
708-515-0600
www.propanegas.com

NATIONAL READY MIXED CONCRETE ASSOCIATION
900 Spring Street
Silver Spring, MD 20910
888-846-7622
www.nrmca.org

NATIONAL RENEWABLE ENERGY LABORATORY
1617 Cole Boulevard
Golden, CO 80401
303-275-3000

NATIONAL ROOFING CONTRACTORS ASSOCIATION
10255 W. Higgins Road, Suite 600
Rosemont, IL 60018-5607
800-323-9545
www.nrca.net

NATIONAL TECHNICAL INFORMATION SERVICE
U.S. DEPARTMENT OF COMMERCE
5285 Port Royal Road
Springfield, VA 22161
800-553-6847
www.ntis.gov

NATIONAL TILE ROOFING MANUFACTURERS ASSOCIATION
P.O. Box 40337
Eugene, OR 97404-0049
888-321-9236
www.ntrma.org

NATIONAL TRUST FOR HISTORIC PRESERVATION
1785 Massachusetts Avenue, NW
Washington, DC 20036
800-944-6847
www.nationaltrust.org

NATIONAL WOOD FLOORING ASSOCIATION
16388 Westwoods Business Park
Ellisville, MO 63021
800-422-4556
www.woodfloor.org

NORTH AMERICAN INSULATION MANUFACTURERS ASSOCIATION
44 Canal Center Plaza, Suite 310
Alexandria, VA 22314
703-684-0084
www.naima.org

NORTH AMERICAN SOCIETY FOR TRENCHLESS TECHNOLOGY
1655N Ft. Myer Drive, Suite 700
Arlington, VA 22209
703-351-5252
www.nastt.org

NSF INTERNATIONAL
P.O. Box 130140
789 N. Dixboro Road
Ann Arbor, MI 48133-0140
800-NSF-MARK
www.nsf.org/wastewater

OAK RIDGE NATIONAL LABORATORY BUILDINGS THERMAL ENVELOPE SYSTEMS AND MATERIALS PROGRAM
P.O. Box 2008, Mail Stop 6070
Oak Ridge, TN 37831-6070
865-574-4345
www.ornl.gov/roofs+walls

PAINTING AND DECORATING CONTRACTORS OF AMERICA
3913 Old Lee Highway, Suite 33B
Fairfax, VA 22030
800-332-7322
www.pdca.com

PLASTICS PIPE INSTITUTE
1825 Connecticut Avenue, NW, Suite 680
Washington, DC 20009
202-462-9607
www.plasticpipe.org

PLUMBING-HEATING-COOLING CONTRACTORS ASSOCIATION
180 S. Washington Street
P.O. Box 6808
Falls Church, VA 22040
800-533-7694
www.naphcc.org

PLUMBING MANUFACTURERS INSTITUTE (PMI)
1340 Remington Road, Suite A
Schaumburg, IL 60173
847-884-9764
www.pmihome.org

POLYISOCYANURATE INSULATION MANUFACTURER ASSOCIATION
1331 F Street NW, Suite 975
Washington, DC 20004
202-628-6558
www.pima.org

PORTLAND CEMENT ASSOCIATION
5420 Old Orchard Road
Skokie, IL 60077-1083
800-868-6733
www.portcement.org

RADIANT PANEL ASSOCIATION
P.O. Box 717
Loveland, CO 80539
800-660-7187
www.radiantpanelassociation.org

REMODELING CONTRACTORS ASSOCIATION
17 S. Main Street
E. Granby, CT 06026
800-937-4722
www.remodelingassociation.com

RESILIENT FLOOR COVERING INSTITUTE
401 East Jefferson Street
Suite 102
Rockville, MD 20850
301-340-8580

ROOF COATINGS MANUFACTURERS ASSOCIATION
Center Park
4041 Powder Mill Road, Suite 404
Calverton, MD 20705
301-230-2501
www.roofcoatings.org

ROOF CONSULTANTS INSTITUTE
7424 Chapel Hill Road
Raleigh, NC 27607
800-828-1902
www.rci-online.org

ROOFING INDUSTRY COMMITTEE OF WIND ISSUES
13303 U.S. 19 N
Clearwater, FL 34624
813-536-0456

ROOFING INDUSTRY EDUCATIONAL INSTITUTE
2305 East Arapahoe Road, Suite 135
Littleton, CO 80122
303-703-9870
www.nrca.net/rlei

SCREEN MANUFACTURERS ASSOCIATION
2850 S. Ocean Boulevard, Suite 114
Palm Beach, FL 33480-6205
773-525-2644
www.screenmfgassociation.org

SEALANT, WATERPROOFING AND RESTORATION INSTITUTE
2841 Main
Kansas City, MO 64108
816-472-SWRI
www.swrionline.org

SHEET METAL AND AIR CONDITIONING CONTRACTORS NATIONAL ASSOCIATION (SMACNA)
4201 Lafayette Center Drive
Chantilly, VA 20151-1203
703-803-2980
www.smacna.org

SINGLE PLY ROOFING INSTITUTE
200 Reservoir Street, Suite 309A
Needham, MA 02494
781-444-0242
www.spri.org

SOCIETY FOR THE PRESERVATION OF NEW ENGLAND ANTIQUITIES
Harrison Gray Otis House
141 Cambridge Street
Boston, MA 02114
617-227-3956

SOCIETY OF PLASTICS INDUSTRY INFO CENTER
1801 K Street NW, Suite 600K
Washington, DC 20006
www.plasticsindustry.org

SOLAR ENERGY INDUSTRIES ASSOCIATION (SEIA)
1616 H Street NW, 8th Floor
Washington, DC 20006
202-628-7745
www.seia.org

SOUTHERN FOREST PRODUCTS ASSOCIATION
P.O. Box 641700
Kenner, LA 70064-1700
504-443-4464
www.sfpa.org

SOUTHERN PINE COUNCIL
P.O. Box 641700
Kenner, LA 70064-1700
504-443-4464
www.southernpine.com

SOUTHFACE ENERGY INSTITUTE
241 Pine Street
Atlanta, GA 30308
404-872-3549
www.southface.org

STEEL DOOR INSTITUTE
30200 Detroit Road
Cleveland, OH 44145-1967
440-899-0010
www.steeldoor.org

STEEL JOIST INSTITUTE
3127 10th Avenue, North
Myrtle Beach, SC 29577-6760
843-626-1995
www.steeljoist.org

STEEL WINDOW INSTITUTE
1300 Sumner Avenue
Cleveland, OH 44115-2851
216-241-7333
www.steelwindows.com

STRUCTURAL BOARD ASSOCIATION
412-45 Sheppard Avenue
Willowdale, Ontario, Canada M2N 5W9
416-730-9090
www.sba-osb.com

STRUCTURAL INSULATED PANEL ASSOCIATION
3413 56th Street NW, Suite A
Gig Harbor, WA 98335
www.sips.org

STUCCO MANUFACTURERS ASSOCIATION
2402 Vista Nobleza
Newport Beach, CA 92660
949-640-9902
www.stuccofgassoc.com

SUMP AND SEWAGE PUMP MANUFACTURERS ASSOCIATION
P.O. Box 647
Northbrook, IL 60065
847-559-9233
www.sspma.org

SUPERIOR GUNITE/SHOTCRETE ASSOCIATION
12306 Van Nuys Boulevard
Lake View Terrace, CA 91342
818-896-9199
supgun@pacificnet.net

SUSTAINABLE BUILDINGS INDUSTRY COUNCIL (SBIC)
1331 H Street, NW, Suite 1000
Washington, DC 20005
202-628-7400
www.sbicouncil.org

SWEDISH COUNCIL FOR BUILDING RESEARCH
P.O. Box 12866, S-112-98
Stockholm, Sweden
46-8-617-73-00
www.bfr.se

TECHNICAL PRESERVATION SERVICE BUREAU
NATIONAL PARK SERVICE
1849 C Street, NW, Suite 200
Washington, DC 20240
202-343-9578
www2.cr.nps.gov

TELECOMMUNICATIONS INDUSTRY ASSOCIATION
2500 Wilson Boulevard, Suite 300
Arlington, VA 22201
703-907-7700
www.tiaonline.org.

TILE COUNCIL OF AMERICA, INC.
100 Clemson Research Boulevard
Anderson, SC 29625
864-646-8453
www.tileusa.com

TIMBER FRAME BUSINESS COUNCIL
c/o Jerry Rouleau
P.O. Box B1161
Hanover, NH 03755
603-643-5033
www.timberframe.org

TRACE RESEARCH & DEVELOPMENT CENTER
5901 Research Park Boulevard
Madison, WI 53719
608-262-6966
www.trace.wisc.edu

TRENCHLESS INFORMATION CENTER
NORTHEAST CONSULTING, INC.
22 Jewelberry Drive
Webster, NY 14580
716-787-1519
www.no-dig.com

TRUSS PLATE INSTITUTE
583 D'Onofio Drive, Suite 200
Madison, WI 53719
608-833-5900
www.tpinst.org

UNDERWRITERS LABORATORIES, INC.
333 Pfingsten Road
Northbrook, IL 60062-2096
847-272-8800
www.ul.com

UNITED LIGHTNING PROTECTION ASSOCIATION (ULPA)
426 North Avenue
Libertyville, IL 60048
800-668-ULPA
www.ulpa.org

U.S. CONSUMER PRODUCT SAFETY COMMISSION
Washington, DC 20207-001
800-638-2772
www.cpsc.gov

U.S. DEPARTMENT OF ENERGY
BUILDING SYSTEMS AND MATERIALS DIVISION
EE-421
1000 Independence Avenue, SW
Washington, DC 20585
202-586-9214
www.energy.gov

U.S. DEPARTMENT OF HOUSING AND URBAN DEVELOPMENT
HUD User
P.O. Box 6091
Rockville, MD 20849
800-245-2691 TDD 800-483-2209
www.huduser.org

U.S. GREEN BUILDING COUNCIL (USGBC)
1015 18th Street NW, Suite 805
Washington, DC 20036
202-828-7422
www.usgbc.org

VINYL SIDING INSTITUTE
1801 K Street, Suite 600K
Washington, DC 20006
888-FORVSI-1
www.vinylsiding.org

WESTERN RED CEDAR LUMBER ASSOCIATION
1200-555 Burrard Street
Vancouver, BC V7X 1S7 Canada
604-684-0266
www.todaycedar.org

WESTERN WOOD PRODUCTS ASSOCIATION
522 SW 5th Avenue, Suite 500
Portland, OR 97204-2122
503-224-3930
www.wwpa.org

WINDOW & DOOR MANUFACTURERS ASSOCIATION
1400 East Touhy Avenue, Suite 470
Des Plaines, IL 60018-3337
847-299-5200
www.wdma.com

WOOD MOULDING AND MILLWORK PRODUCERS ASSOCIATION
507 First Street
Woodland, CA 95695
800-550-7889
www.wmmpa.com

WOOD TRUSS COUNCIL OF AMERICA
6300 Enterprise Lane
Madison, WI 53719
608-274-4849
www.woodtruss.com

INDEX

A

Accessories (cabinets) (*see* Cabinets)
Airborne and impact sound reduction, 278–279
Air conditioning, 457–462
  essential knowledge, 457
  techniques, materials, and tools, 458–462
Air infiltration barriers, 86–89
  essential knowledge, 86–87
  techniques, materials, and tools, 88–89
Alarm and intrusion systems, 419–420
Appliances, 175–178, 350–362
  access and function improvement, 353–357
  energy efficiency and resource improvement, 359–362
  essential knowledge, 350, 353, 357, 359–360
  techniques, materials, and tools, 350–363
    appearance upgrade techniques, 357–359
    replacement techniques, 350–352
  vents and exhausts, 175–178, 353–357
Asphalt shingles, 175–178
Attic ladders, 318–319
  essential knowledge, 318–319
  techniques, materials, and tools, 319
Awnings (*see* Shutters and awnings)

B

Backsplashes (*see* Countertops and backsplashes)
Balloon framing (exterior walls), 57
Balusters (damaged and broken), 315–317
  essential knowledge, 315
  techniques, materials, and tools, 315–317
    replacement techniques, 316–317
    strengthening techniques, 315–316
Baseboard heating (electric), 413–415
  essential knowledge, 413–414
  techniques, materials, and tools, 414–415
    maintenance techniques, 414
    replacement techniques, 414–415
Basement floors, 4–5
Baths (*see* Kitchens and baths)
Below-grade walls (foundations), 24–27
Bibliography (*see* Further reading)
Bidets (*see* Toilets and bidets)
Braced framing (exterior walls), 57
Brick and masonry veneer (*see* Veneer)
Bulkhead doors (*see* Garage and bulkhead doors)

C

Cabinets, 326–339
  accessories, 332–336
    essential knowledge, 332–333
    techniques, materials, and tools, 333–336
      installation techniques, 333–336
      replacement techniques, 333–334
  additional and replacement, 336–339
    essential knowledge, 336–337
    techniques, materials, and tools, 337–339
  hardware, 327–332
    essential knowledge, 327
    techniques, materials, and tools, 327–332
      installation techniques, 329–332
      replacement techniques, 327–332
  surface maintenance and repair, 326–327
    essential knowledge, 326
    techniques, materials, and tools, 326–327
      refinishing techniques, 326–327
      repair techniques, 326–327
      replacement techniques, 327
Cabling (phone, computer, and TV), 416–418
  essential knowledge, 416–417
  techniques, materials, and tools, 417–418
    repair techniques, 417–418
    rewiring techniques, 418
Carbon monoxide detectors, 424–425
Carpets and rugs, 296–298
Carriages (sagging), 314–315
  essential knowledge, 314–315
  techniques, materials, and tools, 315
Casing and trim, 244–47.
  (*See also* Trim)
  essential knowledge, 244–246
  techniques, materials, and tools, 246–247
    installation techniques, 246–247
    repair techniques, 246
Caulking and sealants (*see* Sealants and caulking)
Ceilings, 265–279, 299–310
  finish materials, 299–310
    moldings and trim, 308–310
    paints and coverings, 306–308
    plaster and drywall, 299–306
  structural assemblies, 265–279
Ceramic tile finish flooring, 295–296
Chimneys (*see* Fireplaces and chimneys)

Clay, concrete, fiber-cement, and composite tile roofs, 194–198
essential knowledge, 194–195
techniques, materials, and tools, 195–198
installation techniques, 195–198
replacement techniques, 195
Coatings and finishes, 52–54, 135–139 (*See also* Finishes)
foundations, 52–54
paint, 135–139
Composite tile roofs (*see* Clay, concrete, fiber-cement, and composite tile roofs)
Computer cabling (*see* Cabling)
Concrete roofs (*see* Clay, concrete, fiber-cement, and composite tile roofs)
Controls, 404–412, 475–479
HVAC and plumbing, 475–479
essential knowledge, 475
techniques, materials, and tools, 476–479
installation techniques, 476–479
replacement techniques, 476
lighting, 404–412
essential knowledge, 404–409
exterior, 406–408
interior, 404–406
techniques, materials, and tools, 405–412
Cooling, 457–462
essential knowledge, 457
techniques, materials, and tools, 458–462
Countertops and backsplashes, 340–349
essential knowledge, 340, 342, 345–346, 348
techniques, materials, and tools, 340–349
installation techniques, 345–349
repair techniques, 341–344
replacement techniques, 340–342
Cracks (foundations), 48–52
essential knowledge, 48
techniques, materials, and tools, 48–52
repair techniques, 48, 51–52
stabilization techniques, 49–50
Crawl spaces, 4–5, 27–29
floors, 4–5
unvented, 27–28
vented, 27

D

Dampproofing and waterproofing, 19–23
essential knowledge, 19
techniques, materials, and tools, 19–23
walls, 19–23
exterior, 19–23
interior, 19–20
Decks and porches (wood), 519–529
essential knowledge, 519, 522, 525–527
preservative-treated wood, 525–527
stairs and handrails, 527–529
Decks and porches (*Cont.*):
structures, 519–522
techniques, materials, and tools, 519–529
application techniques, 525–527
installation techniques, 527
repair techniques, 520–525
replacement techniques, 521–525, 528–529
stabilization techniques, 521–522
Detectors, 422–425
carbon monoxide, 424–425
smoke, 422–424
Distribution systems, 442–450
essential knowledge, 442–444
techniques, materials, and tools, 444–450
installation techniques, 446–450
rehab techniques, 444–446
Doors and door frames, 211–217, 236–244 (*See also* Windows and doors)
existing, 211–217
cost-benefit analyses, 217
essential knowledge, 211
evaluation, 211–212
ratings and standards of, 214–216
techniques, materials, and tools, 216–217
types, 214–216
garage and bulkhead, 239–241
essential knowledge, 239
techniques, materials, and tools, 239–240
repair techniques, 239–240
replacement techniques, 240
interior, 241–243
essential knowledge, 241–242
techniques, materials, and tools, 242–243
repair techniques, 242
replacement techniques, 242
primary entry, 236
essential knowledge, 236–237
techniques, materials, and tools, 237–239
repair techniques, 237
replacement techniques, 237–239
storm and screen, 240–241
essential knowledge, 240–241
techniques, materials, and tools, 241
repair techniques, 241
replacement techniques, 241
Drainage, 12–18
foundation, 15–18
essential knowledge, 15
techniques, materials, and tools, 15–18
excavation techniques, 17–18
installation techniques, 15–16
replacement techniques, 16–17
surface and subsurface, 12–15
essential knowledge, 12
techniques, materials, and tools, 12–15

Drain, waste, and vent systems, 502–508
 essential knowledge, 502–503
 techniques, materials, and tools, 503–508
Drywall and plaster, 299–306

E

EIFs (exterior insulation and finish systems), 124–127
 essential knowledge, 124
 techniques, materials, and tools, 124–127
  installation techniques, 126–127
  repair techniques, 125
Electrical and electronic systems, 389–433, 587–591, 641–645
 baseboard heating, 413–415
  essential knowledge, 413–414
  techniques, materials, and tools, 414–415
   maintenance techniques, 414
   replacement techniques, 414–415
 cabling (phone, computer, and TV), 416–418
  essential knowledge, 416–417
  techniques, materials, and tools, 417–418
   repair techniques, 417–418
   rewiring techniques, 418
 further reading, 587–591
 lighting and controls, 404–412
  controls, 408–410
  essential knowledge, 404–409
  exterior, 406–408
  interior, 404–406
  techniques, materials, and tools, 405–412
   installation techniques, 405–412
   rewiring techniques, 405
 overview, 389
 security systems, 419–433
  detectors, 422–425
   carbon monoxide, 424–425
   smoke, 422–424
  essential knowledge, 419, 422–426, 428–432
  garage door openers, 430–433
  intrusion and alarm systems, 419–420
  lightning protection, 425–427
  surge protection, 428–430
  techniques, materials, and tools, 419–433
   inspection techniques, 419–421
   installation techniques, 424–425, 427, 429–430
   maintenance techniques, 423, 427, 432
   repair techniques, 421
   replacement techniques, 421, 423–424, 432–433
 service panels, 390–396
  essential knowledge, 390–393
  techniques, materials, and tools, 393–396
Electrical and electronic systems (*Cont.*):
 wiring and receptacles, 397–403
  aluminum, 399–401
  essential knowledge, 397–398
  receptacles, 401–406
  techniques, materials, and tools, 398–406
   connection techniques, 403
   installation techniques, 398–399, 402–403, 409–412
   repair techniques, 399
   replacement techniques, 400–402
   rewiring techniques, 398, 407
   splicing techniques, 400–401
Energy sources, 439–441
Engineered wood siding, 114–116
Entry doors (*see* Primary entry doors)
Exhausts and vents (appliance), 511–513
Existing foundations (*see* Foundations)
Exterior lighting and controls, 406–408
Exterior walls, 57–139, 566–571, 605–612
 air infiltration barriers, 86–89
  essential knowledge, 86–87
  techniques, materials, and tools, 88–89
 design and engineering, 57–71
 EIFS (exterior insulation and finish systems), 124–127
  essential knowledge, 124
  techniques, materials, and tools, 124–127
   installation techniques, 126–127
   repair techniques, 125
 finishes (paint and other coatings), 135–139
  essential knowledge, 135–136
  techniques, materials, and tools, 136–139
   application techniques, 137–139
   maintenance techniques, 136
   surface preparation techniques, 136–137
 finishes (stucco), 127–128
  essential knowledge, 127
  techniques, materials, and tools, 127–128
   installation techniques, 128
   patching techniques, 127–128
 fire damage (wood-frame), 70–71
  essential knowledge, 70–71
  techniques, materials, and tools, 71
 framing, 57
  balloon, 57
  braced, 57
  platform, 57
 insect damage mitigation, 68–70
  essential knowledge, 68–69
  techniques, materials, and tools, 70
 insulation, 90–94
  essential knowledge, 90
  techniques, materials, and tools, 91–94

Exterior walls (*Cont.*):
masonry wall (existing) reinforcement, 64–66
essential knowledge, 64–66
techniques, materials, and tools, 65–66
moisture deterioration, 66–68
essential knowledge, 66–67
techniques, materials, and tools, 67–68
epoxy techniques, 68
repair techniques, 67
replacement techniques, 67–68
overview, 57–58
sealants and caulking, 131–134
essential knowledge, 131–133
techniques, materials, and tools, 133–134
installation techniques, 134
removal techniques, 133–134
surface preparation techniques, 133–134
types, 132
seismic resistance (wood-frame), 58–62
essential knowledge, 58
techniques, materials, and tools, 58–62
attachment techniques, 58–60
load path (hold-down) securing techniques, 60–62
sheathing, 79–82
essential knowledge, 79
techniques, materials, and tools, 79–82
repair techniques, 79
replacement techniques, 79–82
siding (engineered wood), 114–116
essential knowledge, 114
techniques, materials, and tools, 114–116
siding (fiber-cement), 120–123
essential knowledge, 120
techniques, materials, and tools, 120–123
repair techniques, 120
replacement techniques, 120–123
siding (hardboard), 111–113
essential knowledge, 111
techniques, materials, and tools, 111–113
repair techniques, 111–112
replacement techniques, 112–113
siding (metal), 99–101
essential knowledge, 99
techniques, materials, and tools, 99–101
maintenance techniques, 99
repair techniques, 99
replacement techniques, 99–100
siding (plywood panel), 117–119
essential knowledge, 117
techniques, materials, and tools, 117–119
repair techniques, 117–118
replacement techniques, 118–119
siding (solid wood), 106–111
essential knowledge, 106–107

Exterior walls, siding (solid wood) (*Cont.*):
techniques, materials, and tools, 106–110
repair techniques, 107
replacement techniques, 107–110
siding (vinyl), 95–98
essential knowledge, 95–97
techniques, materials, and tools, 97–98
repair techniques, 97–98
replacement techniques, 98
stain removal techniques, 97
siding (wood shingles and shakes), 102–105
essential knowledge, 102
techniques, materials, and tools, 102–105
trim, 129–130
essential knowledge, 129
techniques, materials, and tools, 129–130
repair techniques, 129
replacement techniques, 130
vapor retarders, 83–86
essential knowledge, 83–84
techniques, materials, and tools, 84–86
application techniques, 84
installation techniques, 84–86
veneer (masonry and brick), 72–78
essential knowledge, 72
overview of, 72
techniques, materials, and tools, 72–78
application techniques, 74
cleaning techniques, 73
repair techniques, 75–78
repointing techniques, 74–75
wind resistance (wood-frame), 62–66
essential knowledge, 62–64
techniques, materials, and tools, 63–64

F

Fences and retaining walls (wood), 525, 527, 529–533
essential knowledge, 529–530
techniques, materials, and tools, 530–533
realignment techniques, 533
replacement techniques, 530–533
strengthening techniques, 530–533
Fiber-cement roofs (*see* Clay, concrete, fiber-cement, and composite tile roofs)
Fiber-cement siding, 120–123
Finishes, 52–54, 127–128, 135–139
foundations (existing), 52–54
essential knowledge, 52–53
techniques, materials, and tools, 53–54
paint and other coatings, 135–139
essential knowledge, 135–136
techniques, materials, and tools, 136–139
application techniques, 137–139
maintenance techniques, 136
surface preparation techniques, 136–137

Foundations (existing) (*Cont.*):
drainage (surface and subsurface), 12–15
essential knowledge, 12
techniques, materials, and tools, 12–15
floors (basement), 4–5
essential knowledge, 4–5
techniques, materials, and tools, 4–5
installation techniques, 4–5
replacement techniques, 4–5
floors (crawl space), 5–6
essential knowledge, 5
techniques, materials, and tools, 5–6
insulation, 25–31
crawl spaces, 27–29
unvented, 27–28
vented, 27
essential knowledge, 24, 27–28
slabs, 28–31
techniques, materials, and tools, 24–28, 31
walls
below-grade, 24–27
exterior, 24–25
interior, floor-to-ceiling, 24–27
overview, 3–4
permanent wood, 7–8
essential knowledge, 7
techniques, materials, and tools, 7–8
prefabricated, 8–11
essential knowledge, 8–10
techniques, materials, and tools, 9–11
shoring, underpinning, and repair, 41–47
essential knowledge, 41
techniques, materials, and tools, 41–47
replacement techniques, 46–47
stabilization techniques, 41–46
underpinning techniques, 45
ventilation, 32–40
crawl spaces, 35–36
essential knowledge, 32–33, 35–36
soil gases, 36–39
techniques, materials, and tools, 33–40
Frames and framing, 57, 144, 224–229, 270–278
exterior walls, 57
floor, 270–278
alternatives, 270–276
fire damage, 277–278
roofs, 144
windows frames and replacement units, 224–229
Fuel supply systems, 509–510
essential knowledge, 509
techniques, materials, and tools, 510
Further Reading, 563–598
electrical and electronic systems, 587–591
exterior walls, 566–571
Further reading (*Cont.*):
foundations (existing), 563–566
HVAC and plumbing, 591–595
kitchens and baths, 582–587
partitions, ceilings, floors, and stairs, 579–582
roofs, 572–576
site work, 595–598
windows and doors, 576–579

G

Garage and bulkhead doors, 239–241
essential knowledge, 239
techniques, materials, and tools, 239–240
repair techniques, 239–240
replacement techniques, 240
Glazing (windows and doors), 218–223
essential knowledge, 218–219
techniques, materials, and tools, 218–223
application techniques, 221–222
cleaning and polishing techniques, 219–220
installation techniques, 222–223
replacement techniques, 222
Gutter and leader systems, 199–207
essential knowledge, 199–200
techniques, materials, and tools, 200–207
installation techniques, 202–207
repair techniques, 200–202
replacement techniques, 206–207
types, 200

H

Handrails and stairs (decks and porches), 527–529
Hardboard siding, 111–113
Hardware, 248–251, 327–332
cabinets, 327–332
windows and doors, 248–251
Heating, 451–456
essential knowledge, 451–452
techniques, materials, and tools, 453–456
installation techniques, 453–456
replacement techniques, 453
Heat pumps, 463–467
Hot-water heating (domestic), 488–495
essential knowledge, 488–489
techniques, materials, and tools, 489–495
efficiency improvement techniques, 489–490
installation techniques, 490–495
HVAC and plumbing, 437–516, 591–595, 646–649
controls, 475–479
essential knowledge, 475
techniques, materials, and tools, 476–479
installation techniques, 476–479
replacement techniques, 476
cooling, 457–462
essential knowledge, 457
techniques, materials, and tools, 458–462

Finishes (*Cont.*):
stucco, 127–128
essential knowledge, 127
techniques, materials, and tools, 127–128
installation techniques, 128
patching techniques, 127–128
Finish flooring, 287–298
carpet and rugs, 296–298
essential knowledge, 296
techniques, materials, and tools, 296–298
installation techniques, 298
maintenance techniques, 296–297
repair techniques, 297–298
restoration techniques, 297–298
ceramic tile, 295–296
essential knowledge, 295
techniques, materials, and tools, 295–296
maintenance techniques, 295–296
replacement techniques, 296
restoration techniques, 295–296
vinyl sheet and tile, 291–295
essential knowledge, 291–292
techniques, materials, and tools, 292–295
installation techniques, 294–295
removal techniques, 293–294
repair techniques, 292
replacement techniques, 292–293
stain removal techniques, 292
wood, 287–291
essential knowledge, 287
technology, materials, and tools, 287–291
maintenance techniques, 287–288
refinishing techniques, 289–291
repair techniques, 288–289
sanding techniques, 289
stabilization techniques, 288–289
Finish wall and ceiling materials, 299–310
moldings and trim, 308–310
essential knowledge, 308
techniques, materials, and tools, 308–310
repair techniques, 308–310
replacement techniques, 308–310
paints and wall coverings, 306–308
essential knowledge, 306
techniques, materials, and tools, 306–308
cleaning techniques, 307
removal techniques, 307–308
repair techniques, 307
stain removal techniques, 306–307
plaster and drywall, 299–306
essential knowledge, 299
techniques, materials, and tools, 300–306
refastening techniques, 300–301
repair techniques, 301–306
replacement techniques, 304
Fire damage, 70–71, 14
floor framing
essential knowledge,
techniques, materials,
roofs, 147–149
essential knowledge, 147-
techniques, materials, and
wood-frame exterior walls, 70–
essential knowledge, 70–71
techniques, materials, and tool
Fireplaces and chimneys, 480–489
essential knowledge, 480–481
techniques, materials, and tools, 481–
installation techniques, 481–487
rebuilding and repointing techniques, 481–482
relining techniques, 482–483
Fire protection systems, 514–516
essential knowledge, 514–515
techniques, materials, and tools, 515–516
Flashing, 153–160, 251–253
roofs, 153–160
essential knowledge, 153
techniques, materials, and tools, 153–160
installation techniques, 154–160
repair techniques, 153
windows and doors, 251–253
essential knowledge, 251–252
techniques, materials, and tools, 252–253
Floors and flooring
foundations (existing) (*see* Foundations)
partitions, ceilings, floors, and stairs (*see* Partitions, ceilings, floors, and stairs)
Floor-to-ceiling interior walls (foundations), 24–27
Foundations (existing), 3–54, 563–566, 601–604
coatings and finishes, 52–54
essential knowledge, 52–53
techniques, materials, and tools, 53–54
cracks, 48–52
essential knowledge, 48
techniques, materials, and tools, 48–52
repair techniques, 48, 51–52
stabilization techniques, 49–50
dampproofing and waterproofing, 19–23
essential knowledge, 19
techniques, materials, and tools, 19–23
walls, 19–23
exterior, 19–23
interior, 19–20
design and engineering, 3–6
drainage (foundation), 15–18
essential knowledge, 15
techniques, materials, and tools, 15–18
excavation techniques, 17–18
installation techniques, 15–16
replacement techniques, 16–17

HVAC and plumbing (*Cont.*):
design and engineering, 437–441
distribution systems, 442–450
essential knowledge, 442–444
techniques, materials, and tools, 444–450
installation techniques, 446–450
rehab techniques, 444–446
energy sources, 439–441
essential knowledge, 439
techniques, materials, and tools, 439–441
fireplaces and chimneys, 480–489
essential knowledge, 480–481
techniques, materials, and tools, 481–489
installation techniques, 481–487
rebuilding and repointing techniques, 481–482
relining techniques, 482–483
heating, 451–456
essential knowledge, 451–452
techniques, materials, and tools, 453–456
installation techniques, 453–456
replacement techniques, 453
heat pumps, 463–467
essential knowledge, 463–464
techniques, materials, and tools, 464–467
hot-water heating (domestic), 488–495
essential knowledge, 488–489
techniques, materials, and tools, 489–495
efficiency improvement techniques, 489–490
installation techniques, 490–495
indoor air quality, 468–474
essential knowledge, 468–469
techniques, materials, and tools, 469–474
overview of, 437–438
plumbing, 496–516
(*See also* Plumbing)
system sizing (replacement), 438–439

I

Impact and airborne sound reduction, 278–279
Indoor air quality, 468–474
essential knowledge, 468–469
techniques, materials, and tools, 469–474
Insect damage mitigation, 68–70
essential knowledge, 68–69
techniques, materials, and tools, 70
Insulation, 24–31, 90–94, 124–127, 163–171
crawl spaces, 27–29
EIFS (exterior insulation and finish systems), 124–127
exterior walls, 24–31, 90–94
roofs, 163–171
slabs, 28–31
Interior doors, 241–243
essential knowledge, 241–242
techniques, materials, and tools, 242–243
repair techniques, 242
replacement techniques, 242
Interior lighting and controls, 404–406
Intrusion and alarm systems, 419–420

K

Kitchens and baths, 323–385, 582–587, 634–641
appliances, 350–362
access and function improvement, 353–357
energy efficiency and resource improvement, 359–362
essential knowledge, 350, 353, 357, 359–360
techniques, materials, and tools, 350–363
appearance upgrade techniques, 357–359
replacement techniques, 350–352
cabinets (accessories), 332–336
essential knowledge, 332–333
techniques, materials, and tools, 333–336
installation techniques, 333–336
replacement techniques, 333–334
cabinets (additional and replacement), 336–339
essential knowledge, 336–337
techniques, materials, and tools, 337–339
cabinets (hardware), 327–332
essential knowledge, 327
techniques, materials, and tools, 327–332
installation techniques, 329–332
replacement techniques, 327–332
cabinets (surface maintenance and repair), 326–327
essential knowledge, 326
techniques, materials, and tools, 326–327
refinishing techniques, 326–327
repair techniques, 326–327
replacement techniques, 327
countertops and backsplashes, 340–349
essential knowledge, 340, 342, 345–346, 348
techniques, materials, and tools, 340–349
installation techniques, 345–349
repair techniques, 341–344
replacement techniques, 340–342
further reading, 582–587
overview, 323–325
product information, 634–641
shapes and design, 324–325
sinks and lavatories, 363–369
access and function improvement, 367–369
essential knowledge, 363, 365–367
techniques, materials, and tools, 363–369
installation techniques, 365–369
surface maintenance and repair techniques, 363–365

Kitchens and baths, sinks and lavatories (*Cont.*):
water, 365–367
conservation, 366–367
purification, 365–366
toilets and bidets, 380–385
access and function improvement, 383–385
essential knowledge, 380–381, 383
techniques, materials, and tools, 380–385
installation techniques, 382–385
maintenance and repair techniques, 380–381
water conservation, 381–383
tubs and showers, 370–379
access and function improvement, 375–379
essential knowledge, 370, 373, 375, 378
techniques, materials, and tools, 370–379
moisture control techniques, 373–375
surface maintenance and repair techniques, 370–373

L

Ladders, attic (*see* Attic ladders)
Landscaping, 553–559
energy efficient and sustainable, 557–559
essential knowledge, 553–554, 557–558
landscape care, 553–557
techniques, materials, and tools, 554–559
Lavatories (*see* Sinks and lavatories)
Leader and gutter systems (*see* Gutter and leader systems)
Lighting and controls, 404–412
controls, 408–410
essential knowledge, 404–409
exterior, 406–408
interior, 404–406
techniques, materials, and tools, 405–412
installation techniques, 405–412
rewiring techniques, 405
Lightning protection, 425–427
Load transfer, 268–270
Low-slope roofs, 179–183
essential knowledge, 179
techniques, materials, and tools, 179–183
repair techniques, 179–182
replacement techniques, 180–183

M

Masonry and brick veneer (*see* Veneer)
Masonry walls (existing) reinforcement, 64–66
essential knowledge, 64–66
techniques, materials, and tools, 65–66
Metal roofing, 184–189
essential knowledge, 184
techniques, materials, and tools, 184–189
repair techniques, 184
replacement techniques, 184–189
Metal siding, 99–101
Moisture barriers (roofs) (*see* Underlayments)
Moisture deterioration, 66–68, 275–277
exterior walls, 66–68
essential knowledge, 66–67
techniques, materials, and tools, 67–68
epoxy techniques, 68
repair techniques, 67
replacement techniques, 67–68
partitions, ceilings, floors, and stairs, 275–277
essential knowledge, 275–276
techniques, materials, and tools, 276–277
repair techniques, 276–277
replacement techniques, 276–277
Moldings and trim, 308–310
(*See also* Trim)

O

On-site waste water treatment, 545–548

P

Paint finishes, 135–139
Partitions, ceilings, floors, and stairs, 265–319, 579–582, 631–634
finish flooring (carpet and rugs), 296–298
essential knowledge, 296
techniques, materials, and tools, 296–298
installation techniques, 298
maintenance techniques, 296–297
repair techniques, 297–298
restoration techniques, 297–298
finish flooring (ceramic tile), 295–296
essential knowledge, 295
techniques, materials, and tools, 295–296
maintenance techniques, 295–296
replacement techniques, 296
restoration techniques, 295–296
finish flooring (vinyl sheet and tile), 291–295
essential knowledge, 291–292
techniques, materials, and tools, 292–295
installation techniques, 294–295
removal techniques, 293–294
repair techniques, 292
replacement techniques, 292–293
stain removal techniques, 292
finish flooring (wood), 287–291
essential knowledge, 287
technology, materials, and tools, 287–291
maintenance techniques, 287–288
refinishing techniques, 289–291
repair techniques, 288–289
sanding techniques, 289
stabilization techniques, 288–289
finish wall and ceiling materials (moldings and trim), 308–310
essential knowledge, 308

Partitions, ceilings, floors, and stairs, finish wall and ceiling materials (*Cont.*):
techniques, materials, and tools, 308–310
repair techniques, 308–310
replacement techniques, 308–309
finish wall and ceiling materials (paints and wall coverings), 306–308
essential knowledge, 306
techniques, materials, and tools, 306–308
cleaning techniques, 307
removal techniques, 307–308
repair techniques, 307
stain removal techniques, 306–307
finish wall and ceiling materials (plaster and drywall), 299–306
essential knowledge, 299
techniques, materials, and tools, 300–306
refastening techniques, 300–301
repair techniques, 301–306
replacement techniques, 304
floor and ceiling structural assemblies, 265–279
essential knowledge, 265–267, 270, 275–278
techniques, materials, and tools, 267–279
airborne and impact sound reduction techniques, 278–279
load transfer techniques, 268–270
reinforcement techniques, 267–268
repair techniques, 270–271, 276–277
replacement techniques, 271–275
restoration techniques, 277–278
salvage techniques, 278
floor framing (alternatives), 270–275
essential knowledge, 270
techniques, materials, and tools, 270–275
repair techniques, 270–271
replacement techniques, 271–275
floor framing (fire damage), 277–278
essential knowledge, 277
techniques, materials, and tools, 277–278
restoration techniques, 277–278
salvage techniques, 278
further reading, 579–582
moisture deterioration, 275–277
essential knowledge, 275–276
techniques, materials, and tools, 276–277
repair techniques, 276–277
replacement techniques, 276–277
overview, 265
sheathing (subflooring), 280–284
essential knowledge, 280–281
techniques, materials, and tools, 280–284
refastening techniques, 280–283
repair techniques, 280–284

Partitions, ceilings, floors, and stairs (*Cont.*):
sound control, 278–279
essential knowledge, 278
techniques, materials, and tools, 278–279
stairs, 311–19
(*See also* Stairs)
structural problems, 265–270
essential knowledge, 265–266
techniques, materials, and tools, 266–268
load transfer techniques, 268–270
reinforcement techniques, 267–268
underlayments, 284–286
essential knowledge, 284–285
techniques, materials, and tools, 285–286
repair techniques, 285
replacement techniques, 285–286
Permanent wood foundations, 7–8
essential knowledge, 7
techniques, materials, and tools, 7–8
Phone cabling (*see* Cabling)
Plaster and drywall, 299–306
Platform framing (exterior walls), 57
Plumbing, 496–516.
(*See also* HVAC and plumbing)
appliance vents and exhausts, 511–513
essential knowledge, 511–512
techniques, materials, and tools, 512–513
design and engineering, 496–497
drain, waste, and vent systems, 502–508
essential knowledge, 502–503
techniques, materials, and tools, 503–508
fire protection systems, 514–516
essential knowledge, 514–515
techniques, materials, and tools, 515–516
fuel supply systems, 509–510
essential knowledge, 509
techniques, materials, and tools, 510
water supply and distribution systems, 498–501
essential knowledge, 498
techniques, materials, and tools, 498–501
distribution improvement techniques, 499
installation techniques, 499–501
Plywood panel siding, 117–119
Porches (wood) (*see* Decks and porches)
Prefabricated foundations, 8–11
essential knowledge, 8–10
techniques, materials, and tools, 9–11
Prefabricated stairs, 317–318
essential knowledge, 317
techniques, materials, and tools, 317–318
Preservative-treated wood, 525–527
Primary entry doors, 236–236
essential knowledge, 236–237
techniques, materials, and tools, 237–239
repair techniques, 237
replacement techniques, 237–239

R

Receptacles (electric) (*see* Wiring and receptacles)
Reference materials, 563–598, 601–655
print resources, 563–598
(*See also* Further reading)
Retaining walls (wood) (*see* Fences and retaining walls)
Risers (*see* Treads and risers)
Roofs, 143–207, 572–576, 613–622
clay, concrete, fiber-cement, and composite tile, 194–198
essential knowledge, 194–195
techniques, materials, and tools, 195–198
installation techniques, 195–198
replacement techniques, 195
design and engineering, 143–149
fire damage, 147–149
essential knowledge, 147–148
techniques, materials, and tools, 148–149
flashing, 153–160
essential knowledge, 153
techniques, materials, and tools, 153–160
installation techniques, 154–160
repair techniques, 153
framing error repairs, 144
essential knowledge, 144
techniques, materials, and tools, 144
further reading, 572–576
gutter and leader systems, 199–207
essential knowledge, 199–200
techniques, materials, and tools, 200–207
installation techniques, 202–207
repair techniques, 200–202
replacement techniques, 206–207
types, 200
insulation, 163–171
essential knowledge, 163–164
techniques, materials, and tools, 164–170
low-slope, 179–183
essential knowledge, 179
techniques, materials, and tools, 179–183
repair techniques, 179–182
replacement techniques, 180–183
metal roofing, 184–189
essential knowledge, 184
techniques, materials, and tools, 184–189
repair techniques, 184
replacement techniques, 184–189
overview, 143–144
seismic resistance (wood-frame), 146–147
essential knowledge, 146–147
techniques, materials, and tools, 147
shakes (wood), 171–174
sheathing, 150–153
essential knowledge, 150–152
Roofs (*Cont.*):
techniques, materials, and tools, 151–152
repair techniques, 151
replacement techniques, 151–152
shingles (asphalt), 175–178
essential knowledge, 175–177
techniques, materials, and tools, 177–178
installation techniques, 178
repair techniques, 177–178
shingles (wood), 171–174
essential knowledge, 171
techniques, materials, and tools, 171–174
installation techniques, 173–174
repair techniques, 171–172
slate roofing, 190–193
essential knowledge, 190
techniques, materials, and tools, 190–193
repair techniques, 190–191
replacement techniques, 191–193
structural decay, 147
essential knowledge, 147
techniques, materials, and tools, 147
underlayments and moisture barriers, 161–162
essential knowledge, 161
techniques, materials, and tools, 161–162
wind resistance (wood-frame), 145–146
essential knowledge, 145
techniques, materials, and tools, 145–146
Rugs and carpets, 296–298

S

Screens
doors (*see* Storm and screen doors)
windows (*see* Storm windows and screens)
Sealants and caulking, 131–134, 254–256
exterior walls, 131–134
essential knowledge, 131–133
techniques, materials, and tools, 133–134
installation techniques, 134
removal techniques, 133–134
surface preparation techniques, 133–134
windows and doors, 254–256
essential knowledge, 254–255
techniques, materials, and tools, 254–256
types of, 254
Security systems, 419–433
detectors, 422–425
carbon monoxide, 424–425
smoke, 422–424
essential knowledge, 419, 422–426, 428–432
intrusion and alarm systems, 419–420
lightning protection, 425–427
surge protection, 428–430
techniques, materials, and tools, 419–433
inspection techniques, 419–421

Security systems (*Cont.*):
installation techniques, 424–425, 427, 429–430
maintenance techniques, 423, 427, 432
repair techniques, 421
replacement techniques, 421, 423–424, 432–433
Seismic resistance (wood frame exterior walls), 58–62
Service panels (electric), 390–396
essential knowledge, 390–393
techniques, materials, and tools, 393–396
Sewer and water lines, 548–551
Shakes (wood), 171–174
essential knowledge, 171
techniques, materials, and tools, 171–174
Sheathing, 79–82, 150–153, 280–284
exterior walls, 79–82
essential knowledge, 79
techniques, materials, and tools, 79–82
repair techniques, 79
replacement techniques, 79–82
roofs, 150–153
essential knowledge, 150–152
techniques, materials, and tools, 151–152
repair techniques, 151
replacement techniques, 151–152
subflooring, 280–284
Shingles
asphalt, 175–178
essential knowledge, 175–177
techniques, materials, and tools, 177–178
installation techniques, 178
repair techniques, 177–178
shakes, 171–174
wood, 171–174
essential knowledge, 171
techniques, materials, and tools, 171–174
installation techniques, 173–174
repair techniques, 171–172
Shoring, underpinning, and repair (foundations), 41–47
essential knowledge, 41
techniques, materials, and tools, 41–47
replacement techniques, 44–47
stabilization techniques, 41–46
underpinning techniques, 45
Showers (*see* Tubs and showers)
Shutters and awnings, 261
essential knowledge, 261
techniques, materials, and tools, 261
Siding, 95–123
engineered wood, 114–116
essential knowledge, 114
Siding (*Cont.*):
techniques, materials, and tools, 114–116
fiber-cement, 120–123
essential knowledge, 120
techniques, materials, and tools, 120–123
repair techniques, 120
replacement techniques, 120–123
hardboard, 111–113
essential knowledge, 111
techniques, materials, and tools, 111–113
repair techniques, 111–112
replacement techniques, 112–113
metal, 99–101
essential knowledge, 99
techniques, materials, and tools, 99–101
maintenance techniques, 99
repair techniques, 99
plywood panel, 117–119
essential knowledge, 117
techniques, materials, and tools, 117–119
repair techniques, 117–118
replacement techniques, 118–119
solid wood, 106–111
essential knowledge, 106–107
techniques, materials, and tools, 106–110
repair techniques, 107
replacement techniques, 107–110
vinyl, 95–98
essential knowledge, 95–97
techniques, materials, and tools, 97–98
repair techniques, 97–98
replacement techniques, 98
stain removal techniques, 97
wood shingles and shakes, 102–105
essential knowledge, 102
techniques, materials, and tools, 102–105
Sinks and lavatories, 363–369
access and function improvement, 367–369
essential knowledge, 363, 365–367
techniques, materials, and tools, 363–369
installation techniques, 365–369
surface maintenance and repair techniques, 363–365
water, 365–367
conservation, 366–367
purification, 365–366
Site work, 519–559, 595–598, 650–655
decks and porches (wood), 519–529
essential knowledge, 519, 522, 525–527
preservative-treated wood, 525–527
stairs and handrails, 527–529
structures, 519–522
techniques, materials, and tools, 519–529
application techniques, 525–527
installation techniques, 527

Site work (*Cont.*):
repair techniques, 520–525
replacement techniques, 521–525, 528–529
stabilization techniques, 521–522
fences and retaining walls (wood), 525, 527, 529–533
essential knowledge, 529–530
techniques, materials, and tools, 530–533
realignment techniques, 533
replacement techniques, 530–533
strengthening techniques, 530–533
further reading, 595–598
landscaping, 553–559
energy efficient and sustainable, 557–559
essential knowledge, 553–554, 557–558
landscape care, 553–557
techniques, materials, and tools, 554–559
underground construction, 544–552
essential knowledge, 544–546, 548–549, 551
storage tanks, 551–552
techniques, materials, and tools, 544–552
installation techniques, 548
remedy techniques, 544–543, 546–548
removal techniques, 551–552
repair techniques, 546–551
replacement techniques, 550–551
waste water treatment (on-site), 545–548
water and sewer lines, 548–551
wells, 544–545
Skylights, 234–235
essential knowledge, 234–235
techniques, materials, and tools, 235
installation techniques, 235
repair techniques, 235
replacement techniques, 235
Slabs, 28–31
Slate roofing, 190–193
essential knowledge, 190
techniques, materials, and tools, 190–193
repair techniques, 190–191
replacement techniques, 191–193
Smoke detectors, 422–424
Soil gas ventilation, 36–39
Solid wood siding, 106–111
Sound control, 278–279
essential knowledge, 278
techniques, materials, and tools, 278–279
Stairs, 311–19.
(*See also* Partitions, stairs, ceilings, and floors)
attic ladders, 318–319
essential knowledge, 318–319
techniques, materials, and tools, 319
balusters (damaged and broken), 315–317
Stairs (*Cont.*):
essential knowledge, 315
techniques, materials, and tools, 315–317
replacement techniques, 316–317
strengthening techniques, 315–316
carriages (sagging), 314–315
essential knowledge, 314–315
techniques, materials, and tools, 315
decks and porches (wood), 527–529
prefabricated, 317–318
essential knowledge, 317
techniques, materials, and tools, 317–318
treads and risers, 311–314
essential knowledge, 311, 313
techniques, materials, and tools, 311–314
repair techniques, 311–312
replacement techniques, 313–314
Storage tanks, 551–552
Storm and screen doors, 240–241
essential knowledge, 240–241
techniques, materials, and tools, 241
repair techniques, 241
replacement techniques, 241
Storm windows and screens, 230–233
essential knowledge, 230–231
techniques, materials, and tools, 231–233
installation techniques, 231–233
repair techniques, 232–233
replacement techniques, 232–233
Structural assemblies (floor and ceiling), 265–279
Structural decay (roofs), 147
Stucco finishes, 127–128
Subfloor sheathing, 280–84
Surface and subsurface drainage, 12–15
Surge protection, 428–430

T

Toilets and bidets, 380–385
access and function improvement, 383–385
essential knowledge, 380–381, 383
techniques, materials, and tools, 380–385
installation techniques, 382–385
maintenance and repair techniques, 380–381
water conservation, 381–383
Treads and risers, 311–314
essential knowledge, 311, 313
techniques, materials, and tools, 311–314
repair techniques, 311–312
replacement techniques, 313–314
Trim, 129–130, 244–247, 308–310
exterior, 129–130
moldings (finish wall and ceiling materials), 308–310
windows and doors, 244–247
(*See also* Casing and trim)

Tubs and showers, 370–379
access and function improvement, 375–379
essential knowledge, 370, 373, 375, 378
techniques, materials, and tools, 370–379
moisture control techniques, 373–375
surface maintenance and repair techniques, 370–373
TV cabling (*see* Cabling)

U

Underground construction, 544–552
essential knowledge, 544–546, 548–549, 551
storage tanks, 551–552
techniques, materials, and tools, 544–552
installation techniques, 548
remedy techniques, 544–543, 546–548
removal techniques, 551–552
repair techniques, 546–551
replacement techniques, 550–551
waste water treatment (on-site), 545–548
water and sewer lines, 548–551
wells, 544–545
Underlayments, 161–162, 284–286
and moisture barriers (roofs), 161–162
essential knowledge, 161
techniques, materials, and tools, 161–162
partitions, ceilings, floors, and stairs, 284–286
essential knowledge, 284–285
techniques, materials, and tools, 285–286
Underpinning (foundations), 41–47
Unvented crawl spaces, 27–28

V

Vapor retarders, 83–86
essential knowledge, 83–84
techniques, materials, and tools, 84–86
application techniques, 84
installation techniques, 84–86
Veneer (masonry and brick), 72–78
essential knowledge, 72, 74
overview of, 72
techniques, materials, and tools, 72–78
application techniques, 74
cleaning techniques, 73
repair techniques, 75–78
repointing techniques, 74–75
Vents and ventilation, 27, 32–40, 511–513
(*See also* HVAC and plumbing)
appliance vents and exhausts, 511–513
crawl spaces, 27
foundations (existing), 32–40
vent systems (*see* Drain, waste, and vent systems)
Vinyl sheet and tile finish flooring, 291–295
Vinyl siding, 95–98

W

Walls
exterior (*see* Exterior walls)
finish materials (*see* Finish wall and ceiling materials)
Waste systems (*see* Drain, waste, and vent systems)
Water, 365–367, 381–383, 498–501, 548–551
conservation, 366–367, 381–383
on-site waste water treatment, 545–548
purification, 365–366
and sewer lines, 548–551
supply and distribution systems, 498–501
essential knowledge, 498
techniques, materials, and tools, 498–501
waterproofing (*see* Dampproofing and waterproofing)
Weather stripping, 256–260
techniques, materials, and tools, 258–260
installation techniques, 258–260
replacement techniques, 258–260
Wells, 544–545
Windows and doors, 211–261, 576–582, 623–631
casing and trim, 244–247
essential knowledge, 244–246
techniques, materials, and tools, 246–247
installation techniques, 246–247
repair techniques, 246
caulking and sealants, 254–256
essential knowledge, 254–255
techniques, materials, and tools, 255–256
types, 255
doors and door frames, 236–243
(*See also* Doors and door frames)
existing, 211–217
cost-benefit analyses, 217
essential knowledge, 211
evaluation, 211–212
ratings and standards, 214–216
techniques, materials, and tools, 216–217
types, 212–214
flashing, 251–253
essential knowledge, 251–252
techniques, materials, and tools, 252–253
further reading, 576–582
glazing, 218–223
essential knowledge, 218–219
techniques, materials, and tools, 218–223
application techniques, 221–222
cleaning and polishing techniques, 219–220
installation techniques, 222–223
replacement techniques, 222
hardware, 248–251
essential knowledge, 248–250
techniques, materials, and tools, 250

repair techniques, 250
replacement techniques, 250
overview, 211
shutters and awnings, 261
essential knowledge, 261
techniques, materials, and tools, 261
skylights, 234–235
essential knowledge, 234–235
techniques, materials, and tools, 235
installation techniques, 235
repair techniques, 235
replacement techniques, 235
storm windows and screens, 230–233
essential knowledge, 230–231
techniques, materials, and tools, 231–233
installation techniques, 231–233
repair techniques, 232–233
replacement techniques, 232–233

weather stripping, 256–260
essential knowledge, 256–258
techniques, materials, and tools, 258–260
installation techniques, 258–260
replacement techniques, 258–260
window frames and replacement units, 224–229

Windows and doors (*Cont.*):
essential knowledge, 224–226
techniques, materials, and tools, 226–229
adjustment techniques, 228–229
installation techniques, 227–228
replacement techniques, 226–228
Wind resistance, 62–66, 145–146
exterior walls (wood-frame), 62–66
essential knowledge, 62–64
techniques, materials, and tools, 63–64
roofs (wood frame), 145–146
essential knowledge, 145
techniques, materials, and tools, 145–146
Wiring and receptacles, 397–403
aluminum, 399–401
essential knowledge, 397–398
receptacles, 401–406
techniques, materials, and tools, 398–406
connection techniques, 403
installation techniques, 398–399, 402–403, 409–412
repair techniques, 399
replacement techniques, 400–402
rewiring techniques, 398, 407
splicing techniques, 400–401

About the Authors

Steven Winter Associates (Norwalk, CT and Washington, D.C.) is one of America's leading architectural consulting firms. Founded in 1972, the firm's areas of expertise include building systems technology, energy-efficient design, sustainable design, and accessibility.

Michael J. Crosbie is a registered architect on the staff of Steven Winter Associates. He is a former editor for *Progressive Architecture* and has authored and contributed to numerous books, including McGraw-Hill's *Time-Saver Standards for Architectural Design Data*, Seventh Edition, and the *Time-Saver Standards for Building Types*, Fourth Edition.